Hormones

Hormones

Fourth Edition

Gerald Litwack, PhD

Emeritus Professor and Chair:

The Geisinger Commonwealth School of Medicine, Scranton, Pennsylvania, United States
The Thomas Jefferson University, Kimmel College of Medicine, Philadelphia, Pennsylvania, United States

ELSEVIER

ACADEMIC PRESS

An imprint of Elsevier

Academic Press is an imprint of Elsevier
125 London Wall, London EC2Y 5AS, United Kingdom
525 B Street, Suite 1650, San Diego, CA 92101, United States
50 Hampshire Street, 5th Floor, Cambridge, MA 02139, United States
The Boulevard, Langford Lane, Kidlington, Oxford OX5 1GB, United Kingdom

British Library Cataloguing-in-Publication Data
A catalogue record for this book is available from the British Library

Library of Congress Cataloging-in-Publication Data
A catalog record for this book is available from the Library of Congress

ISBN: 978-0-323-90262-5

For Information on all Academic Press publications
visit our website at https://www.elsevier.com/books-and-journals

Publisher: Stacy Masucci
Acquisitions Editor: Patricia Osborne
Editorial Project Manager: Pat Gonzalez
Production Project Manager: Swapna Srinivasan
Cover Designer: Mark Rogers

Typeset by MPS Limited, Chennai, India

Dedication

For my family: Ellie, Geoffrey and Susanna, Kate and Andrew, Claudia, Debbie, David and my very young granddaughter, Olivia, who seems to like books.

Also, in remembrance of Anthony (Tony) Norman and his wife, Helen Henry, both of whom contributed to previous editions of this book.

Contents

About the author

Gerald (Gerry) Litwack was educated at Hobart College (BA) and graduated from the University of Wisconsin, Madison (MS and PhD in biochemistry). After spending a summer at the University of Wisconsin as an instructor in a course on enzymes, he sailed to Paris (Liberte) where he spent a year as a postdoctoral fellow of the National Foundation for Infantile Paralysis at the Laboratoire de Chimie Biologique of the Sorbonne. He then became an Assistant Professor of Biochemistry at Rutgers University, later becoming an Associate Professor at the Graduate School of Medicine of the University of Pennsylvania. Future posts took him to the Fels Institute for Cancer Research and Molecular Biology at the Temple University School of Medicine where he was the Laura H. Carnell Professor of Biochemistry and the Deputy Director of the Institute. Later, he was the Chairman of the Department of Biochemistry and Molecular Pharmacology, a Professor of Biochemistry, Vice Dean for Research, and the Director of the Institute for Apoptosis at Thomas Jefferson University Kimmel Medical College with future appointments as a Visiting Scholar at the Geffen School of Medicine at UCLA, the Founding Chair of the Department of Basic Science of the Geisinger Commonwealth School of Medicine, and, as his final position, a Professor of Molecular and Cellular Medicine and the Associate Director of the Institute for Regenerative Medicine at Texas A&M University School of Medicine. Dr. Litwack was a Sabbatical Visitor at the University of California, Berkeley, the University of California, San Francisco, the Courtauld Institute of Biochemistry (London), and the Wistar Institute. He was appointed as the Emeritus Professor and/or Chairman at Rutgers University, Thomas Jefferson University, and Geisinger Commonwealth School of Medicine. He has published more than 350 papers in scientific journals, is named on more than 20 patents, and held numerous editorial positions on biochemical and cancer journals. He has authored 3 textbooks in the areas of biochemistry or endocrinology and has edited more than 65 books. Following his retirement, he lives with his family in Los Angeles where he continues his work as an author and editor and paints in watercolor.

Preface to the fourth edition

The untimely death of Dr. Helen Henry in 2018 followed by the death of her husband, Dr. Anthony (Tony) Norman in 2019, the authors of the third edition of *Hormones*, left vacant an author for the fourth edition. Tony Norman and I wrote the first and second editions of this book. We would take turns in working together, sometimes on the East Coast and sometimes on the West Coast. I could not participate in the preparation of the third edition, published in 2014, because I was involved heavily in authoring the first edition of *Human Biochemistry*, a large textbook that was published by Academic Press late in 2018. This book, *Hormones*, originated from a series of lectures that I presented on the subject as part of the biochemistry course for medical students at the Temple University School of Medicine in the 1980s. Originally, I was to be the first author; however, as the book progressed, it was clear that Tony had much more secondary support (secretarial and other) than I had. For example, his people were instrumental in researching data and preparing the extensive Appendices. For this reason, I decided that it was only reasonable that Tony should be the first author.

After I learned, in April of 2020, about the death of Helen and Tony, I volunteered to be the author of the fourth edition of the book, unless Academic Press/Elsevier had other ideas. They professed their interest in my participation and decided that it was time for the preparation of the fourth edition since nearly 6 years had elapsed from the publication of the third edition.

After reviewing the third edition, it was clear that a great deal of the information and many of the figures were still timely. Accordingly, I have retained many portions of the third edition. I have updated references in all chapters and have elaborated on many subjects, either missing from the third edition or mentioned only briefly. In total, I believe that the fourth edition of the book is up-to-date and more informative on several subjects and stronger than previous editions.

There are a number of changes and additions in this edition. Some of these changes involve up-to-date revisions, enlarged discussions, and, in a few cases, corrections. The most major of these additions concerns a much greater emphasis on hormone receptors and their signaling pathways. A great deal of this information is conveyed in new figures; therefore this edition is longer than previous ones.

I am greatly indebted to Pat Gonzalez, Senior Project Manager at Elsevier, for her ability to cope with my rudimentary method of adding new information to the existing text. She, in addition to Peter B. Linsley of Elsevier, also facilitated the second edition of *Human Biochemistry* (Academic Press) published in 2021. The first edition of *Human Biochemistry* was published late in 2018.

The cover of this edition is graced by a background of molecular structures extending to its entirety. These are the structures of a molecule of follicle-stimulating hormone (FSH) bound to its receptor. This hormone is one of two gonadotrophs (FSH and LH) important in the regulation of development, growth, and reproduction.

Gerald Litwack, PhD
Emeritus Professor and Chair
The Geisinger Commonwealth School of Medicine, Scranton, PA, United States
The Thomas Jefferson University Kimmel College of Medicine, Philadelphia, PA, United States

May 26, 2021

Chapter 1

Hormones: an introduction

I Overview of hormones

A Introduction

The term "hormone" is derived from the Greek *hormon*, the present participle of *impel*, or *set in motion*, an apt characterization of these potent molecules. "Endocrine" is also derived from the Greek: *endo-* for *internal* or *within* and *krinein* meaning *separate*. This term conveys the distance of the site of secretion from the site of action that characterized the systems, such as the pancreas, the thyroid, and the reproductive glands that were studied in the early days of endocrinology.

The cellular constituent that is the immediate recipient of the hormone is the receptor, an entity importance of which is now so dominant in the study of hormones that it is hard to imagine that the existence and nature of these molecules were not appreciated until the early 1970s. The biochemical organization of receptors is diverse but each receptor is structurally organized so that it can specifically recognize and interact with its hormone. Because of the low circulating concentrations of the hormones, the receptor must have a very efficient "capture" mechanism for its hormone. As a consequence of the receptor–hormone interaction (however, transient it may be), signal transduction occurs and a specific biological response(s) is generated within and, in some instances, around the target cell—that is, the cell responds to the presence of the hormone.

The domain of endocrinology includes the study of how, in a higher organism, cell A communicates with cell B by sending a chemical messenger or hormone. A detailed understanding of a particular endocrine system includes an understanding of the following: (1) the anatomical description of cells A and B and their immediate environment, as well as the distance between A and B; (2) the chemical structure of the hormone (H); (3) the details of the biosynthesis of the hormone by cell A; (4) the mode of transfer of H from cell A to cell B; (5) the detailed mechanism by which cell B uses receptors to detect the presence of H; (6) how cell B transduces the presence of H to initiate and sustain a biological response; and (7) how cell B communicates via a feedback loop with cell A to indicate the adequate presence of the hormone.

The study of endocrinology over the past century has been dependent upon the scientific methodologies available to probe the various endocrine systems. Thus in the interval 1900–60, endocrinology was largely pursued at the physiological level. This resulted in the discovery of approximately 25 hormones. The time it took to achieve structural understanding of a hormone was usually inversely proportional to the size of the hormone. For example, the complete structure of thyroxine (molecular weight 770) was defined in 1926, while the sequence and structure of the small protein hormone insulin was not obtained until 1953 (amino acid sequence) and 1969 (three-dimensional structure).

The biochemical era of endocrinology began in approximately 1955–60 and extends to the present time. The availability of radioactive isotopes of carbon (^{14}C), hydrogen (^{3}H), phosphorus (^{32}P), among others, coupled with advances in chemical methodology (chromatography, mass spectrometry, nuclear magnetic resonance spectroscopy, and X-ray crystallography) has led to the detection and chemical characterization of minute quantities (nanograms or picograms) of new hormones and the characterization of many receptors. Now we are experiencing the cellular and molecular biological era of endocrinology. We have an increased ability to visualize how molecules behave in cells, through fluorescent dye tagging, confocal microscopy, and other imaging advances. The sequencing and continuing analysis of the human genome has expanded our knowledge of the molecular players (hormones and receptors) and their evolutionary relationship to each other. The ability to generate mouse models with specific genetic attributes that can be expressed as a function of development stage or other variables helps to fine-tune our functional understanding of hormonal processes. As is always the case in scientific inquiry, new information results in new questions, which leads to new techniques to answer them.

The objective of this book is to provide a status report on the field of human hormones, viewed in the light of our current understanding of cellular and subcellular architecture, as well as the molecular details of their mode of action. In this chapter, some of the first principles of hormone action are presented as a foundation for

Hormones. DOI: https://doi.org/10.1016/B978-0-323-90262-5.00027-5

the more specific considerations of individual hormonal systems in the chapters to follow.

B Review of animal cell structure

To describe the details of the synthesis of hormones and their interactions with target cells in the chapters to follow, it is appropriate to present a brief review of cellular organization. A typical animal cell is shown in Fig. 1.1.

1 Nuclear organization

In eukaryotes the nucleus, containing the chromatin, and the cytoplasm are separated, except during interphase of mitosis. The nuclear envelope, consisting of two membranes separated by a small space, is perforated by nuclear pores through which the transport of macromolecules, proteins, and RNA, between the two major compartments of the cell, takes place. For example, messenger and transfer RNA as well as ribosomal subunits must move from the nucleus to the cytoplasm (Fig. 1.2) and proteins that participate in the synthesis, repair, and transcription of DNA must move into the nucleus from the cytoplasm. The latter include the steroid hormone receptors and other proteins that regulate gene transcription that will be discussed in the following chapters.

Fig. 1.3A shows the fundamental organization of DNA, beginning with the structure of the double helix in the top panel. This is the form in which DNA is found except when it is being transcribed or replicated, at which times the two strands of the double helix are separated. Inside the nucleus is the nucleolus where the DNA-encoding ribosomal RNA is continually being transcribed. The remainder of the DNA in the eukaryotic nucleus is present in a more tightly packed form, arising from an association between the DNA (with its negatively charged sugar–phosphate backbone) and basic, positively charged proteins called histones; the final step in making DNA

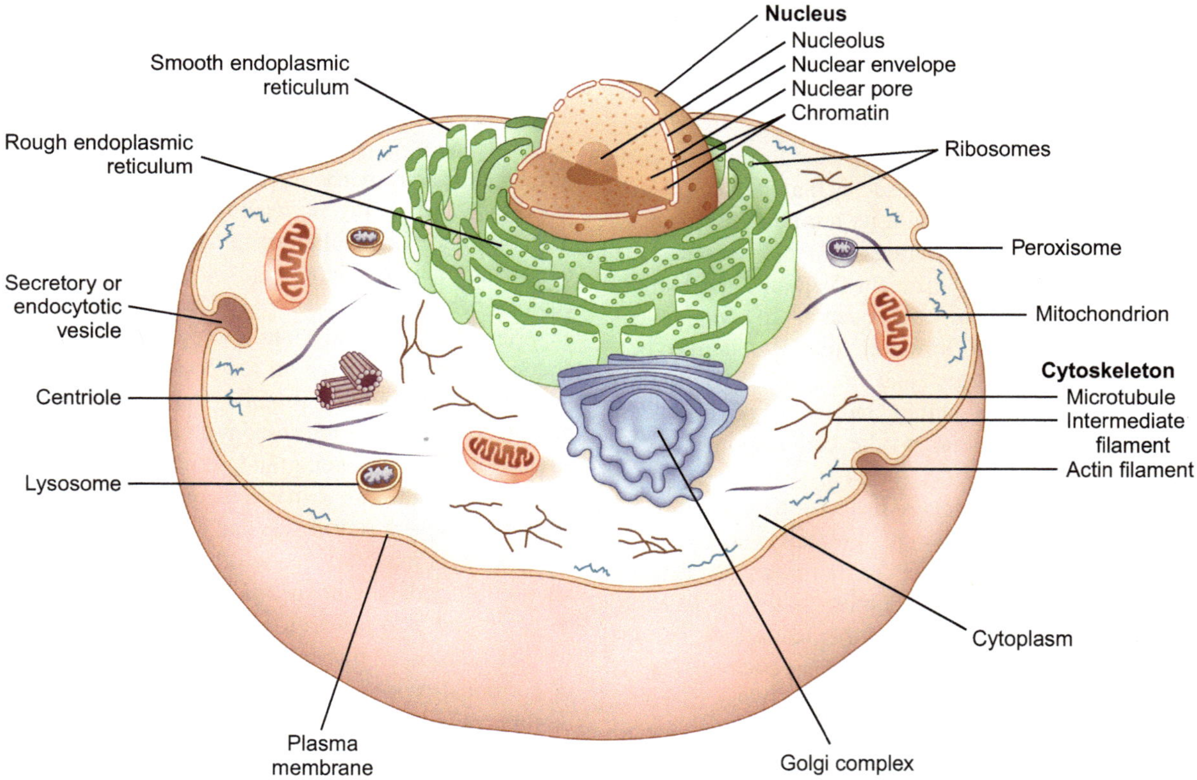

FIGURE 1.1 Structural elements of animal cells. The major features shared by animal cells are shown. In eukaryotes the nuclear membrane separates the genetic material, the chromatin, from the cytoplasm. Molecules move between the two compartments through nuclear pores and a portion of the chromatin (see Fig. 1.3), the nucleolus, is dedicated to the continual production of ribosomes. The cytoplasmic organelles depicted include the smooth endoplasmic reticulum (the microsomes of fractionated cells), which carries out metabolic conversions of carbohydrates and lipids, and the rough endoplasmic reticulum, associated with ribosomes that are synthesizing proteins to be secreted by the cell. These proteins are collected and processed in the Golgi apparatus. Mitochondria generate energy for the cell's function from the products of the metabolism of carbohydrates, fats, and proteins. Peroxisomes and lysosomes degrade molecules that are no longer needed or are potentially harmful. Elements of the cell's cytoskeleton are shown, including the centriole, part of the organizing center for microtubules. The plasma membrane is described in Fig. 1.5.

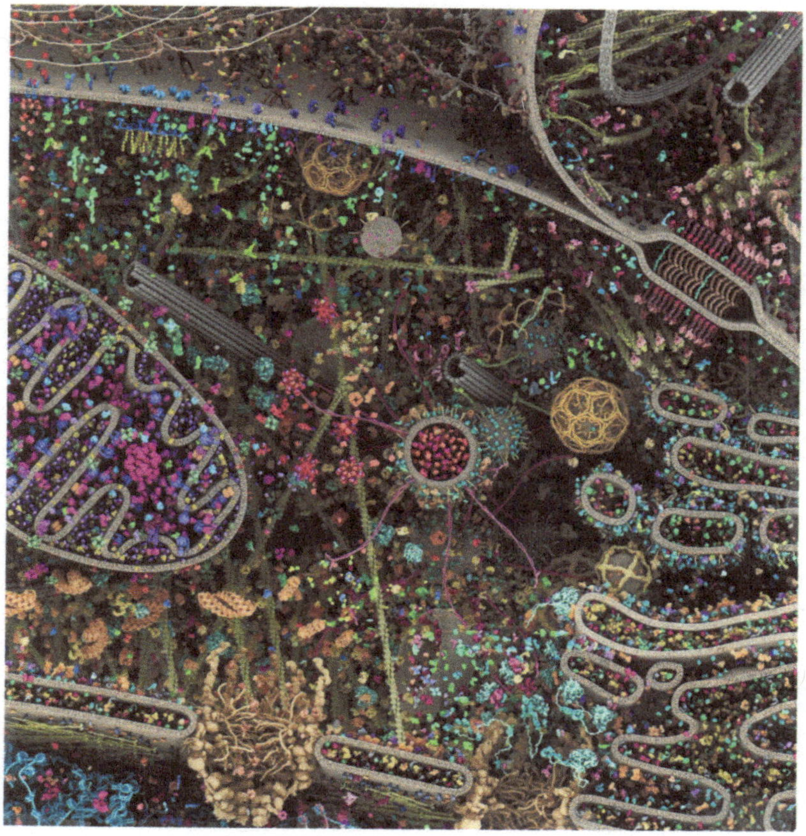

FIGURE 1.2 This is the most detailed image of a human cell. It was achieved by the combined usage of three methodologies: X-ray, nuclear magnetic resonance, and cryo-electron microscopy. The image is described as cellular landscape cross-section through a eukaryotic cell and is the work of Evan Ingersoll and Gael McGill. *Reproduced from https://gaelmcgill.artstation.com/projects/Pm0.Jl.1.*

accessible for transcription involves modification of histone proteins (acetylation) to loosen their association with the DNA. Further steps in the coiling and compaction of DNA are illustrated in Fig. 1.3B. The result of this process is the packing of a linear molecule of DNA that is about 10^5 μm (micrometer) long into a nucleus with a diameter of about 10 μm (Fig. 1.3B).

Fig. 1.4 illustrates details of right-handed double-stranded DNA.

2 The plasma membrane

Although the precise content of substances that comprise the cell membrane differs in different cell types, many components are common to all membranes. These are lipids (including phospholipids, cholesterol, and glycolipids), proteins, and glycoproteins. The cell membrane that encloses the cell resembles the internal membranes, such as those associated with the nucleus, mitochondria, and microsomes.

The primary lipid component of cellular membranes consists of the amphipathic phospholipids containing a polar head group, such as choline, ethanolamine, or serine and a hydrophobic tail consisting of two long-chain fatty acids, usually one saturated and one unsaturated (see Fig. 8.5B). In an aqueous environment, these molecules form a bilayer with the hydrophobic tails on the inside and the polar head groups on the outside, as shown in Fig. 1.5. Cholesterol molecules are interspersed among the fatty acid side chains; the amount of cholesterol, which can have effects on local membrane fluidity, varies with the type of membrane.

Many different types of proteins, seen in Fig. 1.5, are found in or associated with the plasma membrane. In many chapters in this book, we will be considering the actions of membrane hormone receptors that have both extracellular and intracellular domains, and we will look at the structure and function of these integral proteins in some detail. We will also be looking at intracellular proteins that are partially embedded in the lipids of the inner face of the plasma membrane as well as some that are more loosely associated with it.

Complex oligosaccharides may appear on the outer surface as derivatives of sphingosine or other lipids (glycolipids). Proteins may have complex polysaccharides attached to them. Complex polysaccharides that may appear on the surface of the plasma membrane are pictured in Fig. 1.6A. These carbohydrate moieties may play important recognition functions or in the case of

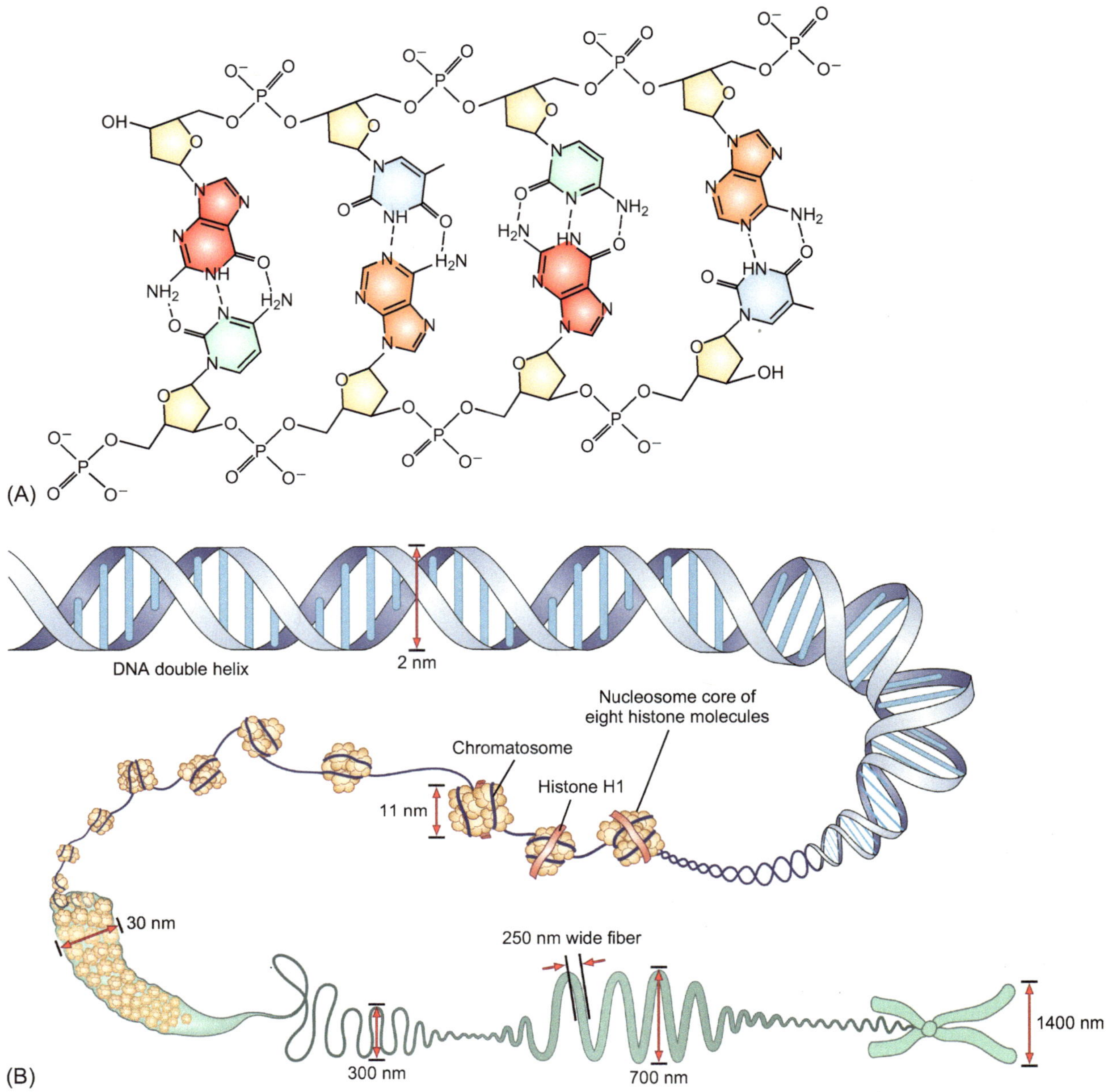

FIGURE 1.3 Organization of DNA. (A) *Double-helical DNA*. The chemical nature of the DNA double helix is shown for a stretch of four base pairs. The negatively charged sugar–phosphate backbone is shown in yellow. Each purine, adenine (A, *orange*), or guanine (G, *red*) pairs with a pyrimidine, thymine (T, *blue*) or cytosine (C, *green*), respectively. The two DNA strands are complementary and antiparallel to one another. The double-helical structure of DNA is stabilized by the hydrogen bonds between the bases on each strand, two for each AT base pair and three for each GC base pair as well as by interactions between the stacked bases in the interior of the helix. (B) *Organization of DNA in chromosomes*. The compaction of double-helical DNA into a chromatid of a chromosome is shown. The first step is the coiling of the double helix of DNA around a core of histone proteins to form the core nucleosome. Histone H1 joins these "beads on a string," 11 nm across, to promote their coiling upon themselves to form a 30-nm fiber. Further structural details are not completely understood, but include 300-nm loops and further coiling of these into the 700-nm chromatid.

membrane receptors for hormones may influence the accessibility of receptors for their ligands.

3 Intracellular organelles

Some of the other intracellular organelles that will be encountered in the cells that make or respond to hormones are illustrated in Figs. 1.1 and 1.2. For example, the synthesis and secretion of protein and peptide hormones depend on the rough endoplasmic reticulum and Golgi apparatus and the specific processing enzymes therein. Lysosomes play an important role in the secretion of thyroid hormones (THs), and both the mitochondria and the smooth endoplasmic reticulum (the microsomal fraction of the cell) are the sites of steroid hormone synthesis in the adrenal gland, gonads, and placenta.

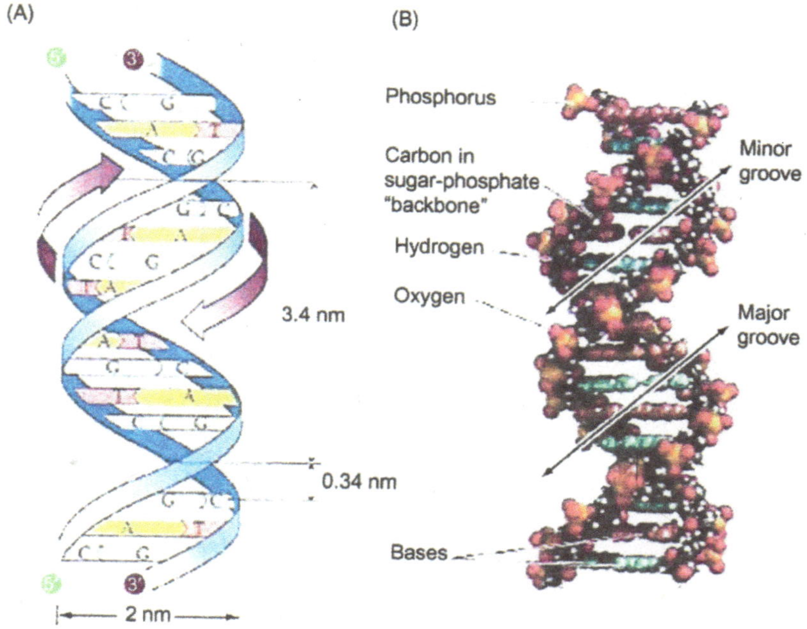

FIGURE 1.4 (A) A right-handed DNA double helix. Down the center of the structure is the axis of symmetry. The diameter of the helix is about 2 nm. One 360-degree turn of the helix takes about 8–10 base pairs and occupies a space of 3.4 nm. The antiparallel character of the two strands is shown by the 5′ end of one strand opposing the 3′ end of the antiparallel strand at the top and the bottom. In (B) the various groupings are indicated. There is a major groove and a minor groove. Many of the transactivating proteins (DNA-binding proteins with a specific function) bind in one of these grooves of DNA to the promoter of a specific gene. The deoxyribose phosphate backbone occurs on the outside of the helix and the hydrophobic bases are located in the interior. The negative charges on the phosphate groups in the backbone attract positively charged motifs on DNA-binding proteins. *Reproduced from Fig. 10.15 of G. Litwack (Ed.), Human Biochemistry, Academic Press, 2018.*

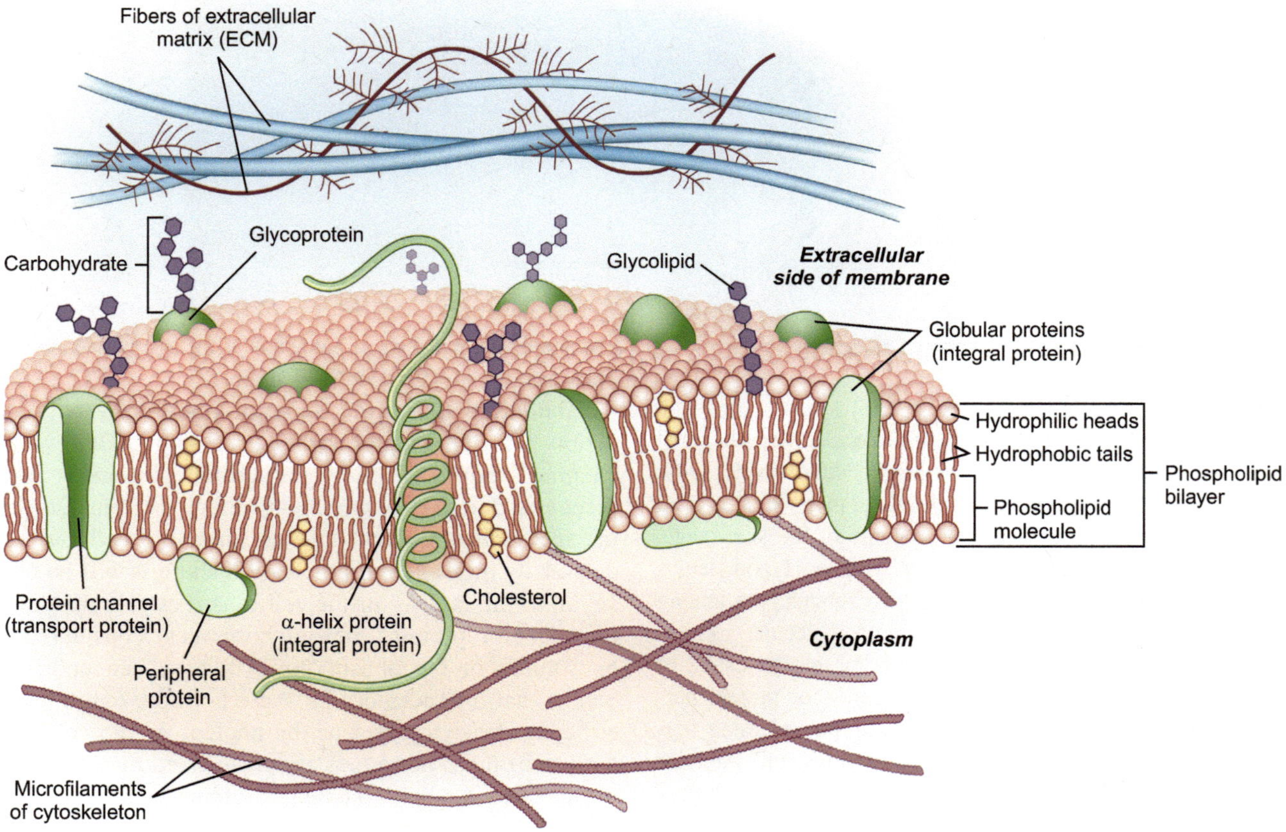

FIGURE 1.5 The plasma membrane. The phospholipids that make up all cellular membranes are oriented with their polar head groups (usually choline, ethanolamine, or serine) facing the aqueous environments inside and outside the cell and their hydrophobic fatty acid side chains toward the interior of the bilayer. Cholesterol molecules (*yellow*) are interspersed among the lipids in both layers. Integral membrane proteins traverse the bilayer as illustrated by the transport protein/channel on the left or the helical protein in the middle; peripheral proteins are embedded in only one side of the bilayer; surface proteins are associated with one face of the membrane but not embedded in it. The exterior surface of the cell displays, in addition to proteins, carbohydrate moieties attached to membrane lipids (glycolipids) and to proteins (glycoproteins).

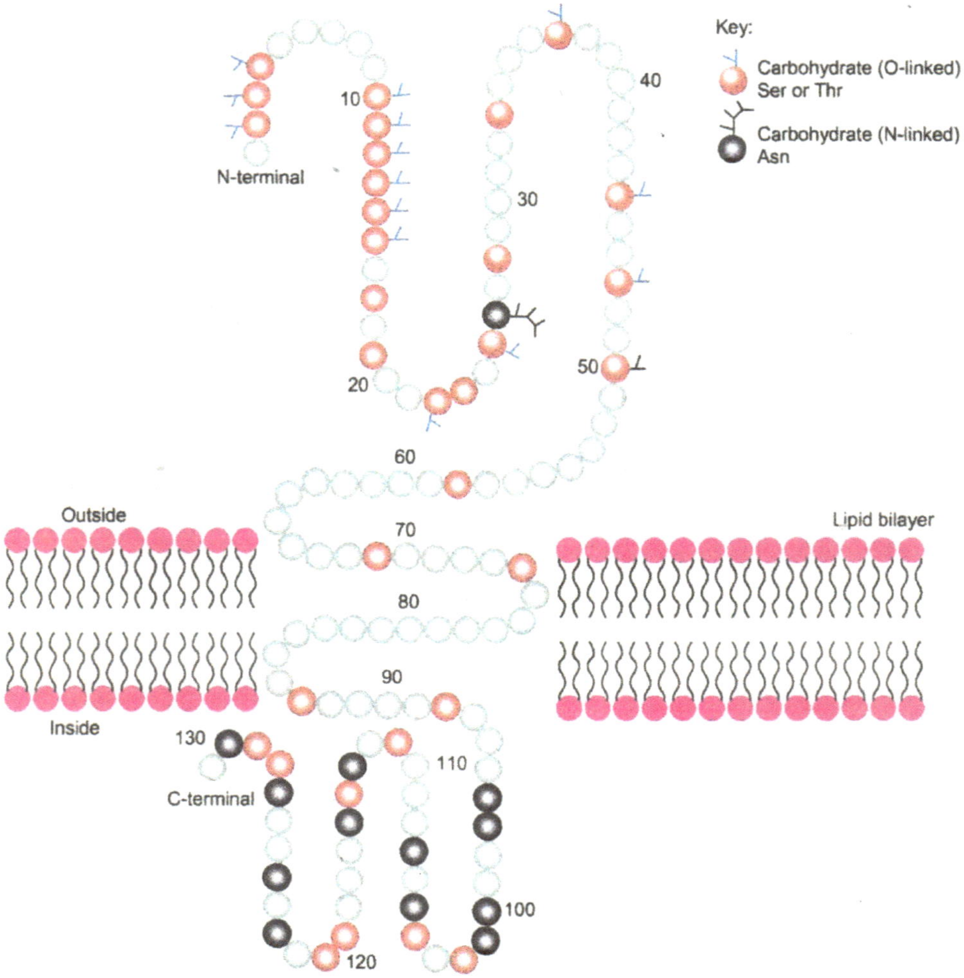

FIGURE 1.6 Diagram of a transmembrane protein with sugar substitutions in the extracellular domain. Sugars that frequently appear in glycoproteins are glucose, galactose, mannose, fucose, and acetylated amino sugars, such as *N*-acetylgalactosamine, *N*-acetylglucosamine, and *N*-acetylneuraminic acid (NANA or sialic acid). Sugars can be O-linked through Ser or Thr or N-linked through Asn. *Reproduced from Fig. 6.29 of G. Litwack, Human Biochemistry, Academic Press, 2018.*

C Hormones and their communication systems

1 Types of hormone molecules

Hormones are heterogeneous in their molecular size, chemical properties, and pathways of synthesis. Nitric oxide (NO; see Chapter 15: Hormones Related to the Kidney and Cardiovascular System) is at one extreme of the size range; the pituitary gonadotropins (Chapter 3: The Hypothalamus and Anterior Pituitary) consisting of two subunits are among the largest of the protein hormones with molecular weights ranging between 25 and 36 kDa, depending on the extent of added carbohydrates (glycosylation). Peptide or protein hormones range from 3 amino acids [thyrotropin releasing hormone (TRH), Chapter 3: The Hypothalamus and Anterior Pituitary] to over 100 per subunit. TH (Chapter 5: Thyroid Hormones) and epinephrine (Chapter 11: Hormones of the Adrenal Medulla) are derived from the amino acid tyrosine. Steroid hormones, and vitamin D and its metabolites are derived from cholesterol or 7-dehydrocholesterol, respectively (Chapter 2: Steroid Hormones: Chemistry, Biosynthesis, and Metabolism). Arachidonic acid, cleaved from membrane phospholipids, is the main precursor of the prostaglandins and other eicosanoids (Chapter 8: Eicosanoids).

The initial step in the action of a hormone, the interaction with its receptor, depends to some extent on its chemical nature. Peptide and protein hormones have receptors that are membrane-spanning proteins so that the molecule does not have to enter the cell but can deliver its message on the outside where it will be conveyed to the interior of the cell by structural changes in the receptor protein. Steroid hormones, considered to be soluble in the phospholipid bilayer, can enter the cell so that the receptors for these hormones are located either in the cytoplasm or the nucleus of the cell. The actions of these hormones are propagated by interaction of the receptor with nuclear proteins and DNA. The amino acid–derived hormones differ from one another: TH has an intracellular receptor similar to those for the steroid hormones and epinephrine interacts with its membrane receptor.

Thus the hormonal messaging systems have evolved using a variety of types of molecules and mechanisms of actions. Understanding these in settings of particular systems is a major focus of this book.

2 Types of hormonal communication systems

Hormones are chemical messengers that send a signal within a physiological system from point A (secretion) to point B (biological action). Three variations on the anatomical and, therefore, functional relationship between point A and point B of these systems are illustrated in Fig. 1.7.

The classic systemic endocrine system is shown in the top panel. The hormone is biosynthesized (and perhaps, but not necessarily, stored) within specific cells associated with an anatomically defined endocrine gland. Upon the receipt of an appropriate physiological signal, which may take the form of either a change in the concentration of some component in the blood (e.g., another hormone, Ca^{2+}, glucose) or a neural signal, the hormone is released into the circulation. It is transported in the bloodstream to one or more target cells, which are defined as targets by the presence of a specific high-affinity receptor, either on the membrane or within the cell, for the hormone. It is what the receptor does after interacting with the hormone that determines the biological response. As will be seen in several of the chapters in this book that many, if not most, hormones have numerous and diverse target cells and the response to the hormone may vary with cell type, indicating that other players in or around the target cell may affect the outcome of hormone–receptor interactions.

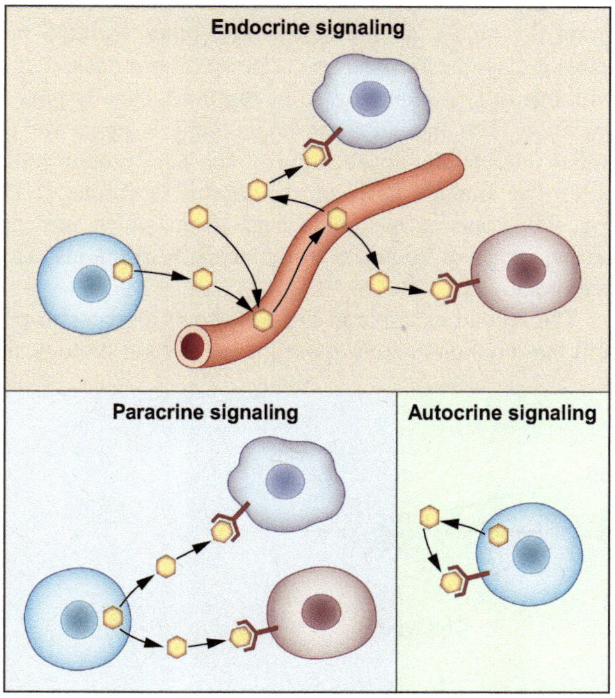

FIGURE 1.7 Types of hormonal signaling. Three of the ways that hormones secreted by one cell can carry out its signaling function are illustrated. The top panel shows the classical endocrine system with a specialized hormone-synthesizing cell secreting its product into the bloodstream. It is carried throughout the body and may interact with one or many distant target cells, which are distinguishable by the presence of a specific receptor for the hormone on its surface (shown) or within the cell (not illustrated). In paracrine signaling (lower left), the signaling cell, which also has many other functions, releases the hormone into the intracellular space and it moves small distances to nearby cells. These target cells also have a specific receptor for the hormone (membrane or intracellular). In paracrine signaling the cells reached by the hormone may be of the same or different types. Finally, cells that secrete the hormone and also have receptors that bind and respond to it are displaying autocrine signaling, as shown on the lower right of the figure.

In some or portions of some endocrine systems, the hormone-secreting cell releases its product not into the general circulation but into a closed system, such as the hypothalamic–pituitary portal system. In this case the hypothalamic-releasing hormones are released into and diluted by a limited volume, ensuring that most of the hormone molecules will be delivered to the anterior pituitary, which contains their target cells (Chapter 3: The Hypothalamus and Anterior Pituitary; Fig. 3.5).

The lower left panel of Fig. 1.7 shows a type of hormonal communication system that does not involve the circulatory system at all. In paracrine systems, hormones secreted from the signaling cell interact with specific high-affinity receptors in neighboring cells that are reached by diffusion: that is, the distance from point A to point B is decreased and dilution in the bloodstream is avoided. As with endocrine systems, the nearby target cells may be all the same type or may differ from each other as illustrated. Most prostaglandins (Chapter 8: Eicosanoids) act through paracrine mechanisms. Several, if not all, of the steroid hormones act by paracrine in addition to endocrine mechanisms. For example, in the testis testosterone is not only released into the blood from the interstitial cells in which it is produced but also diffuses to nearby seminiferous tubules to support the production of sperm (Chapter 12: Androgens). IGF-1 (Chapter 3: The Hypothalamus and Anterior Pituitary and Chapter 17: Growth Factors) is a protein hormone secreted into the bloodstream by the liver in response to growth hormone but is also secreted by other cells to control the growth and differentiation of neighboring cells.

Finally, some cells both produce and respond to the same hormone. This type of system is referred to as autocrine. Examples of these systems involve growth factors and the control (or lack thereof in malignancy) of cellular proliferation.

D Biosynthesis of peptide and protein hormones

The biosynthesis of hormones occurs in specialized cells, usually present in endocrine glands, which express the enzymes that catalyze the steps of their formation and have any other necessary molecules required. Chapter 2,

Steroid Hormones: Chemistry, Biosynthesis, and Metabolism, describes the production of the steroid hormones, while Chapter 5, Thyroid Hormones, Chapter 8, Eicosanoids, and Chapter 11, Hormones of the Adrenal Medulla, describe the biosynthesis of the THs, eicosanoids, and epinephrine, respectively.

Protein and peptide hormones are biosynthesized in specific cells, through the well-known processes of transcription of a specific message encoded in the DNA of the gene for the protein and the translation of the RNA message (mRNA) into a protein. As with other proteins, variations in modifications to the initially produced mRNA and/or protein leads to deviation from the original "one gene, one protein" concept. The biosynthesis of peptide and protein hormones yields many examples of such deviations.

It is now quite well recognized that not only does one gene not lead to a single protein, but also one gene does not lead to a single RNA; that is, two or more RNA transcripts can arise through alternative processing of a single primary transcript. Fig. 1.8 shows schematically how this happens. Exons are joined by splicing them together at very specific sites. Splice site recognition can vary from one cell to another, causing the primary transcript to differ between two cell types. The production of either calcitonin (CT) or CT gene–related peptide was one of the first examples of alternative splicing to be elucidated (see Fig. 9.10). Alternative splicing is by no means an unusual method of generating multiple products of the same gene. While the exact percentage of protein coding genes subject to alternative splicing is not yet known, recent genomic analyses suggest that this number may be as high as 90%.

Another layer of variability in the final product of a gene is the posttranslational processing of the initial protein product. Broadly speaking, this term includes the myriad modifications of the side chains of the amino acids as well as the addition of sugar or lipid moieties to the protein backbone. For this discussion, however, we will confine our attention to alteration of the initially translated protein by proteolytic cleavage, yielding smaller protein or peptide products. These cleavages are catalyzed by one of a family of proprotein convertases (PC1–PC7), serine endoproteases at cleavage sites in the precursor protein that are designated by two basic amino acids (Lys–Lys, Arg–Arg, or Lys–Arg). The reactions take place largely in the rough endoplasmic reticulum and in the Golgi apparatus as the hormone is being prepared for movement into secretory vesicles.

Fig. 1.9 illustrates some examples of the posttranslational cleavage events that yield active hormones. Most simply, virtually all hormones (and other secreted proteins) are synthesized as pre- or pre-pro-hormones, that is, with one or two sequences to be removed, usually prior to secretion. The first of these is generally a signal for the initial intracellular localization of the new protein molecule. The mature form of parathyroid hormone (PTH, Fig. 9.9) contains 84 amino acids from which pre- and prosequences of 25 and 6 amino acids, respectively, have been removed.

The second example in Fig. 1.9 shows a precursor protein that contains within its sequence several biologically

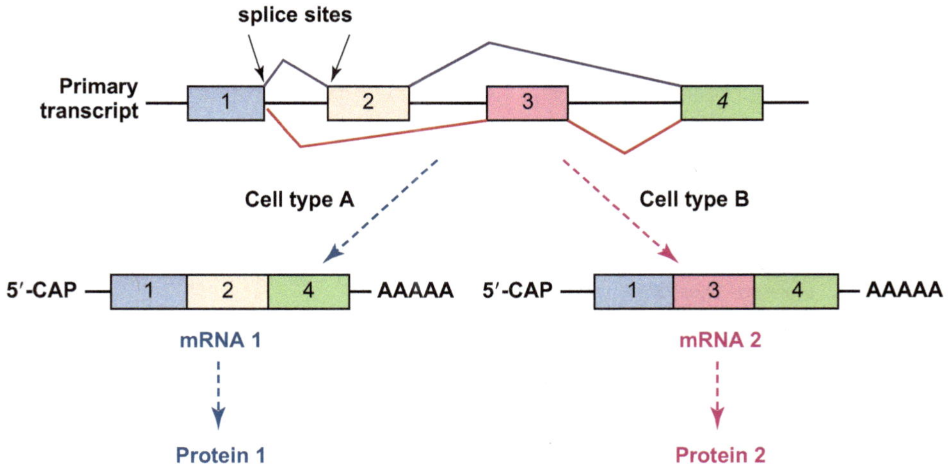

FIGURE 1.8 Alternative splicing of mRNA for hormones. In eukaryotes almost all genes for proteins consist of regions of DNA that carry the code for the protein (exons; *colored boxes*) interrupted by noncoding sequences (*gray line*) in the primary mRNA transcript. Maturation of the primary transcript involves the splicing of these coding regions together as well as the addition of the 5′ cap and the poly A tail typical of eukaryotic mRNA. The splicing of the exons takes place in the nucleus and is carried out by large RNA/protein complexes called spliceosomes. It is the spliceosomes that are responsible for splice site (specific DNA sequences) selection. In the example shown the spliceosomes of one cell type use the splice sites between exons 1, 2, and 4, while those in cell type 2 use exons 1, 3, and 4. The two mature RNAs, thus, encode different proteins. See Fig. 9.10 for the example of calcitonin and calcitonin gene–related peptide. *mRNA*, RNA message.

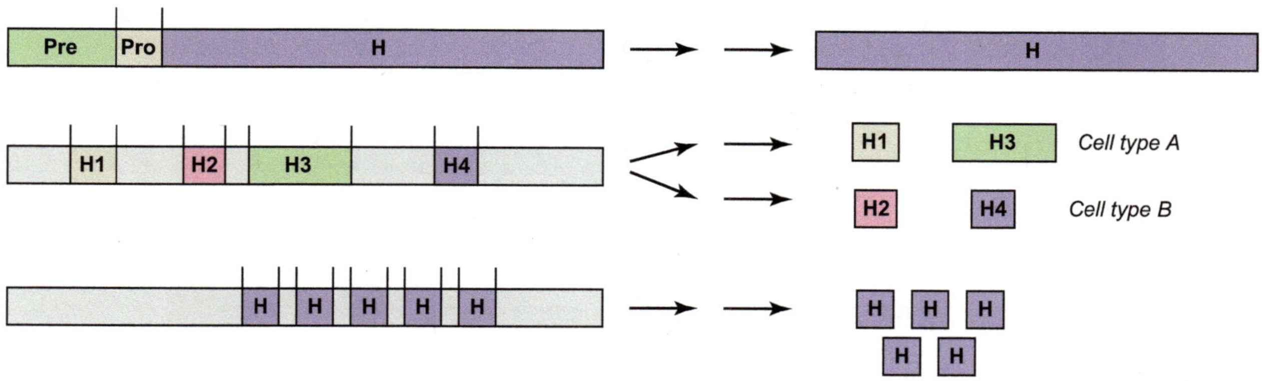

FIGURE 1.9 Processing of pre- and prohormones. Many protein and peptide hormones (right) are synthesized within a larger precursor protein (left), of which three examples are shown here. The process is catalyzed by specific proteases that cleave the protein at specific sites (vertical lines), usually preceded by two basic (Lys, Arg) amino acids. Much of this processing takes place in the endoplasmic reticulum and Golgi apparatus and in secretory vesicles prior to the secretion of the hormones. As illustrated in the top example, many hormones are synthesized with one or two N-terminal portions that are sequentially removed to form the active hormone; see parathyroid hormone, Fig. 9.9. In the second example, several different active peptides are within a single precursor protein, which is processed differently in different cell types; see POMC, Fig. 3.15. Third, a precursor protein can contain several copies of the hormone, each of which is excised at a pair of specific proteolytic sites, as in the case of the tri-peptide, TRH (Fig. 3.9). *POMC*, Proopiomelanocortin; *TRH*, thyrotropin releasing hormone.

active peptides, and which can be differentially processed in different cell types. Such a situation is exemplified by proopiomelanocortin (Fig. 3.15). Adrenocorticotrophic hormone and other hormones are the processing products in pituitary corticotrophs, whereas a different of set of peptides, including β-endorphins, result from the processing of the same precursor in the cells of the intermediate lobe of the brain.

Finally, the precursor protein can contain several copies of a single peptide hormone, as is the case for TRH (Fig. 3.9). This example, as well as that of insulin (not shown in Fig. 1.9 but see Fig. 6.5), played important roles in the establishment of the idea that precursor protein molecules harbor active peptides within their sequences. In the case of insulin, this theory, based on the increasing availability of information from protein sequencing followed by that of DNA sequencing, solved the long-standing question of the origin of the two subunits of insulin. As shown in Fig. 6.5, it is now understood that insulin is synthesized as a single molecule. Disulfide bonds are formed to join two portions of the molecule and proteolytic cleavages release the two joined subunits from the proprotein.

E Regulation of hormone synthesis, secretion, and serum levels

1 Control of synthesis and secretion

The production and/or secretion of most hormones are regulated by the homeostatic mechanisms operative in that particular endocrine system. The secretion or release of the hormone is normally (in the absence of an endocrine disease related to hormone secretion) related to the requirement for the biological response(s) generated by the hormone in question. Once this requirement has been met, the secretion of the hormone is curtailed to prevent an overresponse. Thus a characteristic feature of most endocrine systems is the existence of a feedback loop that limits or regulates the secretion of the hormonal messenger.

Two general categories of endocrine feedback systems are illustrated in Fig. 1.10: those in which the function achieved by the hormone (e.g., elevated serum Ca^{2+} or elevated blood glucose) directly feeds back upon the endocrine gland that secretes the hormone; and those involving the central nervous system (CNS) and hypothalamus. On the left is shown the first case, a simple but effective system, in which changes in the circulating amount of something of physiological importance, in this case serum Ca^{2+}, is both the biological response and the agent that exerts negative feedback inhibition on the gland producing the hormone that caused its increase. Although the actual control of PTH is considerably more complex than shown in this figure (see Chapter 9: Calcium-Regulating Hormones: Vitamin D, Parathyroid Hormone, Calcitonin & Fibroblast Growth Factor-23), the secretion of the hormone in response to low serum Ca^{2+} and its cessation when this cation returns to normal levels is at the heart of the regulation of PTH. Another example of this type of control is the stimulation of insulin by elevated levels of blood glucose and the fall of the hormone when glucose levels fall in response to its actions.

On the right side of Fig. 1.10 is shown a generalized version of a hypothalamic–pituitary–peripheral gland axis, of which several will be encountered in the following chapters. Under the control of numerous areas in the

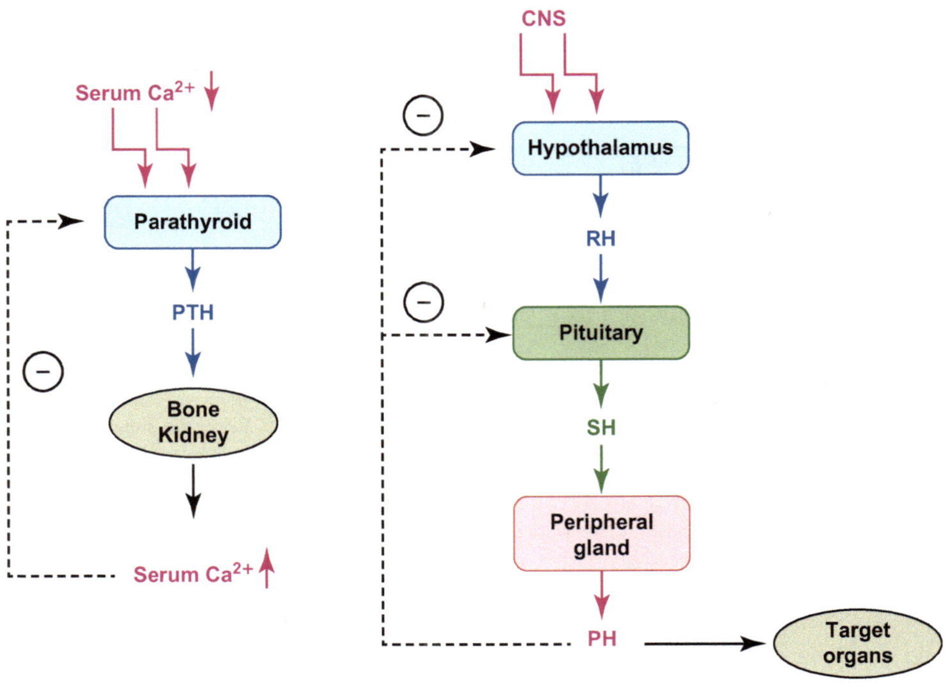

FIGURE 1.10 Models of the regulation of hormone secretion. Two general models of the physiological homeostatic control of hormone secretion are shown. On the left is the type of negative feedback exemplified by the secretion of PTH by the parathyroid gland. The stimulus for the secretion of PTH is a drop in serum Ca^{2+} below the threshold of normality. A calcium sensor in the parathyroid gland cell detects this drop and sets into motion events leading to increased synthesis of PTH and its secretion into the bloodstream. At its target cells, PTH stimulates the movement of Ca^{2+} into the blood and the negative feedback effect of normal circulating levels of the cation result in reduced production and secretion of PTH. See Chapter 9, Calcium-Regulating Hormones: Vitamin D, Parathyroid Hormone, Calcitonin & Fibroblast Growth Factor-23, and especially Fig. 9.12 for more details regarding this system. On the right is a generic version of the hypothalamic–pituitary–peripheral organ axis seen for the hormones of the thyroid gland, gonads, adrenal cortex, and other pituitary hormones. In these systems the hypothalamus receives input from many different areas of the CNS and responds by secreting a hormone (RH, or, in some cases a release-inhibiting hormone) that stimulates specific cells of the pituitary to secrete a peptide hormone that stimulates a peripheral gland (SH). This peripheral gland secretes another hormone, PH, which acts on its target cells to bring about the appropriate biological response and at the same time exerts negative feedback effects on the hypothalamus and/or the pituitary to turn off the system. As will be seen in the chapters devoted to these systems, the actual controls are considerably more complex than depicted here, but the underlying blueprint for them is constant. *CNS*, Central nervous system; *PH*, peripheral hormone; *PTH*, parathyroid hormone; *RH*, releasing hormone; *SH*, secreted hormone.

CNS, specific neurons of the hypothalamus secrete a given hormone (e.g., TRH) that, rather than entering the general circulation, enters the hypothalamic–pituitary portal system and stimulates the secretion of a particular peptide hormone (e.g., thyroid stimulating hormone). This hormone is released into the circulation and travels to its target peripheral endocrine gland (e.g., the thyroid) where it stimulates the release of that gland's hormone (e.g., TH). TH has many target tissues in which it brings about biological responses, but most important in the context of the current discussion are its feedback effects on the hypothalamus and pituitary to shut off the stimulatory hormones from these glands. Again, there are many variations on this basic theme which will be encountered in the consideration of the thyroid gland (Chapter 5: Thyroid Hormones), the gonads (Chapter 12: Androgens and Chapter 13: Estrogens and Progestins), and the adrenal cortex (Chapter 10: Adrenal Corticoids).

The cellular and molecular details of how the synthesis and secretion of hormones are regulated by the players described earlier and others will be covered in the relevant chapters. Here it is important to note that while usually the emphasis is on the increased synthesis of hormones as a point of regulation, there are many other possible regulatory points and these vary with the type of hormone. For example, the steroid hormones (excluding vitamin D metabolites) are regulated primarily at the first step in their synthesis (the cleavage of the side chain of cholesterol; see Chapter 2: Steroid Hormones: Chemistry, Biosynthesis, and Metabolism) and are released as synthesized, not stored in the gland. TH, on the other hand, is stored in large quantities within the thyroid gland. The short-term regulation of its secretion is on the secretory process, while the synthetic process takes place over a longer time frame. Peptide hormones, such as insulin, PTH, and the trophic hormones of the pituitary, are stored in varying amounts in the glands, so the relative roles of synthesis and secretion in the regulatory processes also vary among these hormones.

Two other contributors to the biological availability of hormones deserve mention here. One is the conversion of a relative inactive hormone to an active one in its target glands as occurs with TH and, in some cases, testosterone. Second, the removal of active hormone from the blood must occur as part of the attenuation of its effect (in addition to shutting off the flow of new hormone into the blood). Thus the half-life of an active hormone in the blood, which can vary from seconds to days, is important in understanding its regulatory dynamics.

2 Binding proteins

As discussed in Chapter 2, Steroid Hormones: Chemistry, Biosynthesis, and Metabolism, most steroid hormones have limited solubility in plasma due to their intrinsic hydrophobic character; accordingly, steroid hormones (see Table 2.5) as well as TH (see Table 5.2) are largely (99%) bound to specific plasma transport proteins (PTP), which are synthesized in the liver. Each transport protein has a specific ligand-binding domain (LBD) for its cognate hormone. These ligand domains display little amino acid sequence homology with the ligand binding of the cognate receptors. Nevertheless, the PTP LBD also displays a high affinity (see Section II.D) for its ligand: usually the K_d for the PTP ligand is 10–100 × lower than the K_d of the hormone's receptor.

The current view is that it is the "free" form of steroid hormones and not the complex of the hormone with its PTP that interacts with receptors in or on the target cells to begin the sequence of steps, which results in the generation of a biological response. For some endocrine systems, the concentration of the plasma transport protein can be subject to physiological regulation; that is, the concentration of PTP can be either increased or decreased. Thus changes in the amount of PTP can alter the amount of free hormone in the blood, as well as affect the total amount of hormone in the blood. This role of the binding proteins in the availability of steroid and THs can be of considerable physiological relevance in clinical situations.

II Hormone receptors

A Introduction

When a hormone arrives at a target cell, the first step in delivering its message is interaction with a specific protein receptor. It is the presence of this receptor in the cell that renders it a target for the hormone. All receptors have two key components: (1) a ligand-binding domain that noncovalently but stereospecifically binds the correct hormone for that receptor and (2) an effector domain that responds to the presence of the hormone bound to the ligand domain and initiates the generation of the biological response(s). The interaction between the ligand-binding domain and the effector domain is most likely achieved by a conformational change in the receptor protein so that the effector site may interact with other cellular constituents to initiate the next steps in the signal transduction process (see Section III). In general, steroid hormones and TH interact with receptors that are within the cell, either the nucleus or the cytoplasm (although some form of steroid receptors may reside in the plasma membrane and may show a preference for steroid analogs that differs from the classic cytoplasmic receptor), whereas protein hormones, prostaglandins, and the catecholamines interact with the extracellular ligand-binding domains of plasma membrane–spanning receptors. Exceptions to this generalization exist and will be pointed out when they are encountered. Since mechanisms exist to inactivate and remove both hormone and receptor molecules to curtail their signaling, there is a continuing need for the renewal (biosynthesis and secretion) of the hormone by the endocrine gland and of the receptor by the population of target cells.

B Membrane receptors

Membrane receptors for hormones and other extracellular signals have three clearly identifiable domains: the extracellular component, the membrane-spanning component, and the intracellular component. Each domain has biochemical properties reflecting its location and function. Frequently the membrane receptor comprises a single polypeptide chain where the N-terminus lies outside the cell and the C-terminus lies inside the cell. Others, such as ion channel receptors, are composed of subunits. The diameter of a typical cell membrane is 100 Å, requiring 20–25 amino acid residues organized into an α-helix to cross the membrane once. Since the membrane is hydrophobic, it is not surprising that a receptor's membrane-spanning region consists largely of hydrophobic and noncharged amino acids. Membrane receptors are broadly classified by the number of membrane-spanning regions (for our purposes, one or seven) and by what the cytoplasmic portion of the receptor does when an activating ligand binds. In this section the structures of one type of seven- and one type of single-membrane spanning membrane receptor will be considered. The signaling by these and other receptors will be considered in Section III.A.

1 Protein-coupled receptors

The most frequently encountered class of receptors in the context of hormones is the diverse group of G protein (for guanine nucleotide–binding protein)–coupled receptors for which about 800 genes exist in the human genome. Each of these is specific for ligand and response. G protein–coupled receptors (GPCRs) are found in virtually all eukaryotes and participate in many different cellular functions. About half of the GPCR genes encode receptors that have olfactory functions. About 350 GPCRs have hormones, growth factors, and other small molecules as ligands. Fig. 1.11 shows the fundamentals of the structure of these receptors. The receptor itself has seven α-helical membrane spanning regions. This folding generates three extracellular and three intracellular loops. In some GPCRs, palmitoylation of a cysteine residue in the carboxy region results in another loop. The membrane-spanning helices have been shown by X-ray crystallography to cross at angles to one another as depicted on the right side of panel (A) of Fig. 1.11. The N-terminus of GPCRs is highly variable, as expected from the variety of signals to which these proteins respond. On the right side of panel (A) are shown three examples of types of ligand binding. Small molecules and small peptides have access to a cleft within the helices for binding, whereas larger proteins, such as the

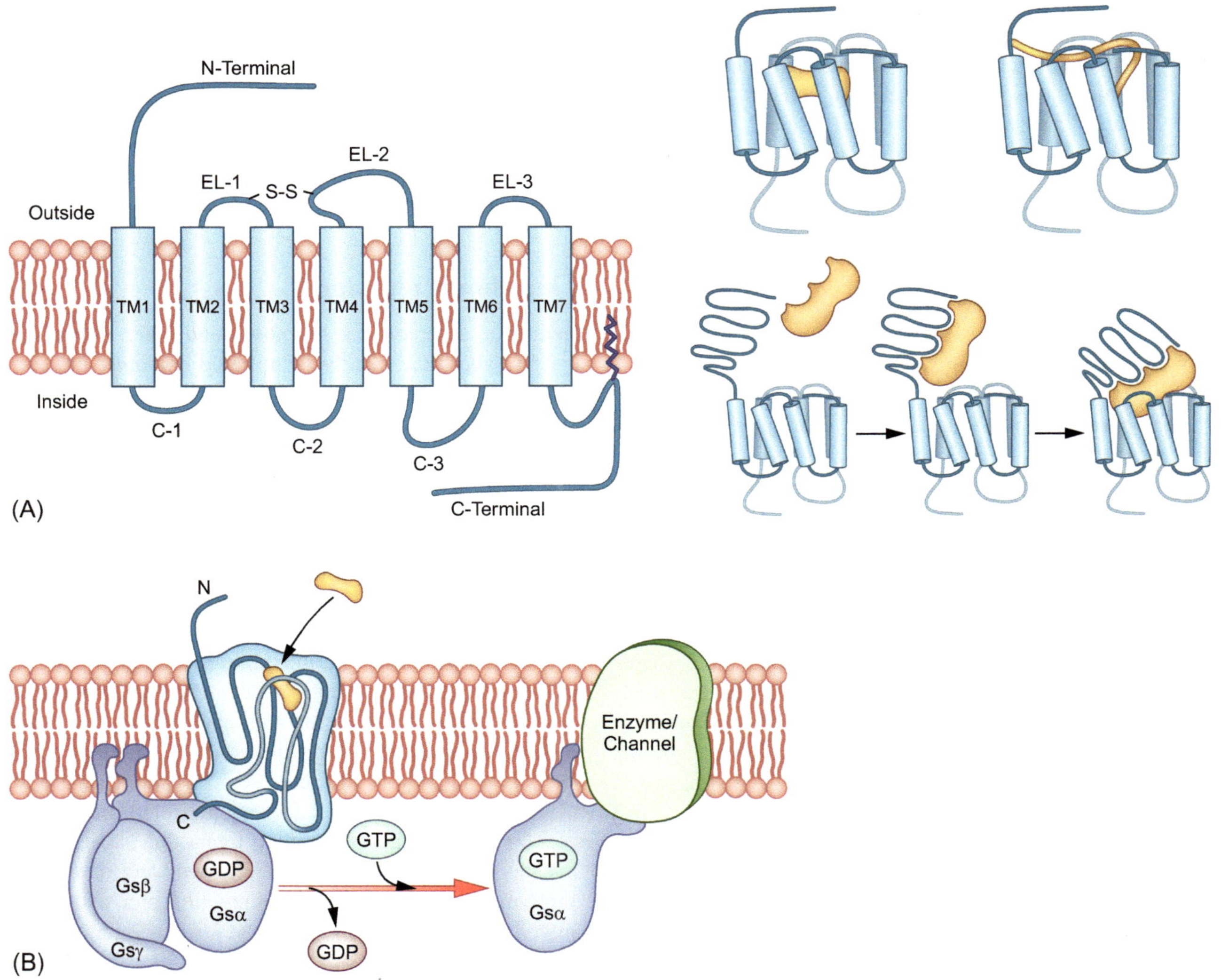

FIGURE 1.11 GPCRs. (A) *General structure of GPCRs.* The GPCRs comprise a large family of proteins that share the main structural features shown in the left side of panel (A). The predominant characteristic of these proteins is the arrangement of their single polypeptide chains into seven membrane spanning regions, creating three extracellular and three intracellular loops. One or more sites on the intracellular C-terminal portion of the cell may be palmitoylated, which plays a role in the receptor's position in the membrane. The right-hand side of panel (A) shows examples of the heterogeneity of the N-terminal portion of GPCRs, reflecting the diversity of ligands for these proteins. Top left: small molecules such as catecholamines or eicosanoids bind to a pocket within the membrane spanning helices; top right, small peptides are partially within a binding pocket but also interact with the extracellular portion of the receptor; bottom, large glycoproteins such as the gonadotrophins or growth factors have binding sites created by the structure of the extracellular portion of the receptor. (B) *Receptor interaction with G-protein.* Inactive G-proteins (left) consist of three subunits in a heterotrimer, α, β, and γ. Two of the subunits, α and γ, have lipid moieties binding them to the membrane and GDP is bound to the α-subunit. When a ligand binds to the receptor and activates it, GDP is replaced with GTP; the α-subunit dissociates from the trimer and moves through the membrane to a nearby protein, an enzyme or ion channel, for example, and activates it, initiating the biological response (see Fig. 1.20). *GDP*, Guanosine diphosphate; *GPCRs*, G protein–coupled receptors; *GTP*, guanosine triphosphate.

glycoprotein gonadotrophins, bind to a site within a longer, more elaborate N-terminus.

The coupling of a GPCR to a G-protein is illustrated in panel (B) of Fig. 1.11. G-proteins are composed of three subunits, α, β, and γ—that is, they are heterotrimeric. The complex is anchored to the membrane by lipid moieties on the α and γ subunits. The contact with the receptor occurs between the cytoplasmic carboxy terminal of the receptor and the α subunit. The α subunit also has a guanine nucleotide–binding site. In the absence of ligand activation, guanosine diphosphate (GDP) occupies this site, and the complex is inactive. When ligand is bound, guanosine triphosphate (GTP) replaces GDP, Gα dissociates from G$\beta\gamma$ and moves through the membrane to a nearby effector protein, such as an enzyme that produces a second messenger or an ion channel, which itself becomes activated upon binding of the α subunit. Gα has intrinsic GTPase activity, which may be aided by nearby GAP (GTPase acceleratory protein) proteins. Gα–GDP quickly finds and binds to a free G$\beta\gamma$ dimer and the inactive heterotrimer is reformed.

As will be discussed in more detail in Section III.A, the biological response that results from a ligand binding to a GPCR depends upon the G-protein attached to the receptor. The human genome encodes about 200 different G-proteins (proteins that bind guanosine nucleotides), a subclass of which are the heterotrimeric proteins ("large G-proteins") described previously.

2 Receptor tyrosine kinases

Receptor tyrosine kinases, or RTKs, are single membrane–spanning receptors and are defined by the presence of tyrosine kinase activity as the main cytoplasmic constituent and initiator of signal transduction. There are 58 receptor tyrosine kinases encoded in the human genome, several of which are important in hormone signaling. Fig. 1.12 illustrates some of the differences in structure seen in this type of receptor. In their monomeric forms, RTKs are single membrane–spanning receptors. However, these receptors dimerize upon the binding of one-ligand molecule, two-ligand molecules, or one-ligand dimer (see Chapter 17: Growth Factors for more details). In some cases, closely related receptors [e.g., epidermal growth factor (EGF) receptor and HER2] in the same cell may heterodimerize. The insulin and IGF-1 receptors, members of the same family, are an exception to this pattern. They exist as dimers of two hemireceptors, each consisting of two subunits, the extracellular α-subunit and the intracellular β-subunit, joined by disulfide bonds. A further set of disulfide bonds joins the two hemireceptors to form the dimerized receptor that then binds one molecule of ligand.

The N-terminal extracellular ligand-binding domains of RTKs consist of one or a few of about 20 structural motifs. In the examples of RTKs in Fig. 1.12, cysteine-rich regions appear in the EGF and insulin/IGF-1 families, whereas the FGFR (fibroblast growth factor receptor) family, along with several others not shown here, consists of a group of immunogammaglobulin-like domains. The carboxy terminals differ primarily in whether the tyrosine kinase catalytic domain is present as a contiguous sequence of amino acids or whether it is interrupted by a stretch of up to 100 non-TK amino acids, as in the insulin/IGF-1 receptors and FGFR. This is referred to as a split tyrosine kinase domain. The insertion has autophosphorylation sites, which suggests that it may be important in interacting with signal-transducing molecules (see Fig. 17.9).

C The nuclear receptor family

The nuclear receptors are a group of ancient evolutionarily related transcription factors. Sequencing of the human genome has revealed 48 members of this class, of which about half appear to be orphans, that is, no activity-modulating ligand has yet been identified for them. We will focus our attention on the 10 separately encoded proteins, activity of which is modulated by the hormones discussed in the following chapters.

The structural organization of the nuclear receptors for the classical steroid hormones, 1α,25$(OH)_2$-vitamin D_3, TH, and retinoic acid is shown in Fig. 1.13A. Each of these proteins functions as a DNA-binding protein, regulating, in a ligand-dependent (and sometimes ligand-independent) way, the expression of genes related to the biological response of the hormone (described in Section III.C).

The receptors for TH (TR) and 1α,25$(OH)_2$-vitamin D_3 (VDR) are typically found in the nucleus of target cells where they (especially TR) may be bound to corepressor molecules which suppress DNA transcription (see Chapter 5: Thyroid Hormones). These receptors form heterodimers with retinoic X receptor (RXR) (also in the nucleus) to bind to specific DNA sequences. The receptors for cortisol, the glucocorticoid receptor (GR) and aldosterone, the mineralocorticoid receptor (MR) are in the cytoplasm prior to ligand binding, where they are bound to chaperone proteins (heat-shock proteins) that maintain them in an inactive state. Upon ligand binding, they undergo nuclear translocation and homodimerization prior to binding to specific DNA sequences. The receptors for progesterone (PR), androgens (AR), and estrogens (ER) also form homodimers and may either be in the nucleus prior to ligand binding or travel between the two compartments.

As shown in Fig. 1.13A, the nuclear receptors consist of a single polypeptide chain divided into six domains. The N-terminal sequence is highly variable in both

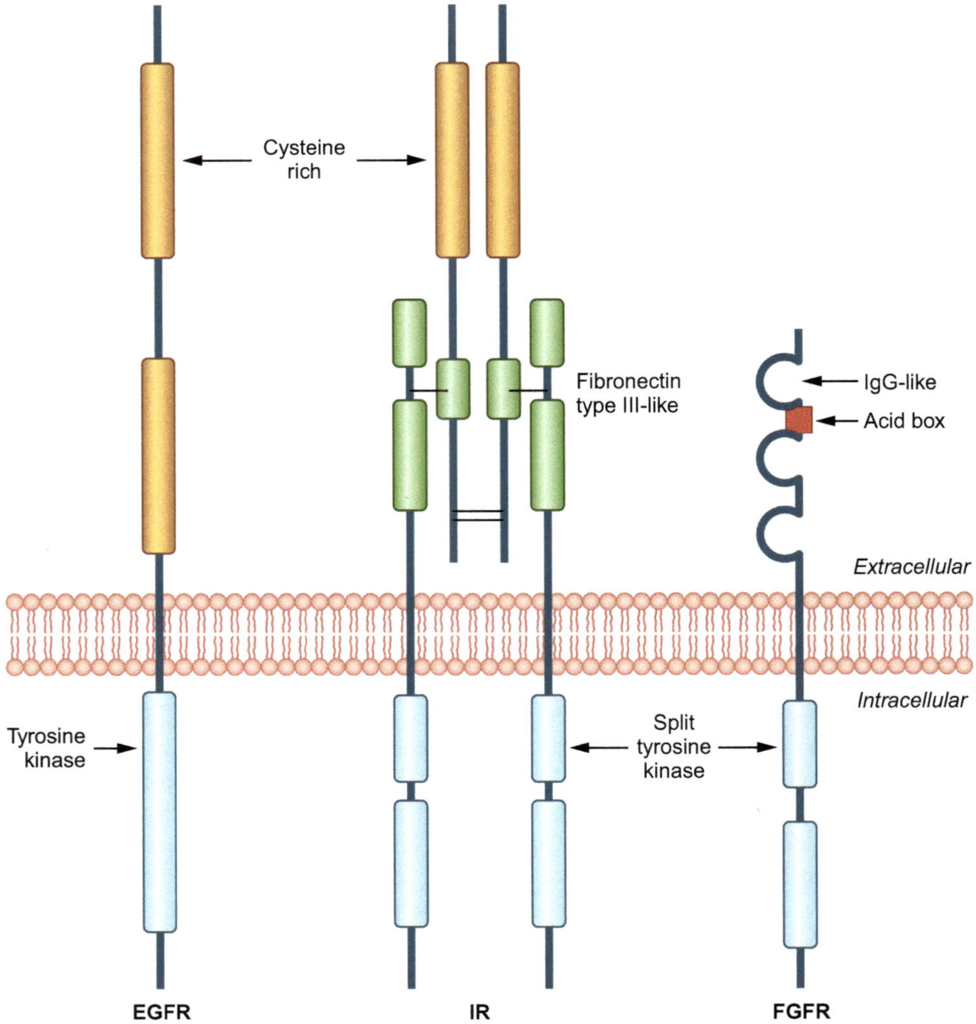

FIGURE 1.12 General structure of receptor tyrosine kinases. The structural features of the RTK are illustrated in three examples: the EGFR; the Ins/IGF-1 receptor; and the FGFR. The RTKs are single membrane spanning proteins with a variable extracellular N-terminal region and a cytoplasmic carboxyl portion that contains the catalytic activity to phosphorylate tyrosines (tyrosine kinase; *blue*) in itself (autophosphorylation) or in nearby proteins. In some receptors (e.g., insulin, FGF), the catalytic domain is split by a nontyrosine kinase sequence. Examples of N-terminal region motifs include cysteine-rich sequences (*gold*), fibronectin type III-like regions (*green*), a series of IgG (*blue*) regions, and the acid box (*red*) as seen between the first and second IgG sequences in the FGF receptor. Most RTKs are monomers that dimerize upon ligand binding. The members of the insulin/IGF-1 family, however, exist as a dimer of disulfide-linked monomers, each consisting of two subunits. See Chapter 17, Growth Factors, for more details on the structures of these and related receptors and their interactions with their ligands. *EGFR*, Epidermal growth factor receptor; *FGF*, fibroblast growth factor; *FGFR*, fibroblast growth factor receptor; *IgG*, immunogammaglobulin; *RTK*, receptor tyrosine kinases.

sequence and length, accounting for the overall difference in size of the receptor proteins. It contains one DNA-binding sequence, termed the AF-1 domain, which not only can regulate gene transcription independently of ligand binding but can also be controlled by ligand binding. This section of the protein can undergo posttranslational modification such as phosphorylation and may also interact with the C-terminal to affect the three-dimensional structure of the protein.

Functionally, the two most critical portions of the receptors are the DNA-binding and ligand-binding (E/F) domains. The three-dimensional organization of both is shown in Fig. 1.13B. The C-domain, on the left, is a highly conserved sequence encoding a two zinc finger motif which is widely used in transcription factors. As seen in Fig. 1.14, the "fingers" resulting from the coordination of a Zn^{2+} atom by four cysteine residues are the contact sites for DNA-binding, one of which carries the amino acid sequence for recognizing the correct specific DNA site (hormone response element, HRE; see Section III.B) for binding.

The E/F domain is composed of 12 α-helices (Fig. 1.13) arranged in approximately the same three-dimensional structure for all the nuclear receptors. It is less highly conserved than the C-domain as it differs

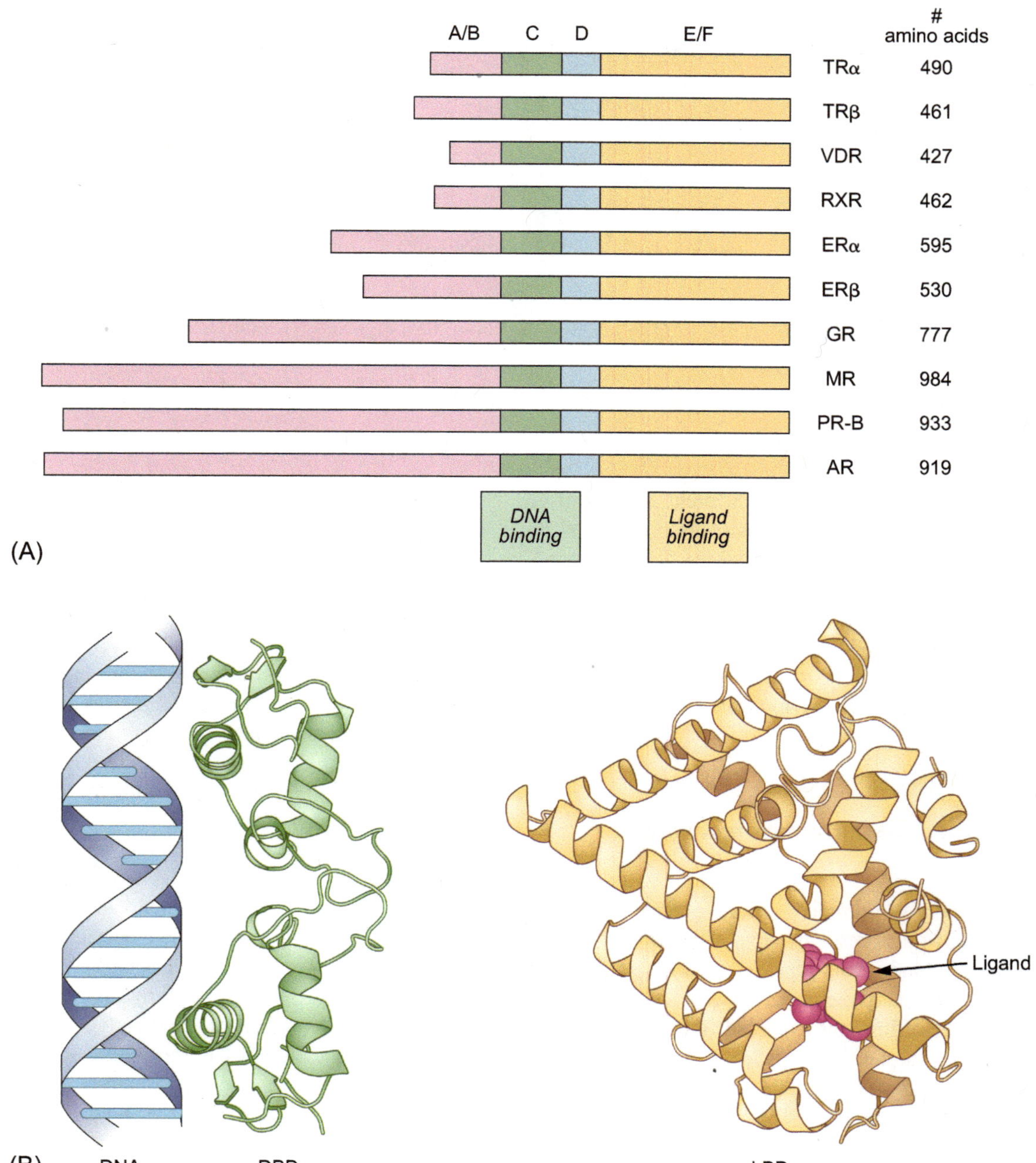

FIGURE 1.13 Nuclear receptor structure. (A) *Primary structural organization*. Shown are the structural features of nuclear receptors for thyroid hormone (TRα and TRβ), 1,25-dihydroxyvitamin D_3 (VDR), the retinoic acid (RXR), estrogen (ERα and ERβ), cortisol (GR), aldosterone (MR), progesterone (PR B), and testosterone/dihydrotestosterone (AR). These receptors share a highly conserved DNA-binding domain (C, *green*) and a short nonconserved region (D, *blue*), which serve as a hinge between the N-terminal and C-terminal portions of the molecule. The ligand-binding domain (*gold*) is less conserved than the DNA-binding domain but is approximately the same size and adopts approximately the same three-dimensional structure in all the receptors. The difference in size between the receptor proteins is the highly variable N-terminal A/B domain (*pink*). Two elements that are necessary for the control of gene transcription, termed activation functions, exist, AF-1 in the A/B domain and AF-2 in the E/F domain. (B) *Three-dimensional structure of the DNA- and ligand-binding domains of the nuclear receptors*. For the DNA-binding domain (left), the interaction of the ER homodimer with DNA is shown. Each DNA-binding site consists of two loops of DNA known as zinc fingers and described more fully in Fig. 1.14. Recognition of the specific DNA sequence to be bound lies within the CI zinc finger (closest to the DNA), whereas CII is involved with receptor dimerization. The ligand-binding domains of the nuclear receptors are less conserved than the DNA-binding domain, but they share many common features. There are 12 α-helices arranged in three layers. The ligand-binding pocket is within the more variable region. In addition to ligand binding, there are sites for a dimerization surface, a coregulator-binding surface, and ligand-dependent transcriptional activation moiety, AF-2. *RXR*, retinoic X receptor; *VDR*, vitamin D_3 receptor.

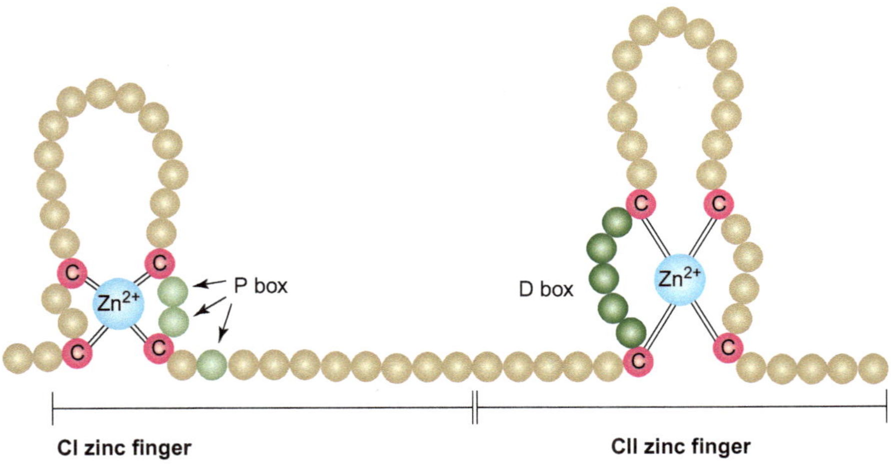

FIGURE 1.14 Steroid hormone receptor zinc fingers. The amino acids in the DNA-binding domain (see Fig. 1.13) of a steroid hormone nuclear receptor are represented by circles. The coordination of a Zn^{2+} atom (*blue*) by four cysteines (*pink*) causes the formation of a loop. One of these, CI, which contains the P-box (*light green*), is involved in binding to the specific DNA-binding site (hormone response element, see Fig. 1.25) and discriminating between closely related sites for different hormones. The D-box (*dark green*) in the second zinc finger, CII, plays a role in receptor dimerization.

among receptors to accommodate different ligands in the ligand-binding pocket. This region also contains the dimerization interface for the formation of both homodimers and heterodimers. Finally, the ligand-dependent transcription function, AF-2, resides in the E/F domain, specifically the C-terminal helix 12. This small helix has a great deal of ligand-dependent flexibility and its position determines the access of nuclear coregulators to the receptor protein (see Section III.C).

D Measurement of hormone–receptor interactions

For every category of hormone (steroid, peptide, or protein) to produce a biological response, it is essential that the hormone physically binds to its partner (cognate) receptor. The receptor is always a protein that has a specific binding domain with high affinity for its ligand. The protein receptor either folds itself around the ligand or the protein's most stable structure/shape is one that is formed as a ligand attempts to bind to its receptor. The binding of a hormone to its receptor is never a covalent linkage; such a bond would be equivalent to turning a light on in a room and forgetting to ever turn it off. The interactions between the receptor and its cognate hormone are facilitated by the formation of a cluster of noncovalent interactions. They can be electrostatic (+ vs. −) or hydrogen bonding (between a hydrogen acceptor and a hydrogen donor) and/or hydrophobic interactions.

$$\text{Hormone} + \text{Receptor} \rightleftharpoons \text{Receptor} - \text{Ligand complex}$$

$$[H] + [R] \underset{k_{-1}}{\overset{k_{+1}}{\rightleftarrows}} [\text{HR}]$$

This reaction can be converted into an equation that mathematically describes the dynamic movement of the forward reaction and the reverse reaction. This is the heart of the Scatchard analysis.

Historically, the principal technique for studying the quantitative interaction of hormones with their receptors in vitro is Scatchard saturation analysis. The technique is dependent upon access to high specific activity radioactive preparations of the hormone under study. Steroid hormones and proteins or peptides can be chemically synthesized to incorporate radioactive carbon (^{14}C; half-life = 5730 years), and/or radioactive hydrogen (^{3}H; half-life of 12.3 years) to make the hormones radioactive for long intervals of time. Neither ^{14}C nor ^{3}H is hazardous to health. There is no radioactive form of oxygen for carboxyl and hydroxyl groups.

Since the cognate receptor has a highly specific ligand-binding domain, under usual incubation conditions, the hormone–receptor complex is formed rapidly, within several minutes at room temperature. The association of H and R to yield HR and the dissociation of HR into H and R are readily reversible processes; that is, it is a dynamic equilibrium since the hormone does not become covalently bound to the receptor. Thus the equilibrium can be expressed in terms of the association constant, K_a, which is mathematically equivalent to 1/dissociation constant (K_d). The individual rate constants k_{+1} and k_{-1} numerically describe the rates of the forward (on-rate) and backward (off-rate) reactions, respectively, as written in the next equation:

$$K_a = \frac{[\text{RH}]}{[\text{H}][\text{R}]} = \frac{k_{+1}}{k_{-1}} = \frac{1}{K_d} \tag{1.1}$$

As shown in Fig. 1.15, hormone bound to receptor is corrected for nonspecific binding of the hormone to other components in the assay. This can be measured conveniently, if the hormone is radiolabeled, by measuring the radioactive receptor without (*brown line* in Fig. 1.15) and with the addition of an excess (100–1000 times) of unlabeled hormone (*blue line*). The excess unlabeled hormone displaces the high-affinity hormone-binding sites but not the low-affinity nonspecific binding sites. Thus when the "radioactive plus

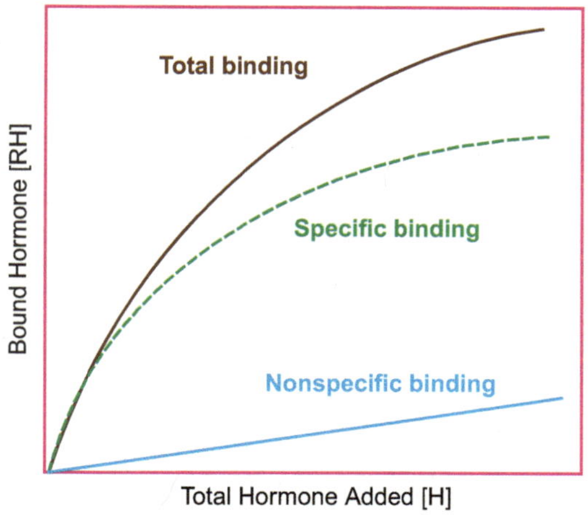

FIGURE 1.15 A typical plot showing the concentration dependency for the hormone [H] under experimental evaluation. The horizontal axis shows increasing concentrations of [H] that generate the data shown on the vertical axis of the increasing amounts of bound hormone [RH]. There are two categories of bound hormone; the brown solid line shows "total binding" and the green dashed line shows the "specific binding." "Total binding" represents the sum of "nonspecific binding" (*bottom solid blue line*) plus "specific binding." The "specific binding" value comes to a horizontal asymptote because there is only a limited amount of receptor. It is crucial to the success of the experiment that there is precisely the same amount of unoccupied receptor in all samples at the start of the experiment. The bottom line labeled "nonspecific binding" increases linearly with the amount of H that was added.

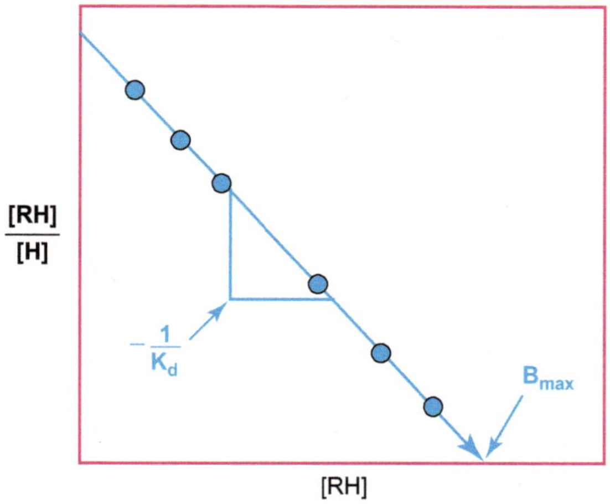

FIGURE 1.16 Typical plot of a Scatchard analysis of specific binding of hormone to its receptor. The vertical axis shows the ratio of bound ligand [RH] to free hormone [H].

nonradioactive" curve is subtracted from the "radioactive" curve, the resulting curve (*green dashed line*) represents specific binding to receptor. This is of critical importance when the receptor is measured in a system containing other proteins. As an approximation, 20 times the K_d value of hormone is usually enough to saturate the receptor.

The Scatchard plot of bound/free = [RH]/[H] on the ordinate versus bound [RH] on the abscissa yields a straight line, as shown in Fig. 1.16. The goal is to determine accurately the numerical value of the dissociation constant, K_d. Most measurements of K_d are made by using the Scatchard analysis, which is a manipulation of the equilibrium equation. The equation can be developed by a number of routes but can be envisioned from mass action analysis of the preceding equation. At equilibrium the total possible number of binding sites (B_{max}) is equal to the unbound plus the bound sites, so that $B_{max} = [R] + [RH]$, and the unbound sites (R) will be equal to $[R] = B_{max} - [RH]$. To consider the sites left unbound in the reaction, the equilibrium equation becomes

$$K_a = \frac{[RH]}{[H](B_{max} - [RH])} \tag{1.2}$$

Thus

$$\begin{aligned} \frac{\text{Bound}}{\text{Free}} &= \frac{[RH]}{[H]} = K_a(B_{max} - [RH]) \\ &= \frac{1}{K_d}(B_{max} - [RH]) \end{aligned} \tag{1.3}$$

When the line in Fig. 1.16 is extrapolated to the abscissa, the intercept gives the value of B_{max} (the total number of specific receptor-binding sites). The slope of the straight line is $-K_a$ or $-1/K_d$.

The K_d values for steroid receptors typically fall in the range of 10^{-10}–10^{-8} M. This very low number is a reflection of how far to the right the reaction between H and R to form RH lies. That is to say, in a mixture of H and R, there is virtually no free hormone and most exists as HR. Also, the low K_d value is a testament to the three-dimensional organization of the ligand-binding domain, which very effectively captures the ligand. These interactions are generally marked by a high degree of specificity so that both parameters describe interactions of a high order, indicating the uniqueness of receptors and the selectivity of signal reception.

From these analyses, information is obtained about the K_d and the maximal number of high-affinity receptor sites (receptor number) in the system. Fig. 1.17 illustrates the regulation of steroid receptor performance for binding the same ligand to three related receptors that have similar but not identical amino acid sequences. The three receptors' K_ds for binding curves *K*, *L*, and *M* are 10^{-10}, 10^{-9} and 10^{-8} M, respectively. If the prevailing hormone concentration bathing the target cell is 10^{-9} M, then depending upon the affinity of the receptor for its ligand, the level of receptor occupancy can vary from 80% (curve *K*) to 50% (curve *L*) to 20% (curve *M*). These differences in receptor performance very likely reflect differences in the amino acid sequence in the interior of the ligand binding domain.

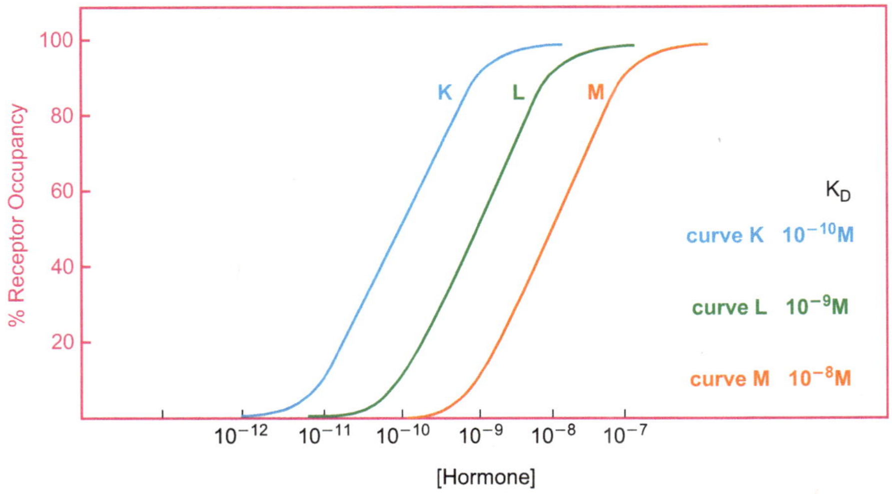

FIGURE 1.17 Regulation of receptor performance, by determination of the K_d. The horizontal line shows the concentration dependency of three different hormones (K, L, and M), which have different affinities for binding to the receptor from 10^{-11} to 10^{-8} M. The specific K_d values for K, L, and M are listed in the right side of the box. It is clear that hormone K has the highest affinity for the binding to the receptor and that hormone M has the lowest affinity for the receptor. This result was obtained by using the Scatchard analysis approach.

III Mechanisms of hormone action

A Cell signaling by membrane receptors

In Section II the structures of some membrane receptors were described, along with the fact that to deliver its message to the cell, the hormone has to cause the receptor to change so that it generates a change within the cell. In this section, we will consider what some of those intracellular changes are and how they initiate a series of events that will bring about a change in the cell's activity, the biological response to the hormone. The realm of intracellular signaling is vast and it will be necessary for us to focus our attention on the portions of systems that are encountered most frequently in the study of hormones.

1 G protein–coupled receptors

As discussed in Section II.B, a heterotrimeric G-protein that has GDP bound to its α-subunit is inactive. As illustrated in Fig. 1.11, when a ligand binds to a GPCR, the receptor changes conformation and interacts with an adjacent G-protein in such a way that the latter exchanges its GDP for a GTP, thus activating the α-subunit. That is, the receptor acts as the guanine nucleotide exchange factor (GEF) that activates this particular G-protein. Different activated α-subunits have different activities. The human genome encodes 16 different α-subunits, along with 5β- and 14γ-subunits. β/γ-subunits appear to be mostly interchangeable with regard to their interactions with α-subunits, but it is now also recognized that these two proteins have some activities of their own, either as a dimer or individually. We will only be concerned with the activities of the α-subunits in the following discussion. Events downstream of the signaling pathways are not presented in detail here but do appear in the chapters dealing with specific hormonal systems.

Fig. 1.19 shows the outcome of receptor-initiated G-protein activation in the case of the three types of α-subunits that will be encountered most frequently in this book. On the left are $G_s\alpha$ and $G_i\alpha$. These proteins interact with the membrane enzyme adenylyl cyclase, which converts one molecule of ATP into one of cyclic AMP (cAMP) (adenosine-3′-5′-cyclic monophosphate) the first intracellular second message to be discovered in the 1970s. For example, Fig. 1.18A shows the interaction of adrenaline binding with the adrenergic receptor (containing G_α, G_β, and G_γ) and its interaction with adenylyl (adenylate) cyclase. Cyclic AMP binds to the regulatory subunits of protein kinase A (PKA) (cyclic AMP-dependent protein kinase) causing the catalytic subunits to become active. At this point the pathway can go in one of several directions, depending on the cell type. All outcomes are dependent on the phosphorylation of protein substrates at specific serine or threonine residues by the activated PKA. Four of these are shown in Fig. 1.19, illustrating the diversity of possible responses to this second messenger, including changes in ion transport, in gene transcription, and in the activity of existing enzyme proteins. The activity of target proteins might be either increased or decreased by phosphorylation. An extracellular signaling agent that triggers the activation of a $G_i\alpha$ protein will have the opposite effect on a pathway that is stimulated by $G_s\alpha$.

On the right side of Fig. 1.19, $G_q\alpha$ is shown interacting with phospholipase C. This enzyme catalyzes the reaction shown in Fig. 1.20, the hydrolysis of the membrane lipid phosphatidylinositol 4,5-bis-phosphate into IP_3 (inositol 1,4,5-triphosphate) and diacylglycerol (DAG), each of which are second messengers. DAG is necessary for the activation of protein kinase C and IP_3 activates the release of Ca^{2+} from intracellular stores in the endoplasmic reticulum. This divalent cation acts as yet another

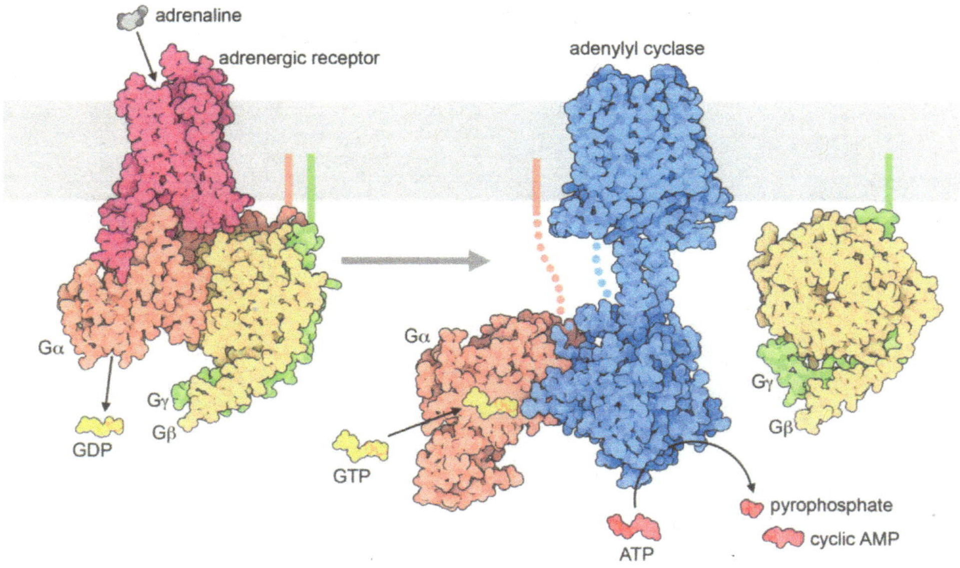

FIGURE 1.18 Signaling with G-proteins. Hormones like adrenaline bind to a GPCR receptor *(left)* which binds to a heterotrimeric G-protein and releases GDP. Then the G-protein separates into two pieces and the G_α-subunit binds to GTP and activates adenylyl cyclase *(right)*. *GDP*, Guanosine diphosphate; *GPCR*, G protein–containing receptor; *GTP*, guanosine triphosphate. *Reproduced from https://pdb101.rcsb.org/motm/58.*

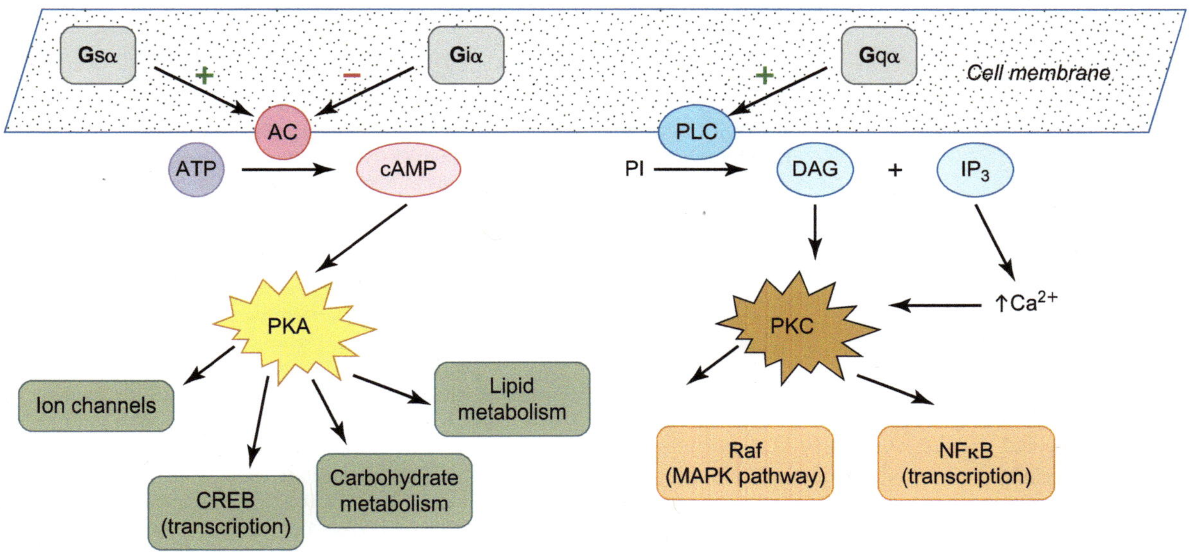

FIGURE 1.19 Hormonal signaling by GPCRs. Two of the most common signaling pathways used by GPCRs are illustrated, those initiated by $G_s\alpha$, $G_i\alpha$, and $G_q\alpha$. These three G-proteins are each the most widely distributed members of the three subfamilies of G-proteins which bear the names $G_s\alpha$, $G_{i/o}\alpha$, and $G_{q/11}\alpha$, respectively. Sometimes the abbreviations of the proteins will have a different order of the α and the subfamily, s, i, or q. In Fig. 1.11 the G-protein molecules are shown after being released from their βγ partners through interaction with a GPCR (see Fig. 1.11). As shown in this figure, $G_s\alpha$ and $G_i\alpha$ stimulate or inhibit, respectively, adenylyl cyclase in the plasma membrane, bringing about an increase or decrease in this second messenger within the cell. When levels of cAMP rise, PKA (cyclic AMP-dependent protein kinase) is activated. Depending on the cell type, one or more steps of activation (or in some cases inactivation) ensue, some of which may involve additional phosphorylation events. Examples of these include the opening of ion channels in the cell membrane, phosphorylation of the transcription factor CREB protein, activation or inhibition of enzymatic steps in the metabolism of glycolysis or lipids. On the right side of the figure, Gqα, also released from a receptor–G protein complex, activates phospholipase C, which catalyzes the release of two second messengers, inositol triphosphate (IP_3) and DAG. See Fig. 1.20 for the details of this reaction. DAG activates protein kinase C, which can activate one of several targets, in this example Raf, which allows entry to the MAPK pathway (see Fig. 1.22) or the transcription factor NFκB to affect gene transcription. *cAMP*, Cyclic AMP; *CREB*, cyclic AMP response element binding protein; *DAG*, diacylglycerol; *MAPK*, mitogen-activated protein kinase; *NFκB*, nuclear factor-κB; *PKA*, protein kinase A.

second messenger with many possible actions in the cell, including the stimulation, along with DAG, of protein kinase C. Targets for protein kinase C include the phosphorylation and augmentation of enzymes in the mitogen-activated protein (MAP) kinase pathway (see Fig. 1.23 in Section III.B) and the phosphorylation of the nuclear transcription factor nuclear factor-κB. The increased intracellular Ca^{2+} may activate exocytosis,

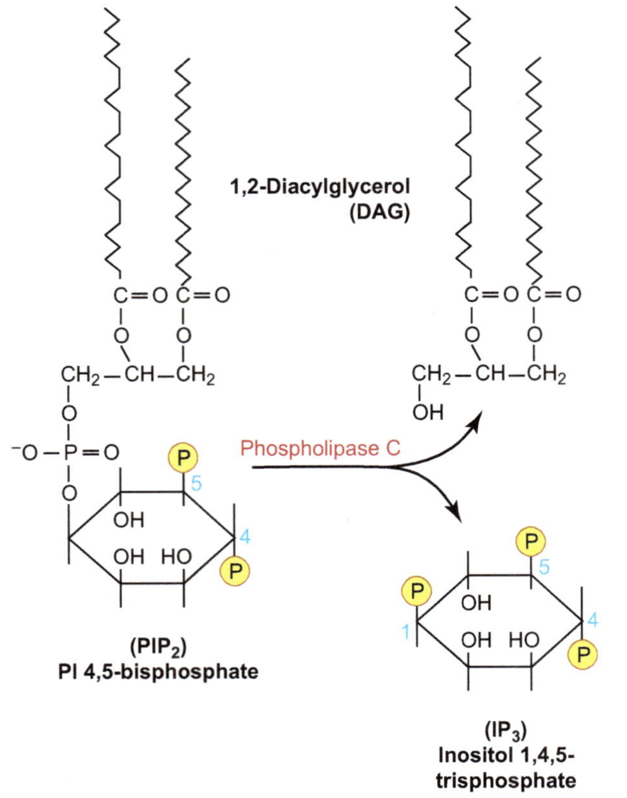

FIGURE 1.20 Phospholipase C reaction. The cleavage by phospholipase C of phosphatidylinositol 4,5-bisphosphonate into DAG and inositol 1,4,5-triphosphate (IP_3) is shown. *DAG*, Diacylglycerol.

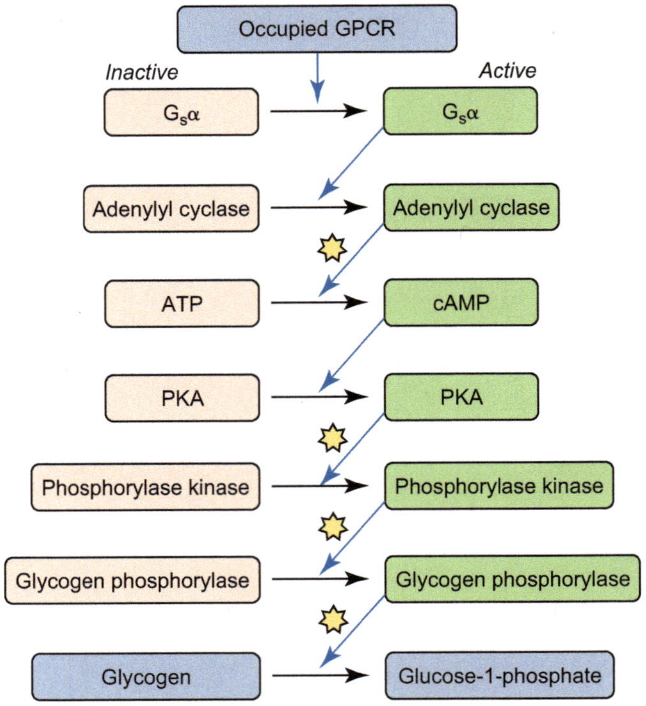

FIGURE 1.21 Amplification of hormonal signaling. The cascade of events from the binding of epinephrine to its G protein–coupled receptor in skeletal muscle to the breakdown of glycogen is shown as one example of the amplification of hormonal signaling. On the left are molecules that are inactive in the cascade prior to the initial binding event and on the right are the active forms of the molecules. As shown in Figs. 1.11 and 1.19, $G_s\alpha$ is activated by hormone binding and, in turn, activates adenylyl cyclase in the membrane, generating the second messenger cAMP. PKA (cyclic AMP-dependent protein kinase) phosphorylates and activates glycogen phosphorylase kinase (phosphorylase kinase), which phosphorylates and activates glycogen phosphorylase. The phosphorylation of glycogen initiates its breakdown into, ultimately, glucose, which is used for the energy needs of the muscle cell. At the steps with a gold star, one activated molecule may generate 10–100 (or more) active molecules. Thus the signal from one occupied GPCR can be amplified several fold, allowing for a large rapid response to a small signal. *cAMP*, Cyclic AMP; *GPCR*, G protein–coupled receptor; *PKA*, protein kinase A.

the basis of the Ca^{2+} secretion coupling mechanism often associated with the secretion of peptide hormones by their secretory glands, for example, pituitary hormones and insulin.

Ca^{2+} may also bind to calmodulin, a 17-kDa calcium-sensing protein. Calmodulin is present in the cytoplasmic compartment of virtually all cells of higher organisms. It binds four Ca^{2+} ions tightly ($K_d \cong 10^{-8}$ M) and then undergoes a conformational change so that it interacts with a number of Ca^{2+}-regulated proteins. These include Ca^{2+}–calmodulin-dependent protein kinases, Ca^{2+}-ATPase (a Ca^{2+} pump), myosin light chain kinase, and phosphatidylinositol-3-kinase, to name a few.

Regardless of the type of activated G-protein α-subunit, the GTP bound to the α-subunit is soon hydrolyzed to GDP. The α-subunit is inactivated and binds again to the βγ-subunits, turning off the signal. The speed with which the signal is turned off may be modulated by nearby proteins called GTPase activating (or accelerating) proteins, or GAPs. These proteins can, in turn, be regulated both positively and negatively by other signaling systems. This is an example of one type of cross talk between signaling pathways.

One important feature of signaling pathways initiated by membrane receptors, their amplification, or cascade property is exemplified by one of the first of the pathways to be elucidated, the cAMP-mediated control of glycogen breakdown in the skeletal muscle. As depicted in Fig. 1.21, several steps in the cascade (indicated by the stars) allow for, at the very minimum, amplification of 1–2 orders of magnitude because of the catalytic nature of the event. Thus a rapid robust response can be obtained from a fairly small change in the concentration of circulating hormone. This is a general feature of membrane-initiated signaling events that involve one or more catalytic events. A higher order cascade is seen in hormonal systems that involve the CNS, hypothalamus, pituitary, and peripheral target organ, as described in Fig. 3.1.

2 Receptor tyrosine kinases

Fig. 1.22 illustrates three of the main signaling pathways used by receptor tyrosine kinases. On the right is the MAP kinase cascade with three tiers of activation

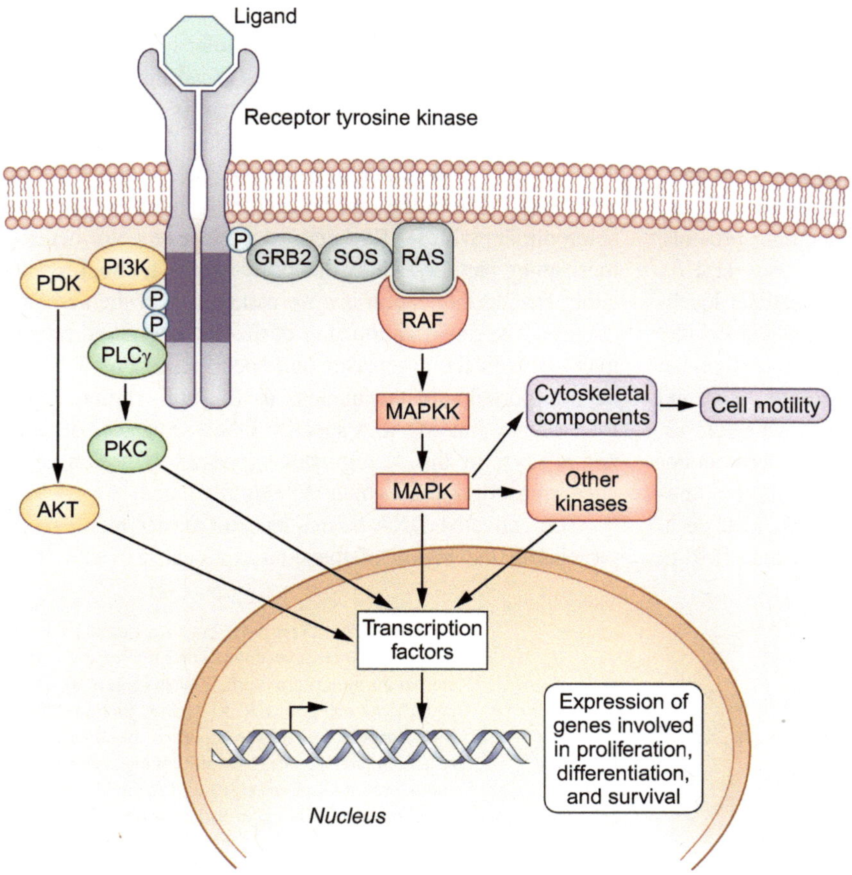

FIGURE 1.22 RTK signaling. The active form of an RTK is a dimer that is formed either before or as the ligand binds. Such an activated form is depicted here (*purple*; see Fig. 1.12 for some specific RTK structures). The extracellular ligand binding/dimerization brings about a conformational change in the receptor protein that conveys the message to the intracellular portion of the receptor. Tyrosine kinase (*dark purple*) is activated and autophosphorylates the cytoplasmic portion of the receptor (*blue circles*). This leads to the interaction of the receptor with other intracellular transducing enzymes such as phospholipase C, leading to the activation of PKC (see Fig. 1.19) or PI3 kinase which phosphorylates and activates AKT. Protein kinases B and C phosphorylate and activate proteins involved in the downstream effects of the hormone, including transcription factors. On the right is shown the activation of the MAP kinase pathway through the adaptor protein GRB2, the guanine nucleotide exchange factor SOS, and the activation of the small G-protein, Ras. This leads to the cascade of the activation of Raf (MAPKKK), MAPKK, and MAP kinase (ERK1/2), which can phosphorylate cytoplasmic proteins and nuclear transcription factors. *AKT*, Protein kinase B; *MAP*, mitogen-activated protein; *MAPKK*, MAP kinase kinase; *PDK*, phosphoinositide-dependent kinase; *PI3K*, phosphatidylinositol-3-kinase; *PKC*, protein kinase C; *SOS*, son of seven less.

beginning with that of the kinase Raf (MAPKKK) by a small G-protein, Ras. Ras, in turn, is activated by the GEF, SOS, recruited to the phosphorylated RTK by the adaptor protein GRB2. MAP kinase has many possible target proteins to alter both events in the cytoplasm and transcription in the nucleus. The exact signaling pathway used will depend on the ligand and the availability of pathway components in the cell being stimulated.

The phosphorylation of the cytoplasmic domain of the RTKs can also activate the PLCγ/protein kinase C pathway described in Fig. 1.19. In addition, phosphatidylinositol (PI) 3-kinase adds a phosphate group to the 3-position of phosphatidylinositol-4,5-bis-phosphonate, resulting in the triphosphorylated polar head group PI-3,4,5-triphosphate, which activates phosphoinositide-dependent kinase. This protein kinase activates protein kinase B, also known as protein kinase B, which phosphorylates nuclear transcription factors.

Note that these pathways share the same cascade property as that described in Fig. 1.21, with amplification at each catalytic phosphorylation step. Also as with the GPCR pathways described earlier, the RTK pathways are subject to termination to limit the timing and size of the signal. Protein phosphatases, often subject to their own regulatory systems, are available to remove the activating phosphate groups to provide this termination function.

Although the GPCR and RTK pathways are responsible for transmitting many of the hormonal signals to be encountered in this book, there are others that are less frequently encountered, but no less important to the system in which they play a role. These include the JAK/STAT pathway, which is described in connection with growth hormone in Chapter 3, The Hypothalamus and Anterior Pituitary; the guanylyl cyclase pathway that mediates the actions of atrial natriuretic factor as described in Chapter 10, Adrenal Corticoids; and the Smad pathway of TGFβ in Chapter 17, Growth Factors.

B Regulation of gene transcription by steroid hormones

The nuclear receptors for the steroid hormones, including $1,25(OH)_2D_3$, and for TH are highly regulated gene transcription factors. As with membrane hormone receptors, the regulation of the activity of these proteins occurs through the binding of ligand and a consequent conformational change that alters the activity of the receptor. Stimulation (activation) or repression of the transcription of specific genes ensues, altering the biological activities of the cell.

One way in which members of the group of hormone-regulated nuclear receptors differ is their location in the cell in the absence of ligand. Originally it appeared that the receptors for the cholesterol-derived classic steroid hormones were localized in the cytoplasm of the cell in the absence of ligand where they are bound to molecular chaperone proteins, HSP90 (heat-shock protein 90) and in some cases HSP70. Fig. 1.23 illustrates this difference between intracellular localization of the two types of receptors. It is now understood that this is not a rigid classification scheme; both forms of the estrogen receptor (ERα and ERβ), the androgen receptor, and one form of the progesterone receptor (see Chapter 13: Estrogens and Progestins) are found in the nucleus and/or travel freely between cytoplasm and nucleus in the unliganded state. Receptors for TH and $1,25(OH)_2D_3$ localize in the nucleus upon completion of synthesis where they bind to other proteins and to DNA. What all these receptors have in common is that, regardless of intracellular localization, in the unoccupied state they are maintained in an inactive state and binding of the ligand activates them and, if necessary, causes or permits their translocation to the nucleus.

The other feature that distinguishes receptors for TH and $1,25(OH)_2D_3$ from the other receptors is that these two form heterodimers with RXR, a general nuclear receptor heterodimerization partner, while the others form homodimers. In either case, dimerization is a necessary step of the activation process. The dimerization step of the glucocorticoid receptor may occur in the cytoplasm (not specified in Fig. 1.23) and be transported into the nucleus as the homodimer. This is followed by binding to a specific DNA sequence, often but not always, in the 5′ regulatory (promoter) region of the gene, transcription of which is being regulated.

The activated DNA-bound receptor dimer recruits other proteins to the region of the transcriptional start site. These

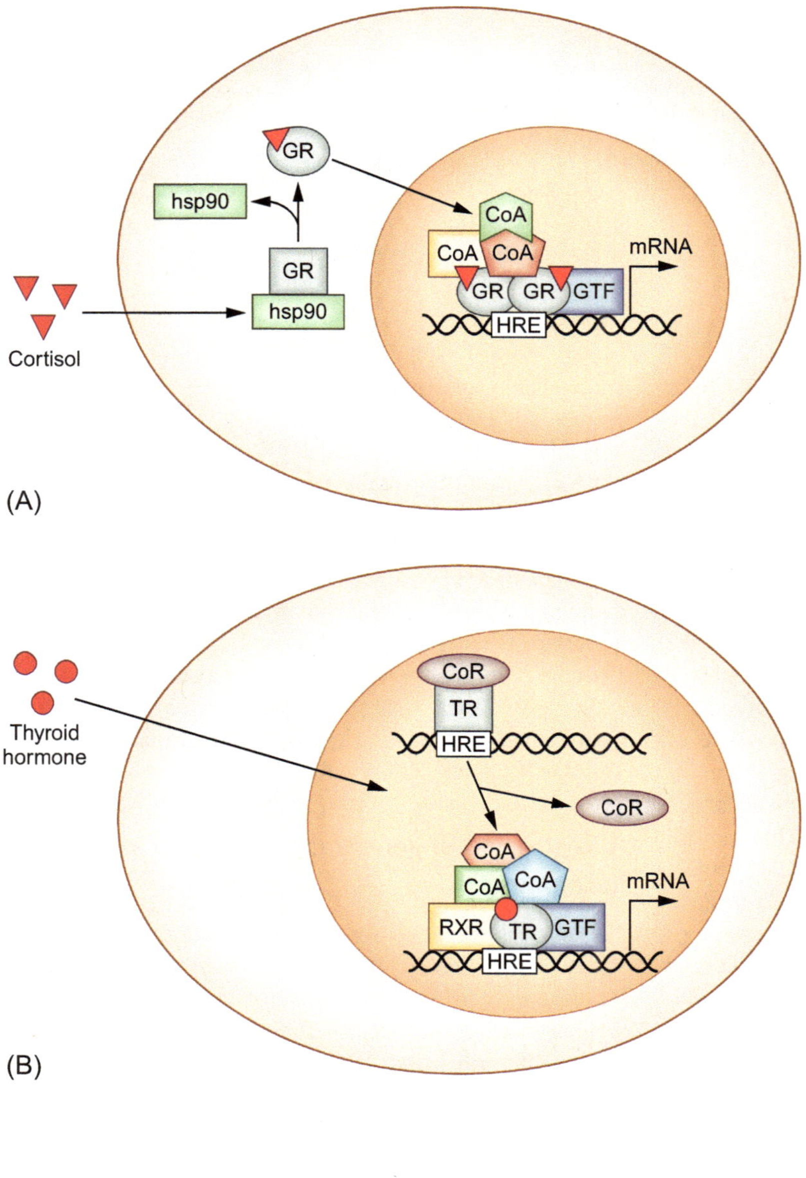

FIGURE 1.23 Transcriptional activation by nuclear receptors. Nuclear receptors for some of the classical steroid hormones are typified in this figure by the glucocorticoid receptor, GR, and its interaction with cortisol (panel A). In the absence of ligand, these receptors are in the cytosol complexed with heat shock proteins (*green*; HSP) that maintain them in an inactive state. The conformational change brought about by ligand binding causes the HSP to dissociate and the receptor translocates to the nucleus. GR, MR, ER, AR, and PR all forms heterodimers prior to DNA binding. Nuclear receptors for $1,25(OH)_2D_3$ (VDR) and TR are in the nucleus (panel B) prior to ligand binding and form heterodimers with the RXR. These are usually maintained in an inactive state by forming a complex with a corepressor (*purple*; CoR), which can bind to DNA and repress its transcription. In this case, ligand binding results in a conformational change that causes the corepressor to dissociate, activating the VDR/RXR heterodimer. The ligand-activated dimer of either type of nuclear receptor (hetero- or homodimer) binds to a specific sequence of DNA, a HRE (see Fig. 1.24). A variety of proteins, termed coactivators, are recruited to the complex to modify chromatin structure and recruit and stabilize the basal transcriptional machinery. This includes the GTF and DNA-dependent RNA polymerase. *CoR*, Corepressor proteins; *GTF*, general transcriptional factors; *HRE*, hormone response element; *HSP*, heat shock protein; *RXR*, retinoic X receptor; *TR*, thyroid hormone receptor; *VDR*, vitamin D_3 receptor.

proteins are termed coactivators. Among the first of these to be recognized were the steroid receptor coactivators, SRC-1, SRC-2, and SRC-3. These proteins bind to the receptor dimer and recruit other coactivator proteins, some of which have the enzymatic activity to modify the structure of chromatin. For example, histone acetylases (p300/CBP) neutralize the positively charged histones forming the nucleosomes (see Fig. 1.3), rendering the gene to be regulated more available for the binding of transcriptional proteins. Other histone modifications catalyzed by coactivators, such as methylation by arginine methyltransferases (PRMT1/CARM) and phosphorylation, play a role in chromatin modification. Still other coactivators recruited by SRC interact with and stabilize the basal transcriptional machinery, including basal transcriptional factors, and RNA polymerase II. Not all coactivators are present in the regulatory complex at the same time, but join, carry out their function, and are released to make room for the next set of coactivators. The combined result of these sequential activities is an increased rate of transcription of the gene that is being regulated.

Fig. 1.24 shows two typical DNA sequences of HREs for ER and VDR. HREs are composed of two hexameric half-site motifs. As can be seen in the figure, the directionality of both the relationship of the monomers to each other and the DNA half sites to each other differ according to homo- versus heterodimerization. Recall from Fig. 1.13 that it is the two zinc fingers of the receptor that are making contact with the DNA.

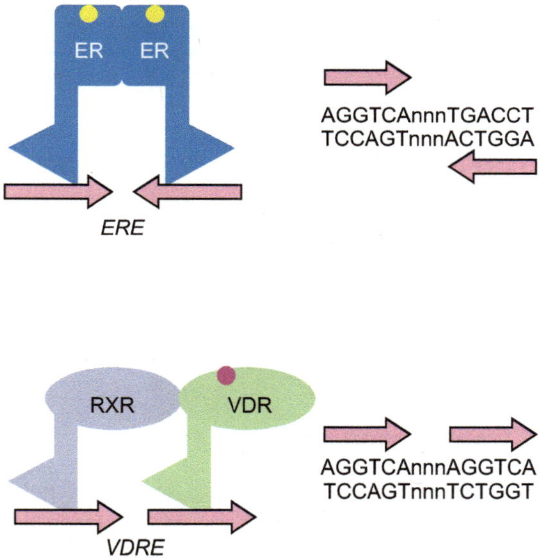

FIGURE 1.24 Nuclear receptor binding to DNA. The binding of nuclear hormone receptors to HRE within DNA is depicted. Monomers in homodimeric receptors are arranged as mirror images (head-to-head) and bind to hormone response elements that are inverted repeats, as shown in the example of an ERE at the right. The heterodimers that the receptors for $1,25(OH)_2D_3$ (VDR) and TR form with RXR are arranged sequentially (head-to-tail) and bind to response elements that are direct repeats as shown for VDR and the VDRE in the lower part of the figure. In response elements for homodimeric receptors, the number of nucleotides between the two repeats (*n*) is 3, while those for the heterodimeric receptors may vary from 3 to 5 (represented by *n*) with the response element. *ERE*, Estrogen response element; *HRE*, hormone response elements; *RXR*, retinoic X receptor; *TR*, thyroid hormone receptor; *VDR*, vitamin D_3 receptor; *VDRE*, vitamin D receptor element.

Nuclear receptors can also bring about repression of gene transcription. In some cases, notably that of the TH receptor (see Fig. 5.15), the unliganded receptor is complexed with the corepressor, silencing mediator for retinoid/TH receptors, which recruits proteins with histone deacetylase activity. The removal of acetyl groups from histones leads to their reassociation with DNA, reformation of the nucleosome, and inaccessibility of the DNA to other binding and activating proteins.

C Membrane-initiated actions by steroid hormones

It has become accepted that a fundamental activity of a steroid receptor with its bound ligand hormone working in a target cell is to either stimulate or repress the expression of genes. The steroid hormone $1\alpha,25(OH)_2D_3$, bound to its receptor, VDR, can activate as many 3000 genes (microarray data) from the ~total 22,000 human genes. It is likely that these numbers are similar for the other steroid hormones as well as TH.

It is also now understood that the hormones that bind nuclear receptors and initiate changes in gene expression also have receptor-mediated effects on processes independent of these nuclear actions. Table 1.1 compares the nuclear and nonnuclear (i.e., membrane initiated) actions of these hormones. Frequently, these rapid nongenomic responses are generated within 1–2 minutes to 15–45 minutes. This contrasts with genomic responses, which generally take several hours to days to be fully functional and apparent and which can be blocked by inhibitors of transcription and translation.

As one example, the steroid hormone $1\alpha,25(OH)_2D_3$ has been shown to initiate many biological responses via rapid response pathways that are independent of parallel genomic responses. Like the other steroid receptors, the VDR has a LBD and a DNA-binding domain that are essential to create a functional nuclear heterodimer with the retinoid X receptor (see Fig. 1.13). However, $1\alpha,25(OH)_2D_3$ is distinguished from the other steroid hormones in its conformational flexibility, as illustrated in Fig. 1.25.

Two characteristic shapes of $1\alpha,25(OH)_2D_3$, a bowl shape and a linear shape, are illustrated. It has been shown experimentally that one of these shapes, the bowl shape, generates genomic responses and the planar shape leads to nongenomic responses. These types of studies clearly support the conclusion that rapid nongenomic responses are functionally unique and different from genomic responses. Several examples of

TABLE 1.1 Nuclear receptors and involvement in genomic and/or rapid biological responses.

(Steroid receptor) and ligand	No. of aa receptor	Steroid nuclear receptors and gene transcription generate GR	Steroid membrane receptor generates RR
(Thyroid β) T_3	461	T_3 binding to its nuclear receptor, in target cells, stimulates dissociation of corepressors, recruitment of coactivators, etc. to complete a GR	T_3 activates PI3 kinases and MAP kinase *RR* pathways which can result in glucose uptake, Ca^{2+}-ATPase, Na^+/H^+ antiporter
(Vitamin D receptor) 1α,25$(OH)_2$-vitamin D_3	427	Both intestinal Ca^{2+} absorption and kidney Ca^{2+} reabsorption requires *GR* for the production of new CaBP	*RR* opening of Cl^- channels in osteoblasts and keratinocytes in 20 min; insulin secretion from β cells, MAP kinase activation in NB4 cells
(Estrogen receptor α) Estradiol	595	ERα *GR* are required for normal ovarian function	ERα activates PI3K and then AKt *RR* stimulates nitric oxide (NO)
(Estrogen receptor β) Estradiol	530	ERβ *GR* are required for ovulation and pregnancy	The cell membrane ERβ bound to caveola has been implicated in RR
(Glucocorticoid receptor) Cortisol	777	KO of the mouse GC receptor is lethal at time of birth	Cortisol stimulates PI3-kinase/Akt to activate in seconds NO release
(Mineralocorticoid receptor) Aldosterone	919	Mr *KO* mice die of Na^+ and H_2O deprivation	Aldosterone activates in 3–15 min the *RR* of Na^+/H^+ exchange in renal cells
(Progesterone receptor) Progesterone	933	The progesterone receptor participates in *GR* sexual differentiation determination	Progesterone stimulates *RR* within seconds to minutes, the acrosome reaction in spermatozoa
(Androgen receptor) Testosterone	919	*KO* of the AR male mouse causes the development of female genitalia	Activation of MAP kinase then activates the ERK pathway via *RR*

AKt, Protein kinase B; *CaBP*, calcium-binding proteins; *ERK*, extracellular signal-regulated kinase; *GR*, genomic responses; *MAP*, mitogen-activated protein; *PI3K*, phosphatidylinositol-3 kinase; *RR*, rapid responses. RR are not dependent on genomic responses. *KO*, knockout renders–affected genes inactive.

VDR-mediated rapid responses include the following: the rapid stimulation of intestinal Ca^{2+} absorption, in vivo; insulin secretion from rat pancreatic β cells; mouse osteoblastic cells in tissue culture that were patch clamped resulted in the rapid opening of voltage-gated chloride channels and also stimulation of exocytosis. And in Sertoli cells, located in seminiferous tubules, 1α,25$(OH)_2D_3$ stimulated the rapid opening of Ca^{2+} and Cl^- channels.

The subcellular location of VDR capable of producing both genomic and rapid responses is shown in Fig. 1.26. The majority of the VDR is in the nucleus (80%–85%) with the rest divided between the cytoplasm/ER (10%–15%) and caveoli (5%). It is the ~5% of caveolae-bound VDR that has been shown to be involved in voltage-gated opening of chloride channels and exocytosis.

The rapid membrane responses are not unique to the VDR receptors. Table 1.1 compares for the following receptors (thyroid β, estrogen α and β, glucocorticoid, cortisol, mineralocorticoid aldosterone, progesterone, androgen receptors, and the VDR). All the listed steroids have been shown to repeatedly participate in both activation and repression of genomic responses.

It has also been established that both the ERα and androgen receptors, in addition to the cell membrane caveolae, specifically bind to the cell's mitochondria. In summary, steroid hormones clearly can stand alone with their receptors and initiate rapid nongenomic responses, genomic effects in the cell nucleus that are mediated by second messenger pathway, and also genomic effects where the steroid receptors function as transcription factors.

IV Clinical aspects

In this chapter, some of the shared features of hormones, their receptors, and the cellular context in which they are found have been presented. In the following chapters, dealing with specific hormones, aspects of these subjects will be developed in more specific detail. In each of these chapters the final section will focus on one or more clinical aspects of the hormone(s) discussed in the chapter. It is this connection between the basic science underlying our knowledge of the structures, synthesis, and actions of the hormones and their receptors and the clinical outcomes of derangements in any of these that makes the study of this area so engaging.

Given the broad range of hormones that utilize GPCRs, it is perhaps not surprising that there are a number of mutations in them that result in either loss (hypo-) or gain (hyper-) of function phenotypes. A few examples

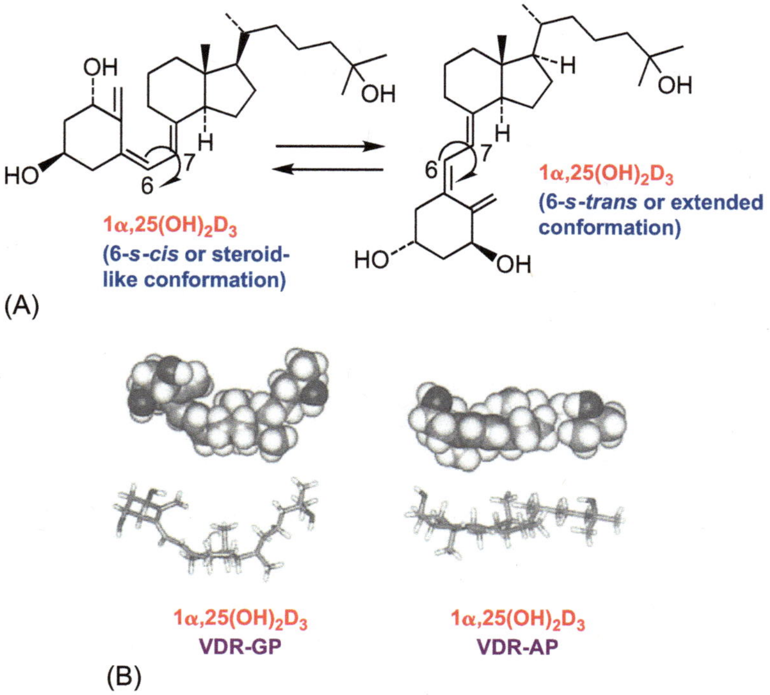

FIGURE 1.25 The steroid hormone $1\alpha,25(OH)_2D_3$ is conformationally flexible and presents different 3D shapes for binding to its receptor, the VDR. Panel (A) presents the structure of two shapes of $1\alpha,25(OH)_2D_3$. The 6-s-*cis* presentation shape on the left side is in rapid exchange with the 6-s-*trans* or extended shape. The change from *cis* to *trans* occurs by a 360-degree rotation around the 6−7 single bond (see the two arrows). This interchange process occurs at the rate of several million times per second. This extreme rotation allows the molecule to assume a very large family of three-dimensional shapes. Panel (B) has two extreme presentations of $1\alpha,25(OH)_2D_3$. The top row is a space-filling three-dimensional representation of 6-s-*cis*-shaped $1\alpha,25(OH)_2D_3$. The two dark images on the left end of the bowl shape represent the space-filling of the 1α-OH group (bottom surface) and the 3β-OH group (upper surface). At the right-hand end of the bowl shape is the space filling view of carbon-25 OH group. On the right side is the three-dimensional planar shape of $1\alpha,25(OH)_2D_3$. The right-hand three-dimensional ligand shows a 6-s-*trans* presentation; overall it has a planar shape. The left dark image is the 1α-OH group, while the 3β-OH group is hidden on the "back" surface. At the right end of the planar shape is the dark image of the 25-OH group. The bottom labels for panel (B) are the VDR-GP (for the VDR genomic pocket engaged in gene transcription; left side) and the VDR-AP (for the VDR alternative pocket engaged in nongenomic responses; right side). In order for the VDR to generate genomic responses, it has been experimentally established that the VDR must have a bound $1\alpha,25(OH)_2D_3$ ligand that is bowl shaped. This can function to turn selected genes either on or off. For the VDR that is localized to the membrane caveolae where it produces rapid responses, it must have a bound $1\alpha,25(OH)_2D_3$ that is in the planar conformational shape. The different shapes of the VDR-GP and the VDR-AP ligands were evaluated by separate determination of the X-ray structure of the VDR-GP with a bound bowl-shape ligand and computer modeling of the VDR-AP with an AP ligand with a planar shape. Panel (B) has two extreme presentations of $1\alpha,25(OH)_2D_3$. The top row is a space-filling three-dimensional representation of 6-s-*cis*-shaped $1\alpha,25(OH)_2D_3$; it has a bowl-like shape. The right-hand three-dimensional ligand shows a 6-s-*trans* presentation; it has a planar shape. The second row is a ball-and-stick presentation of $1\alpha,25(OH)_2D_3$ in the bowl (left) or planar (right) shape. In order for the VDR to generate genomic responses, that is, affect the rate of gene transcription, it has been experimentally established that the VDR must have a bound $1\alpha,25(OH)_2D_3$ ligand in the bowl shape. This is referred to as the VDR-GP. The VDR that is localized to the membrane caveolae and produces "rapid" responses must have a bound $1\alpha,25(OH)_2D_3$ that is in the planar conformational shape. This binding site is referred to as the VDR-AP. The different shapes of the VDR-GP and VDR-AP ligand were evaluated by separate determination of the X-ray structure of each and computer modeling of the AP ligand. *VDR*, Vitamin D receptor.

of each of these types of defects in specific receptors are presented in Tables 1.2 and 1.3. Mutations may render the GPCR unable to bind its hormone with appropriate strength; affect its ability to undergo the appropriate conformational change; or otherwise block its ability to bind to and activate its G-protein. Since most of the losses of function diseases are recessive, it appears that one undamaged allele can compensate for the mutated one.

Loss of function mutations can be nonsense, missense, or frameshift mutations but gain of function mutations are generally missense. The result might be a protein with an incorrect amino acid at a certain position leading, for example, to a receptor that is either overly or constitutively activated. The study of such mutations has yielded important insights into the role of three-dimensional structure and function of GPCRs.

Because of their wide distribution in the regulation of physiological processes, GPCRs make up an important portion of targeted drug development by pharmaceutical companies. It has been estimated that GPCRs constitute as much as

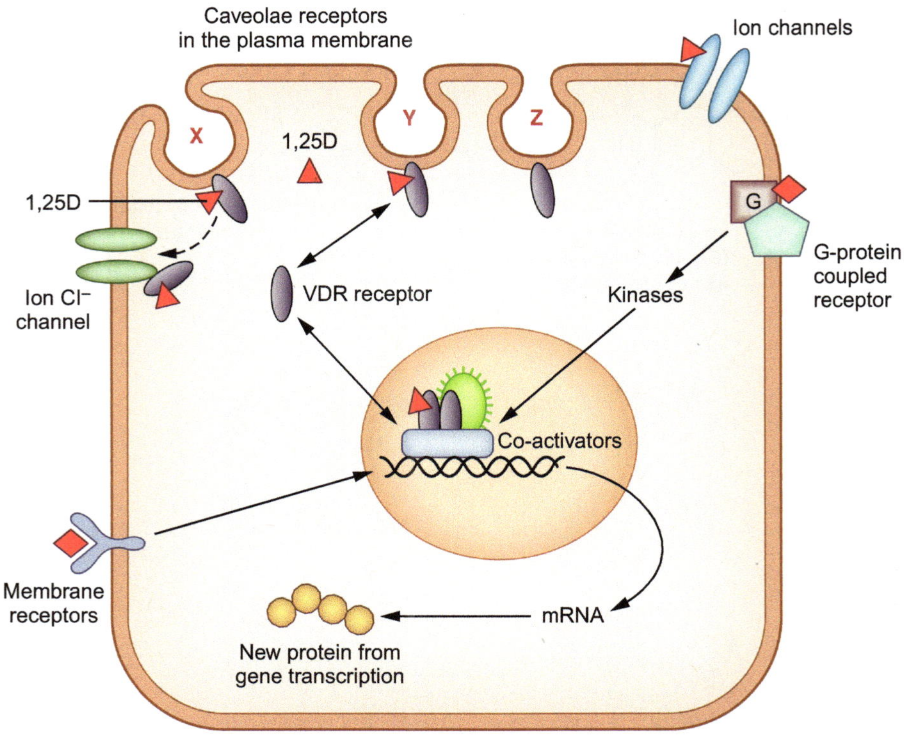

FIGURE 1.26 Distribution of VDR in a target cell. Target cells for the steroid hormone $1\alpha,25(OH)_2D_3$ have VDR distributed between the nucleus ($\sim$80%), the cytosolic compartment ($\sim$15%), and the interior surface of the plasma membrane caveolae ($\sim$5%). The VDR can move by diffusion to these three compartments. The VDR monomer has hydrophobic bonds with the intracellular surface of the caveolae membrane. The X, Y, and Z red letters are outside the cell. The VDR, arriving from the nucleus and cytosol, docks with the surface of the caveolae facing the cytosol of the cell. When the caveolae becomes associated with a monomeric VDR with a bound $1\alpha,25(OH)_2D_3$ (*red triangle*), it moves (in this example) to the nearest chloride channel to activate rapid responses. However, the caveolae-associated VDR must have the $1\alpha,25(OH)_2D_3$ bound to the VDR-AP, and not the VDR-GP for this to happen. This VDR/planar ligand is competent to initiate a variety of rapid responses (see Table 1.1). Illustrated in this figure is the VDR/$1\alpha,25(OH)_2D_3$-initiated opening of chloride channels in the plasma membrane. In the nucleus of the cell, VDR with a bound $1\alpha,25(OH)_2D_3$ forms a heterodimer with the retinoid X receptor to activate gene transcription (see Figs. 1.23 and 1.24). Gene KO of the VDR in mice results in the loss of both $1\alpha,25(OH)_2D_3$-stimulated gene activation and the activation of rapid responses, including the opening of chloride channels, and even exocytosis from the cell. *KO*, Knockout; *VDR*, vitamin D receptor; *VDR-AP*, vitamin D receptor alternate pocket; *VDR-GP*, vitamin D receptor genomic pocket.

TABLE 1.2 Loss of function endocrine diseases resulting from G protein–coupled receptor mutations.

Receptor	Disease	Inheritance
AVP receptor 2	Nephrogenic diabetes insipidus	X-linked R
LH	Leydig cell hypoplasia (males), primary amenorrhea (females)	AR
FSH	Sperm-related hypofertility (males)	AR
GnRH	Central hypogonadotropic hypogonadism	AR
TRH	Central hypothyroidism	AR
TSH	Congenital hypothyroidism	AD
GHRH	Short stature (growth hormone deficiency)	AR
Melanocortin 4	Extreme obesity	AR/codominant

AD, Autosomal dominant; *AR*, autosomal recessive; *AVP*, arginine vasopressin; *FSH*, follicle stimulating hormone; *GHRH*, growth hormone–releasing hormone; *GnRH*, gonadotropin releasing hormone; *LH*, luteinizing hormone; *TRH*, thyrotropin releasing hormone; *TSH*, thyrotropin.

TABLE 1.3 Gain of function endocrine diseases resulting from G protein–coupled receptor mutations.

Receptor	Disease	Inheritance
AVP receptor 2	Nephrogenic syndrome of inappropriate antidiuresis	X-linked D
LH	Precocious puberty (male only)	AD
TSH	Nonautoimmune familial hyperthyroidism	AD
PTH and PTHrP	Jansen metaphyseal hypercalciuria	AD
Ca^{2+}-sensing	Familial hypocalcemic hypercalciuria	AD

AD, Autosomal dominant; *AVP*, arginine vasopressin; *LH*, luteinizing hormone; *PTH*, parathyroid hormone; *PTHrP*, parathyroid hormone–like peptide; *TSH*, thyrotropin stimulating hormone.

65% of prescription drugs. Given that 3%–4% of the human genome codes for these proteins, they will surely continue to be a rich source of information about cellular regulation and of development of drugs to treat GPCR-related diseases.

Further reading

Biosynthesis of peptide hormones

S. Ahn, et al., The RNA-binding protein, HuD, regulates proglucagon biosynthesis in pancreatic alpha cells, Biochimica Biophysica Acta 530 (2020) 266–272.

R. Eggelkraut-Gottanka, A.G. Beck-Sickinger, Biosynthesis of peptide hormones derived from precursor sequences, Ingenta 11 (2004) 2651–2665.

T.D. Madsen, et al., An atlas of O-linked glycosylation on peptide hormones reveals diverse biological roles, Nature Communications 11 (2020) 4033. https:doi.org/10.1038/s41467-020-17473-1.

Membrane receptors

K. Jaqaman, J.A. Ditlev, Biomolecular condensates in membrane receptor signaling, Current Opinion in Cell Biology 69 (2021) 48–54.

I.S. Levina, Y.V. Kusnetsov, T.A. Shchelkunova, I.V. Zavarzin, Selective ligands of membrane progesterone receptors as a key to studying their biological functions in vitro and in vivo, Journal of Steroid Biochemistry and Molecular Biology 207 (2021) 105827. https://doi.org/10.1016/j.jsbmb.2021.105827.

V. Luzovvy, L. Richardson, G. Saade, R. Menon, Progesterone receptor membrane components: key regulators of fetal membrane integrity, Biology of Reproduction 104 (2021) 445–456.

Steroid receptors

S. Saha, S. Dey, S. Nath, Steroid hormone receptors: links with cell cycle machinery and breast cancer progression, Frontiers in Oncology (2021). https:doi.org/10.3389/fonc.2021.620214.

M. Stortz, D.M. Presman, A. Pecci, V. Levi, Phasing the intranuclear organization of steroid hormone receptors, Biochemical Journal 478 (2021) 443–461.

L.S. Trevino, D.A. Gorelick, The interface of nuclear and membrane steroid signaling, Endocrinology 162 (2021) bqab107. https:doi.org/10.1210/endocr/bqab107.

Regulation of gene transcription by steroid hormones

S. Ganguly, D. Naik, A. Muskara, O.Y. Mian, The nexus of endocrine signaling and cancer: how steroid steroid hormones influence genomic stability, Endocrinology 162 (2021) bqaa177. https://doi.org/10.1210/endocr/bqaa177.

A.N. Gerber, R. Newton, S.K. Sasse, Repression of transcription by the glucocorticoid receptor: a parsimonious model of the genomics era, Journal of Biological Chemistry 296 (2021) 100687. https://doi.org/10.1016/j.jbc.2021.100687.

M. Kanki, J. Morgan, P.J. Fuller, M.J. Young, Time of day regulates renal mineralocorticoid receptor transcriptional control of electrolyte balance, Journal of the Endocrine Society 5 (2021) A819–A820.

Membrane actions of steroid hormones

J. Cipolla-Neto, et al., 2021. The crosstalk between melatonin and sex steroid hormones, Neuroendocrinology (2021). https://doi.org/10.1159/000516148.

P.S. Cooke, A.M. Mesa, V.K. Sirohi, E.R. Levin, Role of nuclear and membrane estrogen signaling pathways in male and female reproductive tract, Differentiation 118 (2021) 24–33.

M. Pillerova, et al., On the role of sex steroids in biological functions in classical and non-classical pathways, An update. Frontiers in Immunology (2021). https://doi.org/10.33898/fimmu.2021.674189.

GPCRs in human disease

R.C. Guimaraes, T.T. Goncalves, L.O. Leiria, Exploiting oxidized lipids and the lipid-binding GPCRs against cardiometabolic diseases, British Journal of Pharmacology (2021). https://doi.org/10.1111/bph.15321.

C.-C. Hsiao, et al., GPCRomics of homeostatic and disease-associated human microglia, Frontiers in Immunology (2021). https://doi.org/10.3389/fimmu.2021.674189.

P. Sharma, R.B. Penn, Can GPCRs be targeted to control inflammation in asthma? Health and Disease Volume II (2021) 1–20.

Chapter 2

Steroid hormones: chemistry, biosynthesis, and metabolism

I Introduction

A General comments

This chapter deals with the structural chemistry and biosynthetic pathways of the major classes of steroid hormones. All have a complicated structure of fused rings which can be modified by functional group substitution at many points. Furthermore, the presence of asymmetric carbon atoms introduces steric modifications and isomeric possibilities. The reader will find it prudent to first grasp the essential features of the steroid structures and relationships before attempting to delve into a consideration of their specific hormonal activities in later chapters. Then, when so doing, it may be helpful to turn back to the appropriate portion of this chapter to further heighten the understanding of the structures of the hormones under review.

B Historical perspective

The first steroid hormone, estrone, was isolated in 1929 at a time before the characteristic ring structure of the steroid nucleus had been elucidated. Today well over 230 naturally occurring steroids have been isolated and chemically characterized. In addition, an uncountable number of steroids and steroid analogs have been chemically synthesized and evaluated for their drug properties.

The development of our modern understanding of hormones and the science of endocrinology has closely paralleled studies on the isolation, chemical characterization, and synthesis of steroids and the subsequent elucidation of their pathways of biosynthesis and catabolism. The foundation of many of these developments with steroid hormones is to be found in a lengthy series of papers authored by Professor Adolf O. R. Windaus' chemistry laboratory in Gottingen, Germany (1925–35[1]) that led to the structural determination of cholesterol. This was an extraordinarily challenging problem given the limitation that the techniques of nuclear magnetic resonance spectroscopy, mass spectrometry, and ultraviolet and infrared spectroscopy were not available at that time. Instead, the structure was determined through elaborate classical organic chemistry manipulations, which involved the conversion of the compound under study to known reference compounds. At the present time the application of the powerful separation techniques of high-performance liquid chromatography or gas chromatography, combined with the use of continuous online monitoring by mass spectrometry with computer-assisted data storage and analysis, frequently permits unequivocal structural determinations on impure samples that contain less than 1 ng of the steroid of interest.

C Radioactive steroids

An equally important contribution to our present understanding of the biochemical properties and structure of steroids was the introduction and general availability of radioactively labeled compounds in the 1960s. Radioactive steroids offer two major advantages: first, the presence of the radioactive label provides a significant increase in the sensitivity of detection of the steroid under study in living animals or cells. Prior to the advent of radioactive steroids, investigators relied upon colorimetric or bioassay procedures to quantitate the steroid of interest. Second, the availability of radioactive compounds permitted the investigator to detect, from either in vivo whole animals or in vitro experiments with perfused organs, tissue slices, cell suspensions, cell homogenates, or purified enzyme preparations, the presence of new compounds that in the absence of a radioactive label would otherwise not have been discovered. Thus it was through the application of radioisotope techniques, modern procedures of

1. The Nobel Prize in Chemistry in 1928 was awarded to Adolf Windaus "for the services rendered through his research into the constitution of the sterols and their connection with the vitamins."

Hormones. DOI: https://doi.org/10.1016/B978-0-323-90262-5.00018-4

chromatography, and structure determination that whole categories of new steroid hormones were discovered. For example, research on vitamin D metabolites (Chapter 9: Calcium-Regulating Hormones: Vitamin D, Parathyroid Hormone, Calcitonin, and Fibroblast Growth Factor-23) and catechol estrogens (Chapter 13: Estrogens and Progestins) benefited from preparations of radioactive vitamin D_3 and catechol estrogens.

D Molecular biology contributions

Initially the premise was one step in steroid metabolism was handled by one enzyme. However, the advent of cloning of the cDNAs of each enzymatic step resulted in the discovery that there were fewer steroid proteins than the number of separate enzyme steps. That is to say, one steroid enzyme could carry out two to three quite different catalytic steps. Also it was learned that certain enzymatic steps that occurred in several different tissues were, in fact, carried out by the same enzyme.

II Chemistry of steroids

A Basic ring structure

Steroids are derived from a phenanthrene ring structure to which a pentano ring has been attached; this yields in the completely hydrogenated form, cyclopentanoperhydrophenanthrene, or the sterane ring structure (see Fig. 2.1).

Steroid structures are not normally written with all the carbon and hydrogen atoms as illustrated in the middle panel of Fig. 2.1; instead, the shorthand notation as presented in the bottom panel for sterane of Fig. 2.1 is usually employed. In this representation the hydrogen atoms are not indicated, and unless specified otherwise it is assumed that the cyclohexane A, B, and C rings and the cyclopentane D ring are fully reduced; that is, each carbon has its full complement of carbon and/or hydrogen bonds. Also, indicated for the bottom sterane structure (Fig. 2.1) is the standard numbering system for each of the 17-carbon atoms in the four rings. The three six-carbon cyclohexane rings are designated A, B, and C rings and the five-carbon cyclopentane ring is denoted as the D ring.

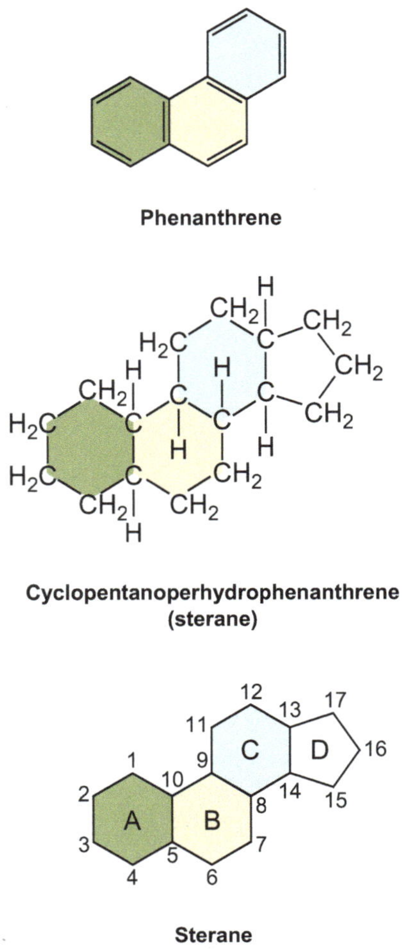

FIGURE 2.1 Parent ring structures of steroids. The creation of the 5-carbon (pentano) ring on phenanthrene followed by the reduction of all the aromatic double bonds creates the foundational, completely hydrogenated, cyclopentanoperhydrophenathrene, also known as sterane (the middle structure). The bottom structure is a different presentation of sterane that illustrates the numbering system or zip code for each of its 17 carbons and the convention for the A, B, C, and D labels for the 4 rings.

B Classes of steroids

In mammalian systems, there are six families of steroid hormones that can be classified on both a structural and a biological (hormonal) basis (see Fig. 2.2). They are the estrogens and progestins (female sex steroids), androgens (male sex steroids), mineralocorticoids (aldosterone), glucocorticoids (cortisol), and vitamin D [$1\alpha,25(OH)_2D_3$]. Also, the bile acids are structurally related to cholesterol and, thus, constitute a seventh member of the steroid family. All of these steroids are biologically derived from cholesterol. Table 2.1 summarizes some fundamental relationships of these principal mammalian classes of steroids.

The parent ring structure for cholesterol is the fully saturated ring structure cholestane (see top row of Fig. 2.3). Cholestane, which has 27 carbons, differs from sterane (Fig. 2.2) by the addition of an 8-carbon side chain on carbon-17 of ring D and the presence of two angular methyl groups at the junctures of the A:B (carbon-10) and C:D rings (carbon-13). The cholestane ring structure also gives rise to the parent ring structures

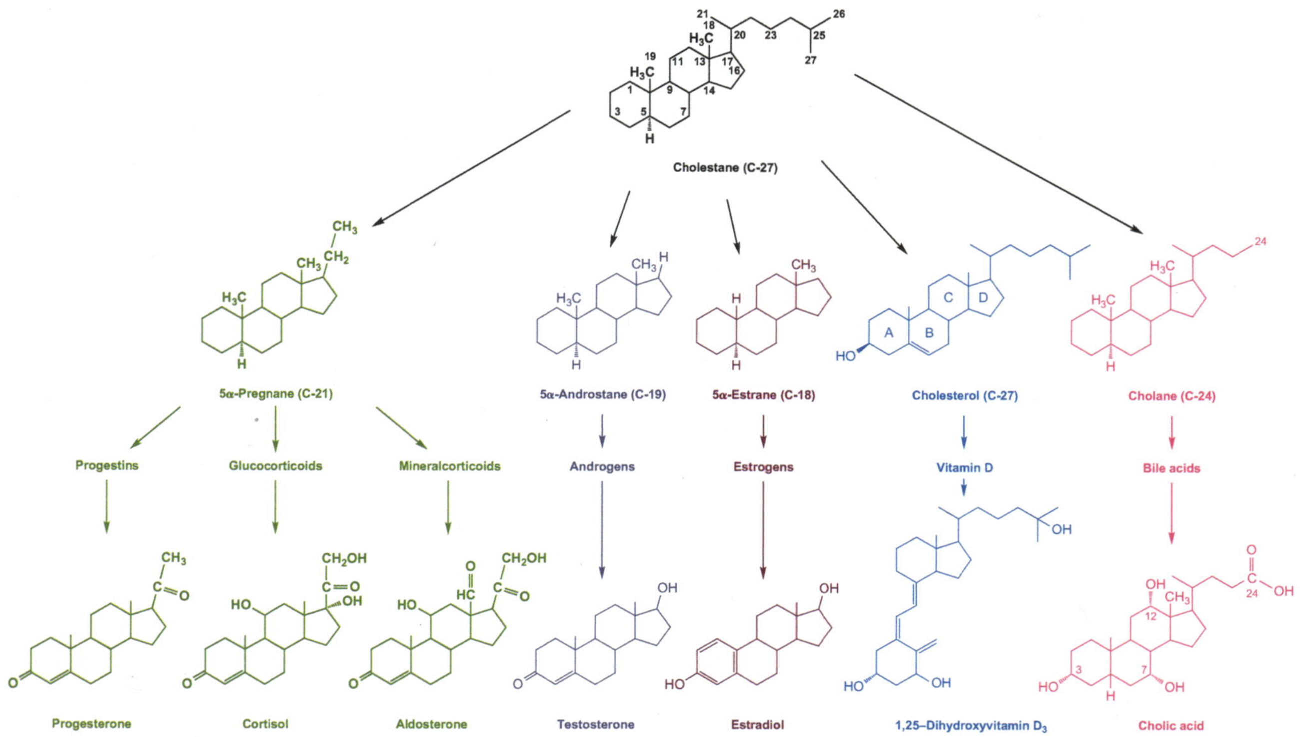

FIGURE 2.2 Family tree of the seven principal classes of steroids (bottom row) that are structurally derived from the parent cholestane (top row). Cholestane has 10 additional carbons added to sterane (see Fig. 2.1); these include two methyl groups, C-18 and C-19, added, respectively, to C-13 and C-10 and an eight-carbon side chain (C-20 to C-27) attached to C-17 of the D-ring.

TABLE 2.1 Classes of steroids.

Steroid class	Principal active steroid in humans[a]	Number of carbon atoms	Parent ring structure[a]
Estrogens	Estradiol	18	Estrane
Androgens	Testosterone	19	Androstane
Progestins	Progesterone	21	Pregnane
Glucocorticoids	Cortisol	21	Pregnane
Mineralocorticoids	Aldosterone	21	Pregnane
Vitamin D steroids	1,25-Dihydroxyvitamin D_3	27	Cholestane
Bile acids	Cholic acid	24	Cholane

[a]*The parent ring steroid structures and active steroid hormone are given in Fig. 2.2.*

for the six classes of mammalian steroids and the bile acids, as summarized in Table 2.1. The parent ring compounds are the completely saturated ring structures pregnane, androstane, estrane, and cholane, which is structurally related to cholestane. These relationships are depicted in Fig. 2.2. The parent ring structures are used as the stem term in constructing the formal nomenclature of any of these steroids.

The biosynthetic pathway of production of each of these general steroid classes will be presented separately later in this chapter. A discussion of their hormonal and biochemical aspects will appear later in individual chapters.

C Structural modification

The basic ring structures presented in Fig. 2.2 can be subjected to a wide array of modifications by the introduction of hydroxyl or carbonyl substituents and unsaturation (double or triple bonds). In addition, heteroatoms such as nitrogen or sulfur can replace the ring carbons, and halogens and sulfhydryl or amino groups may replace steroid

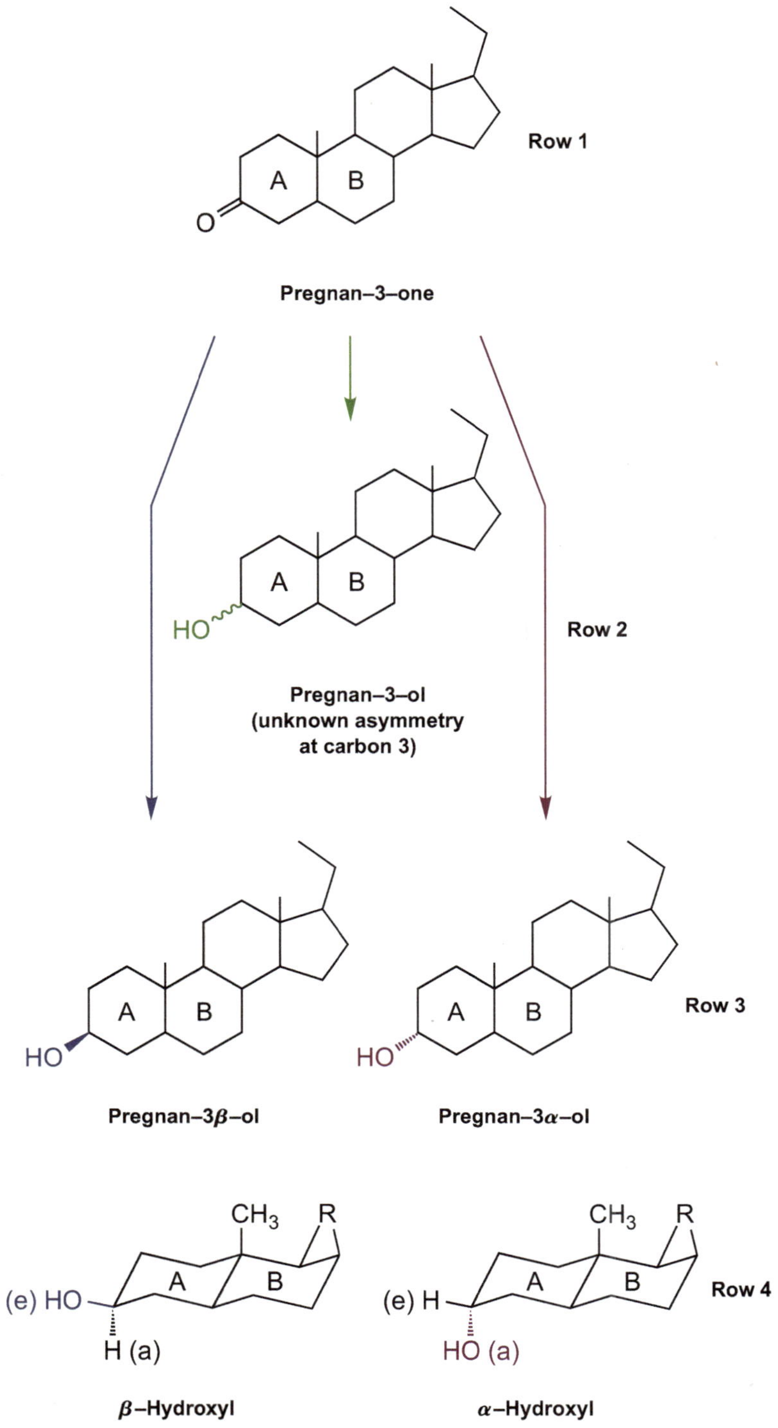

FIGURE 2.3 Structural consequences resulting from the reduction of the 3-keto group of pregnan-3-one to a hydroxyl group (see Row 1). As shown in Row 2, the orientation of the hydroxyl group is not designated (see the green wavy bond of the C3 hydroxyl group). As shown in Row 3, there are two structural options for the orientation of the hydroxyl group on C-3. The hydroxyl can be either "above" the plane of the page (see the *solid* blue hydroxyl group) or below the plane of the page (see the *dashed* red hydroxyl group). Thus in Row 3, application of steroid nomenclature rules designates that the red hydroxyl *below* the plane of the page is a 3α-hydroxyl while the blue hydroxy *above* the plane of the page is a 3β-hydroxyl. Row 4 presents the chair presentation for the A/B rings of the two pregnane isomers; this further emphasizes that the alpha and beta 3-hydroxyl groups each have a distinctly different orientation in space.

hydroxyl moieties. Furthermore, the ring size can be expanded or contracted by the addition or removal of carbon atoms. The consequences of these structural modifications are designated by the application of the standard organic nomenclature conventions for steroids. The pertinent examples of this system are summarized in Table 2.2. Prefixes and suffixes are used to indicate the type of structural modification. Any number of prefixes may be employed (each with its own appropriate carbon number and specified in the order of decreasing preference of acid, lactone, ester, aldehyde, ketone, alcohol, amine, and ether); however, only one suffix is permitted.

Table 2.3 tabulates the systematic and trivial names of many common steroids. All of these formal names are

TABLE 2.2 Steroid nomenclature conventions.

Modification	Prefix	Suffix
Hydroxyl group (OH)	Hydroxy-	-ol
Hydroxyl above plane of ring	β-Hydroxy	–
Hydroxyl below plane of ring	α-Hydroxy	–
Keto or carbonyl group (C=O)	Oxo-	-one
Aldehyde (CHO)	–	-al
Carboxylic acid (COOH)	Carboxy-	-oic acid
Double bond (C=C)	–	-ene
Triple bond (C≡C)	–	-yne
Saturated ring system	–	-ane
One less carbon atom	Nor-	–
One additional carbon atom	Homo-	–
One additional oxygenation	Oxo-	–
One less oxygen atom	Deoxy-	–
Two additional hydrogen atoms	Dihydro-	–
Two less hydrogen atoms	Dehydro-	–
Two groups on same sides of plane	*cis-*	–
Two groups on opposites sides of plane	*trans-*	–
Other ring forms (rings A and B *trans*, as in allopregnane)	Allo-	–
Opening of a steroid ring (as in vitamin D's B ring)	Seco-	–
Conversion at a numbered carbon from conventional orientation (as in epicholesterol or 3α-cholesterol)	Epi-	–

TABLE 2.3 Trivial and systematic names of some common steroids.

Trivial name	Systematic name
Aldosterone	18,11-Hemiacetal of 11β,21-dihydroxy-3,20-dioxopregn-4-en-18-al
Androstenedione	Androst-4-ene-3,17-dione
Androsterone	3α-Hydroxy-5α-androstan-l7-one
Cholesterol	Cholest-5-ene-3β-ol
Cholic acid	3α,7α,12α-Trihydroxy-5β-cholan-24-oic acid
Corticosterone	11β,21-Dihydroxypregn-4-ene-3,20-dione
Cortisol	11β,17,21-Trihydroxypregn-4-ene-3,20-dione
Cortisone	17,21-Dihydroxypregn-4-ene-3,11,20-trione
Dehydroepiandrosterone	3β-Hydroxy-5-androstene-17-one
Deoxycorticosterone	21-Hydroxypregn-4-ene-3,20-dione
Ergosterol	5,7,22-Ergostatriene-3β-ol
Estrone	3-Hydroxyestra-1,3,5(10)-triene-17-one
Progesterone	Pregn-4-ene-3,20-dione
Testosterone	17β-Hydroxyandrost-4-en-3-one
Vitamin D_3 (cholecalciferol)	9,10-Seco-5,7,10(19)-cholesta-triene-3β-ol
Vitamin D_2 (ergocalciferol)	9,10-Seco-5,7,10(19),22-ergosta-tetraene-3β-ol

devised in accordance with the official nomenclature rules for steroids laid down by the International Union of Pure and Applied Chemistry (IUPAC).[2]

D Asymmetric carbons

An important structural feature of any steroid is recognition of the presence of asymmetric carbon atoms and designation in the formal nomenclature of the structural isomer that is present. Thus the reduction of pregnan-3-one to the corresponding 3-alcohol will produce two epimeric steroids (see Fig. 2.3). The resulting hydroxyl may be above the plane of the A ring and is so designated on the structure by a bold solid line; it is referred to as a 3β-ol. The epimer, or 3α-ol, has the hydroxyl below the plane of the A ring and is so designated by a dotted line for the C · OH bond. If the α- or β-orientation of a substituent group is not known, it is designated with a wavy line (C ~ OH).

Another locus where asymmetric carbon atoms play an important role in steroid structure determination is the junction between each of the A, B, C, and D rings. Fig. 2.4 illustrates these relationships for cholestanol and coprostanol. Thus in the 5α-form, the 19-methyl and α-hydrogen on carbon-5 are on opposite sides of the plane of the A:B ring; this is referred to as a *trans* fusion. When the 19-methyl and β-hydrogen on carbon-5 are on the same side of the A:B ring fusion, this is denoted *cis* fusion. In the instance of *cis* fusion of the A:B rings, the

2. The IUPAC definitive rules of steroid nomenclature are presented in full in *Pure & Applied Chemistry* 31 (1972), 285–322 or *Biochemistry* 10 (1971), 4994–4995.

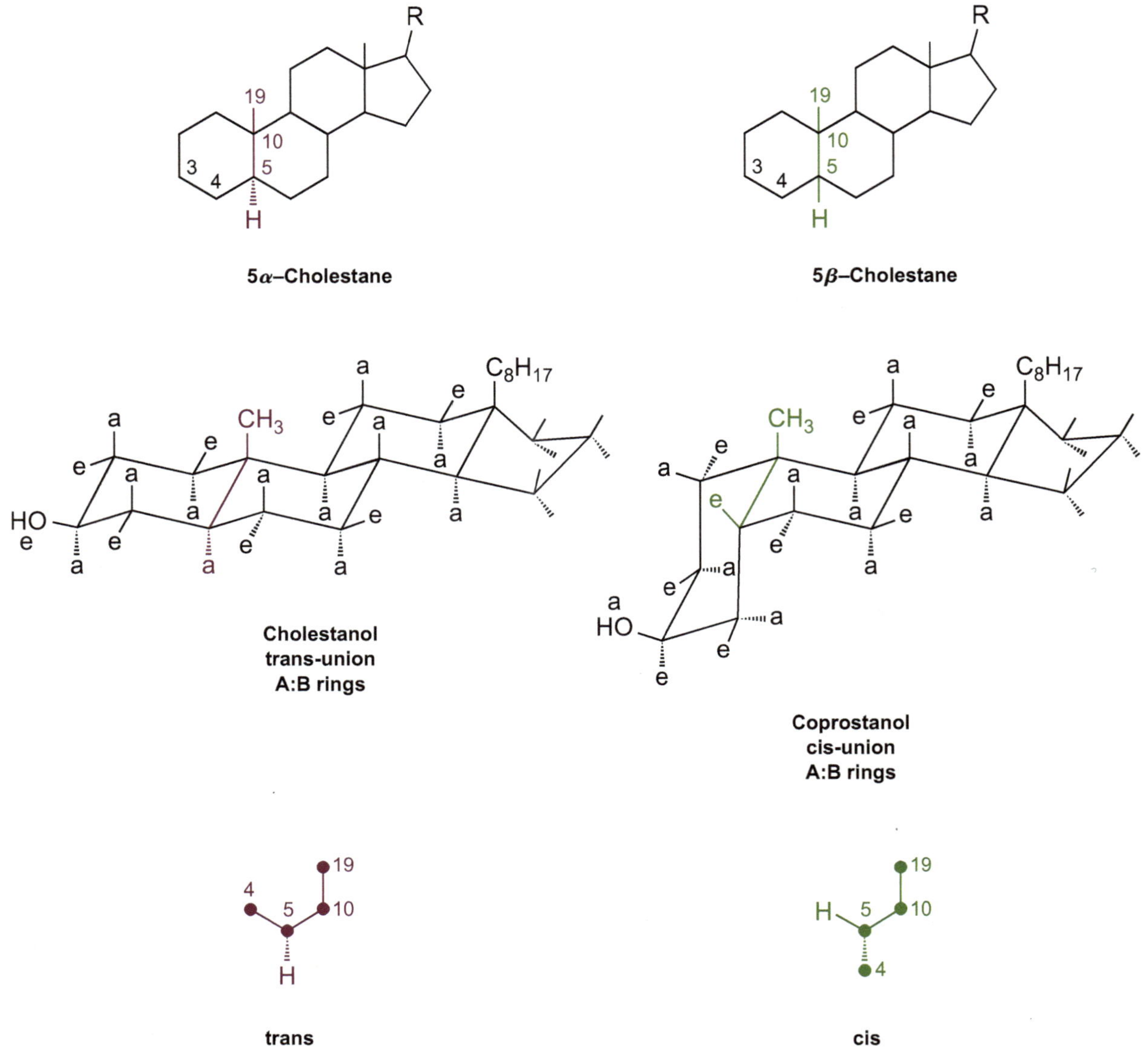

FIGURE 2.4 Comparison of structural relationships resulting from *cis* or *trans* A:B ring fusion in 5α-cholestane versus 5β-cholestane (top row) and cholestanol versus coprostanol (second row). In 5α-cholestane and cholestanol (left side of rows 1 and 2) the A:B ring fusion is *trans* (C-19-methyl-C-5H), while in 5β-cholestane and coprostanol (right side of rows 1 and 2) the A:B ring fusion is *cis* (C-19-methyl-C-5H). The orientation of substituents around carbon-5 for the *cis* and *trans* circumstances are illustrated in the bottom row (●–● indicates carbon–carbon bonds). Finally, with respect to the hydroxyl on C-3, its *beta* orientation is maintained for both cholestanol and coprostanol in spite of their different A/B ring fusions.

steroid structure can no longer be drawn in one plane. Thus in all 5β-steroid structures that have *cis* fusion between rings A and B, the A ring is bent into a second plane that is at approximately a right angle to the B:C:D rings (see Fig. 2.4).

Thus each of the ring junction carbons is potentially asymmetric, and the naturally occurring steroid will have only one of the two possible orientations at each ring junction. Although there are two families of naturally occurring steroids with either *cis* or *trans* fusion of the A:B rings, it is known that the ring fusions of B:C and C:D in virtually all naturally occurring steroids are *trans*. Consideration of the estrogen steroid series (see Fig. 2.2) is a special case in which the A ring is aromatic; thus there is no *cis–trans* isomerism possible at carbons-5 and -10.

The side chain is a third domain of the steroid structure where asymmetry considerations are important. Historically, interest first centered on carbon-20 of the cholesterol side chain, although side chain asymmetry is also now known to be crucial for the insect steroid hormone ecdysterone, for a number of vitamin D metabolites (see Fig. 2.2 and Chapter 9: Calcium-Regulating Hormones: Vitamin D, Parathyroid Hormone, Calcitonin, and Fibroblast Growth Factor-23), and in the production of many bile acids (see cholic acid in Fig. 2.2). While the

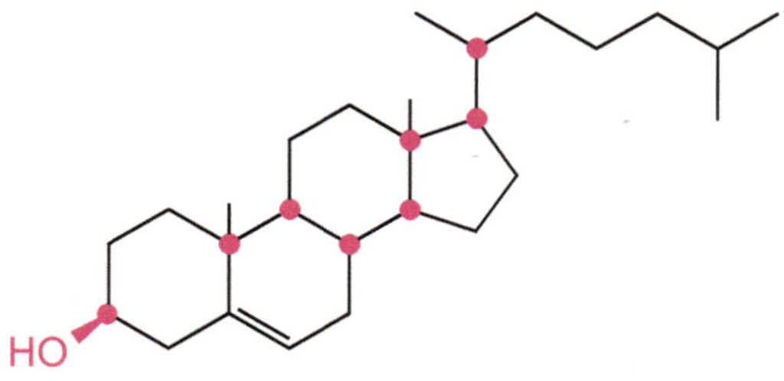

FIGURE 2.5 Asymmetric carbons of cholesterol. The eight asymmetric carbons of cholesterol are indicated by the magenta dots (●). There are 2^8 or 256 structural isomers of cholesterol. Only one structural isomer is produced in higher animals.

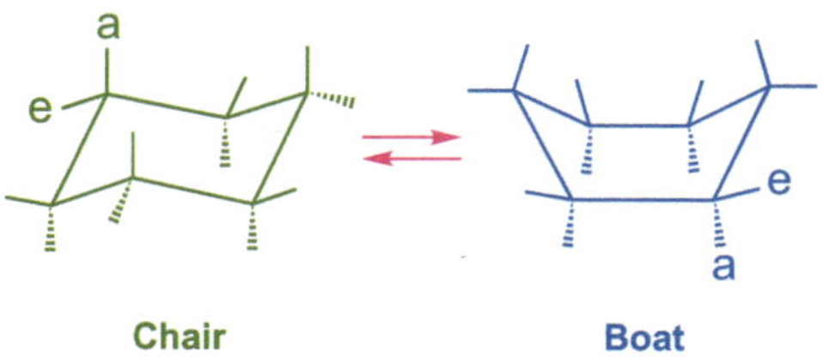

FIGURE 2.6 Principal conformational representations of cyclohexane. The left conformer of cyclohexane is a chair, while the right conformer is a boat. They interchange about a million times per second.

α or β notation is satisfactory for the designation of substituents of the A, B, C, and D ring structures, this terminology is not applicable to the side chain. This is because there is free rotation of the side chain at the carbon-17–carbon-20 bond; thus the side chain may assume a number of orientations in relation to the ring structure.

Fig. 2.5 identifies for cholesterol the eight asymmetric carbon atoms. Introduction of the $\Delta^{5,6}$-double bond in cholesterol deletes one asymmetric center, whereas addition of the eight-carbon side chain adds one asymmetric carbon at position 17. Carbon-20 of the side chain is also asymmetric. Finally, the introduction of a hydroxyl group on carbon-3 creates still another asymmetric center. Thus there are a total of eight asymmetric carbons or 2^8=256 possible structural isomers. Considering that cholesterol is the most prevalent naturally occurring steroid, it is an impressive testament to the precision of evolutionary events and to the specificity of the many enzymes involved in the biosynthesis of cholesterol that only one major sterol product is present in mammalian systems.

E Conformational flexibility of steroids

The steroid nucleus, sterane, is composed of three cyclohexane rings and one cyclopentane ring. The six-carbon atoms of a cyclohexane ring are not fixed rigidly in space but are capable of interchanging through turning and twisting between several structural arrangements in space called conformations.

The two principal conformations of a cyclohexane ring are the *chair* and *boat* forms (see Fig. 2.6). Each of the two substituent groups on the six-carbon atoms of the cyclohexane ring may exist in either the general plane of the ring and are designated as equatorial (*e*) or a plane perpendicular to the ring plane and are designated as axial (*a*). For the equatorial bonds, it is possible to superimpose on the equatorial notation an indication of whether they are below (α) or above (β) the general plane of the ring. Cyclohexane is highly conformationally mobile, interchanging between the boat and chair forms many thousands of times per second. The most stable form of the cyclohexane ring is the chair form; in this conformer, there is a greater interatomic distance between the equatorial and axial hydrogens than in the boat form. Fig. 2.4 illustrates the nature of all of the equatorial (*e*) and axial (*a*) hydrogens on the cholestane and coprostane ring structures.

As indicated, the B and C rings of both cholestane and coprostane are locked into chair conformations (see Fig. 2.4). Although, in principle, the A ring of both these steroids is free to interchange between the boat and chair representations, the chair form is believed to be much more favored. In the case of vitamin D steroids, which do not have an intact B ring due to the breakage of the carbon-9–carbon-10 bond (thus they are termed secosteroids), the A ring is much more conformationally mobile than that of the usual cholesterol-derived steroids (see Fig. 9.6).

An important point for the reader to consider is that the usual structural representation given for steroids (e.g., see Figs. 2.2–2.4) provides no clear designation of either the three-dimensional geometry or the space-filling aspects of the electron orbitals associated with each atom involved in the formation of the requisite bonds required for the full molecular structure. A comparison for cholestanol is given in Fig. 2.7 of the planar representation (A), the planar conformational model (B), a Dreiding three-dimensional model emphasizing bond lengths and angles (C), and a Corey–Pauling three-dimensional space-filling model (D). Certainly the space-filling molecular representation most closely approximates the reality of the three-dimensional shape of the steroid and, thus, provides an insight into the overall molecular shape required to produce a biological response.

The approach of steroid conformational analysis has been of great value to the organic chemist as a tool to predict or understand the course of synthetic organic chemical reactions. It is also known that conformational considerations play an increasingly useful role in the understanding of steroid hormone–receptor interactions. Steroid receptors are known to have very precise ligand specificities; see, for example, Table 2.4, which illustrates the structural preferences of the nuclear receptor for the

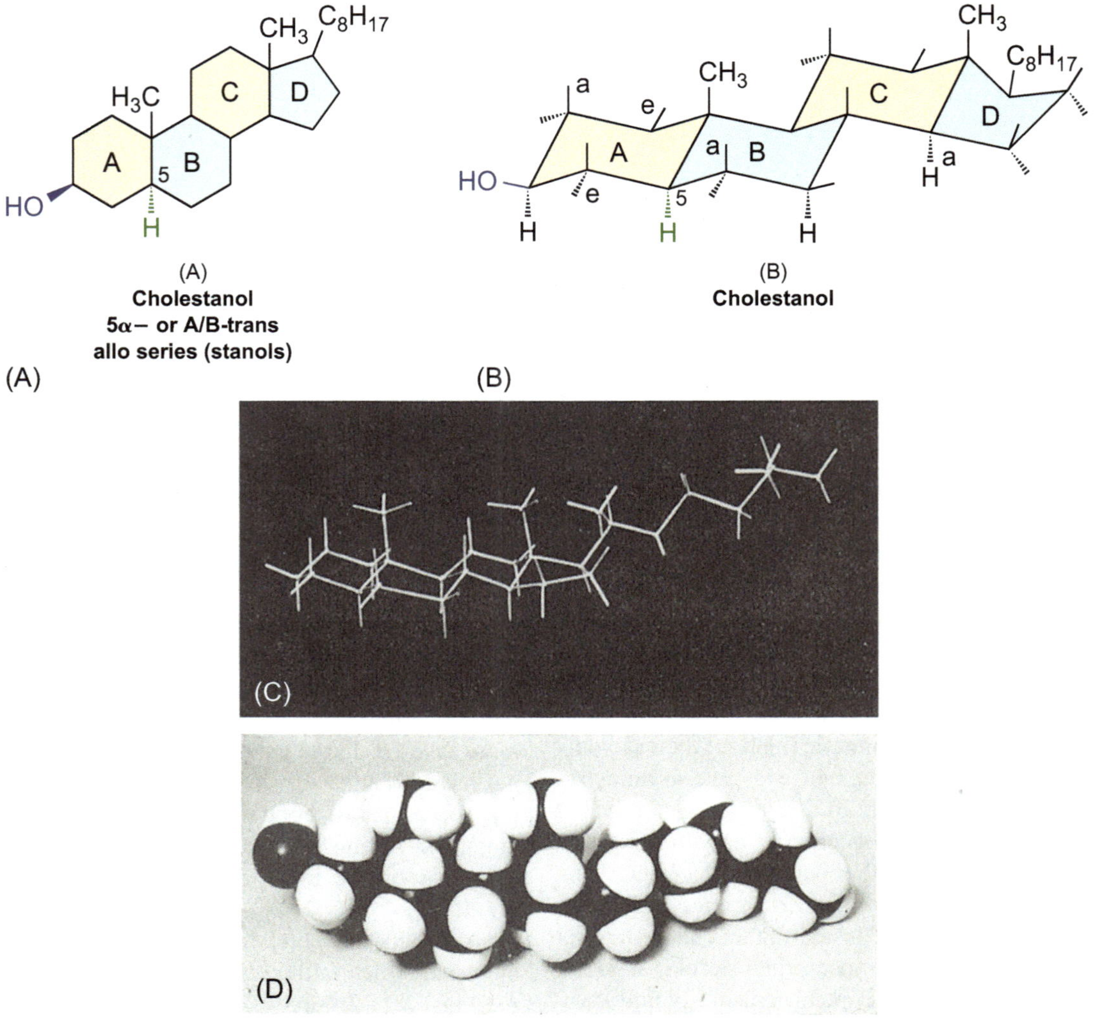

FIGURE 2.7 Four structural representations of cholestanol. (A) A typical two-dimensional structure of cholestanol drawn on a page. (B) A planar conformational model, also drawn on a page. (C) A Dreiding three-dimensional "stick" model of cholestanol emphasizing bond angles and interatomic distances. (D) A Corey–Pauling three-dimensional space-filling model of cholestanol. The C and D models can either be manually assembled from a kit of all possible parts or can be modeled in three dimensions on a computer using a highly sophisticated program.

TABLE 2.4 Ligand specificity of the nuclear 1α,25(OH)$_2$D$_3$ receptor.

Ligand	Structural modification	RCI[a]
1α,25(OH)$_2$D$_3$ See structure in Fig. 2.2	The naturally occurring steroid hormone (see Chapter 9: Calcium-Regulating Hormones: Vitamin D, Parathyroid Hormone, Calcitonin, and Fibroblast Growth Factor-23, for further details)	100%
1α,25(OH)$_2$ − 24-nor-D$_3$	Shorten side chain by one carbon	67
1α,25(OH)$_2$ − 3-epi-D$_3$	Orientation of 3β-OH changed	24
1α,25(OH)$_2$ − 24a-dihomo-D$_3$	Lengthen side chain by two carbons	24
1β,25(OH)$_2$D$_3$	Orientation of 1α-OH changed to 1β	0.8
1α (OH)D$_3$	Lacks 25-OH	0.15
25(OH)D$_3$	Lacks 1α-OH	0.15
1α,25(OH)$_2$ − 7-dehydrocholesterol	Lacks a broken B ring; is not a secosteroid	0.10
Vitamin D$_3$	Lacks both lα, and 25-OH groups	0.0001

RCI, Relative competitive index; *VDR*, vitamin D receptor.

[a]*The RCI is a measure of the ability of a nonradioactive ligand to compete, under in vitro conditions, with radioactive 1α,25(OH)$_2$D$_3$ for binding to the nuclear 1α,25(OH)$_2$D$_3$ receptor (VDR). The RCI is expressed as a percent. Thus the binding of 1α,25(OH)$_2$D$_3$ to its VDR is 100%. When one of the three hydroxyls on 1α,25(OH)$_2$D$_3$ are missing as in 1α (OH)D$_3$, 25(OH)D$_3$, or 3-epi,1α,25(OH)$_2$D$_3$, the RCI values fall to 24% (for the 3βOH) and fall dramatically to 0.15% (for the separate loss of the 1α-OH or the 25-OH).*

Source: Data taken from R. Bouillon, W.H. Okamura, A.W. Norman, Structure–function relationships in the vitamin D endocrine system. Endocr. Rev. 16 (1995) 200–257.

steroid hormone 1α,25(OH)$_2$_vitamin D$_3$. It would be surprising, therefore, if the intimate local structure of the receptor's ligand-binding site did not have the capability of distinguishing and discriminating between the presence or absence of a key hydroxyl or the various conformational forms of the same steroid hormone.

Fig. 2.8 presents a schematic view of the vitamin D receptor (VDR) as it interacts with the three key hydroxyls on carbons 1α, 3β, and 25 of the steroid hormone 1α,25(OH)$_2$D$_3$ to create three anchoring hydrogen bonds (X, Y, and Z) which collectively create a functional ligand–receptor complex (see Fig. 2.8A). The importance of these three hydrogen bonds has been deduced via the evaluation of the three-dimensional structure of the VDR as determined via X-ray crystallographic analysis with its bound ligand the 1α,25(OH)$_2$D$_3$ steroid hormone. As shown in panel B of Fig. 2.8, a modified ligand [25(OH) D$_3$], which is missing the key 1α-hydroxyl group on the

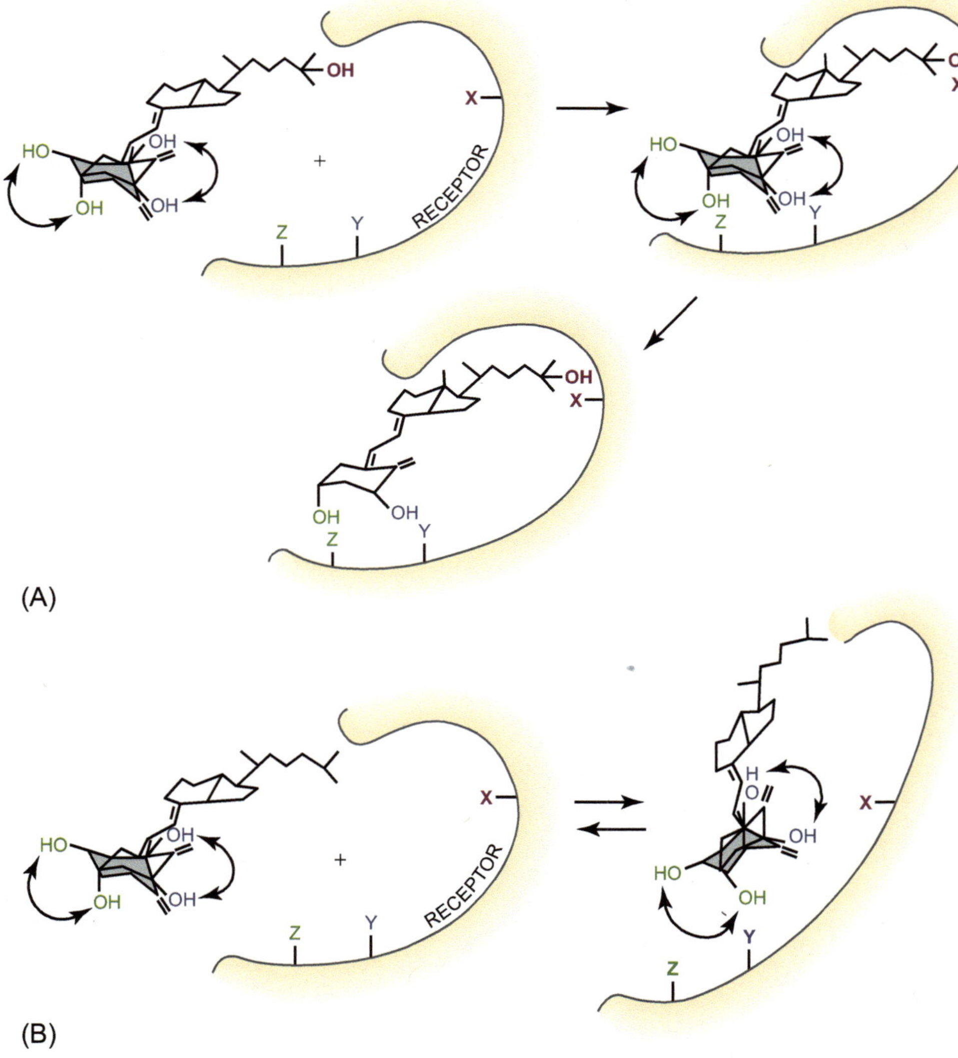

FIGURE 2.8 The absence of the 25-hydroxyl of the steroid hormone 1α,25(OH)$_2$D$_3$ reduces the affinity of 1α (OH)D$_3$ to bind to the VDR by ~660 fold. The schematic model of the VDR illustrates how the receptor for 1α,25(OH)$_2$D$_3$ first "captures" and then forms a stable hydrogen bond (with X) with the 25-hydroxyl group on the end of the conformationally mobile 8-carbon side chain. This is followed by the conformationally mobile A ring's two hydroxyls of 1α,25(OH)$_2$D$_3$ (the 1β and the 3β) to form a stable (using Y and Z) receptor–ligand complex. The X, Y, and Z labels indicate known binding domains on the receptor's interior which each form a stabilizing hydrogen bond with the three hydroxyls of 1α,25(OH)$_2$D$_3$ when the ligand is docked inside the VDR. Panel (A) illustrates the proposed steering effects of the 25-hydroxyl group docking with X which then permits capture of the conformationally active A ring's two hydroxyls by Y and Z. Panel (B) illustrates the consequences of the absence of the side chain 25-hydroxyl group on the poor ability of 1α (OH)D$_3$'s 1α- and 3β-hydroxyls to be captured by the VDR. This is reflected by the fact that 1α (OH)D$_3$ binds to the VDR only 0.15% as well as 1α,25(OH)$_2$D$_3$ (100%; see Table 2.4 and its legend). Thus the following calculation can be made: [RCI for 1α,25(OH)$_2$D$_3$]/[RCI for 1α (OH)D$_3$]=[100]/[0.15]=666-fold difference in relative binding in favor of 1α,25(OH)$_2$D$_3$. *VDR*, Vitamin D receptor.

A-ring, can only bind to the VDR 0.15% as well as the natural hormone, $1\alpha,25(OH)_2D_3$ (see Table 2.4); thus it is unable to form a stable ligand–receptor complex and, therefore, it is not able to produce significant biological responses. Comparable structure–function studies for all the steroid hormones have been carried out by many pharmaceutical companies to identify analogs of the natural steroid hormone(s) that are able to produce useful selective biological responses.

F Other steroid structures

Fig. 2.9 illustrates the diversity of the steroid structures in a wide variety of biological settings ranging from insect

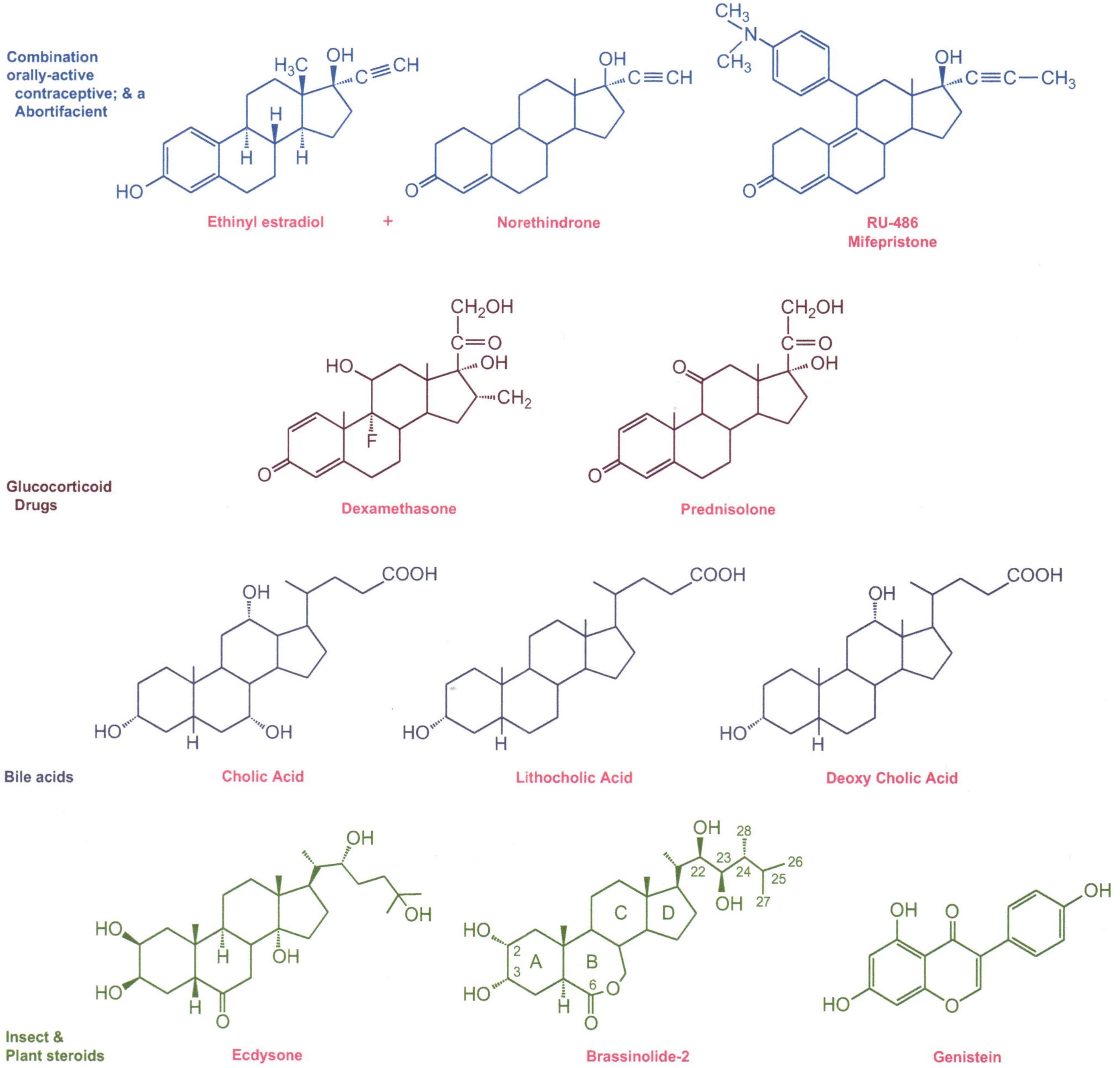

FIGURE 2.9 Structures of biologically significant steroids. Collectively the illustrated steroids emphasize the wide involvement of the backbone cyclopentanoperhydrophenanthrene template (see Fig. 2.1) as a central foundation for the wide diversity of structural additions and modifications that generate unique structures with unique biological functions. The top row (row 1) illustrates two steroids used together as a contraceptive and a third steroid that is an abortifacient or "morning after pill." Row 2 illustrates two widely prescribed drug forms of a glucocorticoid, dexamethasone and prednisolone. Row 3 illustrates three naturally occurring bile acids in man; they are cholic acid, lithocholic acid, and deoxy-cholic acid. Row 4 illustrates an insect steroid hormone, ecdysone, a plant growth regulator-steroid hormone, brassinolide, and a plant member of the family of isoflavonoids, genistein, which have in common phenolic ring structures. Genistein is able to bind to the estrogen receptor. Exposure to genistein occurs principally through foods made with soybeans and soy protein.

steroid hormones (ecdysone) to the world of plant growth regulators (brassinolides) and plant flavinoids (genstein) to functioning as drugs (rows 1 and 2) and as bile acids (row 3). As shown in the top row of Fig. 2.9, a large number of women are able to block their monthly fertile period through the elective use of orally active contraceptive steroids. These are a combination of an estrogen (such as ethinyl estradiol)+a progestin (such as norethindrone). Together these are referred to as a combined oral contraception pill. A controversial, but widely marketed, steroid derivative drug for inducing abortions is RU-486 or Mifepristone; RU-486 is structurally related to progesterone. The second row illustrates two widely prescribed glucocorticoid-related drugs. Dexamethasone is a potent artificial glucocorticoid that is frequently prescribed for the treatment of arthritis, disorders of the immune system, and skin disorders (psoriasis). Prednisolone is a commonly prescribed antiinflammatory steroid. The bile acid structures shown in the third row in Fig. 2.9 are those of the principal forms present in humans. The fourth row illustrates the steroid ecdysone which technically is a prohormone for the major insect molting hormone 20-hydroxy-ecdysone, which binds to its nuclear receptor to modulate gene transcription. The brassinolides are natural plant-growth regulators, some of which function as ligands for membrane receptors. Genistein is a member of the family of plant isoflavonoids that have in common phenolic ring structures. They are produced by the pea family of plants and are biologically active compounds that can bind to the mammalian estrogen receptor.

III Biosynthesis of steroids

A Introduction

1 Profile of bodily cholesterol

The steroid cholesterol (see Fig. 2.5) is the most prevalent steroid in all animals and is essential for animal life. Cholesterol has multiple physiological roles that include its structural presence in all membranes; here it facilitates permeability and fluidity in the various cell membranes as well as the function of clathrin-coated pits and invaginated caveolae. Cholesterol in the caveolae has been shown to be involved in cell signaling processes, by the formation of lipid rafts in the plasma membrane which allows liganded receptor proteins, like the estrogen, progesterone, and $1\alpha,25(OH)_2D_3$ nuclear receptors to be in close proximity with high concentrations of a variety of second messengers which can initiate rapid biological responses.

Cholesterol also is the starting point in the biosynthesis of all steroid hormones, vitamin D_3 and its steroid hormone daughter metabolite, $1\alpha,25(OH)_2D_3$ as well the bile acids.

The level of the total body cholesterol is determined by a complex interplay of dietarily available cholesterol, de novo synthesis of cholesterol, and excretion of cholesterol and bile salts. In an average 70-kg man, 0.8–1.4 g of cholesterol is turned over per day. The daily ingestion of cholesterol in Western countries is 0.5–2.0 g; the efficiency of intestinal absorption ranges from 30% to 50%. However, it is known that dietary cholesterol is not the exclusive source of the bodily pool of cholesterol. The liver and intestine together account for more than 60% of the body's daily biosynthesis of this sterol. When considering the turnover of body cholesterol, it has been suggested that there are two principal components. Pool A, which comprises ~30% of the total body cholesterol, consists of the cholesterol present in liver, bile, plasma, erythrocytes, and intestine. Pool B constitutes the remaining ~70% of the exchangeable body cholesterol; it principally comprises the cholesterol in the skin, adipose tissue, and skeleton.

2 Production of steroid hormones—an introduction

Fig. 2.10 summarizes the six tropic peptide hormones that stimulate the biosynthesis of six essential steroid hormones. The top row identifies the six peptide/proteins that are the stimulatory hormones (chemical messengers). The second row lists six separate endocrine glands each with a specific receptor for the partner tropic hormone (top row). Each endocrine gland is the site of production of a different specific steroid hormone. It is important to appreciate that the production of each steroid hormone is dependent upon the stimulation of its cells of origin by a specific tropic (stimulatory) peptide hormone. For example, adrenocorticotropin hormone (ACTH) that is secreted by the pituitary moves through the circulatory system to the zona fasciculata region of the adrenal cortex where it binds to its partner receptor. This then results in the stimulation of the metabolism of cholesterol to produce the steroid hormone cortisol, a glucocorticoid (see Chapter 10: Adrenal Corticoids) that is released into the circulatory system. The cortisol steroid hormone then delivers its chemical message by travel through the circulatory system to its several target organs, which include the liver, muscle, and adipose tissue.

Fig. 2.11 summarizes both the anatomical sites and major steps of biosynthesis from cholesterol to the final target steroid for each of the six classes of steroid hormones. These include the adrenal cortex (*green panel*), which is anatomically divided into the outer layer zona glomerulosa (aldosterone), the middle layer zona fasciculata (the glucocorticoids corticosterone and cortisol), and an inner layer zona reticularis (the androgen testosterone). The peach-colored panel (Fig. 2.11) focuses on the ovary,

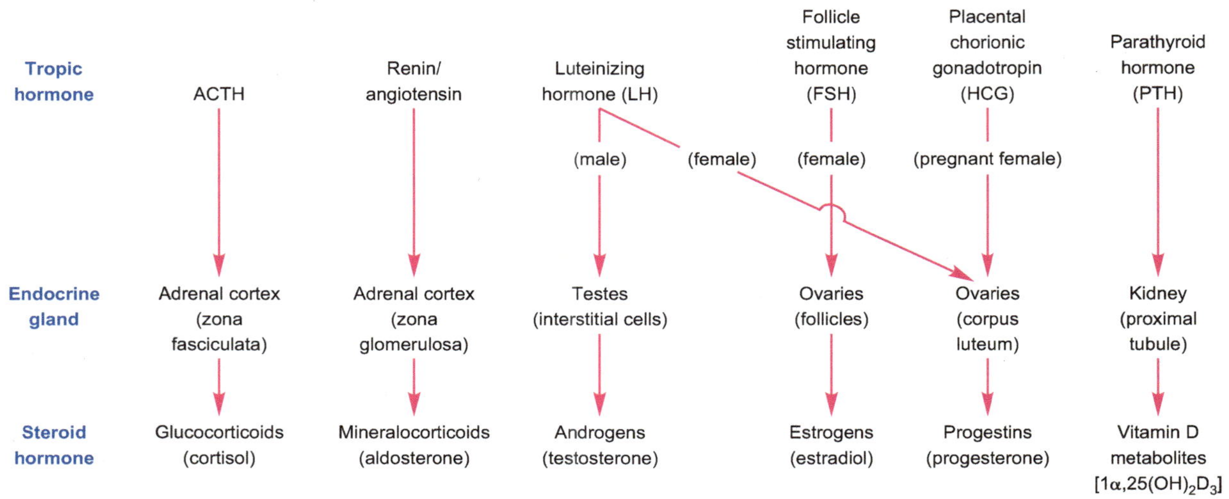

FIGURE 2.10 Relationships between stimulatory peptides or protein hormones and the endocrine glands that are the site of production of six steroid hormones. For each of the six classes of steroid hormones, the specific stimulatory or tropic hormone is indicated (see top row) as well as specification of the organ or cell upon which it acts (see second row), and finally (see bottom row) the newly synthesized steroid hormone. Thus for the tropic hormone ACTH, the newly synthesized steroid is a glucocorticoid (cortisol). The tropic LH has different assignments in males (production of testosterone) and females (production of progesterone). FSH also has different assignments in females (production of estradiol) and in males where it acts upon the Sertoli cells in the seminiferous tubules *not* to increase the production of a steroid hormone but to *increase* the synthesis of the sperm proteins, androgen-binding protein, and inhibin, a peptide hormone. *ACTH*, Adrenocorticotropin hormone; *FSH*, follicle-stimulating hormone; *LH*, luteinizing hormone.

which is anatomically divided into the corpus luteum (progesterone) and the theca and granulosa cells (estradiol). The yellow panel (Fig. 2.11) focuses on the testes' Leydig cells (testosterone). The blue panel (Fig. 2.11) summarizes anatomically the five key steps for conversion of cholesterol into 7-dehydrocholesterol and then vitamin D_3 (in the skin) followed by 25-hydroxylation (in the liver) and followed by the production of the steroid hormone $1\alpha,25(OH)_2D_3$ and the metabolite $24,25(OH)_2D_3$ (both in the kidney). See Fig. 9.7 for further details.

The bile acids, which are the seventh important structural class of mammalian steroids, are principally synthesized in the liver and they function as detergents in vivo to facilitate digestion and absorption of lipids. Also, some orphan nuclear receptors, for example, the farsenoid X receptor and the LXRα receptor, were found to specifically bind cholic and chenodeoxycholic acids (see Fig. 2.9) and to regulate selected gene transcription.

B Properties steroidogenic enzymes

1 Introduction

The role of cholesterol as the starting point for the biosynthesis of all the steroid hormones and bile acids emphasizes the important role that steroid hydroxylation reactions play in the first steps of cholesterol's side chain cleavage. The cleavage of the cholesterol side chain requires the hydroxylation of both carbon-22 and carbon-20 (see Fig. 2.18). Other examples of key hydroxylation reactions include the following: (1) the production of testosterone necessitates hydroxylation of progesterone at carbon-17 followed by the cleavage of the carbon-17—carbon-20 bond aldehyde (see Fig. 2.20); (2) in the biosynthesis of cortisol, three hydroxylases, the 17α-, the 21-, and the 11β-hydroxylases, are required; and (3) the conversion of deoxycorticosterone (DOC) to aldosterone would seemingly involve three separate enzymes, which sequentially hydroxylate carbon-11, then carbon-18, followed by the oxidation of the carbon-18 alcohol to an aldehyde (see Fig. 2.19). It has been learned from molecular biological cloning that one enzyme carries out all three of these steps.

Table 2.5 tabulates the key enzymes concerned with the production of the steroid hormones and enumerates the details of the reactions mediated. The presentation includes long-form name of the enzyme, as well as short name or acronym; a specification of the reaction(s) catalyzed, the subcellular location of the enzymatic activity, and the chromosomal location. Subsequent sections of this chapter focus on a brief overview of particularly important enzymes.

2 Cytochrome P450 steroid hydroxylases

Cytochrome P450 enzymes are responsible for the oxidative metabolism of a diverse family of compounds; these include prostaglandins, fatty acids, biogenic amines, plant metabolites, and steroids, as well as chemical carcinogens,

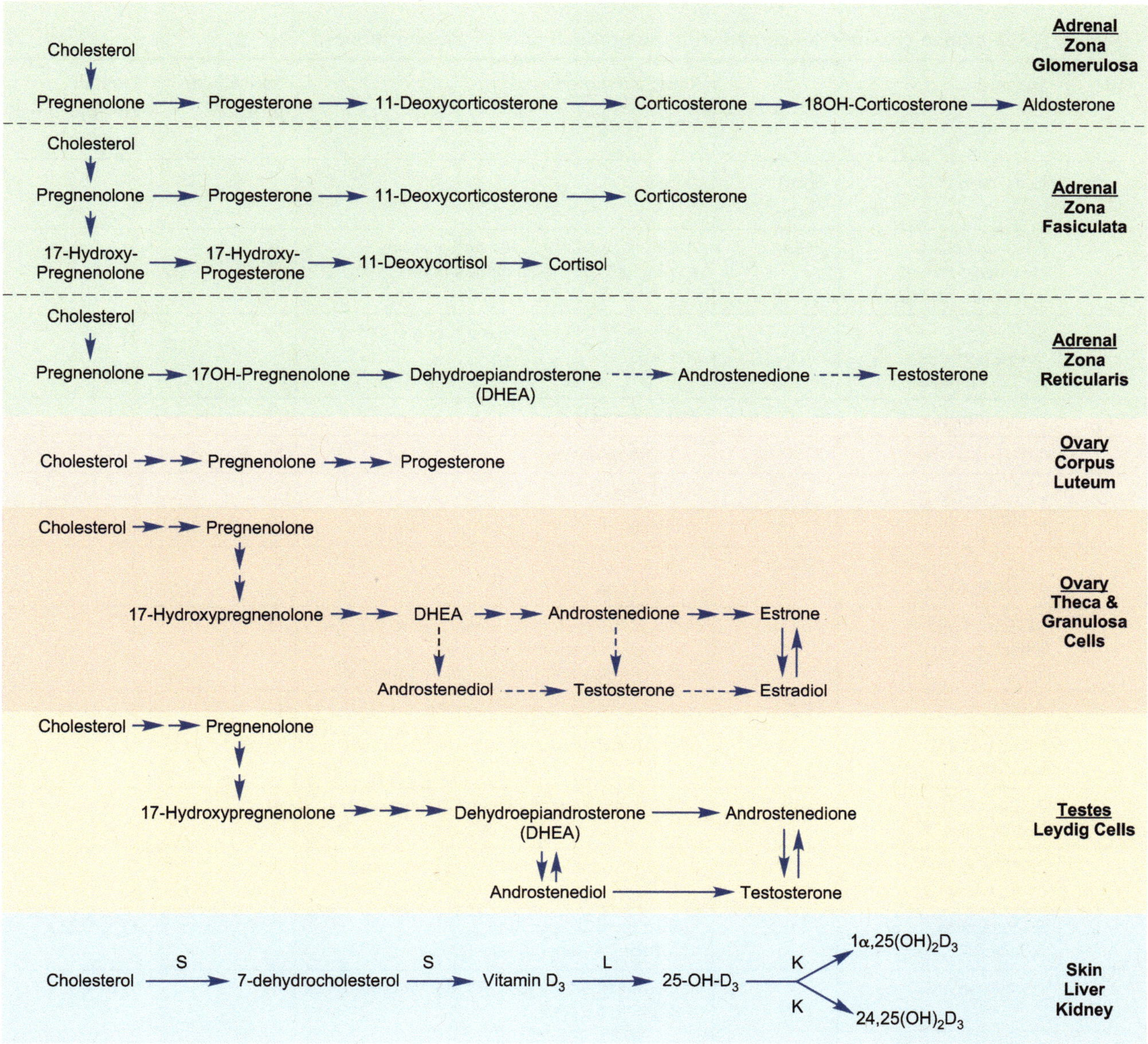

FIGURE 2.11 Pathways of biosynthesis of the principal families of human steroid hormones. Each color panel represents a different anatomical location for the biosynthesis of the designated steroid family. They include the adrenal cortex panel (in *light green*) for its zona glomerulosa (*aldosterone*), zona fasciculata (*cortisol* and *corticosterone*), and the zona reticularis (*testosterone*). Each of the "zona" regions are separated by a dashed line. The sex steroid hormones are shown in the light peach panel for the female ovarian corpus luteum (*progesterone*), the female ovarian theca and granulosa cells (*estradiol*); and in yellow, the male testes Leydig cells (*testosterone*). The bottom blue panel summarizes the biosynthetic steps in the skin, liver, and kidney for the conversion of cholesterol into vitamin D_3 and then into its steroid hormone, 1α,25$(OH)_2$-vitamin D_3. *Modified with permission from W.L. Miller, R.J. Auchus, The molecular biology, biochemistry, and physiology of human steroidogenesis and its disorders, Endocrine Reviews 32 (2011) 81–151.*

mutagens, many drugs, and other environmental pollutants. The following paragraphs present a brief description of the important steroid-related P450 enzymes and their electron transport chain partners.

All of the seemingly diverse steroid hydroxylase reactions are mediated by a family of homologous oxidative enzymes (~57 human enzymes) collectively known as the cytochrome P450 hydroxylases. Each individual P450 enzyme is composed of about 500 amino acids and has a single heme (protoporphyrin ring with a single chelated Fe^{2+} group). The cytochrome moiety is structurally analogous to the hemoprotein cytochromes of the electron transport chain present in mitochondria that are dedicated to the production of ATP. All contain some kind of covalently bound protoporphyrin ring coordinately bound to one atom of iron, which can be reversibly oxidized and reduced. As a class, most of these P450 enzymes are subject to inhibition by the presence of carbon monoxide.

TABLE 2.5 Key human enzymes concerned with the production of steroid hormones.[a]

Entry	Enzyme		Reaction(s) mediated[d]	Subcellular location[e]	Human chromosomal location[f]
	Long name[b]	Short name[c]			
1	P450 side chain cleavage; formerly known as desmolase	P450scc	20*R*-hydroxylation; 22*R*-hydroxylation; scission of C-20–C-22 carbon bond	M	15q23-q24
2	3β-Steroid dehydrogenase/ Δ^5,Δ^4-isomerase	3βHSD1 3βHSD2	Oxidizes 3β-OH to oxo group; isomerizes Δ^5-ene to Δ^4-ene	ER	1p13.1
3	21-Hydroxylase	P450c21	21-Hydroxylation	ER	6p21.1
4	11β-Hydroxylase	P450cllβ	11β-Hydroxylation	M	1q32
			18-Hydroxylation	M	8q21–22
			Oxidation of C-18 hydroxyl to an aldehyde	M	8q21–22
5	17α-Hydroxylase/C-17–C-20 lyase	P450c17*d*	17α-Hydroxylation Scission of C-17–C-20 carbon bond	ER	10q24.3
6	17-Ketosteroid reductase	AKR1C3	Reduces 17-oxo to a 17β-hydroxyl, as in androstenedione→testosterone; estrone→estradiol; DHEA→androstenediol	ER	
7	Aromatase	P450aro	Mediates the aromatization of the A ring	ER	15q 21.1
8	Vitamin D 25-hydroxylase	P450c25	25-Hydroxylation of vitamin D secosteroids	M, ER	
9	$25(OH)D_3$ 1α-hydroxylase	P450c1	1α-Hydroxylation of vitamin D secosteroids	M	
10	$25(OH)D_3$ 24R-hydroxylase	P450c24	24*R*-Hydroxylation of vitamin D secosteroids	M	
11	5α-Reductase-1 5α-reductase-2		Reduction of Δ^5-ene as in testosterone→dihydrotestosterone	NM	5P15 2P23
12	Ferredoxin reductase		In the mitochondria transfers electrons from NADPH to ferredoxin	M	17q24-q25
13	P450 oxidoreductase		In the endoplasmic reticulum transfers electrons from NADPH to a P450 enzyme	ER	7q11.2
14	11β-Hydroxysteroid dehydrogenase	11βHSD1 11βHSD2	Oxidation of 11β-hydroxyl to an oxo as in cortisol→cortisone	ER	1q32-q41 16q22
15	12α-Hydroxylase		12α-Hydroxylation of bile acids	ER	
16	6β-Hydroxylase		6β-Hydroxylation of bile acids	ER	
17	7α-Hydroxylase		7α-Hydroxylation of bile acids	ER	

DHEA, Dehydroepiandrosterone.
[a]*This table summarizes the key enzyme reactions in Figs. 2.11, 2.18–2.21. Note that, in some instances, the same enzyme catalyzes more than one reaction.*
[b]*The full names of the 17 steroid-related enzymes are provided. They belong to either the cytochrome P450 family or the hydroxysteroid dehydrogenases family.*
[c]*A short name in the form of an acronym is provided for many of the enzymes.*
[d]*For each of the enzymes a description of the reaction(s) that are catalyzed by each enzyme is provided.*
[e]*The subcellular location of each enzyme is specified; ER, Endoplasmic reticulum; M, mitochondria; NM, nuclear membrane.*
[f]*For each enzyme the location of its chromosome in the human genome is specified.*
Source: Portions of the data were abstracted from W.L. Miller, R.J. Auchus, Endocrine Reviews 32 (2011) 81–151.

The CO coordinates to the Fe^{2+}, thus blocking the essential cyclical oxidation and reduction that are associated with electron transfer. This inhibition can then be selectively reversed by light with a wavelength of 450 nm. Cytochrome P450 steroid enzymes are known to be present in the liver, adrenal cortex, ovary, testis, kidney, placenta, lungs, intestinal mucosa, and selected regions of the brain.

In each P450 hydroxylase enzyme, the general oxidation–reduction reaction catalyzed is as shown in Fig. 2.12. In principle, the oxygen atom of the hydroxyl group shown as "final hydroxylated product" could be derived from either H_2O or molecular oxygen (O_2). But as indicated in the equation, it is known that the hydroxyl oxygen is derived exclusively from molecular oxygen in all steroid cytochrome P450-containing hydroxylases. As a class, these enzymes are designated as mixed-function oxidases. The term "mixed function" indicates that one atom of the diatomic substrate oxygen ends up in the steroid (as part of the hydroxyl) and the other ends up as part of water. This result has been confirmed by utilization of $^{18}O_2$ and the subsequent appearance of ^{18}O both in the steroid and in $H_2{}^{18}O$.

Virtually all steroid P450 hydroxylases are membrane bound and are present in either the mitochondria (type 1) or endoplasmic reticulum (type 2) of the cell (see Fig. 2.13). Also Table 2.5 tabulates the subcellular localization of the various cytochrome P450 steroid hydroxylases. Depending upon whether the cytochrome P450 hydroxylase is localized in the mitochondria or microsomes, there are two somewhat different electron transport chains that function to transfer a pair of electrons from NADPH to the cytochrome P450 enzyme. These are summarized in Fig. 2.13. In the mitochondrial system,

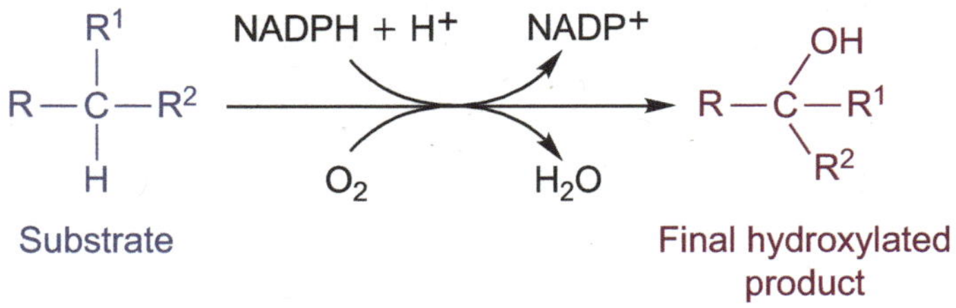

FIGURE 2.12 General reaction of a cytochrome P450-catalyzed steroid hydroxylase. There is a family of related steroid hydroxylase members that utilize NADPH and O_2 as substrates and produce as products a hydroxylated steroid and H_2O.

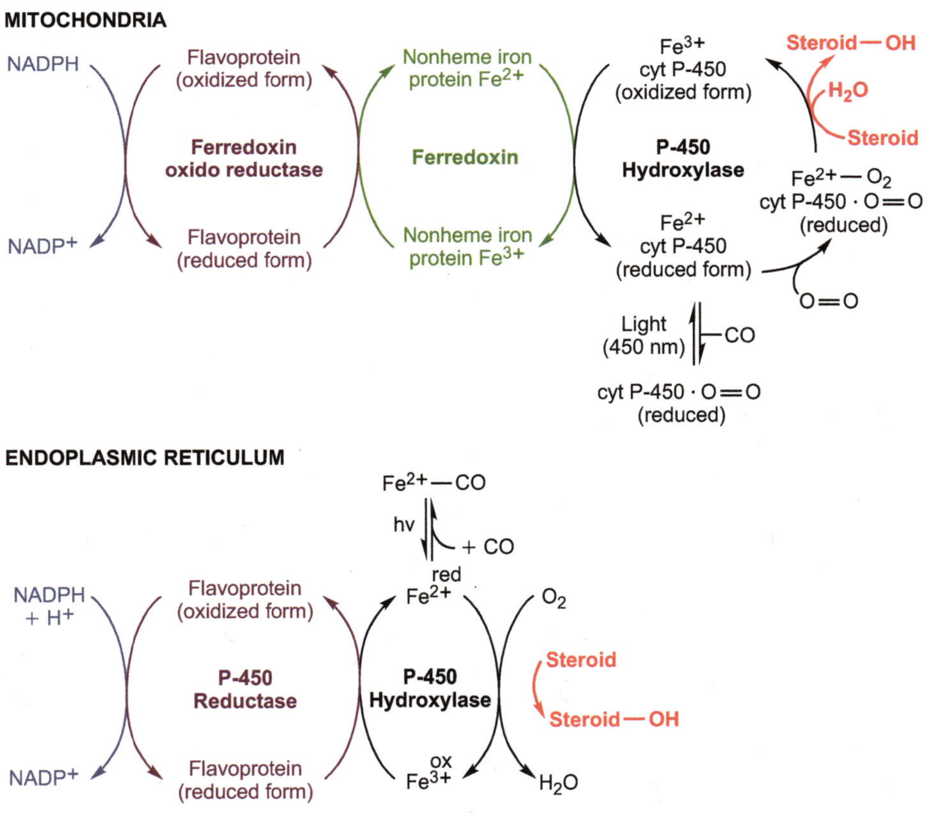

FIGURE 2.13 Electron transport chains for sterol hydroxylases. (A Top) Mitochondrial hydroxylases showing the participation of the three protein components, ferredoxin oxidoreductase, ferredoxin, and a P450 steroid hydroxylase. (B Bottom) Endoplasmic reticulum hydroxylases showing the participation of the two protein components, P450 reductase and a P450 steroid hydroxylase. In both the mitochondrial and endoplasmic reticulum, the inhibitory actions of exposure to carbon monoxide, CO, on the P-450 hydroxylase by the binding of the CO to the oxygenated-P-450 heme group, which displaces and releases the O_2.

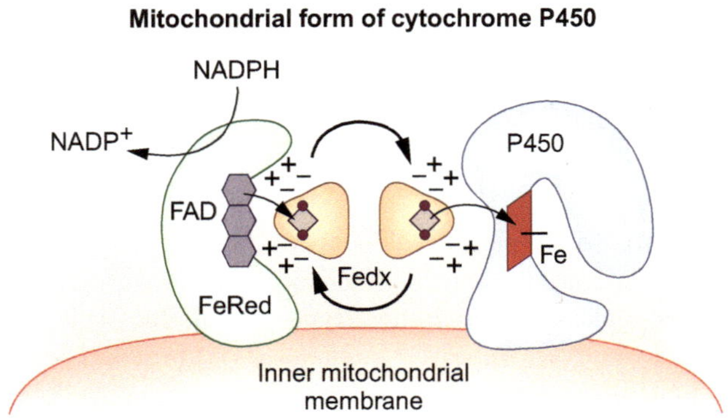

FIGURE 2.14 Electron transport to IMM cytochrome P450 hydroxylases. There are three key mitochondrial proteins that participate in the transfer of electrons from NAPDH to the terminal P450 that generates as a final product a hydroxylated steroid. First the FAD group of FeRed accepts two electrons from NADPH, thus producing the product $NADP^+$. The reduced FAD bound to FeRed then passes the two electrons to an iron–sulfur ferredoxin (Fedx; 12 kDal) protein (see small diamond with two maroon dots). The Fedx then transfers the electrons to the heme of the P450 that is associated with the IMM. To facilitate the two electron transfers, first the negative charges of the $Fedx^{(-)}$ guide the Fedx docking with the positive charges of the $FeRed^{(+)}$ and then after movement to the P450 docking with the positive charges (+) of the P450. For the conversion of one cholesterol into one pregnenolone, three pairs of electrons from NAPH must be transported via the FeRed, Fedx to the P450 to carry out the side chain cleavage process. *FAD*, Flavin adenine dinucleotide group; *FeRed*, ferredoxin reductase; *IMM*, inner mitochondrial membrane. *Modified with permission from W.L. Miller, R.J. Auchus, The molecular biology, biochemistry, and physiology of human steroidogenesis and its disorders, Endocrine Review 32 (2011) 81–151.*

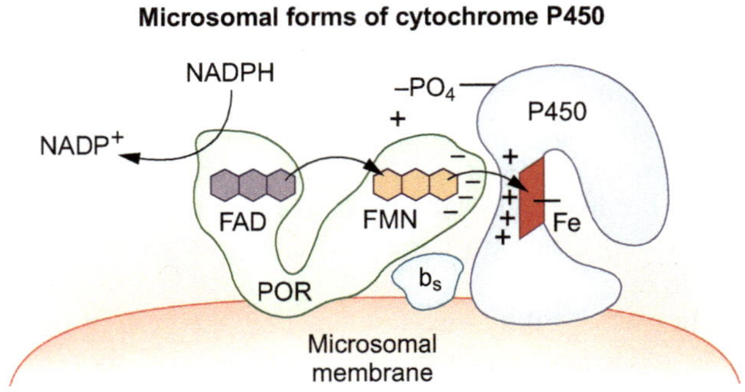

FIGURE 2.15 Electron transport to microsomal forms of cytochrome P450 hydroxylases. There are only two key microsomal proteins that participate to transfer electrons from NAPDH to the terminal P450 that generates a hydroxylated steroid. The POR, which receives a pair of electrons from NADPH, has two bound cofactors that facilitate the transfer of electrons; they are the FAD and the FMN, which work sequentially together to transfer electrons to the second microsomal protein, the P450 oxidoreductase. Receipt of electrons by the POR's bound FAD results in an enzyme conformational change that facilitates the FAD passing its electrons via a redox reaction to the FMN. Then after another conformational change, the FMN domain of the POR protein interacts with the P450 protein's heme group to facilitate (in a redox reaction) the transfer of electrons that will be needed to attach a hydroxyl group on a steroid substrate. The interaction of P450 with the POR is facilitated by the negatively charged acidic residues on the surface of the FMN domain of POR with the positively charged basic residues on the P450 redox-partner. The binding site for the substrate steroid is located on the opposite side of the heme ring (shown as a rectangle) from that interacting with the POR's reduced FMN. *FAD*, Flavin adenine dinucleotide; *FMN*, flavin mononucleotide; *POR*, P450-oxidoreductase. *Modified with permission from W.L. Miller, R.J. Auchus, The molecular biology, biochemistry, and physiology of human steroidogenesis and its disorders, Endocrine Review 32 (2011) 81–151.*

there are at least three separate components: (1) a flavoprotein dehydrogenase known as ferredoxin oxidoreductase, which accepts the electrons from NADPH; (2) a nonheme iron protein, termed ferredoxin[3] (molecular weight 13,000), which accepts the electrons from the flavoprotein and transfers them to (3) the cytochrome P450 hydroxylase protein. The molecular details of the mitrochondrial electron transport process are summarized in the legend to Fig. 2.14. Since the microsomal electron transport system lacks a ferredoxin component, the flavoprotein dehydrogenase transfers electrons directly to the cytochrome P450 enzyme. The molecular details of the microsomal electron transport process are summarized in the legend of Fig. 2.15.

The most important component of both the mitochondrial and microsomal electron transport chain is the cytochrome P450 protein, in that it defines the substrate specificity of the enzyme. Each P450 hydroxylase has a substrate-binding domain that is comparable in its ability to define substrate specificity to that of the ligand-binding domains of steroid receptors and plasma transport proteins for steroid hormones. Thus the three-dimensional structure of the substrate-binding domain of a P450 hydroxylase determines which of the some 22–27 carbons of the substrate will acquire a new hydroxyl group. Comparison of the subcellular localization of the various hydroxylase enzymes with the sequence of steroid movement through a metabolic pathway indicates the important role of cellular compartmentalization. Thus in the conversion of cholesterol into cortisol in the adrenal cortex, the steroid must move sequentially from the mitochondria (side chain cleavage; see Fig. 2.18) to the microsomes (17α- and 21-hydroxylation) and then back to the mitochondria (11β-hydroxylation).

3 3β-Hydroxysteroid dehydrogenase enzyme/ Δ^5,Δ^4-isomerase (3βHSD)

The second most prevalent class of the steroidogenic enzymes are the hydroxysteroid dehydrogenases (HSD). Each HSD enzyme contains 290–380 amino acids (35–45 kDa) that utilize as electron acceptors or electron

3. In the older literature (prior to 1980), this nonheme iron protein was discovered in the adrenals and termed adrenodoxin. With the appreciation of the generic functioning of this heme protein in mitochondrial P450 hydroxylation in many tissues it is now referred to as ferredoxin.

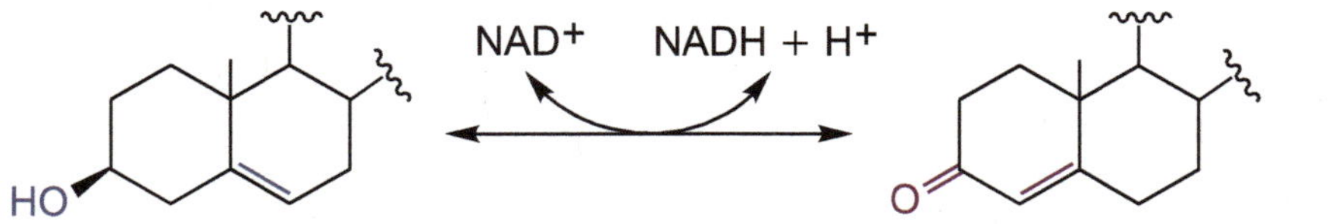

FIGURE 2.16 Reaction catalyzed by the 3β-hydroxysteroid dehydrogenase, Δ^5,Δ^4-isomerase (3βHSD1). Here the dehydrogenation of the 3β-OH is linked by the same enzyme to the isomerization of the adjacent 5,6 double bond to the 4,5 double bond.

donors the nicotinamide adenine dinucleotide (NADH/NAD^+ or NADPH/$NADP^+$) cofactors. Thus a steroid substrate donates two electrons to NAD^+ (see Fig. 2.16; see also Table 2.5).

A key enzymatic transformation in the biosynthetic pathway of five steroid hormones is the cleavage of cholesterol's side chain that is then followed by the conversion of the 5-ene 3β-hydroxysteroid to a 4-ene 3-oxosteroid (see reaction 2 of Table 2.5 and Fig. 2.18). This transformation is mediated by one enzyme located in the endoplasmic reticulum that carries out the two steps. The same enzyme is involved in the adrenals, ovaries, testes, and placenta.

The 3β-hydroxysteroid dehydrogenase Δ^5,Δ^4-isomerase has an obligatory cofactor NAD^+ and catalyzes the reactions shown in Fig. 2.16. The enzyme has been cloned and expressed; evidence indicates the expression of multiple forms of the 3β-hydroxy dehydrogenase/Δ^5,Δ^4-isomerase.

4 Ferredoxin

Ferredoxin, which has ~116 amino acids (14 kDa), is a sulfur/iron electron shuttle protein in the mitochondrial electron transport process associated with steroid hydroxylation (see top panel of Fig. 2.13). It is synthesized as a larger preproprotein having both carboxy-terminal (~60 amino acids) and amino-terminal extensions. The N-terminal extension apparently serves as a leader sequence to facilitate the entrance of the ferredoxin peptide into the mitochondria. The ferredoxin protein acts as a shuttle, accepting electrons from ferredoxin oxidoreductase, which then diffuses in the mitochondrial matrix to a P450 hydroxylase (see Figs. 2.14 and 2.15) where it donates a pair of electrons.

5 Ferredoxin reductase

Ferredoxin oxidoreductase (see Table 2.5) is an inner mitochondrial membrane (IMM)-bound flavoprotein with a molecular weight of 51,100. It is responsible for the transfer of electrons from NADPH to ferredoxin (see Fig. 2.14) and is widely expressed in many human tissues. The details of the electron transfer process are described in Fig. 2.14.

6 Steroidogenic acute regulatory protein

The first step in classical steroid hormone biosynthesis involves the conversion of cholesterol into pregnenolone by the cholesterol side chain cleavage enzyme P450scc (see Fig. 2.18 and Table 2.5, entry 1). Although the side chain cleavage reaction is known to be the rate-limiting enzyme for adrenal and gonadal steroid hormones, it is not the catalytic process of the cholesterol side chain cleavage enzyme that is rate limiting.

A specific cholesterol transport protein, known as the *steroidogenic acute regulatory protein* (StAR), has been implicated as being the essential mediator of cholesterol delivery to the mitochondria and the consequent activation of P450scc (see Fig. 2.17). StAR is biosynthesized as an ~300-amino acid protein with a binding pocket for a single cholesterol molecule. StAR has a mitochondrial targeting sequence of 25 amino acid residues, which is cleaved from the NH_2-terminus after it is transported into the mitochondria. More specifically, the rate-limiting step in classical steroid hormone biosynthesis is the presence and properties of the StAR at the outer mitochondrial membrane (OMM) where it facilitates the movement of cholesterol across the OMM to the IMM site of the P450scc. It is known that the adrenal tropic ACTH hormone (see Fig. 2.10) can rapidly stimulate the gene transcription of StAR in the adrenal cortex which then is the exclusive mediator of the transport of cholesterol from the OMM to the IMM.

The most convincing evidence for the essential nature of StAR in facilitating the actions of cholesterol side chain cleavage and the resultant production in the adrenals of mineralocorticoids and glucocorticoids, and in the gonads, but not the placenta, of estrogens or androgens is that mutations of the StAR gene can result in a defect associated with the disease known as lipoid congenital adrenal hyperplasia (CAH). (See Chapter 10: Adrenal Corticoids, Section V.C, "Congenital Adrenal Hyperplasia and Others.") In lipoid CAH, which is characterized by a deficiency of both adrenal and gonadal steroid hormones, there is a mutation in the StAR gene, so that the StAR protein is nonfunctional due to premature stop codons.

7 Cholesterol side chain cleavage enzyme (p450scc)

The primary enzyme that catalyzes the cleavage of the side chain of cholesterol to initiate the process of steroidogenesis is P450scc that produces as its product pregnenolone (see Fig. 2.18). The p450scc is known as a

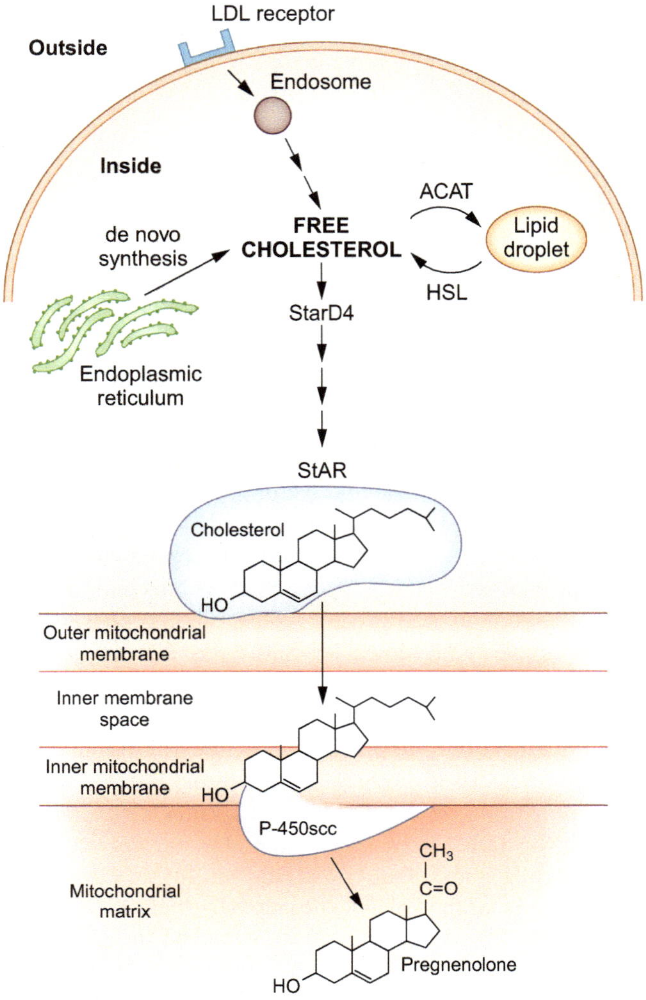

FIGURE 2.17 Model describing the role of the StAR protein in cholesterol side chain cleavage. In humans, cholesterol present in the circulatory system bound to LDL is imported via the plasma membrane LDL-receptor into the adrenal cortex cell's cytosol endosomes where the cholesterol is released. It is also possible for cholesterol to be biosynthesized de novo from acetate in the adrenal cell's endoplasmic reticulum. Free cholesterol can be esterified by ACAT and stored in lipid droplets. In the adrenal cortex, ACTH stimulates a HSL that generates free cholesterol to be used in the mitochondria as a substrate for the cholesterol side chain cleavage enzyme (see Table 2.5, entry 1). There is a family of ~15 StAR cholesterol transport proteins (mol. wt. ~37 kDa) each of which has a single binding site for cholesterol. Some of the StAR family members, for example, StarD4, are believed to assist in transfer of the hydrophobic cholesterol through the cytosol to the mitochondria, but the details are not yet delineated. In the adrenals, StAR facilitates the rapid transfer of the hydrophobic cholesterol molecule across the OMM to the inner mitochondrial membrane where the cholesterol side chain cleavage enzyme (P450scc) is located. Thus with the assistance of StaR, cholesterol is metabolized into pregnenolone (see Fig. 2.17). *ACAT*, Acyl-coenzyme A cholesterol acyltransferase; *ACTH*, Adrenocorticotropin hormone; *HSL*, hormone-sensitive lipase; *LDL*, low density lipoproteins; *OMM*, outer mitochondrial membrane; *StAR*, steroidogenic acute regulatory. *Modified with permission from W.L. Miller, R.J. Auchus, The molecular biology, biochemistry, and physiology of human steroidogenesis and its disorders, Endocrine Review 32 (2011) 81–151.*

lyase. Lyases are enzymes that cleave bonds between carbons bearing vicinal hydroxyls. The properties of these two reactions will be discussed separately.

The cholesterol side chain cleavage enzyme (P450scc) (reaction 1 in Table 2.5) carries out three successive enzymatic steps that are all physically associated with the IMM. These involve the separate 20*R*- and 22*R*-hydroxylations followed by cleavage of the side chain on the two carbons bearing the newly incorporated hydroxyl groups. Each of these steps is an oxidation–reduction reaction requiring a pair of electrons, which is supplied by NADPH. The P450scc protein molecular weight is 49,000. The P450scc from the human adrenal and testes have identical sequences.

8 17α-Hydroxylase/C-17α,20 lyase (P450c17)

Another enzyme that carries out both a hydroxylation reaction and cleavage of a carbon–carbon bond is the P450c17 that is present in three locations. These include in the adrenal zona reticularis (Fig. 2.19), the testicular Leydig cells (see entries 5 and 6 in Table 2.5 and Fig. 2.20), and the ovarian theca and granulosa cells (see Fig. 2.21). The P450c17 enzymes from both the adrenals and testes have been cloned and expressed and found to be identical proteins.

9 11β-Hydroxylase

The 11β-hydroxylase (P450c11) is still another example of one enzyme that can catalyze more than one enzymatic reaction. In cattle, there is one P450 gene with one protein that has 11β-hydroxylase, 18-hydroxylase, and 18-methyl oxidase activities (entry 4 in Table 2.5; see Figs. 2.19). However, in man there are two separate genes *CYP11B1* and *CYP11B2* that produce two separate enzymes, P450c11β (11β-hydroxylase) and P450c11AS (aldosterone synthase), respectively. Both of these enzymes are present in the IMM and donate electrons to ferredoxin and ferredoxin reductase. The existence of two genes is consistent with known human clinical deficiency states associated with CAH (see Chapter 10: Adrenal Corticoids).

10 21-Hydroxylase (P450c21)

The 21-hydroxylation of steroids is essential to the conversion of both progesterone to DOC in the zona glomerulosa (mineralocorticoids) and 17-hydroxy-progesterone to 11-deoxycortisol in the zona fasciculata (glucocorticoid). See Fig. 2.19 for the steps of steroid metabolism in the zona glomerulosa and zona fasciculata, and entry 3 in Table 2.5. The P450c21 is found in the endoplasmic reticulum. Evidence has been presented that humans have two genes coding for P450c21. The details of the expression

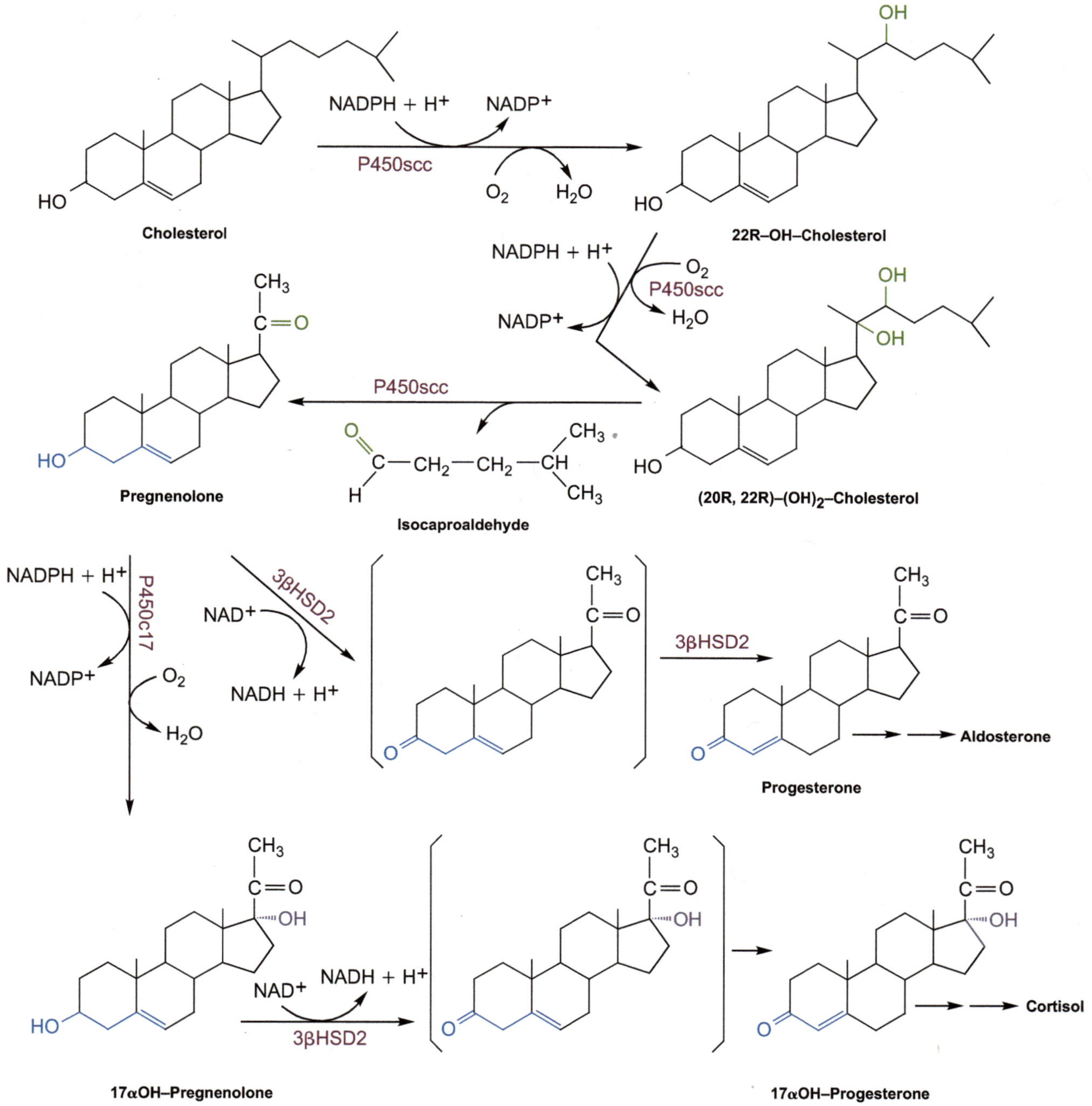

FIGURE 2.18 Cholesterol side chain cleavage to generate the intermediate pregnenolone followed by its separate conversion to either progesterone or to 17αOH-progesterone. The cholesterol side chain cleavage enzyme (csc; see entry 1 in Table 2.5) carries out three distinct enzymatic steps, comprising two successive hydroxylations on C-20 and C-22 (shown in *green*) followed by scission of the C-20–C-22 carbon–carbon bond that releases as products pregnenolone and the rest of the steroid side chain as isocaproaldehyde. Pregnenolone, depending upon the cell type (adrenals, ovary, or testes; see Fig. 2.11), can ultimately be converted into aldosterone, cortisol, estradiol, or testosterone. The two immediate fates of prenenolone are its conversion either by the P450c17 enzyme (see entry 5 of Table 2.1) into 17αOH-pregnenolone or by the 3βHSD2 enzyme into progesterone or 17αOH-progesterone. The 3βHSD2 enzyme carries out two distinct catalytic steps (see entry 2 of Table 2.5): first a dehydrogenation of the 3β-hydroxyl (shown in *blue*) followed by isomerization of the Δ^5,Δ^6 double bond to become a Δ^4,Δ^5 double bond (shown in *blue*) for both progesterone and 17αOH-progesterone.

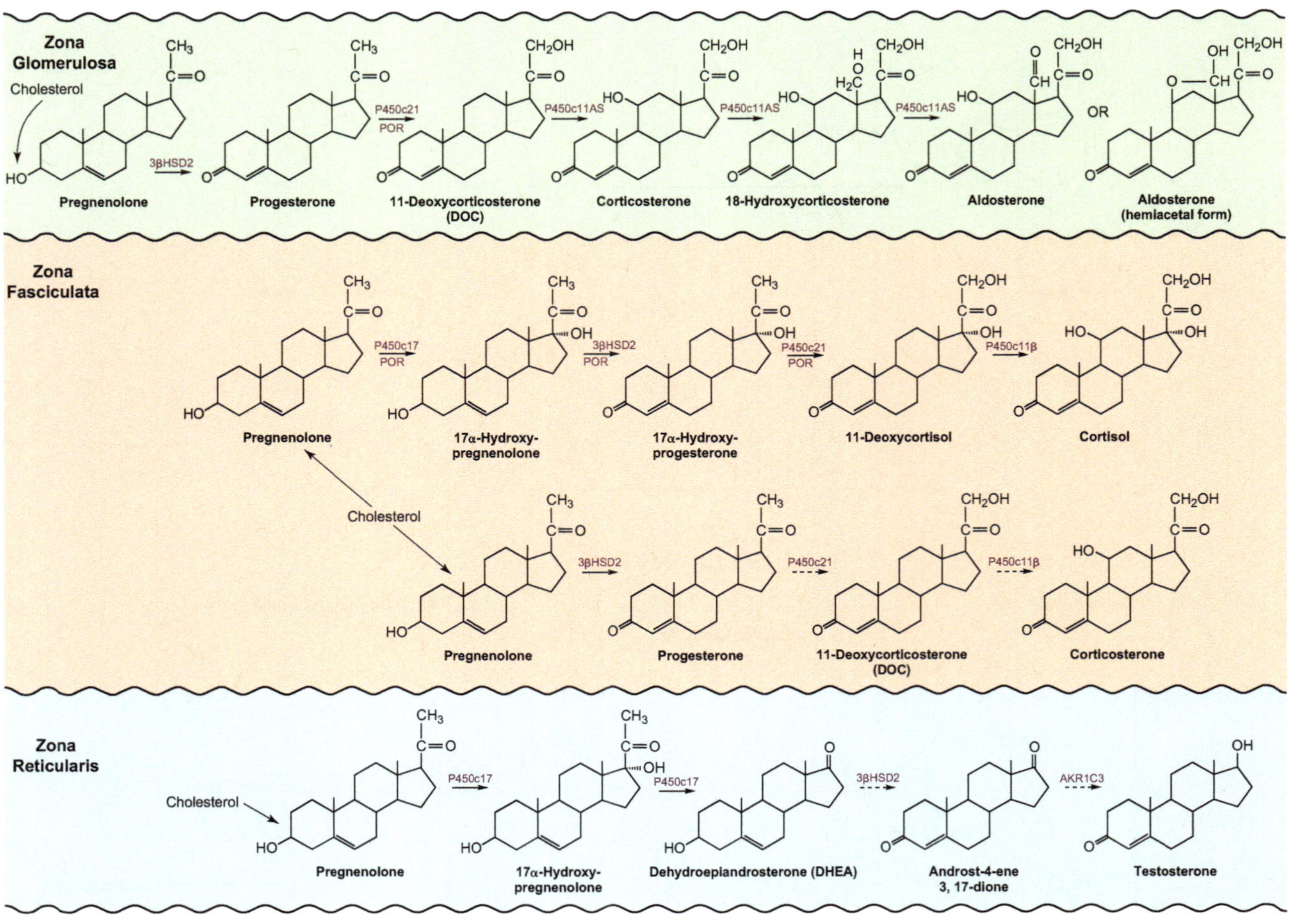

FIGURE 2.19 Adrenal cortex pathways for the production of three classes of steroid hormones. Cholesterol and the presence of the cholesterol side chain cleavage enzyme are the starting point of separate steroid hormone(s) biosynthesis for the three classes. The *zona glomerulosa* (mineralocorticoids) produces aldosterone. A structural hallmark of aldosterone is the presence of both a C-11 hydroxyl and a C-18 aldehyde. The C-18 aldehyde can form either a five member hemiacetal ring, which uses the C-11 hydroxyl group or a six member hemiacetal ring, which uses the C-21 hydroxyl group. These are reminiscent of carbohydrate chemistry. The *zona fasciculata* (glucocorticoids) produces both cortisol and corticosterone. The *zona reticularis* produces limited amounts (represented by the dashed arrow) of the androgens, DHEA, androst-4-ene-3,17-dione, and testosterone. DHEA can be converted to DHEA–sulfate via a sulfotransferase, which forms a mixed ester with the 3β-hydroxyl group. For each of the steroid enzyme transformations, the 3–7 letter/number acronym is shown in magenta next to the arrow. The enzyme's full name for each acronym is tabulated in Table 2.5. *DHEA*, Dehydroepiandrosterone.

of these genes are clinically very important since approximately 1 in 15,000 live births has a 21-hydroxylase deficiency characteristic of CAH; this P450c21 mutation constitutes as many as 90% of all CAH patients. A detailed discussion of this topic is beyond the scope of this chapter.

11 Aromatase (P450aro)

The aromatization of the A ring of either testosterone to yield estradiol or androst-4-ene-3,17-dione to yield estrone is mediated by P450aro. This enzyme is present in the endoplasmic reticulum of the ovary and placenta and also occurs to a limited extent in brain, bone, and adipose tissues. A proposed mechanism for the complex aromatization process has been described (see legend to Fig. 2.22 and entry 7 of Table 2.5).

12 5α-Reductase

The conversion of testosterone to dihydrotestosterone (see Fig. 2.20 and entry 15 of Table 2.5) is mediated by the 5α-reductase. 5α-Reductase is an NADPH-dependent non-P450 enzyme that is present in the nuclear membrane of some target cells for testosterone, for example, seminal vesicles, epididymis genital skin, hair follicles, and liver. It is present in lower concentrations in nongenital skin, testes, and kidney.

13 17-Ketosteroid reductase (AKR1C3)

17-Ketosteroid reductase is a non-P450 enzyme present in the endoplasmic reticulum; it requires NADPH (see reaction 6 of Table 2.5 and Fig. 2.22). The catalyzed reactions of the 17-ketosteroid reductase are readily reversible. It is present in the testes, ovaries, and placenta.

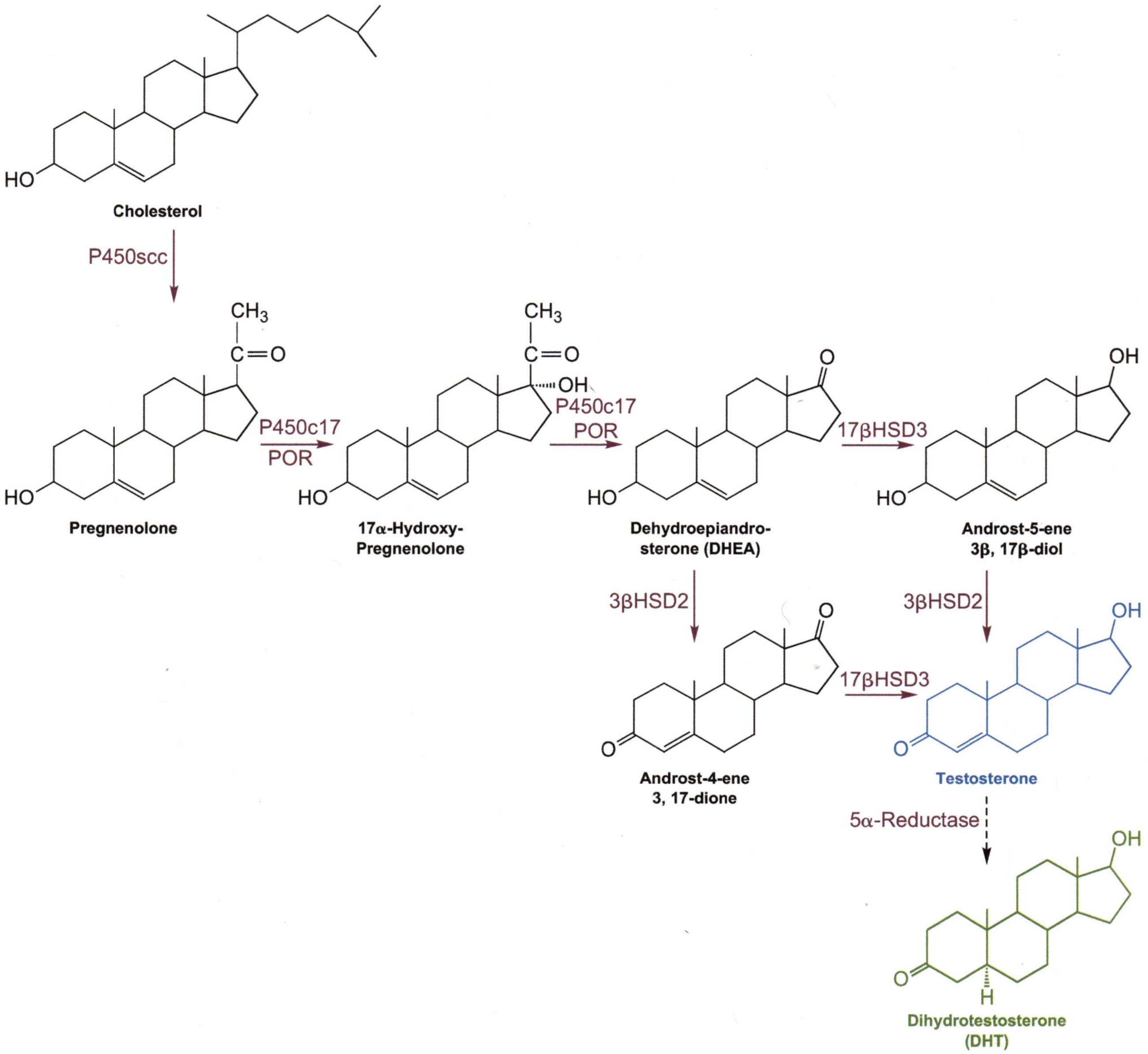

FIGURE 2.20 Pathways of androgen biosynthesis in testicular Leydig cells. Cholesterol is the starting point for the production of the principal androgen, testosterone (shown in *blue*). The conversion of cholesterol to DHEA is in a manner similar to that of the adrenal cortex zona reticularis (see Fig. 2.18). The conversion of testosterone to the more potent DHT (shown in *green*; also see *black dashed arrow*) by the 5α-reductase (see entry 11 in Table 2.5) occurs in androgen target glands such as the prostate, epididymis, seminal vesicles, and certain regions of the brain (see Chapter 12: Androgens). For each of the steroid enzyme transformations, the 3–7 letter/number acronym is shown in magenta next to the arrow. The enzyme full name for each acronym is listed in Table 2.5. *DHEA*, Dehydroepiandrosterone; *DHT*, dihydrotestosterone.

14 11β-Steroid dehydrogenase (11βHSD1)

The enzyme 11β-steroid dehydrogenase (see entry 17 in Table 2.5 and Fig. 2.19) is responsible for the NADPH-dependent oxidation of the 11β-hydroxyl on cortisol to yield cortisone. This enzyme is present in kidney, liver, brain, and lung. One significant contribution of this enzyme in the kidneys is to protect the aldosterone receptor from inappropriate exposure to cortisol, which is an effective ligand. Cortisone, in contrast, is a weak ligand for the aldosterone receptor (see Chapter 15: Hormones Related to the Kidney and Cardiovascular System).

Fig. 2.11 outlines the general framework of the metabolic pathways that are used to convert cholesterol into steroid hormones; the major steps are listed, as are the tissues where the transformations occur. Table 2.5 summarizes the tissue-specific expression of the cytochrome P450 enzymes involved with the production of the classical steroid hormones. A total of six P450 enzymes are

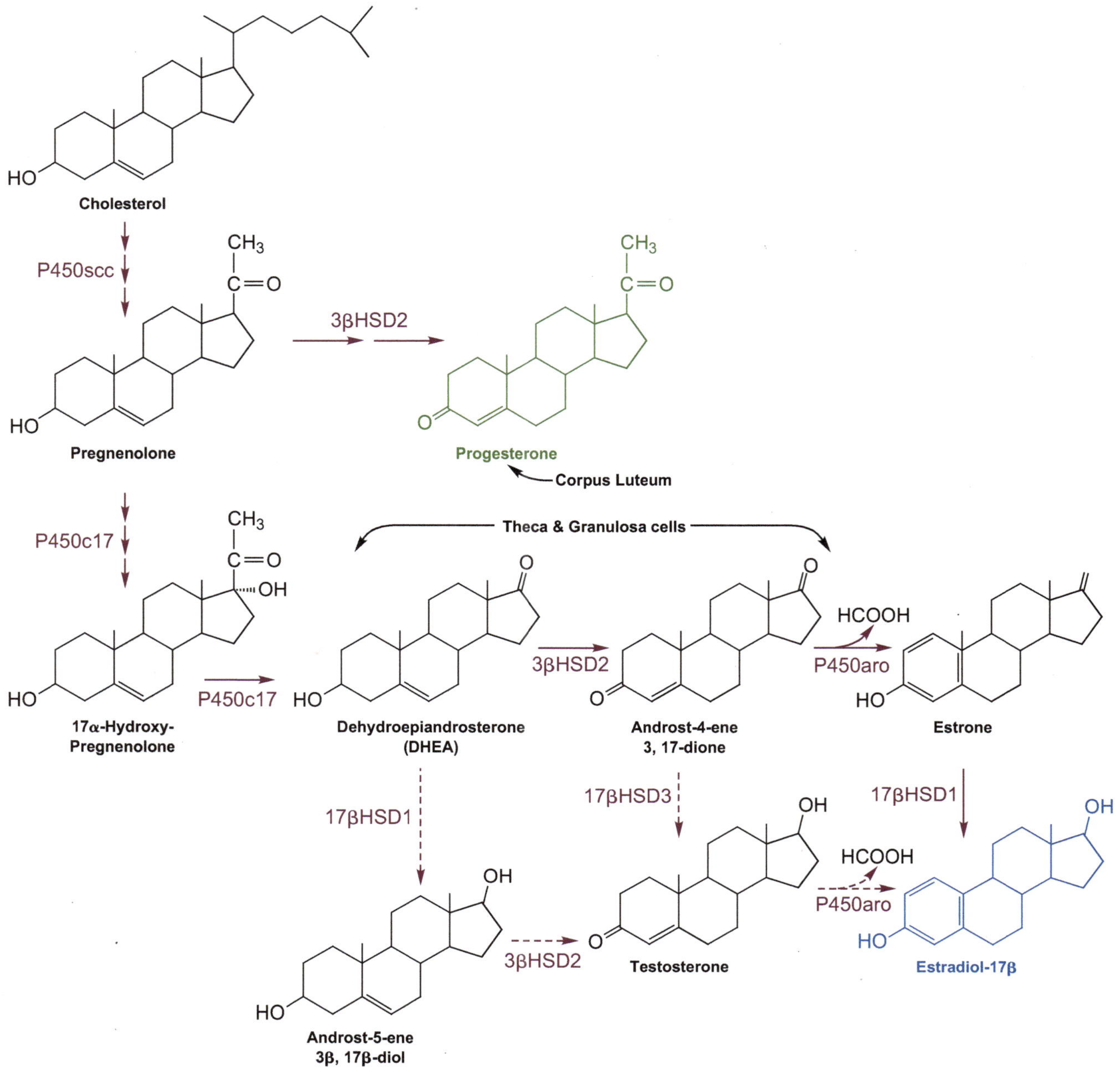

FIGURE 2.21 Pathway of the production of progesterone by the corpus luteum and estradiol by the theca and granulosa cells. Cholesterol is the starting point for the production of both progesterone (shown in *green*) and estradiol (shown in *blue*). There are two pathways from DHEA to estradiol-17. The major pathway is via androst-4-ene-3,17 dione and estrone. The second pathway via androste-5-ene-3β, 17β-diol and testosterone is only a minor pathway (see three *magenta dashed lines*). The chemical details of the aromatization of estradiol are described in Fig. 2.22. For each of the steroid enzyme transformations, the seven letter/number acronym is shown in magenta next to the arrow. The enzyme full name for each acronym is tabulated in Table 2.5. *DHEA*, Dehydroepiandrosterone.

associated with the production of the classical steroid hormones produced by the adrenal cortex, ovaries, and testes.

The purpose of the following sections is to review the metabolic pathways by which the five classes of steroid hormones are produced. All discussions of the endocrine systems for each of the steroid hormone classes as well as of their biological properties and modes of action are deferred to subsequent chapters in this book.

C Biosynthesis of pregnenolone and progestins

Fig. 2.18 illustrates the various steps involved in the production of progesterone and 17α-progesterone. As indicated in Fig. 2.11, the conversion of cholesterol into pregnenolone and then progesterone is a common pathway leading to the production of the five classic families of steroids, for example, mineralocorticoids (aldosterone), glucocorticoids (cortisol and corticosterone),

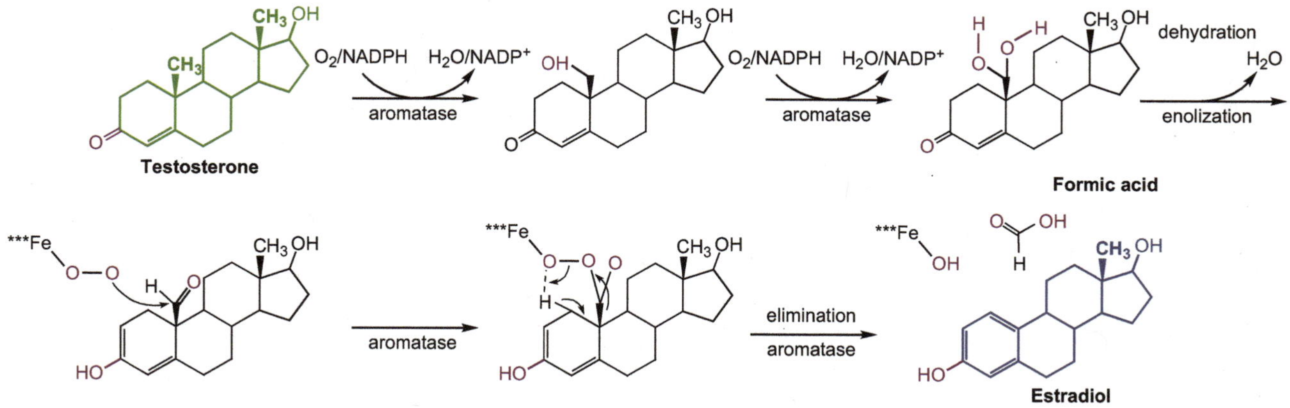

FIGURE 2.22 Proposed mechanism of aromatization of estradiol. The aromatase belongs to the cytochrome P450 super family of enzymes. The aromatase is located in the endoplasmic reticulum and is specifically responsible for the production of the steroid hormone estradiol (shown in *blue*). The reaction requires one molecule of testosterone (shown in *green*), three molecules of NADPH, and three molecules of O_2. The initial step is hydroxylation of carbon 19 (the CH_3) followed immediately by a second hydroxylation of carbon-19 to yield a *gem*-diol, which fragments into a C-19 aldehyde. The third oxidative attack (O_2) results in the conversion of the C-19 aldehyde into a C-19 carboxyl group that is followed by an immediate elimination of the C-19 methyl group as formic acid which completes the aromatization of the A ring. Aromatization of estradiol as described by W. L. Miller, R.J. Auchus in Endocrine Rev. 32 (2001) 81–151.

progestins (progesterone), estrogens (estradiol), and androgens (testosterone).

The conversion of cholesterol into pregnenolone is carried out by only one P450-containing enzyme, the cholesterol side chain cleavage enzyme or P450scc. Once pregnenolone is produced from cholesterol, it may undergo 17α-hydroxylation to 17OH-pregnenolone that ultimately leads in the adrenal cortex to the synthesis of aldosterone and cortisol, in the ovarian *theca* and *granulosa* cells to progesterone, and in the testes into testosterone. The two-step transformation of pregnenolone into progesterone is also catalyzed by one enzyme that carries out two different reactions, the 3β-steroid dehydrogenase oxidation of the 3β-OH and also the movement of the Δ^5 double bond to be a Δ^4-double bond (isomerase).

D Biosynthesis of adrenal cortex mineralocorticoids, glucocorticoids, and some androgens

Fig. 2.19 presents the individual steroid structures that sequentially form the metabolic pathways of the adrenal cortex steroids that describe the biosynthesis of the glucocorticoids, mineralocorticoids, and the androgen dehydroepiandrosterone (DHEA). As discussed in Chapter 10, Adrenal Corticoids and illustrated in Fig. 2.19, the adrenal cortex is anatomically divided into three zones that produce the three different major classes of the adrenal steroids; these are the green panel zona glomerulosa (mineralocorticoids), the peach panel zona fasciculata (glucocorticoids), and the blue panel zona reticularis (androgens). Over 45 steroids have been isolated and chemically characterized from human adrenal gland extracts.

The 21-carbon corticosteroids include the glucocorticoids and the mineralocorticoids; both subclasses are produced by the adrenal cortex (see Chapter 10: Adrenal Corticoids). The corticosteroids are characterized by (1) an oxo group at carbon-3 linked with a Δ4,5 double bond; (2) a two-carbon side chain on carbon-17; (3) an oxo group at carbon-20; and (4) a hydroxyl on carbon-21. Glucocorticoids are characterized by the presence or absence of hydroxyls at both carbon-11 and carbon-17. The principal glucocorticoid in humans is cortisol.

The principal mineralocorticoid in humans is aldosterone. Mineralocorticoids are characterized by a hydroxyl at carbon-11 and an aldehyde on carbon-18 (that is produced by the oxidation of a carbon-18 hydroxyl). As a consequence of its carbon-18 aldehyde moiety, aldosterone can form a five-membered hemiacetal ring by linking the C-18 aldehyde with the carbon-11 hydroxyl or, alternatively, linking the C-18 aldehyde with the carbon-21 hydroxyl to create a six-membered hemiacetal ring.

As indicated in Fig. 2.19, the zona reticularis of the adrenal cortex has the enzymatic capability to produce steroids with mild androgenic activity. The principal adrenal androgens are DHEA and 4-androstene-3,17-dione. The production of these and related steroids may increase in some instances adrenal tumors; women with such tumors may develop secondary male characteristics, including beard growth and hirsutism.

E Biosynthesis of androgens

The androgens are steroids with 19 carbons; see Fig. 2.20. The androgens are characterized by (1) the absence of the two-carbon side chain on carbon-17 and (2) the presence of a keto function on carbon-3. The major naturally occurring steroids with androgenic activity (in decreasing order of relative potency) are 5α-dihydrotestosterone (~150%–200%), testosterone (100%), androst-4-ene-3,17-dione (25%), and DHEA (10%).

The androgens are produced in males in the testes and in females by the ovaries and placenta. Also, as mentioned earlier, the adrenal cortex zona fasciculata can under some circumstances produce steroids with weak, but physiologically significant, androgen activity for the fetus.

Two general metabolic pathways lead from pregnenolone to testosterone; they are, respectively, either the Δ^5- or Δ^4-pathway (see Fig. 2.20). Steroid intermediates on the Δ^5-pathway can be converted to the corresponding steroid on the Δ^4-pathway by the oxidation of the 3β-hydroxyl to a 3-oxo (3β-steroid dehydrogenase) followed by migration of the double bond from C_{5-6} to C_{4-5} (Δ^5,Δ^4-isomerase) (see entry 2, Table 2.5).

The hormonally active form of testosterone in males is believed to be 5α-dihydrotestosterone (5α-DHT). There is evidence for the production of 5α-DHT by the testes, genital skin, liver, and kidney, but it is formed especially in androgen target glands such as the prostate, epididymis, seminal vesicles, and certain regions of the brain. The physiological actions of the androgens are discussed in Chapter 12, Androgens.

In females, in the follicular tissue there is evidence for both the Δ^5- and Δ^4-pathways of androgen production (as outlined in Fig. 2.20). Pregnenolone apparently is a more efficient precursor of carbon-19 steroids than progesterone in ovaries without *corpora lutea*.

In the zona reticularis of the human adrenal cortex (Fig. 2.19), there is also evidence for both the minor production of androgens as indicated by the dashed arrow.

F Biosynthesis of progesterone and estrogens

The principal progestational steroid in females is progesterone. It is produced from cholesterol by both the *corpus luteum* of the ovaries and the placenta (see Fig. 2.21). The physiological actions of progesterone are described in Chapter 13, Estrogens and Progestins.

The several metabolic pathways for the conversion of either androst-4-ene-3,17-dione (25) or testosterone (15) into estrogens are also summarized in Fig. 2.21; a unique feature of this conversion is the loss of carbon-19.

The estrogens are all 18-carbon steroids. They are produced in females in the ovaries (both the follicle and *corpus luteum*) and the fetal–placental unit. In males the testes under some circumstances can produce physiologically significant amounts of both estrone and estradiol (not illustrated). In both males and females the adrenal cortex zona fasciculata can generate small quantities of estrone from androst-4-ene-3,17-dione (not illustrated).

The estrogens in humans are characterized by (1) the loss of carbon-19, (2) the presence of an aromatic A ring, (3) the absence of the 2-carbon side chain on carbon-17, and (4) the presence of a hydroxyl group at both carbon-3 and carbon-17 and, in the instance of estriol, a third hydroxyl at carbon-16. The major naturally occurring steroids with estrogenic activity are estra-3,17β-diol, estra-3,16α,17β-triol.

Fig. 2.22 presents a detailed illustration of the chemistry involved with the conversion of the A-ring of testosterone or androste-4-ene, 3,17 dione into an aromatic ring, which is the hallmark of an estrogen.

Pregnancy is characterized in humans by a massive increase in the production of both progesterone and estrogen. The increase in progesterone and estradiol production occurs only in the placenta.

G Biosynthesis of vitamin D metabolites

Vitamin D_3 is chemically closely allied to the classical steroid hormones. Technically, it is a secosteroid; that is, due to the breakage of the carbon-9, carbon-10 carbon–carbon bond, the B ring is opened up so that only the A, C, and D rings are intact. As discussed in Chapter 9, Calcium-Regulating Hormones: Vitamin D, Parathyroid Hormone, Calcitonin, and Fibroblast Growth Factor-23, there are several families of vitamin D steroids (i.e., vitamin D_2, vitamin D_3, and vitamin D_4) that depend on the structure of the secosteroid side chain. When the side chain is identical to that of cholesterol, then it is the naturally occurring form and belongs to the vitamin D_3 family.

Vitamin D_3 can be produced photochemically by sunlight from a precursor, 7-dehydrocholesterol, present in the skin or alternatively obtained dietarily. Vitamin D_3 is known to be metabolized into a family of over 30 daughter metabolites.

The hormonally active steroid that produces the spectrum of biological responses attributable to vitamin D_3 is 1α,25-dihydroxyvitamin D_3 [1α,25$(OH)_2D_3$]; see Fig. 9.13. The major form of vitamin D_3 present in blood is 25-hydroxyvitamin D_3 [25$(OH)D_3$]. The enzymes in the liver that convert vitamin D_3 into 25$(OH)D_3$ are localized in both the mitochondria and the endoplasmic reticulum, while the enzymes in the kidney that catalyze the conversion of 25$(OH)D_3$ into the steroid hormone 1,25$(OH)_2D_3$ or 24,25$(OH)_2D_3$ are located only in the mitochondria of the proximal kidney tubule. All three enzymes (the 25-, 1α-, and 24- hydroxyl) contain cytochrome P450-type mixed-function oxidases. See Chapter 9, Calcium-Regulating Hormones: Vitamin D, Parathyroid Hormone, Calcitonin, and Fibroblast Growth Factor-23, for more details.

H Biosynthesis of bile acids

The conversion of cholesterol into bile acids takes place largely in the liver. In most mammals, cholic acid and chenodeoxycholic acid (see Fig. 2.9) are the principal bile acids. Before excretion into the bile, the carbon-24 carboxyl of both steroids is conjugated with the amino group of the amino acids taurine or glycine so as to increase the water solubility of the bile acids. The primary function of bile acids is to emulsify or solubilize ingested dietary fats, so as to facilitate their absorption and catabolism.

The steps involved in the conversion of cholesterol into a bile acid include the following: (1) cleavage of the side chain between carbon-24 and carbon-25; (2) microsome-mediated hydroxylations at carbon-7 and carbon-12; (3) epimerization of the $3-\beta$-hydroxyl to a 3α-α orientation (see Fig. 2.9); and (4) the reduction of the Δ^5-double bond so that the A:B ring junction is *cis* (see Fig. 2.4).

In addition to the detergent property of bile acids, some of the bile acids function as a steroid hormone that is a ligand for the nuclear receptor, farnesoid X receptor (FXR). The bile acid–liganded FXR receptor regulates gene transcription in the liver with the consequence that hepatic bile acid synthesis is inhibited. Also the bile acid–liganded FXR may regulate glucose and triglyceride metabolism.

IV Serum-binding proteins for steroid hormones

A Introduction

The hormonally active form of most steroids is generally the molecular species released from the endocrine gland where it was synthesized. The steroid hormone is then transported through the circulatory system by a specific plasma transport protein to various distal target tissues. A target tissue by definition has a receptor for its partner steroid hormone. The hormone and its receptor have a "glove and hand" partner relationship. Thus the protein receptor has a steroid-binding site (usually in the interior of the folded receptor protein) designed to stereospecifically interact with specific functional groups present on the steroid hormone (see Figs. 1.10 and 1.21). A typical ligand K_D for a steroid hormone receptor is in the range of $1-10\times10^{-10}$ M permitting the accumulation of the steroid in the cells of the target tissue. The binding of the steroid hormone to the receptor results in a conformational change of the receptor which enables it to deliver its "message" inside the cell so as to activate a specific second messenger system which, in turn, permits the generation of the appropriate biological response(s) in that target tissue for the specific steroid in question.

Thus a key determinant of the ability of a target tissue to bind the steroid hormone is the hormone's actual blood concentration. The concentration of a steroid in the plasma at any particular time depends on three factors: (1) the rate at which the steroid is biosynthesized and enters the body pools; (2) the rate at which the steroid is biologically inactivated by catabolism and removed from body pools; and (3) the "tightness" of binding of the steroid to its plasma carrier protein. The K_D of a steroid hormone for its plasma transport is always "looser," for example, $1-100\times10^{-8}$ M than that of the K_D of the nuclear receptor. Thus the tighter binding of the steroid hormone to its target receptor when it has arrived at a target tissue allows the hormone to be concentrated inside the target cell.

Previous sections of this chapter have considered in detail the metabolic pathways of production of many steroid hormones; subsequent chapters will describe the physiological properties of many steroid hormones. The remaining section of this chapter will briefly consider the plasma transport proteins for steroids.

B Serum-binding proteins for steroid hormones

The steroid hormones are transported from their sites of biosynthesis to their target steroid hormones, and their transport is facilitated by a family of plasma transport proteins (see Table 2.6). All steroid hormones, except

TABLE 2.6 Plasma transport proteins for steroids, thyroxine, and retinoid hormones.

Plasma protein	Code	Principal steroids bound	Molecular weight ($\times10^3$)
Vitamin D–binding protein	DBP	Vitamin D_3, $25(OH)D_3$, $1\alpha,25(OH)_2D_3$, 24,25 $(OH)_2D_3$	50
Corticosteroid-binding globulin (transcortin)	CBG	Glucocorticoids, progesterone	52 (glycoprotein)
Sex hormone–binding globulin	SHBG	Testosterone, estradiol	A dimer, each subunit 42
Thyroxine-binding globulin	TBG	Thyroxine (T_4), triiodothyronine (T_3)	63 (glycoprotein)
Retinol-binding protein	RBP	Retinol	41

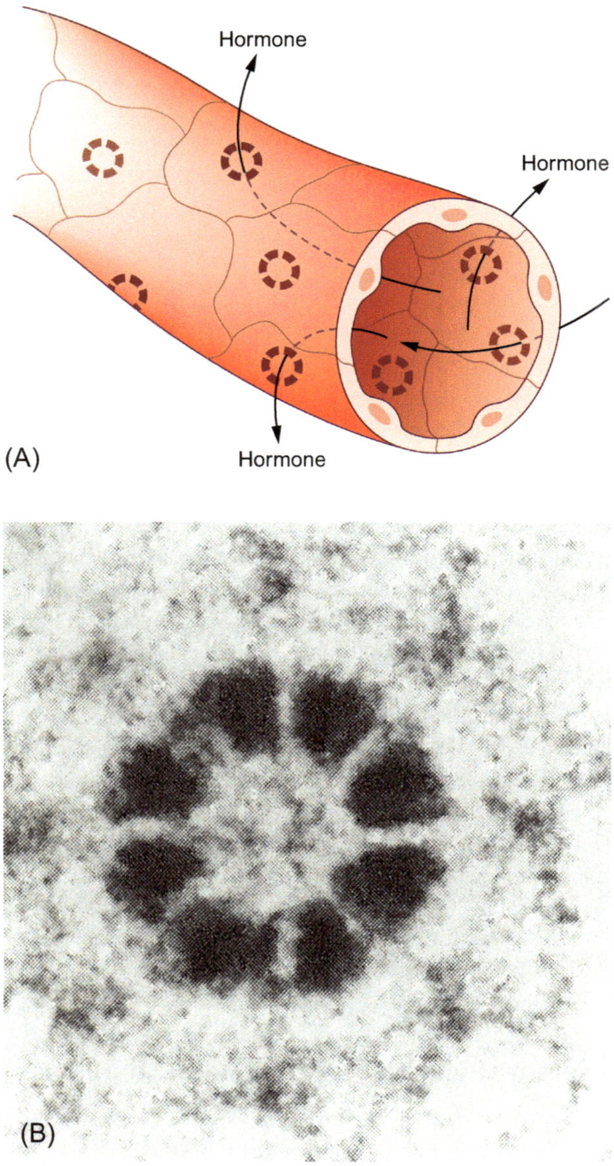

FIGURE 2.23 Examples of capillary wall fenestration. (A) Diagram representing the exit or entry processes of hormones via capillary orifices or fenestrations. (B) A fenestrated diaphragm in the endothelium of an adrenal cortex capillary. Note the eight, dark, wedge-shaped communicatory channels. The existence of fenestrations or pores in the capillary wall allows plasma steroid transport proteins to exit the circulatory system and approach the outer cell membrane of the target cell for the steroid hormone in question. *Adapted with permission from Figs. 2.2 and 2.6 of E. Bearer, L. Orci, Endothelial fenestral diaphragms: a quick-freeze, deep-etch study, Journal of Cell Biology 100 (1985) 418–428.*

one, have their cognate plasma-binding protein. The exception is aldosterone, which is believed to circulate as the free steroid in the plasma compartment. Although the five plasma transport proteins listed in Table 2.6 are all synthesized in the liver, they have no amino acid sequence homology. Also, there is no discernable sequence homology between the ligand-binding domains of the five plasma transport proteins, or the nuclear–cytosol receptor's ligand-binding domain, or the substrate-binding domain of the P450 enzyme(s) that generated the steroid.

In the plasma compartment the steroid hormones move through the circulatory system bound to their partner transport protein. However, an important issue concerns the details of the mode of delivery of steroid hormones to their target cells. Since the "free" form of the steroid hormone is believed to be the form of steroid that moves across the outer plasma membrane of a target cell, it had been postulated that the steroid ligand bound to a plasma transport protein dissociates from its plasma transport protein and then diffuses first through the capillary wall and then through the outer wall membrane of target cells. However, as illustrated in Fig. 2.23 it is apparent that the endothelial wall of capillaries contains fenestrations. Thus it is also possible for the plasma steroid transport protein (with bound steroid hormone) to exit the capillary bed via a fenestration and move to be immediately adjacent to the outer cell membrane of the appropriate target cell for the steroid hormone in question. Here, the steroid hormone will dissociate from the transport protein, diffuse through the plasma membrane, and then bind to an unoccupied partner steroid receptor.

Further reading

Chemistry

L.F. Fieser, M. Fieser, Steroids, Reinhold Publishing Corp, New York City, 1959.

This is a classic presentation of every aspect of steroid chemistry.

Cytochrome P-450

E. Bromek, W.A. Daniel, The regulation of liver cytochrome P450 expression and activity by the brain serotonergic system in different experimental models, Expert Opinion on Drug Metabolism and Toxicity 17 (2021). Available from: https://doi.org/10.1080/17425255.2021.1872543.

Y. Dzichenka, et al., Modified bile acids and androstanes—novel promising inhibitors of human cytochrome P450, The Journal of Steroid Biochemistry and Molecular Biology 205 (2021) 105777. Available from: https://doi.org/10.1016/j-jsbmb.2020.105777.

T. Niwa, et al., Comparison of steroid hormone hydroxylation mediated by cytochrome P450 3A subfamilies, Archives of Biochemistry and Biophysics 682 (2020) 108283.

Steroid metabolism

S. Ganguly, D. Naik, A. Muskara, O.Y. Mian, The nexus of endocrine signaling and cancer: how steroid hormones influence genomic stability, Endocrinology 162 (2021). Available from: https://doi.org/10.1210/endocr/bqaa177. baa177.

L. Schiffer, et al., Human steroid biosynthesis, metabolism and excretion are differentially reflected by serum and urine steroid metabolites: a comprehensive review, The Journal of Steroid Biochemistry and Molecular Biology 194 (2019) 105439.

Reviews on steroid receptors

A. Kobayashi, K. Azuma, K. Ikeda, S. Inoue, Mechanisms underlying the regulation of mitochondrial respiratory chain complexes by nuclear steroid receptors, International Journal of Molecular Sciences 21 (2020) 6683. Available from: https://doi.org/10.3390/ijms21186683.

A.D. Rohira, D.M. Lonard, B. O'Malley, Emerging roles of steroid receptor coactivators in stroll cell responses, Journal of Endocrinology 248 (2021) R41–R50. Available from: https://doi.org/10.1530/JOE-20-0511.

A good source of information

R.S. Kumar, N. Goyal, Estrogens as regulator of hemopoietic stem cell, immune cells and bone biology, Life Sciences 269 (2021) 119091.

G. Litwack, M.L. Sears, T.I. Diamndstone, Intracellular distribution of tyrosine ketoglutarate transaminase and 4-C14-hydrocortisone activities during induction, Journal of Biological Chemistry 238 (1963) 302–305.

W.L. Miller, C.E. Fluck, D.T. Breault, B.J. Feldman, The adrenal cortex and its disorders, Sperling Pediatric Endocrinology, fifth ed., Elsevier, 2021, pp. 425–490. Available from: https://doi.org/10.1016/B978-0-323-62520-3.00014-2.

Steroid acute regulatory protein

M. Galano, et al., Role of constitutive *STAR* in Leydig cells, International Journal of Molecular Sciences 22 (2021) 2021. Available from: https://doi.org/10.3390/ijms22042021.

M. Higa, et al., Novel *STAR* gene variant in a patient with classic lipoid congenital adrenal hyperplasia and combined pituitary hormone deficiency, Human Genome Variation 8 (6) (2021) 2021. Available from: https://doi.org/10.1038/s41439-021-00138-w.

Chapter 3

The hypothalamus and anterior pituitary

I Introduction

A Overview and CNS control of the hypothalamus

The pituitary gland has often been called "the master gland" because so many important functions are governed by the hormones it secretes. In this chapter, we will see that the pituitary hormones are under the control of both the central nervous system, especially the hypothalamus and the secretions of the distal endocrine glands and other internal signals. The hypothalamic releasing hormones function to control the secretions of the anterior pituitary hormones, either positively or negatively. The hormones secreted by the distal endocrine glands exert, in general, negative feedback effects in the brain and in the anterior pituitary.

The hormonal cascade systems involving the central nervous system, hypothalamus, anterior pituitary, and distal endocrine glands are outlined in Fig. 3.1. There are many signals originating either outside or inside the body, which are mediated by the central nervous system. Thus changes in the environment can ultimately stimulate the secretion of hypothalamic releasing hormones, which, through their effects on pituitary hormonal secretion, allow the body to adapt to the change. Likewise, signals from within the organism can trigger this cascade system so that the secretion of hormones from distant target glands can be affected.

Note that proceeding down the cascade shown in Fig. 3.1 from the central nervous system to the peripheral hormone, there is an increasing amount (mass) of the hormone released at each stage: the hypothalamic releasing hormones are secreted in nanogram amounts, the anterior pituitary hormones in microgram amounts, and the terminal hormone in microgram to milligram amounts. In addition, the stability of these hormones increases from a half-life of 3–10 minutes for the releasing hormones to minutes to hours for the anterior pituitary hormones and hours to several days for the distal hormones. These biochemical features of the hormones at different points of the cascade indicate a high degree of amplification stemming from a unique event, the initiating signal from either the external or internal environment. The cascade systems are kept in check by negative feedback loops, long ones from the distal target gland to the hypothalamus, anterior pituitary, or central nervous system and shorter ones from, for example, the pituitary to the hypothalamus.

B Hypothalamic control of the pituitary gland

Signals to the hypothalamus from either outside or within the body may be filtered and transmitted through central nervous system neurons. Primary inputs to the hypothalamus are from the limbic system, brain stem reticular formation, thalamus, subthalamus, basal ganglia, retina, and possibly the neocortex. These neural signals, along with chemical signals (hormones) from the peripheral glands, affect the hypothalamic secretion of either releasing or release-inhibiting hormones of which the primary ones are listed in Table 3.1. The hypothalamic hormonal molecules are transported to the anterior pituitary where they bind to specific receptors on the membranes of their target cells in the anterior pituitary, activating the appropriate intracellular signaling system(s) to bring about the secretion (or inhibition) of the hormone for which that cell is responsible. The secretion pattern of several hypothalamic (and, therefore, pituitary) hormones is pulsatile, as illustrated by the individual examples we will discuss in the following paragraphs.

As shown in Fig. 3.2, the anterior pituitary is the source of several polypeptide hormones, each of which is controlled by at least one hypothalamic input (positive, negative, or both). Taken together, these hormones govern a wide variety of important bodily functions. They are growth hormone (GH, somatotropin), which stimulates growth, particularly that of the skeleton, and often acts through somatomedins (insulin-like growth factors, IGFs) produced in the liver and elsewhere; thyroid stimulating hormone (thyrotropin, TSH), which stimulates the thyroid gland to secrete thyroxine (T_4) and triiodothyronine (T_3) for control of metabolism, development, and differentiation; luteinizing hormone (LH) and follicle-stimulating

Hormones. DOI: https://doi.org/10.1016/B978-0-323-90262-5.00020-2

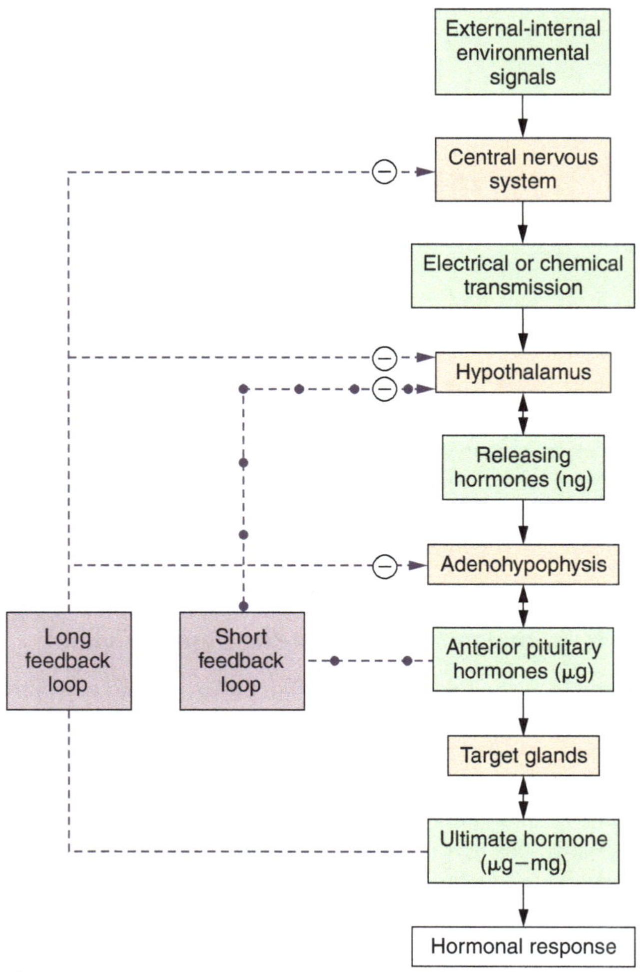

FIGURE 3.1 Overview of hypothalamic–pituitary–peripheral systems. An archetypal cascade of hormonal events is shown, beginning with an external or internal environmental signal. These signals (*green boxes*) are transmitted to the central nervous system and then propagated to the hypothalamus, the anterior pituitary (adenohypophysis), and the peripheral target gland (*pink boxes*) for generating the appropriate biological response. The order of magnitude of the amount of hormone released in each of the three final steps is given in parentheses. The increasing amounts of hormones secreted at each step provide for amplification and the cascade is kept in check by both short (e.g., anterior pituitary to hypothalamus) and long (peripheral gland to hypothalamus) negative feedback loops.

hormone (FSH), which play major roles in reproductive functions; adrenocorticotropic hormone (corticotropin, ACTH), a principal mediator of the stress adaptation through its stimulation of adrenocortical steroid hormones; prolactin (PRL), important for the synthesis of milk constituents during lactation and may have other actions. Not shown in Fig. 3.2 are some important anterior pituitary hormones, which will be discussed later in the chapter: melanocyte-stimulating hormone (melanotropin, MSH), which plays a role in skin-darkening reactions; β-lipotropin (β-LPH), which through its proteolytic products β-endorphin and Met-enkephalin, may promote analgesia in stress and act as a neurotransmitter in signaling the release of other hormones. The sources and plasma half-lives of the trophic anterior pituitary hormones are summarized in Table 3.2.

C Pituitary control of distal endocrine glands

An anterior pituitary hormone is secreted (in microgram quantities) into the circulation from which it reaches its specific receptor in a target cell. A signal transduction process, specific to the hormone, its receptor, and the target cell, ensues which culminates in the release of the target cell hormone, usually in milligram or high microgram quantities. As one example shown in Fig. 3.2, ACTH stimulates the adrenal cortex to secrete the glucocorticoid that is active in humans, cortisol (Chapter 10: The Hypothalamus and Anterior Pituitary). Another example is TSH, which stimulates the secretion of thyroid hormone from the thyroid gland (Chapter 5: Thyroid Hormones). The mechanisms by which the pituitary hormones stimulate the secretion of the peripheral hormones will be discussed in more detail in the chapters that focus on the peripheral hormones.

Finally, it is important to understand that, in several cases, an anterior pituitary (trophic) hormone, in addition to stimulating hormone secretion from its distal target cell, may also be a growth factor for these cells and, therefore, the gland. For example, ACTH, in addition to

TABLE 3.1 Hypothalamic releasing and release-inhibiting hormones.

Name	Abbr.	Hypothalamic neurons	Ant. pituitary hormone
Thyrotropin-releasing hormone	TRH	Paraventricular nucleus	↑ TSH
Gonadotropin-releasing hormone	GnRH	Diffuse	↑ FSH and LH
Corticotropin-releasing hormone	CRH	Paraventricular nucleus	↑ ACTH
Growth hormone–releasing hormone	GHRH	Arcuate nucleus	↑ GH
Somatostatin	SST (SRIH)	Periventricular nucleus	↓ GH
Dopamine	DA	Arcuate nucleus	↓ PRL

ACTH, Adrenocorticotropic hormone; *FSH*, follicle-stimulating hormone; *GH*, growth hormone; *LH*, luteinizing hormone; *PRL*, prolactin; *TSH*, thyroid-stimulating hormone.

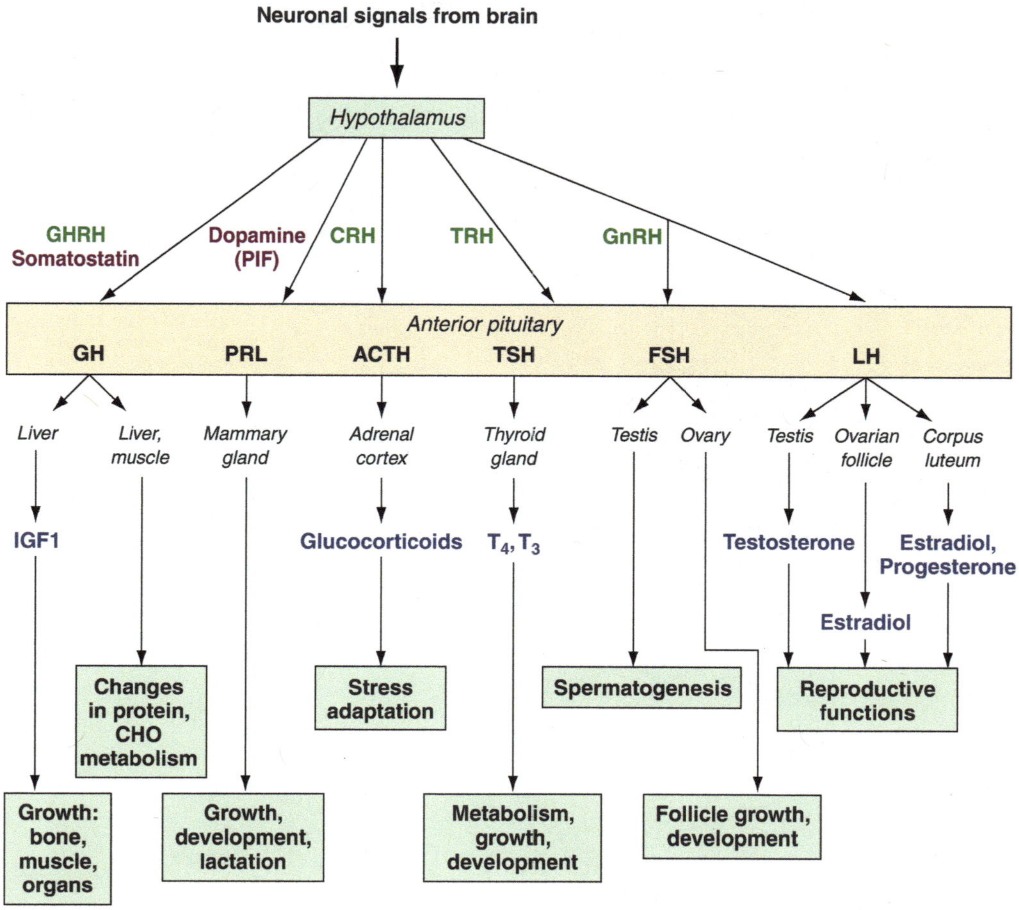

FIGURE 3.2 Summary of hypothalamic–pituitary–hormonal systems. Specific systems that operate according to the principles outlined in Fig. 3.1 are shown. For each the predominant hypothalamic releasing hormone (in *green*) or release-inhibiting factor (*red*) is shown. The main target tissues of the anterior pituitary hormones are indicated, along with the hormones they produce and, in the green boxes, major biological actions.

TABLE 3.2 Secretion of anterior pituitary hormones.

Hormone	Secreting cells	$t_{\{1/2\}}$ in blood (h)
TSH	Thyrotrophs	0.5
FSH	Gonadotrophs	6.0
LH	Gonadotrophs	0.5
GH	Somatotrophs	0.5
PRL	Lactotrophs	3.5
ACTH	Corticotrophs	0.25

ACTH, Adrenocorticotropic hormone; *FSH*, follicle-stimulating hormone; *GH*, growth hormone; *LH*, luteinizing hormone; *PRL*, prolactin; *TSH*, thyroid-stimulating hormone.

causing the secretion of cortisol from the adrenal gland, is also required for the survival of the cells of the adrenal (zona fasciculata cells) that produce cortisol. And, as discussed in Chapter 5, Thyroid Hormones, the growth and development of the thyroid gland is highly dependent on TSH from the pituitary gland.

D Neuroendocrine axis feedback

Each of the cascades initiated in the brain and hypothalamus is subject to negative (and in some rare cases positive) feedback control from the peripheral gland that is affected by the anterior pituitary hormone. Details of these feedback loops vary and will be discussed in the context of regulation of secretion of the particular peripheral hormone. One can note from Fig. 3.1, however, that these feedback effects can be exerted at the anterior pituitary, at the hypothalamus, or in the central nervous system outside the hypothalamus. In addition, short feedback loops, such as the inhibition of the secretion of a hypothalamic releasing hormone by the anterior hormone it controls, play a role in this regulatory network.

II Anatomical relationships

A Hypothalamic–pituitary anatomy

As seen in Fig. 3.3, the hypothalamus is located below the third ventricle of the brain, just above the median

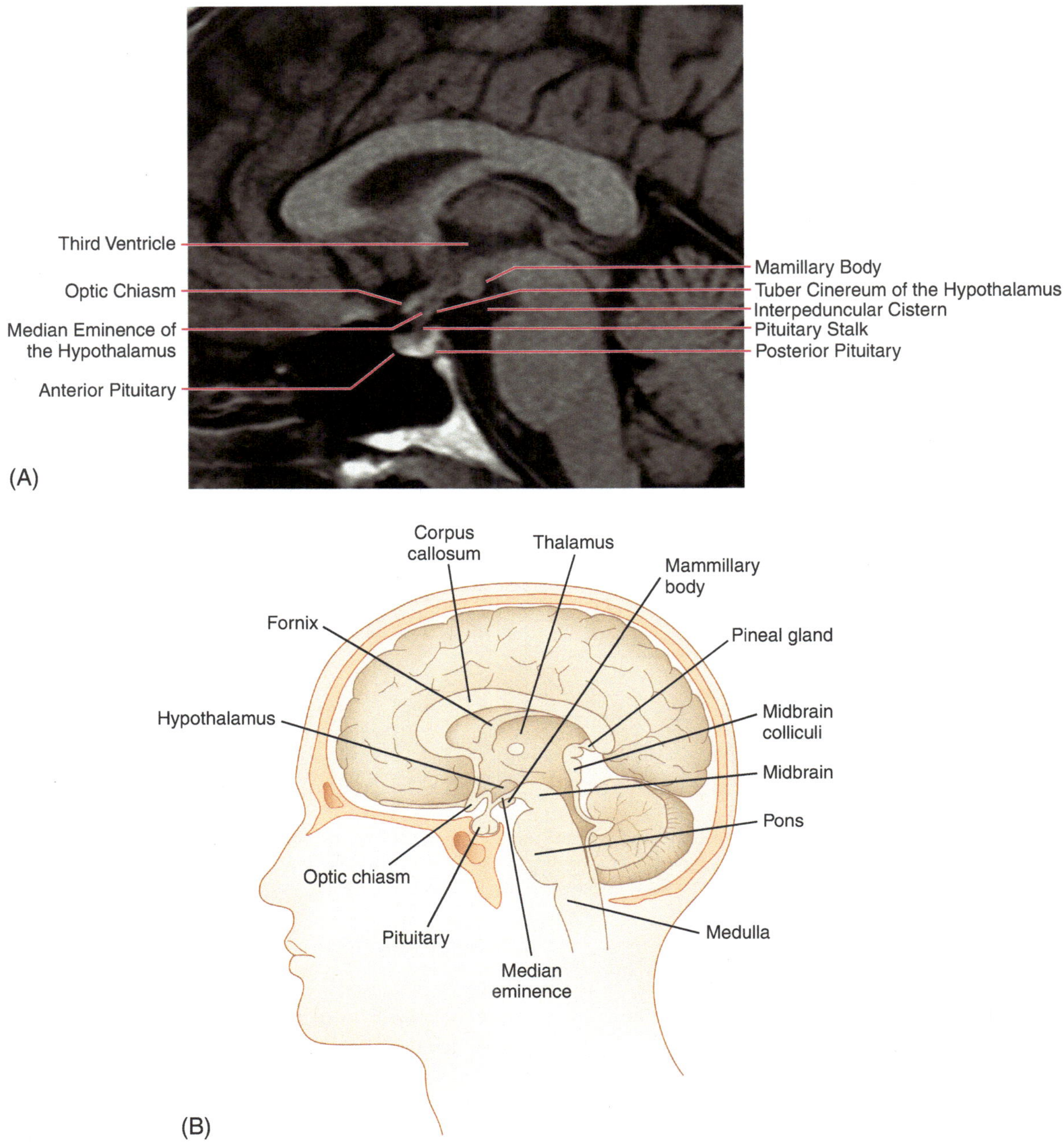

FIGURE 3.3 Anatomy of hypothalamus and pituitary. (A) This lateral view of the human brain shows the anatomical location of the hypothalamus and the anterior pituitary and their relationship to each other and nearby structures. (B) Schematic diagram of the same view as (A). *Adapted with permission from Uygur et al., Chapter 7, Imaging of the sella and parasellar region, Transsphenoidal Surgery, Elsevier, 2010, p. 44.*

eminence. Anatomically, the hypothalamus consists of a collection of neurosecretory centers, including the preoptic area, and the suprachiasmatic, arcuate, supraoptic, and paraventricular nuclei. Some nuclei, such as the supraoptic and paraventricular nuclei, which secrete hormones to be released by the posterior pituitary (see Chapter 4: Posterior Pituitary Hormones) are distinct, while others are less so. These cell groups are not homogeneous; subgroups can be identified within them. Notable is the paraventricular nucleus, which contains cells that secrete one of two pituitary releasing hormones [thyrotropin-releasing hormone (TRH) and corticotropin-releasing hormone (CRH)] as well as the posterior pituitary hormones, oxytocin and vasopressin.

The median eminence is a midline prominence arising at the rostral end of the hypothalamus and continuing into the pituitary stalk. It is the contact between the hypothalamic neurohormones that control anterior pituitary hormone secretion and the hypothalamic/hypophysiotrophic portal circulatory system. The axons terminating in the median eminence contain, in proximity to the portal system, dense vesicles 50–130 nm in diameter, suggesting that they can secrete the contents of the vesicles into the capillaries of the portal system (see next section).

The pituitary is intimately connected to the hypothalamus by the infundibular (pituitary) stalk. The anterior pituitary (adenohypophysis) consists of three parts: the pars distalis, comprising most of the gland and its hormone-producing cells; the pars intermedia; and the pars tuberalis, as shown in Fig. 3.4. Cells of the pars distalis are arranged in anastomosing cords close to fenestrated capillaries in the secondary capillary plexus of the hypophyseal portal system, described later. In humans the pars intermedia is not well defined anatomically in comparison with other species, where it is defined by the residual edge of Rathke's pouch.

B Hypothalamic neural connections

The neuronal circuitry between the rest of the brain and the hypothalamus is extremely complex and a detailed description is beyond the scope of this book. Suffice it to say that many of the electrical signals conducted by the fibers terminating in the hypothalamus will either stimulate or depress the release of releasing hormones from the hypothalamic neurons. Specific instances will be described in the context of specific endocrine systems, but the general organization is depicted in Fig. 3.5.

The hypothalamic neurons have the general structure diagramed in Fig. 3.5. The cell body is the site of synthesis and packaging of the releasing hormones. The secretory granules are transported down the axon, which in some cases may be extremely long, into the nerve ending where a signal will cause the exocytotic release of the granules. The targets of the releasing hormones are the cells in the anterior pituitary, which have cell membrane receptors that are specific for a given hypothalamic factor and, in response activation of the receptor, change the secretion of the anterior pituitary hormone.

C Hypothalamic–pituitary circulatory system

When the hypothalamic neuron receives a stimulatory or inhibitory signal, a specific releasing hormone is secreted from its axon terminal, often in the area of the primary plexus of the closed portal system connecting the hypothalamus with the anterior pituitary as seen in Fig. 3.6. The primary plexus is a group of capillaries with fenestrations (openings in the capillary walls; see Fig. 2.23) that allow the passage of releasing hormone peptides of 3 to more than 40 amino acids into the closed portal circulation. The releasing hormones circulate down the portal vessel to the secondary plexus from which, analogously to the primary plexus, they move via fenestrations into the extracellular space in the vicinity of the anterior pituitary cells containing the anterior pituitary hormones. Most of the blood flow in the portal system is from the

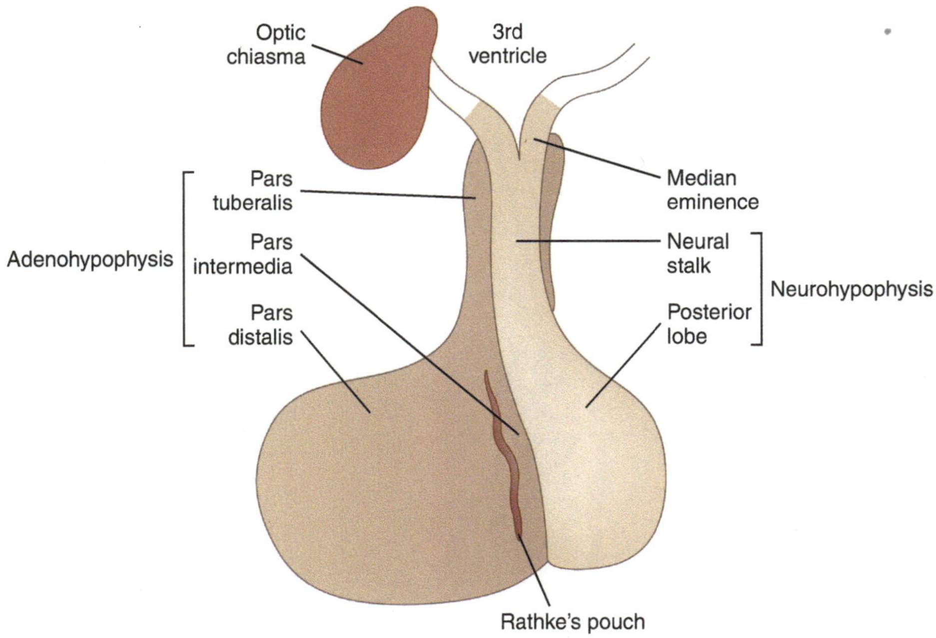

FIGURE 3.4 The pituitary gland in a midsagittal section. The two subdivisions of the pituitary gland are the anterior (darker shade) and posterior (lighter shade) lobes. The anterior lobe (adenohypophysis) is traditionally divided into three components, the *pars tuberalis* (that portion surrounding the stalk), the *pars distalis* (the majority of the anterior lobe), and the *pars intermedia*. In most species, but not in humans, there is a distinct *pars intermedia* or intermediate lobe, defined by the residual edge of Rathke's pouch (*brown streak* in the figure) from which the embryonic gland was derived. In humans the cells in the intermediate lobe are dispersed and this portion of the gland is not anatomically distinct. The neurohypophysis consists of the hypothalamic neurons that produce certain hormones (not shown), the axonal processes of these neurons through the neural stalk and the posterior pituitary gland. The posterior pituitary is the subject of Chapter 4, Posterior Pituitary Hormones.

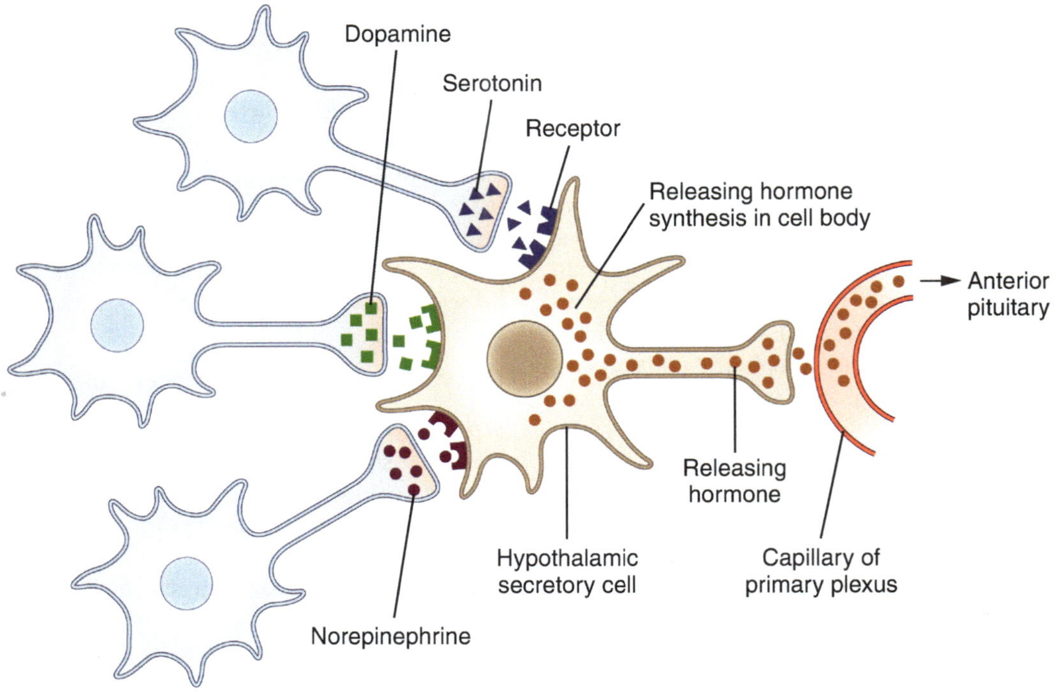

FIGURE 3.5 Hypothalamic neuron receiving and transmitting signals. This schematic diagram shows a hypothalamic neuron receiving multiple signals from other areas in the central nervous system. In response to the signals, the neuron synthesizes and secretes a specific releasing hormone or release-inhibiting hormone into the primary plexus of the hypothalamic–pituitary portal system for transmission to the anterior pituitary.

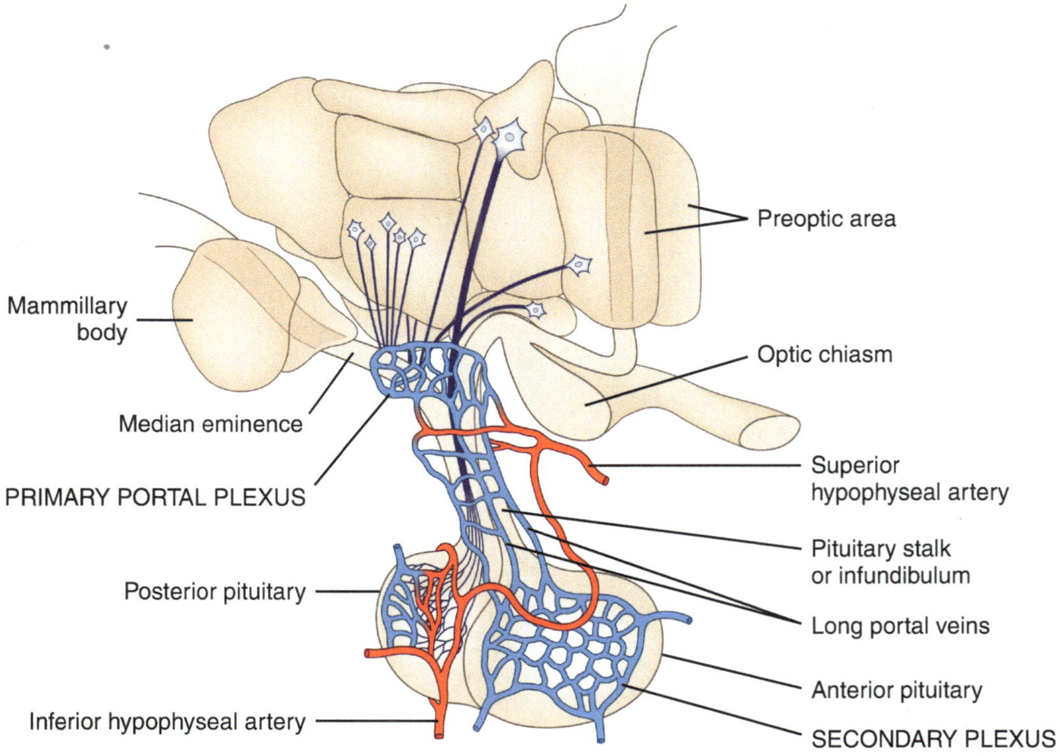

FIGURE 3.6 Portal capillary system of the human hypothalamus and anterior pituitary. Neurohormones are released from the axon termini of the hypothalamic neurons into the primary plexus, transported through the vessels to the secondary plexus from which they move through fenestrations in the capillary walls to interact with specific receptors in target anterior pituitary cells.

hypothalamus to the pituitary, but some may flow in the opposite direction as well.

The closed portal circulation described earlier and in Fig. 3.6 is necessary to maintain the concentration of the releasing hormones, which are secreted in nanogram amounts from the hypothalamus. The appropriate anterior pituitary hormone released in response to the action of the releasing hormone is secreted into the extracellular space and subsequently gains access to the general circulation; here fenestrations in local small vessels are also needed to pass the fairly large hormonal proteins.

III Structure, synthesis, secretion, and target cells of the hypothalamic releasing hormones

The hypothalamic releasing hormones are synthesized as part of larger precursor proteins encoded by messenger RNA (mRNA) transcribed from the gene for the releasing hormone. These preprohormones are proteolytically processed within the same neuron in which they are synthesized to yield the mature peptide. This processing takes place as the secretory vesicles containing the maturing peptide travel from the cell body down the axon to its terminus where the peptide will be released. The structures of the main hypothalamic releasing peptides and one release-inhibiting peptide are shown in Fig. 3.7. The smaller of these peptides, TRH and gonadotropin-releasing hormone (GnRH), have protective modifications at both ends of the molecule, including pyroglutamyl glutamate as the first residue and C-amidation. CRH and GnRH retain the stabilizing C-terminal amidation.

While the focus on the hypothalamic neurohormones in this chapter is on their role in the regulation of pituitary hormone synthesis and secretion, it should be noted that

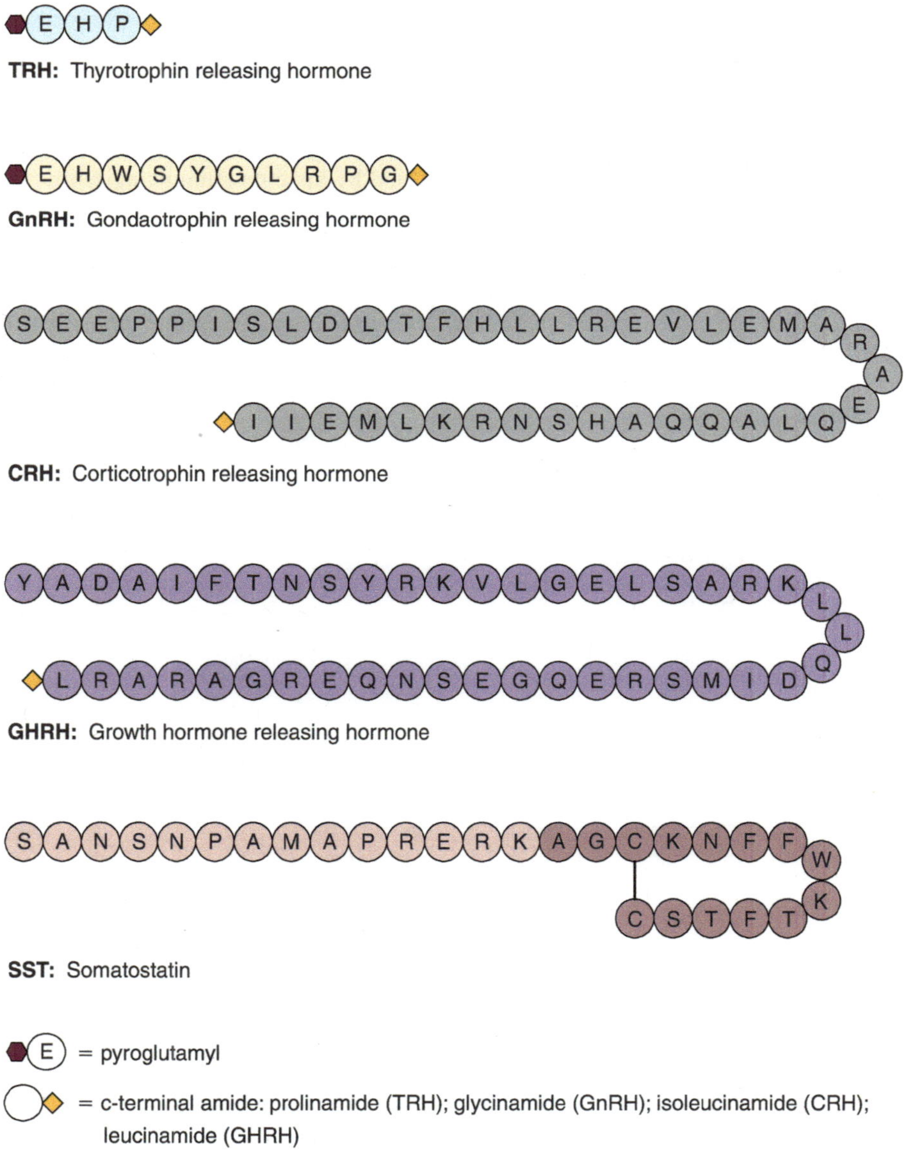

FIGURE 3.7 Hypothalamic peptides that control anterior pituitary hormone secretion. The amino acid sequences of some important hypothalamic regulatory peptides are indicated schematically, reading left to right, N-terminal to C-terminal. TRH and GnRH, the smallest of the peptides, have modifications on both ends, with a pyroglutamyl moiety (*red hexagon*; see Fig. 3.8 for structure) and amidation of the C-terminal carboxyl group (*yellow diamond*). CRH and GHRH have only the N-terminal modification, whereas somatostatin is not modified at either end. The sequence of somatostatin-28 is shown with both lightly and darkly shaded circles. The darker carboxyl terminal 14 amino acids are those of somatostatin-14. *CRH*, Corticotropin-releasing hormone; *GHRH*, growth hormone–releasing hormone; *GnRH*, gonadotropin-releasing hormone; *TRH*, thyrotropin-releasing hormone.

these peptides are also synthesized and play important roles outside the hypothalamic–pituitary system—for example, somatostatin (SST) as 28 (Fig. 3.7, entire sequence) or 14 (Fig. 3.7, darker N-terminal residues) amino acids. SST14 predominates in neural tissue and the pancreas, while SST28 is secreted by the intestine (Chapter 6: Pancreatic Hormones: Insulin and Glucagon and Chapter 7: Gastrointestinal Hormones).

A Thyrotropin-releasing hormone

The chemical structure of the tripeptide *TRH* is shown in Fig. 3.8. Human preproTRH is 242 amino acids and contains six copies of the tripeptide releasing hormone within its sequence. These progenitor TRH sequences are flanked by pairs of basic amino acids (Lys–Arg or Arg–Arg), the signals for the prohormone convertases (PCs) 1 and 2, the proteolytic enzymes responsible for the processing of preproTRH and proTRH as shown on the right side of Fig. 3.9. This arrangement of the preprohormone is similar to that in other vertebrates that have been studied. The processing of TRH is completed by the amidation of glycine at the carboxy terminus and the modification of the N-terminal by glutaminyl cyclase to result in the mature hormone.

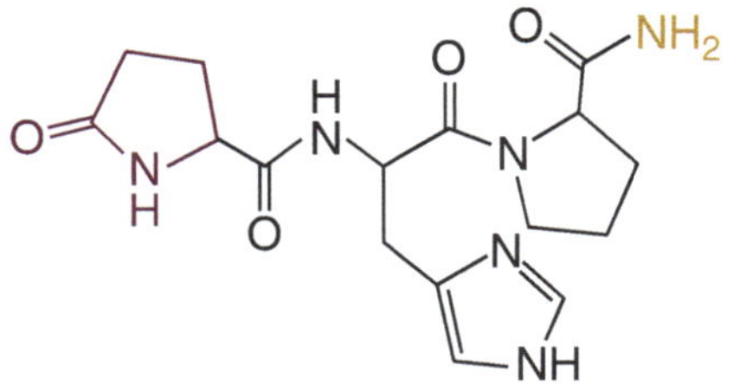

FIGURE 3.8 The structure of thyrotropin-releasing hormone, TRH. The N-terminal pyroglutamyl moiety and the C-terminal amidation of the molecule are shown in red and yellow, respectively. *TRH*, Thyrotropin-releasing hormone.

Fig. 3.9 summarizes some of the regulatory influences on TRH synthesis, through transcriptional effects on the TRH gene expression and on processing through effects on the transcription of the genes for the processing enzymes, PC1 and PC2. The complexity and variety of these signals that affect TRH production are not surprising, given the multiplicity of physiological functions of thyroid hormone (Chapter 5: Thyroid Hormones). Synthesis of preproTRH is stimulated by norepinephrine, enabling the hypothalamic–pituitary–thyroid axis to respond to cold and stress by increasing the rate of metabolism, a hallmark of thyroid hormone activity. The TRH neuron is also activated by appetite-stimulating hormones such as α-MSH and inhibited by anoretic or appetite-suppressing peptides such as Agouti-related protein

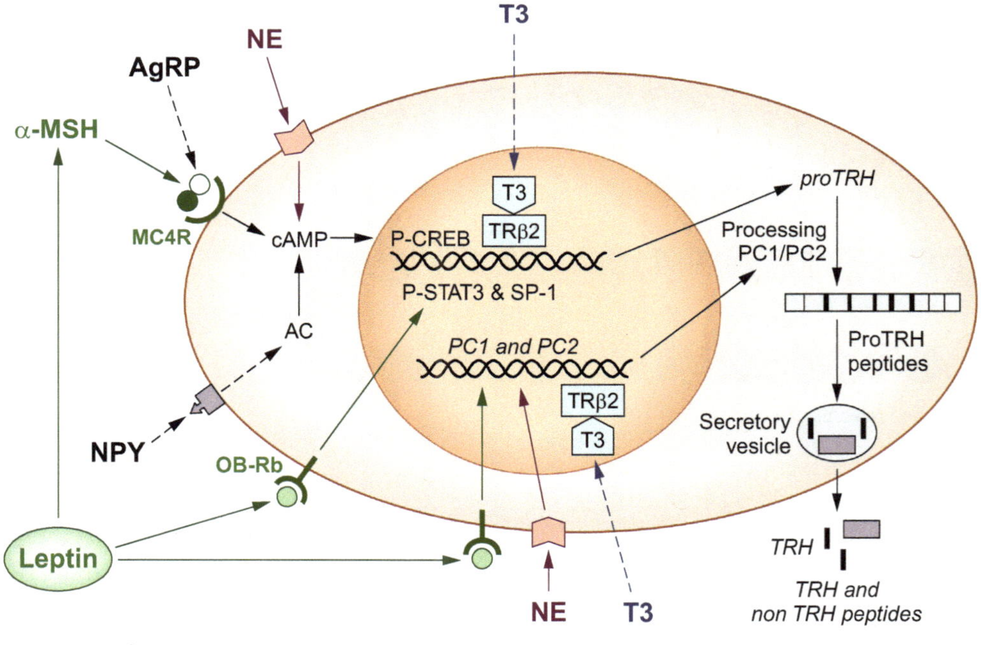

FIGURE 3.9 Control of TRH gene expression and secretion. Signals from the peripheral circulation include the thyroid hormone, T_3, interacting with its nuclear receptor, (TRβ) and leptin interacting with its receptor (OB-Rb). Stimulation is denoted by solid lines and inhibition by dashed lines. Neural input includes NE mediating the response to cold, and the appetite-related peptides α-MSH, its antagonist, AgRP, which both interact with the α-MSH receptor, MC4R; and NPY. On the right side of the figure is shown an overview of the processing of pro-TRH by the enzymes P1 and PC2. At the bottom of the figure are shown the hormonal and neural influences on the transcription of the TRH processing hormones, PC1 and PC2. *AgRP*, Agouti-related protein; *MSH*, melanocyte stimulating hormone; *NE*, norepinephrine; *NPY*, neuropeptide Y; *P1 and PC2*, prohormone convertases I and 2; *TRH*, thyrotropin-releasing hormone.

(AgRP) (which binds to the α-MSH receptor and antagonizes its actions) and neuropeptide Y (NPY). To increase food intake, the adipose hormone leptin stimulates the TRH neuron either directly (through its receptor OB-Rb) or indirectly through the stimulation of α-MSH. Thus the TRH neuron integrates important information about the environment relating to its effects on temperature, food intake, and stress and responds by activating the hypothalamic–pituitary–thyroid axis. The role of peripheral hormones in the control of appetite at the level of the hypothalamus will be discussed in more detail in Section VI and in Chapter 6, Pancreatic Hormones: Insulin and Glucagon.

Negative feedback control of TSH secretion by the peripheral hormone T_3 (see Chapter 5: Thyroid Hormones) is thought to occur primarily at the thyrotrophs in the pituitary rather than the hypothalamus, but thyroid hormones also play a role in the hypothalamic neurons.

As shown in Fig. 3.9, T_3, bound to its nuclear receptor $TR\beta_2$, inhibits the synthesis of mRNAs encoding both prepro-TRH and the enzymes that process it into mature releasing hormone.

The TRH receptor located on target cells in the anterior pituitary (as well as elsewhere in the body) is a typical GPCR (G protein–coupled receptor) of the rhodopsin family with an extracellular amino terminus, three extracellular loops, seven transmembrane regions, three intracellular loops, and an intracellular carboxyl terminus. There are two forms of the receptor, encoded by separate genes, TRH-R1 and TRH-R2. Although the two receptors differ in approximately 30% of their amino acids, three-dimensional studies indicate that the contact points between the ligand, TRH, and its binding site are the same on the two proteins. In the pituitary, TRH-R1 mediates the TRH signal through binding to $G_q/_{11}$ and induction of protein kinase C (PKC)-, phosphophatidyl-inositol- and Ca^{2+}-mediated signaling pathways.

TRH neurons have projections to areas of the central nervous system other than the anterior pituitary. A thorough discussion of these actions is not possible here, but a few illustrative examples are in order. Some axon terminals in the spinal cord have quite high TRH levels and contribute to the regulation of cardiovascular function. TRH from the dorsal motor nucleus of the vagus nerve affects gastrointestinal motility and gastric acid secretion. TRH has been identified in many peripheral tissues such as the retina, the adrenal medulla, and the pancreas, where it plays a role in the specialized functions of these cells.

The function of TRH in the hypothalamus–pituitary–thyroid gland axis is shown in Fig. 3.10. Euthyroidism is maintained by the regulation of TSH by TRH plus the negative feedback of the thyroid hormones (T_3 and T_4). TSH is secreted from the anterior pituitary thyrotrope cell through the action of TRH on its receptor (GPCR type) on the surface of the thyrotrope plasma membrane. TSH induces the secretion of T4 and T3 from the thyroid follicle (see Chapter 5: Thyroid Hormones). T4 (much more concentrated in plasma than T3) inhibits further secretion of TSH through its negative feedback function.

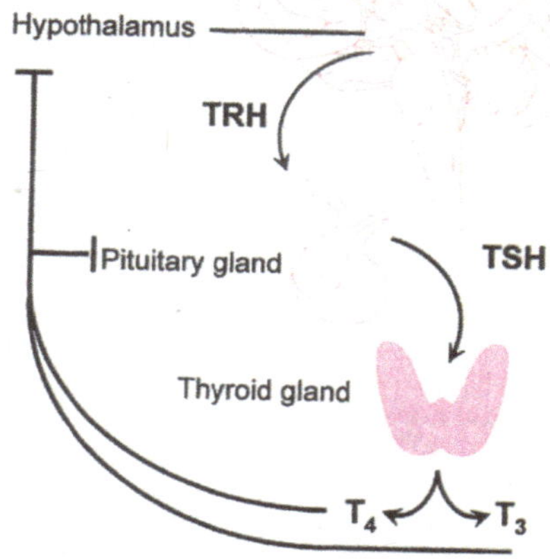

FIGURE 3.10 Normal control path and negative feedback for thyroid hormones. *Reproduced from Fig. 3.9 of G. Halmos et al., Hypothalamic releasing hormones, in: G. Litwack (Ed.), Hormonal signaling in biology and medicine, Academic Press, 2018, p. 49.*

As with all of the hypothalamic releasing hormones, TRH acts on its cognate receptor, a GPCR protein, on the surface of the thyrotropic cell of the anterior pituitary. Activation of this receptor leads to the production of cyclic AMP (cAMP) or to the stimulation of the phosphoinositol system that, in turn, stimulates the phosphorylation of protein kinases, actions of which ultimately lead to the release of TRH. After TRH binds to its receptor, the G-protein (Gq/11) pathway is activated that mobilizes intracellular calcium and the activation of PKC that generates the secretion and synthesis of TSH. The TRH receptor crystal structure is shown in Fig. 3.11.

B Gonadotropin-releasing hormone

The primary structure of the hypothalamic decapeptide *GnRH* (Fig. 3.7) appears to be identical in all mammals, with the exception of the guinea pig, and there is a high degree of conservation of the sequence across the vertebrates. The GnRH precursor protein is also highly conserved across species, consisting of (from the amino to the carboxyl terminus) a 23 amino acid signal peptide, GnRH, a proteolytic cleavage site Gly-Lys-Arg (GKR), and the 56 amino acid GnRH-associated peptide (GAP). GAP is cosecreted with GnRH following the processing of the precursor into the mature peptides in the secretory granules of the GnRH neurons. Other forms of GnRH have been identified in which amino acid variations occur

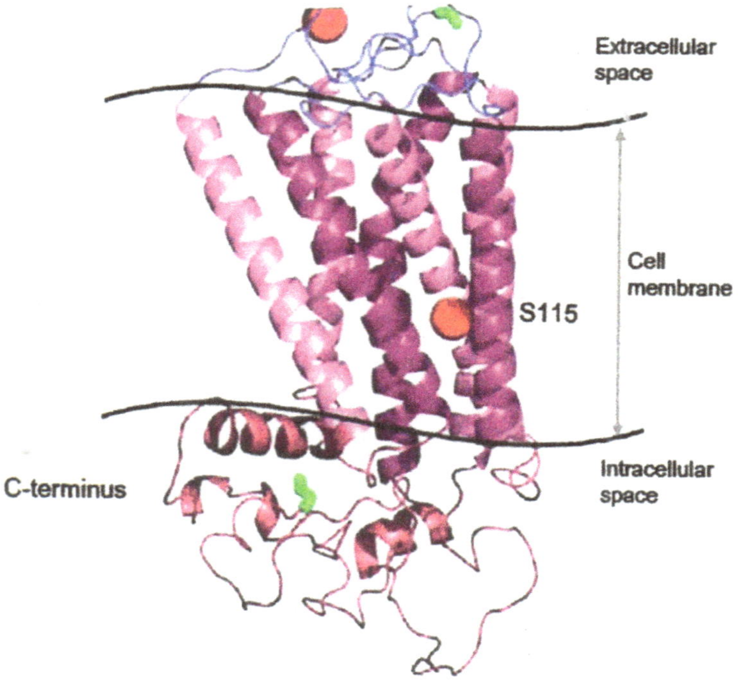

FIGURE 3.11 Crystallographic modeling of the TRHR showing the positions *(red spheres)* of the two mutations associated with central hypothyroidism: R17X truncating the protein in the extracellular domain and an in-frame deletion of three amino acids (Ser115–Thr117) plus a missense change (Ala118; p.S115–T117delCT118) located at the cytoplasmic end of the third transmembrane domain of the receptor. The N-terminal start codon and C-terminal end codon are highlighted in *green*. *TRHR*, Thyroid hormone releasing hormone receptor. *Reproduced from Fig. 3.12 of G. Halmos et al., Hypothalamic releasing hormones, in: G. Litwack (Ed.), Hormonal Signaling in Biology and Medicine, Academic Press, 2018, p. 50.*

at positions 5, 7, and/or 8 and which differ in species and tissue distribution.

GnRH is absolutely required for reproductive function. In its absence, secretion of the two gonadotrophic hormones that depend on it is either completely (LH) or greatly (FSH) diminished.

The secretion of GnRH from the neurons in which it is synthesized is characterized by its pulsatile nature, as shown in Fig. 3.12. In most mammals, including humans, it is difficult to measure GnRH directly due to the small amount secreted and its short half-life. Therefore the measurement of the electrophysiological activity of a group of GnRH neurons has been accepted as a surrogate for the measurement of the actual hormone. In this way the pulses of LH secretion from the pituitary can be correlated as occurring with or shortly following secretion of GnRH from its neurons, as shown by the classical example of such a study shown in Fig. 3.12. Other studies show that when the pulsatility of GnRH secretion is interfered with experimentally both LH and FSH secretion are severely affected. Furthermore, the two gonadotropins show pulse frequency discrimination, with LH responding to faster frequencies of the GnRH pulse and FSH to slower frequencies.

The pulse generator for GnRH secretion is now thought to lie in neurons of the preoptic area that contain a triad of neuropeptides, kisspeptin, neurokinin B, and dynorphin (the so-called KNDy neurons), which project into the cell bodies and terminals of GnRH neurons of the hypothalamus. Inactivating mutations in kisspeptin leads to some types of hypogonadism and infertility in humans

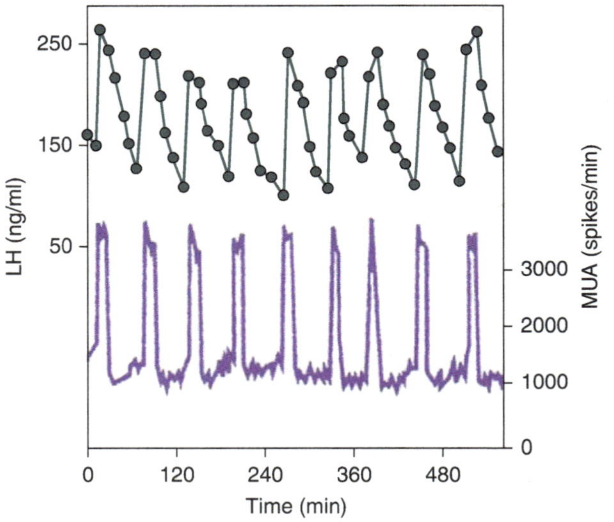

FIGURE 3.12 Demonstration of the pulsatility of GnRH and LH secretion in the monkey. In the lower part of the figure is shown the pulsatile nature of the firing of GnRH secreting neurons given as MUAs. Above are the measured plasma levels of LH in the same animal over the same time course. *GnRH*, Gonadotropin-releasing hormone; *LH*, luteinizing hormone; *MUAs*, multiple unit activity. *Adapted from Cardenas et al., Proceedings of the National Academy of Sciences of the United States of America 90 (9360) (1993).*

and this peptide will be the subject of further discussion in Chapter 12, Androgens and Chapter 13, Estrogens and Progestins.

The human GnRH receptor, like the TRH receptor and other GPCRs, has a seven membrane–spanning domain, and the extracellular domains are qualitatively similar; the intracellular domains are distinct in that the cytoplasmic tail is absent from the GnRH receptor (GnRHR). The two-dimensional structure of the GnRHR is shown in Fig. 3.13. The primary signaling pathway for gonadotropin secretion is through the activation of phospholipase C, release of IP_3 and diacylglycerol, and activation of PKC. These signals are transmitted to the nucleus through the c-junN-terminal kinase pathway to activate transcription of the gene for the β-subunit of either LH or FSH. In addition, elevated cAMP and intracellular Ca^{2+} from the activation of voltage-sensitive calcium channels both contribute to the stimulus of secretion of stored GnRH.

C Corticotropin-releasing hormone

In addition to ACTH, *CRH* also stimulates the secretion of β-endorphin and other proopiomelanocortin (POMC: see Section IV.D)-derived peptides from the anterior pituitary. In other areas of the brain, CRH is an important mediator of responses to stress. It is also an important placental hormone involved in the events of parturition (see Chapter 14: Hormones of Pregnancy, Parturition and Lactation). The primary structure of CRH, which is located in the carboxyl terminus of the 196 amino acid preprohormone, is shown in Fig. 3.7. PreproCRH is encoded by the second of the two exons in its gene. Processing involves the PC-catalyzed cleavage of CRH from the prohormone precursor and the amidation of the carboxy terminal; this amide group is required for biological activity.

The cell bodies of the neurons that synthesize and secrete CRH into the hypothalamic–pituitary portal system are located in the paraventricular nucleus of the hypothalamus. Neuronal input to these neurons is from the limbic system (amygdala and hippocampus) and brainstem regions governing autonomic functions. CRH is found in other areas of the central nervous system, including the cerebral cortex, limbic system, cerebellum, brain stem, and spinal cord.

In the CRH synthesizing neurons functioning as part of the hypothalamic–pituitary–adrenal axis, glucocorticoid receptor activation by the steroid hormone can repress transcription from the *crh* gene as part of negative feedback in the mouse; the precise mechanism by which

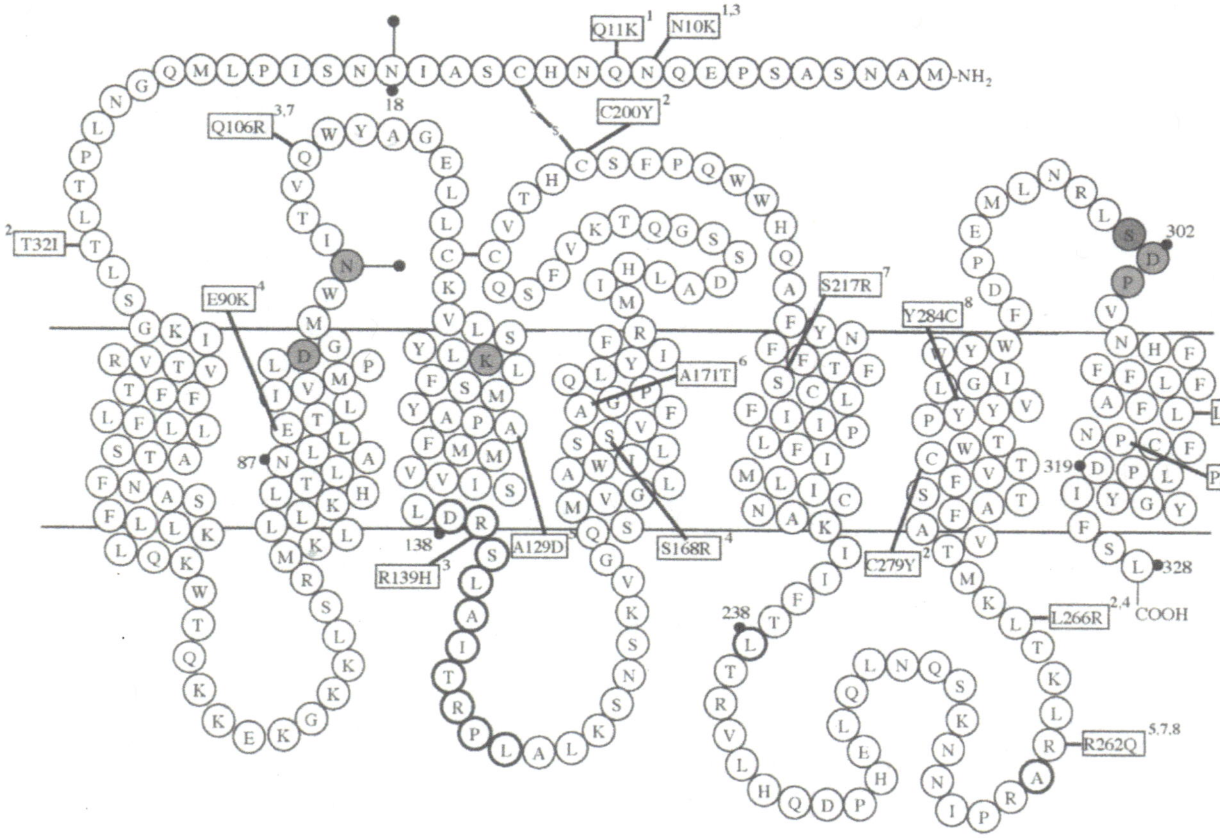

FIGURE 3.13 Two-dimensional representation of proposed topology for the human GnRH-R. Numbering is relative to N-terminal amino acid residue. Known N-linked glycosylation sites are marked by the symbol of a hyphen-type line with a black dot at its end. Shaded residues have been implicated in forming the proposed GnRH-binding pocket. Bolded circles indicate amino acid residues important for G-protein coupling. Common point mutations associated with the human disease hypogonadotropic hypogonadism are denoted in boxes. *GnRH-R*, Gonadotropin releasing hormone receptor. *Reproduced from Fig. 3.1 of L.A. Rispoli, T.M. Nett, Pituitary gonadotropin-releasing hormone (GnRH) receptor: structure, distribution and regulation of expression, Animal Reproduction Science, 88 (2005) 57–74.*

this occurs has not yet been elucidated. On the other hand, stress-related noradrenergic and glutamatergic excitatory signals can activate the gene, in part via the activation of the transcription factor CREB (cAMP response element binding protein). A second protein, transducer of regulated CREB activity, plays a vital role in cAMP-mediated regulation of CRH expression in these neurons. The regulation of CRH gene expression is localization-specific; for example, in other areas of the brain and in the placenta glucocorticoids stimulate, rather than inhibit, CRH gene transcription.

There are two forms of the CRH receptor: CRH-R1, encoded on chromosome 17 in humans, and CRH-R2 on chromosome 7. Both belong to the class II GPCR superfamily and contain seven membrane–spanning regions. The two types of cortocotropin releasing hormone receptors (CRH-Rs) are shown schematically in Fig. 3.14. CRH-R1 is located in the brain and pituitary and CRH-R2 is located in the brain and in peripheral tissues, notably the cardiac myocyte. In addition to CRH, urocortins interact with both types of the CRH receptors and the activated receptors elicit several downstream signaling responses. Cardiac activity of endogenously produced urocortin in the heart is mediated by the activation of cardiovascular CRH-R2 receptor. The genes for each of the two receptors consist of more than a dozen exons and alternative splicing of these leads to several subforms of the receptor with various activities. The two CRH receptors share about 70% amino acid identity, with most of the divergence coming at the N-terminal region of the molecules. CRH-R1 mediates the effects of CRH in the anterior pituitary corticotrophs but is widely distributed elsewhere in the brain. It also occurs in peripheral tissues such as those of the reproductive tract, the adrenal gland, adipose tissue, cardiac muscle, and some cells of the immune system. CRH-R2 has very specific locales in the brain and can also be found in peripheral tissues. In Fig. 3.15 a schematic is presented showing the actions of urocortin in the heart, its binding to the type 2 CRH receptor (CRHR2), and signaling that follows the activation of the receptor. Cytoprotection occurs by the binding of urocortin to CRHR2.

D Hypothalamic control of growth hormone secretion

GH secretion is subject to change in response to a wide variety of physiological situations, some of which are summarized in Table 3.3. Many, but not all, of these are mediated through the two major hypothalamic factors discussed next.

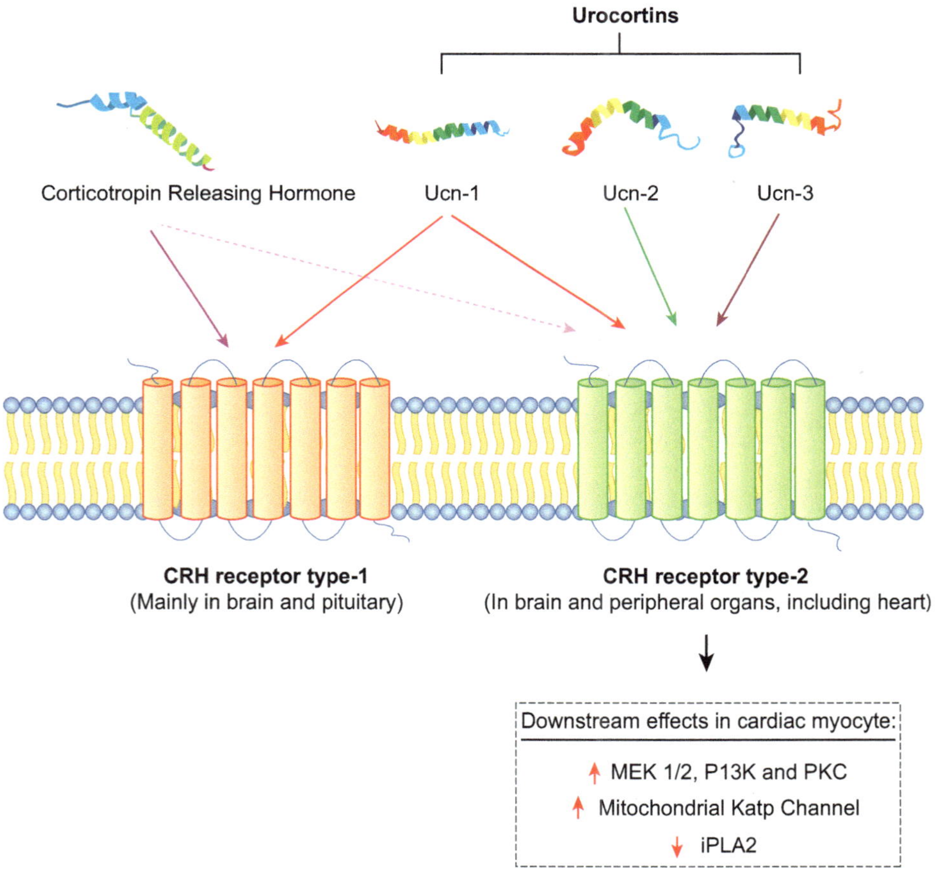

FIGURE 3.14 Schematic diagram depicting interaction of CRH and urocortins with CRH receptors type-1 and type-2, both of which are 7-transmembrane helix receptors. *CRH*, Corticotropin-releasing hormone; *iPLA2*, calcium-independent phospholipase A2; *Katp*, ATP sensitive potassium channel; *MEK ½*, mitogen-activated protein kinase; PI3K, phosphatidylinositol 3-OH kinase; *PKC*, protein kinase C; *Ucn*, urocortin. *Reproduced from Fig. 1 of C. Basman et al., Cardioprotective ability of urocortin in myocardial ischemia-reperfusion injury: where do we stand?, Current Molecular Pharmacology, https://doi.org/10.2174/1874467210666170231 01422.*

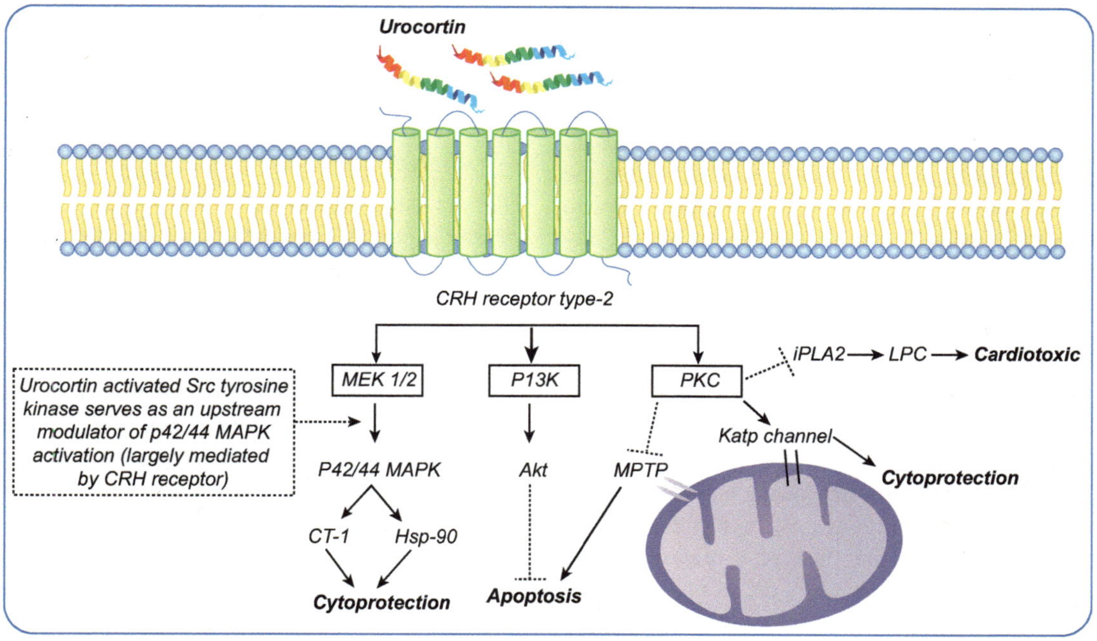

FIGURE 3.15 Multiple mechanisms of action through which urocortin ultimately leads to cardioprotection, subsequent to activation of CRH receptor type 2 in the heart. *Akt*, Protein kinase B; *CRH*, corticotropin-releasing hormone; *CT-1*, cardiotropin-1; *Hsp90*, heat-shock protein 90; *iPLA2*, calcium-independent phospholipase A2; *Katp*, ATP sensitive potassium channel; *LPC*, Lysophosphatidylcholine; *MAPK*, mitogen-activated protein kinase; *MEK ½*, mitogen-activated protein kinase; *MPTP*, mitochondrial permeability transition pore; *PI3K*, phophatidylinositol 3-OH kinase; PKC, protein kinase C. *Reproduced from Fig. 3.2 of C. Basman et al., Cardioprotective ability of urocortin in myocardial ischemia-reperfusion injury: where do we stand?, Current Molecular Pharmacology, https://doi.org/10.2174/187446721066617023101422.*

TABLE 3.3 Factors that affect growth hormone secretion and action.

↑ GH secretion	↓ GH secretion	↓ IGF-1 secretion/action
Deep sleep	β-Adrenergic input	GH receptor deficiency
α-Adrenergic input	IGF-1	IGF-1 receptor deficiency
Cholinergic input	Glucocorticoids	Acute or chronic illness
Fasting	↑Free fatty acids	Undernutrition
Androgens	Hyperglycemia	
Estrogens (paracrine)	Hypothyroidism	
Stress	Obesity	
Hypoglycemia		

GH, Growth hormone; *IGF-1*, insulin-like growth factor-1.

1 Growth hormone–releasing hormone

GH–releasing hormone (GHRH) was originally isolated from ectopic tumors producing it (and thereby causing observable derangements in growth), rather than from the hypothalamus and its structure determined by two laboratories in 1982. GHRH is secreted by the arcuate nucleus of the hypothalamus and is derived from a 108 amino acid preprohormone yielding either GHRH (1–44) shown in Fig. 3.7 or GHRH (1–40), shortened at the carboxyl terminal. Both forms of GHRH are found in the human hypothalamus and since the full biological activity of GHRH lies in amino acid residues 1–29, the physiological differences between the two GHRH forms are likely minimal. Structurally, GHRH belongs to a family of proteins that includes secretin, glucagon, glucagon-like peptides (GLP-1 and GLP-2), and vasoactive intestinal peptide (VIP).

Following secretion from its neurons into the portal circulation, GHRH binds to its receptor, GHRH-receptor (GHRHR), a GPCR on the somatrophs of the anterior pituitary. The schematic linear structure of the GHRHR is shown in Fig. 3.16. Increased cAMP production leads to the increased synthesis of GH. cAMP also stimulates the opening of Ca^{2+} and K^{+} ion channels, which play roles in the secretion of existing GH from the cell, in its characteristic pulsatile fashion. Phospholipid signaling may also be involved in the exocytosis associated with the release of GH by the somatotroph when stimulated by GHRH. In addition to its effect on GH secretion and synthesis by somatotrophs, GHRH stimulates the proliferation of these cells through the activation of the mitogen-activated protein kinase (MAPK) pathway.

GHRH and/or its receptor also occur outside the central nervous system in such tissues as the pancreas where it stimulates insulin, glucagon, and SST release; in the gastrointestinal tract where it stimulates gastrin release and stimulates epithelial cell division; and in tumors of many types.

2 Somatostatin

The stimulatory effect of GHRH on GH release from the anterior pituitary somatotroph is countered by the GH release-inhibiting hormone, SST, also known as somatotropin release inhibiting hormone. The amino acid sequence of the two forms, SST28 and SST14, the forms that play the major role in the regulation of pituitary function in humans, are depicted in Fig. 3.7. The periventricular nucleus (located rostral to the paraventricular nucleus) is the major site of SST-producing neurons in the hypothalamus. In addition to its inhibitory effect on GH secretion by somatotrophs, for which it was named, SST modulates a much broader range of peptides from the pituitary (e.g., TSH), as well as from

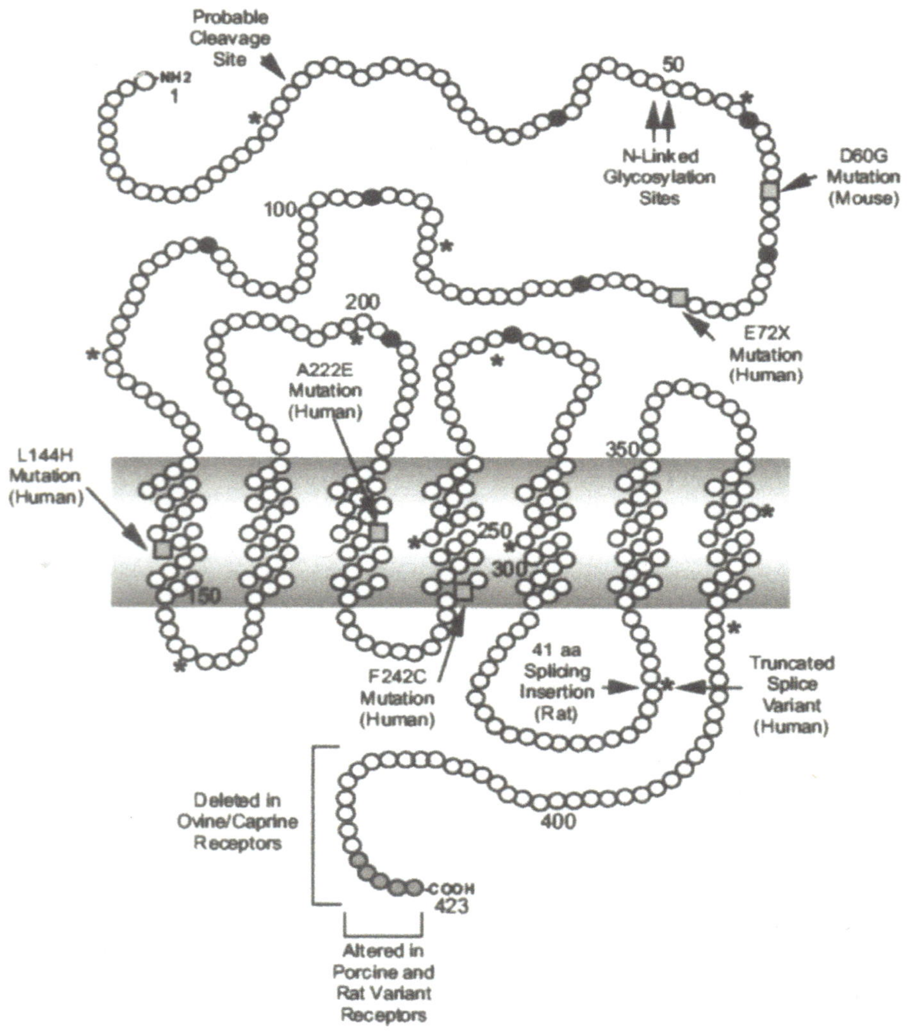

FIGURE 3.16 Schematic structure of the GHRH receptor. The black filled amino acids represent six cysteine residues in the N-terminal extracellular domain conserved in this receptor family, as well as the cysteines in extracellular loops one and two conserved in most GCPRs. Asterisks indicate positions at which introns interrupt the GHRH coding sequences within the gene. Characterized inactivating mutations in the receptor are indicated by the squares. Also indicated are several of the best characterized receptor variants, although others have been reported. *GCPR*, G protein-coupled receptor; *GHRH*, growth hormone–releasing hormone. *Reproduced from Fig. 3.5 of K.E. Mayo et al., International union of pharmacology. XXXV. The Glucagon Receptor Family, Pharmacological Reviews, 55 (2003) 167–194.*

the stomach, brain, intestine, and pancreas (Chapter 6: Pancreatic Hormones: Insulin and Glucagon and Chapter 7: Gastrointestinal Hormones). For example, one of the SST receptors (SSTR4; see later) acts in the brain to decrease Alzheimer's-related amyloid β peptide levels by increasing their rate of degradation.

There are five SST receptors, designated SSTR1–SSTR5. These are encoded by separate genes on separate chromosomes. They differ in their tissue expression patterns, their affinities for various SST agonists, and the G-proteins, and, therefore, the cell signaling pathways, to which they are coupled. Clearly this degree of variation allows SST to have diverse effects in many different target tissues. SST 14 has greatest affinity for SSTR2, which is responsible for the inhibition of the secretion of pituitary hormones, including GH. Upon ligand activation, SSTR2 binds the Gαi/o subunit, which inhibits adenylyl cyclase, diminishing cytoplasmic cAMP levels. SSTR2 activation also initiates opening of K^+ channels and closure of Ca^{2+} channels. Both of these mechanisms contribute to the suppression of pituitary GH secretion. As stated, there are five SSTRs and these are labeled SSTR1 through SSTR5. There are constitutive receptors and ligand-induced receptors and they play various roles in inhibiting hormone secretion, stimulating apoptosis, and inhibition of cell growth plus other effects. Fig. 3.17 shows the SSTRs that are involved in each of these effects. SSTRs function as both monomers and as homodimers or heterodimers as reported in Fig. 3.18.

E Hypothalamic control of prolactin secretion

In mammals the primary hypothalamic control of PRL secretion is predominantly through the inhibitory action of dopamine (DA), secreted by neurons in the dorsomedial area of the arcuate nucleus and the inferior ventromedial nucleus of the hypothalamus. PRL is the only pituitary hormone regulated in this fashion—that is, showing unrestrained secretion in the absence of hypothalamic input. Fig. 3.19 depicts the signaling leading to the restraint of PRL secretion by DA. The most immediate (within seconds) response to DA is increased potassium conductance and inactivation of voltage sensitive Ca^{2+} channels. The resulting diminished intracellular Ca^{2+} leads to reduced exocytosis of secretory vesicles containing PRL. Within a few minutes, the inhibition of adenyl cyclase reduces intracellular Ca^{2+} leading to decreased PRL gene expression. Sustained exposure to DA input leads to decreased lactotroph proliferation.

Fig. 3.19 also shows the receptors for two other hypothalamic peptides, TRH and VIP, which have been shown to be capable of stimulating PRL secretion. These effects are exerted through the signaling pathways shown and result in increased expression of the PRL gene. The physiological role of these stimulatory peptides in humans remains somewhat unclear, however, and by far the most

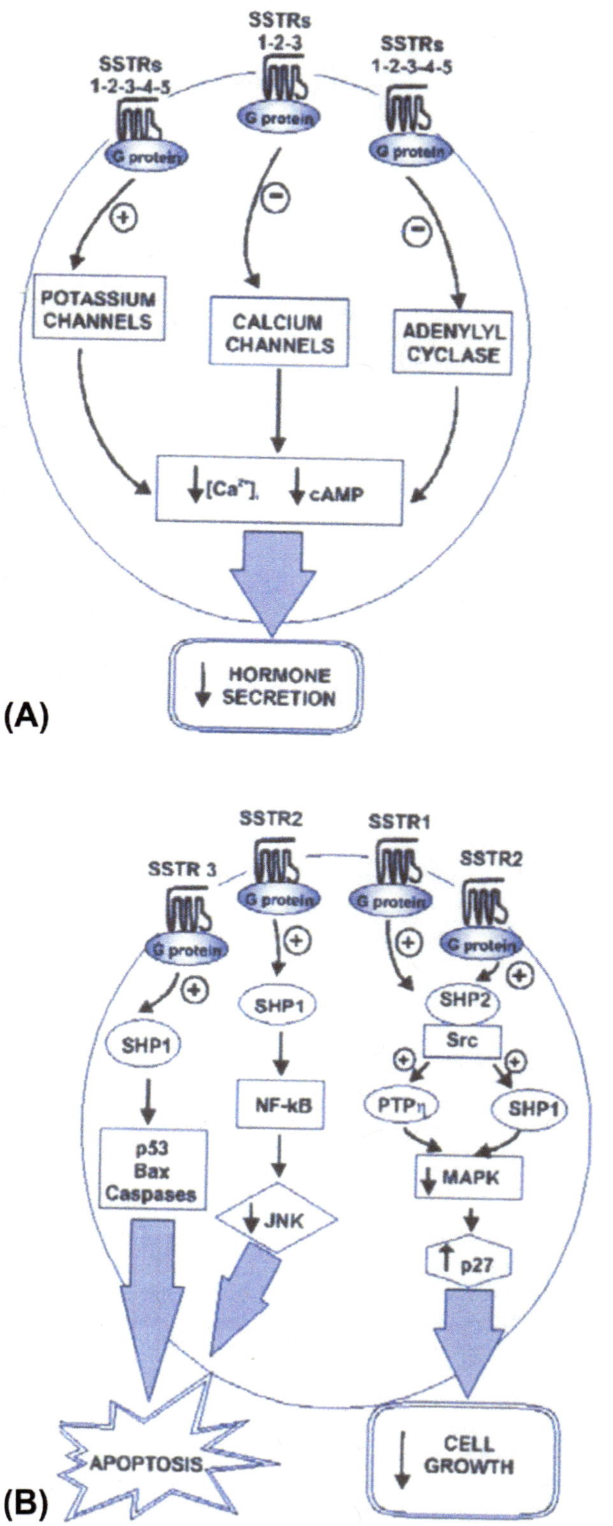

FIGURE 3.17 Diagrammatic representation of the intracellular pathways activated by individual SSTRs to elicit the inhibition of hormone secretion (A) and cell growth arrest or apoptosis (B). Only the best characterized pathways are reported. *SSTR*, Somatostatin receptor. *Reproduced from Fig. 3.1 of T. Florio, Molecular mechanisms of the antiproliferative activity of somatostatin receptors (SSTRs) in neuroendocrine tumors, Frontiers in Bioscience, 13 (2008) 822–840.*

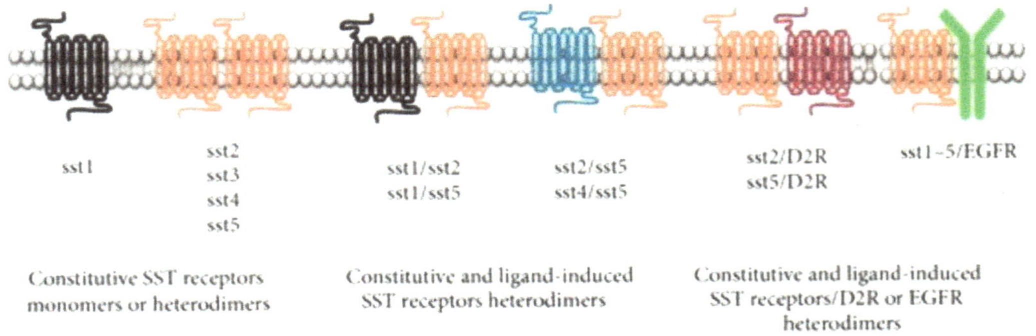

FIGURE 3.18 Dimerization patterns of somatostatin receptors. Each SST receptor subtype shows a different constitutive tendency to dimerize. SST1 exists as a monomer, while SST 2, 3, 4, and 5 homodimerization occurs both constitutively or after somatostatin binding. SST receptors may form heterodimers with other members of the SST receptor family, either in resting conditions or upon ligand binding. SST receptors also dimerize with receptors of other GCPR families (e.g., D2R) or with other receptor families, such as tyrosine kinase receptors (i.e., EGFR), originating heterodimers. *EGFR*, Epidermal growth factor receptor; *SST*, somatostatin; *GCPR*, G protein-coupled receptor. *Reproduced from Fig. 3.2 of F. Barbieri et al., Peptide receptor targeting in cancer: the somatostatin paradigm, International Journal of Peptides, 2013 (2013) 20: article ID926295; https://doi.org/10.1155/2013/926295.*

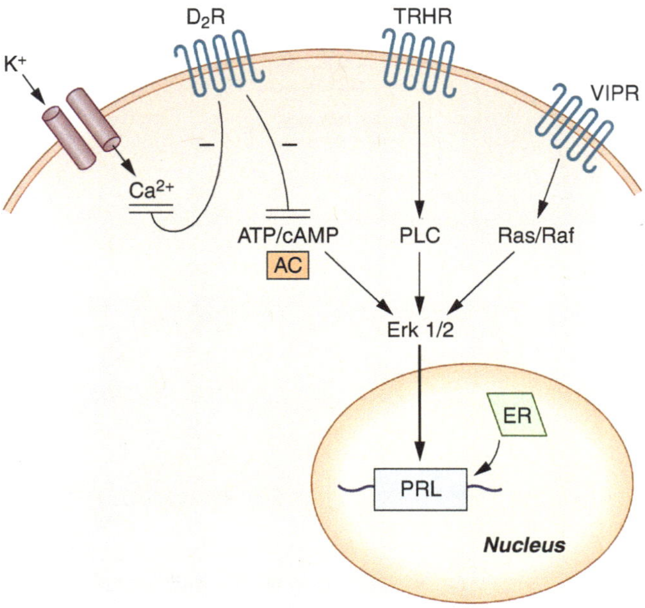

FIGURE 3.19 Hypothalamic control of prolactin secretion. The inhibitory effect of the dopamine receptor, D_2R, the predominant influence on prolactin secretion by the lactotroph, results from the inhibition of cyclic AMP production, which decreases PRL gene expression. The inwardly rectifying potassium channel is also inhibited by D_2R activation, reducing intracellular Ca^{2+} and exocytosis of secretory vesicles containing prolactin. The receptors and signaling pathways for two known hypothalamic stimulators of PRL secretion, TRH and VIP, are also shown, as is the stimulation of PRL gene expression by the activated ER. *AC*, Adenylyl cyclase; *ER*, estrogen receptor; *PLC*, phospholipase C; *PRL*, prolactin; *TRH*, thyrotropin-releasing hormone; *VIP*, vasoactive intestinal peptide.

powerful positive influence on PRL secretion is the neuronal one exerted by suckling as discussed in Section V.B following and in Chapter 14, Hormones of Pregnancy, Parturition and Lactation.

IV Chemistry of the anterior pituitary hormones

A Glycoprotein hormones

There are three anterior pituitary (TSH, LH, and FSH) and one placental (human chorionic gonadotropin, hCG) heterodimeric glycoprotein hormones. These hormones share a common α-subunit and have distinct β-subunits, as illustrated in Fig. 3.20. The α- and β-subunits are each stabilized by disulfide bonds, and disruption of these bonds by reducing agents alters the internal configuration of the peptide chains, causing dissociation of the heterodimer. It is the β-subunit that confers the specificity on the structure of the hormone so that each of the four heterodimers interacts only with its specific receptor. The one α- and four β-subunits that make up the four hormones, each contain either one or two asparagine N-linked oligosaccharides. In addition, hCG has a carboxy-terminal peptide extension, relative to the other β-subunits, that has four serine residues serving as O-linked glycosylation sites. These sites on hCG are not highly important in receptor binding and signal transduction but serve to increase the in vivo half-life of the hormone and, therefore, its bioactivity. The N-linked oligosaccharides in hCG are important for biological activity and their removal results in the conversion of the active hormone into an antagonist that binds to the receptor but does not activate it.

As discussed in some detail in Chapter 5, Thyroid Hormones, the biological role of TSH is to stimulate the synthesis and secretion of the thyroid hormones, thyroxine (T_4) and T_3, from the epithelial cells making up the follicles of the thyroid gland. Similarly, the important biological roles of FSH and LH throughout the reproductive system in males and females will be discussed in Chapters 12–14. Also, hCG, which is initially produced by the human trophoblast upon implantation in the uterus,

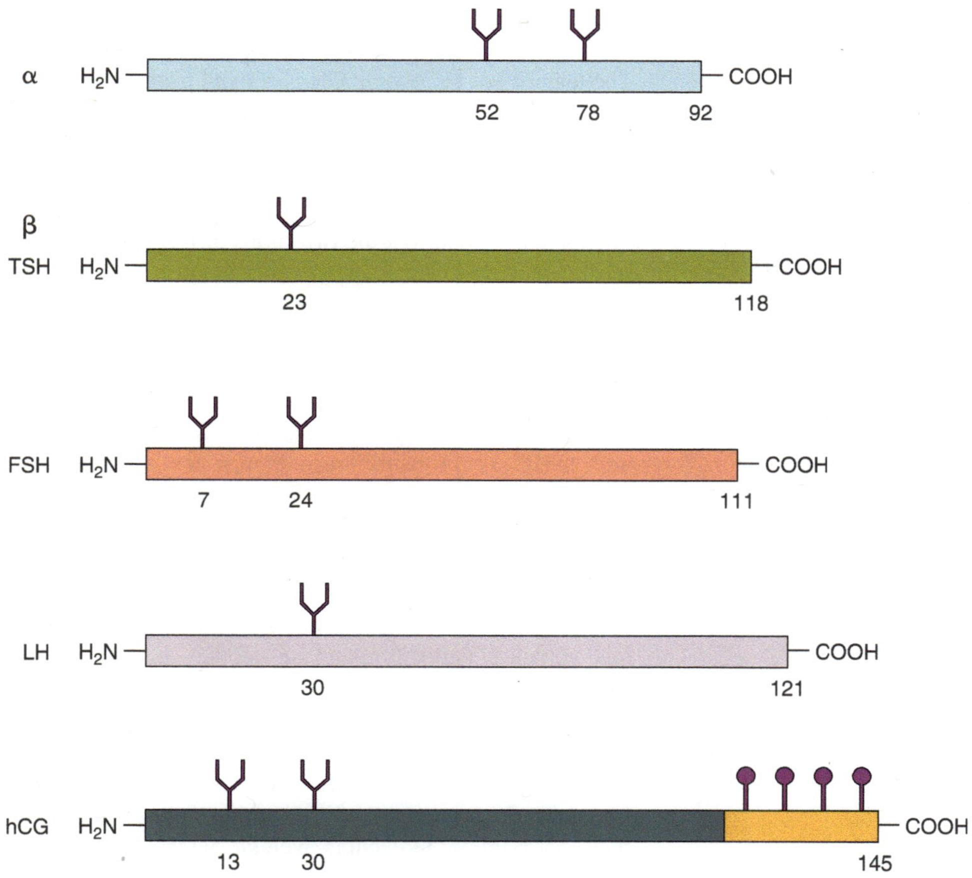

FIGURE 3.20 The subunit structure of the pituitary glycoprotein hormones. The three pituitary glycoprotein hormones, TSH, FSH, LH, and the placental version of LH, hCG, all consist of two subunits. They share a common α-subunit (*blue*), which combines with a unique β-subunit to form each of the four active heterodimeric hormones. The positions of the *N*-glycosylated asparagine sites are indicated by the forks for each of the five subunit types; hCG is characterized by a C-terminal extension with *O*-glycosylated serine sites. *FSH*, Follicle stimulating hormone; *hCG*, human chorionic gonadotropin; *LH*, luteinizing hormone; *TSH*, thyroid stimulating hormone.

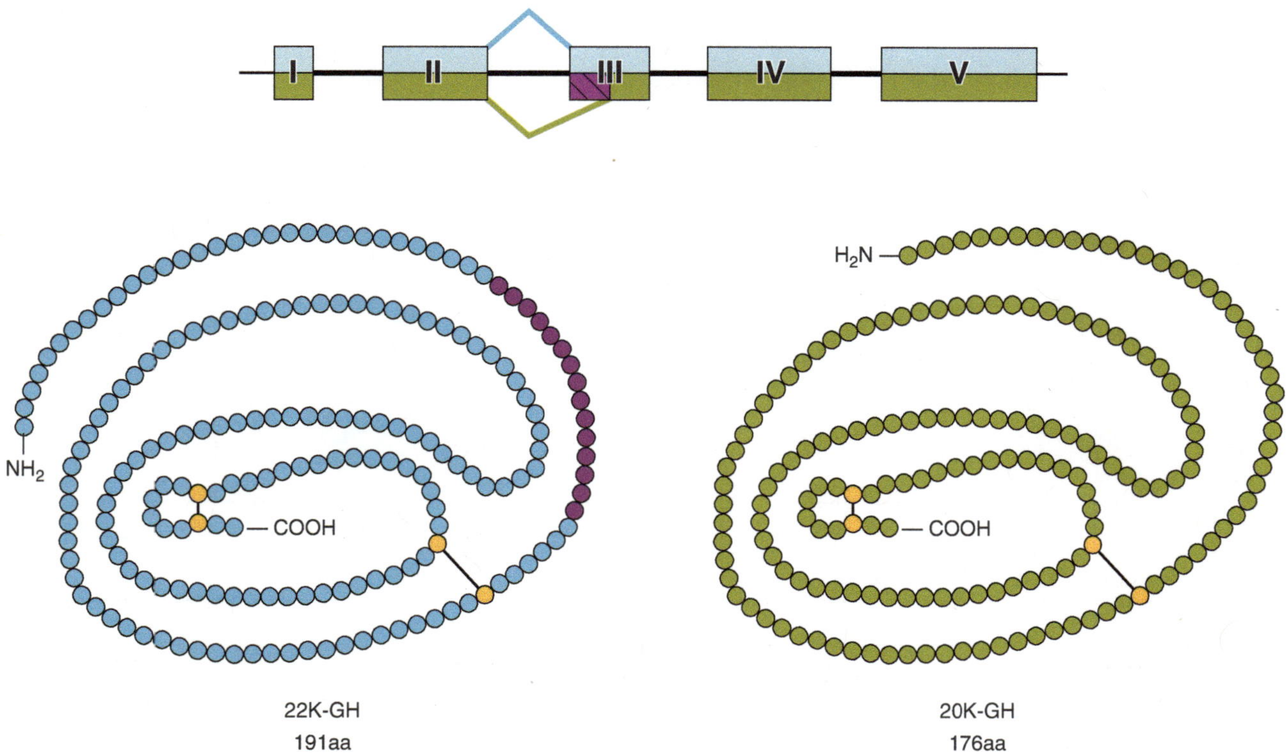

FIGURE 3.21 Gene and amino acid sequences of GH. The two main forms of GH are shown. GH-22k (left) is derived from all five exons, whereas in GH-20k use of an alternative splice site in exon III results in omission of 15 amino acids, which are indicated in pink in GH-22k. *GH*, Growth hormone.

mimics the actions of LH and is essential for the maintenance of pregnancy in its early stages. This will be covered in Chapter 14, Hormones of Pregnancy, Parturition and Lactation.

B Growth hormone

As shown in Fig. 3.21 the gene for human GH (hGH) contains five exons and four introns. GH gene expression in somatotrophs requires the transcription factor Pit-1, which is also involved in the control of the expression of hormone-encoding genes in lactotrophs and thyrotrophs. The majority of circulating GH is derived from the complete set of five exons and is a protein of 191 amino acids with a molecular weight of 22 kD. Approximately 10%–15% of circulating GH is smaller (176 amino acids, 20 kD) due to the omission of 15 amino acids encoded in the first section of exon III. The ratio of the two forms of GH in the blood is relatively constant and is independent of age.

Human placental lactogen (hPL, also known as chorionic somatomammotropin) is produced by the placenta and is very closely related to hGH. Its gene lies in the same cluster on chromosome 17; hPL also contains 191 amino acids, shares greater than 90% amino acid identity with hGH, and probably arose from a new duplication of the primate GH gene or extensive interchange of the GH gene with the PL genes. hPL circulates primarily in the maternal compartment and has little direct effect on fetal growth. Further discussion of the GH gene cluster and the functions of hPL can be found in Chapter 14, Hormones of Pregnancy, Parturition and Lactation.

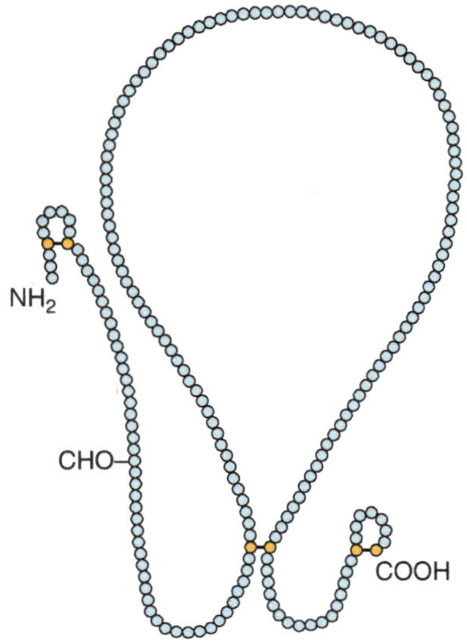

FIGURE 3.22 Amino acid sequence of human prolactin. Prolactin is a single polypeptide of 199 amino acids and 3 internal disulfide bonds as shown.

C Prolactin

The primary sequence of the PRL molecule is shown in Fig. 3.22. PRL consists of 199 amino acid residues, and the approximate molecular weight of the glycosylated form is about 23 kD. It is encoded by a gene on chromosome 6, outside the GH/CS cluster on chromosome 17, but there is about 40% amino acid homology between PRL and GH/CS. The gene duplication event leading to

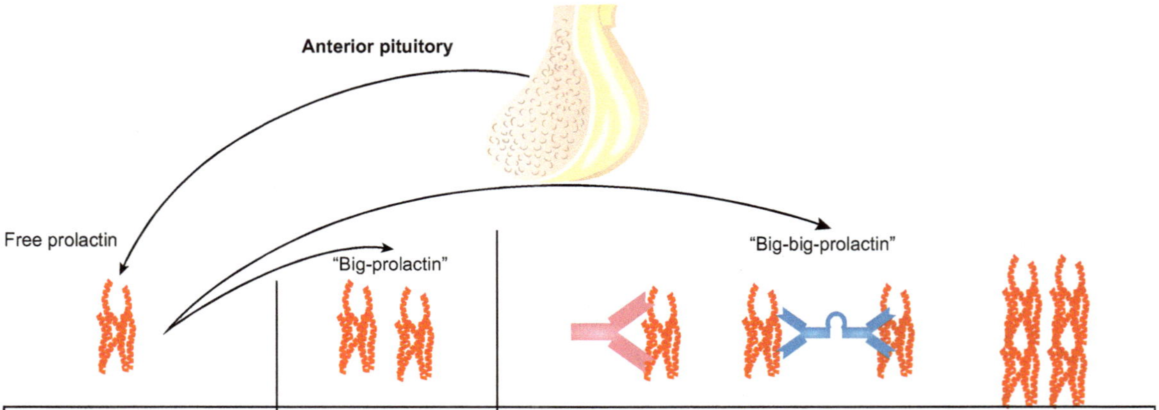

Monomeric	Dimeric	IgG-prolactin	IgA-prolactin	Prolactin aggregrates
23 kDa	40-60 kDa	>100-150 kDa		
Active	Inactive	Inactive		
60%-90%	15%-30%	0%-8%	0%-2%	0%-3%

FIGURE 3.23 Structure of monomeric prolactin, "big prolactin" and "big-big prolactin." *Reproduced from Fig. 3.1 of M. Saleem, M.H. Coates, Prolactin biology and laboratory measurement: an update on physiology and current analytical issues, The Clinical Biochemist Reviews, 39 (2018) 3–16.*

these three genes likely occurred over 400 million years ago. Circulating variants of PRL are generated by post-translational modifications such as proteolysis leading to smaller forms and dimerization or aggregation leading to larger forms referred to as big PRL and macro-PRL, respectively. In Fig. 3.23, are shown the various forms of PRL suggesting that the larger forms beyond the monomer may be inactive. While the biological significance of these larger forms is not fully understood, their presence has importance in the context of clinical measurements of serum PRL levels.

D POMC derivatives: ACTH, α-MSH, β-lipotropin, endorphin

POMC (named for three of the peptide types that are derived from it) was one of the early examples of a propeptide containing more than one biologically active sequence within it. As shown in Fig. 3.24, POMC is encoded by portions of two of the three exons that comprise its gene on chromosome 2.

The processing of POMC differs depending on the cell in which it is synthesized and the PC activities in those cells. In the CRH-stimulated corticotrophs of the anterior pituitary, the major products of POMC processing are an N-terminal fragment, the two active peptides, ACTH and β-LPH, and the joining peptide, which has no known further biological function in humans. A small amount of β-LPH is processed to β-endorphin in these cells. The principal biological action of ACTH is to stimulate corticosteroid production in adrenal cortical cells. This hormone interacts with receptors in membranes primarily of the zona fasciculata and zona reticularis, although there are also receptors in the zona glomerulosa. The linear structure of the 7-transmembrane ACTH receptor (MC2R) on the zona fasciculata cells of the adrenal cortex is shown in Fig. 3.25 showing the primary sequence and indicating the various possible mutations of the human receptor. A further discussion of the role of ACTH in the adrenal cortex appears in Chapter 10, Adrenal Corticoids.

In the intermediate lobe (or in scattered cells thereof in humans) POMC processing is more extensive, with the cleavage of ACTH into α-MSH and corticotropin-like intermediate lobe peptide (ACTH 18–39). Also in these cells, N-POMC is processed into Lys-γ3-MSH, a hormone that, through the activation of hormone-sensitive lipase, may play a role in lipid metabolism in the adipocyte (see Section VI). Finally, β-LPH is cleaved to generate β-endorphin and β-MSH, leaving γ-LPH.

Fig. 3.26 shows the structures of proteins, including the ACTH receptor (MC2R) in the plasma membrane of a zona fasciculata cell. ACTH binds to the MC2R in the extracellular space as long as MC2 accessory protein adopts an antiparallel homodimeric conformation. The Gα subunit of MC2R

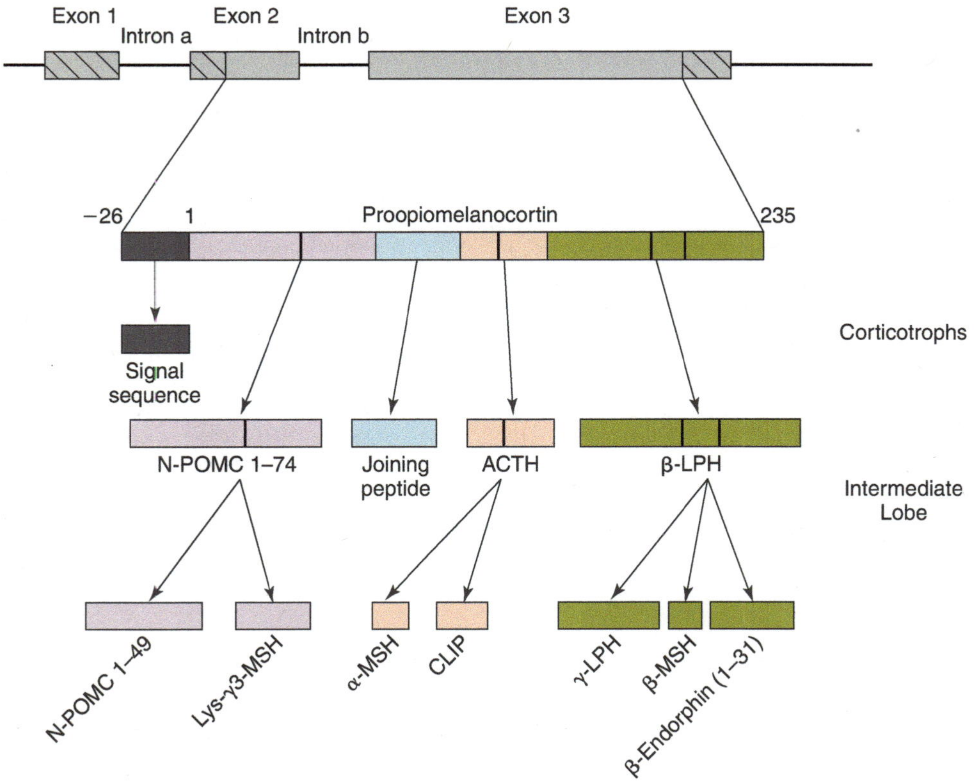

FIGURE 3.24 The POMC gene and its protein products. POMC is encoded by portions of the second and third exons of its gene and consists of a signal peptide and several bioactive peptides. The protein is processed differently in different cell types. In pituitary corticotrophs the proprotein is cleaved into ACTH (*pink*), the main secretory product of these cells, and β-lipotropin (*green*). In the cells of the intermediate lobe, ACTH and β-lipotropin, along with the N-terminal section, N-POMC 1–74 (*light purple*), are further processed as shown. The vertical lines within the proteins represent the dibasic residues at the cleavage sites. *ACTH*, Adrenocorticotropic hormone; *POMC*, proopiomelanocortin.

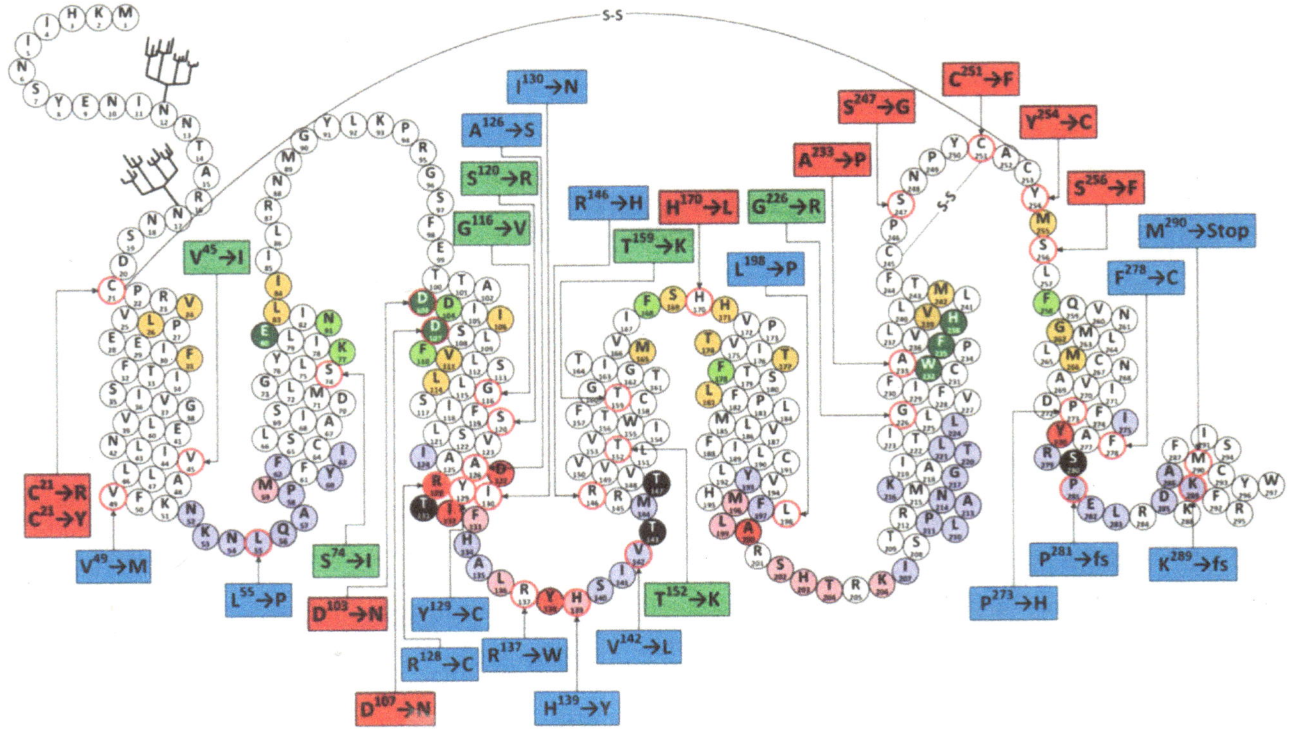

FIGURE 3.25 Snake-like plot of ACTHR. Forked structures represent glycosylation. -S-S- represents disulfide bonds. Circles with red outline represent naturally occurring missense mutations (labels enclose information on the resulting residue replacement). Red labels indicate residues located on the extracellular part of the receptor. Green labels indicate residues located within the central part of transmembrane domains. Blue labels indicate residues located within the intracellular part of receptor. Dark green circles represent residues that form the backbone of the MCR-binding pocket. Orange circles represent residues equivalent to those forming the ligand-binding cavity of MC4R based on previous models. Light green circles represent residues that have been identified as important for ligand recognition. Circles shaded in red, pink and light purple represent residues equivalent to those that directly interact with Gα, are involved in Gα coupling–related intramolecular interactions, and that form the surface of the Gα-binding cavity, respectively. Circles shaded in black represent S/T residues that are crucial for ACTHR expression and functional regulation. Circles with angled cross-hatching represent residues that are unique to ACTHR (yet conserved in other MCRs). *ACTHR*, Adrenocorticotropic receptor; *GH*, growth hormone; *MCR*, melanocortin receptor. *Reproduced from Fig. 3.6 of D. Fridmanis, A. Roga, J. Klovins, Frontiers in Endocrinology (2017); https://doi.org/10.3789/fendo.2017.00013.*

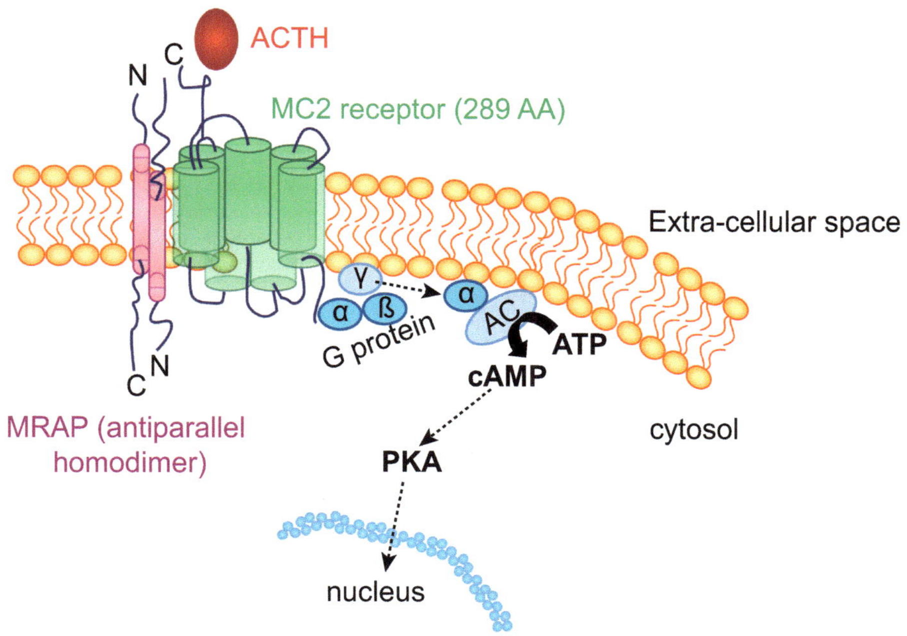

FIGURE 3.26 ACTH receptor protein expression. MC2R mRNA is translated at the endoplasmic reticulum and is unable to traffic beyond this point to the plasma membrane. MRAP mRNA is translated and adopts an antiparallel homodimeric conformation at the endoplasmic reticulum. Only the MRAP–MC2R membrane complex is competent to bind ACTH at physiological concentrations and to generate a steroidogenic signal. *ACTH*, Adrenocorticotropic hormone; *MRAP*, MC2 accessory protein; *mRNA*, messenger RNA. *Reproduced from Fig. 3.4 of A. Angelousi, A.N. Margioris, C. Tsatsanis, ACTH action on the adrenals, in: K.R. Feingold, B. Anawalt, A.S. Boyce (Eds.), Endotext.com.Inc., 2020, 2000–2020.*

dissociates from the receptor and binds and activates adenylate cyclase that forms cAMP from ATP. cAMP activates protein kinase A that enters the nucleus and stimulates the production of proteins involved in the formation of cortisol which is secreted from the cell into the general circulation.

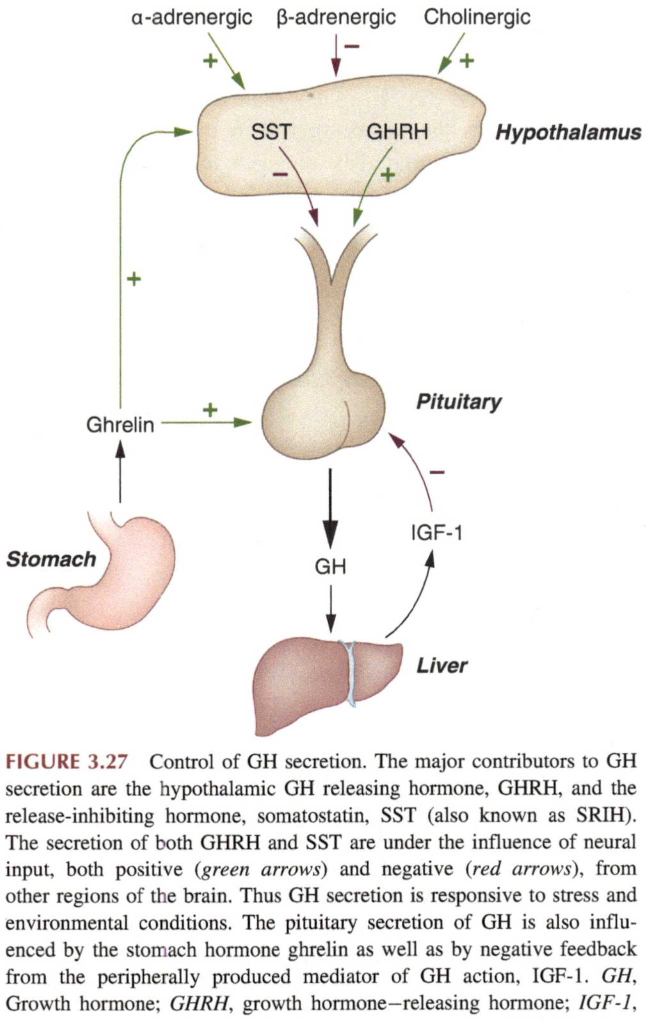

FIGURE 3.27 Control of GH secretion. The major contributors to GH secretion are the hypothalamic GH releasing hormone, GHRH, and the release-inhibiting hormone, somatostatin, SST (also known as SRIH). The secretion of both GHRH and SST are under the influence of neural input, both positive (*green arrows*) and negative (*red arrows*), from other regions of the brain. Thus GH secretion is responsive to stress and environmental conditions. The pituitary secretion of GH is also influenced by the stomach hormone ghrelin as well as by negative feedback from the peripherally produced mediator of GH action, IGF-1. *GH*, Growth hormone; *GHRH*, growth hormone–releasing hormone; *IGF-1*, insulin-like growth factor-1; *SRIH*, somatotropin release-inhibiting hormone; *SST*, somatostatin.

V Regulation and biological actions of growth hormone and prolactin

A Growth hormone

1 Regulation of secretion

As shown in Fig. 3.27, the neuroendocrine control of GH secretion is primarily under the control of the two hypothalamic hormones, GHRH and SST, the release-inhibiting hormone. The synthesis and secretion of these peptides is discussed in Section III.D. The balance between the secretion of GHRH and SST by the hypothalamus is determined by many inputs, including neuronal pathways that allow GH secretion to be responsive to stress and other environmental conditions. GHRH affects both the expression of the GH gene in the somatotrophs and the secretion of existing hormone, whereas SST's effects are predominantly on the secretion of existing GH.

In addition to the hypothalamic hormones, somatotroph secretion of GH is under negative feedback control by IGF-1, a peptide hormone that, as its name implies, shares considerable structural similarity with insulin (see later). The synthesis of IGF-1 (once known as somatomedin) is stimulated by GH in the liver; it mediates GH effects in several peripheral tissues as described further next and exerts negative feedback control on GH secretion at the pituitary.

GH secretagogues (GHSs) have been recognized for more than three decades as small molecules that stimulate pituitary GH secretion independently of GHRH. Originally the two pathways were distinguished by the different second messengers involved in their action, cAMP for GHRH and intracellular Ca^{2+} for GHS. In the mid-1990s the receptor for these synthetic molecules was cloned, named GHS-R, and identified as an orphan GPCR since its natural ligand was unknown. In 1999 the natural ligand for this receptor was identified as a 28 amino acid peptide produced in the stomach. It was named ghrelin and found to play a role in the regulation of appetite (see Section VI). As shown in Fig. 3.27, it stimulates GHRH secretion by the hypothalamus and, in the pituitary, it

FIGURE 3.28 Amino acid sequence of ghrelin. The 28 amino acid sequence of ghrelin is shown, with acylated serine in the third position. The acyl group can be either 8 or 10 carbons long and is required for the growth hormone secretagogue activity of ghrelin, but not for many of its other biological activities.

synergizes with GHRH, stimulating the release of GH in response to a maximal dose of the hypothalamic releasing hormone. In fact the effect of ghrelin on GH secretion requires an intact hypothalamic–pituitary system and it has no effect on patients with hypothalamic lesions.

The structure of mature ghrelin is shown in Fig. 3.28. Its 28 amino acids are derived from a 117 amino acid prohormone and the active peptide is octanylated at the third position. Although only a small proportion of circulating ghrelin is acylated (1.8%), this posttranslational modification is required for its interaction with the GHS-R and stimulation of GH secretion. It is likely that the form lacking this modification, desacyl ghrelin, is an active peptide in ghrelin's impact on metabolism and appetite and in the development of adipocytes.

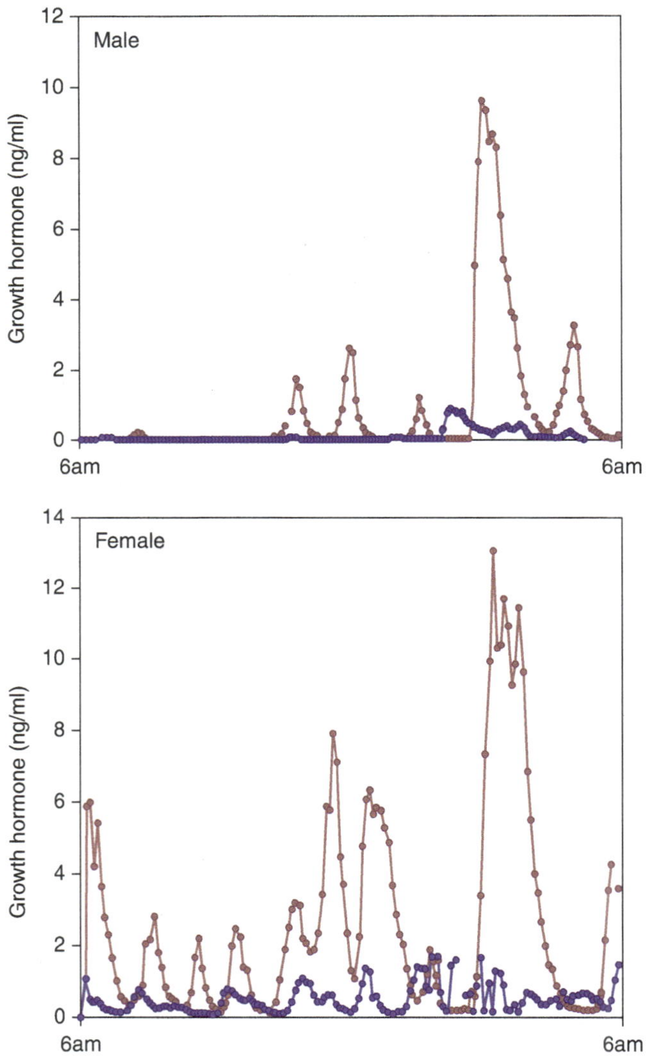

FIGURE 3.29 Patterns of GH secretion. Typical patterns of GH secretion are shown for a young male (top panel) and female (bottom panel) adult humans. The purple line at the bottom of the graph shows the pattern obtained in the presence of a GHRH antagonist which diminishes the amplitude of the GH pulses. *GH*, Growth hormone; *GHRH*, growth hormone–releasing hormone.

Like many of the other pituitary hormones, the secretion of GH from the pituitary is pulsatile and, as shown in Fig. 3.29, peaks of GH secretion occur throughout a 24-hour period. The magnitude of the peaks can vary with a number of factors, one of which is sleep, with which GH secretion is tightly linked, particularly in males. The sexually dimorphic pattern of GH secretion is characterized by the much more predominant nocturnal peak, relative to those during the day, in males than in females. This pattern, as well as the age-related decrease in the amplitude of GH pulses, is related to changes in GHRH secretion, indicating that endogenous GHRH secretion is the main regulator of pulsatile GH secretion in humans. Basal levels of GH are largely dependent on SST's influence on the somatotrophs.

2 GH receptor and signaling

The GH receptor, GHR, belongs to the class I cytokine/hematopoietin receptor superfamily. It contains 638 amino acids that comprise a glycosylated extracellular ligand-binding domain, a single membrane–spanning segment of 24 amino acids and the cytoplasmic signaling component. A soluble derivative of the extracellular domain is known as GH–binding protein, GHBP, which circulates in the blood. GH bound to GHBP is not available to bind to receptors on target tissues or for degradation, so that the binding protein levels can influence the pool of available hormone.

As shown by crystallization and other studies in the early 1990s, the active GH–GHR complex consists of one molecule of GH and two receptor monomers. Two views of the crystal structure of the GH/GHR complex are shown in Fig. 3.30. The binding site between the hGH and its receptor (PDB ID: 1a22 chains A and B) are mapped through clusters of evolutionarily important residues. Initially it was thought that GH binds to a monomer and brings about the dimerization to activate the receptor. More recent work, however, indicates that the binding of GH to the dimerized receptor causes an internal rotation of the receptor, activating the receptor complex so that the intracellular domain initiates the signaling pathway outlined in Fig. 3.31. Each monomer of the GHR dimer binds to a molecule of a tyrosine kinase of the Janus family, JAK2, which then phosphorylates several tyrosine residues on the intracellular domain of the dimeric GHR. These serve as docking sites for at least three different intracellular signaling pathways: STAT (signal transduction and activator of transcription); MAPK; and phosphoinositide-3 kinase. The transcription of multiple genes is affected as a result of the activation of these pathways. One of these genes encodes the peripheral mediator of GH actions, IGF-1, described in the next section. Two others, IGF-binding protein (IGFBP-3) and acid-labile

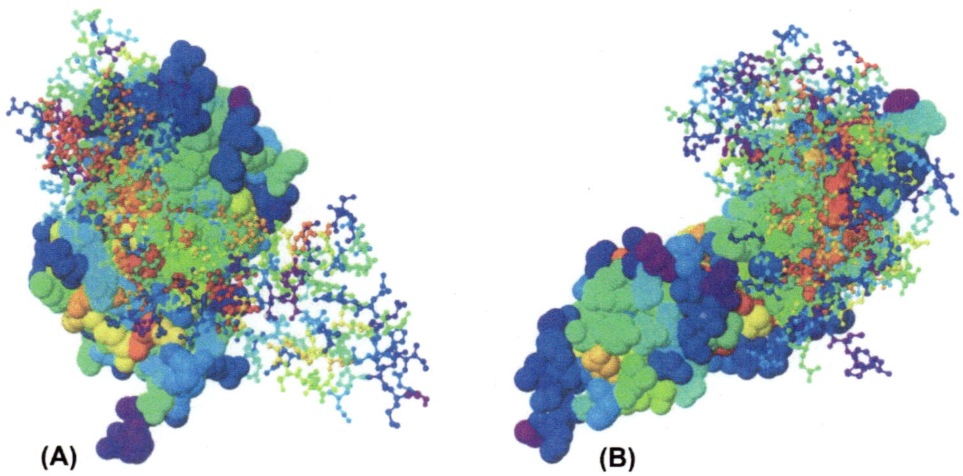

FIGURE 3.30 The human growth hormone in complex with the growth hormone receptor (PDB code: 1a22) with ET analysis. (A) Human growth hormone is shown in spacefill mode, while the human growth hormone receptor is shown as ball and stick. (B) The human growth hormone receptor is shown as spacefill, while the human growth hormone is displayed as ball and stick. *ET*, Evolutionary trace. *Reproduced from Fig. 3.2 of R.C. Lua et al., UET: a database of evolutionarily-predicted functional determinants of protein sequences that cluster as functional sites in protein structures, Nucleic Acids Research, 44 (2015): Database issue; https://doi.org/10.1093/nar/gkv1279.*

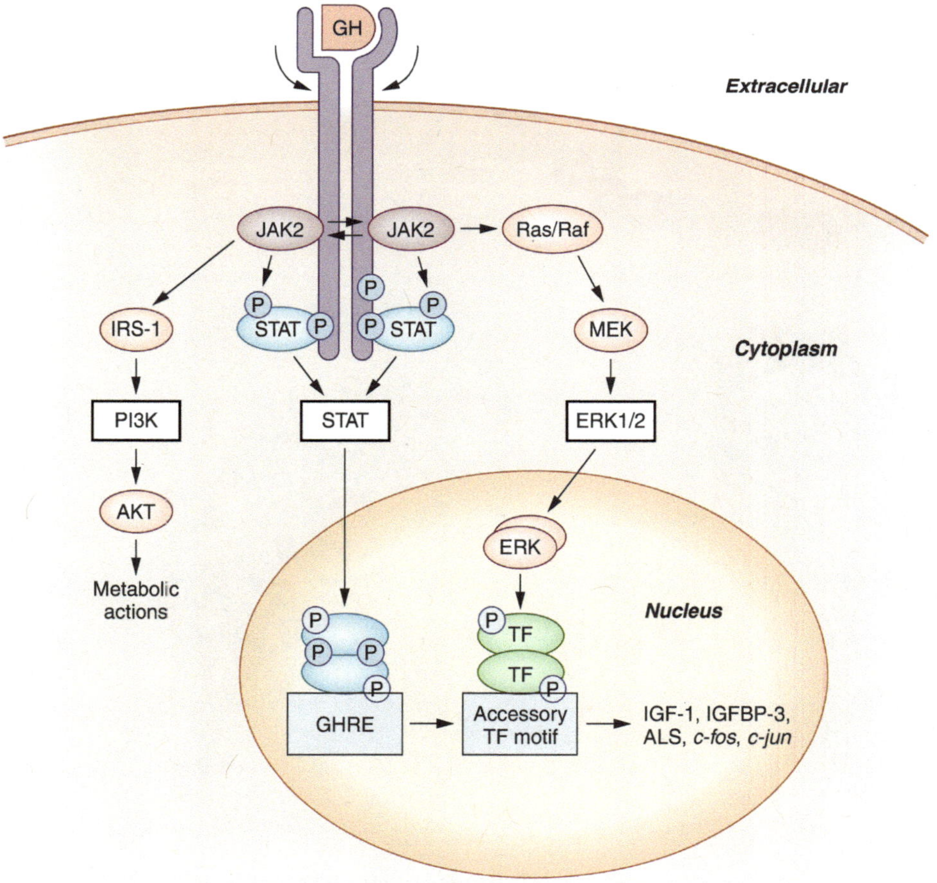

FIGURE 3.31 Growth hormone signaling. The binding of GH to a preformed GHR dimer results in internal rotation leading to the activation of receptor-associated JAK2. Three major downstream signaling modules are shown. Through interaction with the IRS, the PI3 kinase/AKT pathway is activated, leading to changes in metabolic actions. Latent cytoplasmic STAT molecules (*blue ovals*) are recruited to the GHR where they are phosphorylated on tyrosine residues. Phosphorylated STAT molecules move to the nucleus where they bind to specific gene regulatory elements, altering gene expression. Activation of the ERK pathway also leads to changes in gene expression. Some of the genes, expression of which is altered by these two pathways, are indicated. *ALS*, Acid labile subunit (see text); *ERK*, extracellular signal regulated kinase; *GH*, growth hormone; *GHR*, growth hormone receptor; *GHRE*, growth hormone receptor enhancer; *IGF*, insulin-like growth factor; *IGFBP*, insulin-like growth factor binding protein; *IRS*, insulin receptor substrate; *IRS*, insulin receptor substrate; *JAK2*, Janus kinase 2; *STAT*, signal transducer and activator of transcription; *TF*, transcription factor.

subunit, are necessary for maintaining appropriate circulating levels of IGF-1.

3 GH–IGF-1 axis

Many of the peripheral actions of GH are mediated by the peptide IGF-1, the main source of which is the liver. As its name implies, IGF-1 (and its close relative, IGF-2) is structurally related to proinsulin. As shown in Fig. 3.32, the single polypeptide IGF-1 has domains similar to those of proinsulin, including the A and B domains (the two subunits of the mature insulin; see Chapter 6: Pancreatic Hormones: Insulin and Glucagon), a C-domain (similar to the C-peptide released when proinsulin is converted to mature insulin), and a short carboxy-terminal D-domain. In the A (*blue*) and B (*pink*) domains, the amino acids that are identical to those in the A and B subunits of insulin are shown in the darker color to emphasize the structural relationship between the two proteins. IGF-2 is very similar to IGF-1 structurally and in biological activity. The primary difference between the two is that IGF-2 is expressed predominantly in early embryonic and fetal life, and IGF-1 is expressed in the adult. IGF-1,

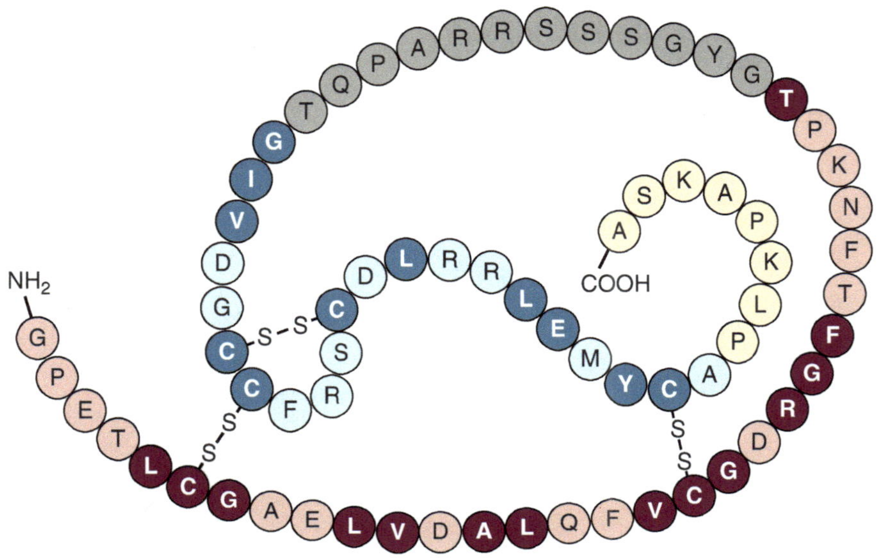

FIGURE 3.32 Amino acid sequence of human IGF-1. The 70 amino acids in IGF-1 occur in four domains. From the N-terminus these are: B, pink; C, gray; A, blue; and D, yellow. Darker amino acids with white lettering are those that are identical in human proinsulin. The A and B domains are homologous to the A and B chains of the insulin molecule and the C domain is the equivalent of the C-peptide in proinsulin. *IGF-1*, Insulin-like growth factor-1.

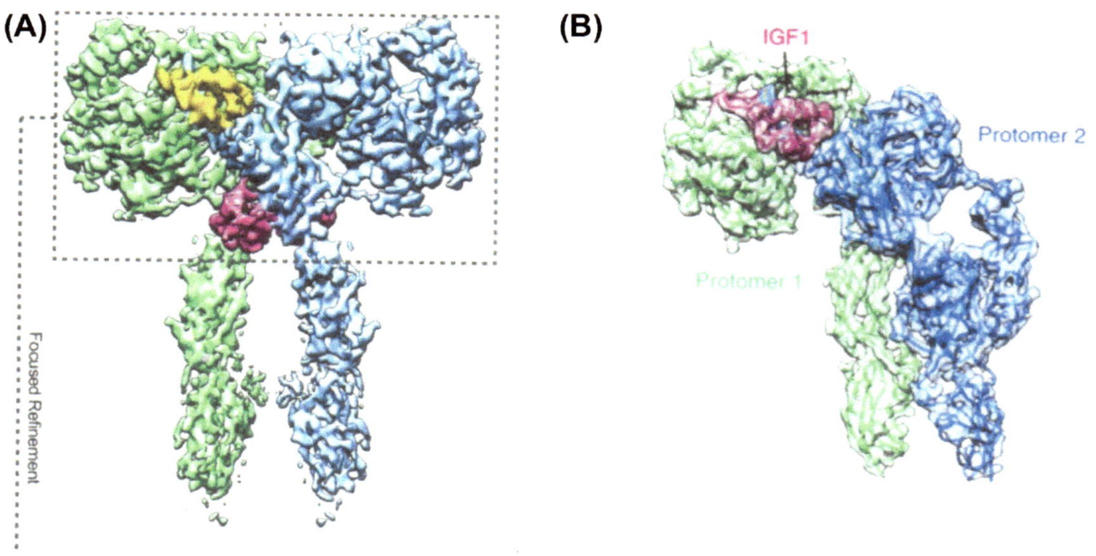

FIGURE 3.33 The cryo-EM structure of the insulin receptor dimer is shown in (A). It has two identical insulin binding sites on each side of the structure. In (B) the IGF1R, a tyrosine kinase, that regulates cell growth and proliferation and can be activated by IGF1, IGF2 and insulin. This is a cryo-EM structure of full length IGF1R-IGF1 complex in the active state. Only one IGF1 molecule binds to the asymmetric IGF1R dimer in contrast to the IR. *cryo-EM*, Cryo-electron microscopic; *IGF1R*, insulin-like growth factor receptor. *(A) Reproduced from E. Uchikawa, G.S.H. Yu, X.-C. Bai, eLife (2019); https://elifesciences.org/aticles/48630. (B) Reproduced from J. Li, E. Choi, Y.X.-C. Bai, Nature Communications (2019), DOI:10.1038/s41467–019–12564–0.*

IGF-2, and insulin diverged from a common ancestral gene. In contrast to insulin the IGFs circulate bound to specific IGF-binding proteins, a group of molecules that play an important role in the availability and plasma half-lives of the IGFs.

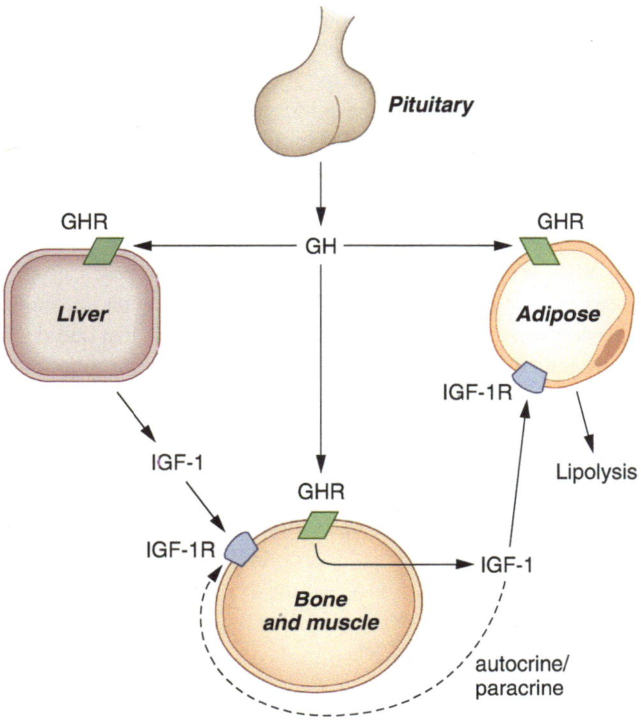

FIGURE 3.34 The GH–IGF-1 axis. In the liver GH interacts with its receptor to stimulate the production of its major mediator, IGF-1 (somatomedin). Through its receptor, IGF-1R (*blue*), IGF-1 mediates effects at several target tissues, including bone and muscle necessary for body growth and adipose tissue. These tissues also respond directly to GH, as indicated by the presence of the GHR (*green*). *GH*, Growth hormone; *GHR*, growth hormone receptor; *IGF-1*, insulin-like growth factor-1.

The Type I IGF receptor, which mediates the mitogenic and metabolic effects of both IGF-1 and IGF-2, is structurally similar to the insulin receptor but distinct enough that there is little physiologically significant crossover in receptor binding between insulin and the two IGFs. A comparison of the structures of the insulin receptor and the IGF-1 receptor (IGF1R) is shown in Fig. 3.33. In both cases the homodimers are depicted. (A) is the cryo-EM structure of the insulin receptor dimer and (B) is the cryo-EM homodimeric structure of the IGF1 receptor. Whereas two molecules of insulin can bind to sites on each insulin receptor monomer, only one molecule of IGF1 binds to the dimeric IGF1R.

As shown in Fig. 3.34, while some peripheral effects of GH in bone, muscle, and adipose tissue (some of the primary targets of GH) are the result of direct action of GH with its receptor in those tissues, IGF-1, interacting with its own distinct specific receptor, is responsible for some of these actions, notably in muscle and liver for linear growth. Most of the circulating IGF-1 that mediates these effects is produced in the liver, although IGF-1 is also produced in target tissues locally and participates in autocrine and paracrine pathways. Both IGF-1 and IGF-2 are important participants in the differentiation and proliferation of many diverse cells in the body. These aspects of IGFs and their serum-binding proteins will be discussed in more detail in Chapter 17, Growth Factors.

Normal growth charts, in terms of both the rate of growth and actual height, are shown in Fig. 3.35. Prenatal bone growth is largely dependent on IGF-1 and IGF-2 and is independent of GH. Postnatal skeletal growth requires GH and IGF-1, both of which have receptors on the epiphyseal growth plate where long bone growth occurs. Thus individuals with a defect in the GH receptor or in its signaling pathway show subnormal growth rates only after

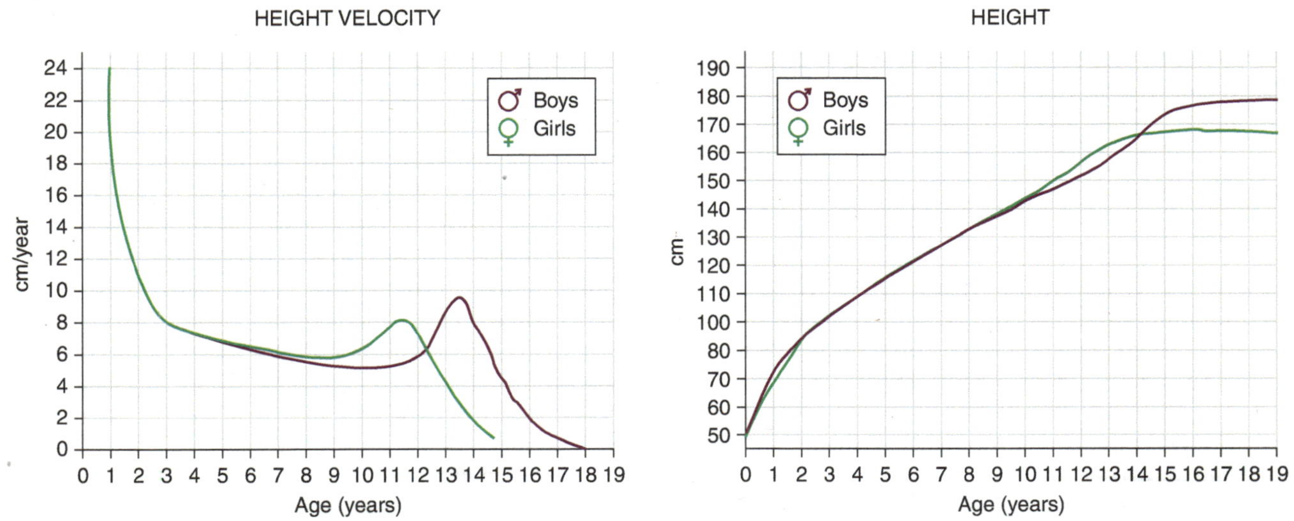

FIGURE 3.35 Growth curves from birth through adolescence. The left panel shows the changes in rate of growth and depicts the prepubertal growth spurt, which begins and ends earlier in girls than in boys. This difference is reflected in the earlier flattening of the absolute growth curve for girls in the panel on the right.

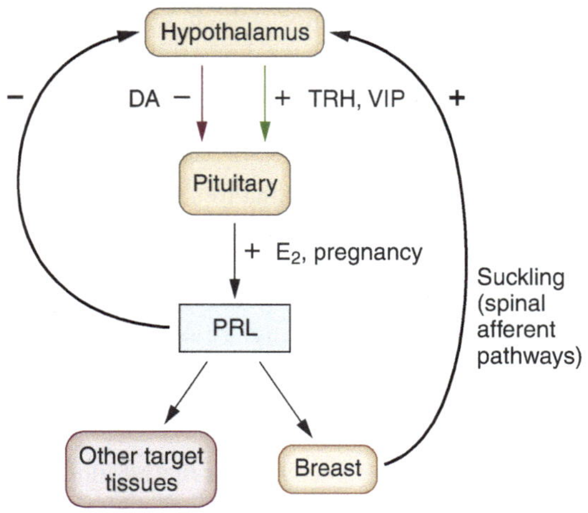

FIGURE 3.36 Feedback regulation of prolactin secretion. The hypothalamic influences on prolactin secretion described in Fig. 3.19 are shown here, as well as the positive effects of estrogen and, during pregnancy, other hormones (see Chapter 14: Hormones of Pregnancy, Parturition and Lactation). The powerful positive neural influence of suckling on prolactin release is also indicated, as is the short-loop negative feedback exerted by prolactin at the hypothalamus.

birth. A defect in either IGF-1 synthesis or in the IGF1R results in growth abnormalities both pre- and postnatally. Both GH and IGF-1 continue to have effects on bone through childhood and during the peripubertal periods of skeletal maturation and acquisition of peak bone mineral density. Some portions of the adult skeleton remains sensitive to GH and IGF-1 in adulthood resulting in the GH excess condition of acromegaly (see Section VII).

In addition to its skeletal effects, GH influences several aspects of fuel metabolism. These effects, which are a direct response to GH or a response to IGF-1, include increased lipolysis in adipose tissue leading to increased circulating free fatty acids; increased triglyceride uptake by liver and muscle through the stimulation of lipoprotein lipase. GH effects on carbohydrate metabolism are, in the short term, insulin-like but chronically oppose insulin action. The overall effect of GH on protein metabolism is an anabolic one.

B Prolactin

1 Regulation of PRL secretion

As discussed in Section III.E and shown in Fig. 3.36, PRL secretion from pituitary lactotrophs is predominantly under the negative control of DA from hypothalamic neurons. The stimulation of release by other hypothalamic factors, including VIP and TRH, also occurs, often in concert with decreased DA levels. This attribute, that is, high constitutive secretion of PRL when not under hypothalamic control, is unique to mammals; in other species the more commonly observed balance between negative and positive regulatory factors is observed.

In mammals, peripheral feedback effects on PRL secretion include the neuroendocrine reflex that is established in response to suckling, a positive feedback loop that serves the lactational function of mammals. In this reflex, impulses resulting from the stimulation of nerve endings in the nipple are transmitted via the spinal cord to the brain stem and hypothalamus (see Fig. 14.16).

Extensive experimental data indicate that in rats, estrogen stimulates PRL secretion at the hypothalamus through effects on DA secretion and at the pituitary through effects on gene expression. In humans the rise in the number of lactotrophs in the pituitary correlates with increased serum estrogen levels and, in hypogonadal situations, a clear effect of estrogen on PRL secretion can be demonstrated. At the same time, a rise in PRL levels does not accompany the mid-cycle estrogen peak in women nor are basal levels of the hormone decreased when ovaries are removed. Thus it is likely that although estrogen stimulation of PRL secretion may be important under some circumstances, in most women estrogens do not play a significant role in the day-to-day control of PRL secretion.

PRL itself can also participate in the regulation of its own secretion through the short feedback loop to the hypothalamus, where it exerts PRL receptor–mediated inhibitory effects on DA synthesis, secretion, and turnover, all leading to decreased levels of DA and increased pituitary secretion of PRL.

2 Prolactin receptor and signaling

The human PRL receptor, PRLR, is closely related to the GHR and both are members of a subfamily of hemapoietic cytokine receptors that includes erythropoietin and most interleukins. The dimeric forms of the PRLR and the GHR are shown in Fig. 3.37. Both receptors show similarities and, notably, the interface of the two monomers is equivalent in both receptors. The monomeric form of this type of receptor is characterized by a single-pass transmembrane region and a cytoplasmic tail that has no intrinsic tyrosine kinase activity but can be phosphorylated by cytoplasmic kinases. The cytoplasmic tail contains two regions, Box 1 and Box 2, which are highly conserved across the cytokine receptor superfamily. The gene for the PRL receptor contains 10 exons, and alternative splicing leads to the presence of isoforms that differ by their abundance and structure in the intracellular domain, within different species. In humans the long form is the most abundantly expressed and mediates many of the biological effects of PRL on its target tissues. Shorter forms have been identified and studied but their physiological roles have not yet been clarified.

As depicted in Fig. 3.38, a PRLR dimer with one ligand bound is the activated form of the PRL receptor. Although evidence exists for involvement of PRL binding in the dimerization, it is now thought that, as with GHR,

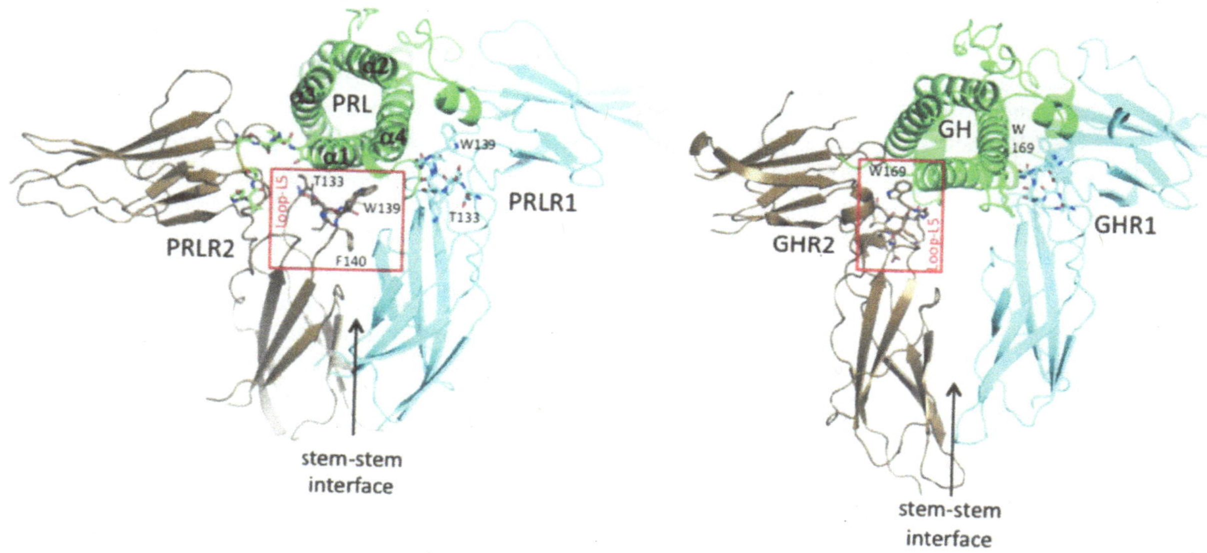

FIGURE 3.37 Comparison of the X-ray crystallographic images of the PRLR on the *left* and the GHR on the *right*. Each figure shows the dimeric forms of the receptors. PRLR1 and PRLR2 are the two monomeric chains; similarly, GHR1 and GHR2 represent the monomers of the dimeric structure. The stem–stem interface is the dimerization interface. The residues that stabilize the receptor dimerization interface are equivalent in the two complexes. *GHR*, Growth hormone receptor; *PRLR*, prolactin receptor. *Reproduced from Fig. 3.1 of J. van Agthoven et al., Structural characterization of the stem-stem interface between prolactin receptor chains compared with the natural hormone, Journal of Molecular Biology, 404 (2010) 112–126.*

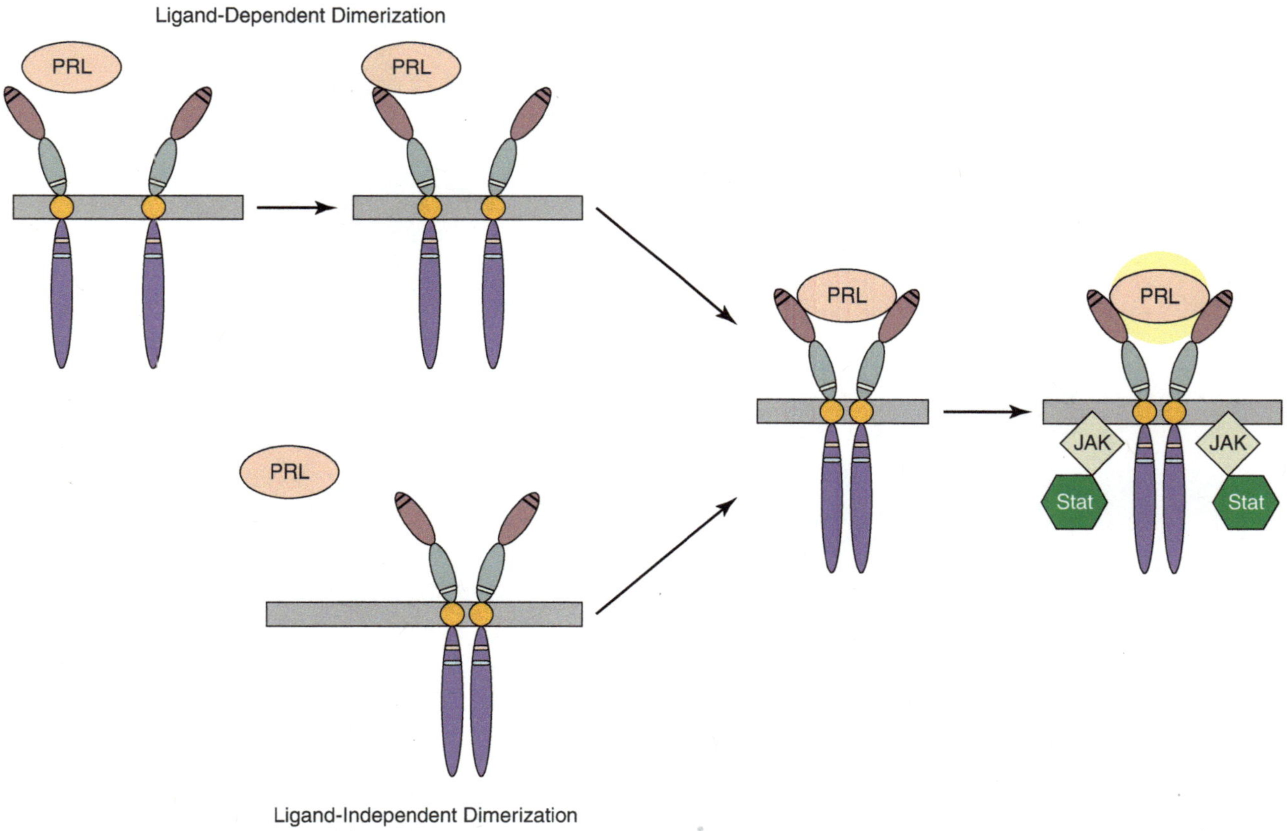

FIGURE 3.38 The hPRLR. The human prolactin receptor monomer (upper right) is characterized by two distinct ECD, a transmembrane domain and an ICD. Each of the domain types has highly conserved motifs, indicated by bands from the top of ECD (*beige* and *blue*) to the bottom of ICD (*purple*). Two segments of the ICD, Box 1 and Box 2, carry out differing signaling functions within the cell. The remainder of this figure depicts the two possibilities of the order of PRL binding and dimerization to form the active receptor. In the upper portion, PRL binds a monomer, then dimerization occurs while in the bottom panel dimerization of two unliganded monomers of hPRLR takes place prior to the binding of one molecule of prolactin. It is now believed that as is the case with the GH-GHR complex, dimerization takes place prior to hormone binding. The JAK/Stat pathway is the major signaling mechanism for PRL in target tissues. *ECD*, Extracellular domains; *GH*, growth hormone; *GHR*, growth hormone receptor; hPRLR, human prolactin receptor; *ICD*, intracellular domain; *JAK*, Janus kinase 2.

the dimer is formed in the absence of ligand. The best understood signaling pathway of the occupied dimerized PRLR is that initiated by JAK2, a nonreceptor tyrosine kinase that is associated with Box 1 of the intracellular domain of the PRLR. Activated JAK2 phosphorylates Stat5, a latent cytoplasmic transcription factor (see Chapter 1: Hormone Introduction) that translocates to the nucleus where it activates the transcription of specific genes, expression of which brings about the biological response of the target cells or tissues.

3 Prolactin biological actions

PRL is present in all vertebrate species and its activities vary widely between classes of vertebrates. The biological actions of PRL can be grouped into six general categories: reproduction; metabolism; water and electrolyte balance; growth and development; immunoprotection; and behavior. For example, in teleosts certain gill ion transport systems are the targets of PRL as is the nasal salt gland in birds. PRL is involved in skin sloughing in reptiles and feather growth in birds.

Many actions of PRL are broadly supportive of successful reproduction. Since reproductive strategies vary with species so do the specific biological actions of PRL. In mammals, PRL's most important actions (and the source of the name of the hormone) have to do with its indispensable role in lactation. This includes the growth and development of the mammary gland and the changes it undergoes during pregnancy; the production of milk; and the delivery of milk to the young. The role of suckling in PRL secretion has already been touched upon and the actions of PRL in mammary gland growth and development and lactogenesis will be discussed in more detail in Chapter 14, Hormones of Pregnancy, Parturition and Lactation.

In rodents, but not in most other mammalian species, including humans, PRL is necessary for the maintenance of the corpus luteum, which secretes the steroid hormones required for the maintenance of pregnancy in mammals. PRL does play a role in human pregnancy, in addition to preparing the breasts for lactation, but it generally acts in concert with other hormones in this regard. It acts alone,

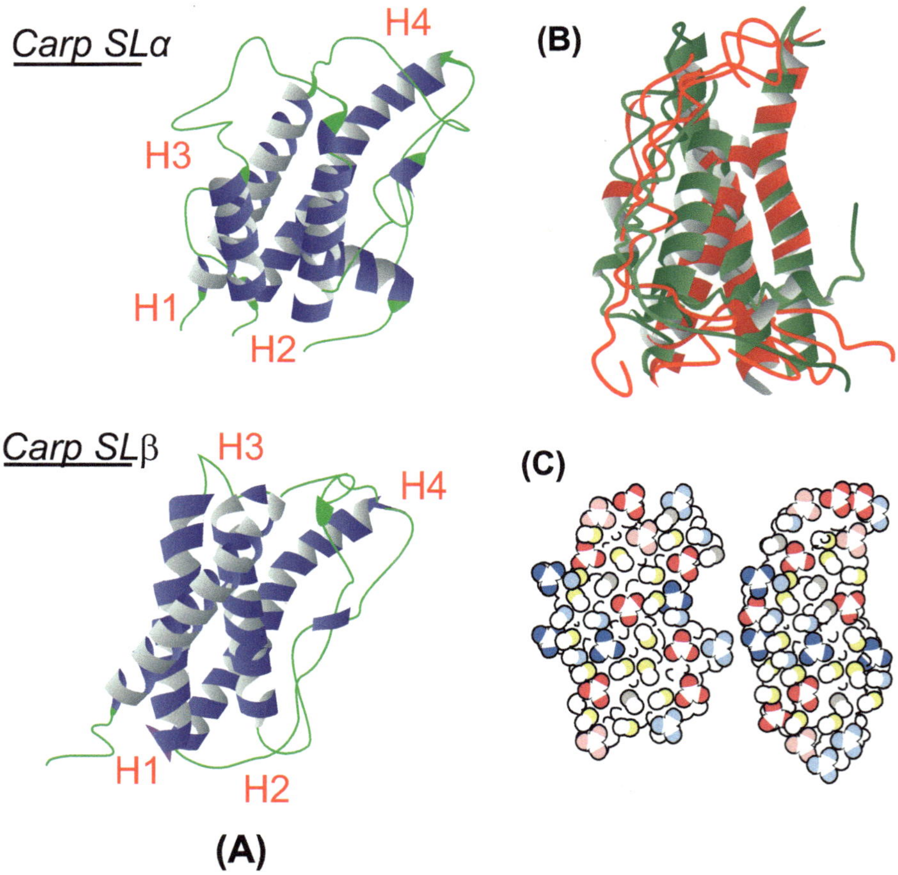

FIGURE 3.39 This figure (A) shows the structures of two forms of somatolactin (α and β) isolated from the carp. (B) shows the structure of prolactin [two separate structures are superimposed (*red* and *green*)]. (C) shows the structure of growth hormone (left) and of prolactin (right). *(A) is reproduced from Q. Jiang et al. (2008), Grass carp somatolactin: I. Evidence for PACAP induction of somatolactin-α and -β gene expression via the activation of pituitary PAC-I receptors, American Journal of Physiology—Endocrinology and Metabolism, https://doi.org/10.1152/ajpendo.90385.2008. (B) Reproduced from K. Tellum et al. (2005), Journal of Molecular Biology; https://www.semanticscholar.org/paper/Solution-structure-of-human-. . .tin.-Tellum-Hoch/b294c3a9ee9a179ae7ab1616beb93c8b10af4b02/figure/4. (C) Reproduced from https://pdb101.rcsb.org/motm/52.*

however, in the stimulation of the size and number of lactotrophs in the pituitary gland which accounts for a large part of pituitary growth during human pregnancy.

High levels of PRL, such as in the hyperprolactemia resulting from PRL secretion from an ectopic or pituitary tumor, inhibit gonadal function in both males and females and can lead to infertility or impotence. This antigonadal effect of PRL is one of the most highly conserved of its actions across classes and species. However, the role of normal levels of PRL in gonadal function in males and nonpregnant females is not yet fully understood.

Aside from its positive role in the mammary gland, PRL is viewed as an inhibitory hormone with respect to its suppression of the secretion of GnRH from the hypothalamus. The mechanism of this suppression is not well understood. PRL actions are mediated by the PRL receptor and by several cytokine receptors. PRL is known to play a key role in maternal behavior.

In the human, as well as in other species, there exists another hormone, somatolactin that is closely related to PRL. Somatolactin was discovered in the flounder/cod pituitary DNA library. Fig. 3.39 compares the crystal structure of two forms of somatolactin to PRL.

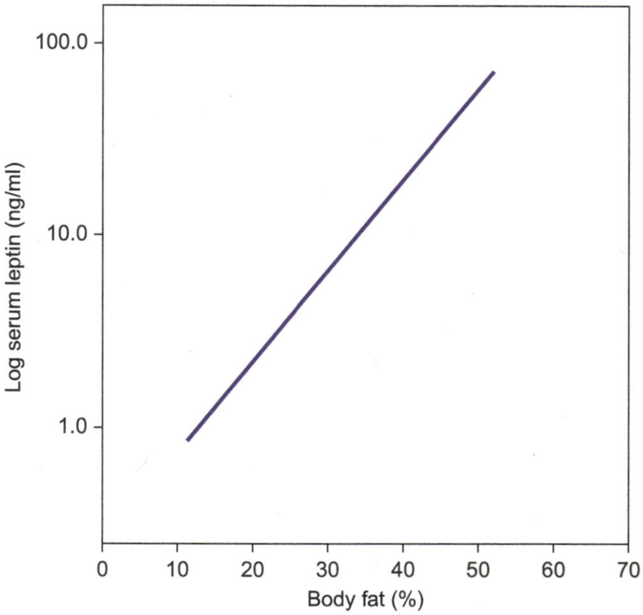

FIGURE 3.40 Relationship between serum leptin levels and body fat. The relationship between serum leptin levels and the percent of body mass that is fat tissue is shown on a logarithmic scale.

VI The hypothalamus and appetite regulation

The involvement of the hypothalamus in the regulation of food intake and energy expenditure has had experimental support for many decades. For example, it was shown in 1940 that hypothalamic lesions in the rat led to morbid obesity. It was also understood that the ability of mammals to maintain fairly stable body weight involves some type of communication between body fuel storage and the control of food intake. The identification of hyperphagic obese mice and the realization that their condition was due to a recessively inherited defect provided a model with which to further study the control of appetite. A series of important experiments with these mice (designated *ob/ob*) using parabiosis (joining two living systems) demonstrated that one or more humoral factors were involved in mediating the *ob/ob* phenotype. More specifically, it was found that the *ob/ob* mice are missing a satiety factor. A related mouse strain with an obese phenotype similar to diabetes mellitus, *db/db* was resistant to the satiety factor. Within the past 20 years, the satiety factor, the product of the *ob* gene, has been identified as the 167 amino acid hormone leptin, from the Greek "leptos" meaning thin. It is produced in the adipose tissue and, as shown in Fig. 3.40 the circulating levels of leptin show a logarithmic relationship to the amount of adipose tissue as a function of total body mass.

Leptin is highly conserved among mammals and has been identified in birds as well. Lack of leptin in both humans and rodents leads to morbid obesity and neuroendocrine disruptions, which can be reversed by exogenous administration of the hormone. Leptin receptors (LEPRs), which transmit their signals through the JAK/STAT pathway, are found in several nuclei in the medial basal hypothalamus. In Fig. 3.41 are shown the crystal structures of the LEPR and the erythropoietin receptor for comparison. The two receptors have similar cytoplasmic domains, both of which interact with JAK2 (A). Later, in (B and C) both receptors interact with JAK2 FERM domain (Protein 4.1R, ezrin, radixin, moesin) via their box 1 domains (top of each figure). The box 2 domain then interacts with SH2. In (D) is shown a version of the dimerized LEPR.

Leptin is not the only regulator of energy homeostasis and some other participants are summarized in Fig. 3.42. In this figure, attention is focused on two important types of neurons in the arcuate nucleus of the hypothalamus. The first synthesizes the neuropeptides AgRP and NPY, which lead to increased food intake. The second type consists of POMC-synthesizing neurons, product of which is α-MSH, leading to the suppression of food intake. Leptin, from adipose tissue, affects both of these pathways, inhibiting the first and stimulating the second for an overall decrease in food intake. In addition to the regulation of appetite by lipid stores, or lipostatic regulation, glucostatic regulation is exerted by insulin, effects of which on the two sets of neurons are the same as those of leptin. Ghrelin, already discussed in this chapter as the endogenous ligand for the GHS receptor, also functions as a "hunger" signal, so that, through action on the NPY/AgRP neurons, feeding behavior increases.

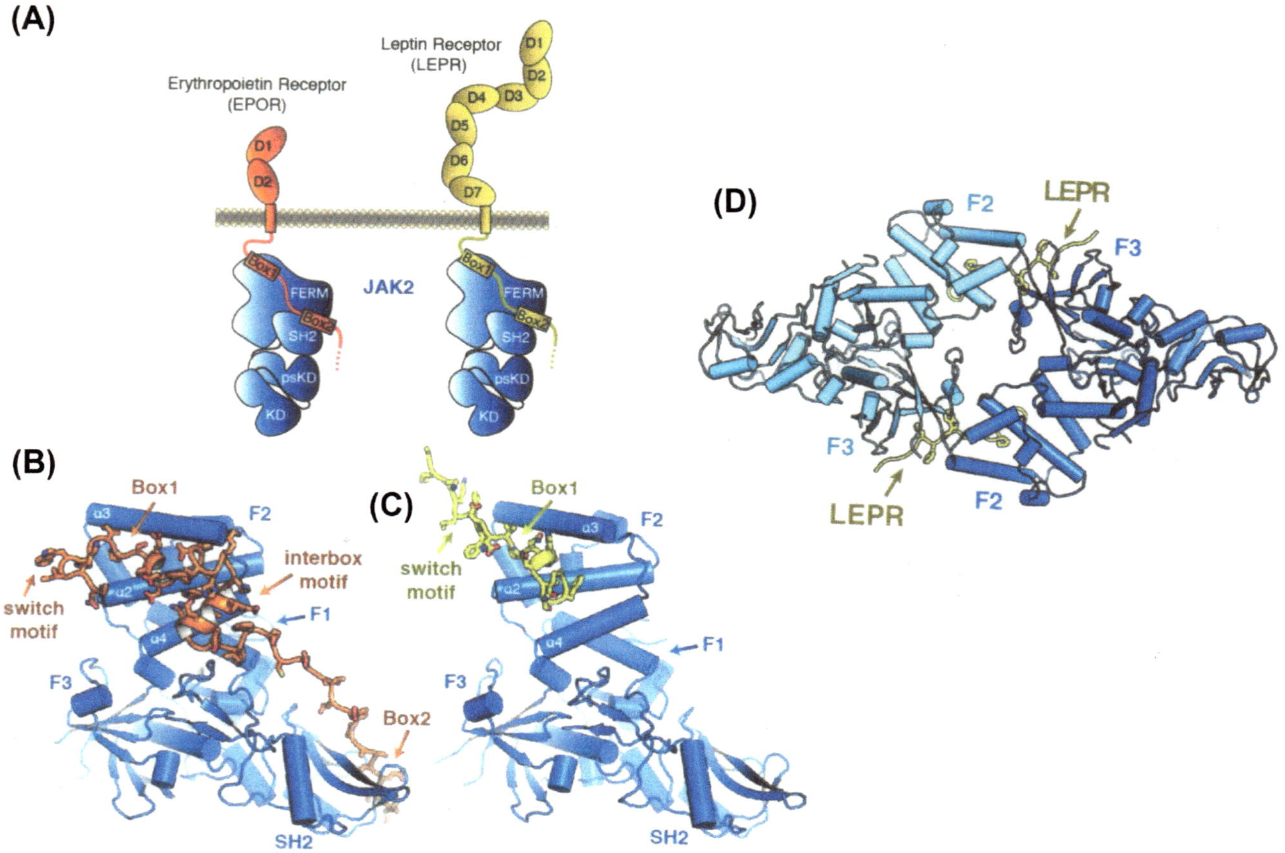

FIGURE 3.41 The structure of the JAK2 FERM-SH2 bound to the EPOR and the LEPR. (A) Schematic representation of EPOR and LEPR bound to JAK2. Each receptor binds to JAK2 via a box 1 interaction with the FERM domain, and a box 2 interaction with the SH2 domain (B and C). The crystal structures of EPOR and LEPR bound to JAK2 are at 2.83 A°, respectively. (B) Cartoon representation of residues 279–335 of EPOR bound to JAK2 FERM-SH2. JAK2 is shown in *blue*, and EPOR in *orange*. (C) Cartoon representation of residues 866–885 of LEPR. (D) A dimerized receptor form of LEPR. *EPOR*, Erythropoietin receptor; *JAK*, Janus kinase 2; *LEPR*, leptin receptor. *Reproduced in part from Fig. 3.1 and Fig. 3.3 of R.D. Ferrao, H.J.A Wallwebber., P.J. Lupardus, Receptor-mediated dimerization of JAK2 FERM domains is required for JAK2 activation, eLife 7 (2018) e38089; https://doi.org/107554/eLife.38089.*

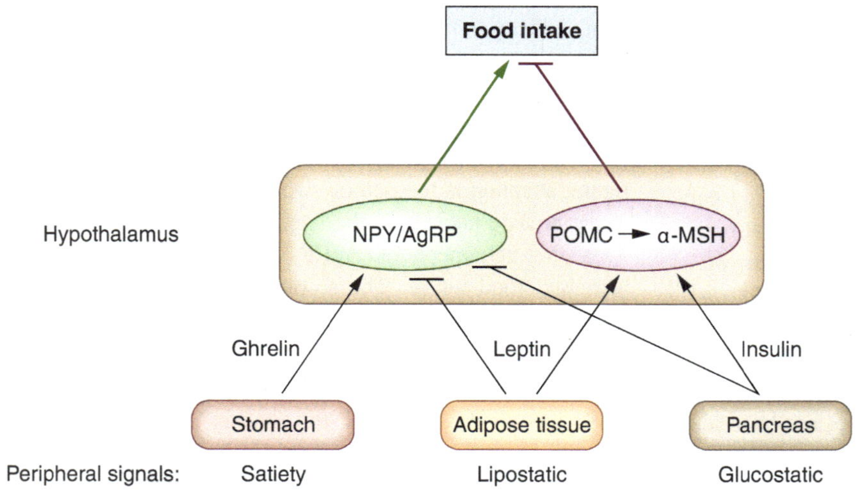

FIGURE 3.42 Pathways of appetite regulation in the hypothalamus. Three peripheral pathways are shown: stimulation of food intake by the stomach hormone, ghrelin and the two' food intake suppressing pathways comprising leptin from adipose tissue and insulin from the pancreas. These peripheral hormones interact with two sets of neurons in the arcuate nucleus of the hypothalamus. On the left is shown the AgRP (NPY) neuron, which acts to stimulate appetite. On the right is shown the POMC neuron which, through production of α-MSH, depresses food intake. *AgRP*, Agouti-related protein; *MSH*, melanocyte-stimulating hormone; *NPY*, neuropeptide Y; *POMC*, proopiomelanocortin.

Understanding the pathways that control appetite, food intake, and energy utilization continues to be the subject of intensive investigation and many aspects remain to be elucidated. Table 3.4 summarizes many of the peptides that are involved in these vital processes. The hormonal control of appetite and energy utilization is also discussed in Chapter 7, Gastrointestinal Hormones.

VII Clinical aspects

A Hypothalamus

Damage to the hypothalamus or to the fragile structures connecting the hypothalamus to the anterior pituitary, the stalk, and the connecting blood portal system can have serious and far-reaching consequences. This kind of damage can happen through trauma to the head or through tumor invasion of the area. These conditions can be collectively listed under panhypopituitarism or hypothalamic hypofunction. Many of the releasing hormones will fail to arrive at their receptors on the plasma membranes of the specific anterior pituitary cell, and the hypothalamic–pituitary–end organ axis will be broken. There is usually an order to the loss of hormonal secretions in hypopituitarism. GH deficiency occurs early followed by LH, FSH, and TSH, and ACTH deficiencies. One rarely sees PRL deficiency; rather since PRL is under predominantly negative control of DA from the hypothalamus, hyperprolactinemia is the result of damage to the connection between the hypothalamus and the pituitary. Diagnosis of disruption of hypothalamic–pituitary communication is now possible through the availability of the releasing hormone assays. Often patients with these syndromes are treated with the hormone of the terminal gland in the pathway, such as the adrenal hormones, gonadal steroid hormones, or thyroid hormones.

The isolated deficiency of GnRH leads to Kallman's syndrome, which refers to anosmic (lacking a sense of smell) patients with hypogonadotropic hypogonadism. There is a cluster of genetic variations that lead to one or more features of this syndrome, with some variants lacking some of the clinical manifestations present in the overt syndrome. The syndrome may originate from an arrest of the migration of the GnRH neurons from the medial part of the nasal epithelium to the hypothalamus/preoptic area at the appropriate time in development. The treatment of individuals with this syndrome usually involves pulsatile administration of GnRH.

Some important disorders of the hypothalamus are manifested not through the effects on the anterior pituitary, but on those of the posterior pituitary. These will be covered in Chapter 4, Posterior Pituitary Hormones.

B Pituitary

1 Mutations in pituitary transcription factors

In recent years a number of specific transcription factors required for the normal development of the anterior pituitary have been identified and studied. It is not surprising that mutations in the genes encoding these proteins can lead to the abnormal development of the cells that secrete the hormones of the anterior pituitary and, therefore, result in isolated, or more commonly, combined, hormonal deficiencies. Some of the best understood of these factors are listed in Table 3.5.

Pit1 is a POU (an acronym for three related transcription factors) domain class 1 transcription factor, which is also known as POU1F1. The protein is found only in the pituitary gland and although it was first identified by its binding to regulatory sites in the GH gene promoter, it is now recognized to also bind to the regulatory region of

TABLE 3.4 Some neuropeptides that regulate appetite.

	Increase (orexigenic)	Decrease (anorexigenic)
Peripheral	Ghrelin	Peptide YY
		Cholecystokinin (CCK)
		Leptin
		Insulin
		Glucagon-like peptide
Central	Neuropeptide Y	α- and γ-MSH (POMC)
	MCH	Glucagon-like peptide
	AgRP	CRH
	Galanin	Serotonin
	Endogenous opioids	Neurotensin
	Endocannabinoids	

AgRP, Agouti-related peptide; *CRH*, corticotropin-releasing hormone; *MCH*, melanin concentrating hormone; *MSH*, melanocyte-stimulating hormone; *POMC*, proopiomelanocortin.

TABLE 3.5 Pituitary transcription factors in hormone deficiency.

Txn factor	Chromosome	Hormone deficiency
Pit 1	3p11	GH, PRL, TSH (variable)
Prop1	5q35	GH, PRL, TSH, FSH/LH, ACTH
Lhx3	9q34	GH, PRL, TSH, FSH/LH
Lhx4	1q25	GH, TSH, ACTH
Tbx19	1q23	POMC

Transcription factors are described in the text. *ACTH*, Adrenocorticotropic hormone; *FSH*, follicle stimulating hormone; *GH*, growth hormone; *POMC*, proopiomelanocortin; *PRL*, prolactin; *TSH*, thyroid stimulating hormone.
Source: Adapted from Endocrinology, (Ed.) Jameson and DeGroot, Chapter 8, p. 148 (2010).

the genes for the GHRH receptor, PRL, and the TSHβ subunit. Patients with Pit1 disorders are usually diagnosed when the failure to grow normally, due to GH deficiency, becomes apparent.

Prop1 (Prophet of Pit1) precedes and induces the expression of POU1F1. In mice, it is expressed exclusively in the embryonic pituitary, where it is implicated in the differentiation of POU1F1-dependent somatotrophs, lactotrophs, and thyrotrophs. The mechanisms for its effects on the gonadotropins and ACTH are not clear and are clinically more variable than those of the other hormones.

Lhx3 and Lhx4 (LIM-homeobox-3 and -4) have many structural features in common but impact different patterns of cell lineages in the developing pituitary. Thus their clinical presentations differ. Patients with Lhx-3 mutations invariably present with combined hypopituitarism with deficiency of GH, TSH, gonadotropins, and PRL. There can occasionally be impaired corticotroph function. Individuals with heterozygous mutations in the gene for Lhx-4 present with a highly variable pituitary hormone deficiency phenotype. All patients demonstrate GH deficiency, and most also have a deficit in TSH secretion, whereas the gonadotroph and adrenal axes are sometimes affected but sometimes normal.

Mutations in Tbx19 (TPIT) have been noted in patients with early onset isolated ACTH deficiency. Severe hypoglycemia, jaundice, and neonatal death often result from such defects. Tbx19 is required for normal development and differentiation of both corticotrophs and melanotrophs and is required for the activation of POMC expression.

2 Growth hormone deficiency in childhood

In childhood, there are many possible causes, both genetic and environmental, of short stature and a slow growth rate. Careful laboratory and clinical evaluations, including measurements of IGF-1 levels, are required to determine whether the cause is a deficiency of GH. If present, GH deficiency may be either congenital (e.g., defective GH or GHRH synthesis or secretion) or acquired (through a tumor or injury to the head). The GH deficiency may be isolated, that is, GH is the only hormone involved, or there may be disruptions in other pituitary hormones as well. GH deficiency untreated in childhood leads to adults with proportionate extreme short stature.

The true incidence of GH deficiency, most of which is idiopathic (cause unknown) is estimated to be around 1 in 3500 children. The child usually has a normal birth weight but exhibits growth failure and abnormal facial development during the first 2 years of life. The objective of treatment with replacement recombinant hGH is to normalize height during childhood. Growth rate is most accelerated during the first year of treatment (the "catch-up" phase) with average rates of 8–10 cm/year. If, thereafter, the growth rate slows too much, other factors such as hypothyroidism or nutritional factors should be considered.

3 Excessive GH in childhood

There are four types of growth patterns that can result in heights more than two standard deviations from the normal mean. These are intrinsic tallness, the expression of the genetic potential of the individual; advanced growth, where growth occurs earlier and stops earlier at normal adult height; prolonged growth, that is, beyond the normal time at the end of puberty, often from a deficiency of sex steroid hormones; and accelerated growth, due to excessive GH levels. This last type is the one we will focus on here.

Hyperproduction of GH can result from a tumor of the somatotroph (pituitary adenoma) during growth. The resulting accelerated growth results in gigantism, which is characterized by a height age that is greater than either bone age or chronological age and a supranormal growth rate. This form of GH overproduction is relatively rare. If left untreated, the tumor destroys the functional gland, impairing the other pituitary hormones and resulting in death. This disease often involves enlargement of the *sella* but is principally manifested by unusually high circulating levels of GH. Therapy involves the removal of the tumor, radiation of the tumor, and/or a GH receptor antagonist.

4 Excessive GH in adults: acromegaly

Excessive GH secretion in adults leads to a constellation of symptoms known as acromegaly and most commonly results from one of several types of GH-secreting pituitary tumors. Hypothalamic tumors that oversecrete GHRH may also be responsible. Acromegaly is quite rare and its incidence is not well established. It generally has a slow progression and can be present for a decade or more before being accurately diagnosed. Since the acromegaly is of adult onset and the epiphyseal growth plates of the long bones close at the end of puberty, the hypersecretion of GH results in excessive growth of only those tissues still capable of responding to it. These include the mandible, bringing about changes of facial structure and the hands and feet, which become enlarged. These changes in appearance are often so gradual that they are not the patient's initial complaint upon presentation. Rather, the headache resulting from the growing tumor is the most common symptom that brings the patient to the clinic. Soft-tissue overgrowth causes a number of symptoms, including neuropathy; cardiomyopathy and cardiovascular disease; debilitating arthritis; sleep apnea due to airway obstruction; respiratory disease and, because of the role of GH in carbohydrate metabolism, carbohydrate intolerance. Untreated acromegaly leads to a two- to fourfold increase in mortality compared to age-matched controls.

The primary goal of treatment of acromegaly is to reduce excessive GH levels. Surgery is one method to achieve this but is accompanied by serious risks, one of which is the difficulty in removing all of the tumors without damaging nearby structures. Radiation of the tumor has also been used, but the response in terms of lower GH levels is much slower than with surgery. Pharmaceutical therapies include GH receptor antagonists and SST receptor ligands that block GH secretion by the tumor cells and reduce tumor size.

Further reading

Thyrotropin-releasing hormone (TRH)

I. Lazcano, et al., Evolution of thyrotropin-releasing factor extracellular communication units, General & Comparative Endocrinology 305 (2021) 113642.

Gonadotropin-releasing hormone (GnRH)

C.A. McArdle, et al., Chapter 7—Gonadotropin-releasing hormone receptors and signaling, Cellular Endocrinology in Health and Disease, second ed., 2021, pp. 149–181.

Corticotropin-releasing hormone (CRH)

H.M. Baumgartner, J. Schulkin, K.C. Berridge, Activating corticotropin-releasing factor systems in the nucleus accumbens, amygdala, and bed nucleus of stria terminalis: Incentive motivation or aversive motivation? Biological Psychiatry 89 (2021) 1162–1175.

Proopiomelanocortin (POMC)

C. Quarta, et al., POMC neuronal heterogeneity in energy balance and beyond: an integrated view, Nature Metabolism 3 (2021) 299–308.

Growth hormone (GH)

J. Donato Jr., et al., Central regulation of metabolism by growth hormone, Cells 10 (2021) 129. Available from: https://doi.org/10.3390/cells10010129.

Prolactin (PRL)

A. Zamora-Morata, E.D. Martin, Prolactin enhances hippocampal synaptic plasticity in female mice of reproductive age, Hippocampus 31 (2021) 281–293.

Hypothalamus and appetite regulation

S. Yousefvand, F. Hamidi, The role of ventromedial hypothalamus receptors in the central regulation of food intake, International Journal of Peptide Research and Therapeutics 27 (2021) 699–702.

Pituitary development

M. Kano, H. Suga, H. Arima, Induction of functional hypothalamus and pituitary tissues from pleuripotent stem cells for regenerative medicine, Journal of the Endocrine Society 5 (2021). Available from: https://doi.org/10.1210/jendoso/bvaa188. bvaa188.

Growth hormone deficiency

F. Scarano, et al., Impact of long-term growth hormone replacement therapy on metabolic and cardiovascular parameters in adult growth hormone deficiency: comparison between adult and elderly patients, Frontiers in Endocrinology (Lausanne) 12 (2021) 635983. Available from: https://doi.org/10.3389/fendo.2021.635983.

Acromegaly

F. Ceccato, et al., Clinical presentation and management of acromegaly in elder patients, Hormones 20 (2021) 143–150.

Chapter 4

Posterior pituitary hormones

I Introduction

Two important hormones are secreted from the posterior pituitary in both males and females. These are vasopressin (VP), the antidiuretic hormone; and oxytocin (OT), which in female mammals is important in parturition (birth) and lactation. The two hormones are structurally closely related nonapeptides that are derived from a common ancestral gene. Each is synthesized in a specific group of cells in the hypothalamus and is released from the axon termini of these neurons in the posterior pituitary. Here the hormones are stored until the appropriate signal brings about their release into the bloodstream. This chapter will focus on the biochemical aspects of the synthesis, secretion, and receptor interactions of the two hormones and the biological actions of VP. The role of OT, a crucial hormone in mammalian reproduction, will be covered in Chapter 14, Hormones of Pregnancy, Parturition and Lactation.

II Anatomy of the posterior pituitary

The posterior pituitary lies below the hypothalamus and, unlike the anterior pituitary, is composed primarily of neural tissue. As shown in Fig. 4.1, it forms a structural and functional unit, the neurohypophysis, with the supraoptic and paraventricular neuron clusters of the hypothalamus. The axons of the neurons traverse the supraoptico-hypophyseal tract and terminate on capillaries, where hormones are released into the circulation. Both the supraoptic and paraventricular nuclei contain magnocellular neurons that produce either VP or OT. The paraventricular nucleus also contains smaller parvicellular neurons that project through the median eminence to the primary plexus of the anterior pituitary and cosecrete VP and corticotrophin releasing hormone (CRH).

The sites of VP synthesis in the hypothalamus appear to be close to the osmoreceptor sites, which sense changes in electrolyte (solute) concentrations in circulation and signal release of the hormone from neuronal terminals in the posterior pituitary. The osmoreceptor is close to the thirst center in the hypothalamus and also interacts with the renin–angiotensin system. Collectively, these systems appear to be the primary elements for the regulation of water balance (see Section IV.B and Chapter 15, Hormones Related to the Kidney and Cardiovascular System).

III Chemistry, biosynthesis, and secretion of the posterior pituitary hormones

A Structures of oxytocin and vasopressin

The primary amino acid sequences of the nonapeptides secreted by the posterior pituitary gland are shown in Fig. 4.2. The disulfide bridge between Cys residues 1 and 6 results in a peptide with a cyclic portion consisting of six amino acids and a tail of three amino acids. These peptides, which share all except two of their nine amino acids, were the first from the pituitary gland, sequences of which were determined.

The broad conservation of the similarities in the structures of the two hormones, as well as mode of their synthesis and secretion and receptor structure, suggest a common ancestral gene for the two peptides. The location and structural organization of the genes for OT and VP, *OT* and arginine vasopressin (*AVP*), are also consistent with the likelihood that they arose from a common ancestral gene. As shown in Fig. 4.3, in humans the two genes appear in tandem on chromosome 20, separated by 8 kilobases, and have nearly identical structure. The three exons of the genes are transcribed and translated into a preprohormone consisting of a signal peptide followed by the amino acids of OT or VP, a peptide-specific neurophysin (NPI for OT, NPII for VP) and, in VP, a terminal sequence encoding a cosecreted peptide known as copeptin. Not surprisingly, the neurophysins also share a great deal of sequence homology.

B Synthesis and secretion of vasopressin and oxytocin

The posterior pituitary hormones are synthesized in the cell bodies of the discrete neurons of origin in the hypothalamus. As is the case for most proteins destined for secretion, this occurs on the rough endoplasmic reticulum, and the translation

Hormones. DOI: https://doi.org/10.1016/B978-0-323-90262-5.00008-1

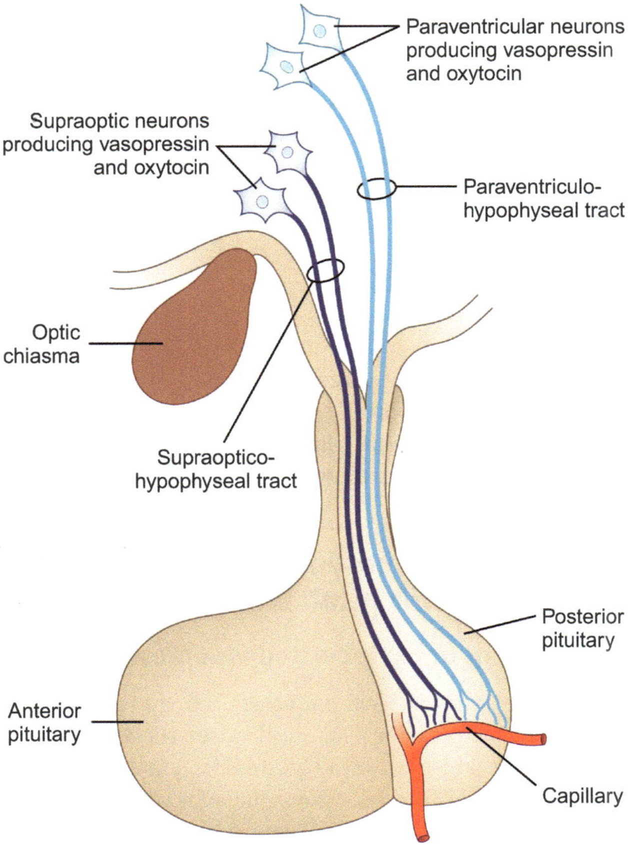

FIGURE 4.1 Anatomy of the neurohypophysis. The neurohypophysis consists of the magnocellular neurons in the supraoptic and paraventricular nuclei of the hypothalamus; the axonal processes of these neurons which form the supraoptico- and paraventriculo-hypophyseal tracts; and the neurosecretory products of these neurons secreted into the capillary system in the posterior pituitary. The cells synthesizing the two hormones can be distinguished from one another by their immunohistochemical and electrophysiological features.

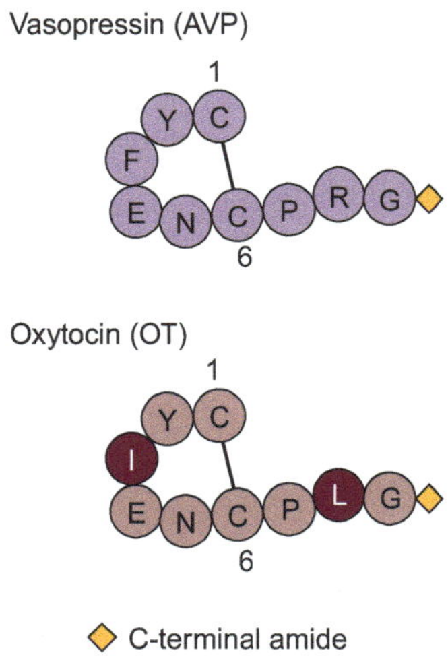

FIGURE 4.2 The primary structures of AVP and OCT. In pigs the amino acid at position 8 is LVP. *AVP*, Arginine vasopressin; *LVP*, lysine-vasopressin; *OCT*, oxytocin.

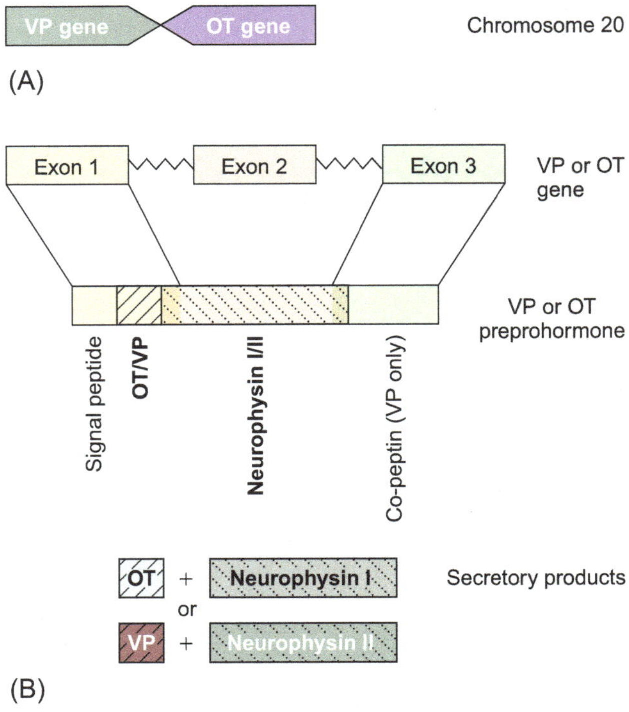

FIGURE 4.3 The location and structural organization of the OT and VP genes. (A) The genes for VP and OT (*purple*) are shown in their head-to-head tandem arrangement on chromosome 20. (B) The exon/intron structure (top) and peptide processing of the VP and OT preprohormones (middle) are virtually identical. The first exon (*yellow*) and part of the second (*orange*) encode the signal peptide and either OT (bottom, *light blue*) or VP (bottom, *magenta*), depending on the gene. The large midportion of the second exon encodes either neurophysin I (OT; bottom, *gray*) or neurophysin II (VP; bottom, *green*), both 93–95 amino acids. The terminal portion of the second exon and the third exon of the VP gene encodes the protein copeptin. *OT*, Oxytocin; *VP*, vasopressin.

product is processed through the Golgi apparatus. As shown in Fig. 4.4, processing of the precursor hormone is begun in the large dense core vesicles of the neuron and continues during transport along the axon. At the nerve termini the secretory granules containing the mature hormone and its associated neurophysin are stored until a subpopulation of neurons is stimulated to release its contents into the adjacent capillaries for transport into the general circulation. The neurophysins are secreted along with, but not bound to, their respective hormones. There is currently no distinct biological activity ascribed to the neurophysins. However, because their synthesis and release reflect that of VP or OT, their measurement can have clinical applications. The same is true of the other peptide coreleased with VP, copeptin, which serves as a biomarker for AVP in a number of clinical settings such as nonspecific stress and certain cardiovascular conditions.

The release of either OT or VP is under the control of separate signals for each hormone. Some of those involved in the regulation of VP release are indicated in Fig. 4.4. The signals impinging on the vasopressinergic neuron may be either chemical or electrical, as depicted in Fig. 4.4.

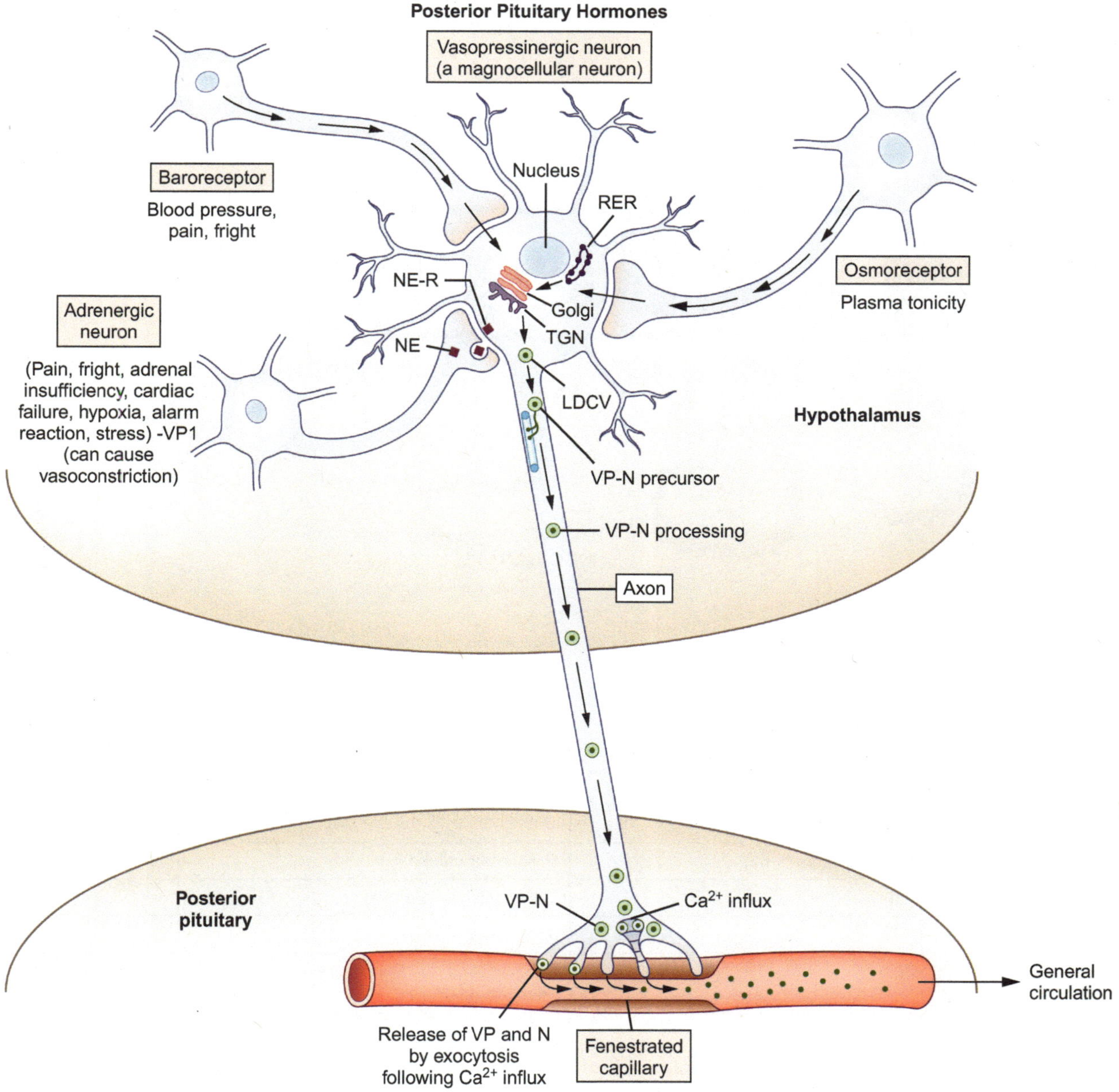

FIGURE 4.4 Pathway of neurohypophysis hormone synthesis and secretion. The neuron (vasopressinergic in this illustration) in the hypothalamus receives input from a variety of sources. Inputs shown are nerves with receptors for sympathetic input (adrenergic neuron); blood pressure input (baroreceptors); and plasma tonicity (osmoreceptors). In response the neuron synthesizes the precursor hormone which, after passing through the Golgi apparatus, is packaged for transportation down the axon to the nerve ending where the hormone and its associated neurophysin are released into the circulation. Processing of the precursor to the mature peptide takes place as the secretory granules move down the axon. The axon terminates on the capillary walls where peptides are released in a Ca^{2+}-dependent manner and taken up through fenestrations in the capillary wall. *NE*, Norepinephrine; *NE-R*, norepinephrine receptor; *RER*, rough endoplasmic reticulum; *TGN*, *trans*-Golgi network; *LDCV*, large dense core vesicle; *VP-N*, vasopressin-neurophysin precursor.

IV Biological actions of AVP

A AVP receptors

There are three AVP receptors, V1a, V1b, and V2, each encoded by a separate gene and each with distinct tissue distribution, signal transduction pathways, and biological consequences. Some of the distinguishing characteristics of these members of the rhodopsin family of the G protein–coupled receptor (GPCR) superfamily are shown in Table 4.1. The secondary structure of VP2, the form in the renal nephron, is shown in Fig. 4.5. Both V1 subtypes stimulate G_{q11} leading to the activation of phospholipase C and breakdown of phosphoinositides, whereas the V2 receptor activates, through G_s, adenylate cyclase and

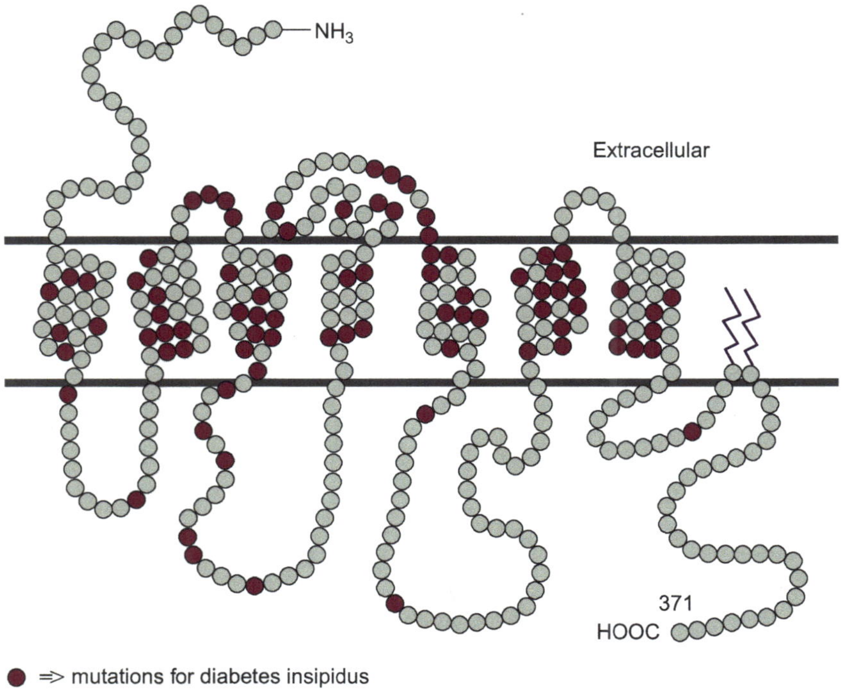

FIGURE 4.5 The secondary structure of the V2 receptor for arginine vasopressin. Residues shown in red are positions of mutations leading to diabetes insipidus, the primary clinical manifestation of lack of AVP. *AVP*, Arginine vasopressin.

TABLE 4.1 Distribution and activity of arginine vasopressin (AVP) receptors.

AVP receptor subtype	Tissue	Physiological activity	2nd messenger system
V1a	Vascular smooth muscle	Vasoconstriction	PLC: Ca^{2+}, IP_3, DAG
	Liver	↑ Glycogen breakdown	
	Platelets	↑ Adhesion	
	Brain	Stress adaptation, social recognition, memory, other	
V1b	Ant. pituitary corticotrophs	↑ ACTH release	Same as V1a
V2	Kidney: distal nephron, basolateral membrane	↑ Aquaporin → ↑ water resorption	cAMP: PKA

ACTH, Adrenocorticotropin hormone; *cAMP*, cyclic AMP; *DAG*, diacylglycerol; *IP_3*, inositol 1,4,5-triphosphate; *PKA*-cyclic AMP-dependent protein kinase.; *PLC*, phospholipase C.

followed by cyclic AMP (cAMP) activation of cyclic AMP-dependent protein kinase (PKA). V1a is fairly widely distributed and has a variety of functions in the vasculature and elsewhere. The localizations of the other two VP receptors are more restricted. V2 in the kidney is responsible for the primary physiological effect of VP, water resorption, giving the hormone its other name, antidiuretic hormone. The V1b receptor, located mainly in the anterior pituitary corticotrophs, is known primarily for its effect on adrenocorticotropin hormone (ACTH) secretion. The V1b receptor has been found to have activity on the vascular system outside the pituitary as shown in Fig. 4.6. In the lower left the V1b receptor, activated by VP, stimulates the release of catecholamines causing vascular contractions that increase blood pressure. This activity is in concert with its effect on the pituitary to release ACTH that stimulates the release of cortisol and aldosterone from the adrenal cortex which also raises blood pressure. These actions of V1b receptor together with those of V1a and V2 receptors are summarized in this figure.

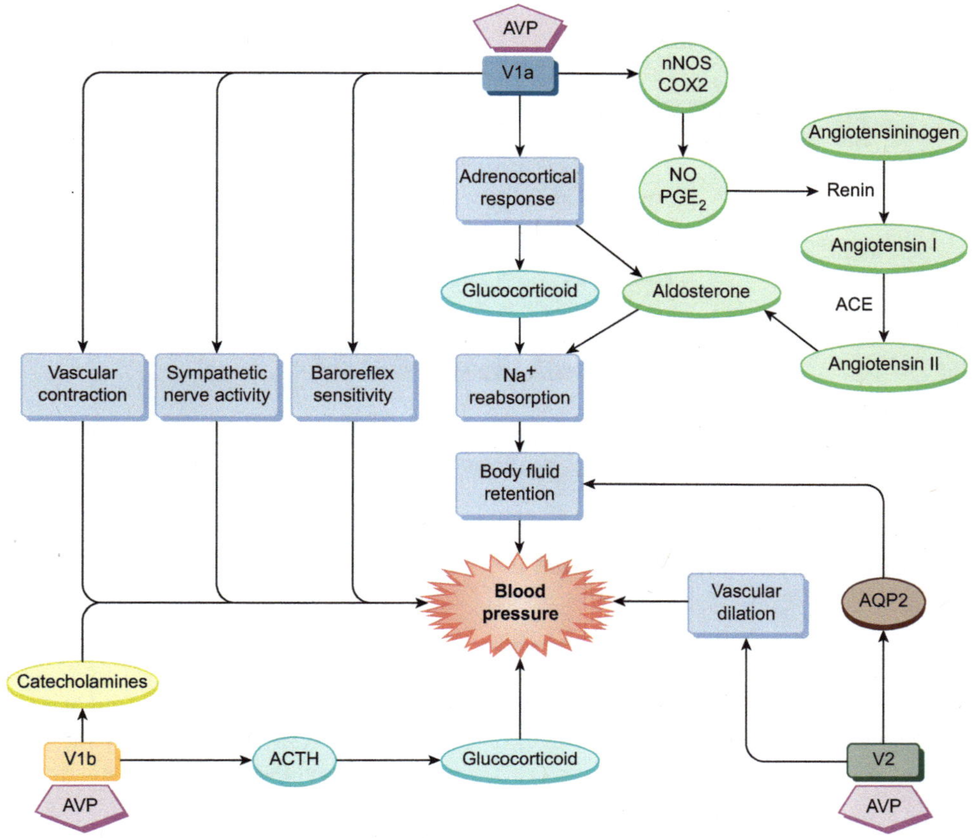

FIGURE 4.6 Contributions of AVP and its receptors to the regulation of blood pressure homeostasis. AVP is involved in the regulation of blood pressure homeostasis via V1a, V1b, and V2 receptors. AVP regulates vascular contraction, aldosterone/glucocorticoid release from the adrenal gland, renin production in the kidney, and sympathetic nerve activity and baroreflex in the CNS, through the V1a receptors. AVP also stimulates ACTH and catecholamine release from the anterior pituitary gland and adrenal gland, respectively, through the V1 receptor. In addition, AVP plays a crucial role in blood pressure control by maintaining body fluid retention via the V2 receptor-AQP2 pathway, while AVP stimulates vascular dilatation via the V2 receptor expressed in the vascular endothelium. Hyperosmolarity and decreased plasma volume lead to the secretion of AVP from the posterior pituitary, which can bind to the V1a receptor in macula densa cells and to the V2 receptor in collecting duct cells. The AP/V1a receptor stimulates the expression of nNos and COX-2, leading to the production of NO and PGE2 by macula densa cells, respectively. Both NO and PGE2 stimulate renin production from granule cells and subsequent increases in angiotensin II and aldosterone levels promote water reabsorption. The AVP/V2 receptor enhances the translocation of AQP2 (aquaporin P2) to the apical membrane from the cytoplasm. Water resorption is further accelerated by the V1a receptor-stimulated renin–angiotensin system. The body fluid is regulated not only by the AVP/V2 receptor system but also by the renin–angiotensin, which might be enhanced by the AVP/V1a receptor. *ACTH,* Adrenocorticotropin hormone; *AVP,* arginine vasopressin; *CNS,* central nervous system. *Reproduced from Figure 3 of Kashimizu, T.-A., et al. (2011), Vasopressin V1a and V1b receptors: From molecules to physiological systems, Physiological Reviews; https://doi.org/10.1152/physrev.00035.2011.*

B AVP regulation of water handling

The primary action of AVP, also known as antidiuretic hormone, is the inhibition of water diuresis (loss of water in the urine), mediated by AVP V2 receptors in the distal nephrons of the kidney. AVP also elevates blood pressure and constricts coronary arteries as described in the following section.

Fig. 4.7 shows an overview of the main features of water balance. The hypothalamic osmoreceptor detects elevations in serum tonicity reflecting low blood volume. The vasopressinergic neuron receives the signal and transmits it to its own nerve ending for the release of vasopressin (AVP) and neurophysin. AVP binds to a membrane receptor located in distal tubular cells of the kidney, and, through cAMP and PKA, aquaporin-2 is phosphorylated and translocated to the cell apical membrane of the distal nephron. The increase in water channels in renal cells results in increased movement of water from urine through the renal cell into blood. The net increase in blood pressure and lowered osmolality results in a quenching of the signal to the vasopressinergic neuron. The increased serum tonicity that activates the AVP pathway also signals the hypothalamic thirst center, which is closely associated with the osmoreceptor. Consequently, more water is consumed, which leads to the dilution of body fluids. Simultaneously, the kidney has sensed increased Na^+ concentration in the blood, which inhibits the Na^+ conservation mechanism (Chapter 15: Hormones

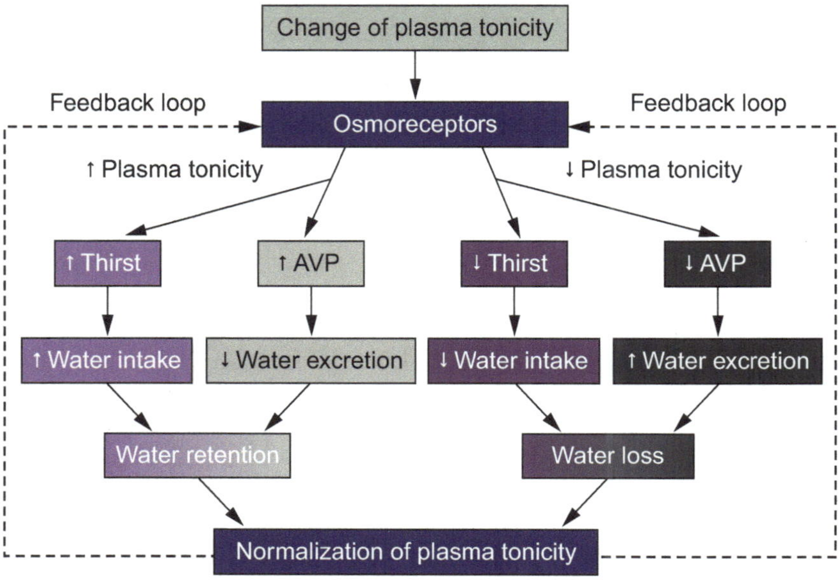

FIGURE 4.7 Role of AVP in the regulation of water balance and plasma tonicity. Changes in plasma tonicity are detected by the hypothalamic osmoreceptors, leading to changes in AVP. The responses to increased tonicity (decreased water and volume) are shown on the left and those in response to decreased tonicity (increased water and volume) are on the right. *AVP*, Arginine vasopressin.

Related to the Kidney and Cardiovascular System) and results in the excretion of Na^+, which, in turn, tends to reduce blood pressure. Thus these two systems function to maintain homeostatic water balance in the body.

C Cardiovascular effects of AVP

AVP acts on the smooth muscle cells of the vasculature through its receptor V1a (Table 4.1). The sensitivity of these cells to AVP, and, therefore, its vasopressor activity, varies with location with splanchnic, hepatic, and renal vessels being the most responsive at physiological concentrations of the hormone.

D Effects of AVP on glucose homeostasis

Glycogenolysis is stimulated by AVP in the liver through the mobilization of intracellular stores of Ca^{2+} (from mitochondria and/or endoplasmic reticulum) by the ligand-activated receptor V1a. Increased cytosolic Ca^{2+} stimulates phosphorylase *b* kinase and, hence, phosphorylase *a* activation through phosphorylation. AVP also alters the phosphorylation and, therefore, the activation state of a dozen or so other intracellular liver proteins, including pyruvate kinase.

E Effects of AVP on ACTH secretion

Subpopulations of the hypothalamic neurons that express and secrete CRH (see Chapter 3: The Hypothalamus and Anterior Pituitary) also synthesize and cosecrete AVP. In these neurons the amounts of both AVP and CRH are inversely related to glucocorticoid levels, suggesting that they are both subject to feedback inhibition. In humans, infusion of the anterior pituitary with AVP alone has little effect on ACTH secretion, but its effect is synergistic with that of CRH. As indicated in Table 4.1, this biological effect of AVP is mediated by the receptor V1b, which is distinct from the CRH receptor.

F AVP and the brain

Central nervous system (CNS) neural networks that are anatomically and functionally separate from the neurohypophysis contain vasopressinergic neurons. In many mammalian species, these central AVP actions play important roles in social and reproductive behavior. In general, AVP tends to mediate male behaviors, while OT mediates those associated with female aspects of reproduction (see Section V). For example, in the monogamous male prairie vole, AVP plays a role in pair bonding. In the Brattleboro rat, which has a spontaneous mutation in the AVP gene, social recognition memory is impaired due to the lack of functional AVP. Information is beginning to emerge to indicate that these pathways, which utilize the V1a receptor, operate in humans as well. For example, genetic association studies suggest a relationship between structural variations in the V1a gene and behavioral patterns such as autism spectrum disorder and social phobias.

V Biological actions of oxytocin

A The oxytocin receptor

The biological actions of OT depend on its high affinity for its receptor, relative to those for VP, V1a, V1b, and V2; the tissue-specific distribution of these receptors; and the regulation of its secretion. Comparison of the oxytocin

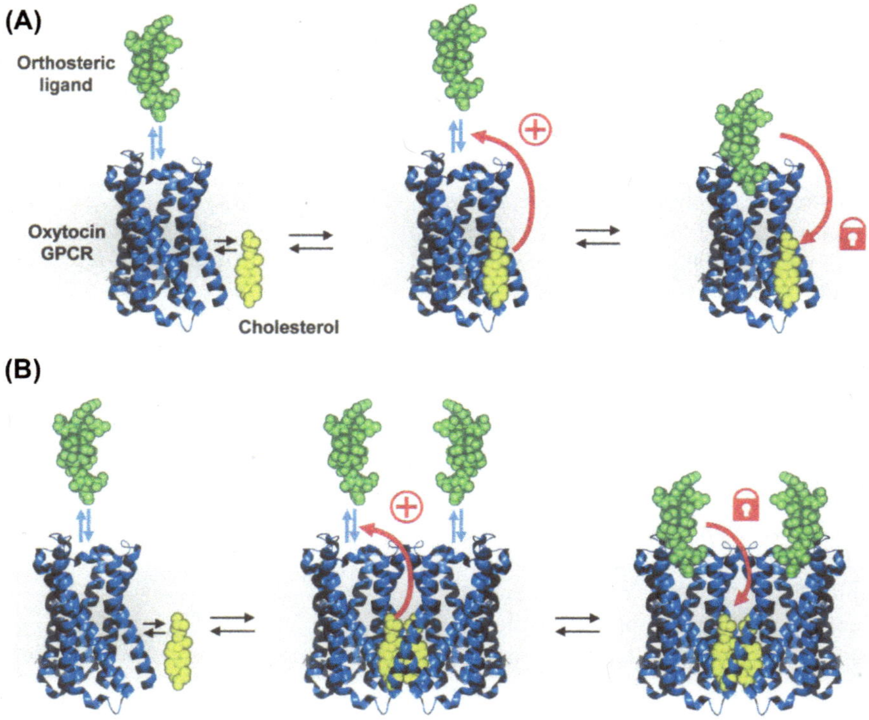

FIGURE 4.8 Monomeric and dimeric models of cholesterol sequestration by ligand-bound OTR. (A) Diagram showing the monomeric model of cholesterol sequestration by ligand-bound OTR. Cholesterol and ligand binding on OTR are in equilibrium between bound and unbound forms. Binding of cholesterol is known to induce a high-affinity state of OTR for its ligand. In this model, binding of ligands will induce conformational changes that stabilize the binding of cholesterol and render it inaccessible to external chelators. In *gray* the lipid membrane is in *blue*, the structure of a GPCR (M2, PDB code: 3uon) representing TR in *green* an external orthosteric ligand (ET, PDB code: 5glh) and in *yellow* a cholesterol molecule (PDB code: 3d4s). (B) The dimeric model showing the cholesterol-induced stabilization of OTR homodimers that represent the high-affinity state of the receptor for its orthosteric ligands. In this model, binding of orthosteric ligands would stabilize the homodimeric form of OTR and embed bound cholesterol molecules at the interface. *ET*, Endothelin; *GPCR*, G protein–coupled receptor; *OTR*, oxytocin receptor; *TR*, thyroid hormone receptor. *Reproduced from Figure 9 of Lemel, L., et al. (2021), The ligand-bound state of a G protein coupled receptor stabilizes the interaction of functional cholesterol molecules, Journal of Lipid Research, online 26 February, 2021, 100059; https://doi.org/10.1016/j.lr.2021.100059.*

receptor (OTR), depicted in Fig. 4.8, with that of V2 shown in Fig. 4.5 reveals that the two are closely related structurally, particularly in the membrane-spanning regions. Like the VP receptor V1a, the OTR is coupled to $G_{q/11}$ through which it activates phospholipase C-β. Downstream events include Ca^{2+} release from intracellular stores, protein kinase C activation, and activation of the mitogen-activated protein kinase cascade. Some OT signaling also involves increasing intracellular calcium through influx from extracellular sources. Finally both within the brain and peripheral tissues OT can, through the activation of PLA_2, stimulate the production of prostaglandins.

B Oxytocin and parturition

In many mammals, OT is an important player in the initiation of parturition (the process of the birth of the young at the end of pregnancy). In others (mice) the lack of OT through genetic manipulation does not impair the birth process, suggesting either that it is not required or that redundant processes are available for this crucial function. In humans the maternal and fetal blood levels of OT are low at the onset of labor and rise progressively during its course. In all species studied the number of uterine OTRs rises during pregnancy, increasing the sensitivity of the myometrium to the hormone. Through its effects on intracellular calcium concentration and prostaglandin production, OT stimulates and coordinates the contractions of the myometrium. As will be discussed in more detail in Chapter 14, Hormones of Pregnancy, Parturition and Lactation, the signals for OT release come from the uterus, but the increase in its secretion as labor progresses is due in part to local positive feedback effects at the hypothalamic neuron.

C Oxytocin and lactation

In mammals, suckling and other stimuli (auditory, visual) related to feeding the infant are transmitted rapidly over a spinal reflex arc to the paraventricular nucleus of the hypothalamus (see Chapter 14: Hormones of Pregnancy, Parturition and Lactation). This stimulus appears to be cholinergic and is transmitted to nerve endings in the posterior pituitary, resulting in Ca^{2+} uptake, depolarization, and exocytosis of OT into the circulation. Negative emotions, such as fear and stress, can inhibit OT release. Only a few seconds are required for OT to reach its receptors in the myoepithelial cells of the mammary gland. Through the activation of phospholipase C and the resulting changes in intracellular Ca^{2+}, the mammary ductules and ducts contract, and milk is ejected from the gland through the nipple into the infant's mouth.

D Oxytocin and the brain

The OTR is expressed in the brain in a region-, sex-, and species-specific manner. This suggests that the biological actions of OT are quite likely to vary by species and by gender and indeed behavioral and neurobiological studies in many mammalian species, including humans, confirm that this is the case. Table 4.2 lists some of the locations of the OTR in the CNS in the human and, for comparison, in the adult rat. The species differences in OT receptor localization are clear from this table. For example, several areas show abundant OT binding in the human and negligible binding in the rat. In most cases the association between binding and physiological activity has not yet been established. The interaction between the OTR and membrane cholesterol has been established, as shown in Fig. 4.9. Cholesterol molecules observed in the structures of OTR and other GPCRs, either in monomeric or dimeric interfaces, after OT binds to its receptor, stabilize the interaction with cholesterol.

TABLE 4.2 Oxytocin receptors in the central nervous system.

Brain region	Human	Rat
Olfactory system		
Islands of Calleja	+	+++
Cortical areas		
Peduncular cortex	(+)	+++
Basal nucleus of Meynert	+++	ND
Basal ganglia		
Ventral pallidum groups	++	+++
Globus pallidus	++	ND
Limbic system		
Lateral septal nucleus	+++	+
Thalamus and hypothalamus		
Paraventricular thalamic nucleus	+	+
Anterior medial preoptic area	++	ND
Posterior hypothalamic area	++	ND
Lateral mammillary nucleus	++	+
Brain stem		
Substantia nigra pars compacta	+++	ND
Hypoglossal nucleus	++	ND
Nucleus of solitary tract	+++	(+)
Trigeminal nucleus	+++	+

OT receptors were determined by binding studies. The number of + signs denotes whether the level is low, moderate, or high; (+) means possible but not definitive presence; ND=Not detectable.
Source: Adapted from Gimpl and Fahrenholz, American Journal of Physiology (2001) 81:630–683.

1 Maternal behavior

In several species, OT availability is associated with maternal behavior characteristic for the species. For example, in rats, pup-grooming and hovering are increased by OT as are reciprocal mother–pup behaviors. In mice lacking the OT gene, some maternal behaviors, such as pup retrieval and licking, are decreased compared to wild-type mice. In humans, several studies suggest that there is a similar positive relationship between OT levels and positive interactions between human mothers and their infants. In particular, higher levels of OT measured during pregnancy are associated with positive maternal behavior (gaze, touch, and vocalization) and other indices of maternal adaptation. The role of OT, which is dependent on a number of other physiological and psychosocial factors, in human parental behavior will become better understood as larger and more detailed studies, as well as those extended to fathers, are carried out.

2 Social behavior: trust and fear

Building trust is a necessary step in relationships and social bonding. Studies in humans suggest that OT increases trust of strangers (and others). Although the role of OT in everyday life is not known, much attention is being paid to its possible involvement in social disorders, particularly isolating ones, such as phobias and autism. Related to its possible effects on social interaction through trust is the role that OT has in modulating the hypothalamic–pituitary–adrenal axis and its stress response.

VI Clinical aspects

A Oxytocin

There are presently no clear clinical disorders that can be attributed to a mutation in the gene for OT or for its receptor in humans, although it is conceivable that some

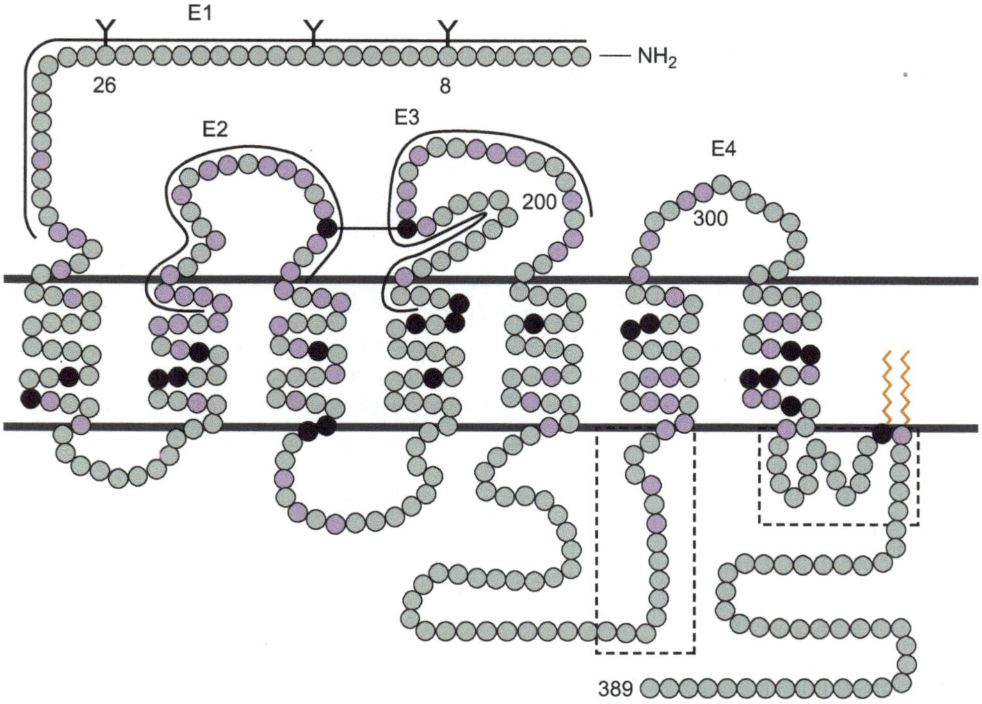

FIGURE 4.9 The structure of the oxytocin receptor. Residues shownin black are conserved in the entire GPCR family (as are the seven trans-membrane regions) and those in purple are conserved among the oxytocin and vasopressin receptors. The extracellular domains are labeled E1–E4 and the solid lines show binding sites for oxytocin and other agonists. The dashed rectangles indicate residues involved in G-protein binding and the palmitoylation sites are shown with orange hydrocarbon chains. *GPCR*, G protein–coupled receptor.

social behaviors discussed earlier could be affected by such changes in OT or the function of its receptor. That is not to say that OT is of no clinical interest, as its use plays a critical role in the management of labor and delivery (Chapter 14: Hormones of Pregnancy, Parturition and Lactation). Of interest is research suggesting that variations in the methylation of the gene for OT, but not any change in the DNA itself, may be related to autism.

B Diabetes insipidus

Lack of VP action, due to damage to the hypothalamus or pituitary or a gene mutation in either the gene for the hormone or for the receptor in the kidney, leads to the condition known as diabetes insipidus (DI) (literally urine that is tasteless). This condition is to be distinguished from diabetes mellitus (urine that is sweet) which is due to the lack of or resistance to insulin (Chapter 6: Pancreatic Hormones: Insulin and Glucagon). DI is characterized by polyuria (excessive urine volume) and polydipsia (excessive thirst/water intake).

The most common cause of hypothalamic DI in both children and adults is a primary brain tumor that impacts the hypothalamic magnocellular neurons that secrete VP. Inherited forms of DI, resulting from defects in VP gene expression, occur in both autosomal dominant and recessive forms; symptoms are generally recognized during the child's second year of life when parents become more aware of the child's need for water. In nephrogenic diabetes insipidus (NDI) the kidney is unresponsive to VP, due to a defect in the receptor or the signaling pathway. In congenital NDI, polyuria and polydipsia must be recognized at birth and treated immediately to avoid life-threatening dehydration. NDI can also be acquired through chronic renal disease, the use of one of several drugs, including lithium and alcohol, and other disease states such as multiple myeloma and sarcoidosis.

Hypothalamic DI can be treated with replacement VP or its therapeutic analog desmopressin, which has some beneficial pharmacokinetic properties. Dosages can be tailored to the individual by monitoring urine volume and osmolality. The treatment of nephrogenic DI is more complex since it is resistant to hormone replacement therapy. Most approaches involve treating the symptoms with low-sodium diet and drugs such as thiazide diuretics.

Further reading

Vasopressin secretion

C.J. Frazier, et al., An angiotensin-responsive connection from the lamina terminalis to the paraventricular nucleus of the hypothalamus evokes vasopressin secretion to increase blood pressure in mice, The Journal of Neuroscience 41 (2021) 1429–1442. Available from: https://doi.org/10.1523/JNEUROSCI.1600-20.2020.

Vasopressin receptor

L. Wang, et al., Cryo-EM structure of the AVP-vasopressin receptor 2-Gs signaling complex, Cell Research 31 (2021). Available from: https://doi.org/10.1038/s41422-021-00483-z.

S. Abraham, et al., Epigenetic regulation of arginine vasopressin receptor 2 expression by PAX2 and PAX transcription interacting protein, American Journal of Physiology — Renal Physiology 320 (2021). Available from: https://doi.org/10.1152/ajprenal.00371.2020.

Vasopressin and water regulation

A. Kim, et al., Neural basis for regulation of vasopressin secretion by anticipated disturbances in osmolality, bioRxiv (2021). Available from: https://doi.org/10.1101/2021.01.27.428388.

Biological actions of oxytocin

K. Chen, Y. Ye, N.F. Troje, W. Zhou, Oxytocin modulates human chemosensory decoding of sex in a dose-dependent manner, Elife 10 (2021) e59376. Available from: https://doi.org/10.7554/eLife.59376.

Oxytocin receptor

B. Chruscicka, et al., Molecular, biochemical and behavioral evidence for a novel oxytocin receptor and serotonin in 2c receptor heterocomplex, Neuropharmacology 183 (108394) (2021) 108394. Available from: https://doi.org/10.1016/j.neuropharm.2020.

Oxytocin and the brain

D. Hiraoka, et al., Epigenetic modification of the oxytocin gene is associated with gray matter volume and trait empathy in mothers, Psychoneuroendocrinology 123 (2021) 105026. Available from: https://doi.org/10.1016/j.psyneuen.2020.105026.

D. Martins, O. Dipasquale, Y. Paloyelis, Oxytocin modulates local topography of human functional connectome in healthy men at rest, Communications Biology 4 (2021). Available from: https://doi.org/10.1038/s42003-020-01610-z. Article number 68 (2021).

Diabetes insipidus

C.M. Mutter, et al., Diabetes insipidus: pathogenesis, diagnosis, and clinical management, Cureus 13 (2021) e13503. Available from: https://doi.org/10.7759/cureus.13523.

Chapter 5

Thyroid hormones

I Introduction

A The thyroid gland and its hormones

The thyroid gland is the largest of the endocrine organs and stores more of its hormone than any of the other endocrine glands. Through the secretion of thyroxine, T_4, and its active derivative, T_3, the thyroid governs myriad physiological events that bring about functional alterations in virtually all metabolic pathways and organs. Thyroid hormone modulates oxygen consumption and basal metabolic rate (BMR), as well as lipid, carbohydrate, and protein metabolism. It has fundamental effects on the nervous system both during development and in childhood and adulthood. Thyroid hormone regulates the synthesis and degradation rates of numerous proteins, including other growth factors and hormones, so that many of its effects emerge as secondary influences on other endocrine pathways.

The effects of thyroid hormone fall into two categories of biological responses: (1) effects on cellular differentiation and development, particularly, but not limited to, the nervous system; and (2) effects on metabolic pathways through which the body uses carbohydrates, lipids, and proteins. These two actions are interconnected in that changes in development and growth are both a cause and a consequence of hormonal modulation of metabolism. The wide range of the biological effects of thyroid hormone is reflected in the fact that virtually every tissue in the body expresses a receptor, TR, for the active hormone, T_3. The presence of different forms of the receptor, TRα or TRβ, in different tissues expands the range of effects this hormone can have in its target tissues.

With an understanding of thyroid hormone secretion, its regulation, and its actions and some of their mechanisms, the stage will be set to observe the results of excess or insufficiency of this hormone. The symptoms of the most common disease states of the thyroid gland are a direct consequence of an interruption of the known actions of its hormones. Furthermore, treatments of these conditions are based directly on our understanding of the mechanisms of the synthesis of thyroid hormone and its regulation.

B Iodine metabolism

The principal iodinated products produced by the thyroid gland are thyroxine (T_4) and triiodothyronine (T_3), which contain four and three atoms of organically bound iodine, respectively. Accordingly, the normal functioning of the thyroid is dependent on adequate and regular dietary intake of iodine.

Iodine is a rare element; although present in ocean water, it is distributed unevenly in the soils of the various land masses of the world and, therefore, the food content of iodine varies widely throughout the world. It is estimated that about 2 billion people suffer from iodine deficiency, which is the leading preventable cause of mental retardation in the world. The recommended daily allowance of iodine established by the US National Research Council is 150 μg/day for adults. In the United States the average daily intake of iodine is in the range of 250–700 μg/day, due largely to iodized salt and iodate in bread. The metabolic fates of this iodine intake are depicted in Fig. 5.1.

In the absence of adequate dietary access to iodine, an individual will adaptively develop iodine deficiency; endemic goiter exists when more than 10% of preadolescent children in a population grouping have enlarged, mildly hypertrophied thyroid glands due to iodine deficiency. In most adults with endemic goiter, iodine intake is below 50 μg/day. Endemic goiter does not exist in the United States but can be found in areas of the world in which there is a suboptimal level of iodine in the soil and in food crops grown thereon, for example, large areas of central Africa, Central Asia, the Andes of South America, and Indonesia. Endemic goiter can be prevented by various interventions, including the oral administration of potassium iodide at 6-month intervals.

II Anatomy of the thyroid gland

Fig. 5.2 shows the location and structure of the thyroid gland in humans. In euthyroid adults the thyroid gland weighs 15–25 g. It consists of two brownish-red lobes that lie on either side of the trachea and are connected by an isthmus that lies across the trachea. In humans and

Hormones. DOI: https://doi.org/10.1016/B978-0-323-90262-5.00028-7

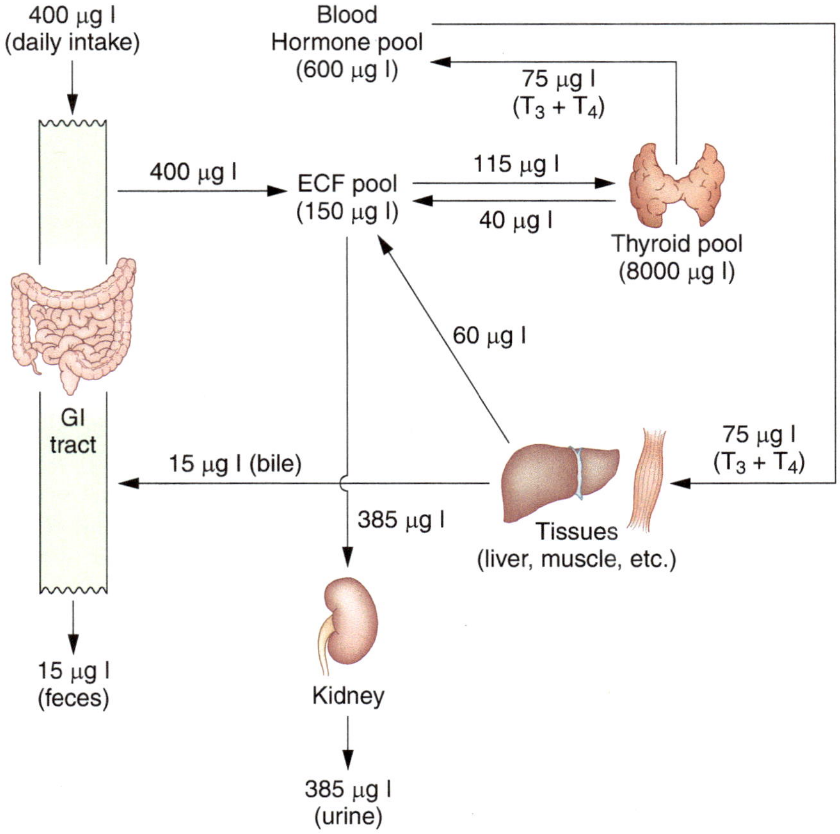

FIGURE 5.1 Physiology and distribution of iodide. In a healthy adult with a dietary intake of 400-μg iodine (I) per day, approximately 115 μg is accumulated across the thyroid epithelial cell membrane. Of this amount 75 μg is incorporated into thyroid hormones. ECF (25 L). Numbers in parentheses indicate the three major pools of I^-: the blood pool as circulating T_3 or T_4; the ECF pool as free I^-, and the thyroid gland pool as protein-bound or free I^-. *ECF*, Extracellular fluid.

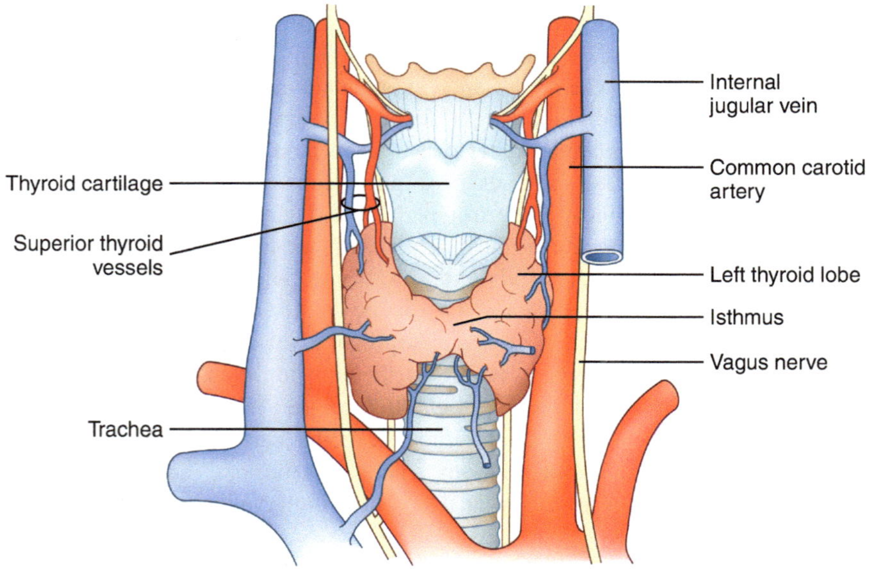

FIGURE 5.2 Anatomy of the adult thyroid gland. The two lobes of the thyroid gland, with the isthmus lying across the trachea, are shown, along with major blood vessels and nearby anatomical landmarks.

many other mammals, the parathyroid glands (see Chapter 9: Calcium-Regulating Hormones: Vitamin D, Parathyroid Hormone, Calcitonin & Fibroblast Growth Factor-23) lie on the surface of the thyroid gland but are not related to it structurally or functionally. Septa extend from the capsule into the thyroid gland, dividing it into lobules, which are, in turn, composed of follicles, the functional units of the thyroid gland. In the human thyroid gland, there are 20–30 million follicles. The thyroid gland has a rich blood supply provided by four arteries that radiate into arterioles, eventually forming capillary beds which surround each follicle as depicted in Fig. 5.3.

Each thyroid follicle consists of a single layer of epithelial cells enclosing a lumen filled with a viscous proteinaceous solution, the colloid. Each follicle is encased by a thin basement membrane. The colloid is clear in appearance and is composed of homogeneous granules that are secretory products of the follicle cell, principally the protein thyroglobulin (Tg), which is the storage form of the thyroid hormones.

III Chemistry of the thyroid hormones

The mammalian thyroid gland biosynthesizes, stores, and secretes two molecular species of thyroid hormone, thyroxine (T_4; 3,5,3′,5′-tetraiodothyronine) and triiodothyronine (T_3; 3,5,3′-triiodothyronine). As detailed next, the synthesis of thyroxine is dependent on the incorporation of iodine atoms into the tyrosine residues of a protein called Tg. The result of these iodinations of Tg is shown in Fig. 5.4A. Fig. 5.4B shows the ring numbering system for the thyronine rings of T_4 and its relatives, including T_3. The ring carbon numbering designations are important because the number and position of iodine atoms on the molecule determine its biological activity, or lack thereof. Thus T_3 is the biologically active form of the thyronines, while thyroxine and reverse T_3 are relatively inert. Fig. 5.4B and C shows two notable deiodination products (rT3 and R2) of thyroxine and T3, respectively.

A detailed study of the structural requirements for a molecule to display thyroid hormone activity indicates that such molecules must have, at a minimum, a central lipophilic core containing bulky ring substituents which are not necessarily iodine atoms. Thus extensive structural modification of the molecule is possible without complete or even major loss of biological activity. Methyl groups, bromine, fluorine, and nitrate are tolerated with decreasing activity in the 3, 5, 3′, and 5′ positions, and the alanine side chain can be replaced by formate, acetate, propionate, or pyruvate groups with some decrease in, but not loss of, activity. The ether link between the two phenolic rings can be replaced by a sulfur or methylene linkage.

IV Synthesis and secretion of thyroid hormones

A The thyroid epithelial cell

Each epithelial cell of the thyroid follicle, or thyrocyte, is specialized to carry out all the steps required for the synthesis and secretion of T_4 and T_3. These are as follows:

1. active transport of iodide into the thyroid gland follicular cells;
2. oxidation of iodide and iodination of tyrosyl residues within the protein Tg;
3. transfer and coupling of iodotyrosines within Tg to form T_4 and T_3;
4. storage of Tg as the colloid in the lumen of the thyroid follicle;
5. endocytosis of the colloid back into the thyroid epithelial cell;
6. proteolysis of Tg with concomitant release of T_4 and T_3 as well as free iodotyrosines and iodothyronines;
7. secretion of T_4 and T_3 into the blood; and
8. deiodination of iodotyrosines within the thyroid follicular cells for the reutilization of the liberated iodine.

To carry out these functions, the thyroid follicular epithelial cell (thyrocyte) has a specialized architecture, reminiscent of other secretory cells, and cell-specific protein components, including Tg, thyroid peroxidase (TPO), and the Na^+/I^- symporter (NIS). The two adjacent thyrocytes depicted in Fig. 5.5 show the subcellular organelles of the thyrocyte and the locations of the proteins involved in thyroid hormone synthesis. Thyrocytes have polarity, with the basal end in contact with the capillary bed surrounding the follicle and the apical end in contact with the lumen of the follicle. The NIS is localized in the basal membrane of the cell and TPO is in the apical membrane. Much of the interior space of the basal and perinuclear portions of the thyrocyte is occupied by rough endoplasmic reticulum (RER). A well-developed Golgi apparatus is located apical to the nucleus as are the associated

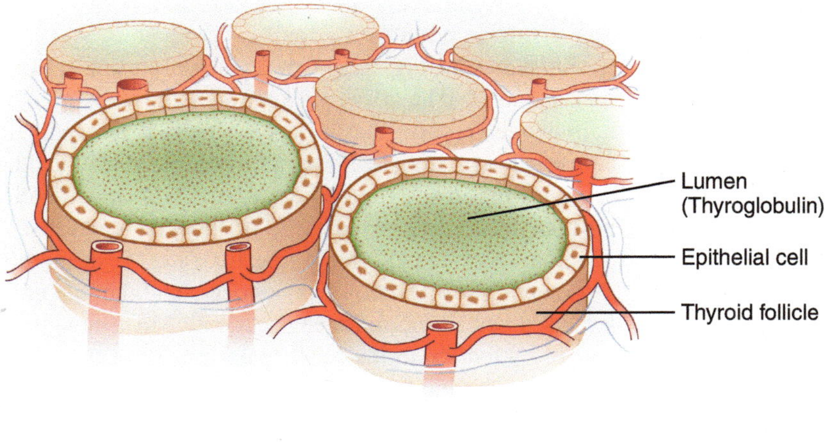

FIGURE 5.3 Thyroid follicles. Each thyroid follicle, the functional unit of the thyroid gland, is composed of a single layer of epithelial cells enclosing a lumen containing a colloidal material (*green*). Each follicle is surrounded by a capillary bed, part of the rich vasculature of the gland.

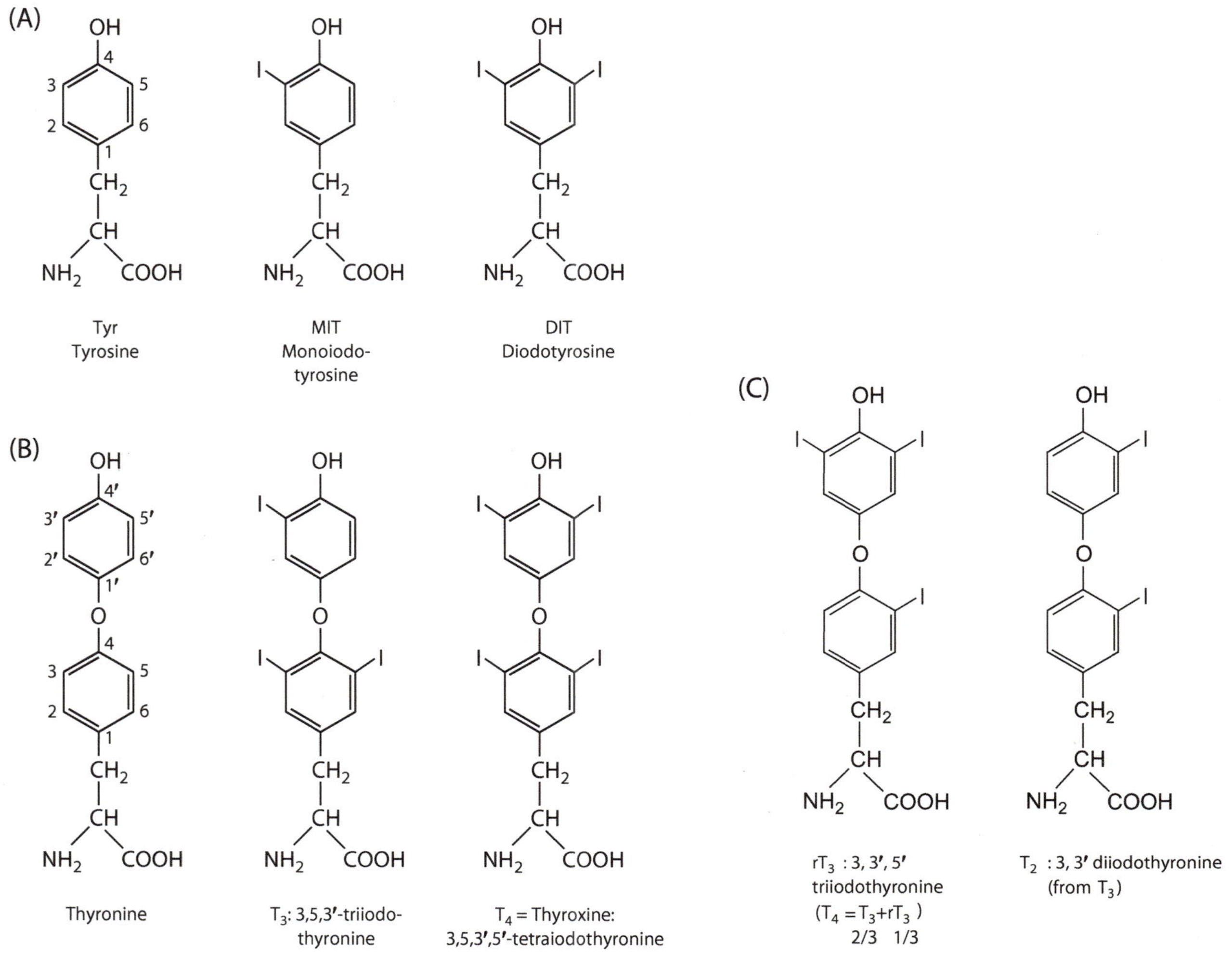

FIGURE 5.4 Thyroid hormone structure. (A) The amino acid precursors to thyroid hormone are shown. (B) The numbering system for the two rings is depicted along with the two main secreted iodinated products, T_3 and T_4. (C) Two deiodination products of T_3 and T_4 are shown.

exocytotic vesicles for the secretion of Tg into the lumen of the follicle. In addition, the cell contains endocytic vesicles and lysosomal structures that are responsible for the breakdown of Tg as discussed later. The apical surface of the thyrocyte is arranged as microvilli and some pseudopods.

B Thyroglobulin

As mentioned previously, the colloid of the thyroid follicle comprises primarily Tg, which is the most abundant protein of thyroid tissue. An understanding of the biochemistry and biological properties of Tg is integral to an understanding of the generation of thyroid hormone.

Tg is a dimeric molecule with a molecular mass of 660 kDa. The monomer is 330 kDa and consists of 2748 amino acid residues, of which 134 (4.5%) are tyrosine. Approximately 10%–11% of Tg's final molecular mass is attributable to carbohydrate (10%) and 5–50 atoms of iodine (0.1%–1%). The human gene for Tg is a single copy located on chromosome 8 that extends more than 300 kb and contains 48 exons. Mature Tg messenger RNA (mRNA) is translated on the RER and secreted into the RER cisternae where it is *N*-glycosylated and undergoes extensive intramolecular disulfide bond formation.

As shown in Fig. 5.6, 4 of Tg's 67 tyrosine residues (per monomer) have been experimentally shown to be hormonogenic sites, that is, a tyrosine that is eventually converted to a thyroid hormone molecule. Several other tyrosines are also iodinated early in the Tg modification process, and a subset of these is thought to consist of donors of the second tyrosyl ring for T_4 or T_3 at the hormonogenic sites. Details of this process and the fate of Tg are discussed in the following sections. Now, though, we turn our attention to the other major player in the formation of the thyroid hormones, iodine.

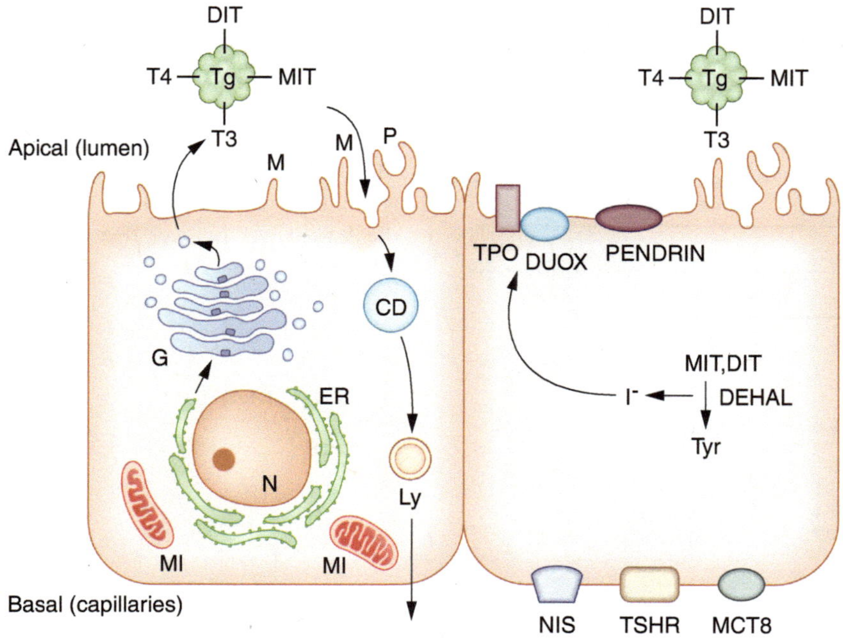

FIGURE 5.5 Thyroid epithelial cell structure and function. In the two thyrocytes shown, the organelles (left) and proteins (right) involved in thyroid hormone synthesis and secretion are depicted. The arrows in the thyrocyte on the left indicate the pathway of Tg from its synthesis in the endoplasmic reticulum through the Golgi apparatus; secretion, accompanied by thyroid hormone synthesis (Tg with T_4, T_3, MIT, and DIT), into the lumen; reuptake into the cell at the M or P to form a CD; breakdown in Ly and release of thyroid hormone into the capillaries. At the basal membrane of the thyrocyte on the right, are shown NIS, the sodium/iodide transporter that is responsible for iodide uptake into the thyrocyte (see Fig. 5.8); TSHR; and MCT8, which participates in secretion of T_4 and T_3 into the capillaries. At the apical membrane is TPO, which catalyzes the steps of thyroid hormone synthesis as Tg moves into the lumen and the dual oxidase, DUOX, which supplies peroxide (H_2O_2) for the reactions. Pendrin, a solute carrier protein, may participate in the movement of I^- from the cell into the lumen of the thyrocyte. Within the cell, unused I^- is removed from MIT and DIT by DEHAL. *CD*, Colloid droplet; *DEHAL*, iodotyrosine dehalogenase; *DIT*, diiodotyrosine; *DUOX*, dual oxidase; *ECF*, extracellular fluid; *Ly*, lysosomes; *M*, microvilli; *MCT8*, monocarboxylate transporter 8; *MIT*, monoiodotyrosine; *NIS*, Na^+/I^- symporter; *P*, pseudopods; *Tg*, thyroglobulin; *TPO*, thyroid peroxidase; *TSHR*, receptor for thyroid stimulating hormone.

C Iodide uptake: the Na^+/I^- symporter

The biosynthesis of thyroid hormone by the thyrocyte is dependent upon the continuous dietary availability of iodine which is absorbed by the intestine after reduction to iodide. The thyrocyte actively accumulates I^- so that the concentration in the thyroid gland is normally 20–30 times that in the blood and this ratio can be as high as 300. Although the active uptake of I^- does occur elsewhere in the body, for example, salivary and mammary glands, 90%–95% of it is normally found in the thyroid gland.

The transfer of iodide from the blood across the basement membrane of the thyrocyte occurs by unidirectional active transport by the integral membrane glycoprotein, NIS. Fig. 5.7 shows the movement of iodide (I^-) from the bloodstream through the NIS symporter in the basal thyrocyte membrane, through the cytoplasm to the pendrin channel in the apical membrane. Pendrin is thought to move I^- from the apical cytoplasm into the colloid where it makes contacts with Tg. Tg becomes complexed with T_3, T_4, monoiodotyrosine (MIT), and diiodotyrosine (DIT). The complexed form of Tg reenters the thyrocyte cytoplasm through the Megalin channel to form endosomes. The subsequent actions of lysosomal proteases release T_3 and T_4 into the basal space and then into the bloodstream. Human NIS is a protein of 643 amino acids, containing 13 membrane spanning regions, and is highly homologous among mammalian species. NIS is a member of the family of solute carriers for which the driving force is an electrochemical sodium gradient generated by Na^+/K^+ ATPase as shown in Fig. 5.8. Certain monovalent anions such as ClO^{4-} (perchlorate), SCN^- (thiocyanate), and TcO^{4-} (pertechnetate) are able to effectively compete with I^- for access to the transport process on the follicular cell and, therefore, inhibit I^- uptake by the thyrocyte.

As discussed in more detail in Section V, TSH (thyroid stimulating hormone; see Chapter 3: The Hypothalamus and Anterior Pituitary) is the most important physiological factor affecting iodide uptake by the thyroid. TSH increases the rate of transcription of the mRNA for NIS and mediates posttranscriptional modifications to bring about the functional maturation and basal membrane localization of the protein. Under certain conditions, also discussed later, iodide itself can also regulate its uptake by the thyroid gland.

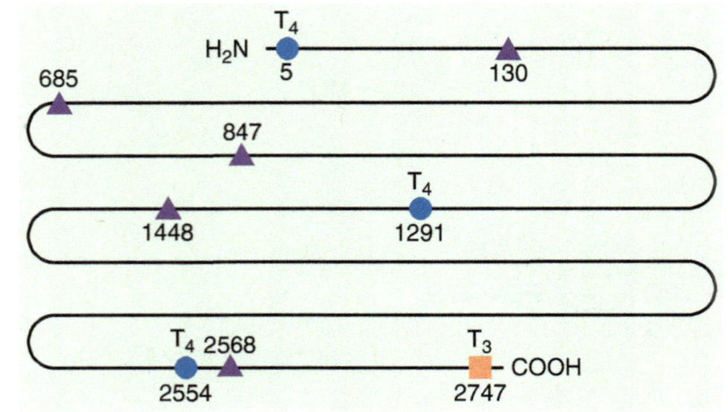

(A)

Residue	Sequence
5	ADYVP
1291	ADYAG
2554	DDYAS
2747	KTYSK

(B)

FIGURE 5.6 Hormonogenic sites of thyroglobulin. The peptide backbone of thyroglobulin is represented by the black line. Certain tyrosine residues are specified by their position from the N-terminal. The three main sites at which T_4 is found are shown by blue circles; the orange square is the site at which T_3 is usually found. These are acceptor sites during the coupling reaction. Purple triangles show the positions of other iodinated tyrosyl residues; these are likely donor sites from which the outer ring of T_3 or T_4 is derived during the coupling reaction (see Fig. 5.9). (B) The amino acid sequences surrounding the four hormonogenic sites in (A) are given.

Once iodide has entered the thyrocyte, a significant proportion moves through the cell and across the apical membrane into the lumen of the follicle, where it is available for incorporation into Tg. There is little free iodide in the thyrocyte and little bound to intrathyroidal proteins. The precise nature of the apical anion transporter(s) responsible for the transfer of I^- from the thyrocyte into the lumen has not yet been fully elucidated. The solute

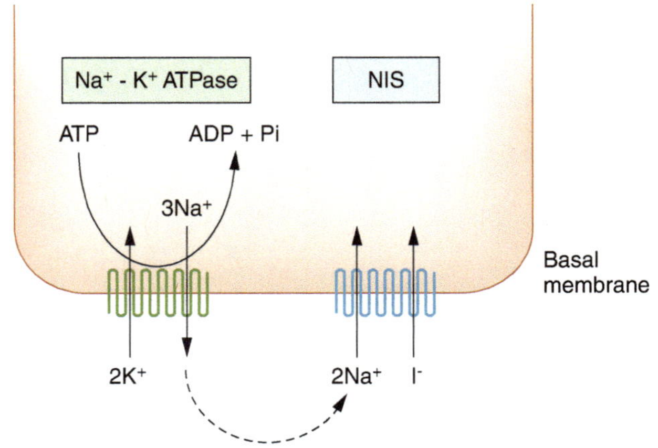

FIGURE 5.8 Iodide uptake by the thyrocyte. Na^+/K^+ ATPase and the sodium/iodine symporter (NIS) are shown in a segment of the basal membrane of the thyrocyte. *NIS*, Na^+/I^- symporter.

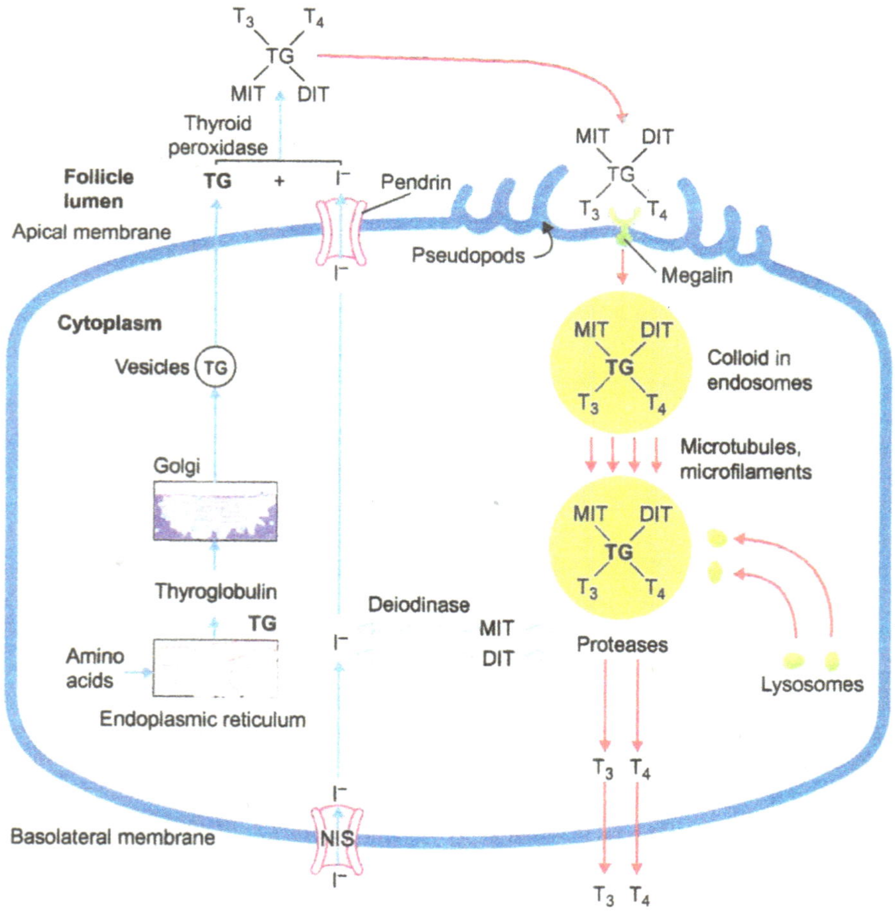

FIGURE 5.7 Summary of the steps from the import of iodide ion through the 2NIS on the basolateral side of the thyrocyte, through the derivatization of thyroglobulin forming the thyroid hormones and intermediates in the follicle, to the entry through the apical membrane of derivatized thyroglobulin through megalin and the release of thyroid hormones from thyroglobulin in the lysosome and their exit from the thyrocyte through the basolateral membrane. *NIS*, Na^+/I^- symporter. *Reproduced from Figure 19.60 of G. Litwack, Human Biochemistry, Academic Press, 2018, p. 640.*

carrier protein, pendrin (Fig. 5.5), has been implicated in this process, because mutations in it lead to Pendred syndrome, an autosomal recessive disorder characterized by impaired iodide organification, goiter, and sensorineural deafness. Experimental studies are consistent with a role, but not an exclusive one, for pendrin in this process.

D Thyroid peroxidase and DUOX: tyrosine iodination and coupling

The process of incorporation of 1 or 2 iodide atoms into the aromatic rings of a tyrosine residue of Tg followed by conversion to a thyroid hormone involves three separate steps, each of which is catalyzed by the enzyme TPO:

1. oxidation of I^- to I^{ox};
2. addition of I^{ox} to specific tyrosine acceptor residues of Tg, particularly at the hormonogenic and donor sites (Fig. 5.7); and
3. coupling of two iodotyrosyl residues to generate one thyroid hormone residue and a corresponding dehydroalanine residue within the Tg molecule.

As with Tg, the enzyme TPO, a heme-containing 933 amino acid member of the peroxidase family, is synthesized on the RER of the thyrocyte. It is processed, including glycosylation, by the Golgi apparatus and is ultimately anchored in the apical membrane by a hydrophobic region in its C-terminal region. Most of the protein, including its catalytic domain, is located in the lumen of the follicle.

Fig. 5.9 shows TPO in the apical membrane of the thyrocyte, along with the enzyme responsible for generating the H_2O_2 required for the peroxidase activity of TPO. This Ca^{2+}/NADPH-dependent oxidase, which is expressed abundantly but not exclusively in the thyroid gland, is called dual oxidase, DUOX, because it contains both an NADPH oxidase domain and peroxidase activity. When there is sufficient I^- available, H_2O_2 production by DUOX is the rate-limiting step in thyroid hormone production. High levels of iodine inhibit the enzyme's activity and TSH stimulates it. There are two forms of the enzyme, DUOX1 and DUOX2, which are encoded on separate genes on chromosome 20. Both DUOX1 and DUOX2 require specific maturation factors, DUOXA1 and DUOXA2, to ensure their proper location and activity in the cell. The genes for these maturation proteins are located in the intergenic region between the genes for the two enzymes. In addition to the thyroid gland, these enzymes appear in other tissues to support varying ion transport functions. Mutations in DUOX2 or DUOXA2 may result in congenital hypothyroidism, but the range of phenotypes observed suggest that the extent of condition may be influenced by other factors.

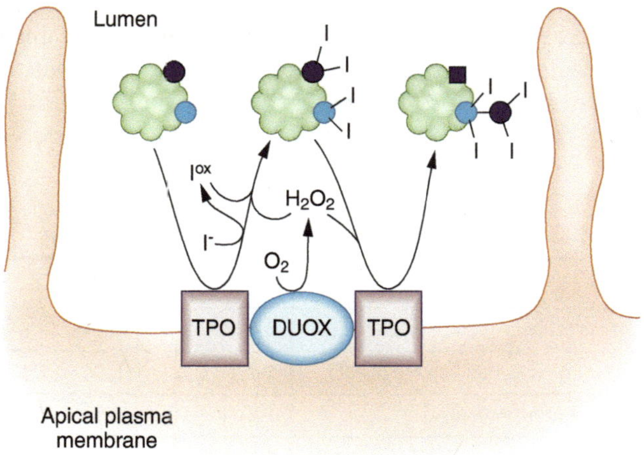

FIGURE 5.9 Thyroid hormone synthesis by TPO. In this schematic diagram, two tyrosine residues are represented by the blue (hormonogenic site) and purple (donor site) circles. In the reactions catalyzed by TPO, H_2O_2 generated by the dual oxidase, DUOX, is required for both the iodination of these residues and for their coupling to form T_4 (see Fig. 5.10). The residue remaining at the latter site after the coupling is dehydroalanine, represented by the purple square. *DUOX*, Dual oxidase; *TPO*, thyroid peroxidase.

In the first step catalyzed by TPO, the H_2O_2 generated by DUOX is used to oxidize I^-, as depicted in Fig. 5.9. This oxidized I^- (I^{ox}), the exact nature of which has not yet been definitively elucidated, is incorporated into a tyrosine ring of Tg to form either a MIT or DIT residue of the protein. One of the major physiological controls of the abundance of iodotyrosine, iodothyronine, and, therefore, thyroid hormone in Tg is the dietary iodide supply. In the absence of ample dietary iodine, the amount of thyroid hormone produced per molecule of Tg falls due to its sparse iodination. Iodination of Tg is also inhibited by excess iodide; this is known as the Wolff–Chaikoff effect, a transient phenomenon to which the thyrocyte adapts by "escaping" the inhibition. Thus proper functioning of this stage of thyroid hormone synthesis, iodination of tyrosine residues, is dependent on adequate but not excessive levels of iodine availability.

Finally, TPO catalyzes the coupling between two iodinated tyrosine residues, also depicted in Fig. 5.9 and in more detail in Fig. 5.10. In this reaction an iodinated (a diiodinated ring is depicted in Fig. 5.9) tyrosyl ring is removed from a donor site (see Fig. 5.7) on Tg and becomes the outer ring of T_4 at the hormonogenic site. This leaves behind dehydroalanine at the donor site. If the tyrosine at the donor site had been monoiodinated, the resulting molecule at the hormonogenic site would be T_3. A typical Tg molecule from an individual with normal thyroid status contains about 5 residues of each MIT and DIT, 2.5 residues of T_4, and 0.7 residues of T_3.

E Thyroglobulin storage, endocytosis, and breakdown

Following the formation of T_4 (and some T_3) as residues of Tg, the protein containing the hormone is stored in the granules comprising the colloid material of the thyroid follicular lumen. In a person with normal thyroid function, the thyroid gland stores approximately 50 days' worth of its hormone, more than any other endocrine gland.

The next step in the ultimate secretion of the thyroid hormones is the reuptake of Tg into the thyrocyte, the breakdown of the protein into its constituent amino acids, and the release of the T_4 and T_3 molecules into the bloodstream. This process is begun by the uptake of Tg-containing granules at the apical end of the cell. In the basal (non-TSH-stimulated) state, uptake is primarily by micropinocytosis involving clathrin-coated vesicles. Mechanisms of sorting Tg molecules according to their state of iodination appear to operate, allowing for recycling of intact molecules bearing few or no hormone residues prior to reaching the degradation stage of the pathway. When the gland is stimulated by TSH, Tg is taken up from the lumen by macropinocytosis involving pseudopod formation. Regardless of the pathway of uptake, endosomes containing Tg destined for breakdown fuse with lysosomes. Lysosomal proteases digest the protein to its constituent amino acids, including MIT, DIT, T_4, and T_3.

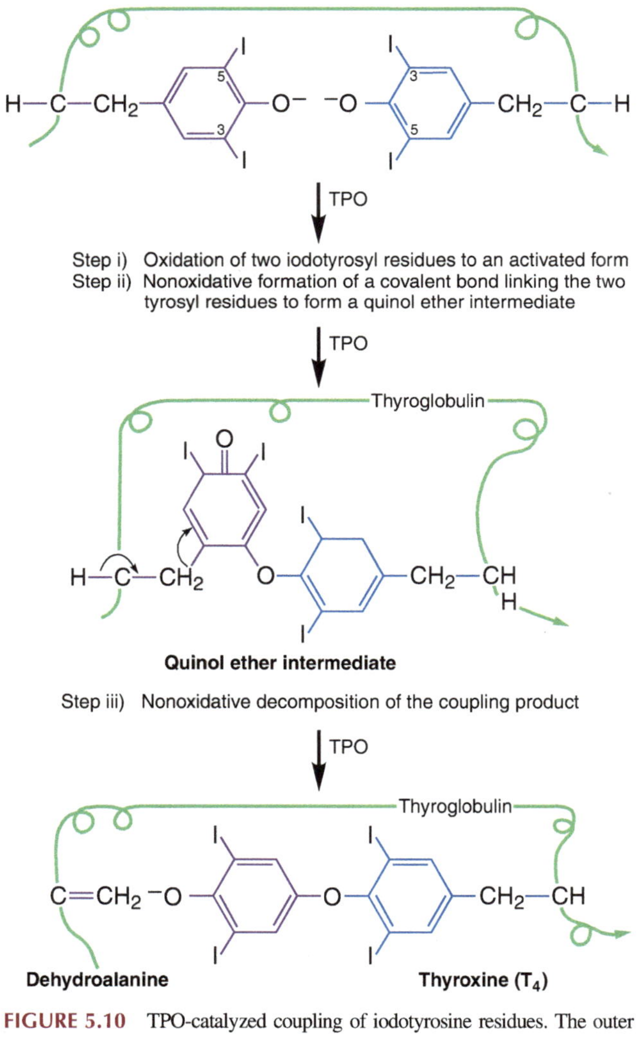

FIGURE 5.10 TPO-catalyzed coupling of iodotyrosine residues. The outer ring of T_4 or T_3 being formed at the hormonogenic site is derived from a donor site (see Figs. 5.6 and 5.8). It has not yet been elucidated whether the intermediate is a free radical or is ionic. *TPO*, Thyroid peroxidase.

F Secretion of T_4 and T_3 and recycling of I^-

The predominant thyroid hormone released from the gland is T_4, with lesser amounts of T_3, a portion of which is formed by intrathyroidal deiodination of T_4. The transport of T_4 and T_3 across the basolateral membrane of the thyrocyte into the circulation was long assumed to take place by passive diffusion. More recently, however, it has become clear that the process is mediated by one or more transport proteins. The monocarboxylate transport protein, monocarboxylate transporter 8 (MCT8) (Fig. 5.5), is currently the prime candidate for this role in thyroid hormone secretion as it is expressed in the basolateral membrane of the thyrocyte and its absence in mice results in lowered secretion of thyroid hormone. This is the same transport protein that participates in bringing T_3 and T_4 into target cells, as discussed in Section VI.A.

The iodine in the MIT and DIT residues of Tg is released through the action of iodotyrosine dehalogenase (DEHAL1) and returned to the intrathyroidal I^- pool. DEHAL is an NADPH-dependent flavoprotein that shows specificity for iodotyrosines relative to iodothyronines. Since the amount of iodide recaptured per day by the thyroid gland is two to three times the amount of "new" iodide transported into the gland from the blood, it is clear that the efficiency of this reutilization pathway is of fundamental importance to the overall iodine economy, and, therefore, thyroid function.

G Transport and metabolism of thyroid hormones

1 Transport

Table 5.1 lists the blood concentrations and some kinetic properties in humans of the two main iodine-containing molecules released from the thyroid gland. Note that the concentration of T_4 is approximately 70 times that of T_3 and that only 0.02%–0.2% of either hormone is

TABLE 5.1 Properties of circulating T_4 and T_3.

Measurement	T_4	T_3
Serum concentration-total (μg/dL)	8.1	0.14
Serum concentration-free (ng/dL)	1.2 (0.02%)	0.29 (0.2%)
Percent from thyroid	100	5
Production ratea (μg/day)	90	32
Serum half-life (days)	7	1

[a]*Production rate of T_3 includes peripheral deiodination of T_4.*
Source: Adapted from J.L. Jameson, L.J. DeGroot, Mechanisms of thyroid action, In: L.J. DeGroot, (Ed.), Endocrinology, vol. 1, W.B Saunders, Philadelphia, 1995, pp. 583–601.

circulating free (i.e., not protein-bound) in the blood. Table 5.1 also gives the half-lives of these compounds which show that T_3 turns over much more rapidly than T_4. Since T_4 is effectively biologically inactive until it is peripherally deiodinated to form T_3 (see the following), its circulating pool can be thought of as a large, slowly turning over reservoir of stored hormone in addition to that which is stored in the lumen of the thyroid gland. T_3, on the other hand, has a small vascular pool with a high turnover rate, as befits the active form of the hormone.

In vertebrates, over 99% of T_4 and T_3, circulate in the blood bound to proteins. The three serum proteins responsible for systemic transport of the thyroid hormones are thyroxine-binding globulin (TBG), transthyretin (TTR; also known as thyroxine-binding prealbumin, TBPA), and albumin. Some of the properties of these proteins, all of which are synthesized in the liver, are shown in Table 5.2. TBG binds about 75% of both T_4 and T_3. Most of the remaining hormone is carried either by TTR or albumin (T_3). Since it is generally held that it is the free fraction of hormone that is available to enter cells and initiate biological responses, the role of the serum-binding proteins in maintaining the equilibrium between the bound and free fractions is an important physiological one.

TBG is a single subunit glycoprotein. It contains four tetrasaccharide moieties which account for approximately 20% of the total molecular mass of the protein. Each molecule of TBG has a single iodothyronine-binding site with a higher affinity for T_4 than for T_3. TBG levels in the blood are increased approximately 2.5 fold during pregnancy due to increased estrogen levels. As TBG increases, T_4 secretion is increased to maintain free T_4 levels, increasing the pool of T_4 in the blood.

TTR is composed of four identical polypeptide subunits, each containing 127 amino acids. Although X-ray structural studies indicate two apparently identical binding sites, only one is usually occupied because the affinity of the second is dramatically decreased by negative cooperativity following binding of the first molecule of T_4. TTR contains a separate specific binding site for retinol-binding protein that binds vitamin A (retinol). The two functions of TTR are independent of one another. Although TTR circulates in much higher concentrations in the blood than TBG, it plays a smaller role in the physiology of the thyroid hormones. Much of it is unoccupied by T_4 and even large changes in its concentration have little effect on T_4 production or secretion. One clinical aspect of the protein, unrelated to the thyroid, is its role in systemic amyloid disorders, particularly in the heart, due to genetic variants of TTR which self-associate and contribute to amyloid fibrils.

Human serum albumin (HSA) comprises more than half of the total serum protein and binds a wide variety of small molecules, mostly with low affinity. The affinity of HSA for T_4 is four orders of magnitude lower than that of TBG, yet the total amount of T_3 and T_4 bound by this protein is a significant, if small, portion of the bound hormone.

TABLE 5.2 Properties of thyroid hormone serum-binding proteins.

	TBG[a]	TTR[b]	Albumin
Molecular weight	63,000	55,000	69,000
2° Structure	Monomer	Tetramer	Monomer
Serum conc. (M)	3×10^{-7}	2×10^{-6}	6×10^{-4}
Half-life (days)	5	2	15
# Binding sites for T_4, T_3	1	2	Several
K_a for T_4 (M^{-1})	1×10^{10}	1×10^{8}	1.5×10^{6}[c]
K_a for T_3 (M^{-1})	5×10^{8}	1×10^{6}	1×10^{5}[c]
Occupancy, T_4	30%	0.5%	<0.1%
Occupancy, T_3	0.6%	<0.1%	<0.1%
Proportion of T_4 carried	75%	10%–15%	10%–15%
Proportion of T_3 carried	75%	<5%	20%

TBG, Thyroxine-binding globulin; *TTR*, transthyretin.
[a] *Thyroxine-binding globulin.*
[b] *Transthyretin; also known as thyroxine-binding prealbumin, TBPA.*
[c] *For highest affinity site.*
Source: Adapted from J. Stockigt, Thyroid hormone binding and variants of transport proteins, in: J.L. Jameson, L.J. De Groot (Eds.), Endocrinology, vol. 2, Elsevier, Philadelphia, PA, 2010, pp. 1733–1744.

2 Metabolism

The blood concentration of the thyroid hormones is dependent not only upon the amount of hormone secreted by the thyroid gland but also the hormone's affinity for its carrier proteins, affinity for target tissues, rate of catabolism, and, finally, its rate of clearance. The major pathways of thyroid hormone deiodination to other thyronines are outlined in Fig. 5.11. The figure also shows (top left) the oxidative deamination of T_4 and T_3, yielding tetraiodoacetic acid or triiodoacetic acid, respectively, as well as the production of DIT through the cleavage of the ether bond between the two rings (top right). In addition to the metabolic steps shown in Fig. 5.11, but also important in the overall thyroid hormone economy of the organism is the conjugation of the hormone with either sulfate or glucuronide.

Finally, it has been recognized that monoiodinated derivatives of T_4, which arise by its deiodination and decarboxylation, exist in human serum and have specific biologic activities. These compounds are referred to as iodothyronamines, the most potent of which is 3-iodothyronamine, seen in Fig. 5.12. This compound and thyronamine have a high affinity for the trace amine receptor (TAR1) and have suppressive effects on heart rate, body temperature, and physical activity. This is currently a very active area of research.

The deiodinases that convert T_4 to either T_3 or rT_3, a biologically inactive metabolite, are an integral part of thyroid hormone physiology and deserve a closer look. They are selenocysteine enzymes, characteristics of which are summarized in Table 5.3. Deiodinase Type I, or D1, is found in the thyroid, pituitary, liver, and kidney. Depending on the substrate, D1 removes the iodine atom from either the outer or inner ring, to activate or inactivate the molecule, respectively. In the liver, D1 provides the majority of circulating T_3 and is also responsible for the deiodination of sulfated T_4 and rT_3. The expression of D1 in the liver is regulated by thyroid status such that T_4 degradation is decreased when the hormone is deficient and increased when it is present in excess. D1 expression in the pituitary is also increased when hormone levels are high, resulting in increased T_3 being made available to suppress TSH secretion (see Section V.A).

D2 has only 5′ deiodinase activity; its main physiological function is thought to be the production of intracellular T_3 in the pituitary, certain localized areas of the brain, and in brown adipose tissue (BAT). D2 activity is inversely related to thyroid hormone status, allowing

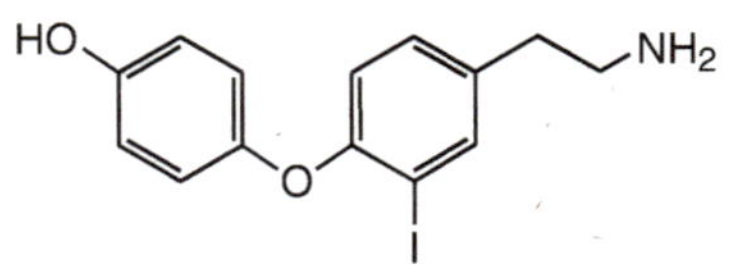

3-Iodothyronamine

FIGURE 5.12 3-Iodothyronamine.

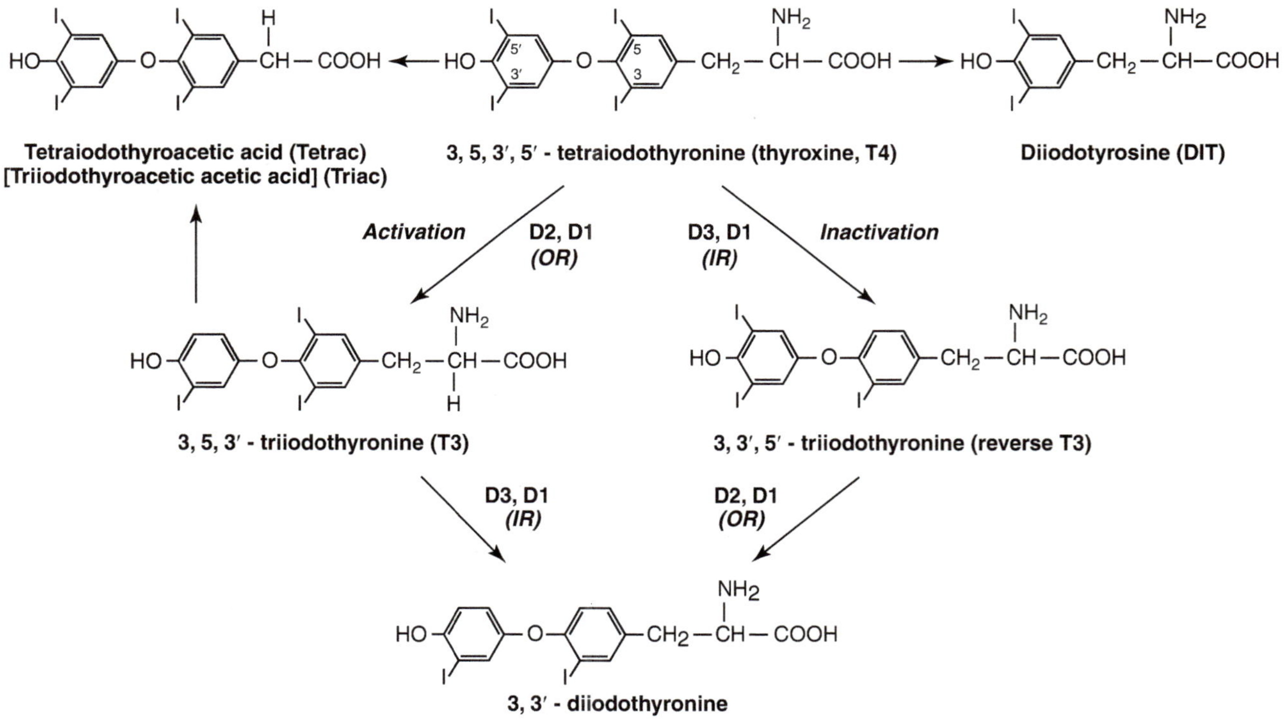

FIGURE 5.11 Peripheral transformations of T_4. At the top are shown the result of oxidative deamination (left) and cleavage of the ether bond between the two tyrosyl rings (top right). The remainder of the figure summarizes the deiodination steps that are catalyzed by one of three deiodinases, D1, D2, or D3. The actions and properties of these important enzymes are more fully described in the text and in Table 5.3. *IR*, Inner ring; *OR*, outer ring.

TABLE 5.3 Properties of iodothyronine deiodinases.

Characteristic	D1	D2	D3
Molecular wt (kDa)	29	30.5	31.5
Deiodination site[a]	IR, OR	OR	IR
Substrate preference	$rT_3 > T_4 > T_3$	$T_4 > rT_3$	$T_3 = T_4$
Inhibition by PTU[b]	High	Low	Low
Hypothyroidism[c]	↓	↑	↓
Hyperthyroidism[c]	↑	↓	↓
Tissues	Liver, kidney, pituitary, and thyroid	CNS[d], pituitary, BAT[d], placenta, muscle, and heart	Placenta, CNS, and fetal liver

[a]*IR, inner (phenolic) ring; OR, outer (tyrosyl) ring.*
[b]*PTU, propylthiouracil.*
[c]*Enzyme activity response (increased or decreased) to either hypo- or hyperthyroidism is shown by the arrow.*
[d]*CNS, central nervous system; BAT, brown adipose tissue.*
Source: Adapted from A.C. Bianco, D. Salvatore, B. Gereben, M.J. Berry, P.R. Larsen, Biochemistry, cellular, and molecular biology and physiological roles of the iodothyronine selenodeiodinases. Endocrine Reviews 23 (2002) 38–49.

some protection against the production of excessive or inadequate intracellular T_3 in the face of hyper- or hypothyroidism.

D3 has only inner ring deiodinase activity and, therefore, is only involved in the degradation of iodothyronines. In adult mammals, D3 occurs primarily in the brain. It is more abundant in fetal tissues, where it may be important in protecting some tissues against high T_3 level concentrations that occur during certain differentiation processes.

H Antithyroid drugs

Specific compounds that interfere with the thyroid follicular peroxidase are used both in research and clinically. The most potent of these compounds contains the thionamide group, of which 6-*n*-propylthiouracil and methimazole are most commonly used in the clinical setting. Fig. 5.13 shows the structures of these drugs. The monovalent cations that block iodide uptake include thiocyanate and perchlorate which may be used in specialized clinical situations. Many other agents used for nonthyroidal purposes or present in the diet or environment can influence thyroid function by affecting thyroid hormone synthesis. These include the inhibition of thyroid hormone synthesis by sulfonamides used for control of bacterial growth, alterations in thyroid hormone transport in the blood caused by anabolic steroids and glucocorticoids, or changes in the peripheral deiodination of thyroid hormones resulting from propranolol or glucocorticoid administration.

V Regulation of thyroid hormone secretion

A The hypothalamic–pituitary–thyroid axis

All of the functions of the thyrocytes described in the previous section are under the control of TSH from the pituitary, which, in turn, is secreted in response to thyrotropin-releasing hormone (TRH) from the median eminence of the hypothalamus. The structures of the tripeptide TRH and the heterodimeric TSH were described in Chapter 3, The Hypothalamus and Anterior Pituitary. Fig. 5.14 outlines this hypothalamic–pituitary–thyroid axis. The physiological stability of the rates of secretion of TSH and the thyroid hormones are primarily due to (1) the stimulation of TSH by TRH from the hypothalamus; (2) the effects of TSH on thyroid hormone production and secretion; and (3) the feedback effects of T_4 and T_3 on the pituitary and hypothalamus.

The TRH that participates in this control system is synthesized in hypophysiotropic neurons of the paraventricular nucleus of the hypothalamus, axons of which project to the anterior pituitary where TSH is released. Destruction of these neurons leads to hypothyroidism, demonstrating their vital role in the tonic secretion of TSH by the pituitary. Additional neuroendocrine interactions take place between the hypothalamus and the pituitary, including the β-adrenergic-mediated increase in TRH secretion in response to cold or stress. Somatostatin and dopamine, two hypothalamic hormones known for their roles in the control of growth hormone and prolactin secretion, respectively (see Chapter 3: The Hypothalamus

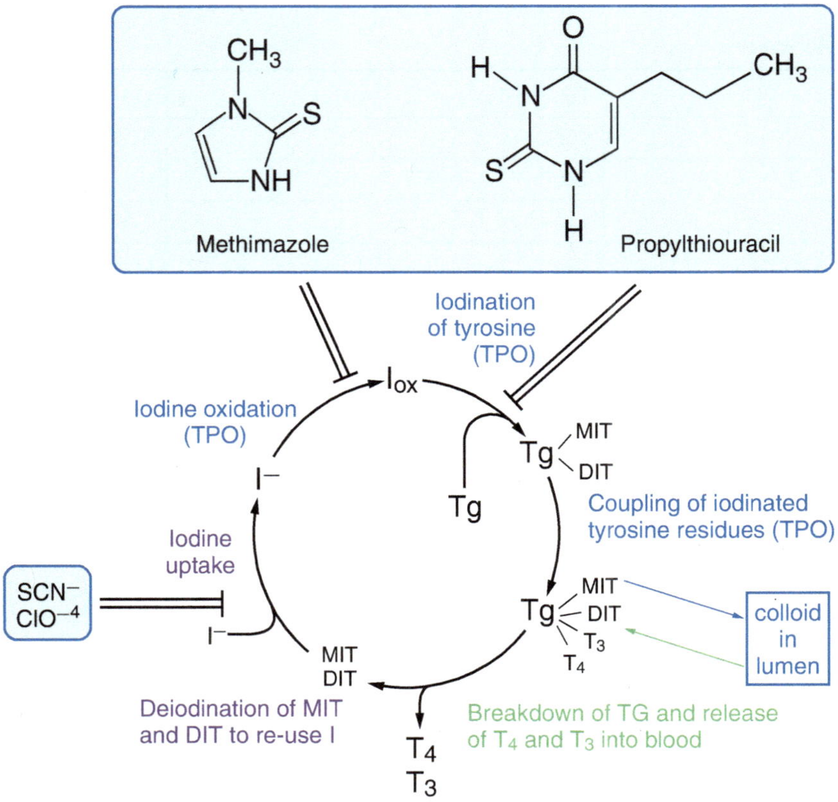

FIGURE 5.13 Inhibitors of thyroid hormone synthesis. The steps of thyroid hormone synthesis are shown clockwise, beginning with iodide uptake at about 7 o'clock. Some common antithyroid agents and their points of interference are shown. *ClO^{-4}*, Perchlorate; *DIT*, diiodotyrosine; *MIT*, monoiodotyrosine; *SCN^-*, thiocyanate; *Tg*, thyroglobulin; *TPO*, thyroid peroxidase.

and Anterior Pituitary), exert an inhibitory influence on TSH secretion.

In the thyrotrophs of the pituitary gland, TRH interacts with its specific G protein–coupled receptor which, through the phospholipase C–catalyzed hydrolysis of phosphatidyl inositol to IP_3 and diacylglycerol (DAG), leads ultimately to the stimulation of the transcription of TSHβ. The production of this subunit of the TSH heterodimer is the rate-limiting step in its assembly for TSH secretion. Protein kinase A-dependent pathways may also play a role in TRH-regulation of TSHβ gene transcription. In addition to its effects on TSHβ gene transcription, TRH also affects the bioactivity of TSH by altering its glycosylation pattern.

Circulating levels of T_3 and T_4 exert classic negative feedback effects on the pituitary secretion of TSH. The thyrotrophs of the pituitary gland contain high levels of deiodinase type II (D2) so that T_4 is readily converted to T_3 which then interacts with the thyroid hormone receptor, TR, to inhibit transcription of both the α and β subunits. This effect on the β-subunit is much more rapid and sustained than that on the α-subunit. In animal models, T_3 also decreases the number of TRH receptors in the thyrotroph.

At the hypothalamus, T_4 and T_3 exert feedback inhibition on TRH secretion, due mainly to decreased transcription of TRH mRNA (see Fig. 3.9). This effect is mediated by the thyroid hormone receptor (TRβ2; see the following). The deiodination of T_4 to T_3 does not appear to take place in the TRH neurons themselves, which lacks the D2 deiodinase, but possibly in nearby glial cells.

B Autoregulation

There are two main ways in which iodide governs the activity of the thyroid gland in processes generally known as autoregulation. Both are thought to be ways in which the gland can respond to wide variations in dietary iodide availability. When iodide intake exceeds what the thyroid requires for normal amounts of thyroid hormone synthesis, Tg iodination and the ensuing hormone production is blocked. In humans, such effects are seen at iodide intake of 500–1000 μg/day. The gland can, over a period of weeks, adapt to these high levels of iodine so that the blockage in hormone production is no longer effective. The iodide inhibition of Tg iodination, and, therefore, of thyroid hormone production, is known as the Wolff–Chaikoff effect and the subsequent adaptation is known as escape from the effect. The mechanisms of the Wolff–Chaikoff effect and the escape from it are not yet fully understood but appear to involve alterations in the thyrocyte's sensitivity to TSH, since signals downstream from the TSH receptor such as intracellular Ca^{2+}, cyclic AMP, and DAG are decreased by inhibitory levels of iodide.

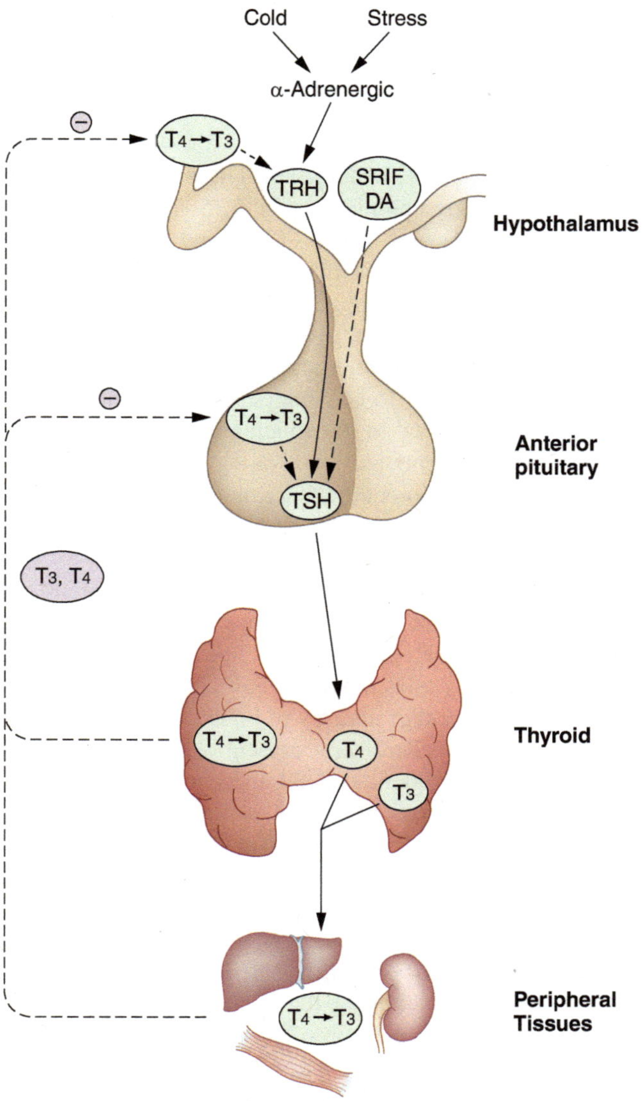

FIGURE 5.14 The hypothalamic–pituitary–thyroid axis. Control of thyroid hormone involves the stimulation of pituitary TSH by TRH from the hypothalamus. Many signals from other areas of the brain, such as cold and stress, influence TRH secretion by the hypothalamus. TSH stimulates synthesis and release of T_4 (and some T_3) from the thyroid gland. T_4 is deiodinated to T_3 (*green ovals*: $T_4 \rightarrow T_3$) in peripheral target tissues as well as in the pituitary and in the hypothalamus. The circulating thyroid hormones (*purple oval*) exert feedback inhibition on TSH secretion in the pituitary and, to a variable degree depending on species, on TRH secretion. SST (SRIF) and DA also influence TSH secretion. Stimulatory and inhibitory effects are represented by solid and dashed lines, respectively. *DA*, Dopamine; *SRIF*, somatotropin release-inhibiting factor; *SST*, somatostatin; *TRH*, thyrotropin-releasing hormone; *TSH*, thyroid stimulating hormone.

Large amounts of iodide can also, independently of TSH, block the release of thyroid hormone from the gland. Various mechanisms for this effect have been proposed but none are yet fully supported by experimental evidence. The effect, however, is useful clinically in blocking hormone release prior to thyroid surgery.

TABLE 5.4 Thyroid responses to thyroid stimulating hormone.

Time frame	Increase in
Immediate: 2–10 min	Cyclic AMP
	Endocytosis of thyroglobulin
	T_4 and T_3 secretion
Intermediate: 2–15 h	Iodide uptake
	Thyroglobulin synthesis
	Thyroid peroxidase synthesis
	Na^+/I^- symporter synthesis
Chronic stimulation: days	Mitogenesis
	Gland size (goiter)

C Effects of TSH on the thyroid epithelial cell

Virtually every function of the cells making up the thyroid follicle is stimulated in response to the activation of the TSH receptor. Some of these responses are summarized in Table 5.4. This G protein–coupled receptor acts through both phospholipase C and cyclic AMP–dependent pathways to increase all aspects of thyroid hormone production. Acutely, that is, in a matter of minutes, these effects include each step of TPO-catalyzed T_4 production as evidenced by increased iodide efflux into the thyroid follicular lumen and the uptake of Tg by macropinocytosis, leading to thyroid hormone secretion. In hours, the transcription of NIS to support increased iodide uptake from the blood is apparent. Transcriptional effects on other proteins such as TPO and Tg are also apparent, depending on the species under investigation. Finally, if TSH stimulation of the thyroid gland is sufficiently strong and prolonged (greater than 24 hours), thyrocyte proliferation ensues, leading over the long term to the goiter associated with hypothyroidism and hyperthyroidism of various etiologies. Any agent that either decreases circulating T_4 and T_3 levels so that circulating TSH increases or otherwise activates the TSH receptors of the thyroid gland will have these same consequences.

VI Biological actions of T_3

The effects of the thyroid hormones can be divided into two broad classes. There are effects on the metabolism and specialized functions of cell typified by, but by no means limited to, the regulation of the BMR. Second, there are effects on cell growth and differentiation, originally elucidated in the context of the metamorphosis of the aquatic legless tadpole into an amphibious four-legged frog. These effects, by definition, occur only during a certain window of time during the developmental process. The majority of effects of thyroid hormones are now believed to be mediated through interactions of T_3 with one of its nuclear receptors resulting in changes in gene expression. It is appropriate, therefore, that we begin this section with a consideration of the nuclear receptor for T_3.

A Thyroid hormone receptor

It is well accepted that the thyroid hormones cross the plasma membranes of target cells, not by diffusion, but with the help of specific transport proteins. Although several proteins have been implicated, the one that has received the most attention is the same protein that was recently shown to be involved in the secretion of T_4 and T_3 from the thyrocyte, the monocarboxylate transporter, MCT8 (see Section IV.F). In humans, mutations in this protein result in neurological symptoms, indicating the importance of its presence in neural tissue in particular. In other target tissues the role of MCT8 and other possible thyroid hormone transporters is currently under investigation. Newer research indicates that there are several known thyroid hormone transporters: eight in the liver; six in the kidney; eight in the brain; two in the cochlea; one in the heart; one in bone; and five in skeletal muscle. MCT8 is one of the transporters in liver, kidney, brain, heart, and skeletal muscle. Members of the OAT (organic anion) family of transporters occur in liver, kidney, brain, cochlea, and skeletal muscle. MCT8 is the sole transporter, to date, in heart, while MCT10 is the sole transporter in bone. Other transporters, such as LAT1 and LAT2, appear in several tissues.

The nuclear receptor for thyroid hormones, TR, belongs to the nuclear receptor superfamily which in humans contains 48 members, including the receptors for the classical steroid hormones, vitamin D, and several nonsteroidal lipids (see Chapter 1: Hormone Introduction). TR affinity for T_3 is 10–15 times greater than for T_4, providing the basis for the designation of T_4 as an inactive prohormone from which is derived the active hormone, T_3.

There are several isoforms of TR, of which the four best studied are represented in Fig. 5.15. The isoforms are encoded by two genes, *THRA* (TRα1 and TRα2) on chromosome 17 and *THRB* (TRβ1 and TRβ2) on chromosome 3. The subtypes of each gene product are the result of alternative splicing during mRNA processing. TRα1 and the two TRβ isoforms have approximately the same affinity for T_3 and ability to activate gene transcription. TRα2 does not bind T_3 due to the splicing of an alternate exon at the carboxy terminus, inactivating the ligand-binding domain; the alteration at the C terminus of TRα2 also disrupts its dimerization properties, leading to changes in its ability to bind DNA.

TRα and TRβ isoforms are expressed differentially during development with TRα1 appearing in early stages and TRβ forms at later stages. The TR isoforms are expressed widely and in overlapping patterns in adult tissues, although neither appears to be expressed in the testis. TRα1 and TRα2 are present in relatively high concentrations in brain, skeletal and cardiac muscle, kidney, lungs, liver, and brown fat, whereas the TRβ2 isoform is predominantly expressed in the anterior pituitary gland, hypothalamus, retina, and the cochlea.

Studies of mice in which one or more of the isoforms has been removed by targeted gene deletion have resulted in complex phenotypes due, in part, to redundancy of the expression and function of the isoforms. It is likely that both TRβ isoforms participate in the regulation of TRH and TSH secretion by T_3 and in cochlear development. Deletions of TRα1 reveal that this isoform is important in cardiac and intestinal function, and in temperature adaptation to cold. When both *Thra* and *Thrb* genes are

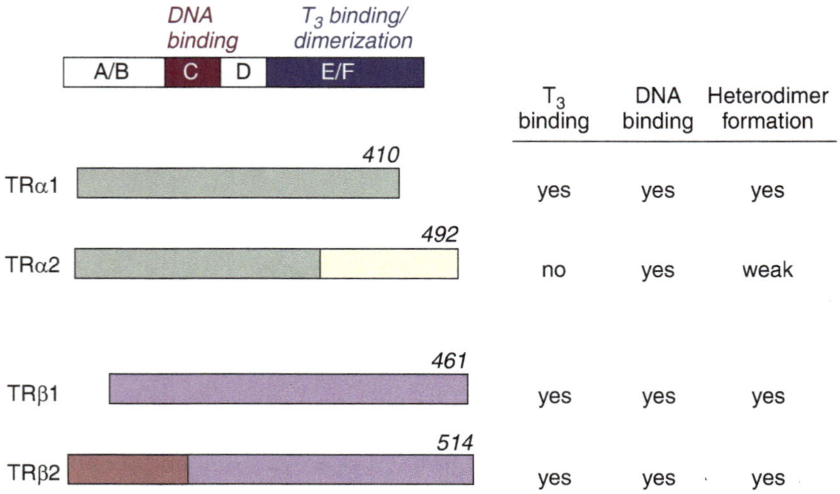

FIGURE 5.15 TR isoforms. At the top of the figure is shown the generic organization of the nuclear receptor family of proteins with the DNA-binding and ligand-binding domains indicated. Four TR isoforms are depicted next with differences in primary structure indicated by different colors. The genes *THRA* (chromosome 17) and *THRB* (chromosome 3) encode TRα and TRβ, respectively. The subtypes of each gene product are the result of alternative splicing during mRNA processing. The functional characteristics of each TR isoform are indicated on the right side of the figure. *mRNA*, Messenger RNA; *TR*, thyroid hormone receptor.

inactivated in mice, the phenotype differs from that in the complete absence of T_3, due to the activity of unliganded TR in repressing expression of certain genes (see next). In general, these studies show that TRα, TRβ1, and TRβ2 not only have specific functions but can also substitute for one another in some cases.

As described in Chapter 1, Hormone Introduction, the ability to bind to regulatory regions (elements) of target genes is a critical component of the action of all steroid nuclear receptors, a class which includes the thyroid hormone receptor even though the hormone itself is not a steroid. One hallmark of the thyroid receptor is that, even in the absence of ligand, the unoccupied receptor is associated with the nuclear chromatin; this is because the receptor is not anchored to cytoplasmic proteins and its nuclear localization signal is operative. Although the TR can bind to DNA as a homodimer, the TR–retinoid-X receptor (RXR) heterodimer is much more common and we will limit our discussion to this form of the receptor complex.

The unliganded thyroid receptor is functionally active and can repress the transcription of genes that are activated by T_3, as shown in the top portion of Fig. 5.16. In the absence of T_3 the unliganded TR–RXR heterodimer is shown binding to a *trans* regulatory element and corepressor proteins. This complex blocks transcription. When T_3 is bound, the heterodimer replaces corepressors with coactivators, allowing recruitment of additional transcription factors to the transcriptional start site and increased transcription. Similar molecular mechanisms operate with genes, expression of which is decreased by T_3.

B Membrane-mediated actions of thyroid hormone

The ability of physiological concentrations of thyroid hormone to bring about cellular changes within seconds to minutes, rather than the hours required for genomic effects, has been the subject of research for many years,

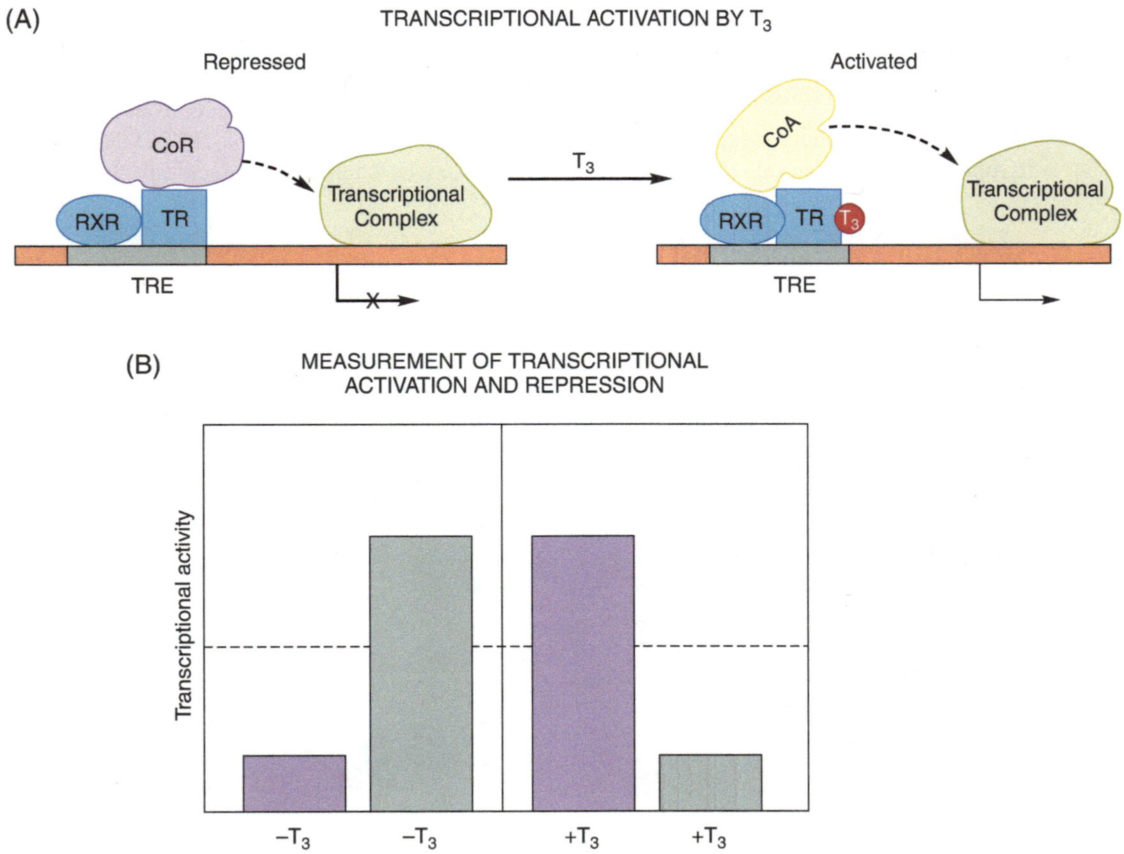

FIGURE 5.16 Transcriptional activation and repression by T_3. A simplified model for the switch from the repressed to the activated state of a gene, expression of which is stimulated by T_3. In the absence of the hormone, the unliganded receptor heterodimer (RXR-TR) binds to a specific TRE upstream of the transcriptional start site (*bold arrow*) and recruits one or more CoR, which inhibit the activity of the transcriptional complex. When T_3 is present, the hormone–receptor complex releases the corepressor(s) and recruits CoA, which stimulate the activity of the transcriptional complex. (B) Measurement of transcriptional activation (*purple bars*) shown in panel (A). The blue bars show the measurement of transcriptional activity of genes, expression of which is repressed by T_3. In this case the unliganded receptor heterodimer, bound to a negative TRE, recruits coactivators and, in the presence of T_3, recruits corepressors, diminishing transcriptional activity [not shown in (A)]. The dashed line represents a "basal state" in which intermediate amounts of T_3 are present. *CoA*, Coactivators; *CoR*, corepressor proteins; *RXR*, retinoid-X receptor; *TR*, thyroid hormone receptor; *TRE*, thyroid regulatory element.

and several high affinity binding sites for the hormone have been studied. Recently, attention has focused on integrin αvβ3, a structural protein of the plasma membrane which has a binding domain for iodothyronines and can initiate events leading to cell division and angiogenesis. Fig. 5.17 shows the functions of plasma membrane avβ3, signal transduction processes of which can lead to the enhancement of the nuclear form of the thyroid hormone T3 receptor and to the formation of mitochondrial ATP synthase which is responsible for respiratory control directly by T3. The activity of avβ3 also could modulate the activity of plasma membrane Na^+/K^+ ATPase as well as other ion pumps.

In other studies, T_3 activates PI3 kinase and mitogen-activated protein kinase pathways in the cytoplasm, affecting the phosphorylation and trafficking of specific proteins and leading in some cases to changes in the expression of certain genes. Rapid actions of thyroid hormone that are manifested at the plasma include those on glucose uptake, Ca^{2+}-ATPase, Na^+/H^+ antiporter, endocytosis, and the activity of the epidermal growth factor (see Chapter 17: Growth Factors) receptor.

C Basal metabolic rate and thermogenesis

The BMR is a measure of the energy expended by an organism at rest, under thermal neutral conditions, and after completion of the intestinal absorption of food. BMR is proportional to both the oxygen consumption and the surface area (body volume) of the organism. In warm-blooded animals the conversion of chemical energy to heat is required to maintain constant internal temperature in response to cold. This is known as adaptive thermogenesis. One of the earliest recognized hallmarks of thyroid hormone action was its ability to stimulate the metabolic rate by increasing oxygen consumption, particularly in the skeletal muscle, cardiac muscle, liver, gastrointestinal tissues, and kidney. Typical data obtained in the rat are shown in Fig. 5.18. In humans the BMR can vary as much as two- to threefold depending on thyroid hormone status. In fact, for the first half of the 20th century, until the development of the radioimmunoassay in the 1960s, BMR was the standard metric for determining thyroid status. Despite this long history, however, it has not yet been possible to identify a unifying explanation to describe the effects of thyroid hormone on BMR or on adaptive thermogenesis, that is, the increase in metabolic rate to maintain constant body temperature in response to environmental cold. It seems likely, however, that both the uncoupling of oxidative phosphorylation (increasing the amount of heat per ATP molecule) and increasing the utilization of ATP, such as through the stimulation of Na^+/K^+-ATPase activity or other ion channels, contribute to increased heat production. T_3

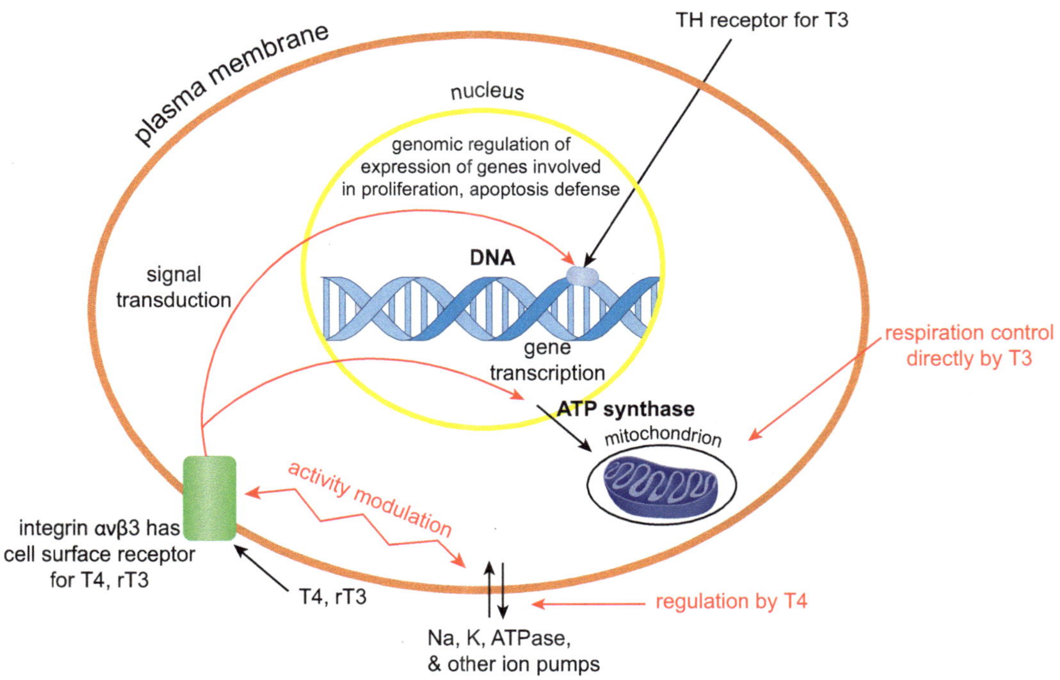

FIGURE 5.17 How the thyroid hormone acts depends on the types of cell or cell-like structure under consideration and that cancer cells, dividing endothelial cells, phagocytes, and platelets respond to the binding of L-thyroxine (T_4) by plasma membrane integrin avβ3. In intact tumor cells, T_4 at the integrin can modulate the transcription of a substantial number of specific genes relevant to cell proliferation and apoptosis. It is possible that in addition to T_4, reverse T_3 (rT_3) may have actions in cancer cells at the thyroid hormone receptor on avβ3. *Reproduced from P.J. Davis, How thyroid hormone works depends on cell type, receptor type, and hormone analogue: implications in cancer growth, Discovery Medicine (2019); https://www.discoverymedicine.com/Paul-J-Davis/2019/02thyroid-hormone-cell-type-receptor-type-and-hormone-analogue-cancer-growth/.*

has wide-ranging effects on intermediary metabolism, as discussed in the following, which also play a role in modulating BMR and mediating adaptive thermogenesis.

BAT is distinguished from white adipose tissue by the nature of its lipid stores, its much larger number of mitochondria, its richer supply of capillaries, and its synthesis of the uncoupling protein, UCP1. These characteristics render BAT, an excellent thermogenic tissue. It also contains a substantial amount of D2 (Type II iodothyronine deiodinase) enabling it to generate T_3 within its cells. BAT has long been recognized as the thermogenic target for T_3 in small mammals and young humans, but its presence and role in adult humans were questionable. Recently, however, the availability of imaging techniques has revealed physiologically important amounts of BAT in adult humans, suggesting that this tissue is the site of T_3-regulated adaptive thermogenesis.

D Metabolic actions

Table 5.5 summarizes some of the major effects of T_3 on the body's metabolic systems. As noted in the previous section, both BMR and adaptive thermogenesis are increased by T_3 and decreased in its absence. Several pathways involved in lipid and carbohydrate metabolism are affected by T_3 both directly and by altering tissue responses to other hormones such as insulin. In addition there is interaction between thyroid hormone and the nutrient signaling through such receptors as peroxisome proliferator–activated receptor (PPAR)α and PPARγ, and the transcription factors carbohydrate response element-binding protein and sterol response element-binding protein. Overall, an elevated level of thyroid hormone is associated with an overall accelerated metabolism, with

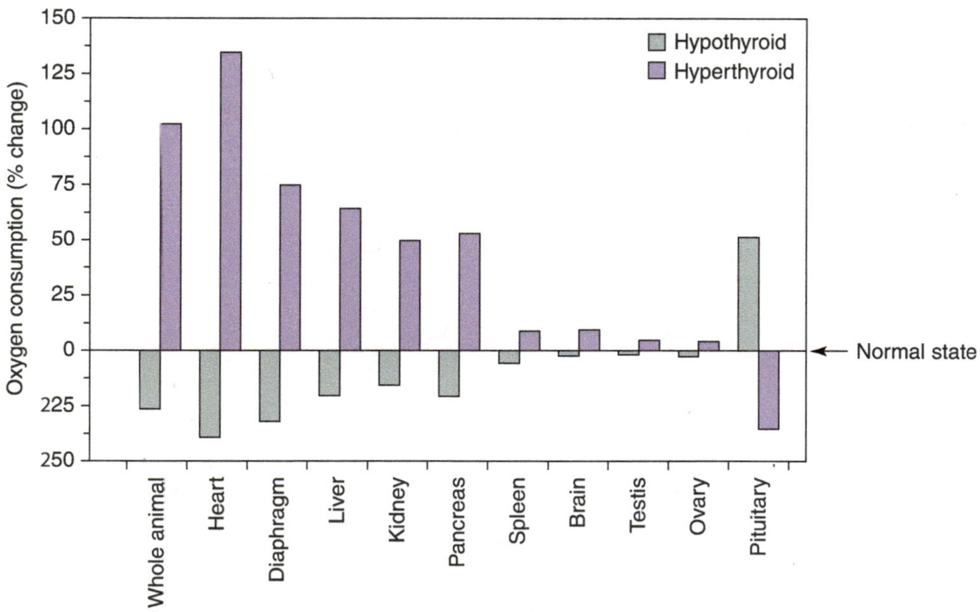

FIGURE 5.18 Thyroid hormone and O_2 consumption. Oxygen consumption in rat tissues was measured in thyroidectomized rats with (*purple*) and without (*green*) treatment with T_4 for 4–6 days. The basis of the paradoxical response of the pituitary has not been elucidated. *Modified from S.B. Barker, Physiological activity of thyroid hormones and analogues, in: R. Pitt-Rivers, W. R. Trotter (Eds.), The Thyroid Gland, Butterworths, London (1964), p. 200.*

TABLE 5.5 Effects of T_3 on metabolic processes.

Process	Effect of T_3	Tissues	Molecular target	Signaling interactions
BMR	↑	Heart, liver, muscle, GI, kidney	Na^+/K^+ ATPase	SNS
Adaptive thermogenesis	↑	BAT, WAT	UCP1, D2	SNS, BA
Fatty acid synthesis	↑	Liver, WAT	ChREBP FAS, ACC	PPAR-α,-γ, LXR, ChREBP, SNS
Fatty acid oxidation	↑	Liver, WAT	CPT	PPAR-α,-γ, LXR, SREBP
Cholesterol synthesis, excretion	↑	Liver, intestine	LDL-R, CYP7A1	PPAR-α,-γ, LXR, SREBP
Gluconeogenesis	↑	Liver, muscle	PEPCK	Glucose, insulin
Insulin secretion: sensitivity	↓	Pancreas	GLUT4	SREBP, LXR PPARα

ACC, Acetyl CoA-carboxylase; *BA*, bile acids; *BAT*, brown adipose tissue; *ChREBP*, carbohydrate response element-binding protein; *CPT*, carnitine palmitoyltransferase; *CYP7A1*, cholesterol 7α-hydroxylase; *D2*, deiodinase type 2; *FAS*, fatty acid synthase; *GI*, gastrointestinal; *GLUT4*, glucose transporter, type 4; *LDL-R*, low-density lipoprotein receptor; *LXR*, liver X receptor; *PEPCK*, phosphoenolpyruvate carboxykinase; *PPARα, PPARγ*, peroxisome proliferator–activated receptor α or γ; *SNS*, sympathetic nervous system; *SREBP*, sterol response element-binding protein; *UCP1*, uncoupling protein type 1; *WAT*, white adipose tissue.

net increased lipid and cholesterol breakdown; low thyroid hormone levels have the opposite metabolic effects.

E Neurodevelopment

Thyroid hormone is necessary for the normal growth and development of many tissues and organs, but nowhere are the effects of a deficiency in hormone availability more profound than in the nervous system. Fig. 5.19 shows the timeline of a number of important processes of neurodevelopment in the human fetus and infant. The fetal thyroid gland appears during the first trimester of pregnancy and hormone secretion is detected during the second trimester. Although the fetal hypothalamic–pituitary–thyroid axis is not active until the third trimester and achieves full function at birth, TRs are expressed in the brain throughout development, suggesting an important role for maternal hormone availability to the fetus. If the maternal supply of the hormone is compromised, the resulting cognitive and motor deficits of the fetus will depend on the timing of the deficit. An adequate exposure to hormones in infancy and early childhood is required for the normal brain development that continues to take place during this period.

VII Clinical aspects

A Clinical features of hyper- and hypothyroidism

As seen in Table 5.7, the clinical features of excess or deficient thyroid hormone are almost a mirror image of one another and reflect known actions of T_3 in various target tissues. Depending on the cause of the condition, both hyper- and hypothyroid individuals often present with an enlarged thyroid gland, known as goiter, due to overstimulation of the TSH receptor.

In instances of thyroid hormone excess, the subject has an accelerated BMR and an elevated cardiac output associated with high body temperature, warm skin, and sweating that is inappropriate for the ambient temperature. Because muscle mass constitutes 50% of body weight, and because muscle tissue has a high rate of oxygen consumption, the increased muscular metabolic activity makes a substantial contribution to the higher BMR. Although hyperthyroid individuals often increase their food intake, there is usually a concomitant weight loss related to the greater gastrointestinal activity and associated diarrhea. Such an individual is often hyperactive,

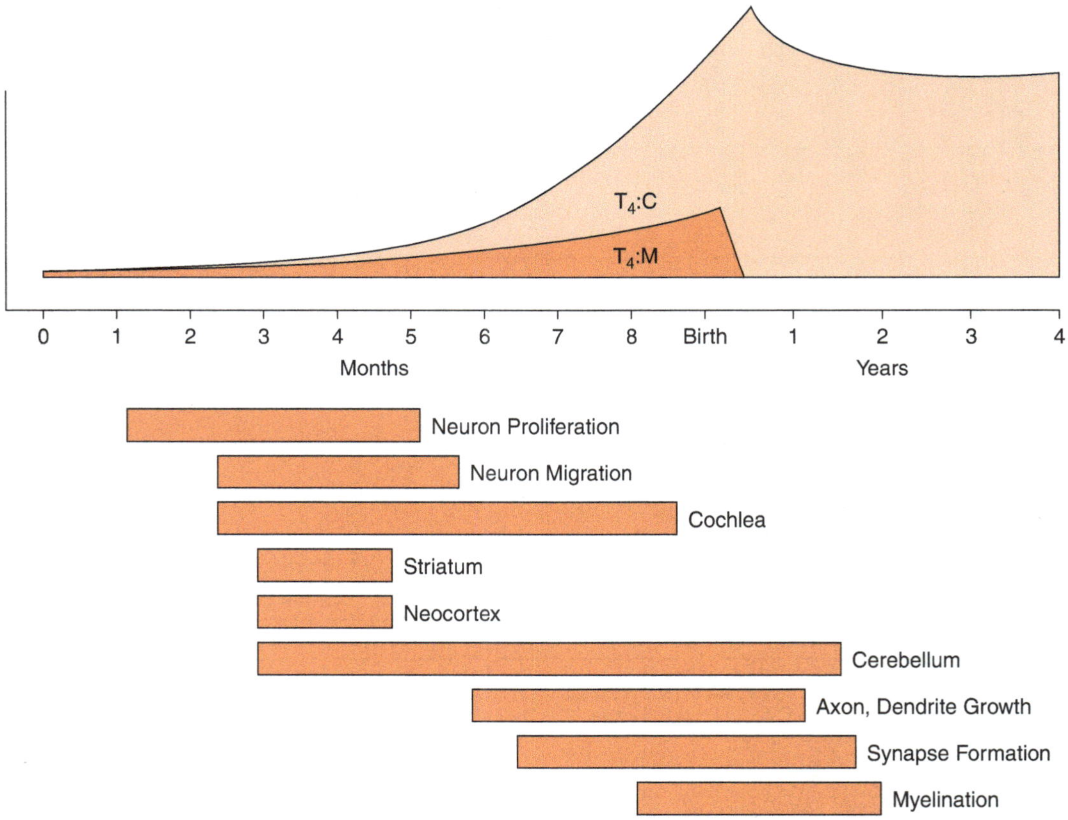

FIGURE 5.19 Thyroid hormone and the development of the nervous system in humans. The top portion of the figure shows the relative contribution by the M and C thyroid glands to the child's total thyroid hormone availability in utero and in early childhood. Below this graph, and using the same timeline, are some major neurodevelopmental events. Deficiency of thyroid hormone during development will have the most adverse effects on those events taking place during the time of deprivation, both before, and after birth. *C*, Fetal/child; *M*, maternal. *Adapted from Jameson and De Groot (Eds.), Endocrinology, sixth ed., 2010. p.1657.*

with rapid movements and exaggerated reflexes and frequently exhibits short attention span. Muscle myopathy occurs and is attributable to both abnormal protein catabolism with negative nitrogen balance and abnormal neuromuscular transmission.

In a hypothyroid subject, both BMR and cardiac output are reduced. Accordingly, the body temperature is lowered and the skin is cool and dry, since sweating is reduced in relation to the ambient temperature. The appetite of a hypothyroid individual is poor and there is reduced gastrointestinal activity. Although the skeletal muscles can be somewhat enlarged, there is no obvious myopathy. The principal external clinical features of hypothyroidism include a myxedemic appearance, described as an accumulation under the skin of mucoproteins and fluids resulting in a puffiness as a consequence of alterations in electrolyte and water balances. In addition to the features mentioned in Table 5.7, a deficit of thyroid hormone during development and childhood results in impaired skeletal growth. The effects of T_3 on the nervous system continue through adulthood so that the hypothyroid individual often shows signs of lethargy, and poor memory. In severe cases of long standing, a precipitating event, such as cold exposure, can result in myxedema coma.

B Hyperthyroidism: etiology and treatment

Thyrotoxicosis is the term used to describe the collection of symptoms that result from an excess of thyroid hormone, from any source. The disease that is responsible for 50%–80% of all cases of thyrotoxicosis was first described in 1825 by Parry but has come to be known in the English-speaking world as Graves' disease (GD) after the eminent 19th century physician, Robert Graves. Its prevalence in women is 1%–2%, about 5–10 times the occurrence in men. There is a strong heredity component to this polygenic disorder, with at least eight genes implicated in the propensity for its development. The genetic effects of Grave's disease are summarized in Table 5.6.

GD is an autoimmune disease in which circulating antibodies directed against the TSH receptor are produced. These antibodies stimulate the TSHR, activating both $G_s\alpha$ and G_q signaling cascades. Thus the thyroid gland responds as if under continual stimulation by TSH (Table 5.4). Since there is no feedback regulation of these antibodies, the result of their production is the unrestrained synthesis and secretion of T_4 (and some T_3) as well as growth of the thyroid tissue. The latter process leads to an enlarged thyroid gland, referred to as goiter.

The treatment of GD is focused on decreasing the high amounts of T_4 being secreted by the overstimulated gland. This can be achieved chemically by inhibiting one or more of the steps of T_4 synthesis and secretion with thionamides such as methimazole or propylthiouracil or iodide uptake by perchlorate, for example (see Fig. 5.12). An additional approach to treating hyperthyroidism is destruction of the excessive thyroid tissue through surgery or administration of radioactive iodide.

The occurrence of ophthalmopathy, the disorder associated with the eye orbital area, depends on the diagnostic procedure used and ranges from 10% to 25% (clinically evident) up to 70% (computed tomography scans) of individuals with GD. The symptoms can range from mild enlargement of the extraocular muscles to vision-threatening disfigurement, with fewer than 5% exhibiting these severe symptoms. While the antigen(s) involved in this part of the autoimmune response in GD has not yet been definitively identified, the TSH receptor is a likely candidate. TSHR is expressed by orbital adipocytes, fibroblasts, and lymphocytes. Ocular fibroblasts participate in the softening and enlargement of

TABLE 5.6 Genetic effects in GD.

Genetic effects are thought to account for up to 79% of the risk for development of GD, based on twin studies
HLA Class II antigens probably account for 20% of the genetic factor. (Specifically DR beta1*0301)
CTLA-4 association may account for up to 30%
CD 40 is a recently noted genetic factor and may (or may not) provide a risk similar to HLA or CTLA-4
Lymphoid tyrosine phosphatase—PTPN22—has also been associated with GD and other endo autoimmunity
TG and TSH-R genes are linked in whole genome screening
Many other genes are probably involved, and their contribution may vary with the population group

CTLA-4, Cytotoxic T-lymphocyte-associated antigen 4; *GD*, Graves' disease; *HLA*, human leukocyte antigen.
Source: Reproduced from L.J. DeGroot, Grave's disease and the manifestations of thyrotoxicosis, in: K.R. Feingold, B. Anawalt, A. Boyce et al. (Eds.), Endotext [Internet], MD Text.com, South Dartmouth (MA), 2015.

the muscles and fat surrounding the eyeball causing its forward protrusion or exopthalmia. However, there is still much that is unknown about the genetic and immunological relationship between GD and the ophthalmopathy associated with it.

There are other causes of hyperthyroidism besides the autoimmune disorder described here. They include mutation in the TSHR gene resulting in increased activity; a functioning thyroid adenoma (toxic adenoma); and a pituitary tumor secreting excess TSH.

C Hypothyroidism: etiology and treatment

Thyroid hormone deficiency can result in a wide variety of clinical and physiological disturbances in virtually every organ system (Table 5.7). The classic disease state is myxedema; the accumulation of mucinous edema results in facial periorbital and peripheral edema involving both the hands and feet. The periorbital edema contributes to the patient's haggard appearance and includes a droop of the upper eyelid. Table 5.8 lists the major causes of hypothyroidism.

TABLE 5.7 Some clinical features of disturbed thyroid hormone status.

System	Hyperthyroid	Hypothyroid
Cardiovascular heart rate	↑: Tachycardia	↓
Gastrointestinal appetite	↑	↓
Intestinal peristalsis	↑	↓
Weight	↑	(↑)
Skeletal muscle	Wasting: weakness	Slow contraction, relaxation: stiffness
Nervous system	↑: Hyperkinesia, irritability, insomnia, and emotional lability	↓: Slow reflexes, memory loss, and lethargy
Skin	Warm, moist	Cool, dry, myxedema
Basal metabolic rate	↑	↓

TABLE 5.8 Causes of hypothyroidism.

Dietary iodine insufficiency
Loss of functional thyroid tissue
Hashimoto's disease (chronic autoimmune thyroiditis)
Treatment with radioactive iodine
Thyroidectomy
Diminished TSH secretion
Low hypothalamic TRH secretion
Defect in TSH secretion from pituitary thyrotrophs
Inherited defects in thyroid hormone synthesis
Na^+/I^- symporter
Thyroglobulin
Thyroid peroxidase
Pendrin
Peripheral resistance to thyroid hormone
Defect in TR

TR, Thyroid hormone receptor; *TRH*, thyrotropin-releasing hormone; *TSH*, thyroid stimulating hormone.

In areas where dietary iodide is sufficient, the most common cause of hypothyroidism is Hashimoto's disease, an autoimmune thyroid disease in which the subject has circulating antibodies against intracellular antigens of the thyroid gland. Hashimoto's disease is characterized by lymphocytic infiltration of the thyroid gland with destruction of the thyroid follicles so that thyroid hormone production and secretion is decreased. The low circulating levels of T_4 and T_3 result in increased TSH secretion from the pituitary, leading to the growth of the thyroid tissue or goiter. The intracellular antigens are most commonly Tg and TPO. Hashimoto's patients may also have antibodies against TSHR and GD patients may have antibodies against TPO and Tg. Thus these two diseases can be considered a single autoimmune entity with Hashimoto's on one end and Graves' on the other end of a spectrum of conditions.

There are a variety of hereditary diseases in which there are defects in one or more of the many steps required for the biosynthesis of thyroid hormone. These include defects in (1) iodide uptake, (2) iodination of Tg, (3) Tg biosynthesis, (4) iodotyrosine dehalogenase biosynthesis, and (5) thyroid gland insensitivity to TSH. Perturbation of the hypothalamic–pituitary–thyroid axis may occur so that there is a deficiency or total lack of pituitary-derived TSH or TRH.

The goal in the treatment of a deficiency of T_4 (and, therefore, T_3) from any of these causes is the replacement of the missing hormone, which can be done quite readily with levothyroxine, the L isomer of the hormone which was first synthesized in 1927. With careful establishment and monitoring of the drug dose, treatment can be highly successful in ameliorating the symptoms of hormone deficiency.

Without question, in many sections of the world, the prime cause of hypothyroidism is dietary insufficiency of iodine. Growth of the thyroid gland is triggered by increased TSH secretion, which attempts to stimulate the iodide-trapping mechanisms and subsequent steps in the intrathyroidal metabolism of iodine. The morphological consequence is the development of a goiter. Endemic cretinism is the most severe outcome of IDD (iodine deficiency disorders). After birth, neither supplying I^- nor replacing T_4 has significant effect on the neurological and skeletal results of I^- deprivation during development.

Further reading

Proteins involved in thyroid hormone synthesis and secretion

G.L. Francis, Solute carrier proteins and their role in thyroid hormone synthesis, Clinical Thyroidology (2021). Available from: https://doi.org/10.1089/ct.2021;33.107-109.

Y.-Y. Liu, A. Milanesi, G.A. Brent, Thyroid hormones, in: G. Litwack (Ed.), Hormonal Signaling in Biology and Medicine, Academic Press, 2020, pp. 487–506.

Metabolism of thyroid hormones

A. Basolo, A. Matrone, R. Elisei, F. Santini, Effects of tyrosine kinase inhibitors on thyroid function and thyroid hormone metabolism, Seminars in Cancer Biology (2020). Available from: https://doi.org/10.1016/j.semcancer.2020.12.008.

M.A. Shahid, M.A. Ashraf, S. Sharma, Physiology, thyroid hormone, StatPearls [Internet], StatPearls Publishing, Treasure Island, FL, 2020. 2021 Jan-. PMID 29763182.

A.H. Van der Spek, E. Fliers, A. Boelen, The classic pathways of thyroid hormone metabolism, Molecular and Cellular Endocrinology 458 (2017) 29–38.

Regulation of thyroid hormone secretion

Y.-Y. Liu, A. Milanesi, G.A. Brent, Thyroid hormones, in: G. Litwack (Ed.), Hormonal Signaling in Biology and Medicine, Academic Press, 2020, pp. 487–506.

Biological actions of triiodothyronine

P.J. Davis, S.A. Mausa, H.-Y. Lin, Nongenomic actions of thyroid hormone: the integrin component, Physiological Reviews 101 (2020). Available from: https://doi.org/10.1152/physrev.00038.2019.

B. Dhole, S. Gupta, A. Kumar, Triiodothyronine stimulates steroid and VEGF production in murine Leydig cells via cAMP-PKA pathway, Andrologia 53 (2021). Available from: https://doi.org/10.1111/and.13972.

Clinical aspects

S. Baksi, A. Pradhan, Thyroid hormone: sex dependent role in nervous system regulation and disease, Biology of Sex Differences 12 (25) (2021) 2021. Available from: https://doi.org/10.1186/s13293-021-00367-2.

F. Subekti, et al., Serum TSH as a prediction of Graves' disease recurrence following antithyroid drug withdrawal: a systematic review, PLos One 16 (2021) e0245978. Available from: https://doi.org/10.1371/journal.pone.0245978.

Van der Spek, E. Fliers, A. Boelen, Thyroid hormone and deiodination in innate immune cells, Endocrinology 162 (2021) 1–15.

Chapter 6

Pancreatic hormones: insulin and glucagon

I Introduction

A Background information

A feature essential for life of higher vertebrates is their ability to maintain a relatively constant blood glucose concentration. In the higher animals, glucose is essential as an energy source for *all cells*. Although some cells can utilize alternate "fuel metabolites," such as amino acids or fatty acids, the brain and its neurons are dependent upon a continuous supply of blood-delivered glucose. Superimposed on the requirement for the maintenance of a constant blood glucose level are the perturbations in blood glucose that may occur naturally as a consequence of ongoing physiological and metabolic events. These include the following seven topics: (1) the efficiency of both the intestinal absorption of foodstuffs [e.g., carbohydrates, proteins, fat, and their concomitant systemic transport to storage depots (liver, muscle, and/or adipose tissue)]; (2) the extent of muscular activity; (3) the thermogenesis and response to environmental extremes of heat and cold; (4) the presence of starvation; (5) the presence of pregnancy and lactation; (6) the individual's age; and (7) the presence and severity of disease or injury states.

The endocrine gland largely responsible for the maintenance of blood glucose levels and the appropriate cellular uptake and exchange of "fuel metabolites" is the pancreas. The principal hormones secreted by the pancreas are insulin and glucagon; in addition, the pancreas also secretes somatostatin, pancreatic polypeptide (PP), and gastrin; see Table 6.1.

Insulin is a pluripotent hormone in that it has a wide sphere of influence; directly or indirectly it affects virtually every organ and tissue in the body. The main function of insulin is to stimulate anabolic reactions to increase the amount of stored carbohydrates, proteins, and fats; this process will have the metabolic consequence of producing a lowered blood glucose level and the storing of energy.

Glucagon can be thought of as an indirect antagonist of insulin. Glucagon stimulates catabolic reactions that ultimately lead to the breakdown of glycogen, proteins, and fat and result in an elevation in blood glucose levels. Thus the pancreas is continuously adjusting the relative amounts of glucagon and insulin secreted, in response to the continuous perturbations of blood glucose and other fuel metabolites occurring as a consequence of changes in the relative amounts of anabolism and catabolism in the various tissues.

B Regulation of blood glucose

The blood concentration of glucose normally lies within the range of 80–110 mg/100 mL (4.4–6.1 mM); see Table 6.2. Reduction in blood glucose levels below 45–55 mg/100 mL (~2.5 mM) for a continued interval of time will lead to an impairment of brain function, tremors, and convulsions due to the activation of the sympathetic nervous system, resulting ultimately in death. Conversely, a prolonged elevation of blood glucose above 130 mg/100 mL generates a state of hyperglycemia, due to a relative lack of insulin. This then leads to a devastating wasting of metabolic energy, osmotic diuresis, and metabolic acidosis. Because glucose is a "small molecule," all blood glucose is completely filtered into urine by the glomerulus of the kidney (see Fig. 15.3B/C). Under normal physiological circumstances, virtually all of this filtered glucose is reabsorbed and placed in the blood compartment. However, in circumstances of excessive hyperglycemia, this reabsorptive process is saturable and it can only accommodate a finite throughput of glucose. Thus a significant fraction of the filtered blood glucose will be lost in the urine. If there is not a balanced increase in the dietary intake of glucose, there will be, of necessity, a compensatory increased catabolism of the storage forms of glucose, named, glycogen. Over an extended interval of time, this can result in a significant loss of "stored" metabolic energy.

Hormones. DOI: https://doi.org/10.1016/B978-0-323-90262-5.00022-6

TABLE 6.1 Hormones secreted by the pancreas.

Pancreatic cell of origin[a]	Hormone	No. of amino acid residues	Function(s)
β (B)	Insulin	51	Reduction of blood glucose
α (A)	Glucagon	29	Elevation of blood glucose
β-cells	Amylin[b]	37	Slows gastric emptying and promotes satiety
Δ (D)	Somatostatin	14	Inhibits secretion of insulin, glucagon, and PP
(F)	PP	36	Stimulates the secretion of gastric acid (HCl) by the parietal cells of the stomach
(G)	Gastrins[c]	34, 17, and 14	Induces secretion of HCl by parietal cells adjacent to the antrum of the stomach

PP, Pancreatic polypeptide.
[a]*See Fig. 6.3A, cellular sources of pancreatic hormones.*
[b]*Amylin is structurally homologous to calcitonin gene–related peptide (CGRP); see Fig. 9.10.*
[c]*Gastrin is predominantly secreted by the antrum of the stomach, the duodenum, and the pancreas. The antrum is at the bottom of the stomach.*

TABLE 6.2 Effects of altered blood glucose levels on several constituents of the blood and urine.

Parameter	"Normal" state	Chronic elevated blood glucose level (diabetes)	Chronic lowered blood glucose level (starvation)
Blood insulin (nmoles/L)[a]	300–600	50–100[b]	30–60
Glucagon (pg/mL)	<50	200–300	150
Glucose (mg/100 mL)	80–110	>130	50–60
Free fatty acids (mM)	0.4–0.7	2.0	4.0
Ketones[c] (mM)	1–2	20	10–20
Urine			
Volume (mL/24 h)	600–2500	3000–3500	600–1200
Glucose (mg/100 mL)	10–20	500–5000	5–10

[a]*1 μ/mL insulin=6 nmoles/L [see Heinemann, L. Diabetes Care 33:33 (2010)].*
[b]*For type 2 or insulin-dependent diabetes.*
[c]*The ketones consist of β-hydroxybutyrate, acetoacetate, and acetone.*

In a normal (70 kg) man the 24-hour urine volume ranges from 600 to 2500 mL, with a glucose concentration of 10–20 mg/100 mL of urine; this results in a daily loss of 600–50,000 mg (6–50 g) of glucose in 1 day. However, in a typical untreated diabetic (without renal complications), the 24-hour urine volume may exceed 3000–3500 mL, with a glucose concentration of 500–5000 mg/100 mL of urine; see Table 6.2.

However, the adverse consequences of a relative lack of insulin (hyperglycemia) are not restricted to a derangement of carbohydrate metabolism. In the absence of insulin, triglycerides and fatty acids will be mobilized from adipose tissue and amino acids from muscle tissue (discussed in detail later). These substances then proceed to the liver, where the fatty acids and branched-chain amino acids are converted into ketone bodies. The ketone bodies include β-hydroxybutyrate, acetoacetate, and acetone. Eventually, this results in progressive ketonemia, and when the renal threshold for acetoacetate and β-hydroxybutyrate is exceeded, both substances will appear in the urine.

In addition to the dominant partnership of insulin and glucagon in maintaining glucose homeostasis, there is an extensive contribution of other physiological factors and hormones to the regulation of blood glucose and glucose homeostasis. These are summarized in Table 6.3.

C Nutritional and metabolic interrelationships

1 Introduction

Any detailed understanding of the integrated actions of glucagon and insulin to effect blood glucose homeostasis cannot be achieved by limiting the assessment of their actions to those on carbohydrate metabolism alone. Due to the ready metabolic interchanges that occur between carbohydrate, protein, and fat constituents, it is essential to have a clear understanding of the intermediary metabolism of all of these substances. In addition, some appreciation of the principles of dietary nutrition is required because many key enzymes of intermediary metabolism of higher animals are adaptively regulated to reflect the current dietary intake of carbohydrate, protein, and lipid. The general design is such that the ingested components are diverted to storage sites during periods of feeding and are later reutilized by the metabolic processes of glycogenolysis, gluconeogenesis, and ketogenesis during intervals of food deprivation.

2 Substrate stores

The three categories of food substances that contribute significantly to caloric intake are carbohydrates, proteins, and fat (as triglycerides). Table 6.4 summarizes the caloric contribution of these prime nutrients when they are metabolized completely to CO_2 and H_2O.

Table 6.5 lists for a normal (70 kg) man the magnitudes of body pools of carbohydrate, fat, and protein and their theoretical caloric yields (in kilocalories), if they were completely metabolized to CO_2 and water.

Virtually the complete body pool of protein is in a continuous state of flux. In the steady state, there is a balance between biosynthesis and degradation. The amino acid efflux from human muscle is 0.5–1.0 g/kg/day. The released amino acids are further catabolized principally by the liver. Physiological amounts of insulin are known to markedly reduce amino acid efflux from muscle. Although a normal 70-kg man has a total of 10 kg of protein, only ~60% of this is theoretically metabolically mobilizable; the remaining 40% represents structural proteins, for example, collagen. In reality, no more than 2 kg of body protein can be metabolically mobilized before there is marked muscular hypertrophy (large cell volume) and myopathy (weakness). In cases of extreme starvation, death ensues not from hypoglycemia but from the loss of respiratory muscle function, which usually leads to terminal pneumonia.

Fat is the body's principal form of stored energy; 76% of the calories derived from the substrate stores tabulated in Table 6.5 come from metabolic oxidation of triglycerides and free fatty acids. Fat or adipose tissue represents the most efficient body storage form of energy. This results not only from the higher caloric yield per gram mass of adipose tissue (see Table 6.4) but also from the fact that the biological "packing" of triglycerides is more condensed

TABLE 6.3 Factors contributing to glucose homeostasis.

Factors resulting in a reduction in blood glucose level	Factors resulting in an elevation in blood glucose level
Insulin	Glucagon actions in the liver to stimulate glycogenolysis
	Epinephrine stimulation of glycogenolysis
Glucose uptake by peripheral tissues	Cortisol stimulation of gluconeogenesis
Glucosuria	Insulin antagonists; growth hormone; cortisol
Exercise	Dietary intake of carbohydrates and proteins; mobilization from storage sites (glycogenolysis)
Stimulation of glucagon catabolism	Stimulation of insulin catabolism

TABLE 6.4 Caloric yield of food substances.

	Respiratory quotient[a]	Energy yield (kcal/g)[b]
Carbohydrates	1.00	4.0
Proteins	0.80	4.1–4.2
Fat (triglycerides)	0.70	9.0

[a]*Respiratory quotient is defined as the molar ratio of CO_2 released to O_2 consumed during the complete oxidation of a food substance to CO_2 and H_2O.*
[b]*The energy yield is expressed as the derived kilocalories that are generated from completely oxidizing 1.00 g of the indicated food component to CO_2 and H_2O. The energy yield is identical irrespective of whether the oxidation occurs in a living organism or in a bomb calorimeter. In the bomb calorimeter, which is submerged in a known volume of water, known amounts of carbohydrate, protein, or fat can be combusted. The energy is released as heat which can be experimentally measured by determining the water temperature before and after combustion.*

TABLE 6.5 Body pools of carbohydrate, fat, and protein in a normal (70 kg) man.

Class	Amount (g)	Caloric yield (kcal)	Total %
Carbohydrate			
Muscle glycogen	400	1600	
Liver glycogen	85	340	2%
Extracellular glucose	20	80	
Fat			
Muscle triglycerides	300	2700	
Adipose triglycerides	15,000	135,000	76%
Plasma free fatty acids	0.4	4	
Plasma triglycerides	4.0	40	
Protein			
Muscle	10,000	41,500	22%
Total		181,000	100%

Source: Data abstracted from N.B. Ruderman, et al. Fuel homeostasis and intermediary metabolism of carbohydrates, fat, and protein. In Becker, K.L. (ed.), Principles and Practice of Endocrinology and Metabolism. Philadelphia, PA: Lippincott, 1992, pp. 1054–1064.

than that of protein or glycogen due to the absence of water molecules in the triglyceride structures. It is the fat depots that allow humans to tolerate prolonged intervals of dietary calorie deprivation. The energy stored in the body fat depots is sufficient to permit the average individual to survive 2–3 months of total food deprivation. Lipolysis or hydrolysis of adipose tissue triglycerides leads to the release of glycerol and free fatty acids into the bloodstream. The free fatty acids are then further catabolized by both muscle and liver tissue.

Approximately 80% of the storage form of carbohydrate, namely, glycogen, is found in muscle; the glycogen concentration is 9–16 g/kg of wet muscle or a total of 350 g (see Table 6.5). The muscle glycogen levels are depleted rapidly during vigorous exercise but are reduced only slowly during prolonged fasting, because muscle, in contrast to the liver, does not contain the enzyme glucose-6-phosphatase, and because normally there is no hydrolysis of the phosphate esters of the various glycolytic pathway intermediates. As a consequence, it is not possible for significant amounts of glycogen (glucose) to leave the muscle. Thus muscle glycogen can only serve as a metabolic fuel in the cell in which it is localized.

The liver glycogen pool (~85 g) plays a key role in facilitating the metabolic adjustment by the body to varying energy requirements. The concentration of hepatic glycogen is 50 g/kg of wet tissue; it represents ~20% of the total body glycogen pool. Liver glycogen may be depleted gradually during periods of fasting. Within 10–12 hours of fasting, the hepatic glycogen is mobilized at a rate of 50 mg/min/kg of liver. In contrast, during 40 minutes of severe exercise, the liver glycogen pool may be depleted by 18–20 g, that is, glycogen is mobilized at a rate of 330 mg/min/kg wet weight. Replenishment of the hepatic levels of glucose occurs only relatively slowly; a period of 12–36 hours is required, depending on the diet and extent of physical activity.

The body pool of extracellular glucose is quite modest by comparison with the muscle and liver stores; only ~20 g of free glucose is present in the extracellular water and the intracellular pool of the liver. The principal function of the blood pool of glucose is to provide a continuous supply of glucose to all of the glucose-dependent tissues. In the basal state the brain is the chief glucose-consuming tissue. During a 24-hour interval the brain requires 125 g of glucose.

The exclusive source of replenishment of the blood glucose pool is the liver. Under usual circumstances, 70% of the hepatic output of glucose is derived from liver glycogenolysis and 30% from gluconeogenesis. During intervals of prolonged exercise (40 minutes), ~50% of the oxidative metabolism of skeletal muscle is derived from glucose uptake from blood.

Ketones or "ketone bodies" accumulate in the blood under normal circumstances, where they function to ensure the survival of the organism, and under pathological conditions, where they can cause coma and death. The ketone bodies are produced almost exclusively by the liver from fatty acids and other carbon fragments derived from amino acids. Acetone is removed from the body via the lungs. The metabolic role of the ketones varies depending upon the nutritional state. Their principal function is to serve under conditions of fasting as a substitute substrate for energy metabolism by muscle, heart, and particularly the brain. Although muscle and heart have the capacity for cellular uptake and oxidation of free fatty acids (derived from lipolysis of adipose tissue), the brain does not have this capacity. Thus under conditions of fasting when the liver glycogen and eventually blood glucose pools become depleted, the physiological significance of the ketone bodies assumes a far greater importance. Under conditions of adequate feeding, the blood levels of the ketones are quite low (see Table 6.2).

The liver is the source of glucose for all tissues of the body. The human liver is the principal site of storage of glucose in its polymeric form known as glycogen. A single molecule of glycogen, when broken down completely by the process of glycogenolysis, can provide 120,000 molecules of glucose. The free glucose is then released into the blood compartment for transport to the brain, muscle, and red and white blood cells. Fig. 6.1 summarizes the balance of glucose production and utilization in humans that ensues over a 24-hour interval after dietary intake and in the absence of excessive exercise. The values for glucose consumption represent the amounts in grams consumed per day by the brain (125 g), red and white blood cells (each 50 g), and muscle (50 g) while in the resting state. Thus in an average day, approximately 275 g of glycogen are broken down by the liver and released as glucose. When starvation extends beyond 12 hours, the hepatic glycogen stores begin to be depleted. This then activates the process of gluconeogenesis in the muscle that results in the release of amino acid fragments from the breakdown of muscle proteins. The amino acid fragments then are transferred via the blood compartment to the liver for conversion by the process of gluconeogenesis into glucose and storage as glycogen.

Maintenance of a constant blood glucose level is governed by the liver and the regulatory actions of the pancreatic hormones. In the absence of hepatic glucose production, the blood level of glucose would fall by 50% in 40–60 minutes.

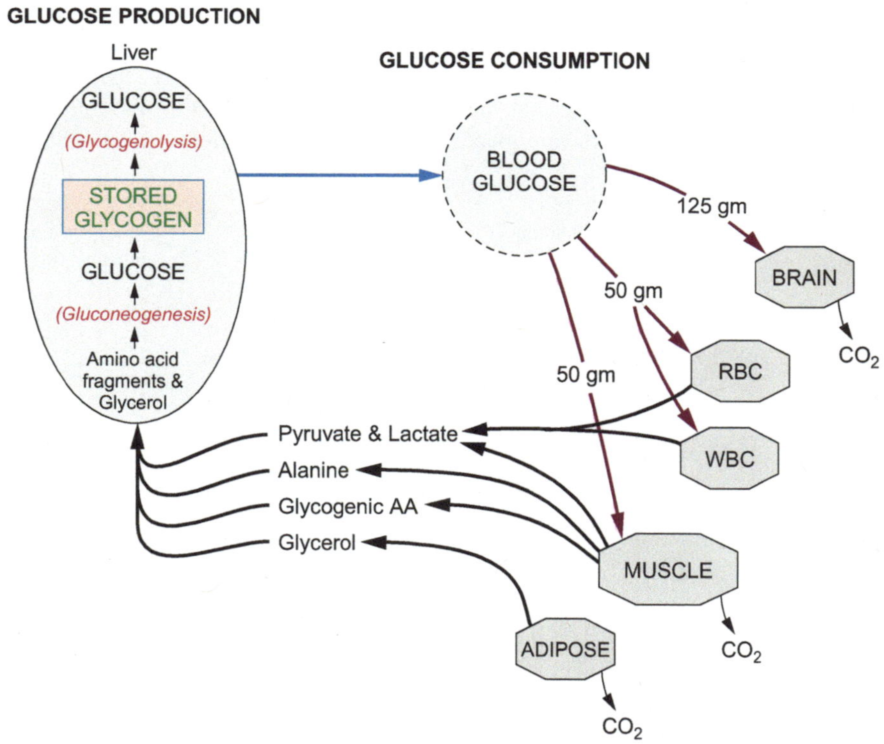

FIGURE 6.1 Glucose production by the liver and its transport as blood glucose to the brain, muscle, and red and white blood cells. This schematic diagram describes the production of glucose from amino acid fragments and its storage as glycogen by the liver. Then in times of need, as judged by a low blood glucose concentration, the liver's stored glycogen is mobilized and released as glucose to be transferred via the blood circulatory system to the brain, muscle, and red and white cells. The liver is the source of glucose for all tissues of the body. The glucose is provided by breaking down the stored hepatic glycogen by the process of glycogenolysis and releasing the glucose into the blood compartment. Gluconeogenesis is the formation of glucose, especially by the liver, from noncarbohydrate sources, such as amino acids supplied by muscle and the glycerol supplied by adipose tissue. *AA*, Amino acid; *RBC*, red blood cell; *WBC*, white blood cell. *Modified with permission from P. Felig, Starvation, in: L.J. DeGroot et al. (Ed.), Endocrinology, vol. 3, Grune & Stratton, New York, 1979, pp. 1927–1940.*

3 Dietary nutritional requirements

The daily caloric requirements can be met by adequate dietary intake of protein, fat, and carbohydrate. One current recommendation is that 15%, 30%, and 55% of the daily energy intake should be derived from protein, fat, and carbohydrate, respectively. Each food substance has its own characteristic energy yield (see Table 6.4), which reflects the relative extent of reduction of the carbon atoms of that food component.

In addition to kilocalories or kilojoules, the diet must supply, on a regular basis, adequate amounts of the various essential dietary constituents; these include water and fat-soluble vitamins, the essential unsaturated fatty acids, the essential amino acids, and both bulk and trace minerals.

II Anatomical, morphological, and physiological relationships

A Anatomy of the hepatopancreatic complex

The hepatopancreatic complex, along with the gallbladder, is responsible for the integrated digestion and subsequent processing of most dietary nutrients (Fig. 6.2). The pancreas contributes to both the digestive and endocrine systems of vertebrates. It is both an endocrine gland that produces several important hormones, including insulin, glucagon, and somatostatin (see Table 6.1), as well as an exocrine gland, secreting pancreatic digestive enzymes (chymotrypsin and trypsin, amylase and lipase, as well as bicarbonate) that pass to the small intestine. These enzymes help in the further breakdown of the protein, carbohydrates, and fat in the chyme.

Anatomically, the pancreas has distinct dorsal (rear side) and ventral (front side) lobes. The pancreas can be divided into lobules that (1) contain the exocrine acinar glands that contain pancreatic digestive enzymes and (2) the endocrine islets of Langerhans that contain insulin, glucagon, and somatostatin for secretion. The endocrine protein of the pancreas constitutes only 1%–2% of the weight of the gland.

Fig. 6.3 lists the major biological function(s) of the five hormones secreted by the pancreas. Also shown schematically is the proportion of the human pancreas islet cells dedicated to the secretion of insulin, glucagon, somatostatin, and the smaller amounts of PP and gastrin.

1 Cells of the pancreatic islets

In both humans and rats, pancreatic islets are composed of at least three major cell types: the α or A (glucagon-secreting) cells, the β or B (insulin-secreting) cells, and the δ or D cells (somatostatin secreting cells). By histological staining with neutral red it has been estimated that there are 13,500 or 890,000 islets, respectively, in the rat and human pancreas. The islets in the rat pancreas range in diameter from 50 to 400 μm. The distribution of cells in a typical rat islet is 15%–18% α, 75%–80% β, and

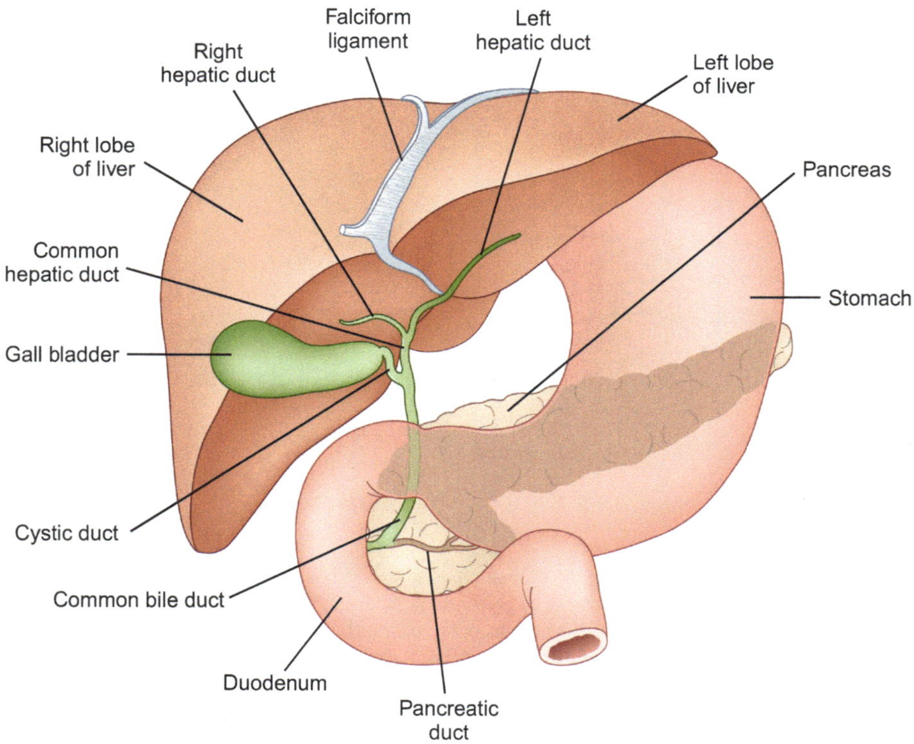

FIGURE 6.2 Gross anatomical features of a human pancreas, liver, biliary system, intestinal duodenum, and stomach.

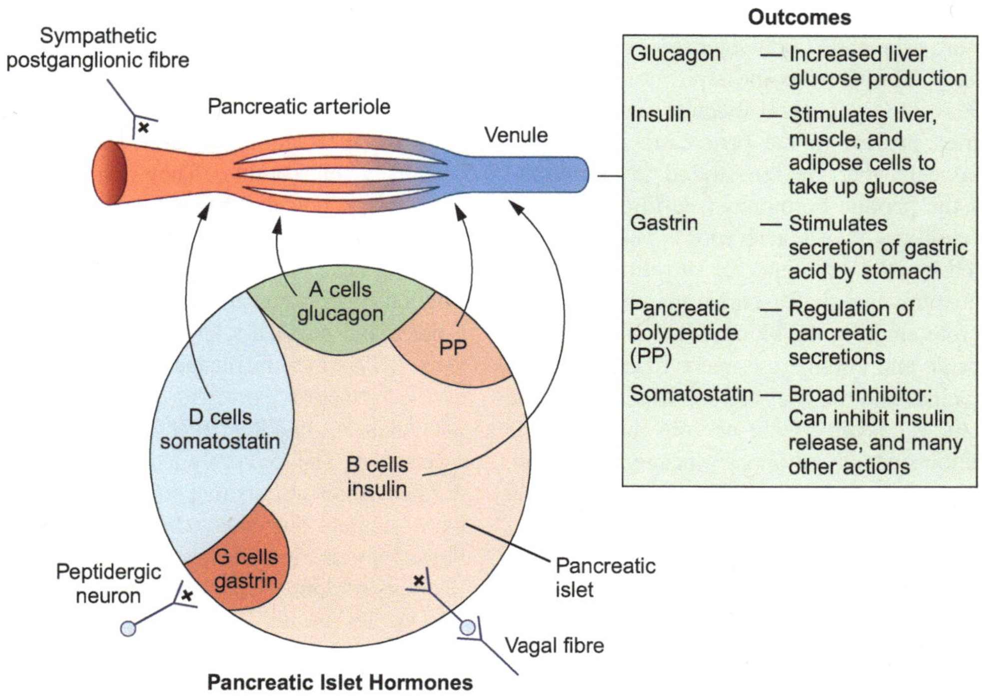

FIGURE 6.3 Schematic diagram of the cell types of a human pancreatic islet. (A) Illustration of the relative proportion of cells that secrete insulin (B cells), glucagon (A cells), somatostatin (D cells), PP cells, and G cells. The hormones are secreted and taken up into the pancreatic arteriole/capillary/venule system that ultimately delivers them to distal target organs (defined by the presence of their cognate receptor) where biological responses are then generated. The major biological functions of each hormone are listed in the green box. *G*, Gastrin; *PP*, pancreatic polypeptide.

2%–10% δ. The islets are surrounded by a basement membrane that encloses all three cell types (see Fig. 6.4).

The pancreatic β-cell also produces a neuropeptide-like molecule known as amylin or islet amyloid polypeptide (IAPP). Amylin is a 37 amino acid peptide with a sequence homology to calcitonin (see Chapter 9: Calcium-Regulating Hormones: Vitamin D, Parathyroid Hormone, Calcitonin; Fibroblast Growth Factor-23). It is cosecreted with insulin in a ratio of 1 amylin to 100 insulin molecules. It was first described as a major protein component of the amyloid deposits that are found in the islets of elderly type I diabetics. The physiological function of amylin is believed to be linked to satiety, which is the quality or state of being fed or gratified beyond capacity.

The pancreas of humans and 13 other species also secretes PP. The physiologic function of PP is to stimulate the secretion of gastric acid (HCl) by parietal cells of the stomach.

2 Vascularization and innervation of pancreatic islets

The arterial supply of the pancreas arises from the splenic, hepatic, and mesenteric arteries, and venous drainage is into the splenic and mesenteric veins. Each islet is normally vascularized by 1–3 arterioles, which abruptly terminate into capillaries, and 1–6 venules, depending upon the size of the islet (see Fig. 6.4).

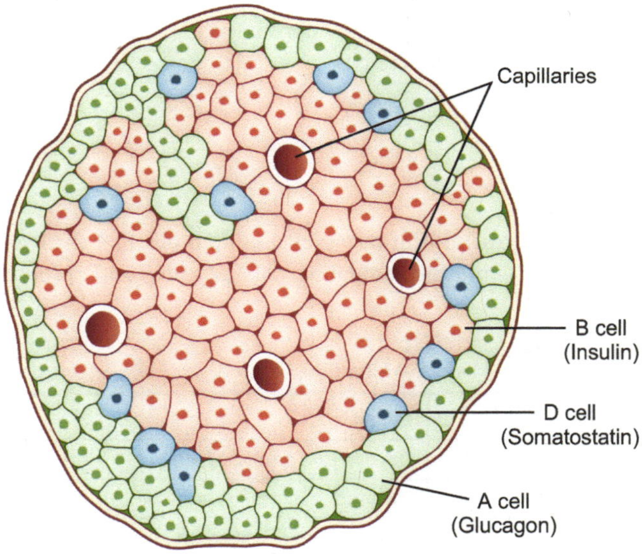

FIGURE 6.4 Schematic diagram of a human pancreatic islet with respect to the relative proportion of cells. The hormone secreting cells are, respectively, insulin (*pink*), glucagon (*light green*), and somatostatin (*blue*). Note that the A cells which secrete glucagon are basically all located on the perimeter of the islet, while the few cells that secrete somatostatin are located adjacent to the A cells. Also shown are the capillaries (*dark red*) that are evenly spread throughout the islet. The capillaries are connected to arterioles, which, in turn, are connected to venules (see Fig. 6.3A) which deliver the released insulin, glucagon, or somatostatin into the general circulatory system.

Morphological studies of the islets indicate that their endocrine cells are arranged so that each islet cell is adjacent to a capillary. This permits the rapid transfer of the secreted hormones into the general vascular system.

The capillaries present in the pancreatic islets comprise endothelial cells that are fenestrated to permit the rapid uptake of the peptide hormones (see Fig. 2.18A for an example of capillary wall fenestration). The hormonal products secreted by islet cells into the surrounding extracellular fluid must traverse the basement membrane of the endothelium before entering the bloodstream.

Both adrenergic and cholinergic nerve fibers are present in both the acinar and the islet regions of the pancreas. Stimulation of the parasympathetic nervous system leads to insulin secretion and inhibition of glucagon secretion, irrespective of whether the stimuli occur at the lateral hypothalamic nuclei, the motor nuclei of the vagus, or the mixed pancreatic nerves. The stimulation of the sympathetic nervous system or application of epinephrine likewise can stimulate glucagon production and inhibit insulin secretion. The hypothalamus appears to play the major integrating role in the balance between sympathetic and parasympathetic regulation of the islet.

III Chemistry, biochemistry, and biological activities of the pancreatic hormones

Table 6.1 lists the peptide hormones and other proteins produced by the pancreas.

A Insulin

1 History

The history of insulin occupies a unique place in the evolution of our understanding of peptide and protein hormones. Table 6.6 summarizes the notable series of "firsts" associated with the development of the chemistry and physiology of this important hormone.

Insulin was discovered in the canine pancreas in early 1921 by F. Banting, C. Best, and J. J. Macleod at the University of Toronto. They then tested their pancreas extract on a dog that had its pancreas removed. Normally, removal of the pancreas is fatal; but they were able to keep their dog alive for the entire summer through regular administration of their pancreatic extract. By early 1922, they had isolated and purified a large amount of the canine pancreas extract. This then facilitated the first administration of their purified pancreas extract to a 14 years old who was dying with diabetes mellitus (type I diabetes). The results were spectacular. This promptly led to the administration of their purified pancreatic extract to many other diabetic children with similar outcomes. In 1923 the Nobel Prize in Physiology or Medicine was awarded jointly to Frederick G. Banting and John J. Macleod "for the discovery of insulin."

Due to insulin's relatively modest size (its molecular weight is 5734 Da), it was an ideal molecule for the pioneering developments of peptide chemistry, amino acid sequencing techniques, determination of its three-dimensional structure via both X-ray crystallography and nuclear magnetic resonance (NMR) spectroscopy and, as well, recombinant DNA cloning of the insulin genome.

2 Insulin amino acid sequence

The primary amino acid sequence of human insulin is detailed in both panels of Figs. 6.5 and 6.6.

Human insulin is a molecule of 5734 Da which falls on the borderline between a large polypeptide and a small protein. It is composed of two separate peptide chains, designated the A-chain and the B-chain. These two chains are joined together by two disulfide bridges. The human insulin A-chain consists of 21 amino acid residues and the B-chain

TABLE 6.6 History of insulin.

Year	Event	Investigator(s)
1869	Discovery of pancreatic islets	P. Langerhans
1921	Discovery and isolation of insulin	F.G. Banting and J.J. Macleod[a]
1955	Determination of primary amino acid sequence of insulin	F. Sanger[b]
1964	X-ray crystallography used in structure determination of important biomolecules	D. Crowfoot Hodgkin[c]
1967	Discovery and structure determination of proinsulin	D.F. Steiner and R. Chance
1969	Determination of three-dimensional structure of insulin	D. Crowfoot Hodgkin
1979	Cloning of the insulin gene	W. Rutter and H. Goodman

[a] *Banting and Macleod jointly received Nobel Prize in Physiology or Medicine in1923 for the discovery of insulin.*
[b] *Received Nobel Prize in Chemistry in 1958.*
[c] *Received Nobel Prize in Chemistry in 1964 for the development of the technique of X-ray crystallography. This enabled the determination of the three dimensions of large molecules, most notably proteins.*

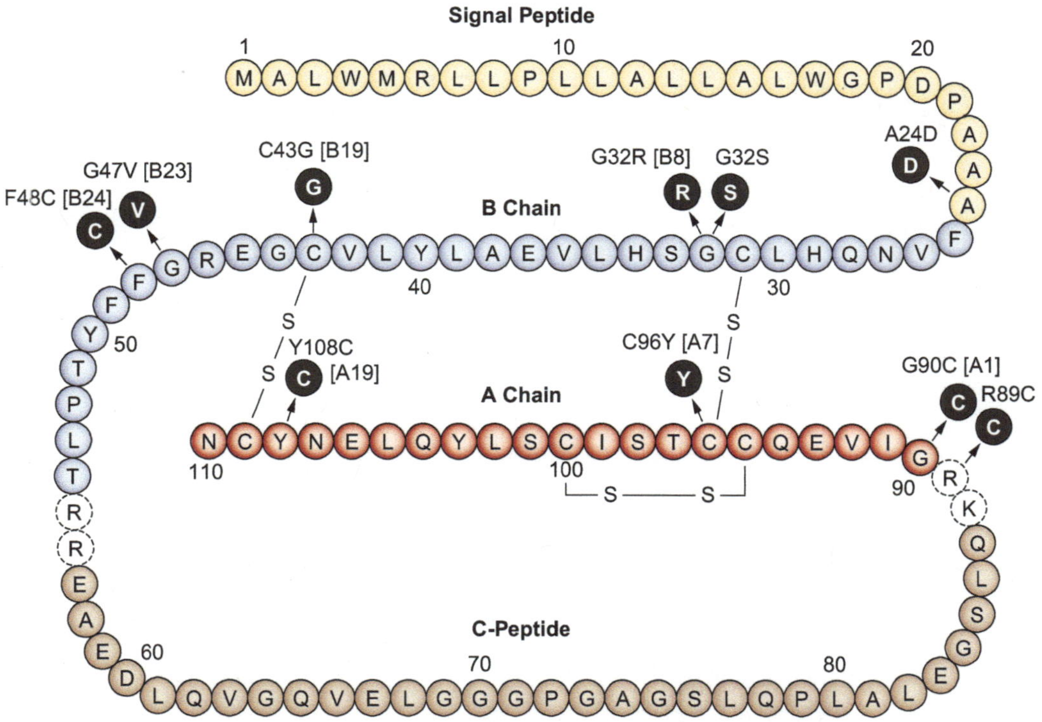

FIGURE 6.5 Amino acid sequence and structure of insulin and human preproinsulin. Amino acid sequence of preproinsulin. The preproinsulin has 110 amino acids. Dashed circles around either an R or K indicate sites of peptidase cleavage that result in the generation of the mature insulin that comprises two separate peptides: the A-chain with 21 and the B-chain with 30 amino acids. A third C-peptide with 21 amino acids served as a linker between the A- and B-chains. The solid black circles represent mutations that result in a disruption of the formation of disulfide bond formation and/or the proinsulin's normal folding; these mutations can lead to the onset of neonatal diabetes. There are also other single amino acid changes that lead to a variety of other insulin-dependent disorders (hyperinsulinemia, hyperproinsulinemia, MODY, and type 1b diabetes) that have clinical consequences. *MODY*, Mature onset diabetes of the young. *Modified with permission from P.E. Oyer, S. Cho, J.D. Peterson, D.F. Steiner, Studies on human proinsulin: isolation and amino acid sequence. Journal of Biological Chemistry 246 (5) (1971) 1375–1386.*

of 30 amino acid residues. A third disulfide exists in the A-chain between the cysteine 6 and 11 (see Fig. 6.6, panel A) or cysteine 95 and 100 (see Fig. 6.5). Note that the differences in numbering of the residues between Figs. 6.5 and 6.6 arise because in Fig. 6.5 both the signal peptide and C-peptide are included in the numbering scheme of the preproinsulin, but they are not included in Fig. 6.6, panel A.

Fig. 6.5 shows the amino acid sequence of the human preproinsulin with its 24 amino acid signal peptide attached (*yellow*). The amino acid residues shown in black circles are a sample of genetic mutations that disrupt either (1) the folding of the proinsulin protein or (2) disulfide bond formation. Either of these outcomes can lead to an appearance of the genetic disease known as permanent neonatal diabetes mellitus.

The primary amino acid sequence of some 72 vertebrate insulins has been determined. Amino acid substitution apparently can occur at many positions without the loss of biological activity; indeed, as many as 34 of the 51 (67%) positions can be replaced.

3 Conversion of preproinsulin into insulin

Fig. 6.7 summarizes the process of conversion of proinsulin, shown at the top of the figure, into mature functional insulin (*red* A-chain and *blue* B-chain connected by two disulfide bonds) shown at the bottom of the figure. Also shown at the bottom is the released C-peptide. There are two pathways from proinsulin to mature insulin. These involve two peptidase enzymes, PC1/3 and PC2; they each have different substrate preferences. Thus PC2 prefers as a substrate the proinsulin intermediate that has already been cleaved at the C-peptide/B-chain (see cleavage of the junction of the *brown* C-peptide-to-*blue* B-chain) with adjacent 31Arg/32Arg labeled as *DES-31,32 Proinsulin* (see Fig. 6.7, panel A ≫ panel B). The enzymatic rate of the PC2+CPE is significantly slower when the proinsulin (panel A) is the starting substrate; here, it effects the proteolysis of the adjacent 64Ly/65Arg labeled as *DES-64, 65 Proinsulin* (see panel A ≫ panel C). Thus the preferred route for the conversion of proinsulin (panel A) into mature insulin and C-peptide (panel D) is the route labeled by panels A ≫ B ≫ D which is significantly more rapid than the route depicted by panels A ≫ C ≫ D.

The C-peptide was first discovered in 1967 at the same time as the insulin biosynthesis pathway. Functionally the C-peptide has a crucial role on the preproinsulin (see Fig. 6.7, panel A) as a stabilizing linker

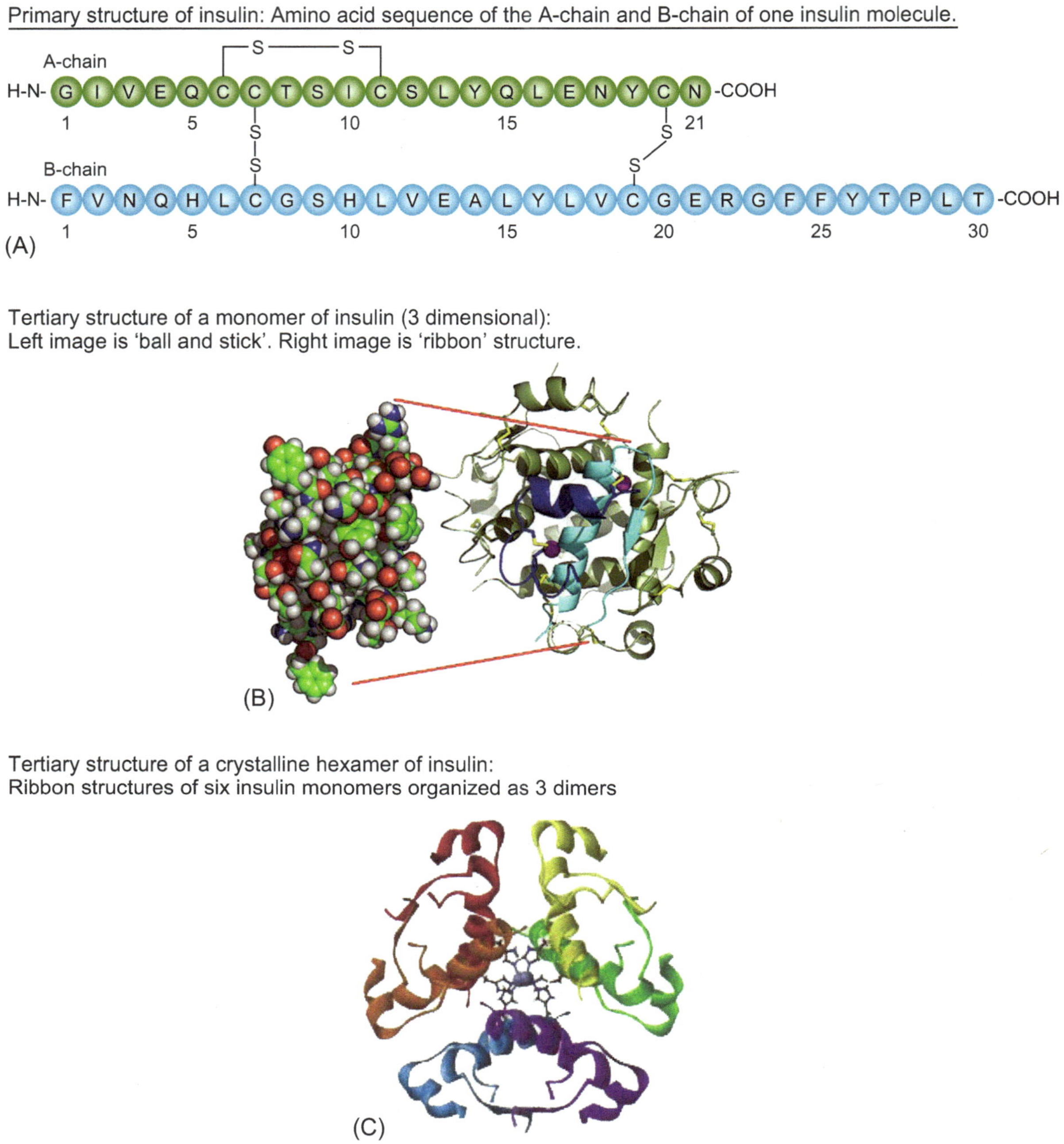

FIGURE 6.6 Three-dimensional orientation of the insulin structure. (A) The primary amino acid sequences of the 51 amino acids (21 from the A-chain and 30 from the B-chain) and 3 disulfide bonds that define insulin. The molecular weight of insulin is 5808 Da. (B) Three-dimensional structure of a monomer of insulin. The left image is a "ball and stick" CPK presentation and the right image is a "ribbon" structure. (C) Insulin is stored as a biologically inactive hexamer. The bottom panel displays the three-dimensional structure of a crystalline construct of six insulin molecules. The ribbon structures of the six insulin monomers are organized as three dimers that are stabilized by six histidines interacting with two zinc ions. *CPK*, Corey, Pauling, Koltun space-filling model. *Panels (B) and (C) were obtained from T.L. Blundell, N. Srinivasan, Symmetry, stability, and dynamics of multidomain and multicomponent protein systems, Proceedings of the National Academy of Sciences of the United States of America 93 (25) (1996) 14243–14428.*

between the B-chain and the A-chain as they undergo the process of efficient folding and processing of the emerging insulin in the endoplasmic reticulum. Not unexpectedly, both the mature insulin and C-peptide are released by the cell in a 1:1 ratio; they are then packaged together in secretory granules of the pancreatic β-cell. Ultimately, the mature insulin and C-peptide are released into the general circulatory system.

There is some evidence suggesting that the C-peptide has a favorable involvement with a number of systems in

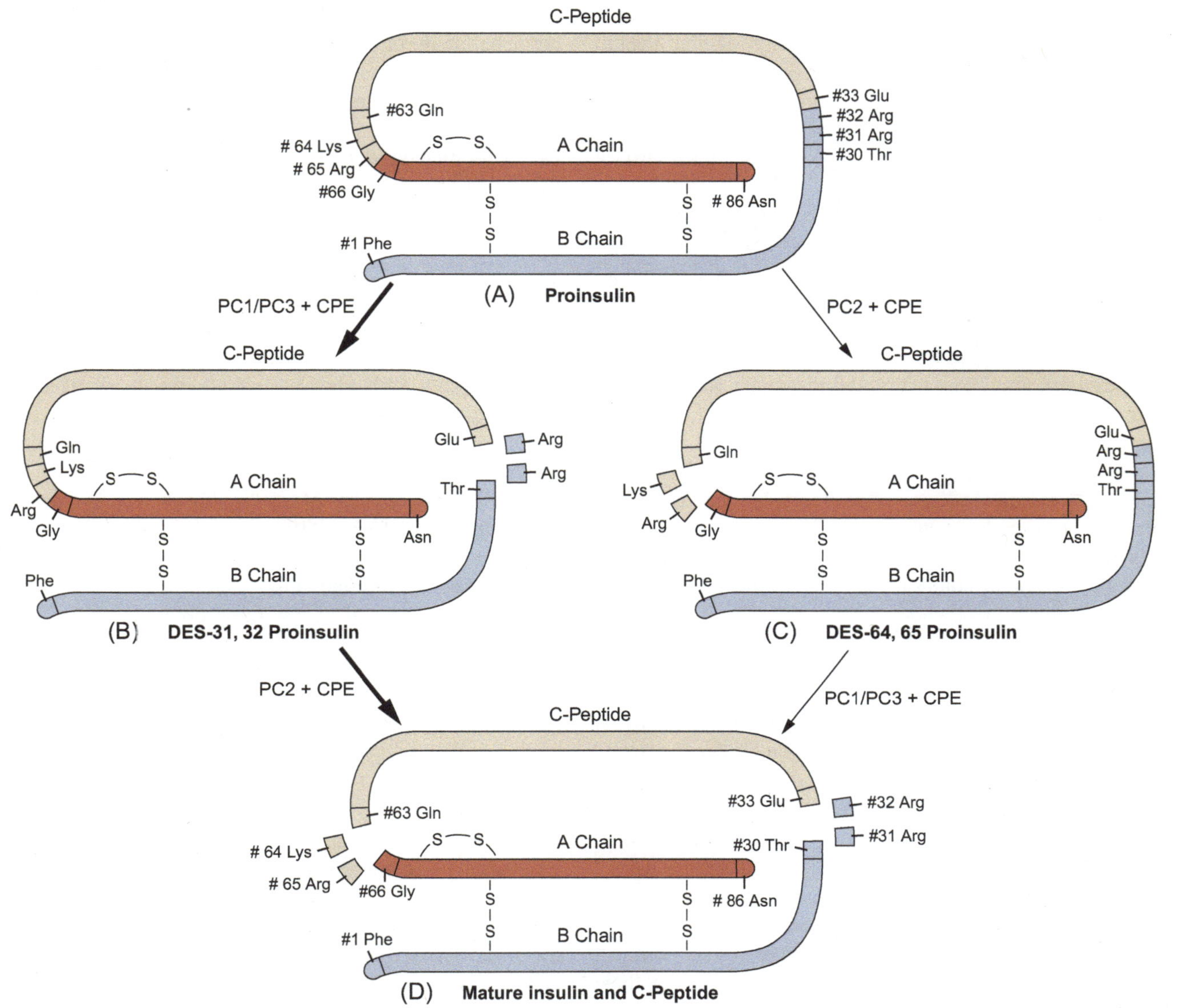

FIGURE 6.7 Steps of proteolytic cleavage of preproinsulin to generate both a mature insulin and a separate C-peptide. (A) Processing of proinsulin by specific endopeptidases, PC1/PC3+CPE (left side) and PC2+PCE (right side) which ultimately generates in (panel D) a mature insulin (a *red* A-chain linked to a *blue* B-chain by two disulfide bonds) and a separate brown C-peptide. The preferred order and role of the enzymatic actions of PC1/3 and PC2 have been determined. PC2 prefers as a substrate the proinsulin intermediate (panel B) that has already been cleaved at the C-peptide/B-chain with adjacent 31Arg/32Arg labeled as *DES-31, 32 Proinsulin* (see panel A ≫ panel B). The enzymatic rate of the PC2+CPE is significantly slower when the intact proinsulin is the starting substrate; here it effects the proteolysis of adjacent 64Lys/65Arg labeled as *DES-64, 65 Proinsulin* (see panel A ≫ panel C). Thus the preferred route, depicted by panels A ≫ B ≫ D, is more rapid (as marked by the *bold black arrows*) than the route depicted by panels A ≫ C ≫ D. The term *DES* indicates cleavage between two adjacent similar basic amino acid residues that define a precise cleavage site. For further details see Y. Rouille, S.J. Duguay, K. Lund et al., Proteolytic processing mechanisms in the biosynthesis of neuroendocrine peptides: the subtilisin-like proprotein convertases. Frontiers Neuroendocrinology 16 (1995) 322–361.

animal models of type 1 diabetes. It has been reported that the presence of C-peptide has favorable effects on kidney function and nerve function.

4 Insulin structure

From careful analysis of structure–function relationships of both naturally occurring and chemically modified insulins, it is apparent that certain structural features of insulin have been conserved throughout evolution. These include the following: (1) the precise positions of the three disulfide bonds; (2) the N- and C-terminal regions of the A-chain; and (3) the hydrophobic residues of the C-terminal region of the B-chain. From biological studies, it has been shown that insulin binds to receptors from tissues of differing species in a way that supports the view that there has been little or no evolutionary drift in the hormone-binding site specificity of the insulin receptor. The insulin receptor specifically recognizes on the insulin molecule an extensive region of ~11 residues around residues A-21 and B-23–B-27. Also insulin binding to its receptor is dependent upon its intact tertiary structure (see next).

Since the amino acid sequence of the evolutionary ancient hagfish insulin stabilized ~300 million years ago and differs from that of human at only 19 of 52 residues (37%), it can be calculated that amino acid substitutions have occurred in insulin at a rate of $\sim 1 \times 10^{-9}$ site/year. This is a typical figure for a highly conserved protein. These findings also imply that the insulin membrane receptor itself has not undergone significant changes throughout the course of vertebrate evolution.

D. Crowfoot Hodgkin, T. Blundell, and associates have determined via X-ray crystallographic techniques the three-dimensional structure of porcine insulin initially at 2.8-Å resolution and then at 1.5 Å. A highly similar structure of insulin in solution has been achieved via NMR spectroscopy.

Human insulin comprises 777 atoms [$C_{254}H_{377}$, $N_{65}O_{75}S_6$] with a molecular weight of 5808 Da. Insulin can exist as a monomer (see Fig. 6.6, middle panel); the left image is a "ball and stick" three-dimensional presentation that includes all of the atoms of the insulin monomer, while the right panel is a ribbon presentation of insulin which highlights the polypeptide backbone. In addition, insulin also can spontaneously organize itself as a dimer, tetramer, and hexamer of the monomer (see Fig. 6.6, bottom panel). The insulin dimers in the crystalline state are held together by hydrogen bonds between the peptide groups of residues B-24 (phenylalanine) and B-26 (tyrosine), which form an antiparallel pleated sheet structure (see one of the three dimer pairs of Fig. 6.6 bottom panel). The hexameric structure of crystalline insulin consists of three dimers ordered around a major threefold axis containing two zinc atoms that are each coordinated at the imidazole groups of three B-10 histidine residues (see Fig. 6.6 bottom panel). Indeed, the propensity for insulin to crystallize into hexamers may be related to the regular arrays of insulin that can be seen via electron microscopic evaluation of the storage granules of the pancreas β-cell.

The biosynthesis pathway for insulin generates monomers; the biologically active form of insulin is the monomeric species. However, monomeric insulin does readily spontaneously polymerize to form hexameric insulin, which is a functional long-term storage form of insulin.

While for many years it was thought that insulin was not a member of any hormone family, either in a structural sense (e.g., oxytocin and vasopressin are structural analogs of one another) or in a protein-processing sense (proopiomelanocortin produces β-lipotropin, adrenocorticotropin hormone, endorphins, etc.), there is now clear evidence that there is a hormone family of homologous growth factors that includes proteins with regions of amino acid sequences identical to that of insulin. These include the following: (1) relaxin, a polypeptide hormone from the *corpus luteum* that is responsible for the dilation of the *symphysis pubis* prior to parturition; (2) insulin-like growth factors I or II; and (3) nerve growth factor.

5 Insulin receptor

Insulin is the chief hormone controlling intermediary metabolism. It affects virtually every tissue in the body, but principally liver, muscle, and adipose tissue. Its short-term effects are to reduce blood glucose and to conserve body fuel supplies. There are also a number of effects of insulin on the regulation of gene transcription and cell replication. The number of insulin receptors in insulin target cells ranges from <100 to more than 200,000 per cell. The highest concentrations of the insulin receptor are found in hepatocytes and adipocytes.

The upper A panel of Fig. 6.8 presents a schematic of the general structure of a single insulin receptor comprising two α and two β subunits that are interlinked by three disulfide bonds, thus creating two ligand-binding domains for insulin on one copy of the receptor. The second half of the schematic of Fig. 6.8 provides a sample of the signal transduction pathways that can be initiated and fully activated in a cell by the binding of insulin to its receptor. In this model, four kinds of biological responses are diagramed: (1) vascular constriction; (2) vascular relaxation; (3) stimulation of glucose uptake followed by glycogen synthesis in the liver, muscle, and adipose tissues; and (4) translocation of glucose transporter 4 (GLUT4) to the cell's plasma membrane. Collectively, this provides an introductory overview of the complexity of the signal transduction events that are available in various cell types with the insulin receptor that can lead to the generation of a wide variety of various biological responses.

The insulin receptor is encoded by a single gene. When it is subjected to alternate splicing during transcription, it generates two long amino acid chains (each ~450 amino acids), IR-A or IR-B. The IR-A and IR-B are then each proteolytically cleaved to generate pairs of α and β insulin receptor subunits. Thus the intact insulin receptor is a heterotetrameric (α/β/α/β) trans-membrane glycoprotein of ~460 kDa with 906 amino acids. Also, a variety of amino acid substitution mutations have been identified in both the α and β subunits of the insulin receptor present in patients with severe insulin resistance. In Fig. 6.9 is shown a model of the insulin receptor in the absence of insulin or in the presence of the insulin molecule. Although the receptor is dimeric with two binding sites for insulin, the second binding site has reduced affinity

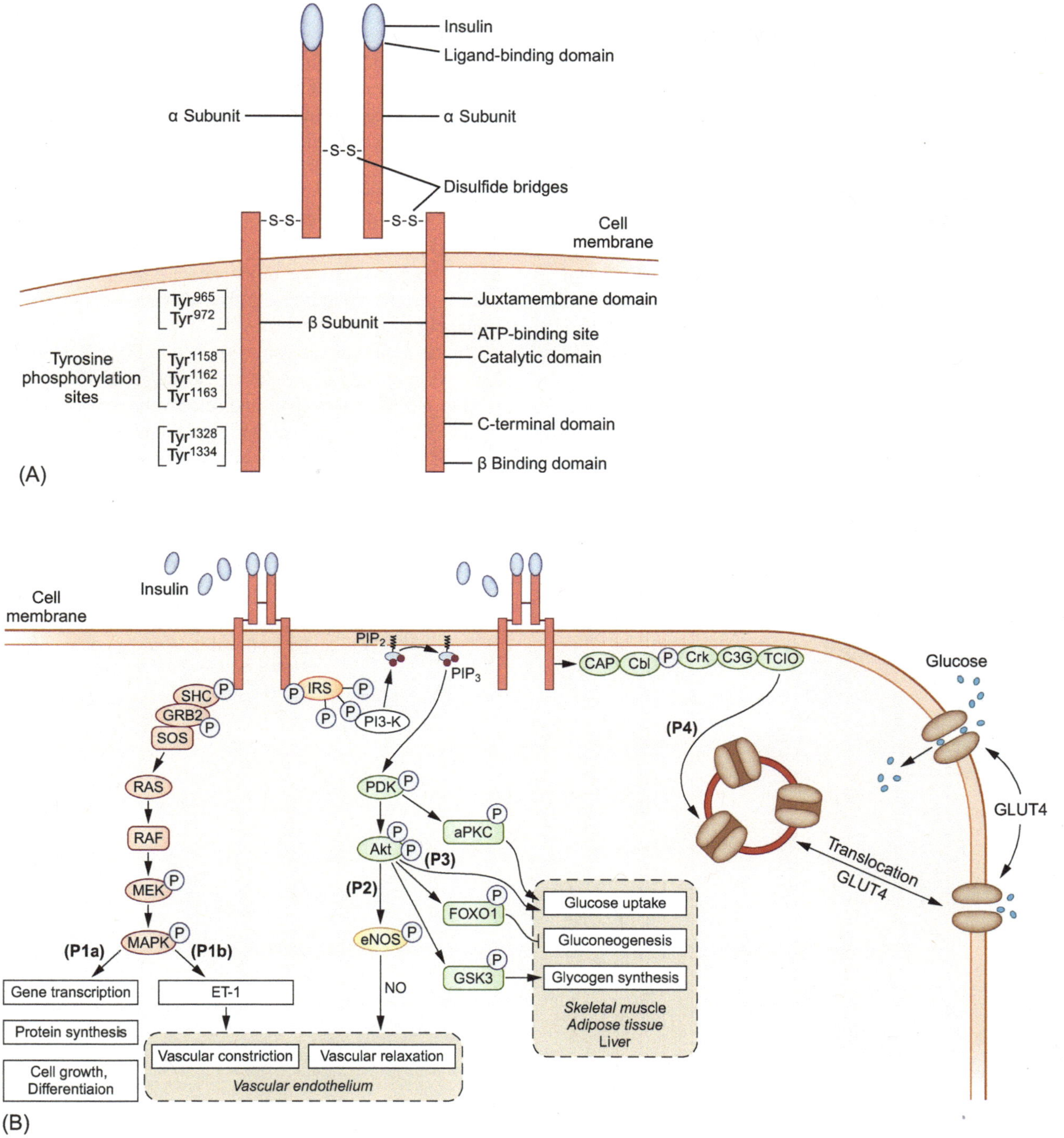

FIGURE 6.8 Schematic diagram illustrating the consequence of insulin binding to its dimeric insulin receptor that spans the cell's plasma membrane. Together the receptor with bound insulin generates an initiating "signal" that results (in this example) in the activation of four signal transduction pathways that generate a variety of biological responses in this composite generic cell. (A) Schematic model of the insulin receptor tetramer in the plasma membrane of a target cell. The two smaller extracellular insulin receptor subunits are stabilized by a disulfide bond. The two larger intracellular receptor subunits are each individually stabilized by a disulfide bond with the extracellular smaller insulin receptor subunit. The process of insulin signaling begins through the binding of two insulin molecules, one to each of the two α subunits of the receptor. This results in a conformational change in the α subunits which is detected by the two intracellular insulin receptor β subunits. (B) Summary of the insulin receptor's stimulation of signal transduction and exocytosis of the GLUT4 glucose transporter. The activated β subunits of the insulin receptor then continue the delivery of the signal transduction message so that one or more of the tyrosine kinases on the β subunit becomes activated. Then, in this composite cell, the activated tyrosine kinase will activate one of five signal transduction pathways designated (P1a, P1b, P2, P3, or P4). The ultimate biological outcome of each of the five pathways is described in panel B. It can range from one of the five following processes: (P1a) activation of gene transcription, and protein synthesis

(*Continued*)

for insulin due to the negative impact of the initial binding of insulin.

Fig. 6.8 presents a schematic of the insulin receptor. The intact insulin receptor is composed of two α subunits (~135 kDa) and two β subunits (95 kDa) that are joined together by three disulfide bonds. The two α subunits are located completely outside of the cell and they are linked to each other by a single disulfide bond. The two β subunits are 10% extracellular and 90% located on the interior surface of the plasma membrane. Each β subunit is linked to an α subunit by a disulfide linkage. As shown in Fig. 6.8, each α subunit appears to contain an insulin-binding domain; however, when the first site is occupied by its insulin ligand, it can induce negative cooperativity, which can reduce the affinity of the insulin ligand for the second insulin-binding site. Occupancy of the insulin receptor initiates a complex series of response cascades that can involve over 50 proteins—enzymes.

The two β subunits that span the plasma membrane both have 28 extracellular amino acids, 23 transmembrane amino acids, and an intracellular domain of 402 amino acids. The extracellular portion of each β subunit is attached via a disulfide bond to the bottom of a partner α receptor subunit. The consequence of insulin binding to its α receptor (*blue ovals*) *outside* the cell results in the activation of the catalytic activity of the β subunit's tyrosine kinase activity on the *inside* of the cell; note the presence of seven tyrosines on the left β subunit. Then one of the insulin receptor's β subunits selectively phosphorylates one of its partner β subunit's seven tyrosines, thereby initiating selective activation of a variety of signal transduction pathways.

In the composite diagram described in Fig. 6.8, physiological concentrations of insulin in the range of 100—500 pM bind to the insulin receptor and selectively activate one of a family of five insulin-dependent signal transduction pathways. Each of these signal transduction pathways mediates unique

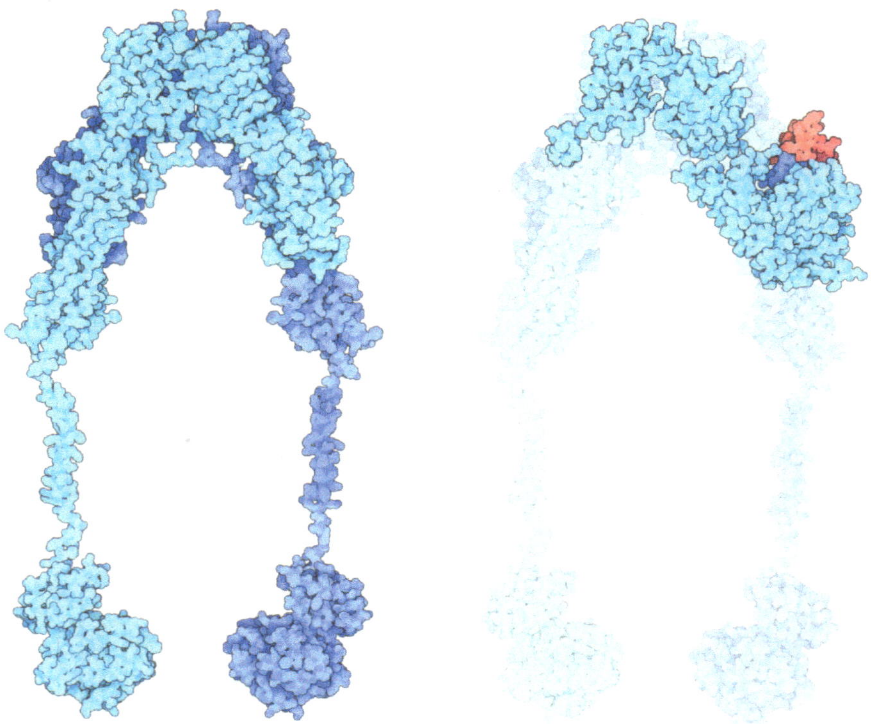

FIGURE 6.9 On the *left* is shown the insulin receptor with the extracellular portion at the top, the intracellular portion at the bottom and the cell membrane shown schematically in gray. On the *right* the extracellular portion of the insulin receptor (*blue*) is bound to insulin (*red*). *Reproduced from Molecule of the Month, https://pdb101.rcsb.org/motm/182.*

that can lead to cell growth and/or cell differentiation via the MAPK pathway; (P1b) vascular constriction of the endothelium via MAPK stimulating blood pressure; (P3) vascular relaxation of the endothelium via PI3-K and Akt stimulating the hormone production of nitric oxide (NO) which causes vascular relaxation; (P2) glucose uptake, activation of gluconeogenesis or glycogen synthesis in liver, skeletal muscle, and adipose tissues by stimulating PKC, FOX1, and GSK3; and (P4) CAP stimulating translocation of GLUT4 to the cell membrane thereby increasing the uptake of glucose into the host cell. Abbreviations for signal transduction proteins: (for the pathways, *P1a* and *P1b*) *SHC*, Src homology 2 domain containing transforming protein 1; *GRB2*, growth factor receptor-bound protein; *SOS*, son of seven less; *RAS*, rat sarcoma oncogene; *RAF*; *MEK*, mitogen-activated protein kinase kinase; and *MAPK*, mitogen-activated protein kinase. (For pathway *P2* greenish eNOS) endothelial nitrogen oxide synthase which secretes the hormone nitric oxide (NO). (For pathway *P3*, orangeish color) starting with *PI3-K*, phosphatidylinositide-dependent protein kinase-1; *PDK*, a constitutive membrane threonine kinase; *Akt*, protein kinase B; *aPKC*, a typical protein kinase; *FOXO1*, forkhead box—containing protein O; and *GSK3*, glycogen synthase kinase-3. (For pathway *P4*) *CAP*, Cbl-associated protein; *Cbl*, Cas-Br-M (murine) ecotropic retroviral transforming sequence; *Crk*, CT-10 factor; *C3G*, guanine nucleotide exchange factor C3G; *TC10*, small GTP binding protein.

biological responses appropriate for the host cell type. The insulin-modulated biological responses are as follows: the *P1a* pathway activates gene transcription that can contribute to growth, mitogenesis, and/or differentiation; in the *P1b* pathway, present in vascular endothelium cells, the secretion of the hormone endothelin-1 (ET-1) is stimulated which causes vascular constriction; the *P2* pathway stimulates the secretion of the hormone nitric oxide (NO) which mediates vascular relaxation; the *P3* pathway is operative in skeletal muscle, adipose tissue, and liver cells where glucose uptake, gluconeogenesis, and glycogen synthesis are stimulated; and the *P4* pathway regulates the uptake of glucose into skeletal muscle, adipose, and cardiac muscle cells.

General features of the insulin signal transduction pathways are described in this hypothetical generic cell (see Fig. 6.8). On the bottom of both of the insulin receptors' β subunit, note the presence of a single phosphorylated tyrosine. In some insulin receptor–containing cells, there can be as many as five phosphorylated tyrosines on a single β subunit. They function as a *switch* that can selectively stimulate (in this model) a menu of five different available signal transduction pathways. (*P1a*) This mitogen-activated protein kinase (MAPK) pathway starts at the bottom left of the phosphorylated β subunit of insulin, which is also covalently linked to the signaling molecule SHC, followed by six additional signaling molecules ending with MAPK. The MAPK can mediate in many cell types the activation and control of general gene transcription, followed by protein synthesis and modulation of cell growth. (*P1b*) When the MAPK signaling pathway is present in the vascular endothelium, its activation can stimulate the secretion of the peptide hormone ET-1. (*P2/P3*) These two signaling pathways start at the bottom of the right-hand β subunit of insulin where the IRS (insulin receptor substrate) is linked via phosphorylation to the insulin receptor and by an additional phosphorylation to the P13-K signaling molecule that leads eventually to the separate activation, depending on the cell type, of either the *P2* or *P3* pathways. The *P2* pathway activates endothelial NO synthase that stimulates the secretion of the only hormone that is a gas named nitric oxide (NO), which then causes vasodilation of the vascular endothelium. The *P3* pathway leads to the stimulation of aPKC, FOX01, or GSK3 in skeletal muscle, adipose tissue, and liver, respectively. The result is stimulation of glucose uptake and an increase in gluconeogenesis and glycogen synthesis.

6 Insulin and its glucose transporter

The stimulation of glucose uptake into selected target cells is the historical standard insulin response. Glucose influx into skeletal muscle, adipose, and cardiac muscle is stimulated by insulin binding to its receptor. When insulin binds to its receptor, it can result in the activation of pathway *P4* (see Fig. 6.8). Thus the six-member signaling pathway starting with Cbl-associated protein and ending with TCIO (small GTP binding protein) stimulates translocation of GLUT4 which is stored in intracellular membranous vesicles in the cell's cytosol to the cell's plasma membrane via an exocytosis process. The final outcome on completion of the exocytosis of the GLUT4 is its functioning as a sodium-independent, facilitated-diffusion GLUT that stimulates the uptake of extracellular glucose and delivers it into the cytosol of the host cell. The principal participant in this pathway is GLUT4, which is 1 of 12 sugar transporter proteins; see Table 6.7. The GLUTs are divided into three classes. Class I members are GLUTs 1–4 and each one is a GLUT. Class II GLUTs are fructose transporters and Class III GLUTs are structurally atypical members of the GLUT family and their functions are not yet well described.

All GLUTs contain 12 membrane-spanning helices with both the carboxyl and amino terminal ends facing toward the cytosol of the cell (Fig. 6.10). GLUT4 can be considered to be an integral membrane protein; it is present in skeletal, cardiac muscle, and adipose tissue. A major effect of insulin acting through its membrane receptor is to stimulate the translocation of GLUT4-containing vesicles in the intracellular cytosol to migrate to the plasma membrane and by the process of exocytosis. Thus the GLUT4 becomes a functional gated channel in the cell membrane. In Fig. 6.11 is shown an open outward conformation of GLUT 3 on the *left*. On the *right*, GLUT 1 is shown in an open inward conformation.

Fig. 6.12 illustrates over a 30-minute interval the translocation of GLUT4 from intracellular storage locations to the plasma membrane using a GLUT4-enhanced green fluorescent protein that was transfected into adipocytes. The translocation of GLUT4 is very dynamic. It has been demonstrated to occur within 30 seconds of the arrival of insulin. Then as the stimulus of glucose absorption dissipates, the decrease in the number of plasma membrane GLUT4 receptors declines. This illustrates the dynamic regulatory properties of the GLUT4 that are required to provide an appropriate number of plasma membrane–associated GLUT4 channels to meet the task of the glucose uptake process. The GLUT proteins have only a single substrate binding site (for glucose) which faces alternatively toward the "outside" or "inside" of the cell. Binding of glucose to one site provokes a conformational change associated with transport, so that glucose is released on the other side of the membrane. Thus GLUT4 is functionally able to both import and export glucose. The glucose-binding sites of GLUT4 are believed to be associated with the transmembrane loops 6, 7, and 8; loop 12 is closest to the GLUT's COOH group.

Fig. 6.13 presents a three-dimensional right-handed α-helix barrel model of GLUT4 that has 12 transmembrane α-helix segments surrounding a central pore. The

TABLE 6.7 Facilitative glucose transporter family: summary of properties.

Protein	No. of amino acids	km (mM)[a]	Tissue and cell locations	Proposed functions
GLUT1	492	3–7	Ubiquitous distribution in cells	Basal glucose uptake
GLUT2	524	17	Liver, β-cells, kidney, small intestine	High capacity, low affinity
GLUT3	496	1.4	Brain and nerve cells	Neuronal transport
GLUT4	509	6.6	Muscle, fat, heart	Insulin-regulated transport in muscle and fat
GLUT5	501	NA[b]	Intestine, kidney, testis	Transport of fructose
GLUT6	507	NA[b]	Spleen, leukocytes, brain	NA[b]
GLUT7	524	0.3	Small intestine, colon, testis	Transport of fructose
GLUT8	477	2	Testis, brain, muscle adipocytes, blastocyst	Fuel supply to spermatozoa insulin-response transport in blastocyst
GLUT9	540	NA	Liver, kidney	Transport of fructose
GLUT10	541	0.3	Liver, pancreas	NA[b]
GLUT11	496	NA	Heart, muscle	Muscle-specific fructose transporter
GLUT12	629	NA	Heart, prostate, mammary gland	NA[b]

Class I GLUTs (1–4) are glucose transporters; Class II GLUTs (5, 7, 9, and 11) are fructose transporters, and Class III GLUTs (6, 8, 10, and 12) are structurally atypical members of the GLUTs family and are not well described.
[a]*Relative net influx of glucose or 2-deoxyglucose.*
[b]*Not available.*
Source: Data were abstracted from F.Q. Zhao, A.F. Keating, Functional properties and genomics of glucose transporters, Current Genomics 8 (2007) 113–128.

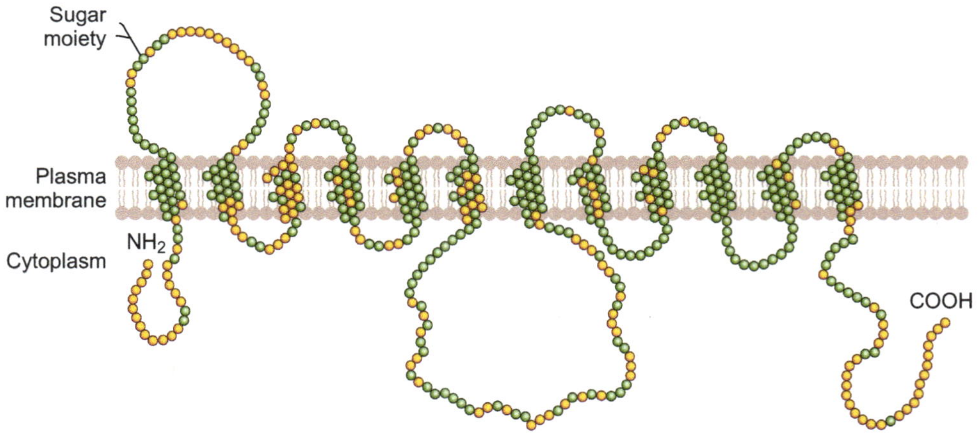

FIGURE 6.10 Schematic presentation of the GLUT4 protein in the cell membrane. GLUT4 is a member of the facilitative glucose transporter family. In each GLUT family member, the peptide chain spans the plasma membrane 12 times; both the amino and carboxyl termini are helices facing toward the cytosol of the cell. The sequence of ~500 amino acids are shown as green or orange spheres that in this example make up the Class 1 amino acids (see Table 6.7); each sphere represents one amino acid residue. A homology plot was made for the entire amino acid sequence for GLUT1 versus GLUT4. The orange spheres indicate residues that are unique to GLUT4, whereas the green spheres indicate amino acid residues that are not conserved for both GLUT1 and GLUT4 proteins. It requires about 20 amino acid residues to span the membrane. Of the 12 sequences that span the membrane, only span 10 which is the third span from the COOH terminus has no conserved residues, that is, in span 10 all the amino acid residues are colored green. Spans 1 (near the NH2 terminus), 7, and 12 each only have 1–2 out of 20 residues conserved (both were *orange* color) and the conserved residues were never in the interior of the membrane. *GLUT*, Glucose transporter. *Modified with permission from N.J. Bryan, R. Govers, D.E. James, Regulated transport of the glucose transporter GLUT4, Nature Reviews Molecular Cell Biology 3 (2002) 267–277.*

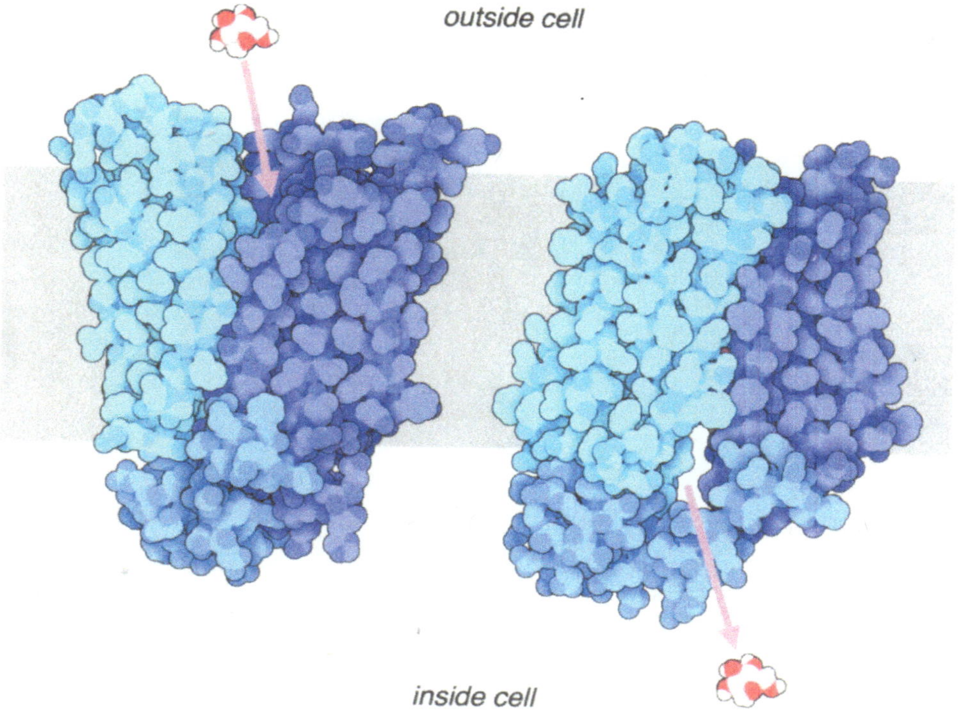

FIGURE 6.11 GLUT3 is shown on the *left* in an open outward conformation and GLUT1 is shown on the *right* in an open inward conformation. The small molecule is glucose. *GLUT*, Glucose transporter. *Reproduced from https://pdb101.rcsb.org/motm/208.*

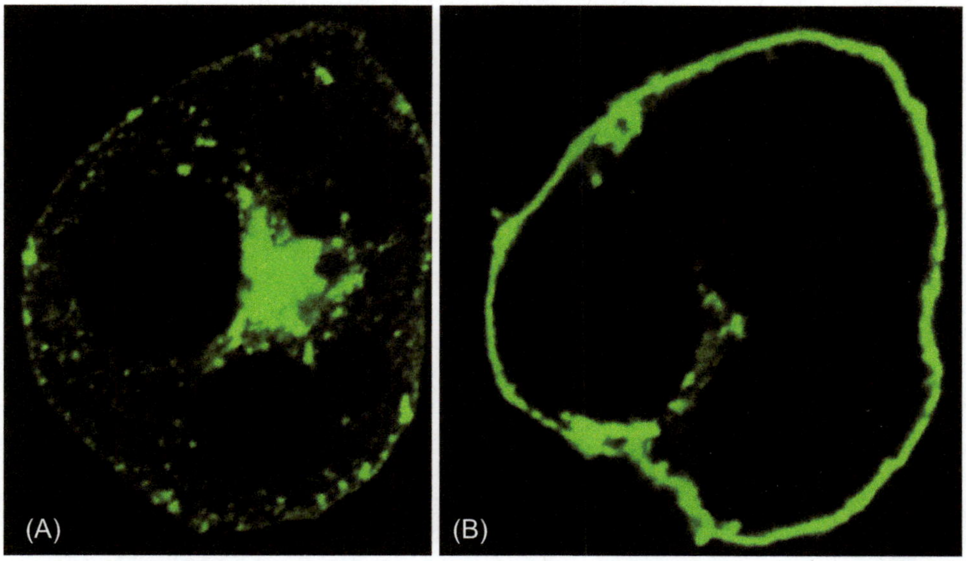

FIGURE 6.12 Insulin binding to its membrane receptor results in the translocation of the GLUT4 transport protein from its intracellular storage sites to the plasma membrane. A GLUT4-enhanced green fluorescent protein construct was transfected into differentiated 3T3L1 adipocytes. They were then incubated in the absence (A) or presence (B) of insulin for 30 min. Next the cells were fixed and the location of the GLUT4-labeled green fluorescent protein determined by confocal fluorescent microscopy. *GLUT*, Glucose transporter. *Figure was provided by R.T. Watson, M. Kanzaki, J.E. Pessin, Regulated membrane trafficking of the insulin-responsive glucose transporter 4 in adipocytes, Endocrine Reviews 25 (2) (2004) 177–204.*

three-dimensional side view (panel A) and top view (panel B) are of the ribbon representations of loop 7 of GLUT4. The key three amino acids, Q (glutamine), L (leucine), and S (serine), are presented as pink CPK (creatine phosphokinase) space filling spheres in both panels A and B. The glutamine, leucine, and serine site is believed to be structurally involved in creating the pore that is involved in the transport of glucose.

The uptake of the glucose present in the blood compartment by the process of facilitated diffusion sends glucose down its concentration gradient. Once the glucose enters into the cell cytosol, it becomes phosphorylated by

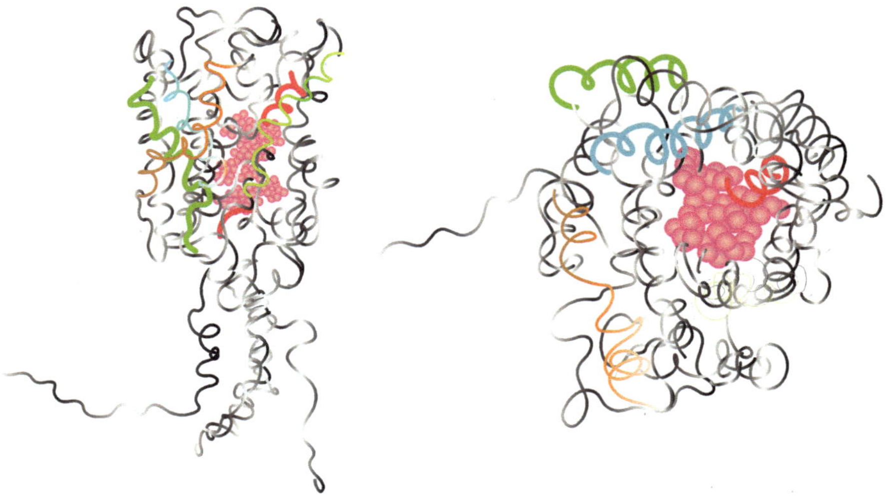

FIGURE 6.13 Three-dimensional model of the GLUT4 glucose transporter. (A) The GLUT4 model describes a right-handed α-helix barrel that consists of 12-membrane segments that surround a central pore. This side view of the GLUT4 shows the extensive ribbon representation of the entire 3D structure. The QLS site, which is believed to be located in the seventh loop of the protein, is colored pink. It is postulated to be structurally involved in creating the pore that is involved in the transport of glucose; it is shown as a CPK space-filling model. The COOH and NH_2 terminal regions are presented as a coil. (B) This is a top view of the 3D presentation of the GLUT4 protein ribbon form. The QLS pink sites are presented as a CPK space filling mode; these residues are believed to participate in the process of glucose transport. The transmembrane helices are colored as follows: helix 2 (*orange*), helix 5 (*yellow*), helix 7 (*red*), helix 10 (*green*), and helix 11 (*cyan*). *GLUT*, Glucose transporter; QLS, glutamine, leucine, and serine; *CPK*, Corey, Pauling, Koltun. *From P. Strobel et al. Myricetin, quercetin and catechin-gallate inhibit glucose uptake in isolated rat adipocytes. Biochemical Journal 386 (2005) 471–478; this paper describes a detailed molecular model of the GLUT4 protein.*

hexokinase to yield glucose-6-phosphate which, because of its negative charge, cannot leave the cell by diffusion through GLUT4. When glucose-6-phosphate is metabolized through glycolysis and the Krebs cycle, the released energy is stored as ATP. In type 2 diabetes mellitus, there is an impairment of GLUT4 trafficking and translocation in the skeletal muscle, which results in the development of insulin resistance.

B Glucagon and glucagon-like peptides

Human glucagon and the glucagon-like peptides, abbreviated as GLPs, are all produced from the same single proglucagon gene. Whereas glucagon is principally biosynthesized in the pancreatic α-cells, the GLPs [GLP-1, GLP-2, and oxyntomodulin (OXY)] are secreted by the intestine's endocrine cells. Besides the pancreas A cells, other sites of expression of glucagon messenger RNA (mRNA) are the endocrine cells of the small and large intestine and also the brain.

1 Glucagon's structure

Glucagon has a molecular weight of 3485 Da and comprises a single amino acid chain of 29 amino acids that is devoid of disulfide linkages. The amino acid sequence of glucagon is presented in Fig. 6.14. The amino acid sequences of human, rat, rabbit, porcine, bovine, and chicken glucagons indicate that there is a high degree of structural conservation for all these species.

The secondary and tertiary structures of glucagon have been studied in solution by optical rotary dispersion-circular dichroism and in the crystalline state by X-ray crystallography (3-Å resolution). The full spectrum of biological responses mediated by glucagon is dependent on the presence of each of the 29 amino acids. This is particularly the case when glucagon binding to its receptor activates adenyl kinase. Fig. 6.15 shows the structure of glucagon in solution (panel A) and an X-ray analysis of glucagon in the crystalline state (panel B). In dilute solutions the glucagon peptide, due to its relatively short chain length, is very unstructured and flexible (panel A). The helical region extends from residues 5 to 25 and can result in the formation of two hydrophobic "sticky" patches that are involved in a self-association process that may result in the presence of glucagon dimers.

Glucagon's N-terminal histidine at position 1 (Fig. 6.15; panel B) is essential for its biological activity. The adjacent residues, 2 (serine) and 5 (threonine), are involved in implementing the adenyl cyclase activity which is key to the initiation of glucagon-mediated biological responses. Residues 19–26 (*blue color*) have an equal potential to stabilize the structure either

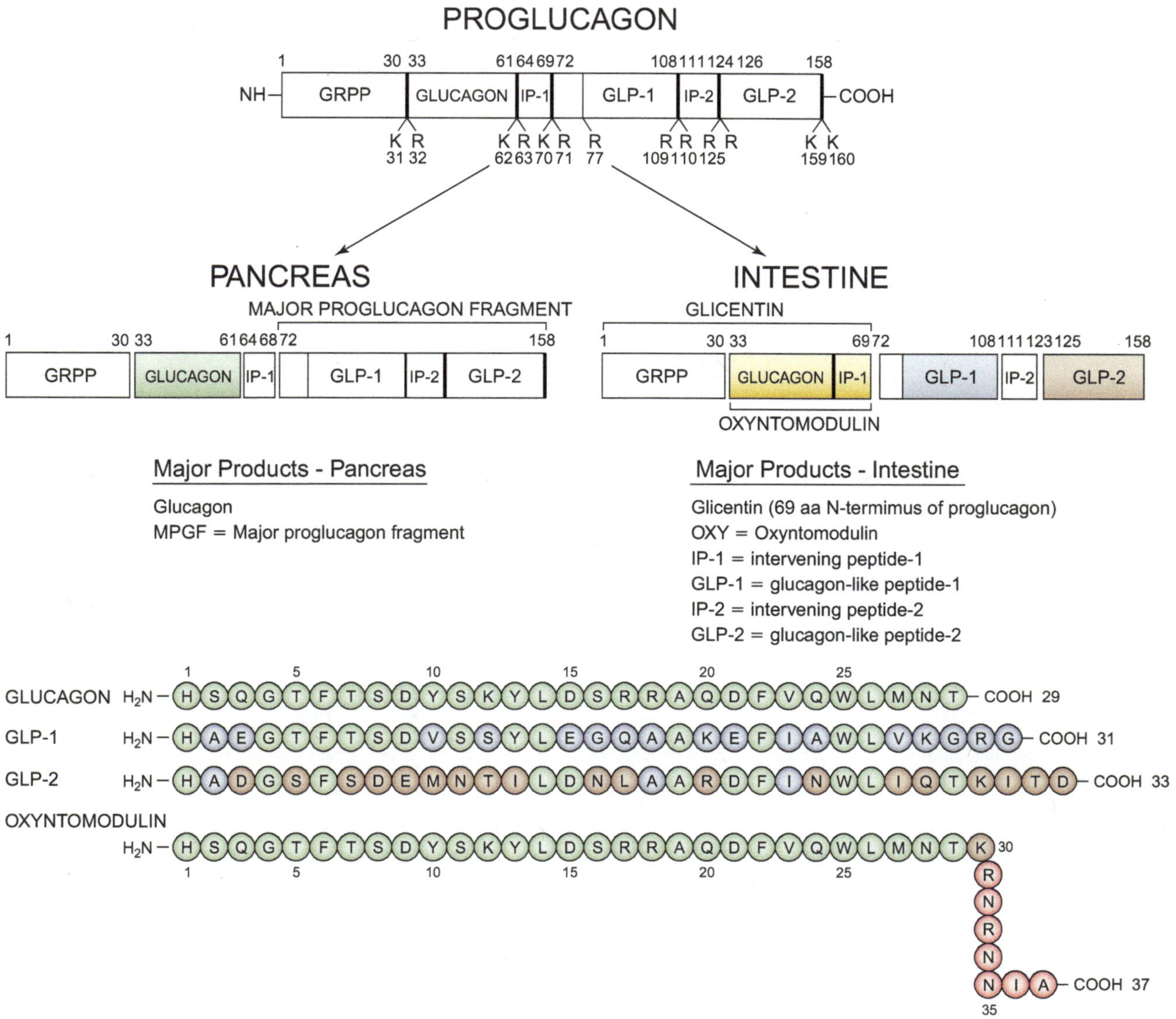

FIGURE 6.14 Schematic diagram of the proglucagon domain organization in the pancreas and intestine and secretion products from the pancreas (glucagon) and small intestine (GLP-1, GLP-2, and oxyntomodulin peptides). In the pancreas the secretion of glucagon is strictly regulated by an array of agents. Examples of agonists include neuropeptides, hormones (epinephrine), and amino acids (arginine, glutamine, and alanine). Antagonists of glucagon secretion include glucose and somatostatin. In contrast, in the intestine both GLP-1 and GLP-2 are secreted by the intestinal L cell. GLP-1 is a potent antihyperglycemic hormone, with a half-life of only 2 min, that stimulates glucose-dependent insulin secretion while suppressing glucagon secretion. The actions of GLP-2 are less clear. In the pancreas, proglucagon is processed to secrete intact full-length glucagon (29 amino acids). The "major" proglucagon fragment (amino acid residues 72–158) has no known biological functions. In contrast, in the intestine, proglucagon is designed to secrete (1) oxyntomodulin, a 37-amino-acid peptide that contains the 29-amino-acid sequence of glucagon followed by an 8-amino-acid carboxy-terminal; (2) GLP-1; and (3) GLP-2. There is also some information suggesting that the intestinal oxyntomodulin displays weak affinity for the glucagon receptor and may mimic glucagon actions in the pancreas and liver. *GLP*, Glucagon-like peptide. *Portions of this figure came from G.I. Bell et al. Nature B304 (1983) 368–371 and Journal of Nutrition, 133 (2003) 3709–3711. The amino acid sequences came with permission from DeGroot, Endocrinology, Chapter 35, 2010, p. 661.*

as β sheets, or as α-helices. The C-terminal two residues, asparagine (28) and threonine (29), increase glucagon's tight binding to its receptor. Several potent antagonists of glucagon have been synthesized by site-directed mutagenesis; thus deletion of the N-terminal histidine, coupled with substitution of 9 Asp by Glu and α-carboxyamidation of the C-terminal Thr, generates a peptide that binds to the glucagon receptor with high affinity but with no capability to stimulate the production of cyclic AMP (cAMP).

Thus the glucagon molecule is unique among medium-sized polypeptides. Although it has a clearly definable

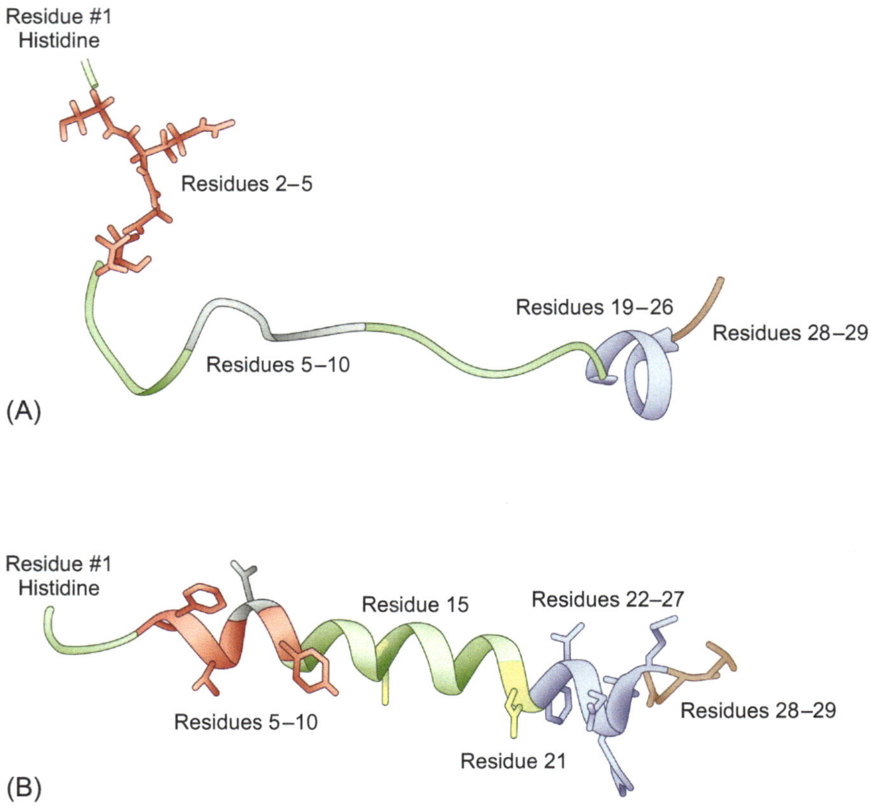

FIGURE 6.15 Presentation of the secondary and tertiary structures/shapes of the 29-amino-acid residues that constitute glucagon. (A) Key secondary structural characteristics of glucagon shapes in solution. The presence of histidine at the H_2N-amino terminus as well as residues 2–5 (*magenta*) are required for the activation of adenyl cyclase, which is key to initiating the biological actions of glycogen in raising blood glucose levels. The position of the amino acid residues shown in black (5–10) and purple (28 and 29) are essential for glucagon to bind tightly to its receptor. Two distinct shapes for glucagon in solution are known. Conformation 1 is made up of both β sheets (from residues 5 to 10) and α-helices (from 19 to 27). Conformational shape 2 comprises all β sheets (5–10) and (19–27). (B) The tertiary X-ray crystallographic shape of glucagon displays the key sites for the binding of glucagon to its receptor. This includes the magenta residues (5–10), the blue residues (22–27), and the orange residues (28–29). *The two figures were modified from originals prepared at Rutgers University Molecular Anatomy Laboratory by Deepthi Mikkilineni in 2008.*

secondary structure (α-helix) and quaternary structure (self-association), it has no definable tertiary structure.

2 Glucagon and glucagon-like peptides—biosynthesis and secretion

Glucagon's stimulation of hepatic gluconeogenesis and glycogenolysis is essential to the liver's maintenance of proper blood concentrations of glucose. Also in adipocytes, glucagon acting through its receptor stimulates lipolysis which converts stored triglycerides into free fatty acids as a substrate for subsequent conversion to glucose.

Glucagon and the GLPs are all produced from a single proglucagon gene; see Fig. 6.14. In the pancreas, only glucagon is expressed and secreted. The secretion of glucagon is strictly regulated by an array of agents. Examples of agonists include neuropeptides, hormones (epinephrine), and amino acids (arginine, glutamine, and alanine). Antagonists of glucagon secretion include glucose and somatostatin.

The proglucagon gene is expressed in the gastrointestinal (GI) tract in the stomach and both the small and large intestine; see Fig. 6.14. It gives rise to several peptides that individually play diverse roles in the control of energy intake and storage, intestinal motility, and nutrient disposal. OXY is derived from the proglucagon gene. It has a 37-amino-acid peptide that includes on its N-terminus the 29-amino-acid sequence of glucagon followed by an 8-amino-acid peptide on its carboxyterminal extension, GLP-1, a 30-amino-acid peptide that participates in the modulation of glucose homeostasis via the control of insulin and glucagon secretion and by the inhibition of gastric emptying and food intake. GLP-1 generates a sensation of increased satiety and likely diminishes appetite via central effects on the regions of the hypothalamus. At times of food intake, GLP-2 is simultaneously secreted with GLP-1 from intestinal endocrine cells.

GLP-2 is trophic to the intestinal tract in both the small and large intestine. The biological activities of GLPs, GLP-1 and GLP-2, are summarized in Fig. 6.16.

OXY binds to both the glucagon and GLP-1 receptor (GLP-1R); however, as yet a separate OXY receptor has not been identified. OXY is released postprandially and has the ability to pass through the blood–brain barrier (BBB) and act as a satiety signal via effects on appetite centers such as the hypothalamus and brainstem.

The glucagon receptor is a member of the superfamily of G protein–coupled receptors; it is a seven transmembrane–spanning G protein–coupled receptor. When it has bound glucagon, it becomes activated and produces cAMP and intracellular Ca^{2+} as second messengers. Its molecular weight is ~62 kDa. The glucagon receptors are principally expressed in kidney and liver and are also found in lower amounts in adipose tissue, heart, adrenal glands, GI tract, and the cerebral cortex. A model of the structure of the glucagon receptor is shown in Fig. 6.17. Glucose is the dominant regulator of glucagon secretion. Falling levels of glucose stimulates glucagon secretion, while elevated levels of blood glucose decrease glucagon secretion.

The GLP-1 receptor also generates its biological responses through the activation of a G protein–coupled receptor that is a member of the glucagon/secretin receptor superfamily. The structure of the transmembrane domain (TMD) of the GLP-1 receptor is shown in Fig. 6.18. The TMD is the site of allosteric modulator binding. The GLP-1-activated second messengers are through calcium and cAMP-stimulated pathways. Also a GLP-2 receptor has a separate G protein–activated receptor, amino acid sequence of which is related to both the GLP-1 and glucagon receptors. As shown in Fig. 6.16, GLP-1 acts on the pancreas to stimulate the release of insulin from the β-cell in the islets of Langerhans. Fig. 6.19 shows a schematic of the action of GLP-1 on its membrane receptor, GLP-1R, and the pathway of signals that culminate in the release of insulin.

C Insulin and glucagon collaborations

The overriding shared responsibility of insulin and glucagon is to maintain blood glucose within normal limits. Shown in Fig. 6.20 are the dose-response curves of (1) how falling glucose levels stimulate glucagon release and (2) how rising blood glucose levels stimulate the secretion of insulin. When the blood glucose concentration rises above ~5 mM then the β-cell begins to secrete insulin. The steep portions

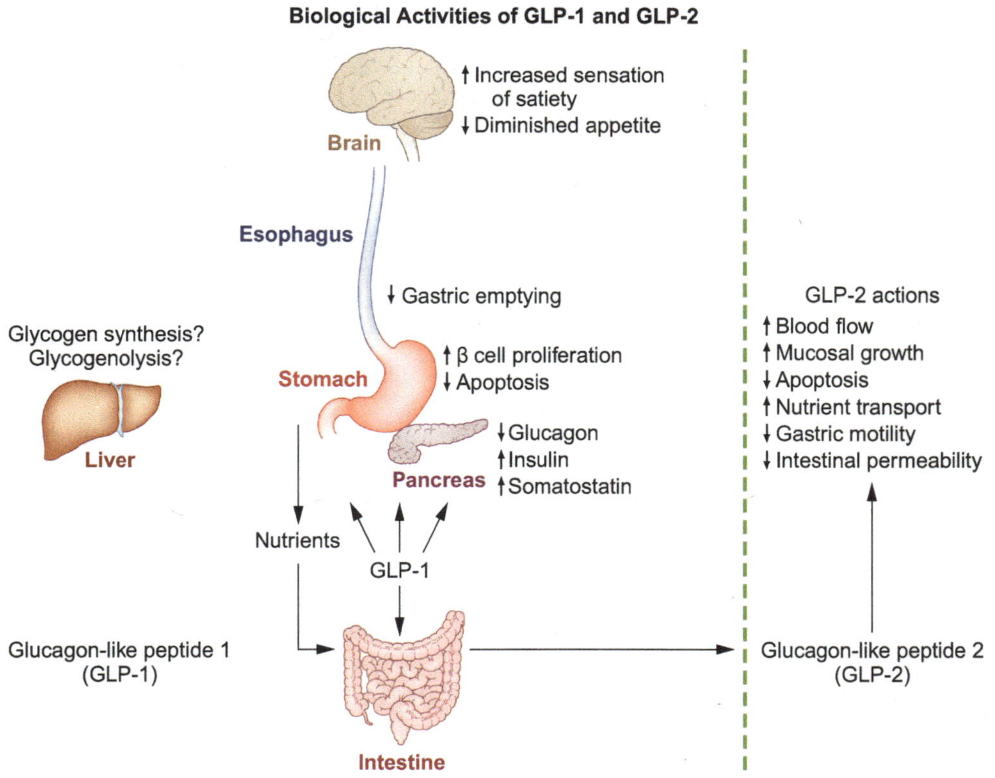

FIGURE 6.16 Biological actions of the GLP. (A) The biological actions of GLP-1 are illustrated in the left panel. They include the following: intestine (secretion of GLP-1), pancreas (reduction in glucagon secretion and an increase in glucose-dependent secretion of insulin and somatostatin), stomach (reduction in gastric emptying), and brain (sensations of satiety and decreased appetite) are illustrated. (B) The biological actions of GLP-2 are tabulated in the right panel. The source of GLP-1 and GLP-2 is illustrated in Fig. 6.14. *GLP*, Glucagon-like peptide.

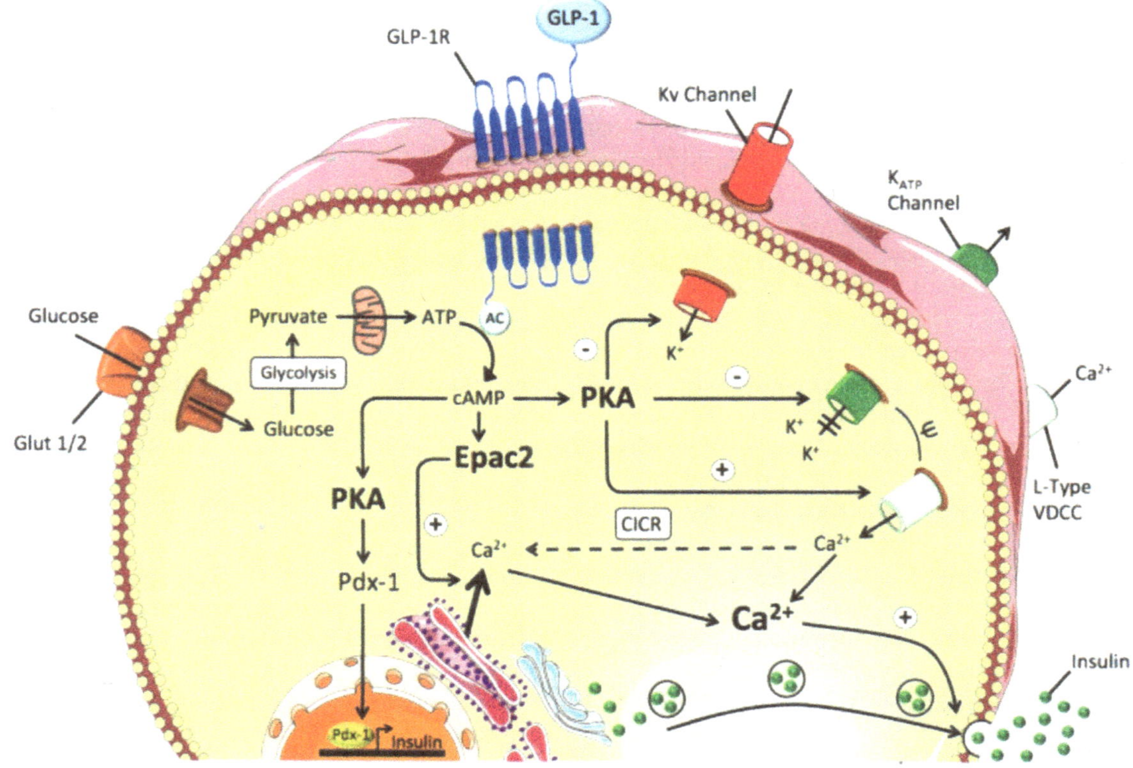

FIGURE 6.17 Schematic on GLP-1-mediated insulin secretion in the β-cell. Calcium increases in the cytoplasm through the uptake of calcium ions from the extracellular space through the calcium channel, L-type VDCC that becomes stimulated by the action of PKA. In the intracellular space, calcium ions are released from its storage in the endoplasmic reticulum through the action of Epac2 activated by cAMP. The accumulation of calcium ions in the cytoplasm results in the release of insulin from its storage sites. *AC*, Adenylate cyclase; *cAMP*, cyclic AMP; *CICR*, calcium-induced calcium release; *Epac2*, exchange protein activated by cAMP; *GLP*, glucagon-like peptide; *GLUT1/2*, glucose transporter *1/2*; *Pdx-1*, pancreatic and duodenal homeobox I; *PKA*, protein kinase A; *VDCC*, voltage-dependent calcium channel. *Reproduced from Fig. 6.5 of T.D. Muller et al., Glucagon-like peptide (GLP-1), Molecular Metabolism 30 (2019) 72–130; doi: 10.1016/j.molmet.2019.09.010; https://www.ncbi.nlm.gov/pmc/articles/pmc6812410/.*

of the insulin release curve occur at blood concentrations of 10–11 mM glucose that occur postprandially. Conversely, the half-maximal glucose-mediated suppression of glucagon secretion occurs at a glucose concentration of 3–4 mM. These two dose-response curves describe the responses of the pancreatic α- and β-cells as they work to effect a stable blood glucose concentration.

The principal biological actions of insulin and glucose are complex and interdependent upon the delicate balance of anabolism and catabolism that occurs throughout the body in response to changes in caloric intake, caloric composition, and the degree of physical activity. In most situations the roles of insulin and glucagon are antagonistic. Tables 6.8 and 6.9, respectively, summarize the actions of insulin and glucagon in their principal target tissues, the liver, muscle, and adipose tissues.

1 Liver

The "normal" blood glucose level is in the range of 80–110 mg/dL (see Table 6.2), whereas the chronic elevated blood glucose level associated with diabetes type 2 is >130 mg/dL. The chronic lowered blood glucose falls in the range of 50–60 mg/dL (see Table 6.2). The American Diabetes Association recommends a post-meal glucose level of <180 mg/dL and a fasting plasma glucose of 90–130 mg/dL.

The actual amount of glucose in the blood and body fluid compartments is normally very small. In a healthy adult of 75 kg with a blood volume of 5 L, a blood glucose level of 100 mg/dL amounts to a total of only ~5 g of glucose. This is slightly less than two packets of sugar for a cup of coffee, but it should satisfy the consumer's sugar receptors.

As shown in Fig. 6.21, an average meal will provide 90 g of glucose or 270 g of glucose per three meals/day. Thus a daily dietary influx will total ~270 g of sugar, resulting in the ongoing elevation of blood glucose levels arising from the ingested three meals. Therefore per one meal, the homeostasis mechanisms summarized in Fig. 6.21 must process 90 g of glucose to lower the blood glucose level back to the normal range of 80–110 mg/100 mL of glucose.

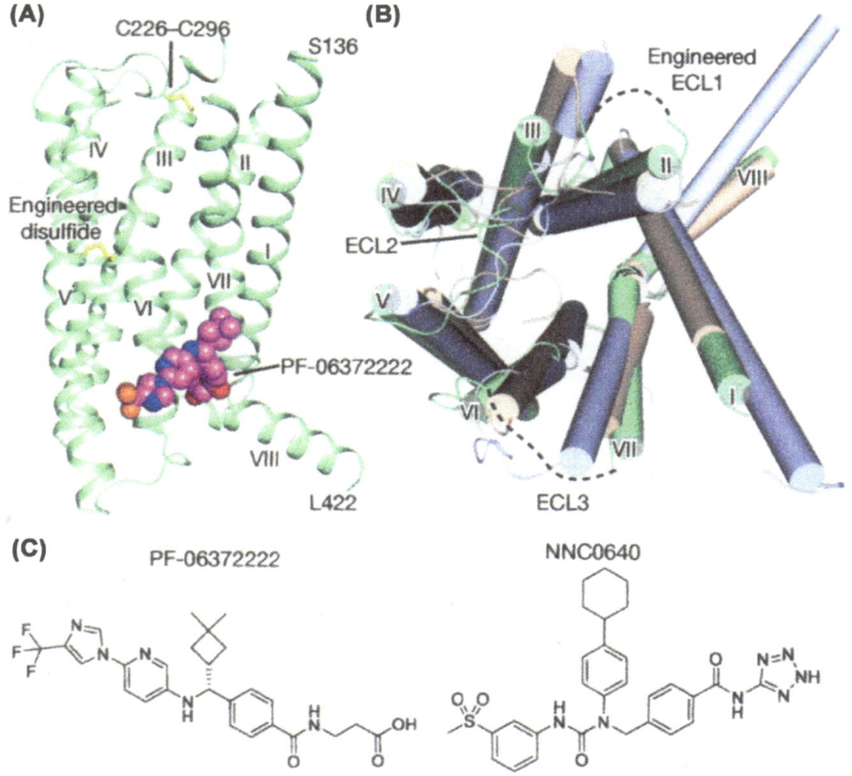

FIGURE 6.18 Overall structure of the GLP-1R TMD. (A) GLP-1R is represented as a green cartoon. Representative ligand PF-06372222 is shown as spheres with purple carbon atoms. Disulfide bonds are shown as yellow sticks and labeled I–VII. (B) Top view of GLP-1R (*green cylinders*) compared with the initial GCGR structure (PDB code 4L6R, *blue*) and the thermostabilized GCGR structure (PDB code 5EE7, *yellow*). (C) Chemical structures of cocrystallized ligands PF-06372222 and NNC0640. *GCGR*, Glucagon receptor; *GLP-1R*, glucagon-like peptide 1 receptor, *TMD*, transmembrane domain. *Reproduced from Fig. 6.1 of G. Song et al., Human GLP-1 receptor transmembrane domain structure in complex with allosteric modulators, Nature Research 546 (2017) 312–315 Letter; doi: 10.1038/nature22378.*

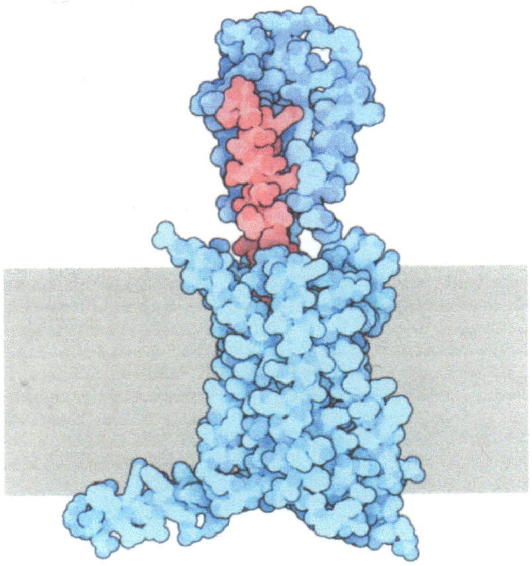

FIGURE 6.19 The glucagon receptor is shown in *blue*, with glucagon in *red*. The plasma membrane is shown schematically in *gray*. *Reproduced from https://pdb101.rcsb.org/motm/184.*

As shown in Fig. 6.21, there are two physiological mechanisms that result in lowering of the blood glucose levels associated with a single meal: (1) one involves storage of ~50% of the excess glucose as glycogen; and (2) the second involves the metabolism of ~50% of the glucose by the process of glycolysis linked to the TCA (tricarboxylic acid) cycle so as to generate substantial amounts of ATP. The elevated blood glucose levels stimulate insulin secretion from the pancreas (see Fig. 6.21). This leads to an increased uptake of glucose and storage over 24 hours as glycogen by the liver (3 × 17 g), muscle (3 × 24 g), and fat/adipose (3 × 2 g). Over the same 24-hour time interval, the brain (3 × 15 g) and muscle (3 × 24 g) metabolize the indicated amounts of glucose via glycolysis linked to the TCA cycle to generate ATP for a wide array of ongoing energy-requiring activities. The kidneys also use 3 × ~8 g/meal lactate as an energy source.

The liver performs an indispensable role in maintaining an adequate blood level of glucose. It possesses the enzymatic capability either to generate glucose from stored glycogen or to store excess glucose as glycogen

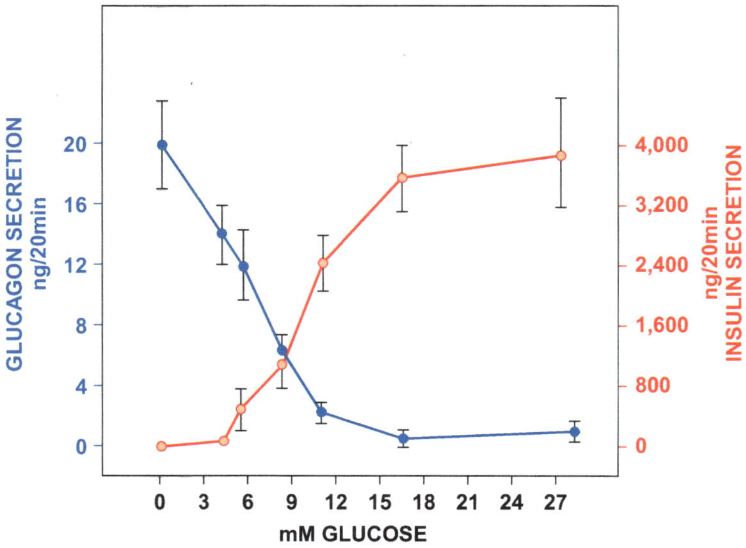

FIGURE 6.20 Effect of changes in blood glucose concentration on the inhibition of secretion of glucagon (*blue*) and stimulation of the secretion of insulin (*red*). These studies were carried out in in vitro perfused rat pancreas. *From R.G. Gosmanov, A.R.Gosmanov, J.E. Gerich, Endotext website: Diabetes mellitus and carbohydrate metabolism—diabetes-manager, Glucagon Physiology 2 (2012); available at http://www.endotext.org.*

TABLE 6.8 Summary of actions of insulin on several tissues.

	Insulin-sufficient state	Insulin-deficient state
Liver	No effect on glucose uptake	Uptake of free fatty acids and conversion to ketones
	Stimulates biosynthesis of hexokinase IV and activates glycogen synthetase	
	Promotes glycolysis and formation of ATP	
Muscle	Stimulation of glucose uptake	Impaired blood glucose
	Stimulates biosynthesis of hexokinase II and pyruvate kinase	
	Stimulates glycolysis and formation of ATP. Increases muscle glycogen levels and creatine phosphate	
Adipose	Stimulation of glucose uptake	Decreases triglyceride synthesis due to a lack of glycerol phosphate
	Enhances glycolysis, which makes available glycerol phosphate, which, in turn, enhances triglyceride synthesis	Stimulation of lipolysis and release of FFA into the bloodstream
	Inhibits lipase activity	
Brain	No direct actions of insulin. Brain is dependent upon blood glucose	None

FFA, Free fatty acid.

TABLE 6.9 Summary of actions of glucagon on several tissues in the glucagon-sufficient state.

Tissue	Effects
Liver	Inactivates glycogen synthetase and activates phosphorylase α, which leads to an activation of glycogenolysis
	Increases the activity of glucose-6-phosphatase
	Enhances synthesis of glucose from pyruvate and lactate as well as amino acids, especially arginine and alanine; that is, it activates gluconeogenesis
Muscle	Muscle does not contain glucagon receptors, and accordingly it has no responses to glucagon
Adipose	Glucagon in large doses can stimulate lipolysis, but under normal circumstances it has little or no effect
Pancreas	Stimulates insulin secretion, particularly after intestinal absorption of amino acids
Brain	None

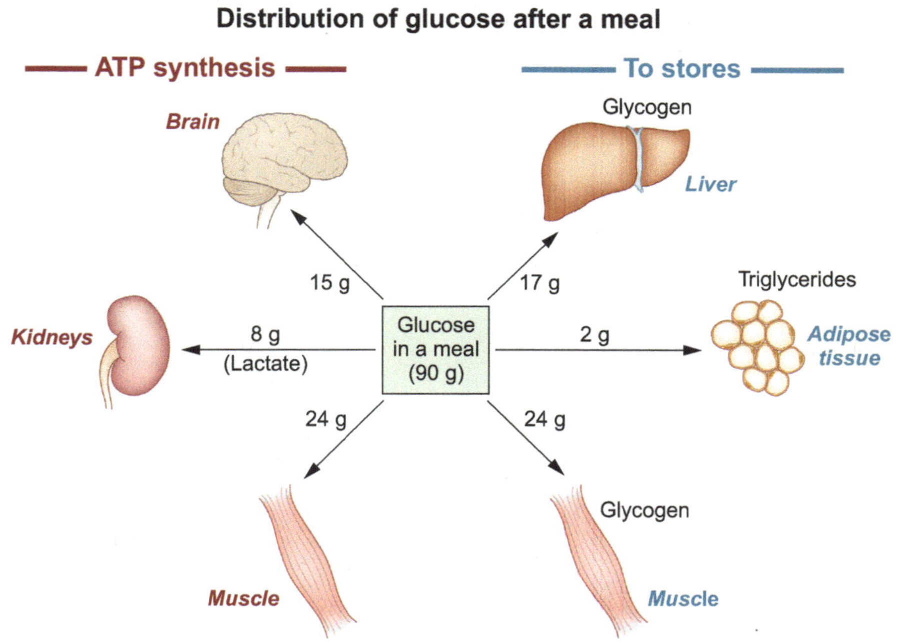

FIGURE 6.21 Distribution of glucose after a meal between storage of glucose as glycogen or metabolism of glucose to CO_2 and production of ATP. An average meal contains ~90 g of glucose. In this example, muscle takes up 50% (~48 g) of the ingested glucose of which half (24 g) is stored in muscle as glycogen, and half (24 g) is metabolized by the glycolytic pathway and the TCA cycle to generate significant amounts of ATP which is utilized as an immediate source of energy. Approximately 17% (~15 g) of the ingested glucose is transported directly to the brain where it is also metabolized to generate ATP to support brain activities. The brain does not store glycogen and, thus, is dependent 24 h/day upon a continuous uptake of blood glucose to support all the cells of the brain. The liver takes up ~20% (17 g) of the meal's 90 g of glucose and stores it as glycogen. This hepatic glycogen is available to be released into the blood so as to maintain the blood glucose concentrations in the normal range, for example, 80–110 g/100 mL (see Table 6.2). The kidneys take up ~10% (8 g) of the glucose meal in the form of lactate which is released from RBCs. Since RBC do not have mitochondria, they must release their anaerobically oxidized glucose as either lactate and/or pyruvate. Finally, in this model, adipose tissue has a modest ~2% (2 g) uptake of glucose which is then converted to triglycerides for long-term storage. *RBC*, Red blood cell. *From E.N. Marieb, K. Hoehn, Human Anatomy and Physiology, seventh ed., Pearson Benjamin Cummings, San Francisco, CA, 2005.*

(Fig. 6.22). The upper half of the figure focuses on the scenario of a "rising" blood glucose level, while the lower half of the figure focuses on the circumstances of a falling blood glucose level. In addition, the liver contains an active gluconeogenesis capability that will permit the production of glucose from three- and four-carbon fragments derived from amino acids. Also, the liver can oxidatively metabolize glucose and other smaller metabolites to H_2O and CO_2, which generates ATP, or convert two-carbon fragments to larger free fatty acids, which then are incorporated into triglycerides and phospholipids.

Several key enzymes in the liver are markedly affected by the presence or absence of glucagon or insulin. Activation of the glycogenolytic system involves the phosphorylation of key enzymes, while inactivation necessitates dephosphorylation. On the one hand, glycogen is phosphorylyzed (broken down) by the cascade of cAMP-governed reactions that convert the relatively inactive phosphorylase *b* into the more active phosphorylase *a*. On the other hand, under conditions where it is appropriate to convert glucose into glycogen, there is activation of a phosphatase that converts the phosphorylated D or *dependent* form of glycogen synthetase into the dephosphorylated I or *independent* form. A key feature of this system is the unique role played by the protein phosphorylase *b* kinase-kinase, which has two distinct catalytic activities. When it is phosphorylated (as a consequence of the presence of glucagon), it initiates the cascade that ultimately leads to the conversion of phosphorylase *b* into phosphorylase *a* and the production of glucose-1-P. Alternatively, when the phosphorylase *b* kinase-kinase is not phosphorylated, it acquires the activity of glycogen-I-synthetase-kinase. Thus the presence of glucagon favors the generation of an active phosphorylase *b* kinase-kinase that leads to glucose formation, while the presence of insulin activates a phosphatase that leads to glycogen formation.

In addition to the glucagon–insulin modulation of glycogen storage and mobilization in the liver, these two peptide hormones also modulate the balance between gluconeogenesis and lipogenesis. Under circumstances of glucose demand, gluconeogenesis will predominate to convert glucose into those carbon skeletons derived from amino acids or glycolytic intermediates produced prior to pyruvate. Under conditions of glucose excess, glycolytic intermediates and free fatty acids can be directed either to (1) storage as triglycerides, (2) to oxidation by the TCA cycle producing ATP, or (3) to production of ketone bodies. Reversal of the glycolytic pathway requires

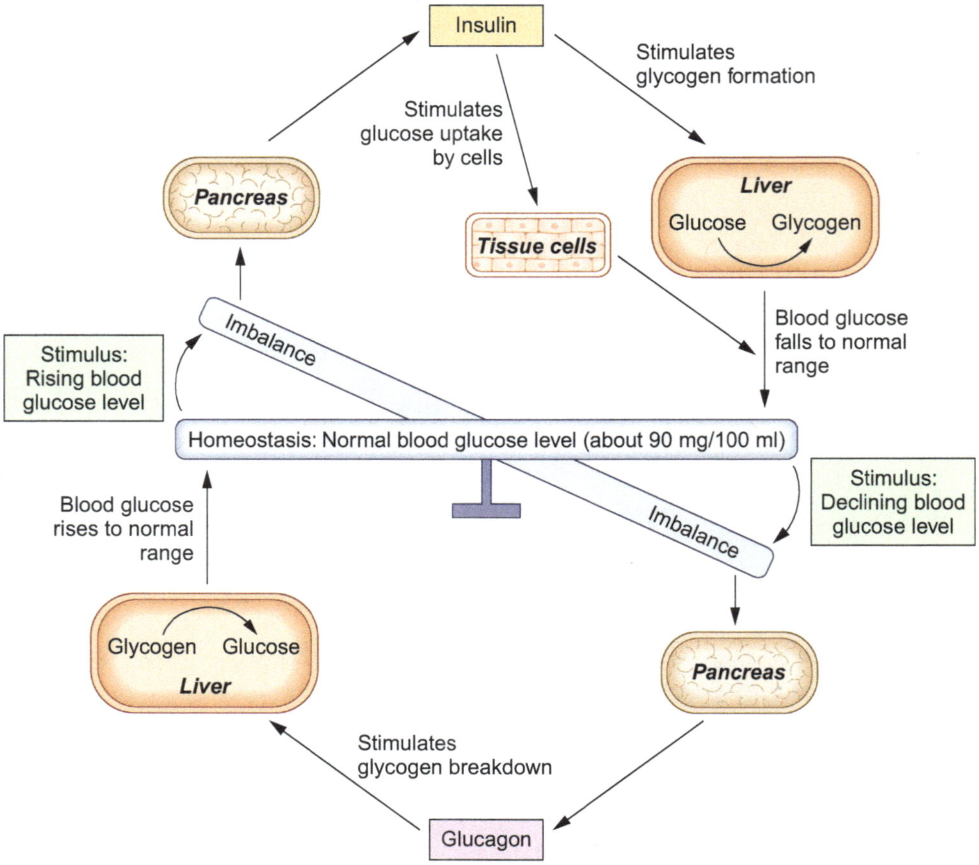

FIGURE 6.22 Regulation of blood glucose levels by insulin and glucagon. Comparison of the relative contributions of insulin and glucagon to the maintenance of normal blood glucose levels in a human. The figure shows the consequences of blood glucose levels deviating from the normal level of about 90 mg/100 mL of blood. The upper half of the figure focuses on the scenario of a "rising" blood glucose level, whereas the lower half of the figure focuses on a scenario of a "declining" blood glucose level. Modest elevation of glucose levels stimulates the pancreas to secrete insulin which in liver and muscle will stimulate the storage of the excess glucose and the metabolic energy that it represents (see Fig. 6.12). In contrast a modest fall in blood glucose levels (bottom half of the figure) stimulates the pancreas to secrete glycogen which stimulates glycogen breakdown to glucose only by the liver. Muscle does not have a glucagon receptor. *Modified from E.N. Marieb, K. Hoehn, Human Anatomy & Physiology, seventh ed., Pearson, Benjamin Cummings, San Francisco, CA, 2005.*

access to NADH (nicotinamide adenine dinucleotide), whereas lipogenesis is dependent upon access to NADPH (nicotinamide adenine dinucleotide phosphate hydrogen) for fatty acid biosynthesis.

In summary, glucagon has been shown to stimulate the conversion of pyruvate, lactate, alanine, and glycerol into glucose. This is accomplished largely by modulating key enzymes of the gluconeogenic pathway. There appear to be no effects of glucagon on stimulating substrate supply or in increasing amino acid uptake by the liver. Insulin appears to exert its inhibitory effects on liver gluconeogenesis by (1) inhibiting or slowing the enzymes of gluconeogenesis and (2) diminishing the flow of amino acids from peripheral tissues, principally muscle, to the liver.

Phosphorylated glycolytic intermediates are not capable of traversing the outer cell membrane of the liver. However, the liver contains an active glucose-6-phosphatase. The activity of this enzyme is increased in the absence of insulin or the presence of cortisol; this ensures the ready conversion of glucose-6-phosphate to free glucose, which may then be exported from the liver cell. By contrast, muscle cells do not have measurable glucose-6-phosphatase activity, and thus the muscle glycogen stores cannot be mobilized for the maintenance of blood glucose levels.

2 Muscle

In the absence of insulin, there is a stimulation of net protein catabolism in muscle. The resulting free amino acids are released into the bloodstream and delivered to the liver where they are oxidatively deaminated. The resulting carbon fragments are then committed to gluconeogenesis or catabolism to yield ketone bodies and/or CO_2. The resulting increase in nitrogen is converted to urea and excreted in the urine.

Muscle tissue is known to contain insulin but not glucagon receptors. Insulin occupancy of these receptors leads to an increased uptake of both glucose and amino acids; there is also an associated stimulation of protein synthesis. After an overnight fast and in the absence of physical activity, muscle tissue is largely dependent upon the oxidation of free fatty acids to meet its energy demands. Under these conditions, there would need to be a significant pool of muscle glycogen. With the initiation of mild-to-moderate exercise, the muscle tissue successively oxidizes its own glycogen, then blood glucose, and finally blood-delivered free fatty acids derived from adipose and hepatic stores.

3 Adipose tissue

Adipose tissue is one of the principal target organs for insulin action. The extensive amount of triglycerides stored in adipocytes serves as an important fuel source under conditions of dietary caloric restriction or prolonged exercise. The mobilized free fatty acids are systematically delivered to a number of key organs (heart and kidney), which can directly utilize them as a substrate for oxidative metabolism. If the free fatty acids were taken up by the liver, they could be esterified into triglycerides or phospholipids, oxidized to CO_2, or more likely converted into ketones, particularly β-hydroxybutyrate and acetoacetate. These ketones, in turn, can also serve as fuels for most extrahepatic tissues, particularly the brain, under conditions of very low blood glucose levels.

4 Pharmacological agents related to the pancreas

Panels A and B of Fig. 6.23 show the chemical structures of two classes of drugs currently used for the treatment of diabetes types 1 and 2. Glipizide and glimepiride (panel A) are second-generation sulfonylureas that are useful in the treatment of type 1 diabetes. They act by stimulating the secretion of insulin from the β-cells in the pancreas. Also there is some evidence that they can limit the production of glucose in the liver.

Metformin (panel B) is clinically useful as a drug for treating type 2 diabetes where it reduces hepatic insulin resistance and, thereby, gluconeogenesis which results in the release of glucose. The basis of metformin's biological actions lies in its ability to activate adenosine monophosphate–activated protein kinase (AMPK), an enzyme that plays an important role in insulin signaling, which leads to favorable effects on body energy balance.

Both alloxan and streptozotocin (panel C) can disrupt the pancreatic islets so that insulin secretion is not occurring.. Streptozotocin and alloxan can enter the β-cells of the pancreatic islets by passage through the GLUT2 receptors present in the islet cell's plasma membrane. This leads to an accumulation in the cytosol of alloxan, which generates reactive oxygen species by a complex series of coupled reactions that ultimately are responsible for the death of the pancreas β-cells.

After the arrival of streptozotocin in the cytosol of pancreas β-cells, it is split into its glucose and methylnitrosourea moieties. Due to streptozotocin's alkylating properties, the latter fragments DNA leading to a modification of many biological macromolecules, which collectively then leads to a destruction of the β-cells. This results in a state of insulin-dependent diabetes.

D Leptin

1 Background

The maintenance of body weight is a complex process involving homeostasis of nutrient intake for both carbohydrate and fat. Insulin and glucagon regulate the metabolic processes of gluconeogenesis and glycolysis in muscle and liver cells (see Figs. 6.1, 6.8 and, 6.15). Normally higher vertebrates maintain the body's energy reserve of fat (triglycerides in adipose tissue) with high precision; thus a typical nonobese, healthy adult would have a body mass containing 15%–20% fat. However, in some vertebrates (including humans), there can be a loss of regulation of the storage of triglycerides in adipose cells, such that the body mass of fat can increase to 30%. Classical parabiotic[1] studies conducted in the early 1970s suggested that genetically obese mice lacked a bloodborne factor (i.e., a hormone) that regulated, via appetite control, the stimulation of glycolysis and thus the extent of body fat deposition. Also, it was known that the mutations in the *obese* gene in the C57BL/6 mouse result in the onset of obesity which includes increased deposition of body fat and hyperglycemia and hyperinsulinemia (similar to type 2 diabetes).

This led in 1994 to the discovery by J. M. Friedman and D. L. Coleman of a new peptide hormone, termed leptin. Leptin is derived from the Greek root *leptos*, meaning thin. Leptin is a 16-kDa protein (167 amino acids) that is the plasma form of the gene product encoded by the *ob* gene. Through the use of selected gene knockout (KO) of the *Ob/Ob* gene in mice, a homozygous wild type (+/+) for leptin could be converted to a homozygous KO mouse (−/−); see

1. Parabiosis: In a parabiotic experiment the circulatory system of one animal is connected to a second animal. Thus D. L. Coleman [*Diabetologia* 9, 294 (1973)] indirectly connected the circulatory system of adult *Ob/Ob* (obese) mice to that of normal mice. This led the *Ob/Ob* mice to lose weight. This suggested the existence of a weight-regulatory substance in the serum of the normal mice, that is, a hormone associated with the maintenance of a normal dietary caloric intake.

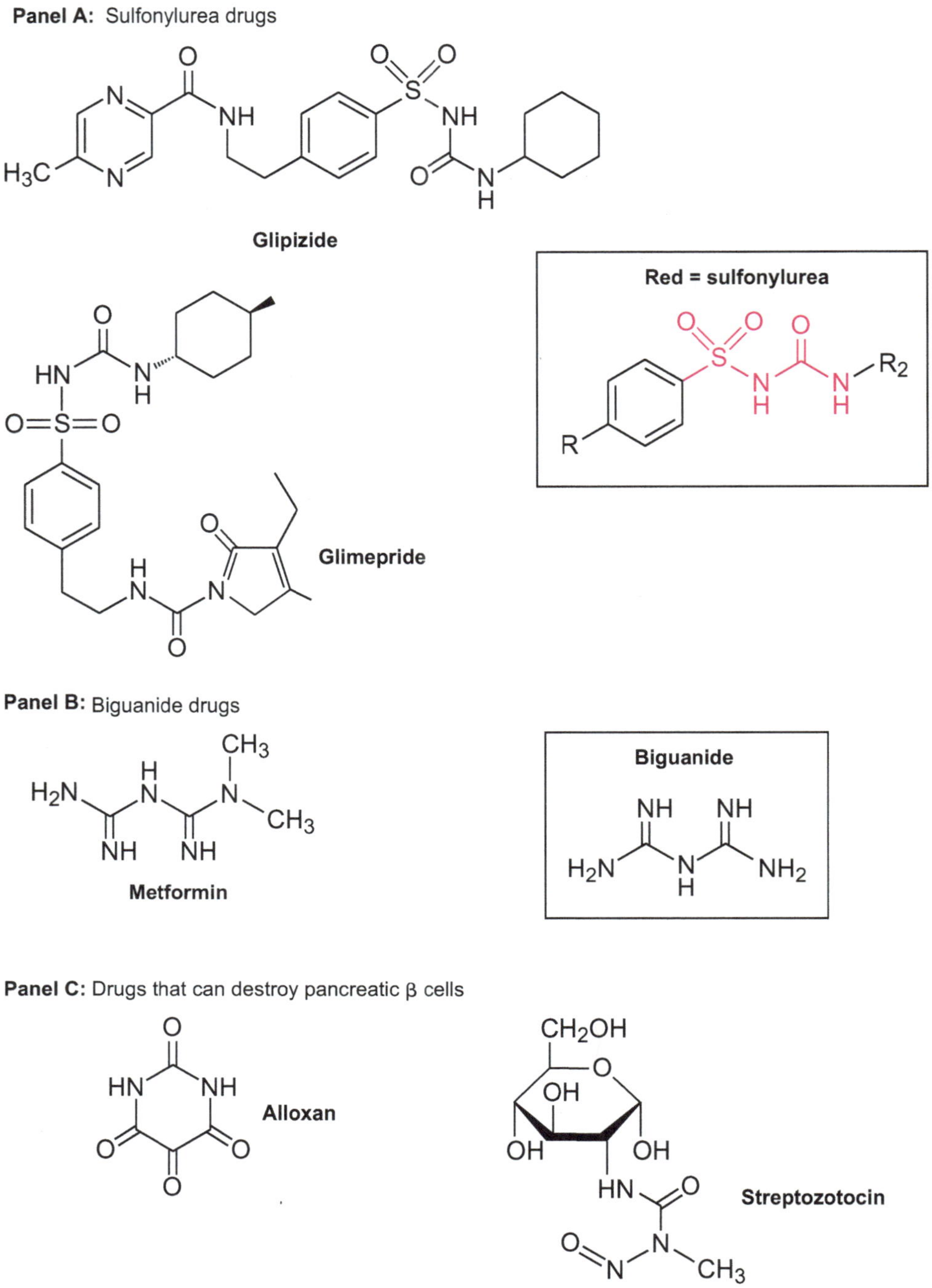

FIGURE 6.23 Chemical structures of some of the most frequently employed oral antidiabetic compounds. (Panel A) The structure of two second-generation drugs, glipizide and glimepiride, that belong to the sulfonylurea class of compounds that have proven to be effective in treating type 2 diabetes. The inset "box" contains the structure of the sulfonylurea radical which is the "active" portion for both glipizide and glimepiride. (Panel B) The structure of one biguanide drug, metformin. The inset box shows the signature structure of the biguanide. Metformin is currently the only biguanide-type drug currently available in the United States. (Panel C) The structures of alloxan and streptozotocin are illustrated. These are both drugs that are employed to induce diabetes by selectively destroying pancreatic β-cells in an experimental animal. Streptozotocin is an antibiotic produced by *Streptomyces achromogenes var. streptozoticus* that has been shown to inhibit the biosynthesis of DNA, including the DNA of pancreas β-cells.

Fig. 6.24. In the absence of the *Ob/Ob* (−/ −) gene, the mouse could not express the gene for leptin. As a consequence, the mouse became obese. The second mouse has a phenotype of *Ob/Ob* (+/+) and is able to produce leptin, thereby controlling the amount of triglycerides that are stored in adipose tissue.

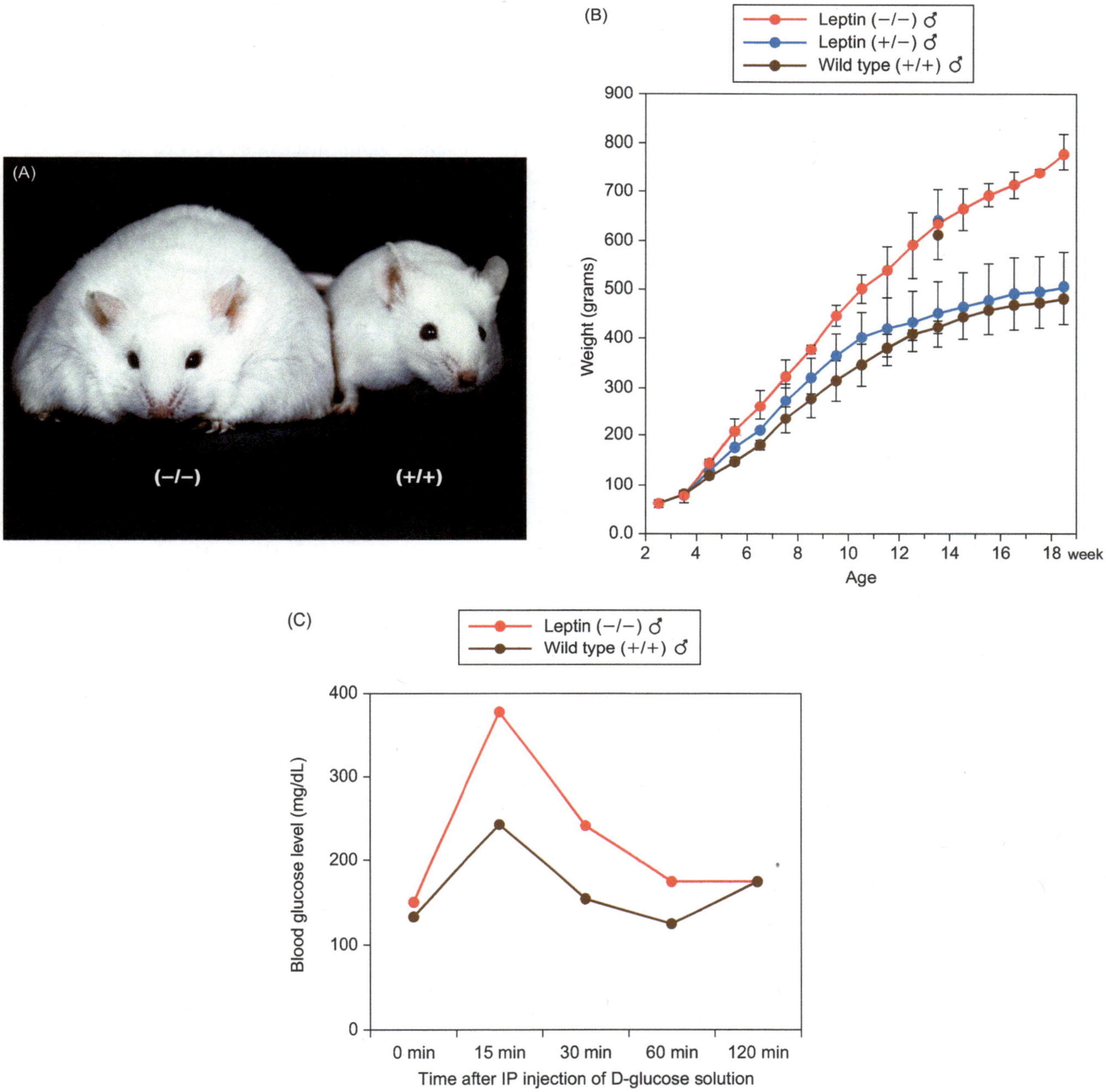

FIGURE 6.24 Importance of leptin and its gene in regulating the body's energy balance. (A) Knockout of the leptin gene. The obese mouse on the left has a knockout mutation on chromosome 7 creating a (−/−) phenotype that has altered the pattern of fat deposition in its body. The mouse on the right has a normal (+/+) phenotype. The "obesity" gene is believed to act in collaboration with other genes that regulate the body's energy balance. Mice with the mutation grow 35%−50% fatter by middle age than a normal mouse, despite eating a low-fat diet. (The mice were photographed at Oak Ridge National Laboratory in Tennessee.) (B) *Effect of the absence of leptin on mouse body weight.* Body weight assessment over 18 weeks for leptin (−/−), (+/−), and (+/+) male rats. In the knockout −/− leptin animal (*red line*), the absence of leptin results in an uncontrolled appetite which over 18 weeks leads to obesity and an abnormal intermediary metabolism. In the heterozygous (+/−) animals (*blue line*), and wild-type (+/+) animals (*brown line*), there is a marked reduction of the obesity due to the presence of leptin. These genetically manipulated animals have been useful in studying diabetes, atherosclerosis, and high blood pressure associated with high serum cholesterol levels. (C) *Effect of the absence of leptin on regulation of serum glucose levels.* Glucose tolerance test in leptin knock out (−/−) and wild-type (+/+) rats. Animals were fasted for 16 h prior to an intravenous injection of a bolus of glucose. Then their blood glucose levels were determined at 15-min intervals over 120 min. In the leptin (−/−) animals, serum glucose levels remained elevated in comparison to the (+/+) wild-type rats that can produce adequate amounts of leptin to control the blood glucose levels. The data shown in panels (B) and (C) were provided by Sage Labs. See the following URL: http://www.sageresearchmodels.com/research-models/knockout-rats/leptin-knockout-rat-%E2%80%94-kilorat%E2%84%A2.

2 Knockout of leptin gene

A follow-up set of experiments with newly born *Ob/Ob* KO mice, (−/ −), heterozygous mice, (−/+), and homozygous wild-type (+/+) mice was evaluated over an 18-week interval (see Fig. 6.24). It is quite apparent that the *Ob/Ob* (−/ −) KO mice that had an absence of leptin were unable to control their body weight. By 18 weeks the heterozygous mice and wild-type mice weighed ~500 g, while the KO mice weighed ~750 g or 150% of the wild type. Thus in the presence of leptin, there was a marked reduction in obesity. In addition, leptin reduced the food intake and increased the *ob/ob* mouse expenditure of metabolic energy.

In a follow-up experiment, a glucose tolerance test was carried out in *Ob/Ob* (−/ −) KO mice and in *Ob/Ob* (+/+) wild-type mice. A glucose tolerance test is a standard clinical assay where glucose is given orally (humans) or intraperitoneally (mice) and sequential blood samples are taken over 1−2 hours so as to determine how quickly glucose is cleared from the blood compartment (see Fig. 6.24). In this experiment the consequences of the presence and absence of leptin on the rapid lowering of blood glucose levels was determined. In the *Ob/Ob* (−/ −) mice the high levels of blood glucose were reduced much more slowly in the absence of leptin than in the *Ob/Ob* (+/+) mice that had normal access to leptin. This clearly demonstrates the modulation of blood glucose levels by the prevailing levels of leptin.

3 Leptin functions as a satiety factor

Leptin is produced primarily in the white adipose tissue; the circulating level of leptin is determined by the total amount of fat in the body. Leptin produced by the adipocytes gene is released into the general circulatory system (see the *red line* in the summary Fig. 6.25) and is delivered to the brain via the circulatory system, where it crosses the BBB by a receptor-mediated transport process. Leptin acts through the leptin receptor (LEP-R), a single-transmembrane-domain receptor of the cytokine receptor family. The LEP-R is present in the neuropeptide Y neurons in the arcuate nucleus and can respond to leptin in such a way as to decrease the activity of these neurons. Thus leptin orchestrates signals in the ventromedial hypothalamic region of the brain where a group of cells in the satiety center suppresses a desire to eat food. The hypothalamus also receives energy signals from ghrelin and insulin which complement the leptin message. Thus leptin can function as a satiety factor as long as there is a functional diabetes *Db/Db* gene that codes for the LEP-R present in the hypothalamus. In summary, leptin signals the brain that the body has had enough food to eat. In Fig. 6.26 is presented a model of LEP-R signal transduction pathways. Activation of the LEP-R by leptin activates Janus-2 kinases, resulting in tyrosine phosphorylation of the receptor and downstream proteins, including STAT3, SHP2, IRS2, and phosphatidylinositol-3-kinase that play roles in regulating the transcription of genes important for food intake and lipid metabolism. In the hypothalamus, leptin inactivates AMPK, increases acetyl-CoA carboxylase (ACC) activity, and decreases food intake. In skeletal muscle, leptin activates AMPK and decreases ACC, and carnitine palmitoyltransferase-1 activity, in turn, increases mitochondrial β-oxidation.

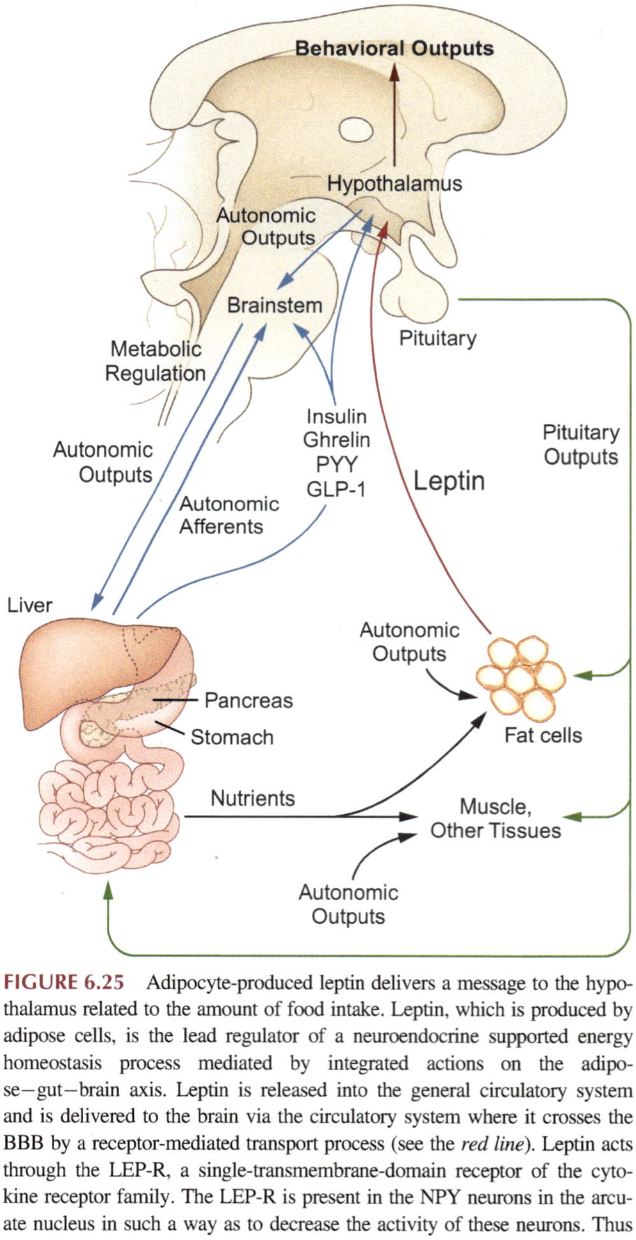

FIGURE 6.25 Adipocyte-produced leptin delivers a message to the hypothalamus related to the amount of food intake. Leptin, which is produced by adipose cells, is the lead regulator of a neuroendocrine supported energy homeostasis process mediated by integrated actions on the adipose−gut−brain axis. Leptin is released into the general circulatory system and is delivered to the brain via the circulatory system where it crosses the BBB by a receptor-mediated transport process (see the *red line*). Leptin acts through the LEP-R, a single-transmembrane-domain receptor of the cytokine receptor family. The LEP-R is present in the NPY neurons in the arcuate nucleus in such a way as to decrease the activity of these neurons. Thus leptin orchestrates signals in the ventromedial hypothalamic region of the brain where a group of cells in the satiety center suppresses a desire to eat food, thereby producing a feeling of satiety. *BBB*, Blood−brain barrier; *LEP-R*, leptin receptor; *NPY*, neuropeptide Y.

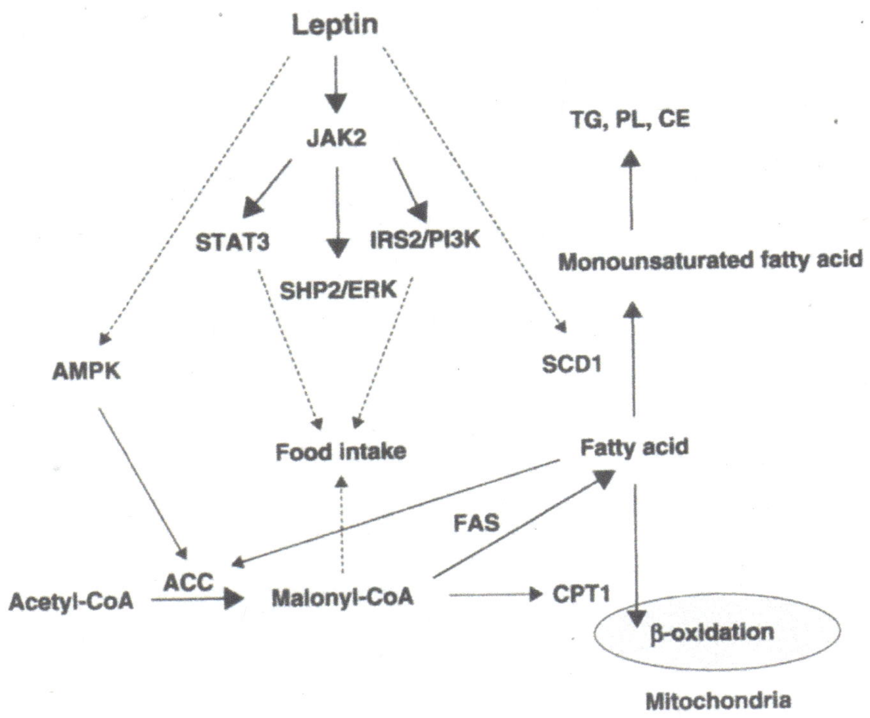

FIGURE 6.26 Actions of leptin. *ACC*, Acetyl-CoA carboxylase; *CE*, cholesterol ester; *CPT-1*, carnitine palmitoyltransferase-1; *FAS*, fatty acid synthase; *PL*, phospholipase; *SCD-1*, stearoyl-CoA desaturase; *TG*, triglyceride. *Reproduced from Fig. 6.4 of F. Zhang, Y. Chen, M. Heiman, R. DiMarchi, Leptin: structure, function and biology, Vitamins & Hormones, 71 (2005) 345–372; https://doi.org/10.1016/S0083-6729(05)71012-8.*

Another component of leptin's metabolic activity is the inhibition of hepatic stearoyl-CoA desaturase-1 activity to regulate lipoprotein metabolism and energy expenditure. Thus leptin acts both centrally and peripherally to regulate energy homeostasis.

The hypothalamus also receives neuronal inputs (*autonomic afferents*) concerning the external environment, including light status and blood glucose levels. This then results in multiple central nervous system actions in the hypothalamus which generate both *autonomic outputs* and *endocrine outputs* (see Fig. 6.24) from the pituitary which feedback to adipose and muscle which collectively results in imposing a decreased food intake. The hypothalamus has responsibility for determining the state of the body's hunger status, fatigue, temperature, thirst, and sleep and circadian cycles. Thus the hypothalamus is essential for life. Besides containing receptors for leptin, the hypothalamus has receptors for other hormones, including insulin, cholecystokinin, ghrelin, peptide YY, and GLP-1; these will be discussed in Chapter 7, Gastrointestinal Hormones.

A number of unexpected actions of leptin have been identified. Leptin inhibits bone formation through a hypothalamic relay which controls bone mass. In rodents, leptin is required for female and male fertility. It has a lesser effect on human fertility. In humans, ovulatory cycles in females are believed to be linked to energy balance (positive or negative) depending on whether a female is gaining or losing weight. The placenta has been shown to secrete leptin.

In summary the essential physiological role of leptin is to signal the presence of a caloric deficiency which mediates changes in metabolism rather than signaling an excess of calories. Also it should be appreciated that much research remains to clearly define the complexity of how energy expenditure contributes to obesity and the details of how hormones may modulate energy expenditure.

E Other pancreatic hormones

1 Pancreatic polypeptide

PP is a 36-amino-acid peptide that is known to stimulate the gastric secretion of HCl and pepsin (see Table 6.1). Most of the PP is expressed and then secreted by the G cells of the pancreas; see Fig. 6.3. PP circulates in the plasma as a dimer with a half-life of 6–7 minutes. The PP binds to its cognate receptor, the Y4 receptor, which is a G protein–coupled receptor. This is linked to inhibiting the production of the second messenger, cAMP. The Y4 receptor mRNA in humans is functionally active in the pancreas, colon, and small intestine. PP can also act as a satiety factor. PP is known to be released after a protein meal. In a recent clinical study where PP was administered to human subjects, the main finding was that PP inhibited gastric emptying which then led to a reduction in food intake over a 24-hour interval.

2 Somatostatin

Somatostatin is derived from a 116-amino-acid preprosomatostatin that results in the generation of somatostatin-28 and somatostatin-14 peptides; both are biologically active. Somatostatin-28 is produced by enteroendocrine D cells that are specialized endocrine cells of the GI tract. Somatostatin-14 has only 14 amino acids and one disulfide bond; it is produced by D cells present in the pancreas and the stomach. Although somatostatin was originally discovered in the hypothalamus, it is also known to be produced by the δ-cells of the endocrine pancreas (see Figs. 6.3 and 6.4) and dispersed cells in the GI tract. Somatostatin, when produced by the hypothalamus, is frequently termed growth hormone release-inhibiting hormone; see Fig. 3.6.

Somatostatin has been classified to be a universal inhibitor. The physiological role of circulating somatostatin is only now becoming clear. It may function in a paracrine fashion to inhibit pancreatic islet functions as well as in the intestine G cells, parietal cells of the stomach that secrete HCl and enterochromaffin-like cells or enterochromaffin-like cells.

3 Amylin

The pancreatic β-cell islets produce the peptide hormone amylin. It has also been formerly known as IAPP. Amylin is a 37-amino-acid peptide that is structurally homologous to the neuropeptide calcitonin gene–related peptide (CGRP). CGRP is discussed in Fig. 9.4. Amylin and insulin are cosecreted from the pancreatic β-cells at the time of food intake. Amylin is also secreted by the intestine, stomach, lung, and the hypothalamus. Amylin functions to slow the gastric emptying of the stomach and to promote satiety.

Knowledge of the amino acid sequence of amylin has permitted, through recombinant DNA techniques, the characterization of both the cDNA and a prepro-amylin, as well as the amylin gene. Amylin appears to have multiple activities. These include negative effects on glucagon secretion and delay of gastric emptying. On the other hand, amylin may act as a satiety factor. Overexpression and oligomerization of amylin in the pancreatic islet β-cells are, in all probability, associated with the pathogenesis of diabetes.

IV Clinical aspects

A Diabetes mellitus

1 Background

The term "diabetes" means "running through"; characteristic features are excess water intake (polydipsia) and water loss (polyuria). *Mellitus* is Latin for honey (i.e., sweet). In a classic sense, diabetes mellitus is not a single disease because it has no single definable and distinct etiology.

The major pathophysiological abnormalities generated by diabetes mellitus include the following: (1) glucose intolerance or inappropriately high blood glucose levels in the presence of elevated or normal insulin concentrations; (2) acidosis and ketosis; (3) reduction of the enzymes for gluconeogenesis; (4) reduced growth; and (5) a microangiopathy, including a thickened basement membrane of capillaries in muscle and retinal blood vessels and of the capillaries of the renal glomerulus.

Diabetes mellitus and type 1 diabetes, or more usually just diabetes, are the most common diseases associated with the pancreatic islets. Of the 313 million residents in the USA in 2011, 25.8 million children and adults or 8.3% had diabetes.[2] Of this cohort, 18.8 million adults and children were formally diagnosed, whereas 7.0 million people were undiagnosed. For persons under 20 years of age, 216,000 or 0.26% in this age group have diabetes. The sex distribution for individuals that were diagnosed with diabetes is that 11.8 million (49%) were men and 12.6 million (51%) were women. Each year approximately 200,000 individuals in the USA are newly diagnosed as diabetic. The lifetime risk in the USA for being diagnosed with diabetes if you were born after 2000 is 33% for males and 38% for females; for Hispanic Americans the risk is 50%. Diabetes is the eighth leading health-related cause of death in the United States.

Clinically, there are two general classifications of diabetes: (1) type 1 or insulin-dependent diabetes mellitus (T1DM), which was formerly designated juvenile-onset diabetes; and (2) type 2 or insulin-independent diabetes mellitus (T2DM), which was formerly designated as maturity-onset diabetes.

2 Type 1 diabetes mellitus or T1DM

It is clear that in T1DM diabetes, the β-cells in the islet are partially or totally destroyed by an aggressive autoimmune process which is mediated by CD4+ and CD8+ cells. The end result is that there is no endogenous insulin secretion. This is determined both by the absence of plasma levels of insulin's C-peptide and the presence of an absolute daily requirement of a drug form of insulin for the maintenance of life. Individuals with T1DM are highly prone to develop metabolic ketosis, and in the untreated state there are the acute symptoms of polyuria, polydipsia, weight loss, and fatigue, all of which are secondary to the absence of insulin. Patients with T1DM require mandatory daily treatment with some form of drug insulin preparation.

2. American Diabetes Association Fact Sheet for 2011; see the following URL, http://www.diabetes.org/diabetes-basics/diabetes-statistics/.

Fig. 6.27 illustrates a timeline of the etiology of type 1 diabetes mellitus that has three chapters. The first is the possible contributions of genetic predispositions (father and mother genes). Also there can be maternal contributions during pregnancy that cannot be avoided. This is followed by the chapter of the prediabetes phase; the length of this time interval can vary widely. Type 1 diabetes is known to occur in genetically predisposed individuals as a consequence of toxic environmental or infectious insult to the pancreatic β-cells coupled with the actions of an aggressive immune system that gradually, but persistently, destroys the β-cells. During this phase, there is onset of the loss of functional β-cells. When the β-cell loss approaches 80%, then the pace of appearance of clinical diagnostics increases which results in a formal clinical diagnosis of T1DM.

3 Type 2 diabetes or T2DM

Type 2 diabetes, also known as T2DM or maturity-onset diabetes, is a form of diabetes mellitus in which there is only a relative lack of insulin. It is frequently not diagnosed until after the individual is of age 40. In this form of the disease, the diabetes is stated to be insulin-independent since the patient will not develop ketosis without insulin therapy. Type 2 diabetics may require only intermittent adjunctive insulin therapy, usually to manage fasting hyperglycemia. In many subjects, adequate glucose and fuel metabolite homeostasis can be achieved by careful dietary management of nutritional and caloric intake.

The most commonly associated physical finding with maturity-onset diabetes is obesity; see Fig. 6.28. Obesity can be defined as an enlargement of the adipose depot and is usually associated with some kind of pancreatic dysfunction. The extent of obesity can be classified by the body mass index (BMI). The equation for determination of BMI is

$$\text{Bodymassindex(BMI)}$$

$$= [\text{weight(kilograms)}]/[\text{height(centimeters)}]2$$

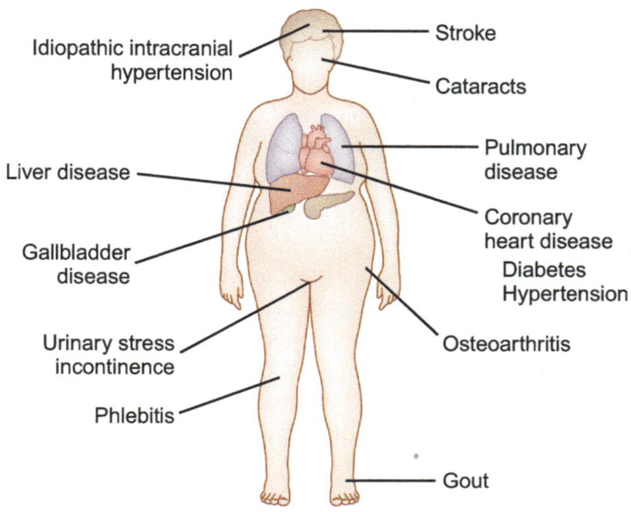

FIGURE 6.28 Overweight older subjects with T2DM have a spectrum of associated medical problems. In overweight older subjects with T2DM typically their life expectancy will be 10 years shorter. About 60% of the reduced life expectancy is linked to blood vessel diseases such as heart attacks and strokes. In the extreme, diabetic older women tend to experience mental and physical disabilities and to have higher rates of premature death. Also women with T2DM have an increase in problems that are associated with the aging process, such as harmful falls, urinary incontinence, osteoarthritis, cataracts, and pulmonary disease, to name a few. *T2DM*, Type 2 diabetes mellitus.

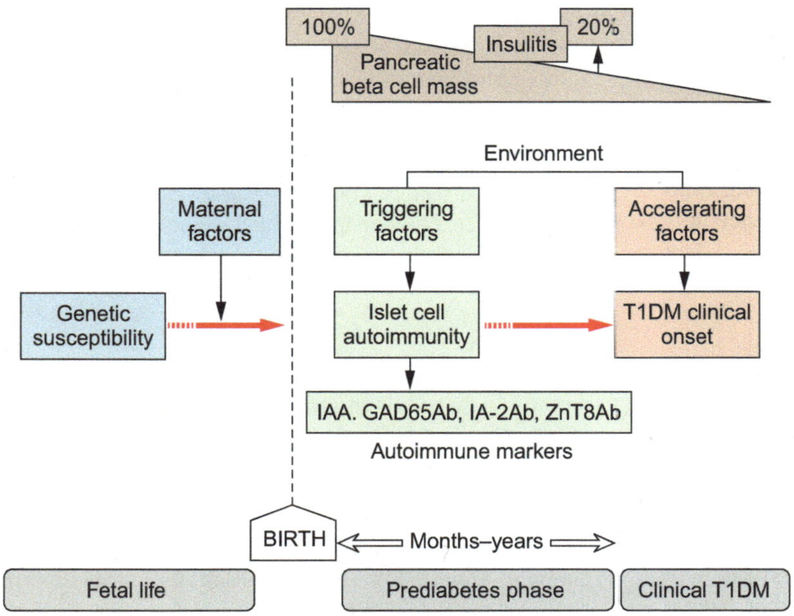

FIGURE 6.27 Timeline for onset of T1DM. T1DM is linked to several autoimmune abnormalities which aggressively carry out the selective destruction of the pancreatic islet β-cells. There are three phases that define the appearance of T1DM: they are (1) fetal life where the new genome (father's+mother's contributions) begins to be expressed as well as some exposure to the mother's immune system; (2) the prediabetes phase that may last months to years with exposure to the environment and triggering factors, which leads ultimately to development of β-cell malfunction; and (3) followed eventually by clinical diagnosis of T1DM. At some point the initiation of β-cell attrition will begin and when it becomes severe enough (loss of ~80% of the pancreas β-cells), some of the classical T1DM symptoms [polyuria (excessive production of urine), polydipsia (the desire to drink), and polyphagia (excessive eating)] will begin to appear, which facilitates the formal clinical diagnosis of T1DM. *T1DM*, Type 1 diabetes mellitus.

The BMI range for a normal healthy adult is 18–25. The BMI values for the following eight categories are shown in parentheses: very severely underweight (<15), severely underweight (15–16), underweight (16–18.5), *normal weight (18.5–25)*, overweight (25–30), moderately obese (30–35), severely obese (35–40), and very severely obese (>40). In the interval of 1980–2002 the prevalence of obesity doubled.

Type 2 diabetes has three metabolic abnormalities that cooccur to create a hyperglycemic state with an unusually high blood glucose concentration that can, in some instances, be >200 mg/100 mL (See Table 6.2 for a summary of the range of blood glucose levels). The three abnormalities are as follows: (1) the liver's basal level of glucose production is significantly increased; (2) healthy skeletal muscle tissue (in the absence of type 2 diabetes) normally takes up 70%–80% of all blood glucose, but in the presence of type 2 diabetes, muscle is resistant to significant glucose uptake; and (3) in D2DM the pancreas has a decreased β-cell function which then stimulates the secretion of glucagon. See Figs. 6.1 and 6.22.

The treatment of type 2 diabetes mellitus with significant obesity is complex, to say the least. There are a host of concurrent diseases that can occur in T2DM; they are hypertension, coronary artery disease, dyslipidemia, congestive heart failure, and stroke.

For the population of hyperglycemic subjects older than 20 years, at least 90% are more than 10% overweight. Frequently after weight loss, these individuals may regain a normal state of carbohydrate metabolism. A significant fact related to insulin-independent diabetes associated with or without obesity is that there is an apparent decrease in the sensitivity of the peripheral tissues to insulin. As a result, obese individuals have an elevated level of insulin secretion.

From the perspective of the insulin receptor present in the various target tissues, it should be emphasized that in the obese form of diabetes, characterized by glucose intolerance, hyperinsulinemia, and resistance to both endogenous and exogenous insulin, there is no evidence for a "nonfunctional form" of the insulin. The circulating ratio of proinsulin to insulin is normal; both cross-react normally in radioimmunoassays for insulin. The defect appears to be both in the number of receptors present in the target cells and in postreceptor events.

The pathogenesis of diabetes mellitus, besides being related to environmental and dietary factors, also clearly contains a genetic or familial component. Although the precise statistics supporting this statement vary with their source, it is estimated that between 25% and 50% of diabetic patients have "family histories" consistent with a genetic component. The reported incidence of diabetes is much higher in children derived from parents who both have diabetes. Also, monozygotic (identical) twins have a significantly higher incidence of diabetes than dizygotic (fraternal) twins. Current evidence favors the position that the inheritance is polygenic or multifactorial; in some specific instances, there is clear evidence of a recessive gene for diabetes.

The Pima Indians have an incidence rate of diabetes that approaches 50% or higher. Also, there are animal models of diabetes (the *Ob/Ob* and *Db/Db* diabetic mouse and the Zucker rat) that clearly reflect genetic transmission.

4 Glucagon diseases

In contrast to the situation relating to insulin, there are relatively few disease states that can be specifically ascribed to dysfunctions of glucagon. The syndrome of hyperglucagonemia was only first described in 1966. This condition arises as a consequence of the presence of a glucagon-secreting tumor of the pancreatic α-cell. This results in a detectable mass in the pancreas, anemia, and the presence of a blistering rash of the buttocks, abdomen, and groin area.

Further reading

Insulin

C.A. Broughton, R. Hovorka, New closed-loop insulin systems, Diabetologia 64 (2021) 1007–1015. Available from: https://doi.org/10.1007/s00125-021-0539-w.

J.E. Campbell, C.B. Newgard, Mechanisms controlling pancreatic islet cell function in insulin secretion, Nature Reviews Molecular Cell Biology 22 (2021) 142–158.

N. Moruzzi, et al., Tissue-specific expression of insulin receptor isoforms in obesity/type 2 diabetes mouse models, Journal of Cellular and Molecular Medicine 25 (2021) 4800–4813. Available from: https://doi.org/10.1111/jcmm.16452.

Glucagon

F.M. Gribble, F. Reimann, Metabolic messengers: glucagon-like peptide 1, Nature Metaolism 3 (2021) 142–148.

B. Jones, et al., Genetic and biased agonist-mediated reductions in β-arrestin recruitment prolong cAMP signaling at glucagon family receptors, Journal of Biological Chemistry 296 (2021) 100133. Available from: https://doi.org/10.1074/jbc.RA1200.016334.

A. Mantovani, et al., Glucagon-like peptide-1 receptor agonists for treatment of nonalcoholic fatty liver disease and nonalcoholic steatohepatitis: an updated *meta*-analysis of randomized controlled trials, Metabolites 11 (2021) 73. Available from: https://doi.org/10.3390/metabo11020073.

S.R. Nason, et al., Glucagon receptor signaling regulates weight loss via central KLB receptor complexes, Journal of Clinical Investigation Insight 6 (2021) e141323. Available from: https://doi.org/10.1172/jciinsight.141323.

M. Wang, et al., Glucagon blockade restores functional β-cell mass in type 1 diabetic mice and enhances function of human islets, Proceedings of the National Academy of Sciences of the United States of America 118 (2021). Available from: https://doi.org/10.1073/pans.2021142118. e2022142118.

Leptin

E. Cecon, et al., Amyloid beta peptide is an endogenous negative allosteric modulator of leptin receptor, Neuroendocrinology 111 (2021) 370–387.

C.L. Farber, et al., Leptin receptor neurons in the dorsomedial hypothalamus regulate diurnal patterns of feeding, locomotion and metabolism, eLife 10 (2021) e63671. Available from: https://doi.org/10.7554/eLife.63671.

K.-P. Huang, et al., Leptin signaling in vagal afferent neurons supports the absorption and storage of nutrients from high-fat diet, International Journal of Obesity (London) 45 (2021) 348–357.

G. Srinivasan, et al., Leptin receptor stimulation in late pregnant mouse uterine tissue inhibits spontaneous contractions by increasing NO and cGMP, Cytokine 137 (2021) 155341.

Diabetes

M.I. Alvelos, et al., Familial neurohypophyseal diabetes insipidus: clinical, genetic and functional studies or novel mutations in the arginine vasopressin gene, Pituitary 24 (2021). Available from: https://doi.org/10.1007/s11102-020-01119-y.

R.,G. Miller, et al., Insulin resistance-associated genetic variants in type 1 diabetes, Journal of Diabetes and Its Complications 35 (4) (2021) 107842.

G. Priya, S. Kalra, A. Dasgupta, E. Grewal, Diabetes insipidus: a pragmatic approach to management, Cureus 13 (2021) e12498. https://doi.org/107759/cureus.12498.

GLUT 4 glucose transporter

C.,R. Gregori, C.,P. Laurindo, U.F. Machado, Estrogen and glycemic homeostasis: the fundamental role of nuclear estrogen receptors ESR1/ESR2 in glucose transporter GLUT4 regulation, Cells 10 (1) (2021) 99. Available from: https://doi.org/10.3390/cells/10.3399.

J. Gu, W. Zhang, L. Wu, Y. Gu, CFTR deficiency affects glucose homeostasis vis regulating GLUT4 plasma membrane transportation, Frontiers in Cell and Developmental Biology 9 (2021) 630654. Available from: https://doi.org/10.3389/fcell.2021.630654.

D. Leto, A.R. Saltiel, Regulation of glucose transport by insulin: traffic control of GLUT4, Nature Reviews Molecular Cell Biology 13 (2012) 383–396.

Chapter 7

Gastrointestinal hormones

I Introduction

A Background

The oral cavity, stomach, intestine, colon, liver, gallbladder, and pancreas operate as an integrated physiological unit to effect the digestion and absorption of dietary-acquired nutrients. A discussion of the principal pancreatic hormones, insulin and glucagon, is presented in Chapter 6, Pancreatic Hormones: Insulin and Glucagon. This chapter will focus only on the gastrointestinal (GI) hormones.

There are ~30 peptides known to be expressed within the digestive tract, thus making the stomach, intestine, and colon the largest endocrine organ in the body. The regulatory peptides synthesized by the gut include hormones, growth factors, and peptide neurotransmitters. The GI hormones are a family of polypeptides produced by specialized endocrine cells present in the stomach, intestine, and colon which function both as traditional hormonal messengers and as neurotransmitters. These hormones mediate a wide variety of specific biological responses carried out by the stomach, intestine, and colon, as well as the endocrine and exocrine pancreas, liver, and gallbladder, which, when collectively integrated, optimize the physiological conditions necessary to permit the efficient digestion and absorption of protein, carbohydrates, and fat as well as minerals and vitamins from the lumen of the intestine. There is also increasing evidence of neuroendocrine communication between the brain and gut that provides an additional level of integration and complexity to these digestive processes.

The first appreciation of gastroenterology's link to endocrinology was an observation by Bayliss and Starling in 1902. They discovered that acid stimulates a chemical messenger from the cells of the duodenal and jejunal mucosa that moved through the circulatory system, resulting in the excitation of the pancreas. They named this intestinal secretion *secretin*. It subsequently provided the impetus to establish the field of endocrinology as a specialist field in medicine. Also it established the special contribution of the GI domain to the early history of endocrinology; see footnotes 1 and 2 for the citation of the landmark Bayliss and Starling article as well as a perspective of their work from the view of 2005 concerning the discovery of secretin.[1,2]

B Resume of the gastrointestinal hormones

Table 7.1 tabulates the six GI hormone families that collectively secrete ~20 hormones. A "family" designation is based on the similarities of the amino acid sequences for each family. Table 7.2 summarizes 10 neurotransmitter hormones produced by or active in the GI tract.

C Problems of food processing and digestion

All animals require continued and regular intake of food so that they can meet their bodily nutrient requirements. Briefly, the diversity of the food components includes the following: adequate caloric intake of carbohydrate, fat, and protein to provide the spectrum of substrates required for the generation of the biosynthesis or modification, and repair of tissue components, and also the metabolic energy needed for the generation of mechanical (muscle contraction) and electrical energy (nerve impulse). Also there must be an adequate intake of essential substances not capable of being biosynthesized by the animal in question. These include the essential amino acids, essential fatty acids, and all the vitamins. And finally there must be an adequate intake of macro and trace minerals. Thus the term "food" is a general label to include the variety of chemical substances, including proteins, carbohydrates, fats, minerals, and vitamins, required to provide nourishment (i.e., maintenance of life).

However, the simple process of swallowing food does not imply that these food substances are actually available to sustain life. It is only after the swallowed and digested food has left the stomach and passed through the intestine

1. W. Bayliss, E.H. Starling, The mechanism of pancreatic secretion. The Journal of Physiology (London) 28 (1902) 325–353.
2. J. Henderson, Ernest starling and "Hormones": a historical commentary. Journal of Endocrinology 184 (2005) 5–10.

Hormones. DOI: https://doi.org/10.1016/B978-0-323-90262-5.00024-X

TABLE 7.1 Gastrointestinal hormone families.[a]

Hormone family	Code	Examples of hormones[b]
Gastrin, cholecystokinin	G C	Gastrin, cholecystokinin
Secretin	S	Secretin, VIP, GIP, GLP-1, GLP-2, glicentin/enteroglucagon, glucagon
Pancreatic polypeptide	PP	Pancreatic polypeptide (PP), PYY, NPY
Bombesin and related peptides	BRP	Bombesin and GRP
Tachykinin	T	Substance P, neurokinins A and B
Neurotensin and related peptides	N[c]	NT
Calcitonin gene–related peptides	CGRP	CGRP, amylin
No formal family	See[d]	Ghrelin, motilin, somatostatin

GI, Gastrointestinal; *GIP*, gastric inhibitory peptide; *GLP*, glucagon-like peptide; *GRP*, gastrin-releasing peptide; *NPY*, neuropeptide Y; *NT*, Neurotensin; *PYY*, peptide YY; *VIP*, vasoactive intestinal peptide.

[a]*This table identifies the "families" of gastrointestinal hormones. Each family is characterized pertaining to the similarities of their amino acid sequences of the biologically active peptide; see the many Chapter 7, Gastrointestinal Hormones, figures containing GI hormone amino acid peptide sequences.*

[b]*Most of the hormones listed in the right column are evaluated in Table 7.4 with respect to the following: (1) their amino acid sequence (see figure legends); (2) the stimulatory scenario that results in their secretion; and (3) some of the main biological responses that are modulated by the GI hormone in question (see Table 7.4).*

[c]*Table 7.2 lists neurotransmitters produced by the gastrointestinal tract.*

[d]*The three entries at the bottom of the table do not formally belong to any specific gastrointestinal hormone family. However, each hormone provides an important contribution to the realm of gastroenterology endocrinology.*

TABLE 7.2 Neurotransmitters produced by the gastrointestinal tract.a

Neuropeptides as neurotransmitters	Actions
GRP; bombesin[b]	GRP is a 27 aa peptide, Fig. 7.11.[c] The G cells of the stomach are innervated by the vagus nerve that stimulates the stomach to release GRP. The GRP stimulates the release of gastrin from the G cells of the stomach which, in turn, stimulates HCl production by parietal cells of the stomach.
CGRP	CGRP is a 37-amino-acid peptide that stimulates somatostatin release and smooth muscle contraction.
Endothelin	Endothelin is a potent vasoconstrictor; it acts locally in the GI tract.
Galanin	Galanin is a neuropeptide that is widely expressed in the brain, spinal cord, and gut of humans. It has 30 aa in the human peptide and is found in many subsets of neurons.
NPY	NPY is a 36 aa peptide that is the most prevalent neuropeptide in the brain (see Fig. 7.10). It also functions as a potent vasoconstrictor and inhibits ions and fluid secretion in the small intestine and descending colon. NPY is believed to stimulate food intake.
Neuromedin B, Neuromedin C	Neuromedin B is a 10 aa peptide that is highly conserved across many species. It stimulates gastrin release and inhibits hypothalamus TSH release. Neuromedin C is a 10 aa peptide neurotransmitter that can cause full stimulation of amylase release from the exocrine pancreas.
Neuromedin U	Neuromedin U is a 25-amino-acid peptide that has two receptors: one in the peripheral nervous system, with highest levels in the GI tract and a second receptor in the central nervous system, with the greatest expression in the medulla and hypothalamus.
SP and the tachykinins	SP is an 11-amino-acid peptide that functions as a neurotransmitter especially in the transmission of pain impulses from peripheral receptors to the CNS. In the small intestine, SP can mediate smooth muscle contractions.

aa, Amino acids; *CGRP*, calcitonin gene–related peptide; *CNS*, central nervous system; *GI*, gastrointestinal; *GRP*, gastrin-releasing peptide; *NPY*, neuropeptide Y; *SP*, substance P; *TSH*, thyroid stimulating hormone.

[a]*Table 7.2 lists mostly amino peptide neurotransmitter hormones produced by the gastrointestinal tract. Neurotransmitters are endogenous molecules [e.g., peptide hormones, amino acids (glycine, glutamate), or acetylcholine] that transmit nerve impulses across a synapse to a postsynaptic element, which may be another nerve, or a muscle, organ, or other tissue which has a receptor for the neurotransmitter; see Fig. 7.4B.*

[b]*Bombesin is a member of the bombesin-like family of peptides found in the skin, gut, and brain of frogs; it shares biological activity with the mammalian gastrin–releasing peptide (GRP), the tachykinins, and neuromedin B. There is amino acid sequence overlap shared by bombesin and gastrin-releasing peptide (GRP); see Fig. 7.11.*

[c]*GRP and bombesin each have an identical C-terminal amidated heptapeptide.*

and appeared in the blood or lymph that they can be said to be "in the body." Food substances, such as minerals and free amino acids, can be absorbed with little or no change in their chemical form at the time they were eaten or drunk. Other food components (protein, carbohydrate, or fat) must be subjected to extensive physical and chemical modification prior to their entry into the lymph or bloodstream. Thus proteins, polysaccharides (both are macromolecules), fats (triglycerides), and phospholipids must all be degraded to their constituent components amino acids, mono- and disaccharides, free fatty acids, and glycerol, respectively, before they can be efficiently absorbed.

The overall physical and chemical properties of the various foodstuffs (e.g., water solubility, net charge), as well as the nature of their structural units and bonding (e.g., amide, glycoside, and ester), pose diverse biochemical challenges to be resolved by the digestive system. The system of digestion, then, is the processing of the ingested bulk food into molecular forms capable of participating in the intestinal absorptive process resulting in the successful delivery to the circulatory system.

Table 7.3 summarizes the steps and processes associated with the digestive system. The digestive process involves the complex integration of voluntary and involuntary muscle contractions, parasympathetic and sympathetic neural actions, release of GI and other hormones, and biosynthesis and release of a host of digestive enzymes (e.g., pepsin, trypsin, chymotrypsin, and amylase) and digestive reagents (e.g., HCl and HCO^-), along with digestive detergents (e.g., bile acids). These substances all collectively work together to process the food and present it to the intestinal absorptive process in an optimal form for absorption. The intestinal absorptive process depends upon the integrated functioning of the mouth, stomach, small and large intestine, and the pancreas, liver, and gallbladder.

As a consequence of evolutionary pressures, a number of hormonal systems have emerged to participate in the regulation of digestion and absorption of key dietary nutrients. These include the GI hormones (this chapter), insulin and glucagon (see Chapter 6: Pancreatic Hormones: Insulin and Glucagon), and vitamin D_3 and its steroid hormone form $1\alpha,25(OH)_2D_3$ (see Chapter 9: Calcium-Regulating Hormones: Vitamin D, Parathyroid Hormone, Calcitonin; Fibroblast Growth Factor-23). In addition, many other hormones are known to exert effects on either the digestion process or the GI tract to modulate the absorption of various nutrients.

Many digestive processes, including motility, absorption, and ion transport associated with secretion and also the GI blood flow, are influenced by the nervous system. Some of this control emanates from connections between the digestive system and the central nervous system (CNS). However, the digestive system is endowed with its own, local nervous system referred to as the intrinsic or enteric nervous system (ENS). The ENS is large and complex, with as many neurons as the entire spinal cord.

The ENS, along with the sympathetic and parasympathetic nervous systems, comprises the autonomic nervous system. The ENS governs the activities of the GI system. The functioning of the ENS depends on $\sim$100 million neurons. The principal components of the ENS are two networks, which are both incorporated in the wall of the digestive tract and extend from esophagus to anus: the myenteric plexus is a network of nerve fibers in the circular and longitudinal layers of the esophagus, stomach, and the entire intestine and governs the motility of the digestive tract. The submucous plexus, as its name implies, is located in the submucosa (see Section II.C). Its principal role is in sensing the environment within the intestinal lumen, regulating GI blood flow, and controlling epithelial cell function.

II Anatomical and physiological relationships

A Gastroenteropancreatic system

The gross anatomical organization of the digestive tract is shown in Fig. 7.1. It includes the oral cavity with its salivary glands, the esophagus, the stomach, small intestine, and the large intestine and the rectum and anus of humans. Fig. 7.1 also illustrates the anatomical relationships between the stomach, liver, pancreas, gallbladder, and the common bile duct.

The digestive tube or alimentary canal in humans is an $\sim$8-m long muscular tube extending from the lips to the anus. Fig. 7.1 identifies the individual components of the alimentary canal; each part makes an essential contribution to the overall process of making ingested food ultimately available as relatively small molecules, for example, amino acids, glucose, glycerol, fatty acids, for all the cells of the body. The operations within the digestive tube are supplemented by several accessory organs, including the teeth, tongue, salivary glands, and pancreas, and liver secretions.

B Stomach

The human stomach is a muscular organ that is located on the right side of the upper abdomen. The stomach is connected to the bottom of the esophagus and to the top of the small intestine. The stomach receives partially processed food from the mouth via the esophagus. The arrival of the partially digested food in the stomach initiates the secretion of acid and enzymes that further digest the food components. The interior ridges of muscle tissue, called rugae, line the interior surface of the stomach. The total interior surface area of the human stomach is $\sim 800\ cm^2$. In adult humans, who have not eaten for some time, their "empty" stomach volume will be in the range of 50–75 mL. In contrast, after consuming a large meal, the interior volume of the stomach can readily hold about 1 L. It is possible for the human stomach to further expand so that it can hold up to 2–3 L. The stomach muscles contract periodically, churning food to enhance the

TABLE 7.3 Components and steps of the digestive process.

Process	Anatomical location	Purpose
Food entry		
Consumption of food/mastication	Mouth	Support bodily nutrient requirements
Salivary secretion (stimulated by the nervous system): ~1000–1500 mL is secreted per day	Salivary glands	Conversion of starches into dextrins and maltose
Voluntary swallowing of food	Tongue/pharynx/esophagus	Entry of food into the stomach
Involuntary peristaltic muscular contraction	Esophagus	To move food into the stomach
Digestion in the stomach		
Secretion of gastric juice (HCl+pepsin) as a consequence of neural, mechanical, and hormonal (gastrin) stimuli (~2000 mL secreted/day)	Stomach	Conversion of proteins to polypeptides
Involuntary mechanical contractions	Stomach	Mixing of intestinal contents and when the pyloric sphincter opens, the transfer of stomach contents (chyme) into the top of the small intestine (duodenum)
Digestion in the small intestine		
Neural and hormonal (secretion of pancreozymes); mediated secretion of pancreatic juice (~1200 mL secreted/day)	Exocrine or acinar pancreas secretions are ducted into the duodenum	
Bicarbonate secretion		HCO_3 neutralizes HCl from stomach and creates a favorable duodenal pH (6.0–7.0) so that pancreatic enzymes may operate efficiently
Amylase, trypsinogen, and chymotrypsinogen secretion and conversion to "active" enzymes		Hydrolysis of amide, glycoside, and ester bonds of peptides, carbohydrates, and lipids
Neural stimulation and duodenal muscular distension and stimulation of intestinal juice (~4000–5000 mL secreted/day)	Duodenum	
Enterokinase (activates trypsinogen conversion into highly active trypsin)		Further processing of proteins→peptides
Proteases (trypsinogen, chymotrypsinogens, lipases, amylases)		Further processing of proteins→peptides
Sucrase, maltase, lactase; specific transport systems (all bound to brush border membranes)		Further processing of dextrins and other partially hydrolyzed substances
Synthesis and secretion of bile acids (by the liver) followed by storage in the gallbladder; hormonally stimulated secretion (by cholecystokinin and secretin) of the gallbladder to release bile acids (~600–800 mL secreted/day)	Liver (site of production), gallbladder (storage), transport (through bile duct) into duodenum	Emulsify fats to facilitate absorption of fatty acids
Involuntary mechanical activity	Duodenum≫≫jejunum≫ileum	Squeezing and mixing of intestinal contents (8–10 contractions/min in duodenum)
Peristalsis	Duodenum≫≫jejunum≫ileum	Onward movement of intestinal contents
Chemical activity	Duodenum, jejunum, and ileum	Continued saponification of fat and hydrolysis of proteins→peptides to amino acids and carbohydrates-dextrins to mono- and disaccharides followed by cellular absorption and transport to the blood or lymph compartments
Digestion in the large intestine involuntary mechanical activity	Large intestine	Mixing movements that promote reabsorption of H_2O and compaction of intestinal contents (~7000–8000 mL of H_2O reabsorbed/day)
Bacterial action	Large intestine	Acts on undigested residues to effect the fermentation of carbohydrate and further degradation of proteins as well as biosynthesis of vitamin K by resident bacteria in the colon

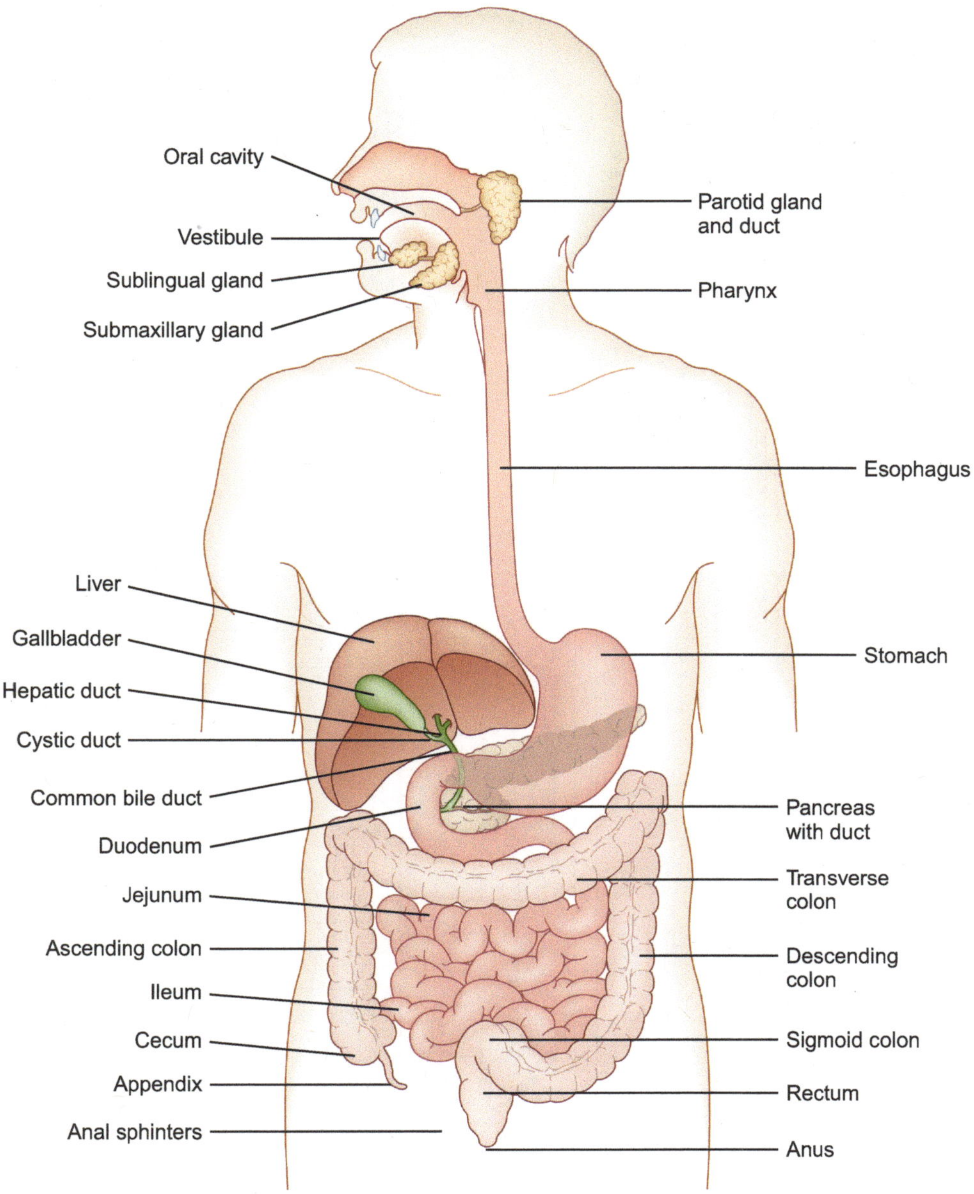

FIGURE 7.1 Diagram of the human digestive system from the port-of-entry oral cavity to the exit at the anus. The total length of the digestive system in an adult is ~27 ft, or 4.5 × the average height of 6 ft of an adult. The esophagus is ~10 in in length. The small intestine is ~22–23 ft in length and the large intestine (the colon) is ~5 ft in length.

digestive process. The pyloric sphincter at the bottom of the stomach is a muscular valve that opens to allow food to pass from the stomach to the top of the small intestine.

Fig. 7.2 presents a cross-sectional view of the human stomach. In panel A the interior gastric mucosa is divided into six anatomical regions. Panel A illustrates, by color, the functionally distinctive regions of the stomach. The *cardia* (1, *light red*) is attached to the bottom of the esophagus through a one-way valve termed the cardiac sphincter so that the stomach can receive the liquids and solids that were recently chewed and swallowed. It is followed by the *fundus* (2, *blue*) at the top of the stomach and the *corpus* (3, *green*) which is the largest component of the stomach.

The fundus (*blue*) has ~20% of the total stomach volume. Food acquired from the esophagus and cardia is ground, mixed, and treated with some digestive enzymes. After about 50 minutes, this *chyme* (acidic digested food) is passed on from the antrum to the corpus for further processing.

The corpus (*green*) constitutes ~50% of the total stomach volume and is the location of vigorous digestion which involves the secretion of hydrochloric acid and pepsinogen. Gastric acid which is secreted by the corpus consists mainly of hydrochloric acid which acidifies the stomach contents to a pH of 1–2. The HCl has two roles. First, this strong concentration of HCl results in the denaturation of the three-dimensional structures of most proteins. Second, the HCl activates the proenzyme, pepsinogen, to release catalytically active pepsin. Pepsin has its highest enzymatic activity at an ambient pH of 2 and a temperature between 37°C and 40°C. The pepsin denatures and degrades many

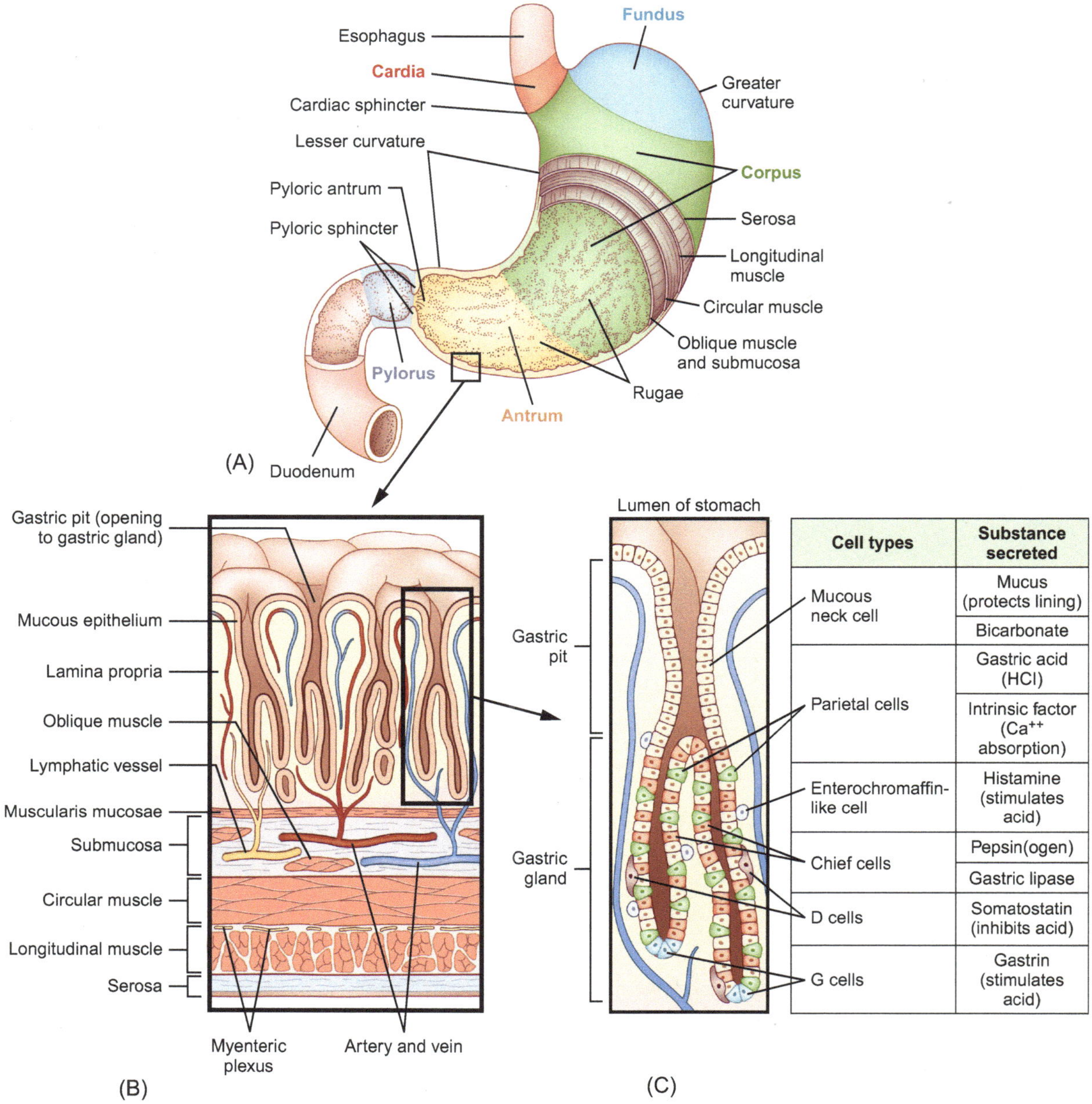

FIGURE 7.2 Longitudinal sections of the human stomach (A). The stomach has five distinctive regions based on physiological function. 1/Cardia (*light red*) is attached to the bottom of the esophagus so that it can deliver to the stomach the liquids and solids that were recently chewed and swallowed. 2/Fundus (*blue*) is at the top of the stomach. 3/Corpus (*green*) is the largest component of the stomach and is followed by 4/antrum (*gold*), which is the second largest component. The bottom of the stomach's antrum merges to create 5/the pyloric antrum (also *gold*), which is connected to the 6/pyloric sphincter, or valve, which is also associated with the pyloric canal (*gray*). The pylorus is a strong ring of smooth muscle at the end of the pyloric canal (about 1-in long) which controls the careful release of acidic digested food (chyme) into the duodenum where it will be subjected to further processing. (B) Blueprint of the five tissue layers that consist of two connective tissue layers (serosa and submucosa) and three muscle layers (circular muscle, longitudinal muscle, and muscularis mucosae). These three muscles can contract and are responsible for some of the many movements of the stomach. The serosa that lies under the longitudinal muscle is the outermost layer that surrounds and encloses the entire stomach organ. The submucosa and the muscularis mucosae constitute a boundary that lies above the circular muscle. (C) As shown in the table portion of panel (C), the gastric pits and gastric glands include many cell types (parietal cells, EC cells, chief cells, D and G cells, and mucous cells), which collectively secrete useful reagents for the digestive process. *EC cells*, Enterochromaffin-like cells.

food proteins to peptides. Pepsin cleaves peptide bonds between aromatic and hydrophobic amino acids. Also, the enzymes trypsinogen and chymotrypsinogen are secreted by the corpus and, after their activation, they too can break down food proteins.

The *antrum* (4, *gold*), which has the second largest stomach volume ($\sim$30%), is located at the bottom of the stomach. It carries out the same digestive process as the corpus. However, the antrum anatomically evolves to create the *pyloric antrum* and the *pyloric sphincter* (5; also *gold*) at the very bottom of the antrum that leads to the pyloric canal (6, *gray*). The pyloric sphincter (valve) is a strong ring of smooth muscle about 1-in long, which controls the careful release of chyme from the stomach into the intestinal duodenum, where it will be subjected to further processing.

Fig. 7.2, panels B and C, illustrate the contributions of gastric pits and gastric glands from the antrum toward providing hydrochloric acid to facilitate food processing into chyme. This is a highly regulated process. The human stomach has several million of these pits, which cover the interior surface of the stomach lining and are known as the rugae. Also, there is a wide variety of specialized cell types in the gastric gland which provide key contributions to the digestive process. See the table in panel C of Fig. 7.2, which lists the antrum's cell types and the substance(s) secreted. These include parietal cells, enterochromaffin-like (ECL) cells, chief cells, D and G cells, and mucous cells, which collectively secrete useful reagents for the digestive process.

C Small intestine and colon

For an overall view of the intestinal tract see Figs. 7.1 and 7.3. The small intestine of humans is divided anatomically into three segments. At the top the duodenum

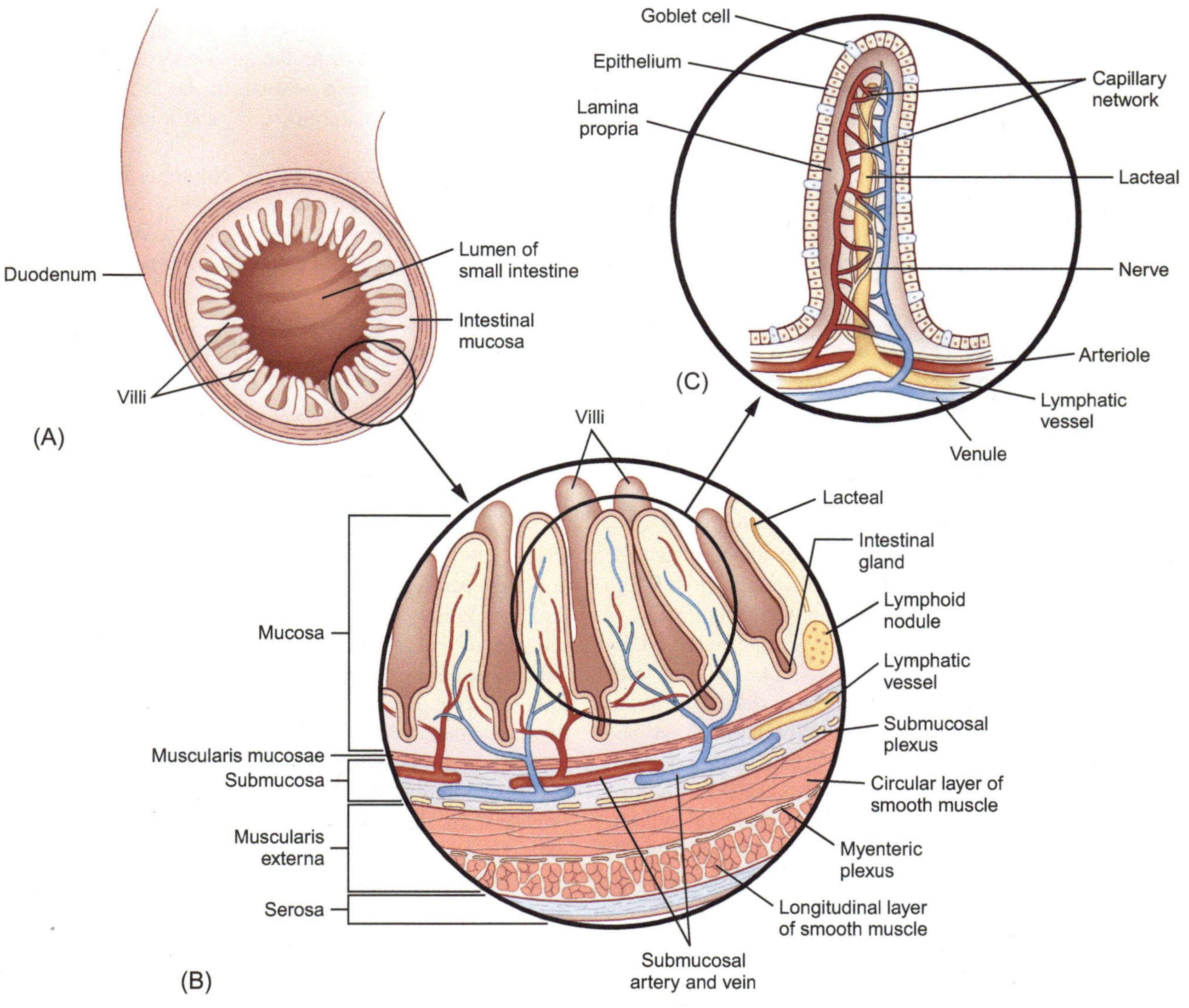

FIGURE 7.3 Organization of the small intestinal duodenum from the macro level down to just one intestinal villus. (A) Macro details of the cross section of the duodenum. The outer width of the duodenum is ~2 cm and the inner width of the lumen is ~1 cm. The predominant cell type is the abundant villus; the villi face the interior of the duodenum, jejunum, and ileum and are responsible for nutrient absorption. (B) Close-up of the intestinal columnar villi that line the inner surface of the duodenum. The lumen of the small intestine is composed of a multitude of these finger-like projections, which are termed villi. See the black circle in panel A illustrating the source of the highly magnified image of the intestinal columnar villi. (C) Cross-sectional highly magnified view of one villus. Typically, the villi range from 0.5 to 1.5 mm in length.

($\sim$25 cm in length and 3.75–5 cm in diameter) is connected to the pylorus and the pyloric sphincter of the stomach and at the bottom to the jejunum. The middle jejunum is $\sim$275-cm long and $\sim$4 cm in diameter and is connected to the ileum. The ileum is $\sim$400 cm in length and $\sim$3.7 cm in diameter. The ileum connects to the colon. The colon is $\sim$180 cm in length and is sequentially divided into the cecum, followed by the ascending colon, the transverse colon, and the descending colon, which connects to the sigmoid colon, which attaches to the rectum. The appendix is attached to the cecum. The human appendix is approximately 11 cm in length by 7–8 mm in diameter. The length of the appendix can range from 2 to 20 cm.

The morphology and cellular organization of the intestine are superbly adapted to efficiently effect the absorption of dietary constituents; see Fig. 7.3A–C. The predominant cell types are the abundant villi (see next paragraph) that face the interior cylindrical surface of the duodenum, jejunum, and ileum and that are responsible for nutrient absorption. Panel A illustrates the details of the cross section of the duodenal wall ($\sim$2 cm in height with a lumen down the center that is $\sim$1 cm in width). See the black circle in panel B; it illustrates the intestinal columnar villi, which line the inner surface of the lumen of the duodenum. The wall of the small intestine is composed of four layers: (1) the mucosa or mucous membrane; (2) the submucosa which has both arteries and veins; (3) the muscularis externa, which is a thin layer of smooth muscle at the bottom of the villi; and (4) villi at the bottom which have the muscularis externa and the serosa.

The lumen of the small intestine is composed of a multitude of finger-like projections that are termed *villi*; see Fig. 7.3, panel C. Typically the villi range from 0.5 to 1.5 mm in height. As a consequence, the functional surface area of the interior duodenum is greatly increased, by about 600×, compared to what its total surface area would be without microvilli. Thus the surface area available for the absorption of partially digested food and micronutrients is greatly enhanced. The villi comprise two kinds of cells: they are the predominant columnar epithelial cell and the intermittent goblet or mucus-secreting cell. The mucous secretion of the goblet cells lubricates the surface of the lumen of the GI tract to facilitate and enhance the movement of digested food along the full length of the small intestine and colon. Beneath the columnar epithelial cells lie the lymphatic and blood vascular systems, which project into each villus and, thus, provide an effective mechanism for the translocation of the various substrates once they exit from the epithelial cells.

The mucosal cells originate in a progenitor population existent at the base of the villus in the crypt of Lieberkühn. As they differentiate, the cells begin to migrate up the villus. Ultimately, after migration up the villus, the cells are sloughed off into the lumen of the intestine. Only after epithelial cells enter the villus do they differentiate into fully functional absorptive cells. Intestinal epithelial cell turnover times of 70–100 hours have been reported in mice, rats, and humans. The rate of epithelial cell renewal increases when there is an increased demand for new cells, as in the recovery from radiation injury, response to parasite infestation, or adaptation to partial intestinal resection.

The apical surface of these cells, which is exposed to the lumen, is referred to as the brush border or the microvillar membrane. A variety of carbohydrate hydrolases and an alkaline phosphatase are specifically associated with this brush border region of the intestinal mucosal cell. This is believed to be the site of localization of permeases that have been shown to be involved in the absorption of food materials, Ca^{2+} and phosphate into the cell.

The outermost surface of the columnar epithelial cell in the microvilli or brush border region is covered with a filamentous, mucopolysaccharide coat termed the glycocalyx or "fuzzy coat." The function of the glycocalyx is not specifically known.

The myenteric plexus (Fig. 7.3, panel B) is a network of nerve fibers in the muscular lining of the esophagus, stomach, and intestines.

D Hormone-secreting cells: their distribution in the gastroenteropancreatic complex

Table 7.4 lists 11 different endocrine-related cell types and their partner hormones that are present in the GI system. A variety of the endocrine cells are dispersed throughout the gastric mucosa, small intestine, and colon. Two basic kinds of cells are identifiable by classical histological procedures of staining with chrome and silver: they are enterochromaffin cells and agyrophil cells or "silver-loving cells." The enterochromaffin cells are sparsely present in the gastric (stomach) mucosa but are fairly prevalent in the small and large intestines. In humans there are approximately twice as many agyrophil cells as enterochromaffin cells in the intestine.

By using immunofluorescent techniques, it has been shown that in the gastroenteropancreatic complex, there may be as many as 11 different types of hormone-secreting cells. These cells are capable of secreting gastrin, motilin, secretin, cholecystokinin (CCK), somatostatin, histamine, GLP-1 (glucagon-like peptide), GLP-2, PYY (peptide YY) GIP (gastric inhibitory peptide), GIP, and the pancreatic polypeptides (PPs). In many cases, these cells have microvillar borders facing the lumen of

TABLE 7.4 Endocrine-related cell types and their hormones that are present in the gastrointestinal system.

Cell name	Location	Function of cell
Chief cell	Small intestine	Stimulate propepsinogen and secretin secretion[a]
D cell	Stomach	Somatostatin secretion
ECL cell[b]	Antrum of stomach; duodenum	Histamine secretion
G cell	Stomach and duodenum	Gastrin secretion as a messenger
I cell	Duodenum and jejunum	CCK secretion
K cell	Intestine	GIP secretion
L cell	Ileum, colon, rectum	GLP-1, GLP-2, and PYY secretion[c]
M cells	Duodenum and jejunum	Motilin secretion
N cells	Small intestine and ileum	Neurotensin
S cells	Duodenum and jejunum	Secretin secretion
Parietal cell	Stomach	Stimulate hydrochloric acid secretion; and intrinsic factor for Ca^{2+} absorption

CCK, Cholecystokinin; *ECL*, enterochromaffin-like cells; *GIP*, gastric inhibitory peptide; *GLP*, glucagon-like peptide; *PYY*, peptide YY.
[a]*The propepsinogen becomes catalytically active in an acid environment. In addition, the hormone secretin stimulates the pancreas to secrete bicarbonate and send it to the small intestine to neutralize the gastric HCl.*
[b]*ECL cell is an enterochromaffin-like cell that secretes histamine.*
[c]*GLP-1 and GLP-2 are glucagon-like peptides.*

the stomach or the intestine so that they can respond to chemicals in the stomach or intestinal contents. Most of the cells appear to produce a single peptide hormone, but in some instances they also produce biogenic amines such as histamine or 5-hydroxytryptamine.

There are four general pathways by which peptide hormone–secreting cells, also known as enteroendocrine cells, of the GI tract can make their chemical messengers available; see Fig. 7.4. The enteroendocrine cells are located in the intestine, the stomach, and the pancreas. These pathways include the following models: (1) paracrine delivery where the hormone (H) or the neurotransmitter (NT) moves from the endocrine cell source to immediately adjacent cells; (2) "neurocrine" delivery where there is a release of an NT by the nerve to the adjacent environment; (3) "open" endocrine cell delivery, where there is a release of the H from an endocrine cell that is "open" to the intestinal lumen so that the H can diffuse out and enter adjacent cells where there should be a cognate receptor; (4) "closed" endocrine cell delivery in which the hormone-producing cell does not face the brush border surface and also cannot engage in paracrine delivery. The release of the hormone H is directly to the blood compartment where it can be transmitted to cells that have a receptor for the hormone (see Fig. 7.4D). Examples of GI hormones that function via these pathways are as follows: paracrine fashion [PP, vasoactive intestinal peptide (VIP)], neurocrine [neuropeptide Y (NPY), calcitonin gene–related peptide (CGRP)], open endocrine delivery (gastrin), and closed endocrine delivery (secretin, motilin, CCK). Fig. 7.19 in Section III. L provides a description in which stomach G cells and the hormone gastrin collaborate as both neurotransmitters (NTs) that interact with a nerve for delivery and which then send an NT hormone to a specified receptor in a target cell.

E Pancreatic, biliary, and intestinal secretions

1 Pancreatic secretions

Exocrine pancreas secretion occurs after the ingestion of a meal and is associated with the appearance of the acidified chyme from the stomach in the duodenum. There is a need to neutralize the pH of the incoming acid. As a consequence, there is a coupled secretion by the exocrine pancreas of the pancreatic juice, which is a rich source of HCO_3^- (bicarbonate, pH ~7) and also of many enzymes. The composition of the human pancreatic juice is given in Table 7.5.

Many of the potent enzymes are secreted in inactive pro forms and must be converted separately to active enzymes in the lumen of the intestine. The pancreatic secretion is mainly dependent upon the actions of CCK; in addition, the stimulation of appropriate cholinergic or peptidergic nerves leading to the pancreas can stimulate exocrine pancreatic secretions.

Pancreatic secretions of electrolytes and H_2O from the ductular and centroacinar cells are primarily under the influence of secretin. Secretin is known to stimulate

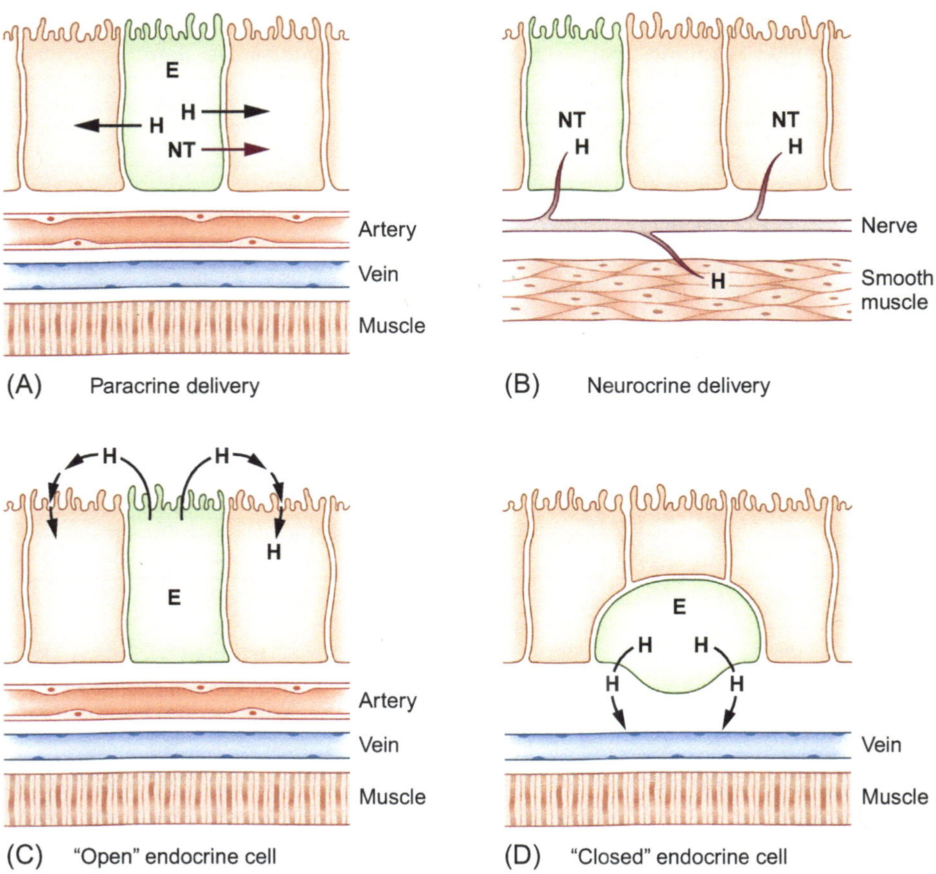

FIGURE 7.4 Schematic models describing pathways by which gastrointestinal hormone–producing cells make their chemical messengers either available locally or deliverable through the circulatory system or via the activation of a neuron. (A) Paracrine delivery, where the hormone (H) or NT moves from an endocrine cell to immediately adjacent cells. For some examples of neurotransmitters that are amino acid peptides see Figs. 7.10 and 7.11 and the neurotransmitters (NPY), and GRP. (B) Neurocrine delivery, where there is release of an NT by the nerve to the local environment, which can include mucosa, other nearby endocrine cells, or smooth muscle cells. (C) Release of a hormone(s) from an endocrine cell that is "open" via the secretion of the H to the intestinal lumen and adjacent cells. (D) Release or secretion of hormones to the blood from an endocrine cell that is "closed" to adjacent cells and does not have an opening to the intestinal lumen. *GRP*, Gastrin-releasing peptide; *NPY*, neuropeptide Y; *NT*, neurotransmitter.

TABLE 7.5 Composition of human pancreatic juice.

Components	
Cations Na^+, K^+, Ca^{2+}, Mg^{2+}, pH=8.0	
Anions HCO_3^-, Cl^-, SO_4^{2-}, HPO_4^{2-}	
Digestive enzymes	**Substrate**
Trypsinogen*b*	Proteins, polypeptides
Chymotrypsinogen*b*	Proteins, polypeptides
Prolactases	Elastin, other proteins
Procarboxypeptidase A	Proteins, polypeptides
Procarboxypeptidase B	Proteins, polypeptides
Prolipase	Fat droplets
Pancreatic lipase	Triglycerides
Phospholipase A_2	Phospholipids
Cholesterol–ester–hydrolase	Cholesterol esters
α-Amylase	Starch
Ribonuclease	RNA
Deoxyribonuclease	DNA

Source: Information in this table was abstracted from W.F. Ganong, Review of Medical Physiology, Chapters 25–26, Appleton-Lange, New York, 1995.

intracellular cyclic AMP (cAMP) in these cells, which activates unknown factors necessary to alter the secretory cell membrane permeability to sodium and hydrogen ions. Thus there is an increased exchange of external Na^+ ions for internal H^+ ions. The increased concentration of extracellular H^+ lowers the local pH, which then increases the production of CO_2 from circulating bicarbonate. This CO_2 then diffuses into the cells, where it combines with water to form carbonic acid (mediated by carbonic anhydrase) and results in the production of HCO_3^-, which can be utilized for secretion.

The pancreatic secretion of digestive enzymes that occurs in the acinar cells of the exocrine pancreas is under the influence of CCK. In the pancreatic acinar cells, CCK interacts with a CCK-A receptor present on the outer cell membrane. This activates a G protein–coupled signal transduction pathway to ultimately stimulate the secretion of amylase, trypsinogen, and chymotrypsinogen stored in secretory granules. In contrast, when CCK binds to the CCK-B receptor it has stimulatory effects on the vagus nerve and has functions relating to anxiety and the perception of pain.

2 Biliary secretion

The liver produces 500–1500 mL of bile per day. Bile canaliculi also known as bile capillaries are thin tubes that receive bile secreted by *hepatocytes*. The bile canaliculi eventually merge and form bile ductules. The bile passes through canaliculi to the hepatic bile ducts and then into the common hepatic duct which drains directly into the duodenum. Table 7.6 summarizes the major organic components present in human bile.

As bile flows through the bile ducts, it is diluted by the addition of a watery, bicarbonate-rich secretion from ductal epithelial cells. Then the bile passes through small ducts that empty directly into the upper portion of the duodenum. This results in the release of secretin from S cells in the small intestine. Secretin stimulates the pancreatic secretion of bicarbonate and water.

TABLE 7.6 Major organic components of human bile.

Components	Bile (mg/mL)
Dihydroxycholanic acids	49
Cholic acid	32
Phospholipid (mostly lecithin)	27
Protein	5
Cholesterol	3
Bilirubin	3
Hexose and hexosamine	1

Source: Abstracted from F. Nakayama, H. Miyake, Species differences in cholesterol-complexing macromolecular fractions in bile in relation to gallstone formation, Journal of Laboratory and Clinical Medicine 67 (1966) 78–86.

Similarly, the exocrine pancreas has ducts that are arranged in clusters called *acini* that merge into the main pancreatic duct which also drains directly into the duodenum. This results in the secretion of CCK by the I cells in the duodenum and jejunum, which results also in the stimulation of secretion of acinar pancreatic amylase, trypsin, and chymotrypsin secretions directly into the duodenum and separately stimulation of the contraction of the gallbladder. CCK mediates the contraction of the gallbladder to release bile into the intestine. The EC_{50} (the concentration of CCK that gives 50% of the maximal response) for this contraction is in the range of 10–50 nM CCK which is within the range of the affinity of CCK1R (or CCKαR) for CCK. Coincidentally, CCK2R (CCKβR) requires a different concentration range of CCK and also is not expressed by gallbladder smooth muscle, whereas gallbladder smooth muscle expresses only CCK1R. Accordingly, CCK binds to its CCK1 receptor (CCKαR) on the gallbladder smooth membrane which signals gallbladder contraction and subsequent emptying of bile into the intestine. Fig. 7.5 summarizes, in addition to the action of CCK, the mechanisms by which extrinsic and intrinsic nerves mediate gallbladder mobility.

Bile is a watery mixture of inorganic and organic compounds that are synthesized by the liver and then stored in the gallbladder. Table 7.6 presents the major organic components of human bile. These constituents are secreted into the intestinal lumen in response to the presence of dietary fat and they act as detergents to disrupt and disperse the oil droplets.

3 Intestinal secretion

Several gut hormones, including VIP, GIP, secretin, CCK, and glucagon, and prostaglandins E_1, E_2, and $F_2\alpha$ have been shown to affect the small intestine and colon by either inhibiting the active absorption of electrolytes and H_2O or stimulating the secretion of H_2O and electrolytes. Both VIP and the prostaglandins are potent stimulators of the adenylate cyclase-cAMP system, and it is probable that this is the basis for their actions on the intestinal secretion of H_2O and electrolytes. The cellular mechanisms of the other gut hormones in this system are not yet known.

F Coordination of gastroenteropancreatic hormone release

A number of GI hormones can only be secreted by their cells of origin as a consequence of the active digestion and absorption of nutrients in the stomach and duodenum.

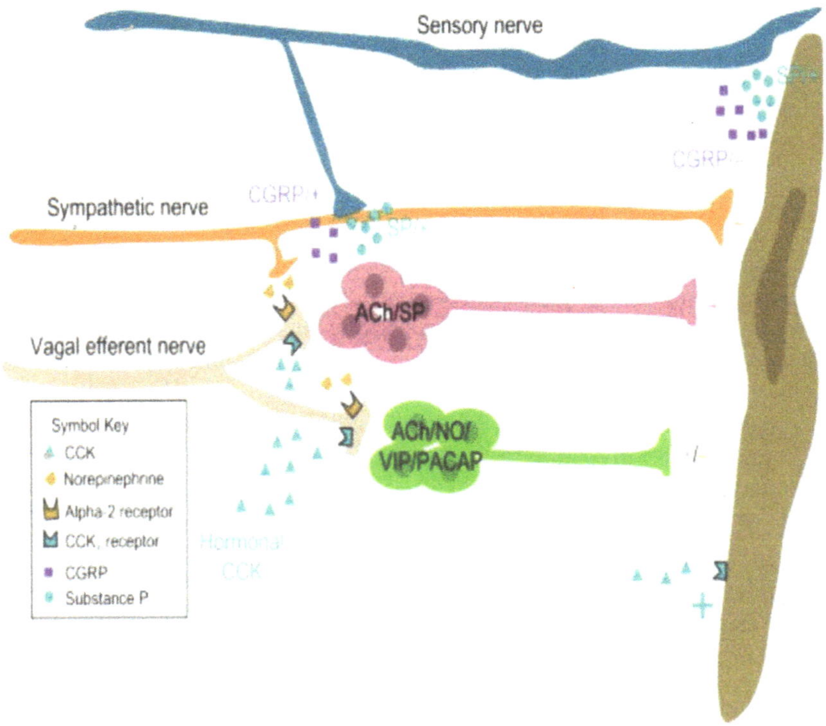

FIGURE 7.5 Schematic diagram demonstrating the mechanisms by which intrinsic and extrinsic nerves can modulate gallbladder motility. *Reproduced from Fig. 30.4F of G.M. Mawe et al., Neuromuscular function in the biliary tract, in Macsween's Pathology of the Liver, seventh ed., Elsevier, 2017.*

The stomach has been recognized to be the major control center to coordinate digestion. Table 7.4 summarizes 11 types and sources of secretory cells of the stomach and intestine. The ingestion of nutrients *into the duodenum* is postulated to be the initiating signal for gastroenteropancreatic hormone release. Ingestion is accomplished by drinking or eating food through the mouth and delivering it via the esophagus and stomach to the GI tract. Carbohydrate ingestion leads to the secretion of GIP and enteroglucagon, while fat ingestion stimulates CCK, GIP, neurotensin, and possibly motilin release.

Because these hormonal responses occur within 15 minutes of ingestion, it seems probable that cellular absorption of the dietary components cannot have played a significant role in stimulating the prompt but selective release of the GI hormones. It is now known that gastric hormones or releasing factors function as control signals for hormone release from the duodenal, jejunal, and ileal cells, as well as from the pancreas. A gastrin-releasing protein has been isolated from the fundic portion of the stomach.

Pepsin is the most important protease in initiating the breakdown of proteins in the stomach. Pepsin can break down any protein in the diet. Pepsinogen (Mol.Wt 40,554) is an inactive enzyme (zymogen) in the stomach and has an additional 44 amino acid residues more than the catalytically active daughter pepsin. The vagus nerve and gastrin hormone facilitate the release of both HCl and pepsinogen from the stomach lining when food is present.

The zymogen pepsinogen is activated solely by the presence of hydrochloric acid at a pH of $\sim$2.5. The presence of the acid permits the pepsinogen to partially unfold and use its autocatalytic properties to cleave off its own 44 amino acid N-terminal peptide, thereby creating active pepsin (Mol.Wt. 34,614). Pepsin then processes the incoming proteins in the stomach and cleaves at any peptide bonds on the amino-terminal side of one of the aromatic amino acid residues (Phe, Trp, and Tyr), creating large peptides.

The challenge is that while the stomach is extraordinarily well organized, it does *not* have the responsibility of converting *all* the food components into free amino acids, and peptides of various sizes, etc. That responsibility falls principally to the proteases available in the duodenum, jejunum, and ileum with the assistance of the pancreas. Accordingly, the pH$\sim$2.5 environment work-product of the stomach's fundus, corpus, and antrum (large peptides, carbohydrates, and fat) are sent through the pyloric sphincter to the duodenum. This promptly signals the release of the GI hormone, secretin, into the blood for the delivery of the secretin messenger to the pancreas. This activates the production of the pancreatic secretion containing two different proteases, zymogen granules that contain either catalytically inactive trypsinogen or chymotrypsinogen. Also at the same time, the procarboxypeptidases A and B are also packaged as zymogen granules. Collectively, they are all delivered to the duodenum via the pancreatic duct. Secretin also stimulates the transport of bicarbonate from the pancreas via the pancreatic duct to its release in the duodenum. The presence of the bicarbonate has the desired consequence of neutralizing the gastric HCl to ~pH of 7.

When the stomach sends partially ingested food nutrients to the duodenum, they become candidates for further breakdown by proteases present in the duodenum, jejunum, and ileum. The ultimate goal is to have small peptides that can be readily absorbed by the duodenal intestinal lining. Enterokinase, also known as enteropeptidase, is secreted from the intestinal glands of the duodenum; see Fig. 7.3, panel B. The enterokinase enzyme converts the inactive zymogen form of trypsinogen (received from the pancreas) into a catalytically active trypsin.

The duodenum secretes enterokinase promptly after the passage through the pyloric sphincter and arrival in the duodenum of digested proteins and other food components. Newly secreted enterokinase consists of a disulfide-linked 82–140 kDa heavy chain which holds the enterokinase in the intestinal brush border and a 35–62 kDa light chain that contains the catalytic subunit that will convert principally trypsinogen (also a zymogen) into a catalytically active trypsin protease. Enterokinase is a serine protease enzyme that preferentially cleaves after a lysine if the following amino acid sequence (Asp–Asp–Asp–Asp–Lys) is present in the candidate zymogen. Trypsinogen is virtually the only protein which has this necessary signal sequence. Trypsin is the principal protease present in the intestine.

Fig. 7.6 summarizes the activation process for six zymogens (chymotrypsinogen, proelastase, procarboxypeptidase, prophospholipase, procolipase, and kallikreinogen); see the left and right green boxes. Colipases are enzymes that break down fats to fatty acids and glycerol. Kallikreins are enzymes that cleave peptide bonds in proteins. Each of these zymogen enzymes are activated in the duodenum by trypsin to produce six active enzymes that also collectively contribute to further maximizing the breakdown process of all the ingested food particles.

The physiological importance of enterokinase (also known as enteropeptidase) is emphasized in the second pink arrow of Fig. 7.6. As emphasized in Section I.C, "Problems of Food Processing and Digestion," it is essential that the GI digestion system be very efficient in processing the food intake to provide maximal amounts of amino acids and carbohydrates which can all be reutilized for ultimate transfer to the circulatory system for delivery to the liver and muscle to participate in gluconeogenesis and/or glycolysis.

G Motor functions of the intestinal tract

The GI hormones play an important physiological role in the regulation of the motor activity of the GI tract. This includes effects on the stomach, small intestine, colon, gallbladder, and bile ducts. The hormones may have indirect effects (neurally mediated) or direct actions (muscular) via smooth muscle motor activity. The GI hormones that primarily stimulate motor activity are gastrin, CCK, and motilin, while the peptides that inhibit motor activity are secretin, VIP, glucagon, and enteroglucagon. Gastrin exerts its effects partly via direct interaction with intestinal or stomach muscle cell receptors and partly via postganglionic cholinergic fibers; it is effective at very low concentrations (e.g., $1-100 \times 10^{-12}$ M).

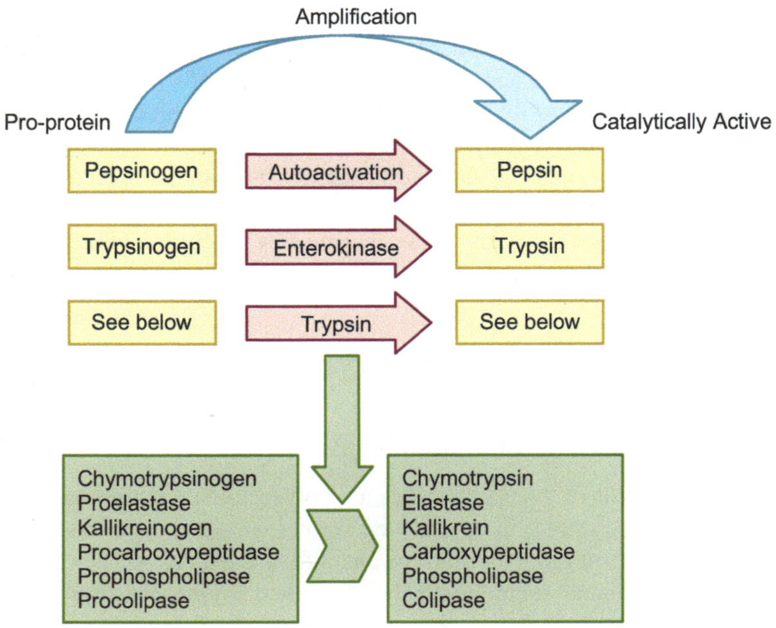

FIGURE 7.6 Activation of both the duodenal preprotease enzymes (enterokinase and trypsinogen) and the stomach's pepsinogen are required to convert an inactive enzyme (zymogen) into a catalytically active useful enzyme. The pair of top yellow boxes describes the autoactivation in the stomach of pepsinogen into active pepsin. Pepsinogen that is secreted into the stomach's corpus and antrum where it is quite acidic (pH 2–3) becomes autoactivated. The second pair of yellow boxes shows the amplification by enterokinase of trypsinogen ≫ trypsin. Trypsinogen becomes activated by the duodenal peptidase, enterokinase (also known as enteropeptidase), by the cleavage of trypsinogen's lysine residue 15. This fully activates the trypsin (right box). The bottom pair of yellow boxes describes the active trypsin activating each of the six proenzymes (see left *green box*) into an active enzyme (right *green box*). The mechanism of pepsinogen's self-activation is described under Section II.E, "Coordination of Gastroenteropancreatic Hormone Release."

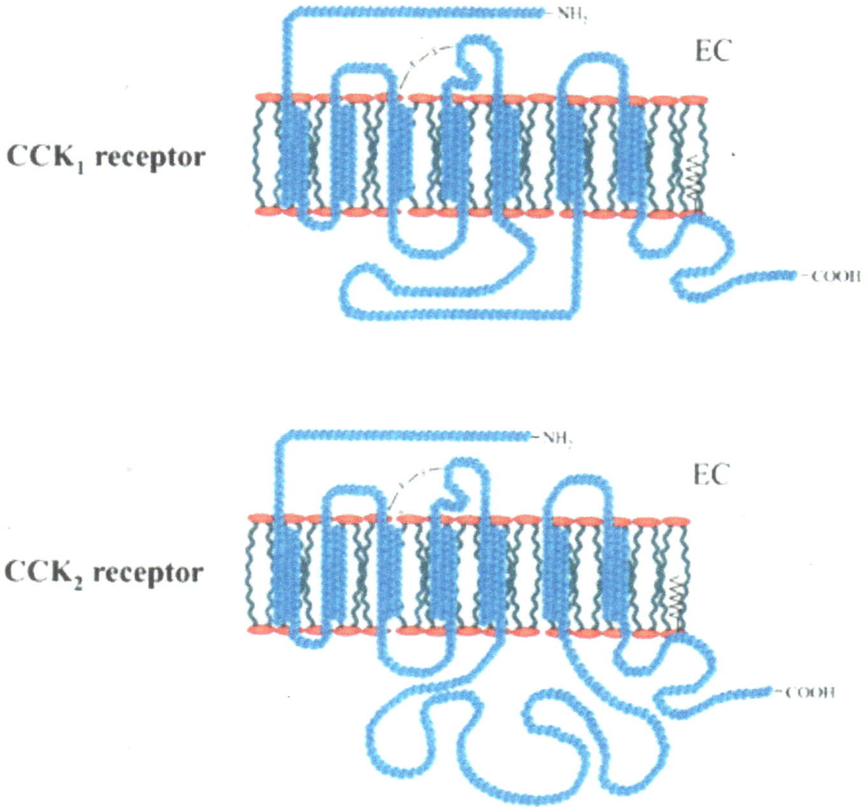

FIGURE 7.7 Predicted membrane topology of CCK$_1$ (CCK-A) and CCK$_2$ (CCK-B) receptors. Both belong to the family of G protein–coupled receptors with seven transmembrane domains. A large portion of the amino acid sequence is conserved. EC, extracellular side of the membrane. Differences in the two structures occur mainly in the cytoplasmic domains. *CCK*, Cholecystokinin. *Reproduced from Fig. 7.2 of G. Varga, A. Balint, B. Burghardt, M. D'Amato, Involvement of endogenous CCK and CCK$_1$ receptors in colonic motor function, British Journal of Pharmacology 141 (2009) 1275–1284; https://doi.org/10.1038/sj.bjp.0705769.*

The actions of CCK binding to its CCK-B receptor (Fig. 7.7) have stimulatory effects on the vagus stimulation of gallbladder contraction. Emptying of bile is the result of four successive events. First, there is a progressive increase in the tension of the gallbladder wall. Then an intermittent opening of the sphincter-like cholecystocystic junction occurs. Next, there are phasic contractions along the common axis of the bile duct that finally result in the sequential opening and closing of the choledochoduodenal sphincter. This results in the periodic discharge of bile into the duodenum. As yet no molecular mechanisms of hormone action are available to describe these complex and critical physiological processes.

H Brain–gut axis

There is now good neurophysiological and neuroendocrine evidence describing the brain–gut axis. A surprising number of gut hormones have also been found in the brain (these include CCK, gastrin, ghrelin, secretin, VIP, glucagon, PP, PYY, and NPY). Further, some peptide hormones that clearly function in the GI system were originally found in the brain [e.g., enkephalins, somatostatin, substance P (SP), and CGRP]. It is also known that neurons of the CNS interact with neurons of the ENS via both afferent and efferent pathways to influence digestive processes. Finally, evidence is now emerging that GI hormones and peptides provide a feedback signal to the CNS to regulate satiety and thermoregulation. There are two enkephalin peptides (isoleucine–enkephalin and methionine–enkephalin) known to be present in the brain, pituitary gland, and also the GI tract. These enkephalins function as neuromodulators or neurotransmitters to inhibit neurotransmitters in the pain perception pathway. The amino acid sequence of leu-enkephalin is Tyr–Gly–Gly–Phe–Leu and the amino acid sequence of met-enkephalin is Tyr-Gly-Gly-Phe-Met. Both of these pentapeptides are known to bind to morphine receptors in the CNS.

Table 7.7 provides a list of 10 questions related to the entire human digestive process. It offers the reader the opportunity to assess the mastery of their understanding of the size and interplay of the 10 digestive components.

III Biochemical properties and molecular actions

A General relationships

All of the GI hormones are polypeptides, and they are each produced in special endocrine cells that are dispersed throughout the stomach and intestine and colon. The GI hormones were originally discovered during the search for physiological agents that regulated and stimulated the

TABLE 7.7 Questions related to the human digestive process.[a]

Question	Answer
What is the distance from mouth cavity to the anus?	Female: 8.5 m ~27 ft; Male: 8.0 m ~26 ft
What is the length of the duodenum+jejunum+ileum?	Female: 7.1 m ~23 ft; Male: 6.9 m ~22 ft
What is the length of the colon (large intestine)?	1.5 m ~4.9 ft
What is the length of the duodenum?	25 cm or ~10 in
Could the absorptive area of the small intestine cover a 9 m × 9 m=81 m^2 floor area, or a tennis court?	YES! It is ~275 square meters, which is approximately the playing area of a tennis court. http://www.chp.edu/CHP/organs+intestine
What is the volume of the stomach?	Empty=50 cm^3; full=~1500–2000 cm^3
After eating a normal meal, how long does it take to empty the stomach of its contents?	The total emptying time of the stomach is 3–5 h
After entry of digested food into the colon, what is the transit time (in hours) for complete transit through the colon?	Transit time through the colon ranges from 30 to 40 h
What is your average body surface area in square meters?[a]	For a person 5′10″ (1.79 m) and 162 lbs (73.6 kg), their surface area is 1.95 square meters
What is the ~weight (in pounds) of water you would drink over 1 year? The RDA[b] for water is 2.2 L for adult females and 3.0 L for males	Male: ~2400 lbs or 1.2 t! Female: ~1750 lbs or 0.875 t!

[a]*The Dubois equation to determine the body surface area of humans is as follows:* $BSA = 0.007184 \times W^{0.425} \times H^{0.725}$. *The weight should be in kilograms, and the height in centimeters.*
[b]*RDA, Recommended dietary allowance, an amount that should be the average intake.*

exocrine secretions and activities of the gut. The application of molecular biology techniques resulted in significant advances with respect to the elucidation of the gene, cDNA, and, therefore, amino acid sequence of many GI hormones. Also, for many of these hormones there is an emerging understanding of their membrane receptors and the details of their signal transduction. As a consequence of these studies, it has been possible to identify structural homologies between the GI hormones and group them into seven families (see Table 7.1) and also to identify neurotransmitters produced by the GI tract (see Table 7.2).

Each of these GI hormones has a specific function to contribute to the selective regulation of the process of digestion. A GI hormone has a specific physiological responsibility conferred upon it by the precise ordering of the amino acid sequence that comprises the single chain of its polypeptide. The diverse amino acid sequences of seven GI families are shown in Figs. 7.8–7.14.

Every GI hormone has its own "mechanism of action" to produce its biological response(s). Thus those hormones with a short half-life, <5 minutes, will rapidly become inactivated by the actions of peptidases that disrupt their amino acid sequence so that the hormone fragment is unable to bind to its partner receptor and produce useful biological responses. This is mediated by circulating peptidase enzymes secreted by the kidney and liver.

B Cholecystokinin/gastrin family

In Fig. 7.8A/B, each of the hormone amino acid residues is given a different color. The CCK peptide hormone is colored blue and gastrin is colored green. The amino acid sequences of CCK and gastrin are presented in Fig. 7.8A/B. A 115-amino-acid prepro form of CCK is cleaved to a 94 amino acid prohormone that is posttranslationally transformed to peptide sizes 83-, 58-, 39-, 33-, 22-, 8-, and 5-amino-acid species. See Fig. 7.8A/B. All of the biological activity of CCK is contained in the octapeptide (amino acids 52–58), and it is likely that the other larger peptides represent intermediates in processing. To achieve full biological activity of CCK, the tyrosine at position 7 must be sulfated.

To identify regions of identical amino acid sequences for CCK and gastrin and its various smaller fragments, the test peptides are aligned so that each of the carboxyl-amide amino acid residues is directly even; see Fig. 7.8A. By general convention for all peptides and proteins, the amino terminus is normally assigned position 1. Thus for CCK-58, gastrin-17, and gastrin-34, this alignment identifies that the position and ordering of the 5 amino acids starting with the carboxyl-amidation of the three peptides are *identical* and have been colored gold. For CCK the minimum peptide with full biological activity has only 7

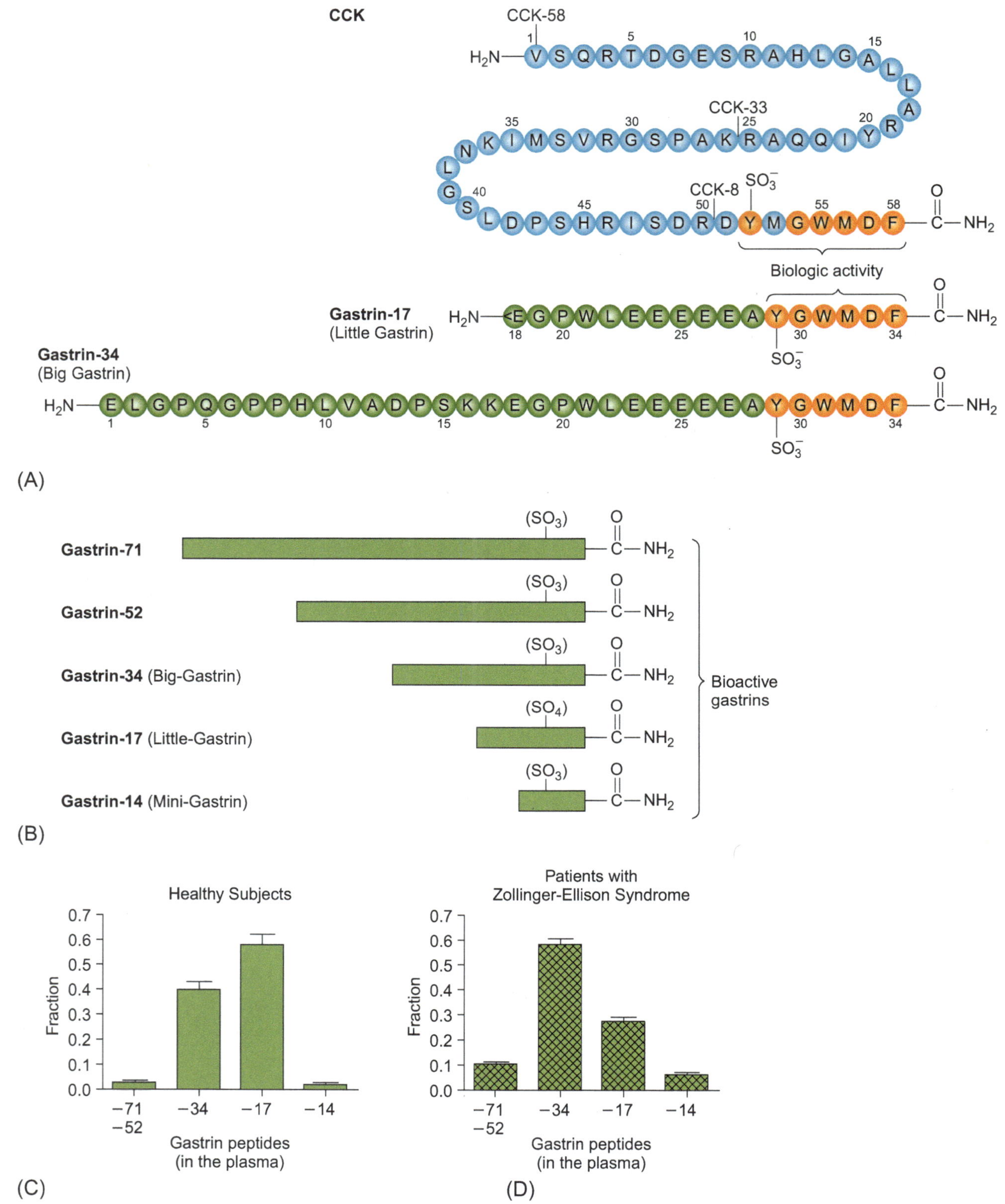

FIGURE 7.8 Identification of the minimal amino acid sequences of CCK and gastrin that are required for biological activity; and for gastrin, a clinical assessment of the size pattern of gastrin peptides required for biological activity in healthy subjects and patients with the ZES. (A) Each of the amino acid residues for the CCK peptide hormone is colored blue, and for gastrin each amino acid residue is colored green. The CCK peptide and the gastrin peptides are aligned such that carboxy terminus amides of the phenylalanines are all aligned together. This result emphasizes that the final six amino acids of CCK, Gastrin-17, and Gastrin-34 are exactly identical and thus they are each colored orange. (B) The 101-amino-acid progastrin is

(Continued)

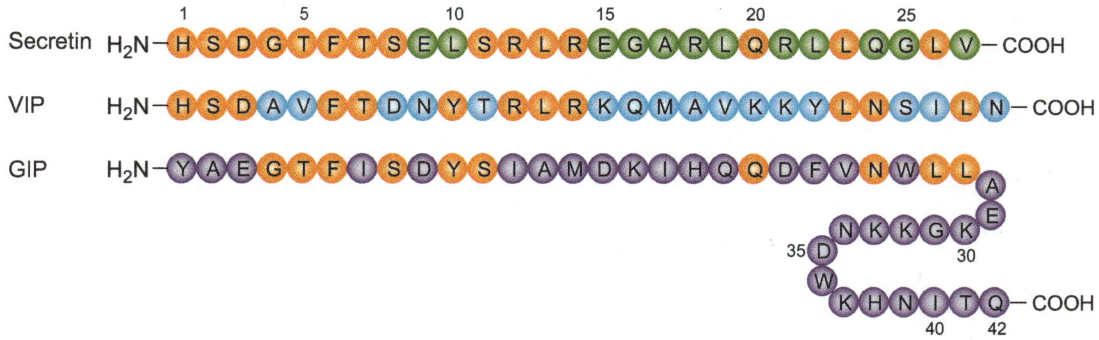

FIGURE 7.9 Amino acid sequence comparison of the secretin peptide hormone family. These include secretin (*green*), VIP (*blue*), and GIP (*purple*). The three peptides are aligned on the basis of each of the NH_2-terminus residue being assigned 1. Each of the 27-amino-acid residues of secretin was separately compared with the first 27 residues of VIP and GIP; all amino acid matches between the three peptide hormones were colored gold. *GIP*, Gastrointestinal inhibitory peptide; *VIP*, vasoactive intestinal peptide.

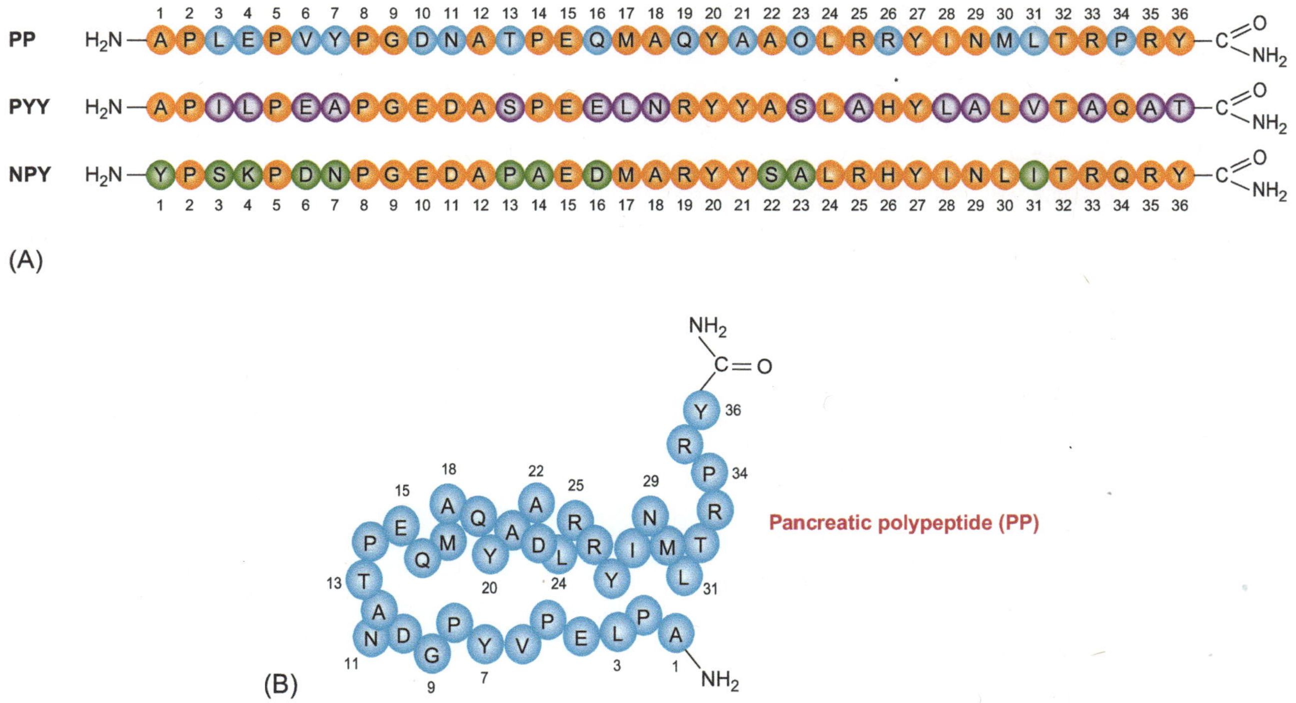

FIGURE 7.10 Amino acid sequence comparison of the PP hormone family. These include PP, PYY, and NPY. (A) The 36 amino acids of each of the hormones are colored differently; PP is blue, PYY is purple, and NPY is green. The three peptides were aligned on the basis of each of the NH_2-terminus residues being assigned 1 and the carboxy-amide being numbered 36. Each of the 36-amino-acid residues of PP was separately compared with the 36 residues of PYY and NPY. All exact matches with PP were colored gold. Also, the amino acid sequences of PYY and NPY were compared and all identical amino acid identities were also colored gold. These matches may reflect evolutionary changes in the order of amino acids. (B) Illustration of the PP fold as determined via X-ray crystallography and also NMR. The PP fold comprises residues 3–31 of the 36-amino-acid peptide. The formation of the "fold" is believed to be facilitated by the four proline residues at peptide positions 2, 5, 8, and 14. *NMR*, Nuclear magnetic resonance; *NPY*, neuropeptide tyrosine; *PP*, pancreatic polypeptide; *PYY*, peptide tyrosine tyrosine.

---

enzymatically processed into at least 5 major peptides that are shown in this panel. The three principal biologically active circulating forms are G-34 (big gastrin), G17 (little gastrin), and G14 (mini-gastrin). Other low concentration gastrins include gastrin-6 (see the 6 *gold* residues of G-17 and G-34 in panel A), and gastrin-52 and gastrin-71. (C and D) The relative amounts (fraction) of five blood-circulating gastrin peptides are shown for healthy subjects and for comparison to patients with the ZES. The suggestion is that the disease process associated with ZES significantly changes the processing of the 101-amino-acid precursor form of progastrin into the family of gastrin peptides. *CCK*, Cholecystokinin; *ZES*, Zollinger–Ellison syndrome. *Data in panels (C) and (D) were abstracted from the following citation: J.R. Rehfeld et al., The Zollinger–Ellison Syndrome and mismeasurement of gastrin. Gastroenterology 140 (2011) 1444–1445.*

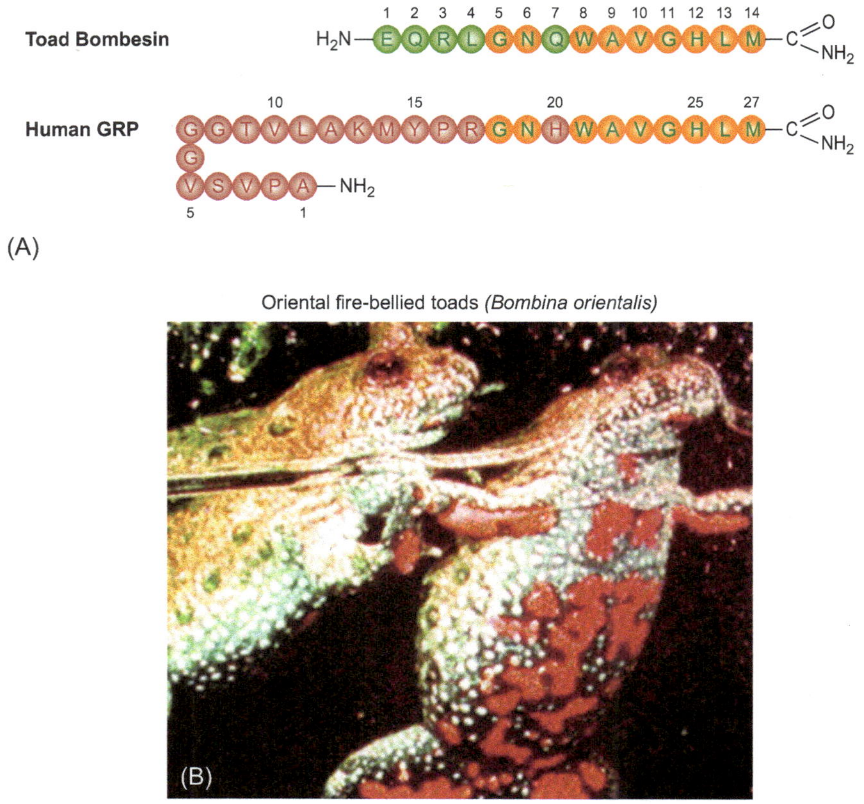

FIGURE 7.11 A toad's bombesin peptide and a human's GRP share identical amino acid peptide sequence. (A) A toad's bombesin peptide (*green*) has 14 amino acids, while the human GRP (*red*) has 27 amino acids. Both of these peptides have a C-terminal amidation terminus. When the sequence of identical amino acids in the two peptides was identified, they are colored gold. (B) Photograph of an oriental fire-bellied toad (*Bombina orientalis*). It is one of ~3500 species of toads. *GRP*, Gastrin-releasing peptide.

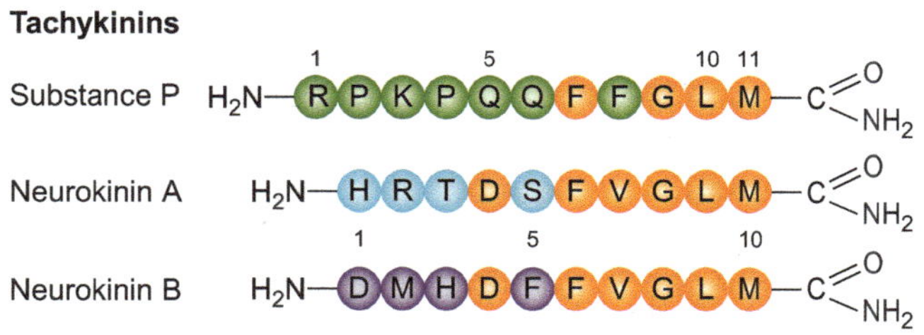

FIGURE 7.12 The amino acid sequences of three tachykinin family members are illustrated. This includes SP (*green*), neurokinin A (*blue*), and neurokinin B (*purple*). Neurokinins A and B have a total peptide sequence of only 10 amino acids; of these six share identical amino acid sequences for residues 4 and 6, 7, 8, 9, and 10. SP has a peptide sequence of 11 amino acids and shares four identical residues (see 7 and 9, 10, and 11 *gold residues*) with the neurokinins A and B. Each of these three peptides has a C-terminal amidation. *SP*, Substance P.

amino acids (see underlining braces); however, CCK peptides with 8, 33, or 58 residues (see labels on the peptide chain) are known to be present in vivo in the circulatory system, which suggests that they may have useful but not yet understood biological responses.

For gastrin the three principal biologically active circulating forms are G34 (big gastrin), G17 (little gastrin), and G14 (mini-gastrin). Other low-concentration gastrins include gastrin-6 (see the six gold residues of G-17 and G-34 in Fig. 7.8A), gastrin-52, and gastrin-71. The minimum gastrin peptide with full biological activity has only 14 amino acids; in each instance the amino terminal residue is the 21 tryptophan. For CCK, sulfation of the hydroxyl of tyrosine 52 is mandatory for biological activity; note that residue 52 is only 7 residues from the phenylalanine carboxyl-amidate terminus of the CCK. Both CCK and gastrin have an identical sequence for the last five residues ending with the carboxyl-amidate. For the gastrins, sulfation of the 29 tyrosine is known to occur but is not considered to be mandatory for full biological activity.

CCK, as determined by immunohistochemical techniques, is found principally in the duodenum and jejunum and also in the CNS, as well as in the nerves of the myenteric plexus of the intestine and the nerves of the urinary bladder and uterus.

The two major biological responses initiated by CCK are stimulation of the acinar cells of the pancreas to secrete pancreatic enzymes and stimulation of gallbladder contraction. The release of CCK from the endocrine cells in the intestinal mucosa is stimulated by the breakdown products of food in the duodenum, particularly 10–18-carbon fatty acids and L-amino acids.

The biological actions of CCK result as the consequence of interactions with two classes of receptors; the CCK-A receptor is predominantly present in the pancreas and gallbladder, while the CCK-B receptor is found in the CNS. There is some evidence that the occupancy of CCK

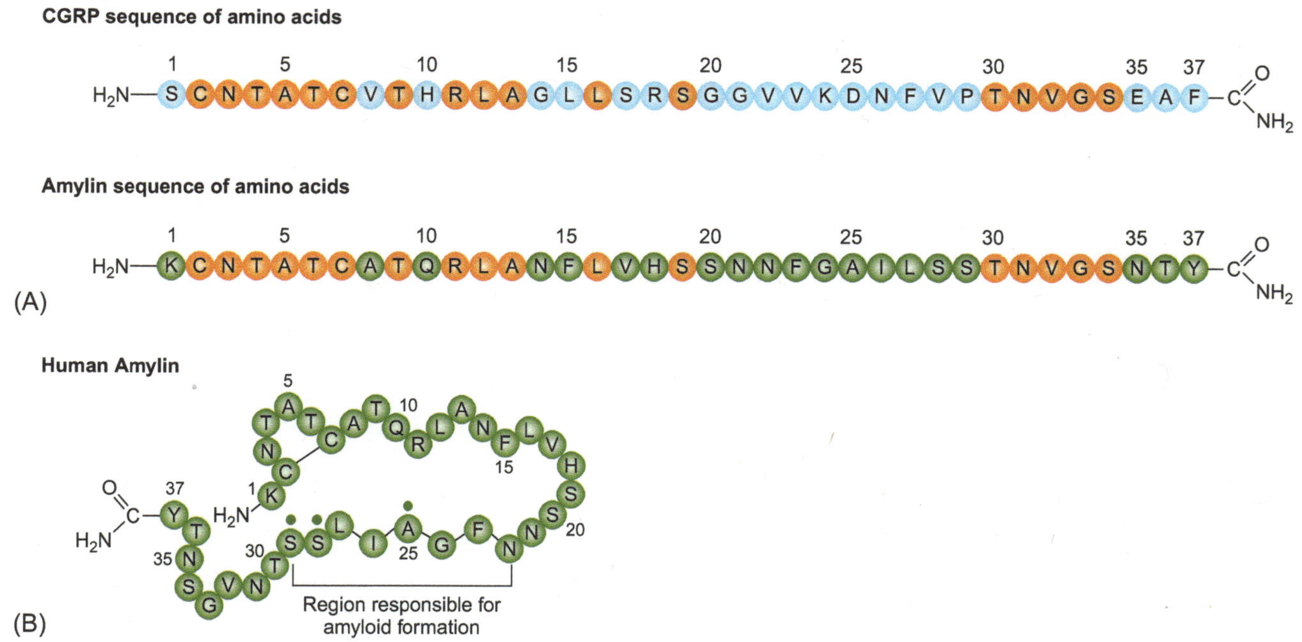

FIGURE 7.13 Amino acid sequences of CGRP and amylin. (A) The CGRP and the related peptide, amylin, are both members of the family of calcitonin gene–related peptide and related peptides. Both peptides, CGRP and amylin, have 37 total amino acids. Of the 37 amino acids of CGRP and amylin, 17 (45%) are colored gold marking the identical amino acid positions in these two peptides. (B) Both peptides have a 6-amino-acid ring structure created by the disulfide bond occurring between cysteine 2 and 7. Both peptides have a 6-amino-acid ring structure created by the disulfide bond occurring between cysteine 2 and 7. *CGRP*, Calcitonin gene–related peptide.

receptors by CCK may contribute to a meal termination signal, that is, to function as a satiety factor.

Both the CCK-A and CCK-B receptors are highly glycosylated. The CCK-B receptor has an amino acid sequence that is 48% identical to that of the CCK-A receptor. Both receptors contain the seven transmembrane domains characteristic of G protein–coupled receptors. In the pancreatic acinar cell, CCK stimulates amylase release via a complex signal transduction system involving IP_3, nitric oxides, cyclic guanosine monophosphate, and Ca^{2+} channels.

The bulk of GI gastrin is localized within the G cells of the stomach antrum; the G-17 peptide form of gastrin predominates here. Gastrin has also been localized in the brain. The principal stimulants for gastrin release are partially digested proteins, peptides, and amino acids. Gastrin secretion can also be stimulated by vagus nerve stimulation and the condition of insulin hypoglycemia. Gastrin secretion is inhibited by antral acidification to pH ~ 1.0–2.5. Gastrin release is also stimulated by bombesin and inhibited by secretin, GIP, VIP, and somatostatin.

The principal biological action of gastrin is to stimulate the release of HCl by the parietal cells. These responses are achieved via direct interaction of gastrin with a cell-surface membrane receptor and via indirect stimulation of the release of histamine from ECL cells present in the gastric mucosa.

The disease known as Zollinger–Ellison syndrome (ZES) is characterized as an advanced ulcer proliferation of the upper GI tract; see Section IV.C at the end of this chapter. The ulcer interferes with the production and actions of gastrin. A study involving humans has been completed so that a clinical assessment of the size pattern of gastrin peptides present in healthy subjects can be compared with the size pattern of patients with the ZES; see Fig. 7.8C/D. A postulate is that the disease process associated with the ZES will make significant changes in the processing of the 101-amino-acid precursor form of progastrin into the family of gastrin peptides. The gastrin peptides were isolated from the blood and separated on the basis of their size via column chromatography so that the relative amounts of the five gastrin peptides could be quantitated. As shown in Fig. 7.8D, the subjects with ZES clearly have a different peptide fragment profile from that of the normal subjects (panel C). For patients with ZES the proportion of gastrin-34 was increased from 38% (healthy control) to 58% (ZES patients). In contrast, gastrin-17 was decreased from 56% (healthy control) to 27% (ZES patients). This suggests that some aspect of the ZES is capable of changing the size pattern of gastrin fragments.

C Secretin family: secretin, vasoactive intestinal peptide

Fig. 7.9 compares the amino acid sequences of secretin (*green*), VIP (*blue*), and GIP (*purple*). They are all members of the secretin family; see Table 7.1. Unlike most other

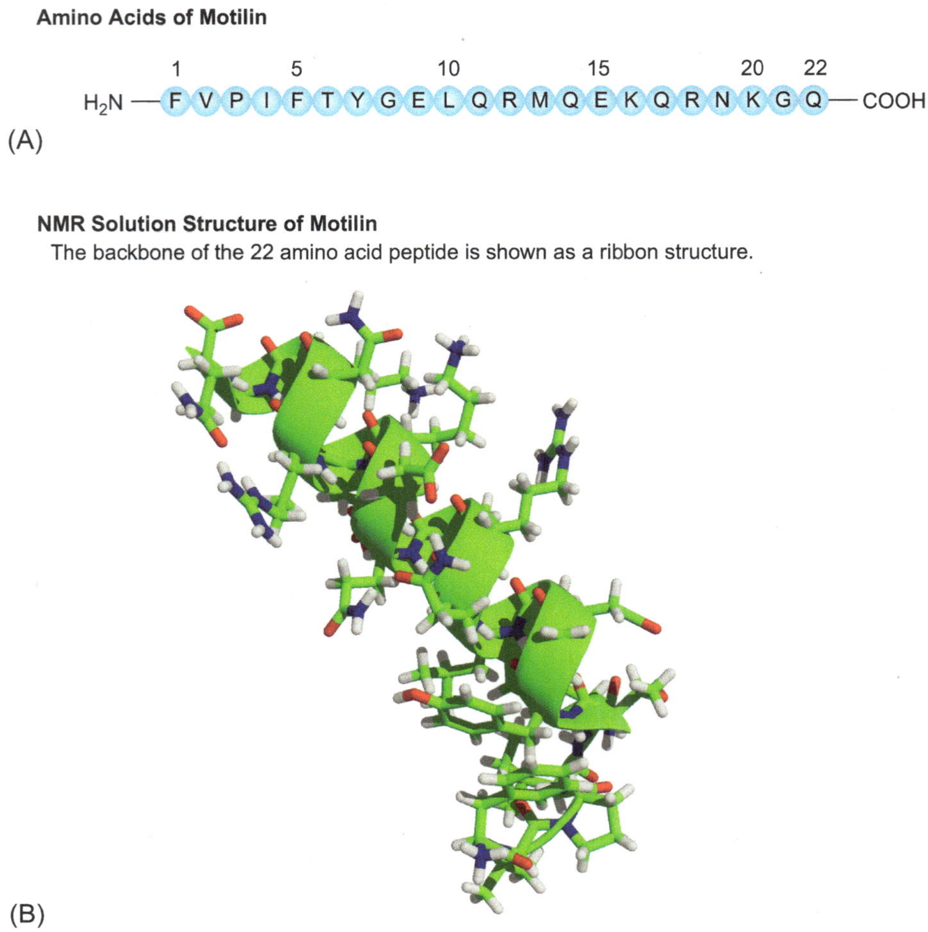

FIGURE 7.14 Motilin's amino acid sequence and its three-dimensional ribbon structure. (A) Amino acid sequence of motilin. (B) NMR solution structure of motilin. The 22-amino-acid sequence of motilin in solution assumes a ribbon structure. *NMR*, Nuclear magnetic resonance.

peptide hormones, only one form of secretin is known. Secretin has only 27 amino acids. The entire molecule of 27 amino acids is required for full biological activity. Secretin is localized to the upper duodenal and jejunal S cells. VIP is a peptide that has 28-amino-acid residues and is widely expressed in the central and peripheral nervous systems and also in the digestive tract. GIP (*purple*), also known as glucose-dependent insulinotropic polypeptide, has 42 amino acids. Total 10 of VIP's 28 amino acids (*gold*) are identical to secretin and are colored gold. On the other hand, 4 of its 28 amino acids are identical to GIP and are colored gold. Secretin has only 7 of its 28 amino acids identical to GIP. When VIP and GIP amino acids are identical, they are also colored gold. These matches may reflect evolutionary changes in the order of amino acids.

Within the GI tract, secretin is found in the S cells of the duodenum and jejunum. Secretin secretion is stimulated by unknown mechanisms when the acidity of the duodenal contents falls below pH ~ 4.5. The principal biological response of secretin is to stimulate the secretion of bicarbonate and water by the acinar cells of the pancreas. This response is stimulated by ligand occupancy of the secretin receptor, which is located on the outer cell membrane of the pancreatic acinar cells. The human secretin receptor has been cloned and found to be a 419-amino-acid peptide (47,900 Da). It belongs to the family of G protein–coupled receptors. Stimulation of adenylate cyclase leads to the secretion of bicarbonate and H_2O.

VIP is a 28-amino-acid peptide secreted by the small intestine, pancreas, and by the suprachiasmatic region of the hypothalamus in the brain. VIP functions as a neurotransmitter from the esophagus to the rectum and also in the salivary gland. VIP promotes smooth muscle relaxation in the trachea, esophageal sphincter, and the stomach. GIP is also known as the glucose-dependent insulinotropic peptide. GIP can induce insulin secretion, which is stimulated primarily by the presence of hyperosmolarity of glucose in the duodenum. GIP is biosynthesized largely by K cells present in the small intestine. K cells and L cells secrete, respectively, the incretins, GIP, and GLP-1. GIP is released into the circulatory system from the duodenum and jejunum following a mixed meal. Receptors for GIP have been identified in selected regions of the GI tract as well as the brain.

GIP, which is a member of the secretin peptide, shares a high sequence homology with both glucagon (15 amino

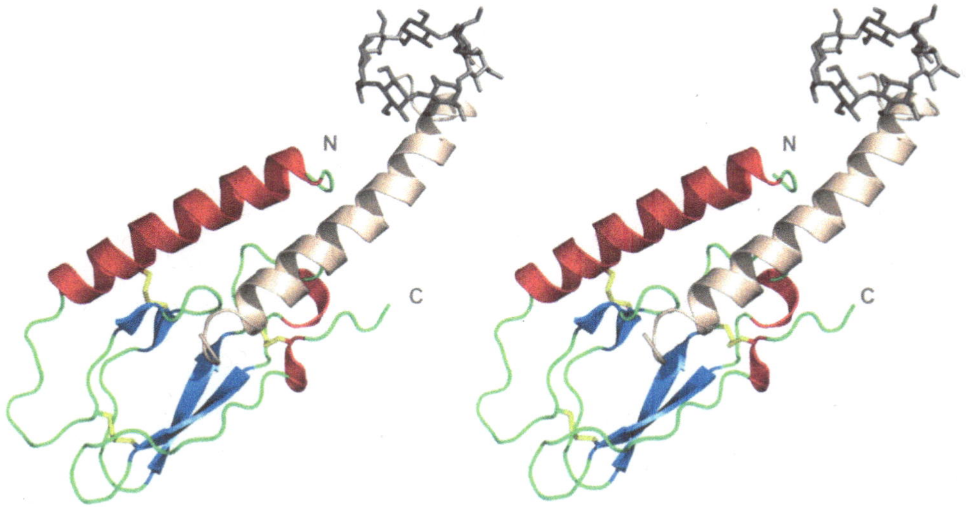

FIGURE 7.15 Overall structure of the GIPR ECD in complex with GIP_{1-42}. Stereoview of the peptide hormone GIP_{1-42} (residues 1′ − 32′ beige; residues 33′−42′ were not visible in the electron density) bound to the ECD of GIPR [colored according to secondary structure, residues 29(N)−122 (C) visible]. Disulfide bridges in the GIPR ECD are shown as yellow sticks. The N terminus of the ligand is bound by a methyl-β-cyclodextrin molecule (*gray sticks*). *ECD*, Extracellular domain; *GIPR*, gastric inhibitory peptide receptor. *Reproduced from Fig. 7.1 of C. Parthier et al., Crystal structure of the incretin-bound extracellular domain of a G protein-coupled receptor, Proceedings of the National Academy of Sciences of the United States of America 104 (2007) 13942–13947.*

acids identical) and secretin (6 amino acids identical). Fig. 7.15 shows the crystal structure of the GIP receptor in complex with GIP_{1-42}. Little information is available concerning the mechanisms by which GIP elicits biological responses. Two categories of biological responses are known: (1) inhibition of gastric acid secretion and gastric motility and (2) potentiation of pancreatic insulin secretion after oral as opposed to intravenous glucose administration.

D Pancreatic polypeptide family: peptide YY and neuropeptide Y

All three members of the PP family have the same number of 36 amino acids and have an amidated C-terminal tyrosine residue; see Fig. 7.10. PP is colored blue, PYY is colored purple, and NPY is colored green; each of the 36-amino-acid residues of PP was separately compared with the 36 residues of PYY and NPY. All exact matches were colored gold. Also the amino acid sequences of PYY and NPY were compared and all identical amino acid identities were also colored gold. These matches may reflect evolutionary changes in the order of amino acids. PP has 13 of its 36 amino acids identical to PYY and 15 of its 36 amino acids identical to NPY. PYY has 17 of its 36 amino acids identical to NPY. PP is secreted by pancreatic islet cells (see Fig. 7.10) and its receptor is present in the small intestine, colon, and the pancreas. Fig. 7.16 shows the linear structure of the seven transmembrane Y4 human PP receptor. Also shown in this figure are a homology model of the human Y4 receptor and the locations of variant amino acid residues. Twelve naturally occurring variants of the human PP receptor are known. PYY functionally is an enterogastrone, which describes any hormone secreted by the mucosa of the duodenum as a response to food intake. NPY is widespread in the peripheral and CNSs and also has important inhibitory actions in the intestinal tract.

The three-dimensional structure of each of the three members of the PP family of hormones includes a hairpin-like motif known as the PP fold, illustrated in Fig. 7.10B as determined via X-ray crystallography and also nuclear magnetic resonance (NMR). The PP fold comprises residues 3−31 of the 36 amino acid peptide. The formation of the "fold" is believed to be facilitated by the four proline residues at peptide positions 2, 5, 8, and 14. It has been proposed that the tightly organized tertiary structure of the PP fold is required for the interaction of the PP family member hormones with their specific cognate G protein−coupled receptors.

E Bombesin and related peptides: gastrin-releasing peptide and neuromedin C

The bombesins include structurally related peptides that were first isolated from amphibian toad skin. Bombesin is a 14-amino-acid peptide originally isolated from the skin of *Bombina orientalis*. The human gastrin-releasing peptide (GRP) gene has been found to encode a number of bombesin-like peptides. Additional members of the bombesin family have been isolated directly from the gut, for example, GRP and neuromedin C. Neuromedin C is a 10-

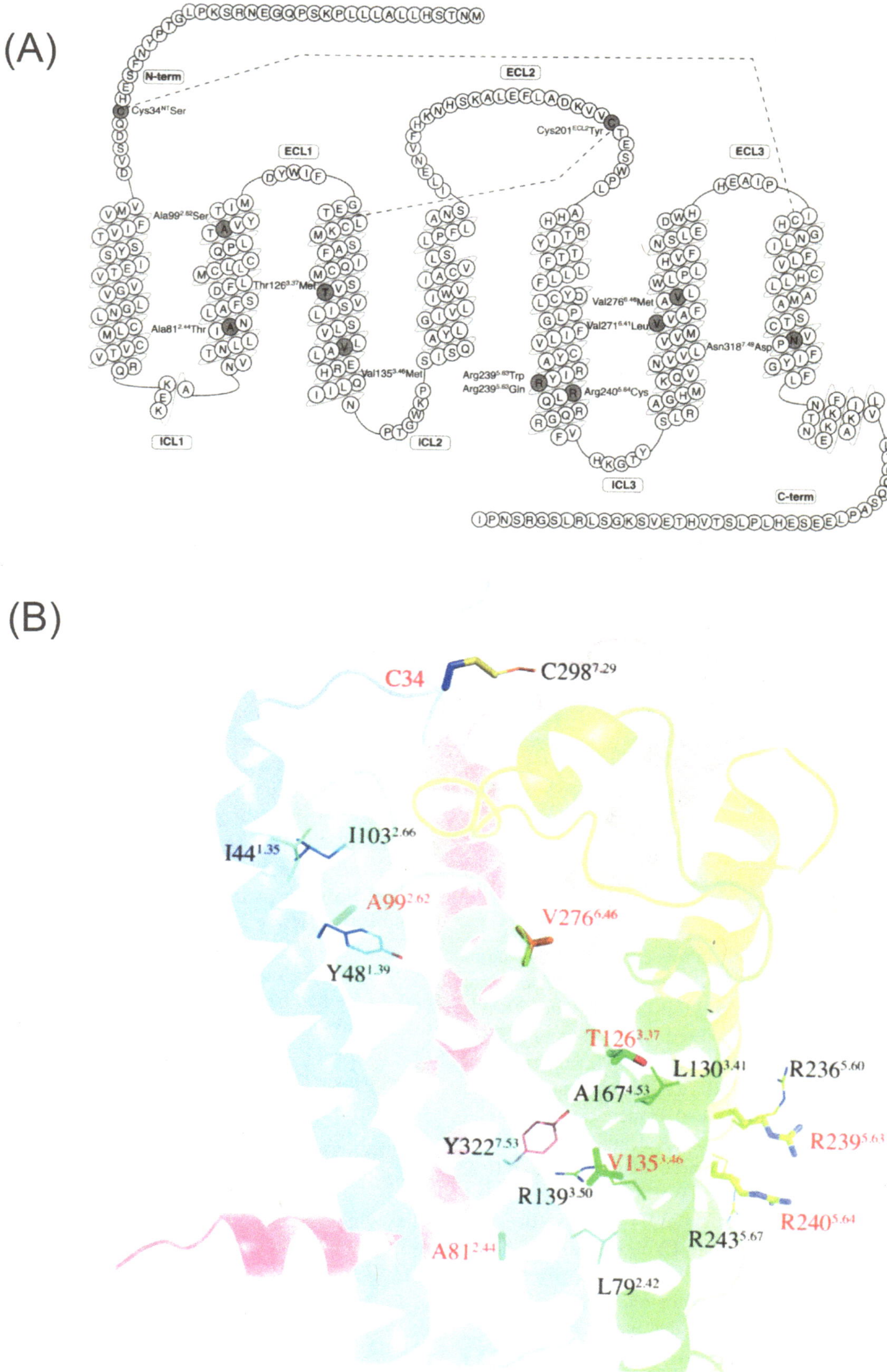

FIGURE 7.16 (A) A schematic representation of the human Y4 receptor. Nonsynonymous mutations were identified and marked with a gray color. (B) Homology model of the human Y4 receptor and the locations of the variant amino acid residues. The three variants that completely lost receptor response are not shown. The mutated residues are depicted with thicker sticks and red label, while their surroundings are indicated with black labels. *Parts (A) and (B) are Fig. 7.1 and Fig. 7.4, respectively, reproduced from K. Shebanits et al., Fnctional characterization in vitro of twelve naturally occurring variants of the human pancreatic polypeptide receptor NPγ4R, Neuropeptides 76 (2019) 101933; https://doi.org/10.1016/j.npep.2019.05.004.*

amino-acid peptide; its aa sequence follows immediately: (G-N-H-W-A-V-G-H-L-M--NH2). This sequence results from enzymatic degradation of the 17 amino terminal end of GRP.

A toad's bombesin peptide and a human's GRP share nine identical amino acid peptides sequence; see Fig. 7.11A. The bombesin peptide sequence is colored green, while the GRP is colored red. The carboxyl ends of both peptides are identical over the last seven residues and are colored gold. From the entire sequence of 14 amino acids of bombesin, 9 amino acids are identical to the human GRP's C-terminus of 14 amino acids. This suggests that there is a GRP genetic linkage between toads and man.

Bombesin was originally discovered and isolated from the skin of the oriental fire-bellied toad (*Bombina orientalis*); see Fig. 7.11B. The bombesin peptide sequence has been reported to occur in GI nerves of mammals, while in toads, frogs, and birds it occurs in endocrine cells.

The bombesin present in the skin of frogs has been shown to stimulate smooth muscle contraction. Also bombesin was found to interfere with the metabolic processes of the toad's body. This results in an increase of bodily waste products that accumulate in the frog's glands and that could be released as a defense mechanism when the frog is threatened.

Also the potential medical–pharmaceutical significance of products derived from amphibian skin secretions have been extensively explored by pharmaceutical companies. Molecular biology studies in toads revealed that the *identical* messenger RNA is expressed in brain, skin, and stomach to the bombesin peptides. Interestingly in skin, it is processed only to bombesin (14 amino acids), whereas in the stomach it is also processed to a large form, similar in size to mammalian GRP (27 amino acids).

The presence of human GRP in the stomach's fundus and antrum and the jejunum, ileum, and colon as well as with the peripheral and CNSs has been determined. In all these systems, GRP functions largely as a neurotransmitter where it is a strong stimulator of other gut hormones, including gastrin, CCK, enteroglucagon, motilin, PP, PYY as well as somatostatin, insulin, and glucagon.

F Tachykinin family: substance P and neurokinins A and B

The tachykinins (rapid kinins), which mediate rapid contractions of smooth muscle, are so named to distinguish them from bradykinins, which have a much slower action on smooth muscle. SP (*green*) has a peptide sequence of 11 amino acids; see Fig. 7.12. Neurokinin A (NKA) (*blue*) and neurokinin B (NKB) (*purple*) each have 10 amino acids. Each of these three peptides has a C-terminal amidation. SP shares four identical residues 7, 9, 10, and 11 (gold residues) with the NKA and NKB. NKA and NKB have a total peptide sequence of only 10 amino acids and share 6 identical amino acid sequences (4, 6, 7, 8, 9, 10) that are colored gold.

Tachykinins have been shown to function both as a vasoconstrictor and as a vasodilator, depending upon the circumstances. In the intestines the tachykinins are expressed largely in extrinsic afferent nerves fibers and, also, in the intrinsic enteric neurons. The biological responses of the three tachykinins, SP, NKA, and NKB are mediated through binding to one of three G protein–coupled receptors.

The pattern of biological responses generated by the tachykinins is complex and includes smooth muscle contraction, hyperalgesia, release of histamine from salivary cells, and involvement with asthma, hay fever, migraine headache, and inflammatory bowel disease.

G Neurotensin

Neurotensin (NT) is a 13-amino-acid peptide. Its bovine sequence is as follows:

$$\text{pyroG} - \text{L} - \text{Y} - \text{E} - \text{N} - \text{K} - \text{P} - \text{R} - \text{R} - \text{P} - \text{Y} - \text{I} - \text{L} - \text{COOH}$$

Neurotensin (NT) and a related neuropeptide, neuromedin N, are both biosynthesized from an amino acid precursor protein of 169–170 residues. Neurotensin NT has been detected in the ileum. The intestinal tract has most of the hormone, while the rest of the NT is present in the hypothalamus. It has been shown that dietary intake of particularly fat stimulates the release of NT. The secretion of NT can be inhibited by somatostatin. Blood levels of NT are known to rise in patients who have the *dumping syndrome* (rapid emptying of the stomach wherein food moves too rapidly into the duodenum and leads to digestive discomforts).

H Calcitonin gene–related peptide family: amylin

The CGRP (*blue*) and the related peptide, amylin (*green*), each has a total of 37 amino acids; see Fig. 7.13A. Of the 37 amino acids of CGRP and amylin, 17 are colored gold, marking the identical amino acid positions in these 2 peptides.

CGRP stimulates, in the pancreas, an $8\times$ increase in amylase secretion. CGRP is primarily released from sensory nerves. This can then result in smooth muscle relaxation.

Amylin is cosecreted with insulin from the pancreatic β-cell. The insulin/amylin secreted ratio is $\sim$100:1. Amylin plays a role in glycemic regulation by slowing gastric emptying and thereby increasing satiety. Amylin is also biosynthesized in the intestine, stomach, lung, and hypothalamus.

Amylin is related to the disease state of amyloidosis that shares in common the deposition of insoluble fibrillar proteins outside of cells. These are part of a growing group of diseases now thought to be caused by misfolding of proteins. It has been found that many amyloidoses have deposits that have a common β-pleated sheet structural conformation.

As shown in Fig. 7.13B both peptides can form a 6-amino-acid ring structure created by the disulfide bond occurring between cysteine 2 and cysteine 7. Human amylin carries out inappropriate formation of amyloids. Residues 22–29 are responsible for the formation of the amyloid fibrils. The amyloids are aggregates of a variety of insoluble proteins which arise from an inappropriate folding of the affected proteins. They also can have a deposit of common β-pleated sheets. Their presence has been associated with diseases because of the accumulation of amyloid fibrils in a variety of organs, including the pancreatic islets, which can lead to a variety (~20 versions) of neurodegenerative disorders.

I Enteroglucagon and oxyntomodulin

The topics of enteroglucagon and oxyntomodulin are covered in Chapter 6, Pancreatic Hormones: Insulin and Glucagon. The same proglucagon (160 amino acids) in the pancreas secretes the classical 29-amino-acid pancreatic glucagon. However, in the intestine it secretes the 69-amino-acid glicentin that includes the 29-amino-acid sequence of pancreatic glucagon; see Fig. 6.8. Enteroglucagon is biosynthesized in the intestinal mucosal L cells of the ileum and colon. It has been postulated to be a trophic factor for the intestinal mucosa.

Oxyntomodulin is a 37-amino-acid peptide that is produced by the stomach's fundic cells (also known as oxyntic cells) that, like parietal cells, secrete gastric acid. The oxyntomodulin 37-amino-acid peptide includes the exact amino acid sequence of pancreatic glucagon. The oxyntomodulin is also found in intestinal and colon cells. Oxyntomodulin and peptide tyrosine–tyrosine (PYY) are secreted from intestinal enteroendocrine cells in response to a meal. These circulating hormones are considered to be satiety signals, as they have been found to decrease food intake, body weight, and adiposity in rodents. It has been proposed that the effects on energy homeostasis are mediated by the brainstem and hypothalamus.

J Motilin

Motilin is secreted by M cells present in the duodenum and proximal jejunum. Motilin has 22 amino acids and has no identifiable sequence homology with other GI hormones; see Fig. 7.14A. Fig. 7.13B illustrates the 22 amino acids of motilin in solution in a ribbon structure as elucidated by NMR analysis. Motilin does not have a long-enough amino acid sequence to utilize the PP fold. The PP fold of the 36-amino-acid PP is shown in Fig. 7.10.

A major role of motilin in the stomach is to stimulate the production of pepsin which, in turn, will stimulate the contraction of the fundus and antrum so as to accelerate gastric emptying. Additionally, motilin stimulates gallbladder contractions. Also, motilin is believed to play a role in the stimulation of small intestinal motility during fasting. Its release from the endocrine ECL cells of the duodenum is cyclical, occurring approximately every 2 hours. Significantly, most of motilin's actions do not coincide with dietary intake, but with fasting. It has been suggested that this enhances the movement of digested food particles in preparation for the next meal. The physiological role of motilin remains to be clarified. The human motilin receptor is shown in primary sequence in Fig. 7.17. Some sections of the receptor have been experimentally deleted (as indicated in the diagram) and the products have been tested for motilin binding and intracellular calcium responses. Both motilin and erythromycin and erythromycin derivatives bind to different sites on the receptor.

K Somatostatin

Somatostatin has several names. It is also known as growth hormone (GH)–inhibiting hormone or somatotropin release–inhibiting factor or somatotropin release–inhibiting hormone. Somatostatin is a cyclic peptide with two naturally occurring forms: either as a 28 (SS-28)-amino-acid peptide or as a 14 (SS-14)-amino-acid peptide (see Fig. 3.7). Somatostatin was first isolated from the hypothalamus, where it functions to inhibit the secretion of GH.

Somatostatin is present functionally in the pancreas D cells (see Fig. 6.3) and as well in the endocrine D cells of the intestine and the stomach; see Fig. 7.2C. It is also found in extrinsic and intrinsic neurons of the myenteric as well as the submucosal plexuses. In each of these settings, somatostatin blocks the release of many GI hormones. They include the following: CCK, gastrin, motilin, secretin, gastric inhibitory polypeptide (GIP), VIP, enteroglucagon, and in the pancreas where SS-14 blocks the release of both insulin and glucagon.

While the principal form of somatostatin is the SS-14 species, both SS-14 and SS-28 display overlapping but distinct biological activities and act through different receptors. No less than five unique somatostatin receptors have been isolated and characterized; they all are G protein–coupled, seven membrane–spanning receptors. It is

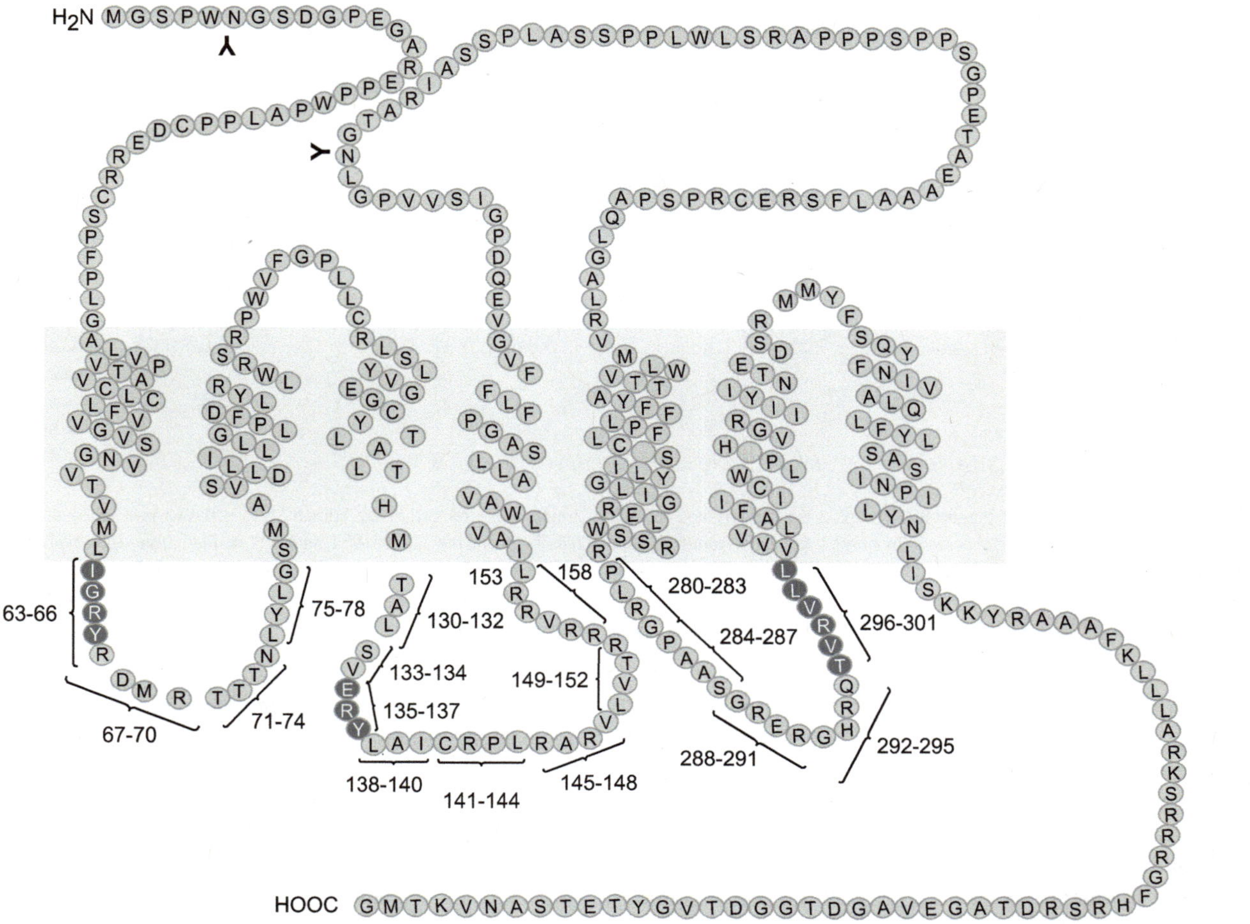

FIGURE 7.17 Primary structures of the human motilin receptor experimental constructs shown in the schematic diagram of the amino acid sequence and possible membrane topology (based on hydropathy) of the motilin receptor, along with the design of the segmental deletions (in numbered brackets) and Ala- or Phe- or His-replacement constructs (areas of dark circles with white lettering). *Reproduced from Fig. 7.1 of H. Tokunaga et al., Mutational analysis of predicted intracellular loop domains of human motilin receptor, Hormones and Signaling 294 (2008) G460–G466; https://doi.org/10.1152/ajpgi.00244.2007.*

also known that the 14 amino acids at the N-terminus of SS-28 have no intrinsic biological activity. The primary and secondary amino acid structure of the human somatostatin receptor (SST_1) is shown in Fig. 7.18. Various motifs and substitution sites are denoted.

L Gastric acid secretion

Gastric acid is a digestive fluid that is produced in the corpus and antrum zones of the stomach; see Fig. 7.2. The gastric acid has a pH of 1.5–3.5 and is composed of hydrochloric acid (HCl) and NaCl and KCl salts ~7 mM with only a trace of other electrolytes. The concentration of hydrogen ions is ~1 million times higher (pH ~ 1) than that of plasma. The function of gastric acid is to denature the organized three-dimensional properties of proteins that were included in a recent meal. The HCl also participates in solubilizing and absorbing Ca^{2+}, Fe^{2+}, and also vitamin B12.

Fig. 7.19 provides a description of how human parietal cells in the stomach are able to secrete a strong concentration of hydrochloric acid (HCl), thereby acidifying the stomach to a pH of 1.5–3.5 which facilitates the digestion of proteins. The model shown in Fig. 7.19A illustrates how the stomach's D and G cells collaborate to secrete the gastrin hormone and have it available in the capillary bed which supplies the stomach's fundus where there are gastrin receptors. Panel B of the figure illustrates the parietal cell's plasma membrane receptor (CCK2R); it is specific for gastrin.

Fig. 7.19B also illustrates three stimulants necessary to orchestrate the production of HCl which is secreted inside the stomach. The stimulants are gastrin, acetylcholine (Ach; a neurocrine messenger), and histamine (a paracrine messenger from an adjacent stomach cell). Histamine (orange balls) (see panel B) is released from the ECL cells followed by binding to its HRH2 receptors on the surface of parietal cells and also on the plasma

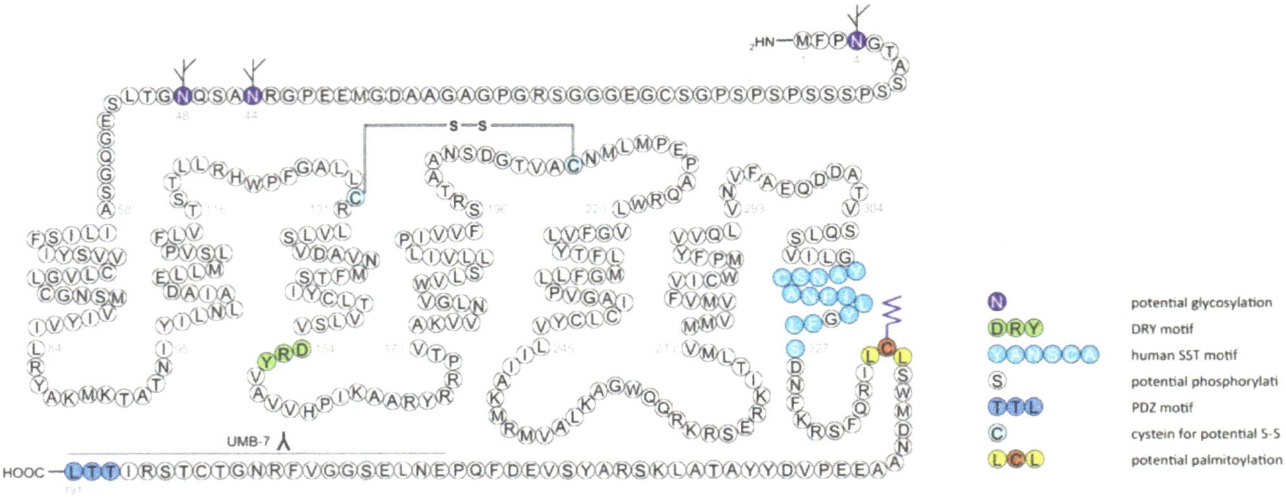

FIGURE 7.18 Structure of human SST receptor, the primary and secondary amino acid structure of the human SST_1 is shown in a schematic serpentine format. Glycosylation sites are colored in purple; the DRY motif is highlighted in green; the human SST motif is in light blue; the disulfide-forming cysteines are in pale blue; and the potential palmitoylation site is in orange. UMB-7 is a rabbit monoclonal antibody, which detects the carboxy terminal tail of SST_1, in a phosphorylation-independent manner. *SST*, Somatostatin. *Reproduced from Fig. 7.3 of T. Gunther et al., International union of basic and clinical pharmacology.CV. Somatostatin receptors: structure, function, ligands, and ne nomenclature, Pharmacological Reviews 70 (2018) 763–835.*

membrane of the D cell. This activates the adenylate kinase that generates the second messenger, cAMP. The GI hormone, gastrin (*blue balls*), is produced in the G cells (see panel A) and then binds to its CCK_2R receptors also associated with the partner D cell (panel A). In addition, the gastrin, after transport through the capillary bed, arrives and enters (see panel B) in both the ECL and parietal cells of the fundus cells. The gastrin then binds to the CCK2R gastrin receptor that is present in the plasma membrane of the parietal cells and also the ECL cells. As a consequence, the parietal cells engage in the production of hydrochloric acid. Separately, the ECL cells initiate a release of histamine (*orange balls*) that moves to the HRH2 plasma membrane receptor of both the parietal cell and D cells, thereby activating phospholipase C to mediate the release of cytosolic Ca^{2+}. Also, Ach that has been released from neurons binds to the M_3 plasma membrane receptor (see Fig. 7.19B) on both the parietal cells and D cells that are linked to stimulating intracellular Ca^{2+}. Thus in the parietal cells, the combination of calcium-dependent and intracellular cAMP signaling systems stimulates downstream protein kinases that lead to the stimulation of H^+K^+-ATPase proton pumps, resulting in the secretion of HCl into the stomach cavity.

The G cell with microvilli on the luminal surface is a classic gut endocrine cell (see examples in Fig. 7.4C). The G cell responds to the presence of food within the stomach. The G cell is the principal site of synthesis of the hormone gastrin. A typical gland is shown in the small black box in the upper middle panel of Fig. 7.19A. The antrum and fundus have a receptor for the G cell's GRP (GRP; see upper right panel) that is present in the plasma membrane of both G cells and D cells. The GRP receptor is located in the plasma membrane where it binds the extracellular GRP hormone (*small yellow balls*) as a ligand in response to a GRP nerve signal (see upper right panel). The GRP, which is present in the neuronal fibers innervating the antral mucosa, functions to stimulate gastrin secretion by acting as a secretagogue of gastrin.

As shown in the lower right panel, the parietal cell can be further stimulated by the genomic expression actions of gastrin on the production of histidine decarboxylase, which has been shown to stimulate the production of the histamine. See the binding of histamine (*orange balls*) to the parietal cell's plasma membrane receptor HRH2 for gastrin, which further stimulates the secretion of hydrochloric acid.

The hormone somatostatin is also involved in the D cells of both the antrum (upper right panel) and the oxyntic cells present in the fundus (bottom right panel). Some background information on somatostatin is provided in Figs. 3.7 and 6.3A. Both the antrum's and fundus's D cell somatostatin has been shown to modulate the gene transcription of gastrin so as to secrete more or less gastrin by adjacent G cells, depending on the physiological circumstances. As shown in Fig. 7.19A/B, somatostatin (*orange circles*) is secreted from the D cells and binds to the SSTR2 receptor for somatostatin in the basal lateral membranes of the adjacent G cells (upper right panel) or adjacent parietal cells (lower right panel). In both instances the presence of somatostatin in the G cells or parietal cells creates an inhibitory effect. In the parietal cells, less HCl is secreted (lower right panel) and in the G cells (upper right panel), less gastrin is produced and available for export by secretion and transport through the capillary

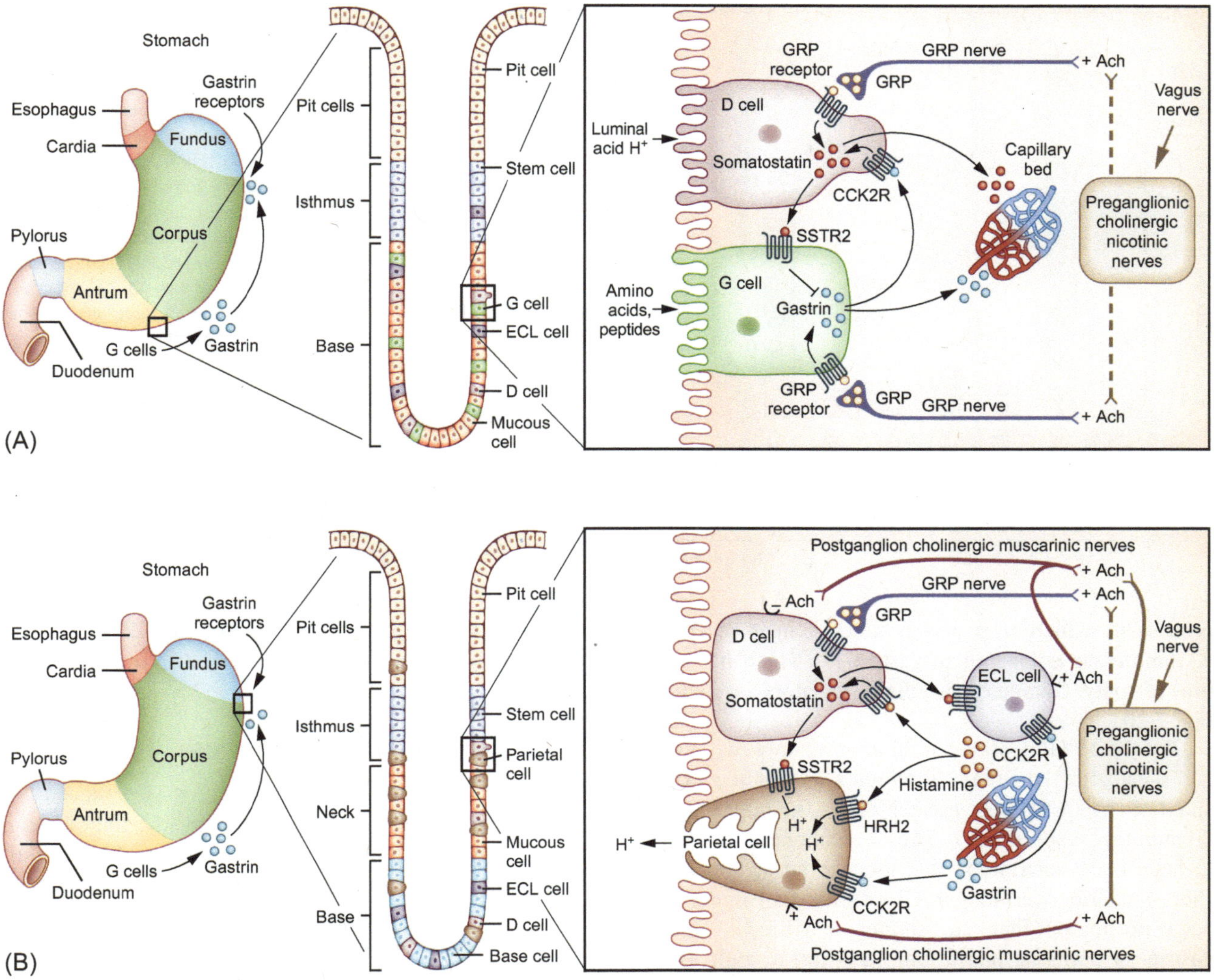

FIGURE 7.19 Integrated signal transduction by D cells, G cells, ECL cells, and parietal cells in the antrum and fundus of the stomach, which results in the secretion of hydrochloric acid in the stomach so as to facilitate food denaturation. The left side of both panels (A) and (B) illustrates the various regions of the stomach. At the bottom of the antrum (left of panel A) and the bottom of the fundus (top left of panel B), there is a small black box. Each box identifies the zone of cells that are enlarged in the center of both panels (A) and (B). In the center panel of this figure, a second pair of black boxes identifies the single cells illustrated in the center and right sides of panels (A) and (B), respectively. In the center of both right panels, a capillary bed is illustrated in red and blue. In the right side of panel (A), the combined partnership of the D and G cells' GRP regulates the production in the G cell of the hormone gastrin (*blue balls*) which is released to the capillary bed for transport to the parietal cells in the stomach's fundus region (see bottom right of panel B). The antrum's receptor for the G cell's GRP (see upper right panel) is present in the plasma membrane of D cells (at the top) and G cells (at the bottom). The GRP receptor binds the GRP hormone (small orange balls) as a ligand in response to a GRP nerve signal (see upper right panel B). As shown in the bottom right panel, the fundus' oxyntic (acid secreting) parietal cells and the ECL cells receive the hormone gastrin from the capillary bed (*small blue balls*) which binds to their CCK2R plasma membrane receptor. This ligand–receptor action results in the production and secretion of essential hydrochloric acid. Following is a key to the abbreviations employed in this figure: G cells are a source of the endocrine hormone, gastrin; D cells are a source of the hormone somatostatin; ECL are enterochromaffin-like cells that are a type of neuroendocrine cells found in the stomach's antrum in the vicinity of parietal cells; parietal cells are a source of hydrochloric acid which is secreted into the stomach; CCK2R is a plasma membrane receptor for gastrin; SSTR2 is a plasma membrane receptor for somatostatin; Ach is messenger produced by a postganglion cholinergic muscarinic nerve; histamine is an endocrine messenger that is derived from the amino acid histidine. *Ach*, Acetylcholine; *CCK*, cholecystokinin; *ECL*, enterochromaffin-like; *GRP*, gastrin-releasing peptide; *SST*, somatostatin.

bed. The GRP nerve generates in the D cells of both the antrum (upper right panel) and fundus (bottom right panel) the production of the GRP hormone (GRP; *yellow balls*), which binds to the plasma membrane receptor for GRP (see 1 *yellow ball*), thus leading to the immediate secretion of somatostatin, which will inhibit the secretion of gastrin (upper right panel) and inhibit the production of HCl by the parietal cells (bottom right panel).

Gastrin is synthesized as a large precursor molecule of 101 amino acids that is relatively slowly processed to

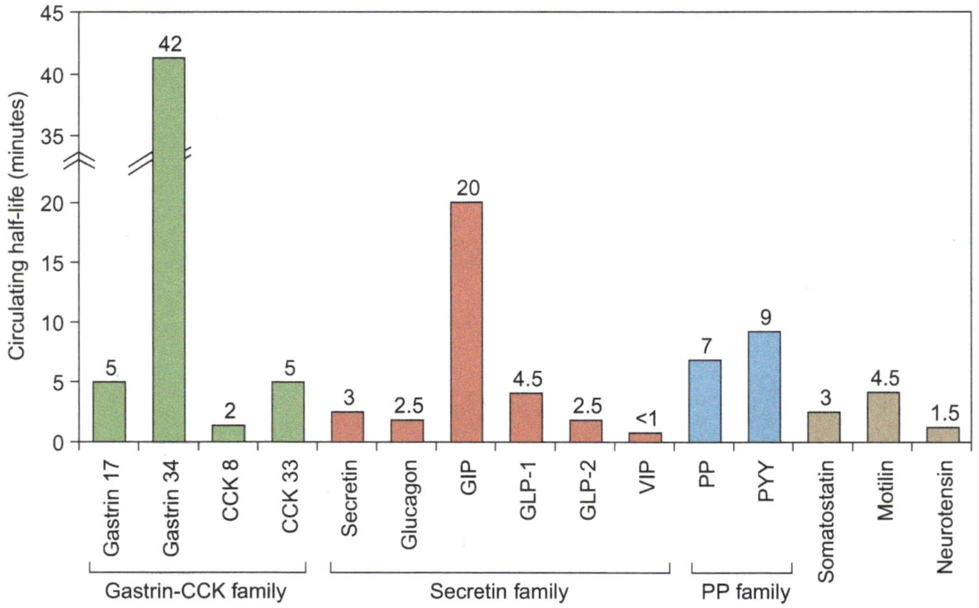

FIGURE 7.20 The circulating half-lives of the blood levels of 15 gastrointestinal peptide hormones are compared by their hormone family. The gut hormones are generally released in response to the intake of meals. This figure illustrates the circulating half-life for 15 GI peptide hormones from three GI families (the gastrin-CCK, secretin, and pancreatic polypeptide families) ranging from <1 min to an average of 3.8 min with only the GIP and Gastrin-34 (big gastrin) having a half-life significantly longer than 10 min. *CCK*, Cholecystokinin; *GI*, gastrointestinal; *GIP*, gastric inhibitory peptide.

gastrin-34, which then has a much longer half-life of 42 minutes as compared to gastrin-17, which has a half-life of only 5 minutes in the circulatory system; see Fig. 7.20.

For the 15 GI peptides in this figure, only two peptides had a half-life less than 10 minutes and only gastrin-34 and GIP had half-lives greater than 10 minutes (42 and 20 minutes, respectively). Thus evolutionarily it appears to have been determined that there is an advantage to a short half-life over longer GI peptide half-lives. It has been postulated that the shorter half-life is safer. If gastrin-34 hormone had a long half-life, it could be producing large quantities of HCl even when there was little food intake to process.

As discussed earlier, Fig. 7.8C/D presents results from a clinical trial that evaluates the relative amounts (fraction) of five blood-circulating gastrin peptides from healthy subjects (panel C) for comparison with the results obtained from patients with the ZES (panel D).[3] The suggestion is that the disease process associated with ZES significantly changes the processing of the 101-amino-acid precursor form of progastrin into the family of gastrin peptides; see Fig. 7.8B. The gastrin peptides can be isolated from the blood and separated by their molecular weight via chromatography so that the relative amounts of the five gastrin peptides can be quantitated (see the histogram). The subjects with ZES clearly have a different peptide fragment profile from normal subjects. For patients with ZES, the gastrin-34 was increased from 38% (healthy) to 58% and the gastrin-17 was decreased from 56% (healthy) to 27%. This suggests that ZES is capable of changing the half-life pattern of gastrin fragments.

M Ghrelin, leptin, and energy use

Ghrelin is a 23-amino-acid peptide that is produced both by the antrum of the stomach and by the hypothalamic arcuate nucleus. Evidence has been presented that ghrelin has effects on lipid and glucose metabolism, gastric acid secretion, and cardiovascular, immunologic, and autonomic nervous system effects. Some proportion of these effects is the consequence of the presence of ghrelin in the hypothalamic arcuate nucleus.

Several different forms of ghrelin peptides are now known to be present in the circulation. Fig. 7.21A illustrates the 117-amino-acid prepropeptide that is derived from ghrelin's gene via the successive processes of transcription and translation and the subsequent posttranslational steps that lead to the final two peptide products, the 24-amino-acid ghrelin and the 23-amino-acid obestatin.

In Fig. 7.21A the four amino acid sequences of the dominant ghrelin are labeled as peptides *a, b, c*, and *d*, while in panel B the peptide *e* is obestatin. Ghrelin peptides *a* and *b* each have 28 amino acids, whereas peptide *c* has lost the C-terminal arginine 28, and peptide *d* has lost residue 14, a glutamine. Ghrelin has a unique partnership with octanoic acid, forming an ester with the 3 serine amino acid; see Fig. 7.21C. The enzyme that acylates ghrelin is a member of the membrane-bound *O*-acyltransferases. When the substrate is ghrelin, the acronym is

3. *Zollinger–Ellison Syndrome* is caused by a nonbeta islet cell, gastrin-secreting tumor of the pancreas that stimulates the acid-secreting cells of the stomach to maximal activity, with the appearance of a plethora of gastrointestinal mucosal ulcers.

changed to GOAT (ghrelin *O*-acyl transferase). A knockout (KO) of this single GOAT enzyme in mice results in a loss of any detectable acyl-ghrelin biological activity in vivo. As seen in peptides *a, c,* and *d*, ghrelin's amino acid 3 (a serine) is acetylated by an 8-carbon fatty acid and on occasion peptides *b* and *c* are acetylated with a 10-carbon saturated fatty acid, or in peptide *b*, a 10-carbon saturated or unsaturated fatty acid. It is known that the serine residue 3 must be acetylated for ghrelin to be biologically active to stimulate GH, to elevate intracellular Ca^{2+}, and to stimulate appetite. The octanoic–ghrelin conjugate is essential for ghrelin to interact with its receptor (GHS-R1a) (see Fig. 7.22) so as to generate GH release, calcium mobilization, and modulate appetite effects. In contrast, ghrelin does not need to be acylated with the octanoic acid to produce cardiovascular and lipogenic effects.

Table 7.8 tabulates factors that either increase or decrease ghrelin's circulating concentration. It is clear from the entries in Table 7.8 that ghrelin has a wide array of activities/responsibilities. Thus short-term fasting elevates total body ghrelin levels; however, ghrelin levels will fall promptly when there is food intake. A variety of human and animal studies have described ghrelin-inhibiting insulin levels. This has the consequence that blood glucose levels become increased. It is to be anticipated that future studies will be designed to sort out the details.

Obestatin was first discovered in 2005 and it was proposed to be concerned with regulating food intake and energy use in a brain–gut network that focuses on hunger or satiety. Obestatin has 23 amino acids; see peptide *e* (*green*) in Fig. 7.21B. It is the second biologically active peptide produced from the ghrelin gene. Obestatin has been found in the GI tract, spleen, mammary gland, and in breast milk. Some recent studies suggest that obestatin can reduce food intake. Further studies are needed to clearly define the physiological role of obestatin.

As described in Fig. 7.23A, ghrelin that is biosynthesized in the stomach can be delivered via the bloodstream to the arcuate nucleus of the mediobasal region of the hypothalamus using active transport to pass through the blood–brain barrier (see A1). Ghrelin generated in the periphery also stimulates vagal currents, which eventually stimulates the *N. tractus solitarius* (see A2) which then communicates with the NPY/Agrp neurons (see A3) in the hypothalamus. The NPY/Agrp neurons are believed to stimulate an increased food intake. Some data have supported the view that ghrelin can be directly produced in the hypothalamus, where it could also stimulate food intake.

The effects of intravenous ghrelin or saline infusion on appetite and food intake have been evaluated in a randomized double-blind crossover study in nine healthy volunteers. These results are shown in Fig. 7.23B. The left and right panels report the mean energy intake from standard meals. The red (saline infusion) and blue (ghrelin infusion) bars report the results for each group. Ghrelin treatment resulted in a clear-cut increase in energy consumed by every individual; the mean increase for the group was $28 \pm 3.9\%$, $P < .001$ higher during the perfusion of ghrelin as compared to saline infusion. The results in the right panel report the individual result for each of the nine participants. All patients that received ghrelin showed a clear increase in their energy intake.

A schematic model of the integrated biological properties of ghrelin and leptin with respect to energy balance is shown in Fig. 7.24. Leptin is a 16-kDa protein (167 amino acids) (see Fig. 6.18), whereas ghrelin is a 28-amino-acid GI peptide hormone (see Fig. 7.21). As described in the figure, ghrelin produced by the stomach sends signals to the hypothalamus, which results in an increased appetite that, over time, can increase body weight. Leptin interacts with its receptor in adipose tissue and also sends a signal to the hypothalamus, which results in a loss of appetite, which can then cause a reduction in body weight. Fig. 7.25 shows the structure of the leptin–leptin receptor dimer complex and the specific leptin-binding sites on the receptor dimer molecules. Table 7.8 presents a summary of the various parameters that modulate ghrelin secretion.

Based on the wide distribution of receptors for ghrelin, it can be extrapolated that there is a broad variety of ghrelin-mediated responses. The presence of the ghrelin receptor has been determined via using classical and real-time reverse transcription and polymerase chain reaction methodology. The ghrelin receptor was reported to be present in 31 tissues, including the following: adrenal gland, atrium, breast, buccal mucosa, esophagus, fallopian tube, fat tissue, gallbladder, human lymphocytes, ileum, kidney, colon, liver, lung, lymph node, muscle, myocardium, ovary, pancreas, pituitary, placenta, prostate, right colon, skin, spleen, testis, thyroid, and veins.

The ghrelin receptor, known as the GH secretagogue receptor, is a G protein–coupled receptor. Intriguingly, a KO of ghrelin receptors was more effective than KO of ghrelin itself in interfering with the generation of biological responses. When challenged by fasting and exposure to cold, only the mice without GHS-R maintained a normal body temperature, thus suggesting the existence of additional unidentified regulators that contribute to the biological effects of ghrelin and/or GHS-R.

IV Clinical aspects

A Peptic ulcer disease

The term *peptic ulcer* refers to an open sore on the inside lining of the esophagus, stomach, and/or the upper portion

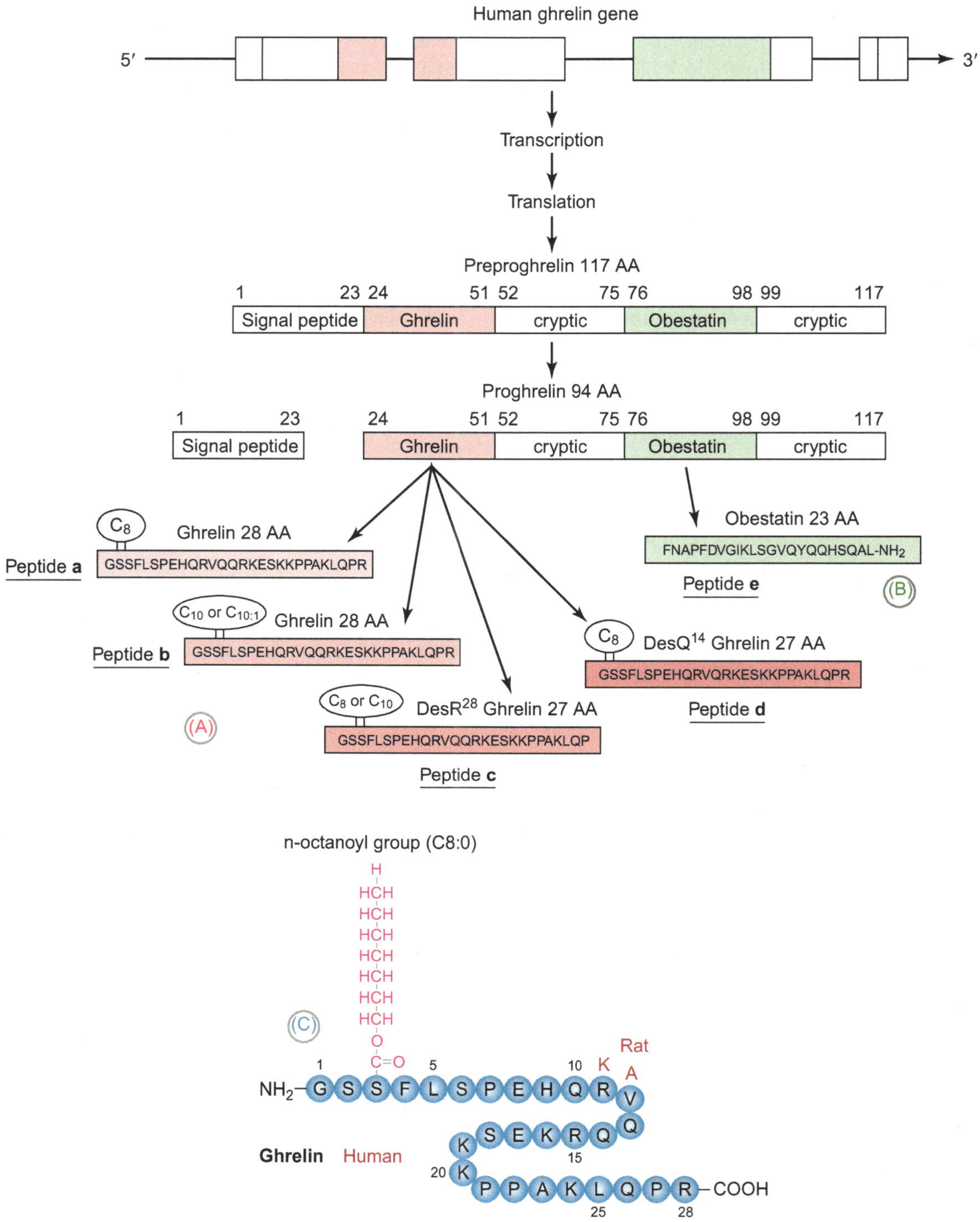

FIGURE 7.21 A schematic presentation of the human ghrelin gene and the resulting preproghrelin. (A) The 117-amino-acid prepropeptide and the subsequent posttranslational steps that lead to the 94-amino-acid proghrelin and the final two products, the 24-amino-acid ghrelin (*pink*) and the 23-amino-acid obestatin. The amino acid sequences of the dominant ghrelin are labeled as peptides a ≫ d and peptide e for obestatin (*green*). Ghrelin peptides a and b have 28 amino acids, whereas peptides c and d have lost the C-terminal residue 28, an arginine. As seen in peptides a, c, and d ghrelin's amino acid 3 (a serine) is acetylated by an 8-carbon fatty acid and on occasion (peptides b and c) are acetylated with a 10-carbon saturated fatty acid, or in peptide b, a 10-carbon unsaturated fatty acid. It is known that the serine residue 3 must be acetylated for ghrelin to be biologically active to stimulate GH, to elevate the intracellular Ca^{2+}, and to stimulate appetite. (B) Obestatin (see *green* peptide e), with 23 amino acids, is the second peptide produced from the ghrelin gene. Some recent studies suggest that obestatin can reduce food intake. (C) The active form of ghrelin. The 28-amino-acid sequences of human and rat ghrelin are shown; residue 1 is the N-terminal glycine. Two amino acids at position 11 and 12 are changed from arginine (R) and valine (V) for the human sequence to lysine (K) and alanine (A) for the rat sequence. Intriguingly, the serine residue in position 3 must be esterified to octanoic acid which is necessary for ghrelin to have any biological activity. Possibly the hydrophobic tail on the octanoic acid is used to tether ghrelin to a plasma or endoplasmic reticulum membrane. *GH*, Growth hormone.

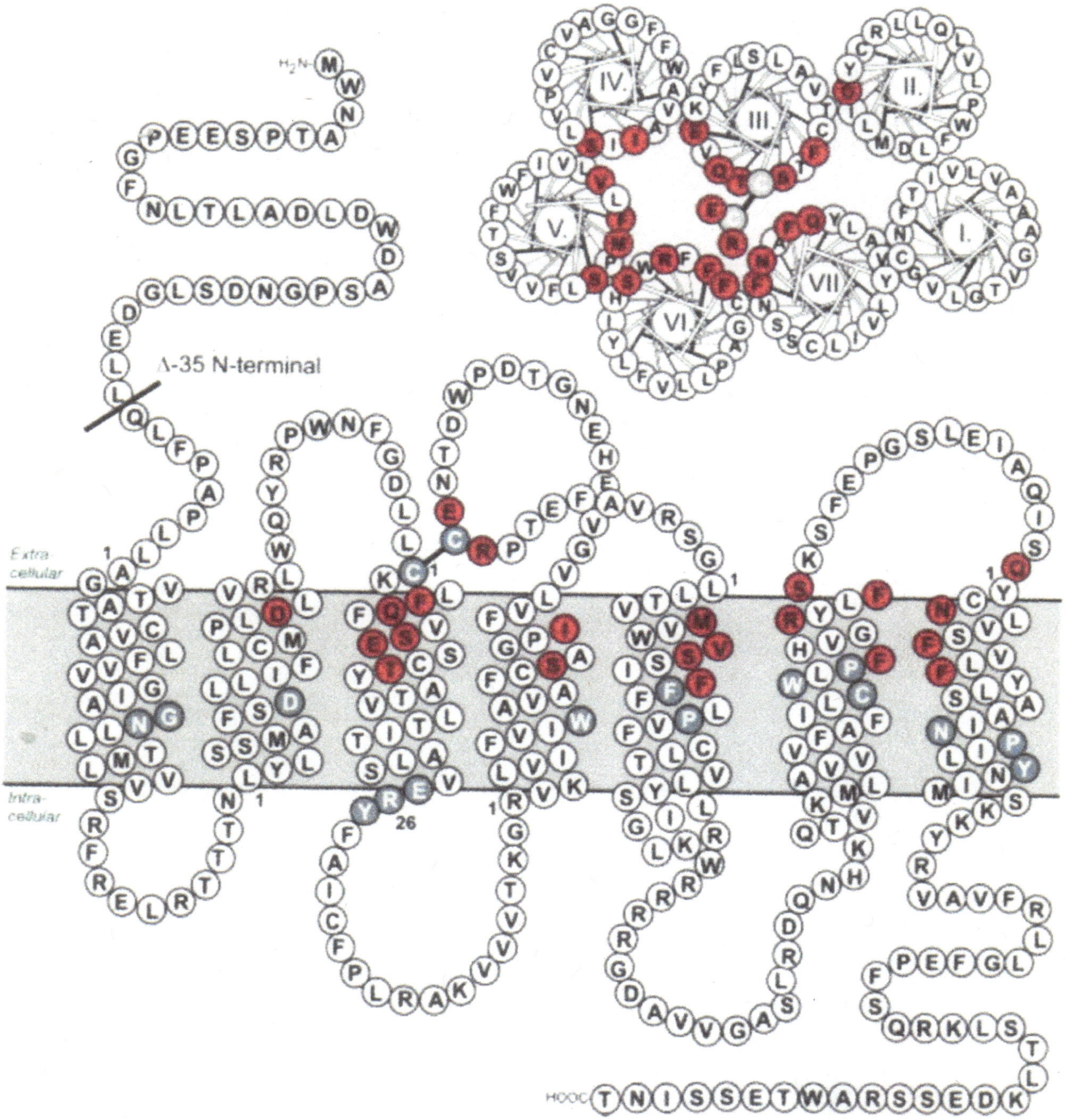

FIGURE 7.22 Schematic figure of the ghrelin receptor. Helical wheel and serpentine diagram of the ghrelin receptor. Mutated residues are highlighted in red. Schwartz/Baldwin's generic numbering system for TTM receptor, which is based on the actual location of the residues in each transmembrane helix, is used. For reference the numbering of highly conserved residues in each helix is marked (highlighted in *white* on *gray*) AsnI:18, AspII:10, ArgIII:26, TrpIV: 10, ProV:16, ProVI:15, and ProVII:17, and the proposed first residue of each transmembrane helix is indicated by "1." The position of the deletion of the amino terminal is indicated by a line, Δ − 35 N-terminal. *Reproduced from Fig. 7.1B of B. Holst et al. Overlapping binding site for the endogenous agonist, small-molecule agonists, and ago-allosteric modulators on the ghrelin receptor, Molecular Pharmacology 75 (2008) 44–59.*

of the small intestine, particularly the duodenum. The most common symptom of a peptic ulcer is abdominal pain. The reason(s) why peptic ulcers develop is not well understood. It has been postulated that there is a shift in the balance of mucosa-protecting (secretion of bicarbonate and mucus) and mucosa-damaging (secretion of acid and pepsin) mechanisms. A common cause of ulcers is the bacterium *Helicobacter pylori*. Usually, *H. pylori* causes no problems;

TABLE 7.8 Modulators of ghrelin secretion.[a]

Ghrelin is decreased	Ghrelin is increased
Glucose	Leptin
Insulin	Growth hormone–releasing hormone
Food consumption	Short-term fasting
Elevated body mass index	Low body mass index
Somatostatin	Thyroxine

[a]*Short-term starvation or food restriction leads to an increase in ghrelin's blood concentration. Ghrelin levels are higher during the night.*

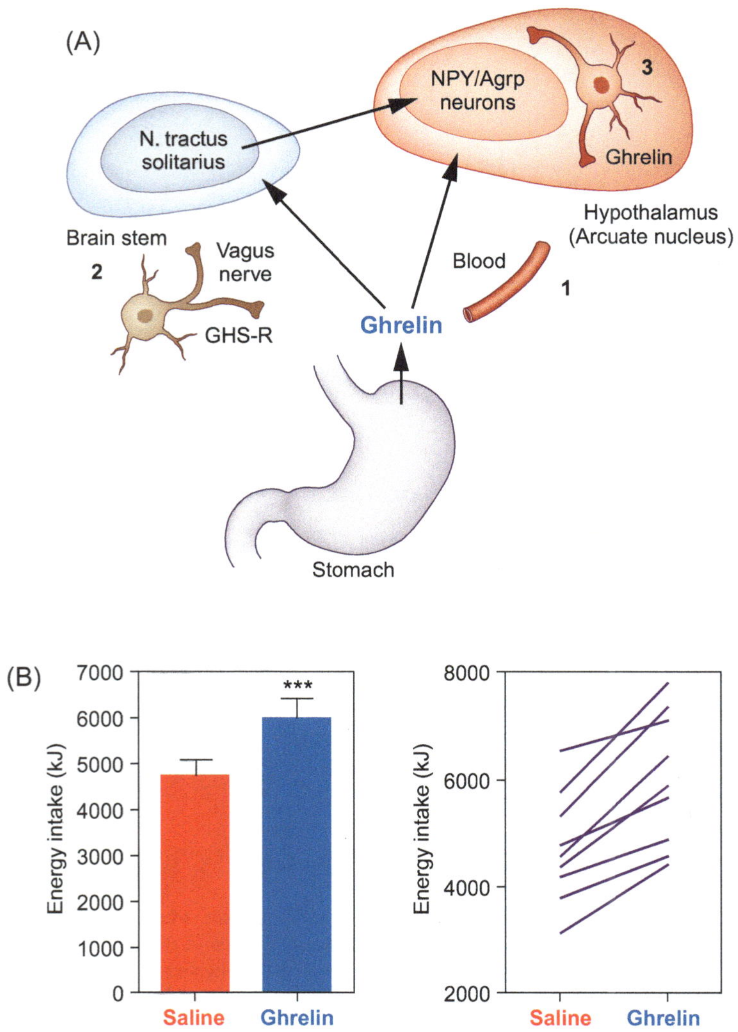

FIGURE 7.23 (Panel A) Ghrelin exerts its biological effects in the hypothalamus via three different signaling pathways (see panels A 1, 2, and 3). (1) Ghrelin biosynthesized in the stomach is delivered to the arcuate nucleus of the hypothalamus via the bloodstream (A1). (2) Ghrelin generated in the periphery stimulates vagal currents, which eventually stimulates the N. *tractus solitarius* (see panel A2) (3) which then communicates with the NPY/Agrp neurons in the hypothalamus (A3). (Panel B) The effects of intravenous ghrelin or saline infusion on appetite and food intake were evaluated in a randomized double-blind crossover study in nine healthy volunteers. The left and right panels report the mean energy intake from standard meals. The red bar (saline infusion) and blue bar (ghrelin infusion) report the energy intake of the group results. The mean increase in energy intake mediated by ghrelin infusion for the group (left *red* panel) was $28 \pm 3.9\%$, $P < .001$ higher compared to saline infusion. The results in the right panel report the individual result for each of the nine participants. All patients who received ghrelin showed a clear increase in their energy intake. *NPY*, Neuropeptide Y. *The data were published by A.M. Wren et al. Ghrelin enhances appetite and increases food intake in humans, The Journal of Clinical Endocrinology and Metabolism 86 (2001) 5992–5995.*

however, it sometimes can disrupt the mucous layer so that the lining of the stomach or duodenum develops inflammation, thereby producing an ulcer.

About 5 million people in the United States currently have peptic ulcer disease. About 25 million people in the United States are expected to develop peptic ulcers at some point in their lives. Peptic ulcer disease affects all age groups but is rare in children. Men have twice the risk of ulcers as women. The risk of duodenal ulcers tends to rise, beginning around age 25, and continues until age

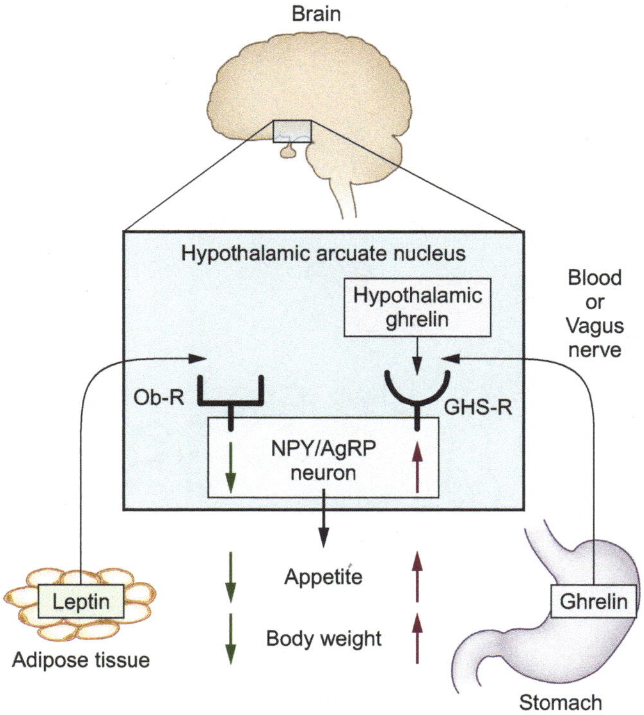

FIGURE 7.24 A schematic model of the biological properties of ghrelin and leptin with respect to energy balance. Leptin is a 16-kDa protein (167 amino acids) (see Fig. 6.18), whereas ghrelin is a 27/28-amino-acid peptide gastrointestinal hormone; see Fig. 7.21. As described in the figure, ghrelin produced by the stomach sends signals to the hypothalamus, which results in an increased appetite that over time can increase body weight. Leptin interacts with its receptor in adipose tissue and also sends a signal to the hypothalamus that results in a loss of appetite, which can then cause a reduction in body weight.

75. The risk of gastric ulcers peaks at ages 55–65. Although individuals with duodenal ulcer disease clearly have increased rates of gastric acid secretion, and although gastrin is a potent secretagogue of HCl secretion, it has not been possible to directly implicate gastrin in the etiology of duodenal ulcer disease.

H2 blockers interfere with acid production by blocking histamine, a substance produced by the body that encourages acid secretion in the stomach. Four H2 drugs have a proven value in the treatment of duodenal ulcer disease; they are famotidine (Pepcid AC), cimetidine (Tagamet), ranitidine (Zantac), and nizatidine (Axid). All have good safety profiles and few side effects.

B Carcinoid syndrome

Carcinoid is the term currently used to describe the major tumor of the neuroendocrine tumor family. The term "carcinoid" was first used about a century ago. It was applied by a German pathologist Otto Lubarsch in 1888 to a group of tumors which he studied and which develop in the intestine. They are visually quite different but functionally similar in their properties to ordinary carcinomas. In addition, the carcinoid tumors are not quite benign. Therefore he coined the term "carcinoid" as describing an intermediate type of malignancy that grows more slowly than the more common intestinal carcinoma and which does not have the classical functional appearance of a carcinoma. Today, the term carcinoid is employed to describe a group of intestinal carcinoid tumors that are derived from ECL, which are cytochemically indistinguishable from many of the GI hormone-secreting cells of the intestinal tract. The most frequent sites of their location are the ileum and the appendix. The chief clinical signs of carcinoid syndrome are extensive body flushing notable on the face and arms and also diarrhea.

Carcinoid tumors are now known to be found most commonly in the GI tract and lungs. Some rare examples of carcinoid sites include the heart, testes, and ovaries, and they have the ability to metastasize to the liver. Carcinoid tumors normally range from 1 to 2 cm in diameter. The clinical "label" of *carcinoid syndrome* was introduced in 1954.

The chief biochemical lesion is an elevated production of 5-hydroxytryptamine, which results from the conversion of tryptophan to 5-hydroxytryptophan followed by decarboxylation to yield 5-hydroxytryptamine. Elevated blood levels of 5-hydroxytryptamine and an increased urinary excretion of 5-hydroxyindoleacetic acid are often diagnostic of the carcinoid syndrome. Also, the tumor contains the enzyme kallikrein (see Chapter 15: Hormones Related to the Kidney and Cardiovascular System), which leads to increased plasma levels of bradykinin.

Originally carcinoid was an extremely rare tumor; however, in the last 15 years it has been diagnosed with greater frequency. Presently, there are about 3000–4000 new carcinoid cases appearing and being reported in the United States every year. There are now ~50,000 subjects with carcinoid of clinically significant degree. Of course, this rate of frequency is still rather uncommon when viewed from the perspective of the more frequently encountered growths like cancer of the breast (230,000 per year) or prostate (240,000/year), both in 2012. Both pharmacological and surgical treatments are employed to treat carcinoid syndrome.

C Zollinger–Ellison syndrome

The ZES was first described in 1955 by Drs. Zollinger and Ellison, who described two patients. It is characterized by an advanced ulcer disease of the upper GI tract (the duodenum), with hypersecretion of gastric acid, and nonbeta islet cell tumors of the pancreas islets. The principal symptoms include the following: abdominal pain, nausea, heartburn, chronic diarrhea, vomiting blood that has been digested, and malnourishment. The ZES may occur intermittently or as

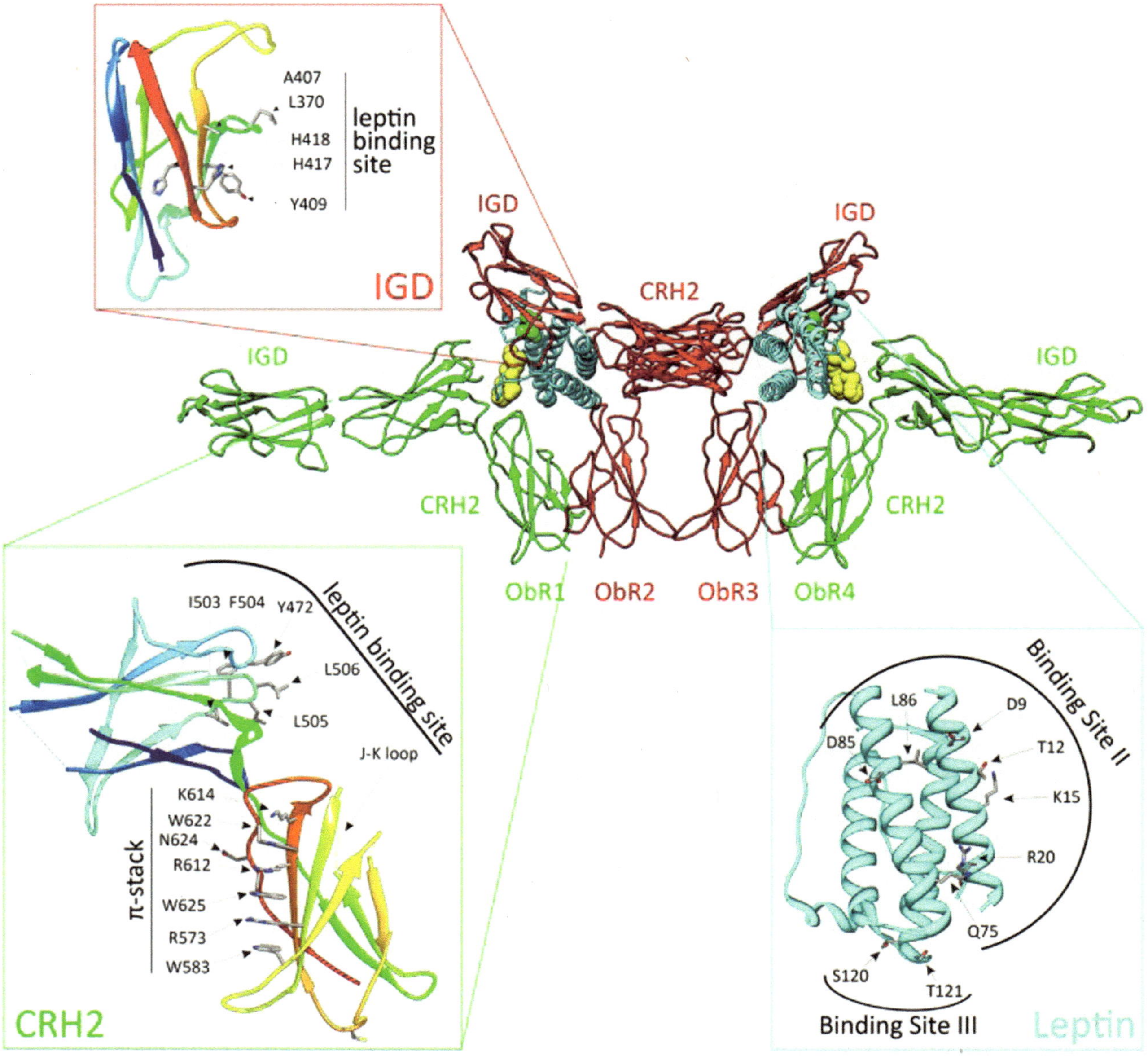

FIGURE 7.25 The activated leptin:ObR complex. Leptin clusters two preformed ObR dimers to form an activated 2:4 leptin:ObR complex. In this model the ObR's are colored green and red, leptin molecules cyan. The hormone binds with its binding site II to the CRH2 domain of the receptor, while site III residues interact with the IGD of a second receptor. These residues are colored yellow and green and defined in the cyan panel. Receptor residues involved in these interactions are shown in red and green panels. Only the CRH2 and ICD domains of each receptor are shown (for clarity). *ICD*, Intracellular domain; *IGD*, immunoglobulin-like domain; *ObR*, leptin–leptin receptor. *Reproduced from Fig. 1 of J. Wauman, L. Zabeau, J. Tavernier, The leptin receptor complex: heavier than expected? Frontiers in Endocrinology 8 (2017) 30; https://doi.org/10.3389/fendo2017.00030.*

part of an autosomal dominant familial syndrome termed multiple endocrine neoplasia, type 1.

Most patients are diagnosed between the ages of 20 and 50. The male–female ratio ranges between 1.5:1 and 2:1. The annual occurrence is estimated at 1–2 cases per million adults. It is estimated that ~1% of operative cases of peptic ulcer disease are individuals with ZES. The syndrome results from the release of large quantities of gastrin by the pancreatic islet tumors, probably D cells. This then results in increased secretion of HCl by the parietal cells of the stomach.

Further reading

Gastrointestinal hormones and obesity

H. Alhabeeb, et al., Gut hormones in health and obesity: the upcoming role of short chain fatty acids, Nutrients 13 (2021) 481. Available from: https://doi.org/10.3390/nu/3020481.

M.L. Ricardo-Silgado, A. McRae, A. Acosta, Role of enteroendocrine hormones in appetite and glycemia, Obesity Medicine 2021 (2021) 100332. Available from: https://doi.org/10.1016/j.obmed.2021.100332.

Pancreatic polypeptide PP and peptide YY

R.A. Lafferty, P.R. Flatt, N. Irwin, Established and emerging roles peptide YY (PYY) and exploitation in obesity-diabetes, Current Opinion in Endocrinology and Diabetes and Obesity 28 (2021) 253–261.

Obestatin

B.G. Malikarjuna, U.V. Manjappara, Obestatin and rosiglitazone differentially modulate lipid metabolism through peroxisome proliferator-activated receptor-γ (PPARγ) in pre-adipose and mature 3T3-Li cells, Cell Biochemistry and Biophysics 79 (2021) 73–85.

Gastrin

G.D. Fang, et al., Gastrin via activation of PPARalpha, protects the kidney against hypertensive injury, Clinical Science (London) 135 (2021) 409–427.

J.G. Prosapio, P. Sankar, I. Jialal, Physiology, Gastrin, StatPearls [Internet], StatPearls Publishing, Treasure Island, FL, 2020.

Gastric inhibitory peptide

K. Gupta, A. Raja, Physiology, Gastric Inhibitory Peptide, StatPearls [Internet], StatPearls Publishing, Treasure Island, FL, 2020. 2021 Jan 2020 Oct 1.

F. Yang, S. Dang, H. Lv, B. Shi, Combined treatment with a gastric inhibitory polypeptide receptor antagonist and a peptidyl peptidase-4 inhibitor improves metabolic abnormalities in diabetic mice, Journal of International Medical Research 49 (1) (2021). 3000060520985664.

Secretin

Z. Cruz-Monserrate, et al., Delayed processing of secretin-induced pancreas fluid influences the quality and integrity of proteins and nucleic acids, Pancreas 50 (2021) 17–28.

C.G. Dengler, et al., Discovery of small molecule positive allosteric modulators of the secretin receptor, Biochemical Pharmacology 185 (2021) 114451. Available from: https://doi.org/10.1016/j.bcp.2021.114451.

Chapter 8

Eicosanoids

I Introduction

Eicosanoids are a class of molecules derived from 20-carbon ("eicosa" is Greek for 20) polyunsaturated fatty acids, most frequently arachidonic acid. The eicosanoids include the prostaglandins (PG), thromboxanes (TX), leukotrienes (LT), and lipoxins (LX). These molecules almost always act on the cells that produce them or on neighboring cells, that is, over short distances and time periods, and, therefore, can be classified as autocrine/paracrine hormones. They are widely distributed in the cells and tissues of the body and, as indicated in Table 8.1, have wide-ranging biological actions.

The existence of the PG was first hinted at in the 1930s in studies demonstrating that lipid extracts of semen could stimulate uterine smooth muscle contractions. Believing (incorrectly) that the active substance was derived from the prostate gland, it was called "prostaglandin." It was another 30 years before the active substances were sufficiently purified by scientists in Sweden to determine the structures of the first two members of the family. PG E and F were named for their solubility in either ether (E) or phosphate (F, "fosfat" in Swedish). PG A and B (not shown) were named for their stability in acids and bases and the remaining letters were attached to the compounds as they were structurally characterized in the mid-1960s.

During the 1960s and 1970s, the widespread nature of the PG became appreciated and it became clear that most if not all nucleated cells produce these and/or other eicosanoids. As shown by some examples in Table 8.1, the eicosanoids play important roles in endocrine systems. For example, in the brain PGs influence the secretion of pituitary hormones that govern endocrine tissues such as thyroid, adrenal, ovary, and testis. PGs affect exocrine hormone targets such as pancreas and gastric mucosa, and endocrine target tissues such as renal tubules, bone, and adipocytes. PGs and other eicosanoids act on the smooth

TABLE 8.1 Some biological actions of eicosanoids.

Tissue	Eicosanoid	Action
Hypothalamus–pituitary axis	PGE_1, PGE_2	Hormone secretion
Ovary	PGE_2	Ovulation
	$PGF_{2}\alpha$	Luteolysis
Uterus	PGI_2	Implantation
	PGE_2, $PGF_{2}\alpha$	↑Contraction
Kidney	PGH_2, PGE_1, PGI_2	↑Blood flow, ↑Filtration
Stomach	PGE_2, PGI_2	↓Gastric acid secretion
Intestine	PGE_1, $PGF_{2}\alpha$	↑Motility nausea, diarrhea
Bronchi	PGE_2, PGI_2	Bronchodilation
	$PGF_{2}\alpha$, TXA_2, LTC_4, LTD_4	Bronchoconstriction
Platelets	TXA_2	↑Aggregation
	PGE_1, PGI_2	↓Aggregation
Blood vessels	PGI_2	Vasodilation
	$PGF_{2}\alpha$, TXA_2, LTC_4, LTD_4	

See Figs. 8.1–8.3 for names and structures of the eicosanoids. *LT*, Leukotrienes; *PG*, prostaglandins; *TX*, thromboxanes.

Hormones. DOI: https://doi.org/10.1016/B978-0-323-90262-5.00005-6

muscles of the reproductive, alimentary, and respiratory tracts and on cardiovascular smooth muscles. Red blood cells, leukocytes, and platelets are targets of eicosanoids in inflammatory reactions, including those of asthma. PGs are important mediators of pain perception.

Virtually every endocrinological/physiological system discussed in the other chapters of this book also involves the local production and action of one or more of the eicosanoids. This chapter will first focus on the structure and nomenclature of the eicosanoids, the pathways by which they are synthesized and the important enzymes involved in these pathways. The current understanding of the cellular receptors for the eicosanoids will be followed by a few examples of their biological activity and current areas of their clinical importance.

II Structure and nomenclature of eicosanoids

The prostanoids, PGs and TXs, can structurally be considered relatives of prostanoic acid (Fig. 8.1A). They are derived from polyunsaturated fatty acids, most commonly

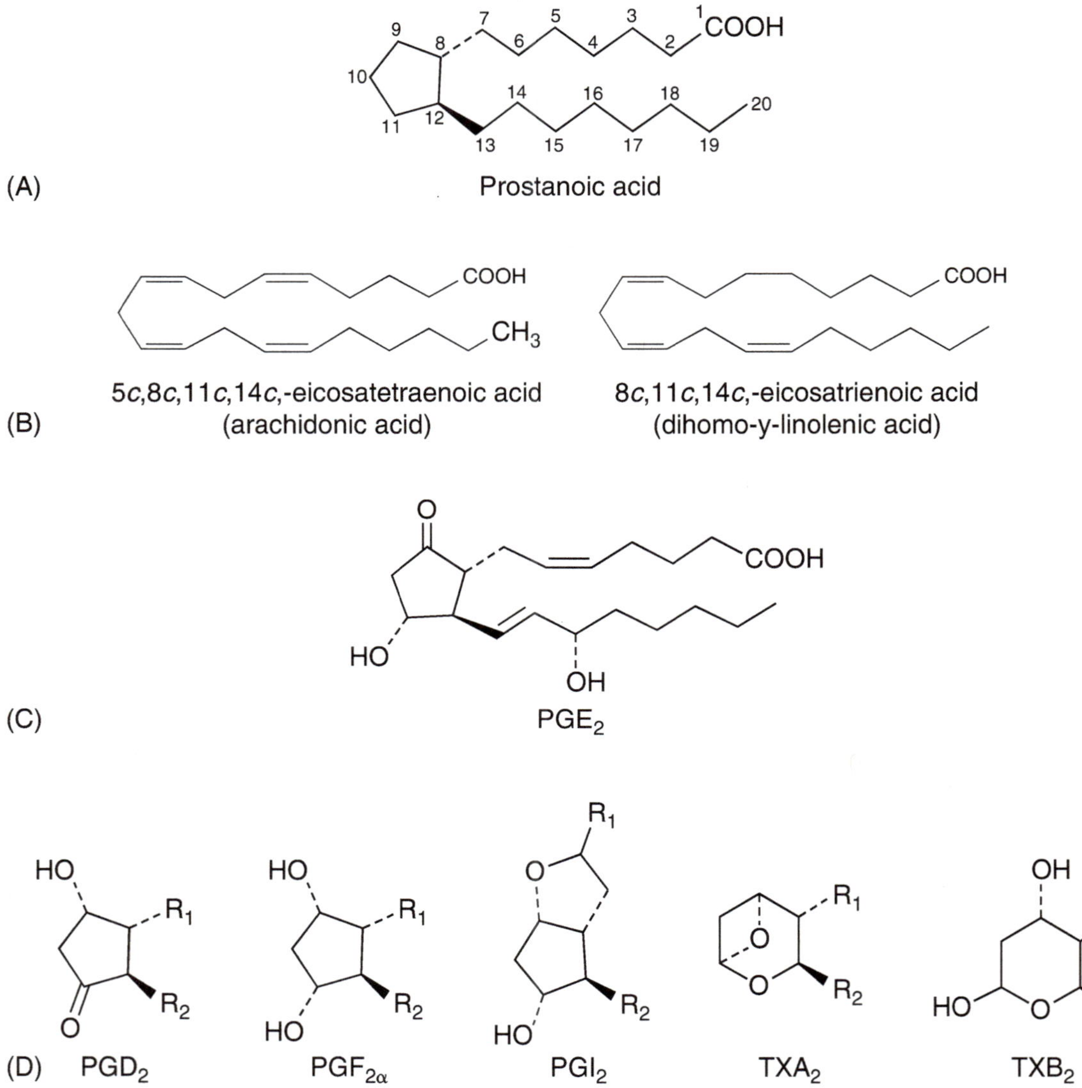

FIGURE 8.1 Prostanoid structures. Prostaglandins D, E, F, and I (prostacyclin) and the thromboxanes are based on that of prostanoic acid, shown in panel (A) with the numbering of the carbon atoms. (B) The prostanoids are derived from 20-carbon polyunsaturated fatty acids. The most abundant of these is arachidonic acid that gives rise to prostanoids of the 2 series. Also shown is eicosatrienoic acid, which can be converted to prostanoids of the 1 series. (C) Prostaglandin E_2, derived from arachidonic acid, is named for the family of prostanoids into which it falls (PG), the substituents on the ring portion of its structure (PGE), and the number of bonds in its two side chains (PGE_2). Had it been derived from eicosatrienoic acid, it would have a single double bond and would be designated PGE_1. (D) The ring structures of other biologically important prostanoids are shown, with the double bonds in the side chains, R1 and R2, assumed to be the same as in PGE_2. PGF2α designates the stereochemistry of the hydroxyl group at C-9. PGI_2 is also referred to as prostacyclin. Also shown are the structures of the TX, of which TXA_2 is the biologically active form. *PG*, Prostaglandins; *TX*, thromboxanes.

arachidonic acid (Fig. 8.1B), taking the form typified by PGE_2 (Fig. 8.1C).

Most of the PG have a five-membered ring and the substituents on it determine the subclass, designated by a single letter, such as D, E, or F (Fig. 8.1D). One prostaglandin, PGI_2, also known as prostacyclin, has two adjacent rings, one of which contains an oxygen. TXs have a six-membered ring with one or two oxygen associated with it.

The subscripted 2 following the third letter designates the number of double bonds in the two side chains of the molecule and depends on the polyunsaturated fatty acid from which it is derived. Arachidonic acid gives rise to prostanoids of the 2 series, whereas dihomo-γ-linoleic acid (Fig. 8.1B), with its three double bonds, is the substrate for prostanoids of the 1 series. 5*c*,8*c*,11*c*,14*c*,17*c*-Eicosapentanoic acid, with five double bonds (Fig. 8.2), is converted to molecules with three double bonds in the two side chains. The hydroxyl group at carbon 9 in PGF is in the α configuration; thus for this one PG, this Greek letter traditionally follows the number.

Representatives of the open chain eicosanoids, the LT and LX, are shown in Fig. 8.3. LTs are essentially modified open chain fatty acids that may be conjugated to

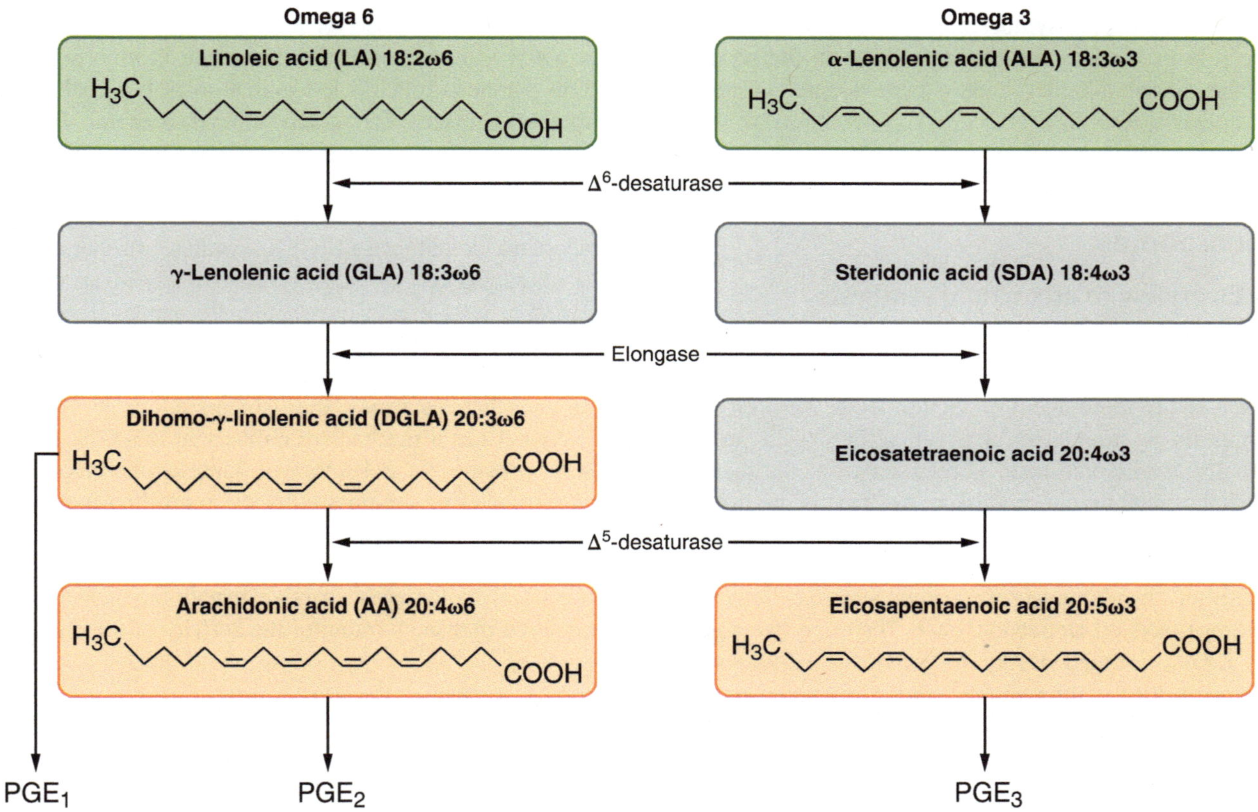

FIGURE 8.2 Dietary sources of eicosanoid precursors. Two dietarily EFA (*green boxes*) give rise to the 20-carbon polyunsaturated fatty acids required for the prostanoids (here typified by PGE) of the 1, 2, and 3 (*orange boxes*) series as well as that of the open chain leukotrienes and lipoxins. The EFAs and the pathways that convert them are referred to as omega 6 or omega 3 depending on the number of the carbon at which the first double bond begins, counting from the terminal methyl group, the omega carbon. The names of the types of enzyme that catalyze these reactions are shown between the two pathways. *EFA*, Essential fatty acids.

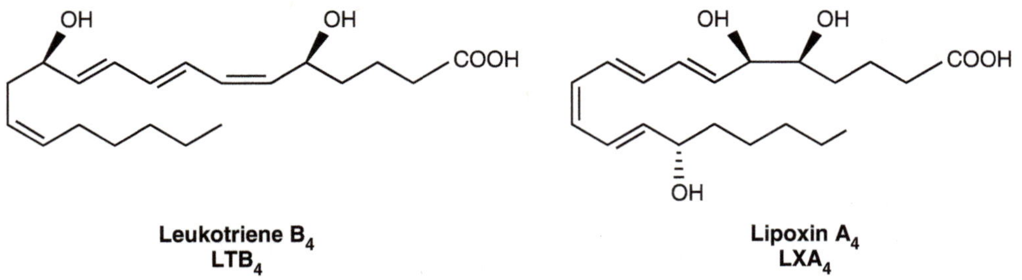

FIGURE 8.3 Other eicosanoids. The structures of representative nonprostanoid biologically important derivatives of arachidonic acid are shown.

glutathione or glutathione degradation products (see Fig. 8.8 later, in Section III.E). LX are open chain fatty acids. As with the prostanoids, the third letter denotes the substituents on the carbon atoms and the number is the number of double bonds, indicating the fatty acid from which the molecule was derived.

Humans are quite limited in their ability to produce fatty acids longer than 18 carbons and to introduce double bonds beyond C-9. An elongase adds two carbons to palmitate (16C) to form stearate (18C), and there are three desaturases to introduce double bonds at the $\Delta 5$, $\Delta 6$, and $\Delta 9$ positions. These enzymes are also able to use the essential fatty acids, linoleic and linolenic acids, to generate the 20-carbon polyunsaturated fatty acids from which the eicosanoids are synthesized, as shown in Fig. 8.2. It is the role of linoleic and linolenic acids in the production of the eicosanoids that renders them essential components of human diets.

III Synthesis and inactivation of eicosanoids

A Overview of eicosanoid synthesis

The pathways for the synthesis of PGs and relatives from membrane lipids are summarized in Fig. 8.4. As will be discussed in more detail in Section III.B, when eicosanoid synthesis is stimulated in a cell a fatty acid precursor, usually arachidonic acid, is released from the membrane phospholipids by phospholipase A_2. This is the initial and often rate-limiting step in the synthesis of eicosanoids.

In cells that produce LT, the first step in the pathway is catalyzed by lipoxygenase, producing the intermediate hydroperoxyeicosatetraenoic acid. There are two pathways to LXA_4, through LTA_4 followed by 12-lipoxygenase action or directly through the action of 15-lipoxygenase.

In all cells that produce any of the prostanoids, arachidonic acid (or other polyunsaturated fatty acid) is first acted on by the membrane PGH synthetase complex (PGHS), also known as cyclooxygenase (COX), producing the cyclic endoperoxide intermediates, PGG_2 and PGH_2. The conversion of PGH_2 to any of the PGs (including prostacyclin), or the TX, will depend on which specific synthase(s) are present in the cell. As indicated earlier, if 20-carbon polyunsaturated fatty acid substrate for COX has either 3 or 5 double bonds, PG of the 1- or 3 series, respectively, will result.

Each of the enzymes that convert PGH_2 to the other prostanoids shown in Fig. 8.4 catalyzes a specific reaction and has its own distinguishing characteristics. Prostacyclin synthase (PGIS) is a cytochrome P450 enzyme that catalyzes the isomerization of PGH_2–PGI_2. This is a nonoxidative reaction so, in contrast to other cytochrome P450 enzymes, neither molecular oxygen nor NADPH is required.

There are two forms of PGD synthase (PGDS), one in mast cells and microglia which is glutathione-dependent (H-PGDS) and the other found in brain, male genital organs, and cardiovascular tissues (L-PGDS). L-PGDS is glutathione-independent but uses other sulfhydral compounds.

PGH_2 is converted to PGE_2 by a glutathione-dependent microsomal PGE synthase (mPGES-1). mPGES-1 is inducible in several tissues and constitutively expressed in some; it is the form of the enzyme involved in PGE_2's actions in inflammation, pain, and cancer (see Sections V and VI). Two other proteins capable of synthesizing PGE_2 have been isolated and characterized: a second membrane bound enzyme, mPGES-2; and a soluble protein, cPGES. These activities are both constitutively expressed. Studies with mice in which the gene for mPGES-2 has been inactivated do not reveal any role for this enzyme in vivo relative to PGE_2 synthesis. cPGES is identical to p23, a cochaperone protein necessary for glucocorticoid receptor function. Whether it also plays a role in normal PGE_2 synthesis is not yet clear.

$PGF_2\alpha$ can arise either from the action of PGH_2 9,11-endoperoxide reductase ($PGF_2\alpha$ synthase) or that of PGE_2 9-ketoreductase. The pathway used in any given cell will depend at least in part on whether the same cell also contains PGE_2 synthase.

Thromboxane synthase (TXAS) is a cytochrome P450 enzyme that catalyzes the isomerization of PGH_2 to TXA_2 but has a mechanism unlike that of PGIS. Each of these enzymes is a subject of intense pharmacological interest, offering the possibility of controlling the production of the prostanoids individually. TXA_2 is quite unstable and rapidly breaks down to TXB_2, a more stable and biologically inactive metabolite which can be measured in urine to monitor the activity of this pathway.

B Phospholipase A_2

The phospholipids that make up most of the lipid component of the plasma and internal cellular membranes serve as substrates for several types of phospholipases, as depicted in Fig. 8.5A, depending on where the cleavage of the phospholipid takes place. Phospholipases C and D are involved in cell signaling events that require the polar head group (such as inositol triphosphate, IP_3). The two types of phospholipase A cleave the fatty acid from either the *sn*-1 or *sn*-2 position, the latter being the position in which arachidonic acid or another polyunsaturated 20-carbon fatty acid is found. Subtypes of PLA_2 include the 14-kDa secreted forms of PLA_2 ($sPLA_2$), such as those found in venoms of snakes and other creatures as well as mammalian pancreatic enzymes; these enzymes are dependent on high (millimolar) concentrations of Ca^{2+}. A second subtype, Group IV, are the cytosolic PLA_2s that are dependent on micromolar concentrations of Ca^{2+}.

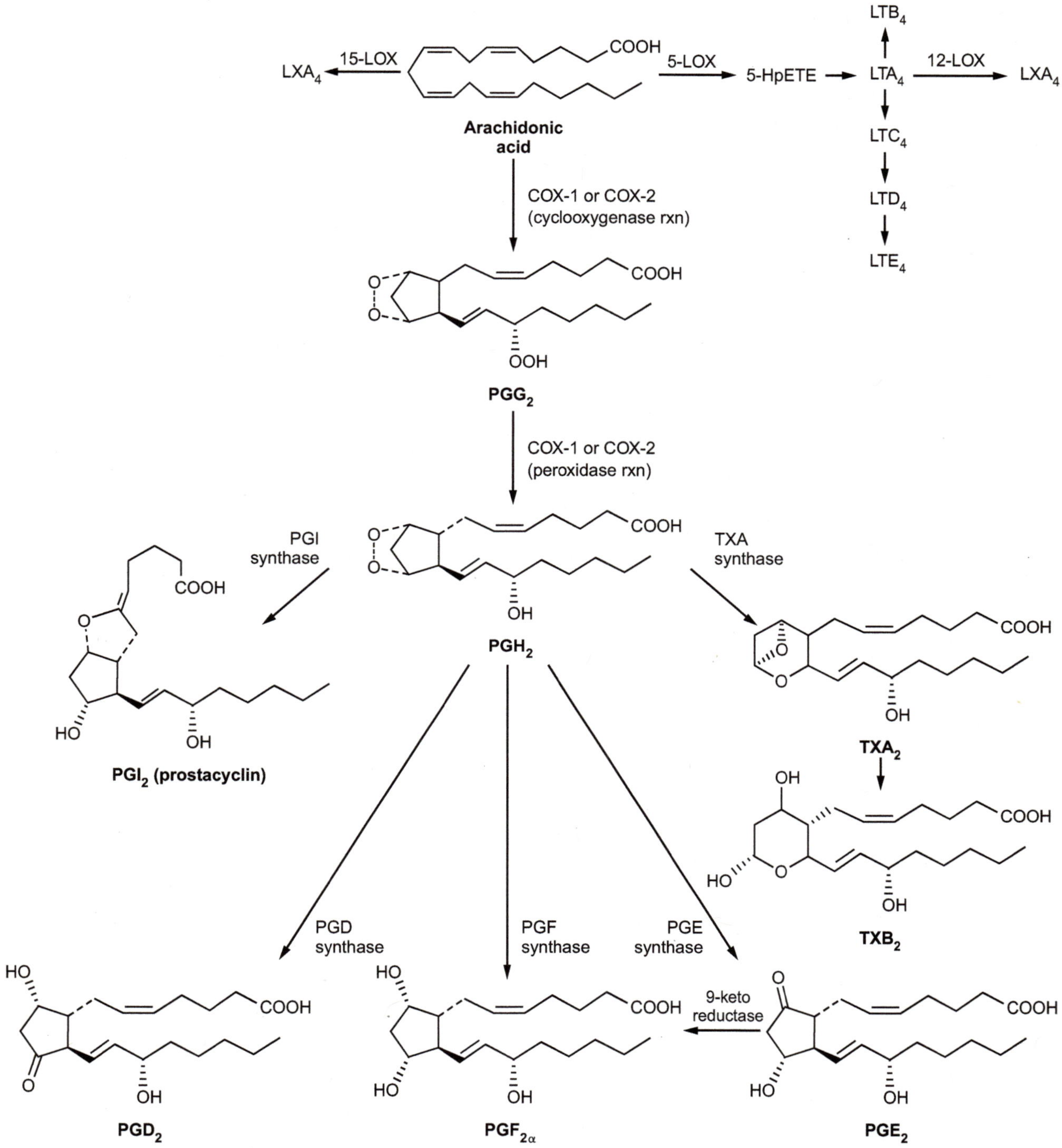

FIGURE 8.4 Eicosanoid biosynthesis. The pathway of the biosynthesis of the main prostanoids from the most abundant substrate, arachidonic acid, is shown. The 5-LO pathway leads to the LT through HpETE. The action of 12-lipoxygenase on LTA_4 leads to lipoxin A_4 (see Fig. 8.3 for structures). Arachidonic acid can also be converted directly to LXA_4 through the 15-lipoxygenase pathway. The COX-1 and COX-2 pathway gives rise, through the COX product PGH_2, to the prostanoids, prostaglandins, and thromboxanes. Each biologically active prostanoid is produced from PGH_2 by a specific synthase. The synthase(s) expressed by a given cell determine which prostanoid(s) the cell will make. TXA_2 has a particularly short half-life and its inactive breakdown product, TXB_2, is shown. *5-LO*, 5-Lipoxygenase; *COX*, cyclooxygenase; *HpETE*, hydroxyperoxytetraenoic acid; *LT*, leukotrienes; *LT*, leukotrienes; *LX*, lipoxins.

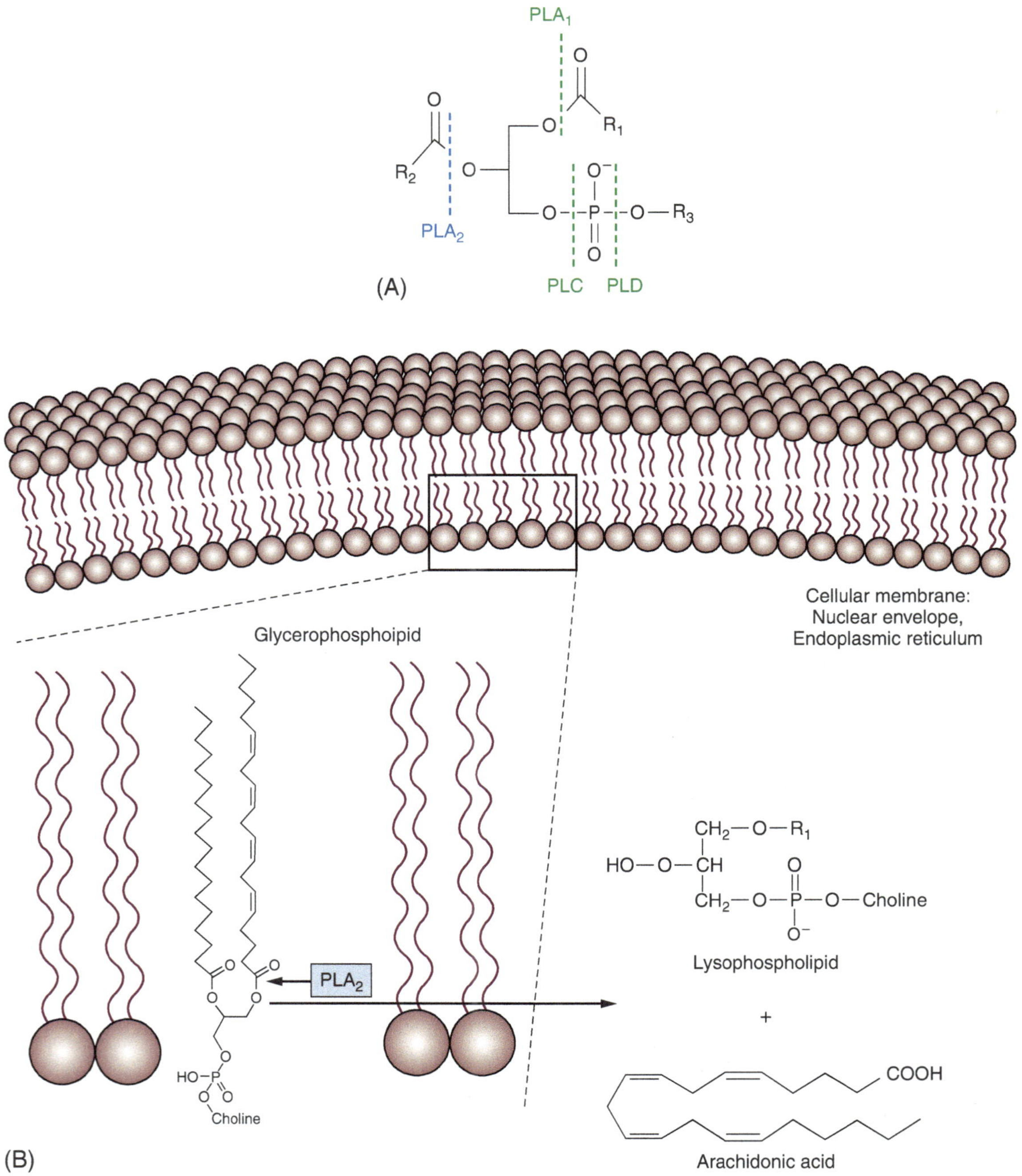

FIGURE 8.5 Phospholipase A2. In panel (A) a glycerophospholipid is shown in which R_1 and R_2 are long-chain fatty acids. R_1 is generally a 16- or 18-carbon saturated fatty acid, while R_2 is usually an 18- or 20-carbon polyunsaturated fatty acid, such as arachidonic acid (see Fig. 8.1 for structure). R3 is a polar head group such as choline. The catalytic activities of different types of phospholipases on this molecule are shown. Phospholipase A_2 releases the long polyunsaturated fatty acid from the second position of the glycerophospholipid. (B) When activated, phospholipase A_2 moves to the phospholipid bilayer of the nuclear envelope or endoplasmic reticulum and removes a molecule of arachidonic acid from a molecule of, in this illustration, phosphatidyl choline. The nearby enzymes of eicosanoid biosynthesis convert arachidonic acid to the appropriate ones for the cell.

Member A of this group is the 85-kDa protein that catalyzes the release of arachidonic acid from intracellular membrane phospholipids for eicosanoid synthesis. This enzyme is sometimes referred to as GIVA PLA_2 as recognition of its place in the classification scheme, but here it will continue to be referred to as PLA_2.

As shown in Fig. 8.5B, the reaction that PLA_2 catalyzes takes place at intracellular membranes, notably those of the nuclear envelope and the endoplasmic reticulum. The products of the reaction are arachidonic acid and lysophospholipid. Like many proteins that move from the cytoplasm to a membrane and as depicted in Fig. 8.6, PLA_2 has a Ca^{2+}

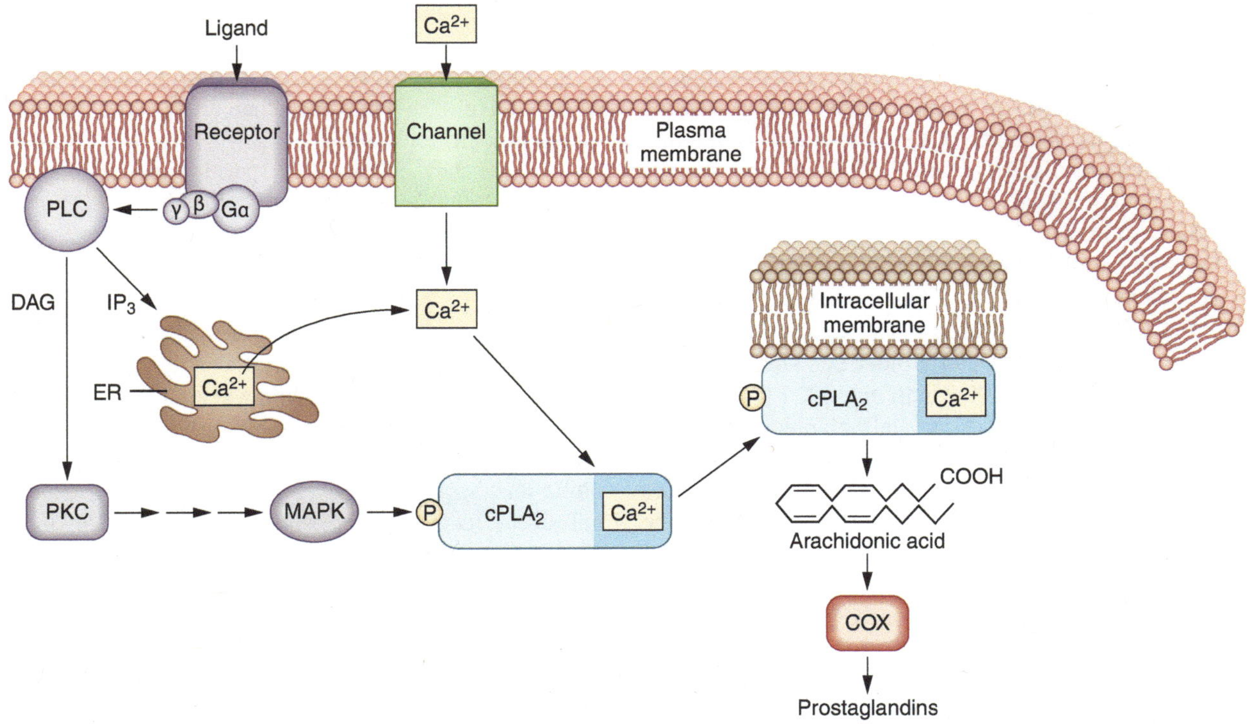

FIGURE 8.6 Activation of phospholipase A2. In this schematic diagram of a cell, the contributors to activation of $cPLA_2$ are shown. Two events bring about the movement of $cPLA_2$ to internal membranes, such as the nuclear envelope or the ER, and are required for the maximum stimulation of PLA_2 activity. These are the phosphorylation (*circle*) of specific serine residues in the catalytic domain (*light blue*) and the binding of Ca^{2+} (box) to specific binding sites (*dark blue*) on the protein. External signals can initiate these events through ligand–receptor stimulation of intracellular signaling pathways. The one illustrated involves PLC that generates both an increase in intracellular Ca^{2+} through inositol triphosphate (IP_3) and activation of a phosphorylation pathway through DAG. Intracellular Ca^{2+} can also be increased through the activation of calcium channels. *COX*, Cyclooxygenase; *$cPLA_2$*, cytosol phospholipase A_2; *DAG*, diacylglycerol; ER, endoplasmic reticulum; *MAPK*, mitogen-activated protein kinase; *PKC*, protein kinase C; *PLC*, phospholipase C.

phospholipid-binding domain. Studies have shown that this domain functions in targeting the enzyme to the membrane area that contains the COX and lipoxygenase. The exact site and how the targeting occurs remain to be determined. Phosphorylation of specific serine residues increases PLA_2 catalytic activity. Thus these two intracellular messages, which can be stimulated by any one of a number of extracellular signals, bring about the maximal activation of phospholipase A_2, initiating the eicosanoid cascade in a particular cell. Part of the antiinflammatory action of glucocorticoids is due to the inhibition of phospholipase A_2, shutting off substrate availability for either COX or lipoxygenase.

The ability of a cell to produce a given prostanoid depends on the availability of the substrate, for example, arachidonic acid, as well as that of COX and the specific synthase required. Part of the antiinflammatory action of glucocorticoids is due to the inhibition of phospholipase A_2, shutting off substrate availability for either COX or lipoxygenase.

C Prostaglandin H synthase/cyclooxygenase

The step that commits arachidonic and other polyunsaturated 20-carbon fatty acids to prostanoid synthesis is catalyzed by one of two enzymes: prostaglandin endoperoxide synthase-1 or -2 (PGHS-1 and -2), also known as COX-1 and COX-2. The two forms catalyze the same two reactions at two different sites on the protein: the insertion of two O_2 molecules into the fatty acid chain called the COX reaction to form PGG_2, followed by the peroxidase reaction at carbon 15 to form PGH_2 (see Fig. 8.4). In vivo other peroxidases may also carry out this second step to convert PGG_2 to PGH_2.

COX-1 and COX-2 are encoded by separate genes; the two isoforms from the same species share about 65% amino acid identity. Each isoform shares about 85% identity with its counterpart from another species. Both isoforms are synthesized with a signal peptide to direct it to the endoplasmic reticulum where it is bound to a single face of the phospholipid bilayer. The mature forms of COX-1 and COX-2 are 576 and 587 amino acids long, respectively.

Recent evidence indicates that both COX-1 and COX-2 exist as homodimers in the cell, the two identical monomers binding to one another over a large shared interface of the proteins. When one monomer binds a ligand, either a substrate molecule (i.e., fatty acid) or a regulatory molecule, it is not catalytically active, but its dimer partner becomes catalytically active. The activity of the catalytic partner can be affected by the molecule bound by the first, or allosteric, monomer. Thus the homodimer actually functions as a heterodimer. These details of structure and function of COX-1 and COX-2 are important in the development of specific inhibitors of the isoforms, particularly COX-2.

One structural and, therefore, functional difference between COX-1 and COX-2 lies in the catalytic sites where the COX reaction takes place. Crucial differences in the amino acids within and around this site result in a more capacious site in COX-2 than in COX-1. This difference, determined by X-ray crystallographic studies, is important to the effort discussed later, to develop drugs that are specific for COX-2, since most COX inhibitors function by preventing access of arachidonic acid to the active site of the enzyme.

The primary physiological difference between COX-1 and COX-2 lies in their regulation. COX-1 is generally considered to be expressed constitutively, with possible increases of two- to fourfold in response to stimuli such as hormones or growth factors. On the other hand, in many tissues and cell types, there is little or no expression of COX-2 under basal conditions but it is inducible by agents signaling such events as inflammation, including lipopolysaccharides, cytokines, and certain growth factors. The concept of COX-1 as the constitutive "housekeeping" form and COX-2 as the inducible one is not absolute: in some cells, for example, in brain and kidney, COX-2 is present at low levels all the time (constitutive) and has "housekeeping" functions of its own such as in the maintenance of cardiovascular homeostasis and of uterine contractions during labor.

D Cyclooxygenase inhibitors

Antiinflammatory steroids are natural (cortisone) or synthetic (prednisone, dexamethasone) glucocorticoids. They are antiinflammatory due to their inhibition of phospholipase A_2, blocking the production of the substrate for COX action and by blockage of the induction of COX-2. However, as discussed in Chapter 10, Adrenal Corticoids, glucocorticoids have many other physiological functions and their use, particularly long term, is, therefore, limited because of many untoward side effects.

COX inhibitors are among the most commonly used drugs in the world. Aspirin (acetylsalicylic acid) has been used to treat fever, pain, and inflammation for over a century, that is, since long before the discovery of PG. As the role of PGs in these processes began to come clear, the role of aspirin as an inhibitor of PG production emerged and it became the founding member of what would be the nonsteroidal antiinflammatory drugs (NSAIDs) which now include several other drugs, examples of which are shown in Fig. 8.7A. Aspirin is unique among these drugs in irreversibly inactivating COX-1 through the acetylation of the active site, while the other drugs block the access of arachidonic acid to it.

Because they inhibit COX-1, which functions in many tissues and physiological systems, the side effects of nonspecific NSAIDs are considerable. Among the most serious are the effects on the gastrointestinal system so that using them to treat chronic pain and inflammation as in arthritis, for example, can lead to potentially life-threatening ulcers and bleeding. Other side effects include kidney and liver damage. Thus when COX-2, an enzyme induced upon inflammation, was discovered in 1989, there was much excitement about the possibility of inhibiting this enzyme selectively, thus avoiding gastrointestinal side effects. This is especially important in treating the pain and inflammation of chronic diseases such as rheumatoid arthritis.

The difference in the size of the active site mentioned earlier is the basis for some of the selectivity of the drugs between the two enzymes. Some of these drugs, shown in Fig. 8.7B, have led to increased thrombolytic and adverse cardiovascular events and it is now clear that developing and prescribing these drugs has become more complex than originally anticipated. While there is still interest in COX-2 selective inhibitors for certain applications, attention has shifted somewhat to the specific synthases in an effort to pinpoint more precisely the active molecule that is being targeted.

E Lipoxygenase

The open chain eicosanoids are synthesized from arachidonic acid released from membrane glycerophospholipids by PLA_2. In this case the first committed step of the pathway is not catalyzed by COX but by a lipoxygenase. There are at least three of these enzymes that are named for the carbon on which they bring about oxygenation. Detail regarding all of them is beyond the scope of this book, but a look at the 5-lipoxygenase (5-LO) pathway will serve as a good example of such a pathway that leads important arachidonic acid derivatives.

Arachidonic acid 5-LO is a 78-kDa nonheme iron protein that requires another protein, 5-LO activating protein, for its activity. The 5-LO catalyzes the addition of molecular oxygen to the substrate, followed by the removal of water to produce LTA_4. This unstable leukotriene can be enzymatically converted to the biologically active leukotriene B_4 (LTB_4) by LTB_4 hydrolase, as depicted in Fig. 8.8. LTA_4 can also be converted to LTC_4 by conjugation with glutathione, catalyzed by the glutathione-*S*-transferase, LTC_4 synthase. The enzymatic shortening of the glutathione moiety results in the biologically active

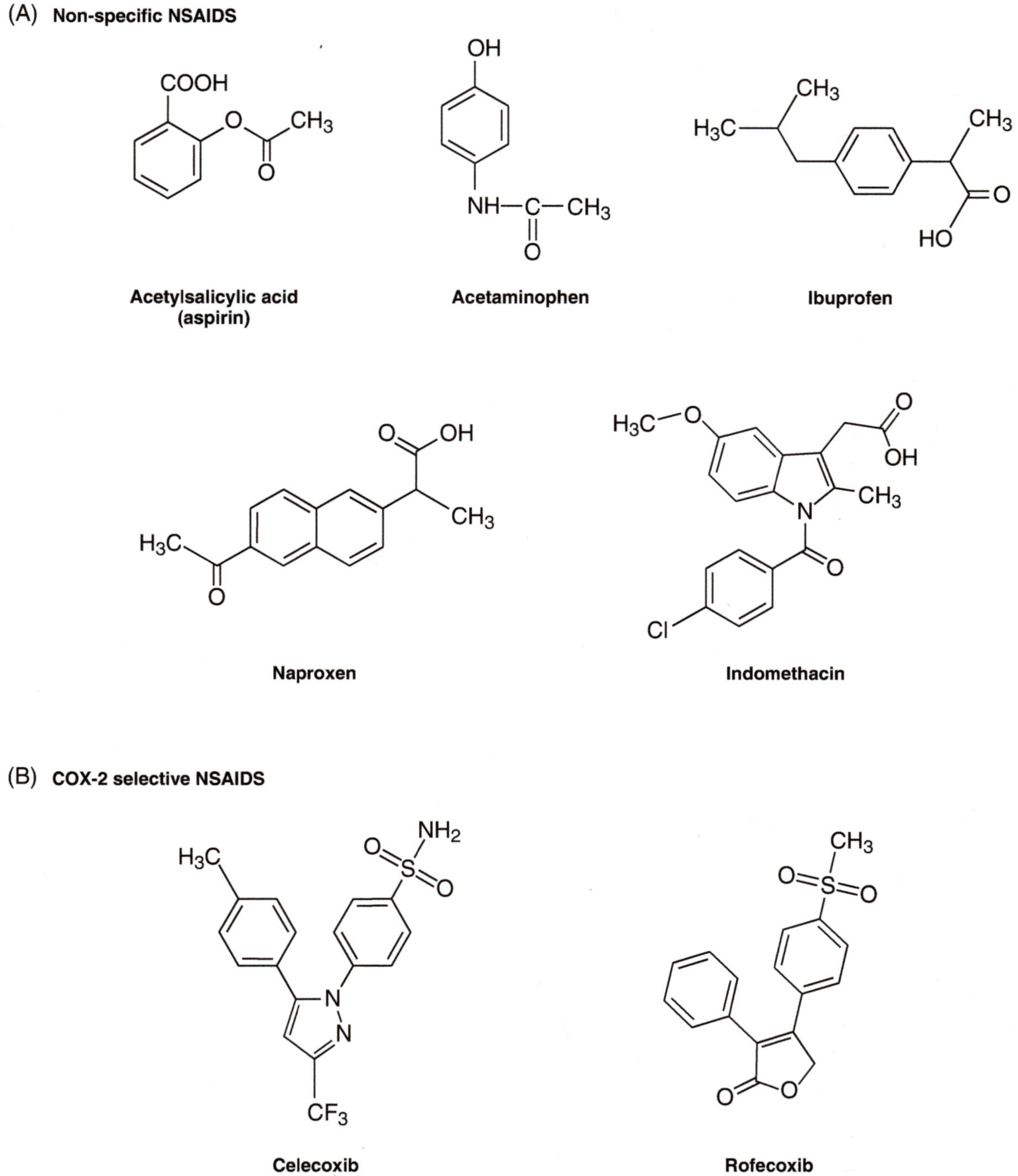

FIGURE 8.7 Cyclooxygenase inhibitors. The structures of some commonly used nonspecific NSAIDS are shown in panel (A). These compounds inhibit both COX-1 and COX-2. (B) Two examples of specific COX-2 inhibitors are shown. Celecoxib is currently on the market, while Rofecoxib was withdrawn in 2005. *COX*, Cyclooxygenase; *NSAIDs*, nonsteroidal antiinflammatory drugs.

sulfated LT, LTD_4, and LTE_4 (along with LTC_4, the cysteinal LTs). The activity of 5-LO is regulated by Ca^{2+} and phosphorylation of specific serine residues. In addition, the high concentrations of the product, LTA_4, irreversibly inhibits the enzyme so that its continued production requires the synthesis of new 5-LO.

F Transport and inactivation of prostanoids

Prostanoids are synthesized by nearly all tissues in the body and are released to act on the cells that made them or on those very close by. Because they are charged anions, these molecules do not diffuse well through the

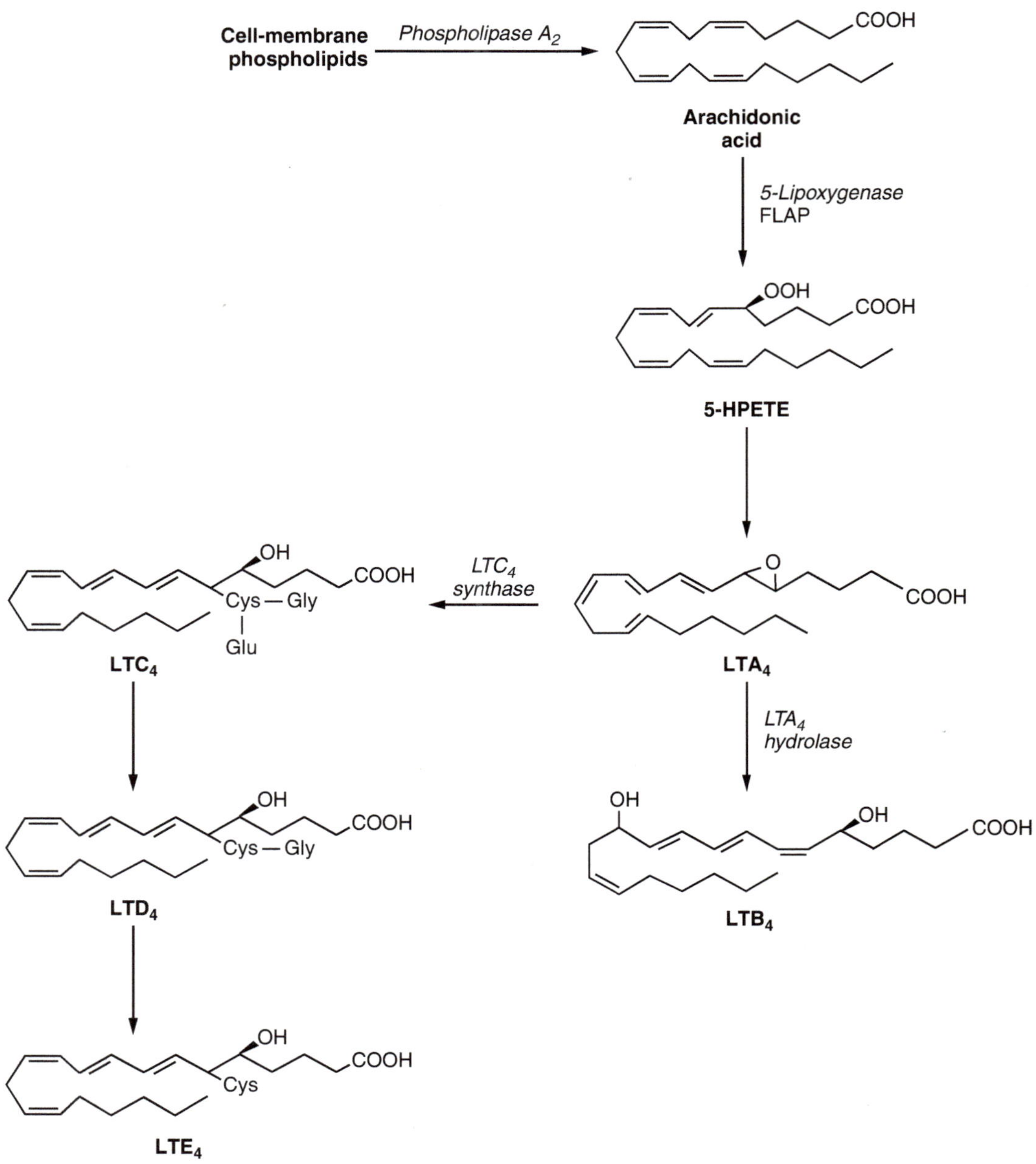

FIGURE 8.8 Leukotriene biosynthesis. The pathway from arachidonic acid through 5-LO is shown. 5-LO requires a membrane protein, FLAP for the two reactions it catalyzes to form LTA_4, which is converted to LTB_4 through the removal of a water molecule by LTA_4 hydrolase. LTC_4 arises from the addition of glutathione to carbon-6 by LTC_4 synthase. The sequential removal of glutamate and glycine leads to the active leukotrienes D_4 and E_4. These three LTs are referred to as the cysteinal leukotrienes or cys-LTs. *5-LO*, 5-Lipoxygenase; *cys-LTs*, cysteinyl LTs; *FLAP*, 5-lipoxygenase activating protein; *LT*, leukotrienes; *LTA_4*, leukotriene A_4; *LTB_4*, leukotriene B_4; *LTC_4*, leukotriene C_4.

plasma membrane. The prostaglandin transporter (PGT) has been identified as a protein containing 12 membrane-spanning regions. Fig. 8.9 shows the details of the operation of the 12-transmembrane PGT in the bovine uterus. PG are shown to operate on receptors of the cell that produces them (autocrine action). The PGT is thought to be the major transporter of PG. It is effective in bringing extracellular prostanoids into the cell for inactivating, thereby playing an important role in signal termination. The exit of newly synthesized prostanoids from the cell may also be carried out by PGT, but a second protein, multidrug-resistance protein 4, a member of a subfamily of ATP-binding cassette transporters is very effective in this regard. Further details about the transport of PG, TX, and other lipid signaling molecules remain to be elucidated.

The most important catabolic step in the removal of the active prostanoids is the oxidation of the hydroxyl group at carbon 15 to a ketone by prostaglandin-15-dehydrogenase, 15-PGDH, which completely inactivates

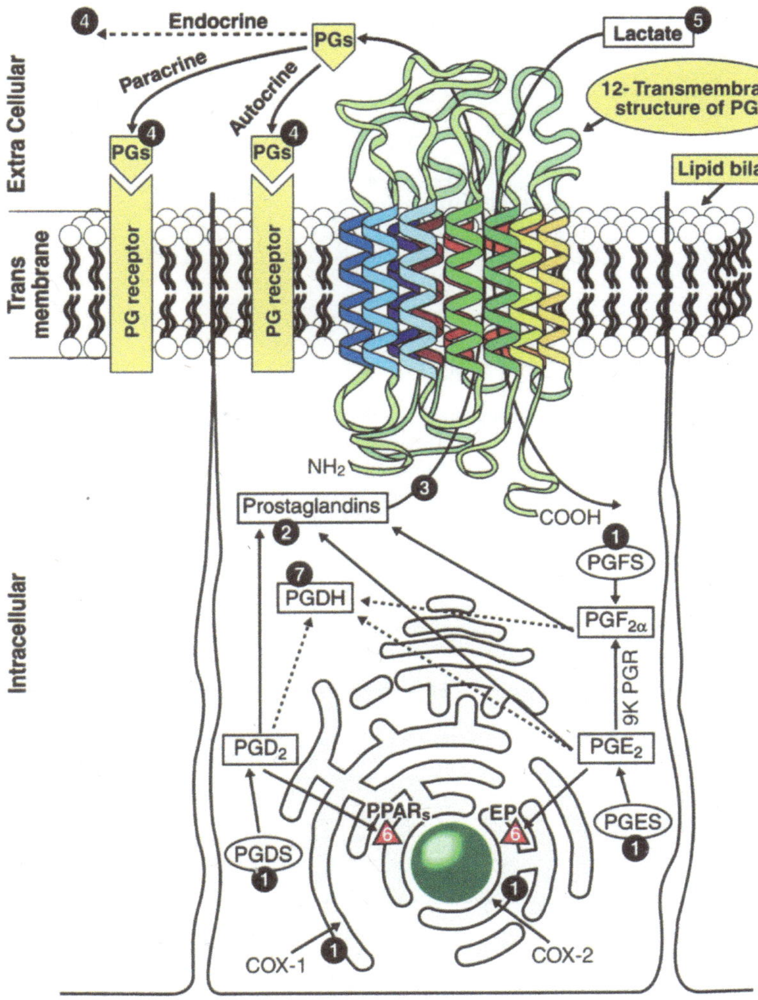

FIGURE 8.9 Proposed hypothetical model of PGT-mediated transport in bovine uterus. (1) Sites of PG-biosynthetic enzymes. (2) The presence of PGs in the cytosol. (3) Transport of PGs across the plasma membrane through PGT. (4) Autocrine or paracrine or endocrine action of PGs on the membrane PG receptors. (5) Obligatory influx of lactate across the plasma membrane through PGT. (6) PG receptors in the nuclear membrane. (7) Cellular oxidation of PGs through catabolic enzyme. *9K PGR*, 9-keto PG reductase; *EP*, PGE receptor; *PG*, prostaglandins; *PGDH*, PG dehydrogenase; *PGDS*, PGD synthase; *PGES*, PGE synthase; *PPAR*, peroxisome proliferator–activated receptor. *Reproduced from Fig. 8.6 of S.K. Banu, J.A. Arosh, P. Chapdelaine, M.A. Fortier, Molecular cloning and spatio-temporal expression of the prostaglandin transporter: A basis for the action of prostaglandins in the bovine reproductive system, Proceedings of the National Academy of Sciences of the United States of America, 100 (2003) 11747–11752.*

the prostanoid. This enzyme is highly expressed in the lungs, placenta, kidney, and spleen. The reduction of the double bond between carbons 13 and 14 by prostaglandin reductase, which has a tissue distribution similar to that of 15-PGDH, is a further catabolic step for these compounds.

Some cancer cells, such as those found in the colon, exhibit high levels of COX and protumorigenic PGE_2. In this and similar contexts, 15-PGDH is considered to be a tumor suppressor because of its ability to inactivate PGE_2 (see Section VI.A).

IV Eicosanoid receptors and signaling

Following their release from the cells in which they are synthesized, eicosanoids travel only very short distances, if at all, before interacting with specific receptors on this cell (autocrine actions) or a neighboring one (paracrine actions). The receptors for the biologically active PG and for thromboxane A_2 are depicted in Fig. 8.10. Table 8.2 gives similar information about the receptors for the LT. Each of these receptors is a seven-transmembrane spanning G protein–coupled receptor and all are encoded by separate genes. Thus far, the only prostanoids that have multiple receptors encoded by multiple genes are those for PGE_2, the most abundant and ubiquitous of the PG, and PGD_2. Splice variants exist on several of the other receptors, including EP3, FP, and TP. DP2 is structurally distinct from the other eight receptors and shares less than 20% homology with them; between the remaining eight receptors, the extent of homology ranges between 20% and 30%. Homology between the same receptor in different species is 70%–90%.

The primary signaling system used by each receptor is shown in Fig. 8.10, but at least five of the nine receptors use more than one pathway, depending on the cellular context. The receptors can be sorted into three classes based on primary signaling pathways and effect on smooth muscle contraction. For example, EP2, EP4, IP, and DP1 signal through cyclic AMP (cAMP) and have a relaxant effect on smooth muscle cells. EP1, FP, and TP stimulate intracellular Ca^{2+} mobilization and cause

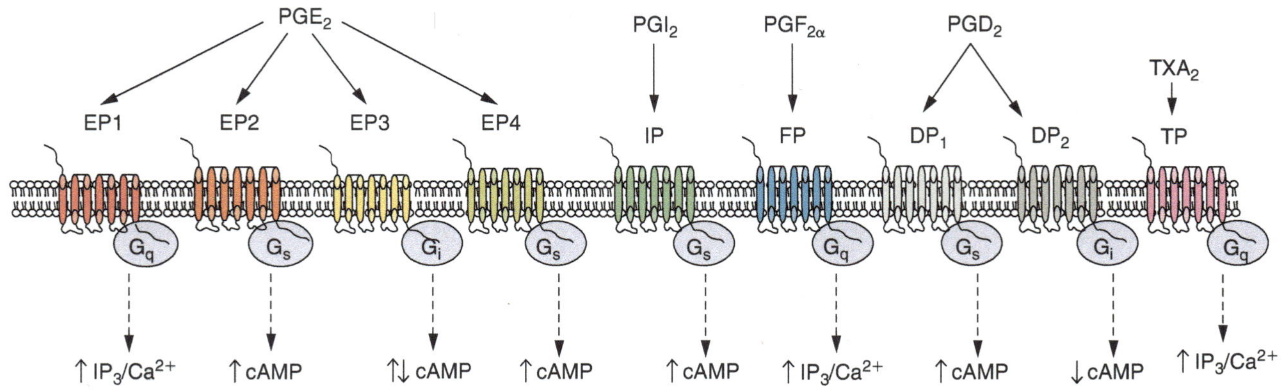

FIGURE 8.10 Prostanoid receptors. The known prostanoid receptors are seven-transmembrane G protein–coupled receptors. The receptors are named for the prostanoid ligand that binds to each; for example, the receptor for PGI_2 is IP, etc. There are multiple receptors for PGE_2 (EP1–EP4) and PGD_2 (DP1 and DP2). DP2 is also known as CRTH cells. The primary regulatory subunit (Gq, Gs, and Gi) for each receptor is shown as the gray oval on the intracellular face of the membrane, along with the signaling pathway initiated by the receptor. Some receptors use different G proteins and signaling pathways in different cells (not shown). *CRTH*, Chemoattractor receptor of Th2 cells; *PG*, prostaglandins.

TABLE 8.2 Leukotriene receptors and signaling.

Ligand	Receptor	Signaling	Location
LTB_4	BLT1	↓cAMP, ↑Ca^{2+}	Leukocytes
	BLT2	↓cAMP, ↑Ca^{2+}	Wide distribution
LTD_4	Cys-LT1	↑Ca^{2+}	Lung SMC, macrophages
LTC_4, LTD_4	Cys-LT2	↑Ca^{2+}	Spleen, heart, and endothelial cells

cAMP, Cyclic AMP; *cys-LTs*, cysteinyl LTs; *LT*, leukotrienes; *SMC*, smooth muscle cells.

smooth muscle cells to contract. EP3 usually decreases cAMP and, therefore, is referred to as inhibitory, but its effects can vary with splice variant and cell type.

Studies with the prostanoid receptors have led to insights into the complexity of signaling networks involving them and other molecules such as hormones and neurotransmitters. Many if not most physiological processes are governed by such networks rather than by individual molecules operating one at a time.

V Examples of biological actions of prostaglandins

As indicated by Table 8.1, prostanoids play roles in numerous physiological settings. In many cases the nature of these roles and their mechanisms remain to be elucidated. Next are some features of a few well-studied examples chosen to illustrate the local nature of the actions of these molecules, to appreciate their importance to the overall physiological processes in which they participate, and to recognize the basis for the side effects of the drugs that inhibit their production.

A Prostacyclin and thromboxane in the vasculature

The roles of TXA_2 and PGI_2 in platelet aggregation and vascular tone are shown in Fig. 8.11. Platelets are enucleated cell fragments formed from bone marrow megakaryocytes. They retain many cytoplasmic components, including mitochondria, granules-containing platelet-specific proteins, coagulation factors, and the enzymes (phospholipase A_2, COX-1, and TXAS) to produce TXA_2. When an injury to the vasculature occurs, platelets are activated through the detection of exposed collagen in the wall of the vasculature. A rise in intracellular Ca^{2+} in the platelets leads to the activation of phospholipase A_2 and COX. The resulting TXA_2 is released and acts on the platelets to promote aggregation through interaction with its receptor, TP, and the reduction in intracellular cAMP levels. TXA_2 also acts on nearby smooth muscle cells of the vasculature, constricting them to prevent blood loss. A platelet plug is formed at the site of the injury, setting the stage for clot formation.

Under normal conditions, that is, when no injury to the vasculature is detected, prostacyclin, PGI_2, is produced by COX-2 and PGI synthase and released by the endothelial cells lining the vasculature. PGI_2 acts on platelets through its receptor, IP, and increased cAMP production to inhibit aggregation. PGI_2 also promotes vasodilation of the smooth muscle cells. Thus balance between the actions of TXA_2 and PGI_2 in the blood vessels is critical in maintaining vascular homeostasis. Since platelets contain COX-1 and endothelial cells contain COX-2, inhibitors selective for the latter enzyme (Section III.D) upset the balance between the two prostanoids and can, therefore, lead to serious cardiovascular side effects.

B Prostaglandins in the kidney

PG participate in several functions of the kidney. Fig. 8.12 depicts the distribution along the renal tubule of the four receptors for PGE_2 and the one for $PGF_2\alpha$. Precise functions for each of the receptors in the different locations have not yet been clearly mapped out, but some have been deduced from experimental observations. For example, inhibition of prostaglandin production by COX

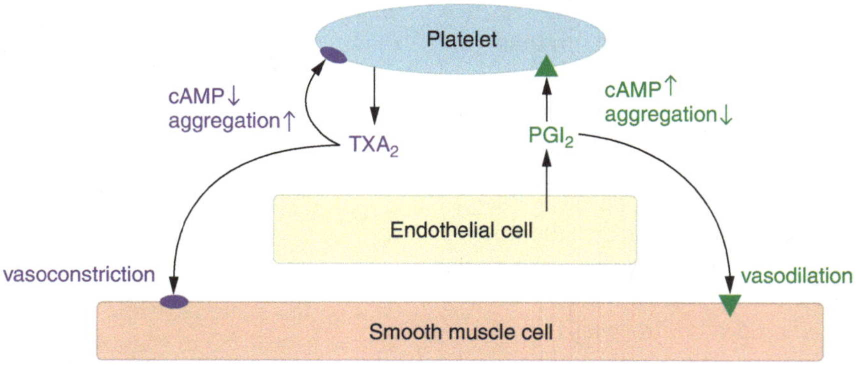

FIGURE 8.11 PGI2 and TXA2 in the vasculature. The balance between PGI_2 and TXA_2 is important in cardiovascular homeostasis. The endothelial cells of the blood vessels produce and release PGI_2. Through interaction with its receptors (IP, *green triangles*) on blood platelets and on the smooth muscle cells of the blood vessel walls, PGI_2 decreases platelet aggregation and promotes vasodilation, respectively. Platelets produce and release TXA_2 that acts through its receptor, TP (*purple ovals*), in an autocrine fashion to promote platelet aggregation and through paracrine action to stimulate vasoconstriction. *PG*, Prostaglandins; *TX*, thromboxanes.

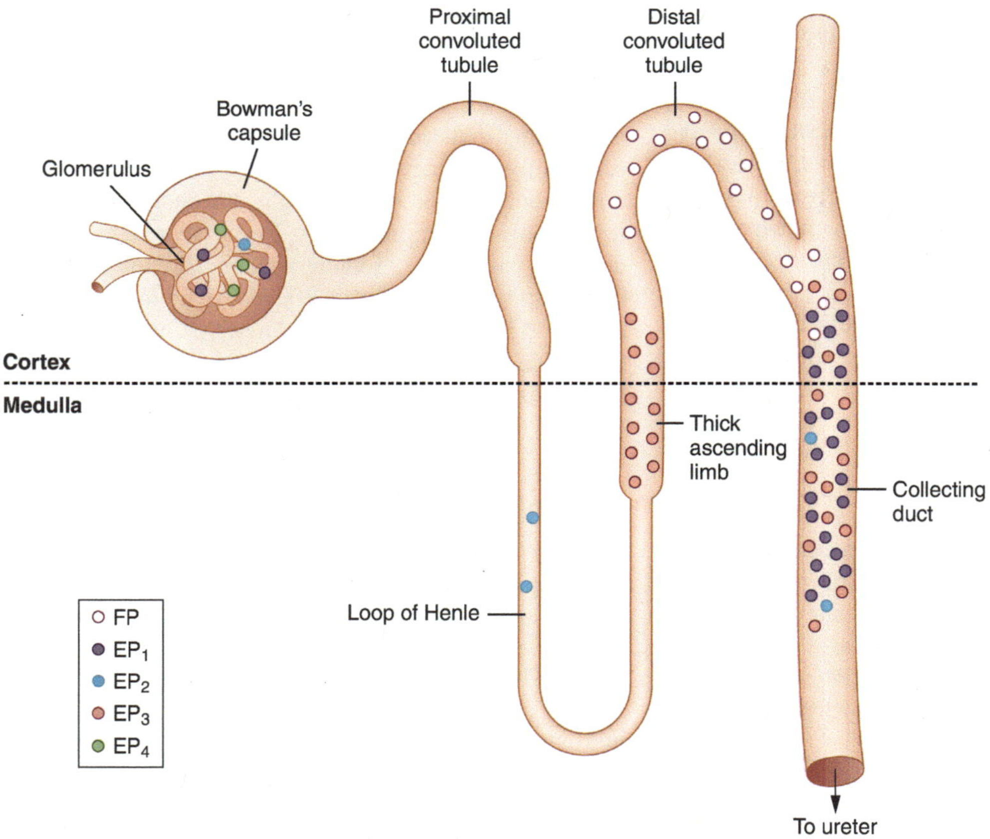

FIGURE 8.12 Prostaglandin receptor distribution in the renal tubule. The distribution of receptors for the main prostaglandins produced by and active on the epithelial cells of the renal tubule is shown. See inset at lower left for legend of symbols used.

inhibitors leads to salt and water retention and the resultant high hypertension. This fact, in the context of studies with mice that have been genetically altered to lack the gene for EP3, FP, or EP1, indicates that each receptor plays a role in NaCl transport sequentially along the tubule, from the thick ascending limb through the distal collecting duct. EP2 is also thought to play a role in salt excretion, but the receptor is present in low levels and its precise location is not yet certain. EP4 has recently been shown to be necessary for the salt deprivation–induced stimulation of renin secretion, the first step of the renin–angiotensin–aldosterone system (see Chapter 15: Hormones Related to the Kidney and Cardiovascular System).

The hemodynamics of blood flow through the kidney, and, therefore, the glomerular filtration rate, is also subject to control by prostanoids. TXA_2 produced in the glomerulus and PGI_2 in the vasculature of the arterioles in the glomerulus and other renal blood vessels influence the state of vasoconstriction or vasorelaxation, respectively, as in the vasculature in other parts of the body.

C Prostaglandins and pain perception

The pain mechanism is essential for survival, since acute pain is a warning mechanism for threatening conditions. The peripheral nociceptors, when excited by potentially harmful stimuli, cause pain by the way of their afferent nerve fibers, as shown in Fig. 8.13. Probably every organ in the body contains these receptors. There are two classes of nociceptive afferent nerves: those found among thin myelinated Aδ fibers, associated with sharp focused pain, and those among the nonmyelinated C fibers and associated with dull, burning, and diffuse pain. In response to an injury, chemical substances released from the injured tissue excite nociceptors or sensitize them to other stimuli (Fig. 8.13A) resulting in the generation of pain. These substances, referred to as algesic because they produce the sensation of pain, include bradykinin which stimulates the intracellular production of PG, leading to the firing of the nerve as described in the following. The afferent signals are sent through the spinal cord in ascending fibers to pain centers in the brain.

PG produced in the nerve terminals are released to act on their membrane receptors on this or nearby termini, as described in Sections III.F and IV. The role of PGE_2 will be discussed here, although PGI_2 also functions in pain perception pathways through similar mechanisms. Furthermore, although only EP1 and EP4 are discussed here, EP2 and EP3 have also been implicated in pain perception in some tissues. As shown in Fig. 8.13B, PGE_2, acting through either protein kinase A (EP4) or protein kinase C (EP1) brings about the phosphorylation of ion channels in the nerve terminus. Inhibition of either COX-1 or COX-2 would lead to analgesia, the effect long known for aspirin and the basis for the intense interest in other COX inhibitors (Section III.D).

The efferent function of the nocireceptor is also depicted in Fig. 8.13A. This includes the release of neurotransmitters such as substance P and CGRP (calcitonin gene–related peptide) which activate other nearby cells, such as mast cells and neutrophils. These cells secrete substances such as histamine into the inflammatory site. The release of substance P and CGRP also results in vasodilation and in the leakage of blood plasma proteins and fluid.

D Prostaglandins in reproduction

The processes of reproduction and their hormonal control in the female are covered in Chapter 13, Estrogens and Progestins and Chapter 14, Hormones of Pregnancy, Parturition and Lactation. It may be useful to make reference to this material while reading the following regarding the role of PG in some of these processes.

1 Ovulation

Much of the information gathered about the role of prostaglandin synthesis and action during ovulation has come from the examination of mice genetically altered to be deficient in either COX-1 or COX-2. The cascade of cellular events that follows the midcycle surge of LH and leads to the release of the ovum shares several characteristics with the process of inflammation. Thus it is not surprising that induced COX-2 in granulosa, rather than the constitutively expressed COX-1 in the thecal cells, is the critical enzyme for the prostaglandin pathway in ovulation. In mice lacking the COX-2 gene, in which ovulation does not take place, follicle rupture and release of the ovum can be achieved most effectively by the administration of PGE_2. In nonhuman primates, it has been shown that PGE_2 is specifically involved in the regulation of plasminogen activator–mediated proteolysis required for follicule rupture.

2 Luteolysis

In nonprimate mammals, such as rodents and domesticated species, the regression of the corpus luteum (luteolysis) at the end of a nonfertilization reproductive cycle is brought about by $PGF_{2}\alpha$ produced by the uterus. In primates, including humans, the corpus luteum can undergo regression in the absence of the uterus although $PGF_{2}\alpha$ is synthesized by the human corpus luteum and FP receptors are found there. This and other evidence suggest that locally produced $PGF_{2\alpha}$ participates in primate luteolysis.

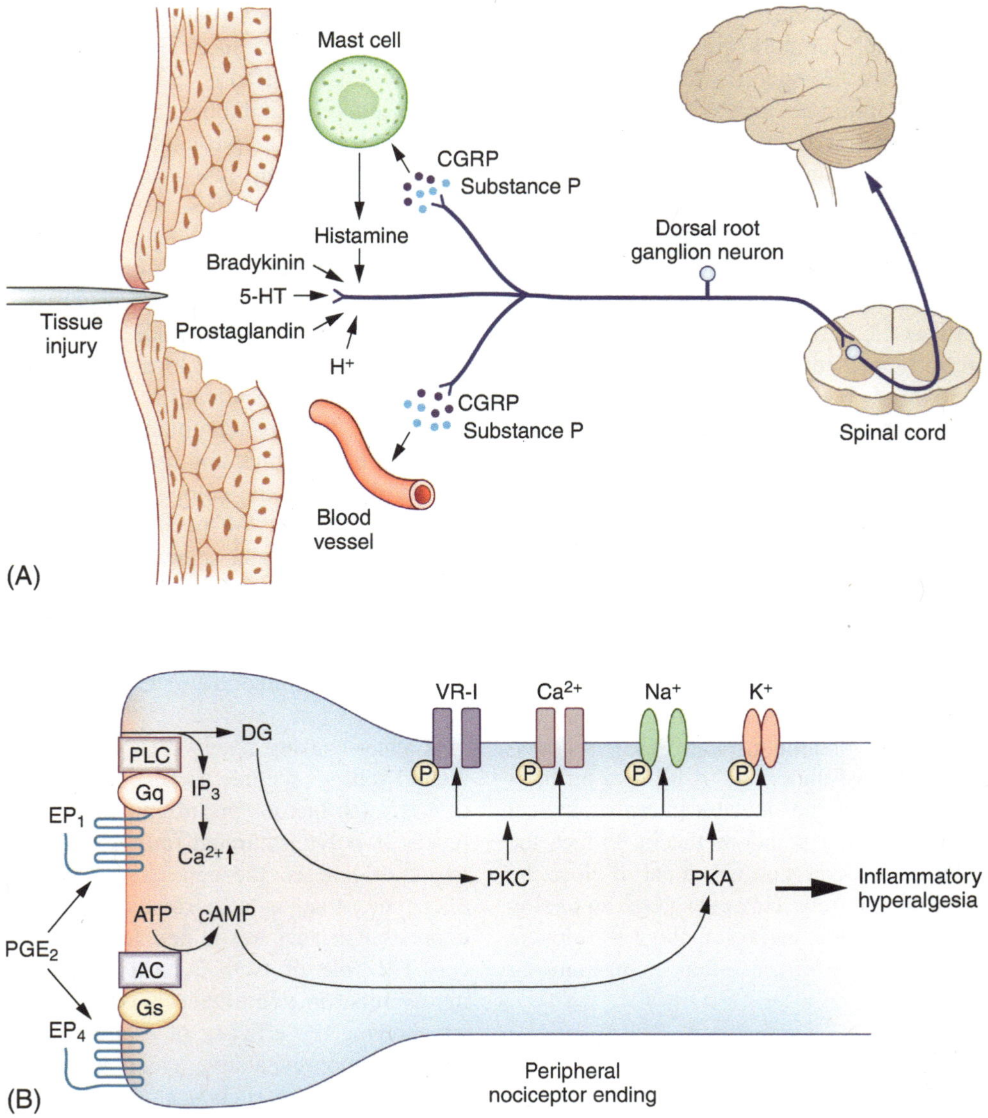

FIGURE 8.13 Prostaglandins in pain perception. (A) When a tissue injury occurs, cells at the site release several substances, including bradykinin, serotonin (5-HT, hydroxytryptamine), and prostaglandins, primarily PGE_2, into the acidic environment of the inflammation. All of these act on the terminus of the nociceptor, the afferent neuron, cell body of which lies in the dorsal root ganglion that will carry the pain signal through the afferent fibers in ascending tracts in the spinal cord to pain perception centers in the brain. At the same time, the efferent function of the nociceptor engages, releasing the neurotransmitters substance P and CGRP (see Chapter 9: Calcium-Regulating Hormones: Vitamin D, Parathyroid Hormone, Calcitonin; Fibroblast Growth Factor-23) leading to the activation of nearby nonneuronal cells, which contribute other molecules, such as histamine, to the inflammatory milieu. (B) Inside the nerve terminal, PGE_2 acting through either EP1 or EP4 (depending on the tissue and species under study) activates PKC or cyclic AMP-dependent PKA, respectively. Phosphorylation leads to the opening of Ca^{2+} and Na^+ channels, including the vanilloid receptor, VR-1 (a mono- and divalent cation channel), and the closing of K^+ channels. Collectively, these events lead to membrane depolarization and transmission of the neural signal to the brain. *CGRP*, Calcitonin gene–related peptide; *PKA*, protein kinase; *PKC*, protein kinase C.

3 Cervical ripening

A critical step in the birth of the newborn is the softening (or ripening) of the uterine cervix, which has functioned to retain the fetus throughout pregnancy, so that the fetus can be expelled when gestation is concluded. Several PG are produced in the cervix and the tissue contains both EP and FP receptors. It is likely that the mechanism of prostaglandin action in the cervix includes the induction of enzymes responsible for the remodeling of collagen and proteoglycans that occurs during cervical softening. The local administration of PGE_2 is a common way to stimulate the process, particularly when labor is being induced, and brings about the same changes as those seen in nontherapeutically assisted softening.

4 Parturition/preterm labor

The biological effect of the first PG studied was their powerful ability to contract uterine smooth muscle. PGE_2 is a potent abortifactant and is used in the early termination of pregnancy. It has been known for three decades that aspirin and indomethacin, both COX inhibitors that, at low doses, are specific for COX-1, delay parturition (birth) in humans and other animals. At higher doses, these NSAIDs both inhibit COX-2 as well. The COXs occur in the placenta and fetal membranes and during normal labor, prostaglandin production by COX-1 is regulated by other hormones involved in parturition as shown in Fig. 14.13. Preterm labor, on the other hand, which shares physiologic features with inflammatory processes, is mediated largely by COX-2. Clinically, preterm labor can be curtailed with systemic or vaginally delivered local doses of indomethacin, attesting to the importance of PG in parturition. However, a side effect of this approach, in addition to possible renal and cardiac damage in the fetus, is the effect of indomethacin on the remodeling of the ductus arteriosus, or DA.

During fetal development the DA carries deoxygenated blood away from the fetal pulmonary circulation to the umbilical–placental circulation to be reoxygenated in the maternal blood pools (Fig. 14.4). Immediately upon birth, this shunting vessel must close so that the fetal circulation is adapted to air-breathing. Fetal PGE_2 is necessary for the maintenance of the open DA prior to birth, as indicated by the closure of the duct by pharmacological dose of COX inhibitors, such as indomethacin. In fact, the use of such inhibitors is a common treatment to close the DA in infants born prematurely. However, if given to stop premature labor, indomethacin can bring about the closure of the DA prior to the birth of the infant, compromising its circulatory health.

VI Clinical aspects

A Prostaglandins and cancer

Inflammation is now understood to play a critical role in the development and progression of many types of cancer. Disruption of the normal role of PG, especially PGE_2, in tissue growth and response to inflammation, is now considered to be among the hallmarks that distinguish cancerous from normal tissue. Others include evasion of the normal apoptotic process that multicellular organisms use to control the number and quality of cells in a tissue; self-sufficiency in growth signals, that is, dysfunction of normal signaling pathways that stimulate and restrain cell growth; unlimited potential for replication, a property shared with stem cells; sustained production of new from existing vascular tissue (angiogenesis); and ability to invade other tissue and metastasize. These are some of the functional consequences of the unrepaired mutations that underlie the conversion of a normal cell to a cancerous one.

Signs of inflammation are present in many cancerous and precancerous tissues. Tumor growth is complex and involves many different cell types, including associated noncancer cells and bone marrow–derived cells which contribute to the growth and angiogenic capacity of cancer cells. Blocking the proinflammatory cytokines or other mediators from these cells can slow cancer progression. On the other hand, infiltration of the tumor site with lymphocytic cells can inhibit tumor growth, so the inhibition of this process is counterproductive. Thus the effects of the mixture of eicosanoids to which tumor cells are exposed may be difficult to unravel and even more difficult to target therapeutically.

PGE_2 is the most common and abundant prostaglandin found to be elevated, and associated with a poor prognosis, in tumors of the colon, breast, lung, and esophagus. From many studies of models of colon cancer, it is clear that COX-2 is a major source of the elevated PGE_2 levels, although in tumors in other tissues, COX-1 may also be involved. PGE_2's actions include inhibition of apoptosis, stimulation of proliferation, and angiogenesis; in addition, it acts as an immunosuppressant and proinflammatory agent.

Cancer-causing agents that induce COX-2 in experimental settings include tobacco smoke, ultraviolet irradiation, growth factors, proinflammatory cytokines, and high levels of polyunsaturated fatty acids. Because of its role in tumorigenesis, the gene for COX-2 is considered an oncogene, that is a normal cellular gene that, when expressed in an unrestrained fashion, contributes to cancer. The role of COX-2 as a pivotal enzyme in PGE_2 involvement in tumorigenesis is supported by many studies showing the efficacy of aspirin and other NSAIDs to act as chemopreventative agents in colon and gastrointestinal tract cancers, such as stomach and esophagus. A significant effort with regard to selective COX-2 inhibitors has been directed toward a more specific chemopreventative as well as therapeutic role in colorectal cancer, but problems remain with their use due to adverse cardiovascular effects (Section III.D).

Not all PG are tumorigenic and not all tumors respond to NSAIDs as well as does colon cancer, suggesting that other elements of the prostaglandin pathway may also be involved. In fact, mPGES-1 is overexpressed in several cancers, including lung, breast, and colorectal, and is currently under study as a cancer therapeutic target. In addition, it has been recognized that high levels of PGE_2 may be due not only to its production but also to the reduction of its inactivation by 15-PGDH (Section III.F). Markedly decreased levels of 15-PGDH mRNA have been seen in cancers of the bladder, colon, stomach, small intestine, lung, skin, kidney, pancreas, and liver. Although such a decrease in 15-PGDH in tumors is not universal, its frequent appearance suggests that reduced removal of PGE_2

contributes to its elevation in a tumorigenic setting. From this standpoint, it is clear that in many situations, 15-PGDH may be considered a tumor suppressor.

B Leukotrienes in human disease

Table 8.3 summarizes some aspects of the involvement of LT in human physiology, particularly in inflammatory settings. As shown in Fig. 8.8, the 5-LO pathway can lead to the synthesis of either LTB_4 through the LTA_4 hydrolase or to the cysteinyl LTs (cys-LTs) (LTC_4, LTD_4, and LTE_4) through the LTC_4 synthase. The top portion of Table 8.3 shows the cell types in which one or the other LTA_4 metabolizing enzyme predominates to produce one of the two types of LT. The middle portion of Table 8.3 lists major target cells of the two leukotriene types, along with an example or two of the response of the target cells to either LTB_4 or a cys-LT. The bottom portion of Table 8.3 lists some human diseases associated with each type of LT, a few of which are discussed in the following. For other diseases, such as chronic pulmonary obstructive disease, pulmonary fibrosis, and irritable bowel disease, circumstantial evidence has implicated LT in the disease process but a clear dominant role has not yet been established.

1 Asthma and other upper respiratory conditions

Asthma is a complex disease resulting in part from narrowing of the airways. For decades the agent that causes the bronchoconstriction of asthma was known as the slow-reacting substance of anaphylaxis (SRS-A) but was not structurally characterized until the early 1980s. The identification of SRS-A as a mixture of LT led to the elucidation of the 5-LO pathway and, in particular, the role of the products of LTC_4 synthase, the cys-LTs, in the pathogenesis of bronchial asthma. Linking the cys-LTs to asthma were the observations that, as bronchoconstrictors, they are more potent than histamine on a molar basis by

TABLE 8.3 Leukotrienes: biological actions and human disease.

	LTB_4	cys-LTs
	LTA_4 hydrolase	LTC_4 synthase
Production from LTA_4	Macrophages	Macrophages
	Neutrophils	Eosinophils
		Mast cells
		Dendritic cells
Target cells	*Neutrophils*	*Endothelial cells*
↑Activity	Chemotaxis, aggregation, and degranulation	Microcirculation permeability and leakage
	Mast cells	*Mast cells*
	Recruitment of immature mast cells	IL-5, IL-8, TNF-α production
	Smooth muscle cells	*Smooth muscle cells*
	Proliferation, migration	Vaso- and broncho-constriction
	Dendritic cells	*Dendritic cells*
	Chemotaxis, accumulation in lymph nodes	Migration from epidermis to lymph nodes
	Macrophages	*Macrophages*
	IL-6, MCP-1, TNF-α production	TNF-α, MMP-9 production
	Lymphocytes	*Lymphocytes*
	T-cell recruitment to peripheral tissue	Allergen sensitization in pulmonary tract, Th2 immune reaction
Associated diseases	Arthritis	Asthma
	Atherosclerosis	Allergic rhinitis
	Cancer	Aortic aneurysm
	Dermatitis	Ischemia/stroke

COPD, Chronic obstructive pulmonary disease; *IBD*, irritable bowel disease; *IL*, interleukin; *MCP*, monocyte chemoattractant protein; *MMP*, matrix metalloproteinase; *TNF*, tumor necrosis factor.

three orders of magnitude. In the 1990s the efficacy of the 5-LO inhibitor zileuton in human asthma was demonstrated and effective antiasthma drugs became available.

In addition to asthma, reactions such as immediate hypersensitivity to allergens and hyperactivity in response to cold and exercise are mediated by LT. For example, cys-LTs are elevated in allergen-challenged subjects and the 5-LO inhibitor zileuton decreases these responses.

2 *Atherosclerosis*

Atherosclerosis is a chronic inflammatory vascular disease. There is much evidence supporting a role for the 5-LO and, in particular, LTB_4 in the development of human atherosclerosis. Blocking the pathway with a BLT1 (see Table 8.2) antagonist protected against atherosclerosis in a mouse model. Also in mice, genetic removal of 5-LO pathway decreases the size of atherosclerotic plaques and specific inhibition of 5-LO reduces monocyte adhesion and infiltration. In humans, atherosclerotic plaque levels of 5-LO correlate positively with disease stage. Genetic polymorphism studies in humans suggest that individuals with some variants of 5-LO show a greater risk of myocardial infarction and stroke. Clinical trials of inhibitors of the 5-LO pathway are underway in patients with cardiovascular disease.

3 *Arthritis*

There is considerable evidence from various mouse models that LTB_4, working through both the BLT1 and BLT2 receptor (see Table 8.2), plays a critical role in autoantibody- or collagen-induced arthritis pathology. Interventions such as blockage of the BLT1 receptor with antagonists, deletion of the gene for the 5-LO activating protein (*ALOX5AP*), or deletion of the genes for BLT1 and/or BLT2 demonstrate that interference with the pathway for LTB_4 production and action leads to diminished arthritic inflammation, cell infiltration, and loss of bone and cartilage resulting from the disease. Synovial fluid from patients with rheumatoid arthritis has LTB_4 and studies with human cells from arthritic sites confirm that the LTB_4 pathway operates and has effects on the arthritic process in humans.

Further reading

Phospholipase A_2

M. Murakami, Y. Taketomi, Y. Miki, H. Sato, T. Hirabayashi, K. Yamamoto, Recent progress in phospholipase A2 research: From cells to animals to humans, Progress in Lipid Research 50 (2011) (2011) 152–192.

Cyclooxygenases 1 and 2

L. Dong, A.J. Vecchio, N.P. Sharma, B.J. Jurban, M.G. Malkowski, W. L. Smith, Human cyclooxygenase-2 is a sequence homodimer that functions as a conformational heterodimer, Journal of Biological Chemistry 286 (2011) (2011) 19035–19046.

W.L. Smith, Y. Urade, P.-J. Jakobsson, Enzymes of the cyclooxygenase pathways of prostanoid biocynthesis, Chemical Reviews 111 (2011) (2011) 5821–5865.

Cyclooxygenase inhibitors

S.H. Hwang, A.T. Wecksler, K. Wagner, B.D. Hammock, Rationally designed multitarget agents against inflammation and pain, Current Medicinal Chemistry 20 (2013) (2013) 1783–1799.

Eicosanoid receptors

T. Hirata, S. Narumiya, Prostanoid receptors, Chemical Reviews 111 (2011) (2011) 6209–6230.

Biological actions

G. Petho, P.W. Reeh, Sensory and signaling mechanisms of bradykinin, eicosanoids, platelet-activating factor, and nitric oxide in peripheral nociceptors, Physiological Reviews 92 (2012) (2012) 1699–1775.

A. Poschke, N. Kern, T. Maruyama, H. Pavenstadt, S. Narumiya, B.L. Jensen, R.M. Nusing, The PGE2-EP4 receptor is necessary for stimulation of the renin-angiotensin-aldosterone system in response to low dietary salt intake in vivo, American Journal of Physiology-Renal Physiology 303 (2012) (2012) F1435–F1442.

Prostaglandins and cancer

I. Cebola, M.A. Peinado, Epigenetic regulation of the COX pathway in cancer, Progress in Lipid Research 51 (2012) (2012) 301–313.

E.R. Greene, S. Huang, C.N. Serhan, D. Panigrahy, Regulation of inflammation in cancer by eicosanoids, Prostaglandins Other Lipid Regulation 96 (2011) (2011) 27–36.

H.-H. Tai, Prostaglandin catabolic enzymes as tumor suppressors, Cancer and Metastasis Reviews 30 (2011) (2011) 409–417.

Leukotrienes and human disease

A. Di Gennaro, J.Z. Haeggstrom, The Leukotrienes: Immune-modulating lipid mediators of disease, Advances in Immunology 116 (2012) (2012) 51–92.

Chapter 9

Calcium-regulating hormones: vitamin D, parathyroid hormone, calcitonin, and fibroblast growth factor 23

I Introduction

A Background information

This chapter will focus on four hormones: parathyroid hormone (PTH); calcitonin (CT); a steroid hormone metabolically produced from vitamin D_3, $1\alpha,25(OH)_2$-vitamin D_3 [$1\alpha,25(OH)_2D_3$]; and fibroblast growth factor 23 (FGF23). Historically, the first three have been intimately involved in calcium and phosphate homeostasis. However, as a consequence of major advances in biological research over the past 30 years, these four hormones have been discovered to have many additional assignments.

Calcium and phosphorus are the most abundant of the inorganic elements in humans; together they are the key structural minerals of mammalian bone. A 70-kg man contains about 1000 g of calcium and 700 g of phosphorus present as phosphate. Calcium and phosphate are also essential to a great number of cellular processes (see Table 9.1).

For optimal growth and function, organisms with a skeleton require an adequate supply of calcium and phosphate to meet both their structural and metabolic needs. These higher organisms must be dietarily supplied with calcium and phosphate on a regular basis. The plasma concentrations of these substances are maintained within a surprisingly narrow limit by an endogenous hormonal control mechanism.

The three primary target tissues involved in calcium and phosphate homeostasis in humans and higher animals are the intestine, bone, and kidney. The intestine is the primary location of the absorption process for calcium and phosphate, which delivers these two ions to the general circulatory system. Functionally, the bone serves two needs. The three-dimensional organization of the precipitated calcium phosphate defines the skeleton, which provides the structural support for the weight of all the soft tissues. At the same time, modest portions of the bone calcium and phosphate are utilized for a minute-by-minute regulation of the serum levels of these two ions. This process is mediated by the dynamic, hormone-regulated process of bone formation and bone resorption. Third, the kidney functions as a gatekeeper for calcium and phosphate (and other ions) where their extent of excretion and renal tubular reabsorption can be monitored and regulated. The maintenance of calcium and phosphate homeostasis thus involves the delicate and coordinated interrelationships of absorption by the intestine, accretion by bone osteoblasts and reabsorption by bone osteoclasts, and renal tubular reabsorption and/or urinary excretion of calcium and phosphate by the kidney.

The total concentration of calcium in plasma is normally in the range of 8.5−10.5 mg/100 mL (1.2−1.3 mM). Calcium in the plasma exists in three forms: free ionized (Ca^{2+}), Ca^{2+} complexed (chelated) by organic ions such as citrate, and as protein-bound Ca^{2+} (see Table 9.2). Under usual circumstances, ~45% of the total plasma calcium is bound to protein, primarily to albumin, and the remaining 55% is "free" and, therefore, ultrafilterable by the kidney glomerulus. There are two basic physiological mechanisms for changing the concentration of ionized calcium in the plasma: one is an endocrine mechanism involving hormones and the second occurs by changing the concentration of plasma protein. All evidence indicates that the biologically functional and, therefore, regulated form of calcium is the "free" ionized species.

In contrast, the plasma concentrations of the several ionic species of phosphate are not regulated so stringently. Under normal circumstances, phosphate's normal range is 2.5−4.5 mg/100 mL (0.8−4.5 mM). Approximately 10% of the plasma phosphate is protein bound, and the remainder exists as free phosphate, either as HPO_4^{2-} or $H_2PO_4^{-}$ (Table 9.2). The relative proportion of these two species is dependent upon the plasma pH. There are many biological functions of phosphate extending from its role as an inorganic buffer ion to multiple involvements as organic phosphates in various enzymatic and structural proteins, phospholipids, and nucleic acids (see Table 9.1).

Hormones. DOI: https://doi.org/10.1016/B978-0-323-90262-5.00002-0

TABLE 9.1 Biological calcium and phosphorus.

Calcium	Phosphorus
Body content	Body content
70-kg man has 1000 g of Ca^{2+}	70-kg man has 700 g of P
Utilization	Utilization
Structural: bone has 98% of body Ca^{2+}	Structural: bone has 90% of body P_i
Plasma $[Ca^{2+}]$ is 8.5–10.5 mg/100 mL (2.12–2.62 mM)	Plasma $[P_i]$ is 2.5–4.5 mg/100 mL (0.81–1.45 mM)
Cell division	Intermediary metabolism (phosphorylated intermediates)
Cell adhesion	Genetic information
Protein secretion	DNA
Plasma membrane integrity	RNA (mRNA)
Nerve pulse transmission	Phospholipids
Muscle contraction	
Blood coagulation	Enzyme–protein components (phosphohistidine, phosphoserine)
Glycogen metabolism	Membrane structure
Enzyme cofactors (amylase, lipases, and ATPases)	
Eggshell (birds)	
Daily requirements (70-kg man)	Daily requirements (70-kg man)
Dietary intake: 600–1600	Dietary intake: 600–2000
Fecal excretion: 300–600	Fecal excretion: 200–600[a,b]
Urinary excretion: 100–300[a,b]	Urinary excretion: 400–1400[a,b]
Sweat: 100–200[a,b]	

mRNA, messenger RNA.
[a]*Values in milligrams per day for an adult.*
[b]*Based on the indicated level of dietary intake.*

TABLE 9.2 Distribution of calcium and phosphate in normal human plasma.

State	Plasma concentration		Percentage of total
	mmol/L	mg/100 mL[a]	
Calcium			
Free Ca^{2+}	1.22	4.88	49
Protein-bound	1.18	4.72	47
Complexed	0.08	0.30	3
Total	2.42	9.90	
Phosphorus			
Free HPO_4^{2-}	0.50	1.55	44
Free $H_2PO_4^-$	0.11	0.34	10
Protein-bound	0.14	0.43	12
$NaHPO_4^-$	0.33	1.02	28
$CaHPO_4$	0.04	0.12	3
$MgHPO_4$	0.03	0.10	3
Total	1.15	3.56	

[a]*For plasma inorganic phosphate ($H_2PO_4^-$, HPO_4^{2-}, and PO_4^{3-}), the values are expressed as milligrams of elemental phosphorus or P per 100 mL of plasma, in accordance with long-standing clinical practice.*

Source: Adapted from M. Walser, Ion association. VI. Interaction between calcium, magnesium, inorganic phosphate, citrate and protein in normal human plasma, Journal of Clinical Investigation 40 (1961) 723.

B Calcium and phosphorus homeostasis

The principal organs of the body involved in the maintenance of calcium and phosphate homeostasis are the intestine, bone, and kidney. It is here that the four calcium-regulating hormones, PTH, CT, $1{,}25(OH)_2D_3$, and FGF23 (a phosphate-regulating hormone), initiate an integrated set of biological responses that results in the maintenance of calcium and phosphorus homeostasis. The steroid hormone $1\alpha{,}25(OH)_2D_3$ is the primary stimulator of the intestinal absorption of both Ca^{2+} and $H_2PO_4^-$. The calcium uptake process is regulated according to the needs of the animal. Once the absorbed Ca^{2+} and $H_2PO_4^-/HPO_4^{2-}$ from the intestine arrives in the plasma, a delicate hormonally mediated balancing of the concentrations of Ca^{2+} and $H_2PO_4^-/HPO_4^{2-}$ occurs in both the skeleton, between bone accretion and bone mobilization, and in the kidney tubules, between urinary excretion and urinary reabsorption. In the event that the dietary availability of Ca^{2+} and HPO_4^-/HPO_4^{2-} is diminished, the balance is tipped in favor of increased bone mobilization to release small quantities of both Ca^{2+} and phosphate to meet the stringent requirement of maintaining a constant serum Ca^{2+} level. Thus the serum Ca^{2+} concentration may become elevated by the stimulation of intestinal Ca^{2+} absorption, bone Ca^{2+} mobilization, or stimulation of the tubular reabsorption of Ca^{2+} in the kidney. Conversely, the serum Ca^{2+} can be lowered by increasing bone formation and inhibiting the renal tubular reabsorption of Ca^{2+}. It is clear that, in the event that there is a dietary shortage of calcium and/or phosphate, the bone is the central organ in calcium and phosphate metabolism, acting as a replacement source to elevate serum calcium and/or phosphate concentrations.

The underlying foundation of the aqueous solubility product, K_{sp}, for $[Ca^{2+}] \times [H_2PO_4^-/HPO_4^{2-}]$ strictly governs the permissible changes in blood $[Ca^{2+}]$ and blood $[H_2PO_4^-/HPO_4^{2-}]$. The value of the K_{sp} is $\sim 1 \times 10^{-7}$ M. Thus if the K_{sp} is exceeded in vivo, then the precipitation of the excess calcium and phosphate will likely occur in the kidney and/or heart and arteries. This soft tissue calcification can damage normal tissue by large amorphous calcium deposits or by the formation of kidney stones. As a general rule, when the hormonal regulation of serum calcium is elevated, there is an associated fall in serum phosphate concentrations, or vice versa.

The US Institute of Medicine (2010) Recommended Dietary Allowance (RDA) average daily intakes for calcium vary from 700 mg/day for 1–3 years old to 1200 mg/day for >70 years of age.

Fig. 9.1 is a schematic diagram illustrating the 24-hour "metabolic balance" of calcium and phosphorus

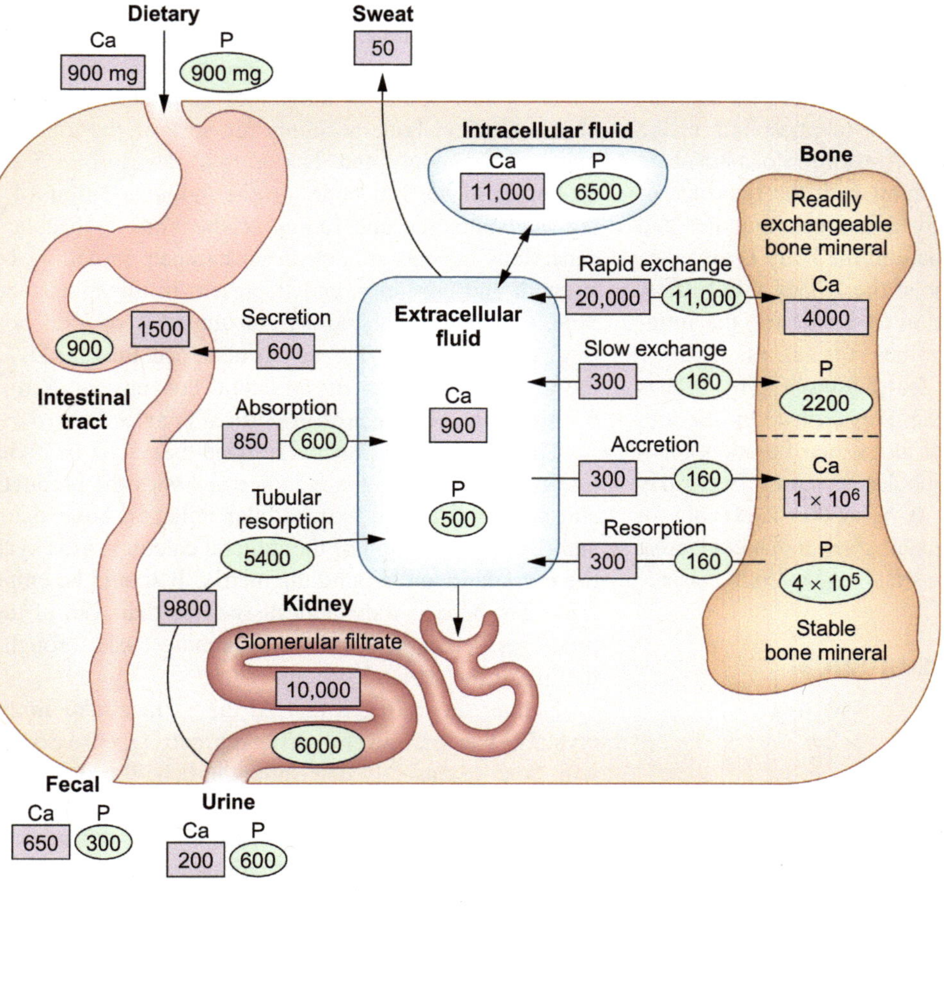

FIGURE 9.1 Schematic model of calcium and phosphorus metabolism in an adult man having a calcium intake of 900 mg/day and a phosphorus intake of 900 mg/day. All numerical values are milligrams per day. All entries relating to phosphate are calculated as phosphorus and are enclosed in ovals. Entries related to calcium are enclosed in rectangles.

metabolism in a normal adult male. Calcium and phosphorus (as phosphate) are both absorbed into the body primarily in the duodenum and jejunum regions of the intestine. In addition to the ~900 mg/day calcium ingested from the diet (for this example), ~600 mg is added to the intestinal contents by pancreatic and intestinal secretions. Of the ~1500 mg of total calcium present in the lumen of the intestine, ~850 mg is absorbed by the intestinal epithelial cells and transported to the blood compartment, leaving the remaining ~650 mg to be excreted in the feces.

After the newly absorbed Ca^{2+} has entered the extracellular pool, it is in constant exchange with the Ca^{2+} already present in the extracellular and intracellular fluid compartments of the body and in certain compartments of the bone and the kidney's glomerular filtrate. The entire extracellular pool of 900 mg of Ca^{2+} turns over approximately 12 times per day. Thus the glomerulus of the kidney filters ~10,000 mg of Ca^{2+} per day, but the renal tubular reabsorption of this ion is so efficient that under normal circumstances only ~200 mg of Ca^{2+} appears in the urine. In the event of hypercalcemia the urinary excretion of Ca^{2+} rises in a compensatory fashion; however, it rarely exceeds a value of 400–600 mg/day. The renal tubular reabsorption of Ca^{2+} is stimulated by the separate actions of PTH and $1\alpha,25(OH)_2D_3$ in the distal nephron of the kidney. Also, depending on the ambient temperature, an additional 50–200 mg of Ca^{2+} may be lost per day through the skin via sweating.

The dynamics of phosphate metabolism are not particularly different from those of calcium. Under normal circumstances, approximately one-third of phosphate in the diet is absorbed by the intestine. Absorption of phosphate is interrelated in a complex fashion with the presence of Ca^{2+} and can be stimulated by a low-calcium diet and also by $1\alpha,25(OH)_2D_3$. Phosphate in the body is also partitioned among three major pools: the kidney ultrafiltrate, the readily exchangeable fraction of bone, and the intracellular compartments in the various soft tissues.

The major excretory route for phosphate (Fig. 9.1) is through the kidney. The handling of phosphate by the kidney is determined by the rates of glomerular filtration, tubular reabsorption, and possibly tubular secretion. Every day the kidney glomerulus filters some 6000–10,000 mg of phosphorus. A normal 70-kg person, given a diet containing 900 mg of phosphorus, excretes ~600 mg/day in the urine.

II Anatomical and physiological relationships

A Intestine

The morphology and cellular organization of the intestine are superbly adapted to efficiently carry out the absorption of dietary constituents, including calcium and phosphate. The anatomical organization of the intestinal mucosa, which optimizes the absorptive surface/volume ratio of the cell thereby facilitating the intestinal absorptive processes, is discussed in Chapter 7, Gastrointestinal Hormones (see Fig. 7.3).

B Bone

1 Bone organization

Bone is a complex tissue made up of cells and extracellular organic and mineral material. The cells are of a wide variety of morphological and functional types, but all have a common origin in the mesenchymal stem cells. The principal cell types are (1) chondrocytes or cartilage cells, which secrete the collagen matrix of the cartilage region; (2) osteoblasts or bone-forming cells; (3) osteoclasts, which are multinucleated giant cells responsible for bone resorption; and (4) osteocytes, individual functional osteoblasts that are trapped in the mineralized bone matrix. On a dry weight basis, bone consists of 65%–70% inorganic crystals of calcium hydroxyapatite and 30%–35% organic matrix known as osteoid. Of the osteoid, 40% is composed of the extracellular protein collagen.

There are two major categories of bone which are differentiated on the basis of density; these are the cortical bone and the trabecular or cancellous bone. See Fig. 9.2, which illustrates an interface region with both cortical and cancellous bone. Cortical bone is composed of densely packed columns of mineralized collagen laid down in layers and is the major component of tubular bones. Cortical bone accounts for 80% of the total bone mass of an adult and has a total area of ~3.2 m^2. Examples of cortical bone are the femur and tibia of the leg, and humerus and radius of the arm. Trabecular or cancellous bone is spongy in appearance, providing both strength and elasticity, and is present in the spinal vertebrae, thin bones (the skull), and the ends of long bones (e.g., the hip bones). Trabecular bone accounts for 20% of the total bone mass of an adult but has ~16 m^2 or 5 × times the surface area of cortical bone.

The formation of functional bone tissue can be divided into two phases: (1) that is concerned with the production and secretion of the extracellular collagen bone matrix; and (2) the deposition of the mineral calcium hydroxyapatite crystals in and around the matrix. It should be emphasized that bone is a dynamic tissue and that both of these processes occur on a minute-by-minute basis throughout the life of the skeletal system.

Neither the collagen matrix, the extracellular mineral crystals, nor the several different cell types associated with bone exclusively determine the behavior of the bone tissue. It is a unique combination of the extracellular proteins and the inorganic phases, as well as the particular biochemical properties of the various bone cell types, that

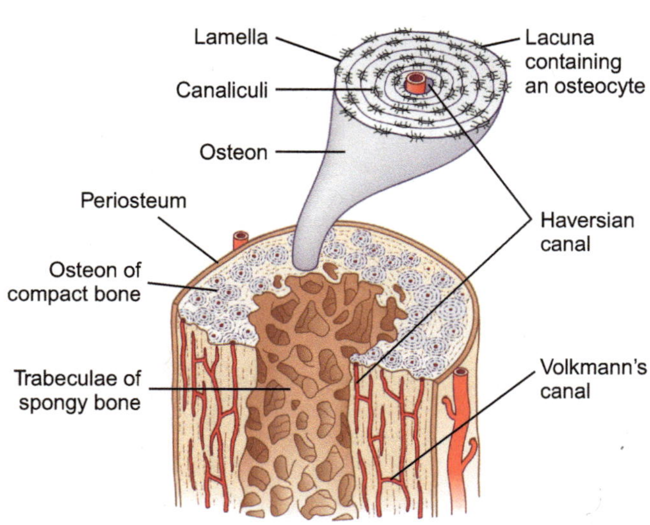

FIGURE 9.2 Schematic representation of the main features of the structure of bone. This presentation illustrates both the transverse (top) and longitudinal sections of an interface region between both cortical and cancellous or spongy bone. Osteons are the fundamental building blocks of cortical bone. Bone comprises concentric layers of osteons along the long axis longitudinally from the top to the bottom of a long bone. Each osteon consists of concentric layers of lamellae of compact bone tissue that consist of parallel layers of collagen sheets that surround a central canal, the Haversian canal. Haversian canals are tiny, interconnecting, parallel channels along the long axis of the bone tissue through which a blood capillary and a nerve pass. Volkmann's canals are small ones that transmit blood vessels from the periosteum (a membranous outer layer of bone) into the bone; they lie perpendicular to the Haversian canals and their capillaries connect with the capillaries of the Haversian canals. Lacunae cavities lie between the lamellae and are connected to one another by a series of small canals termed canaliculae. Each lacuna is occupied by one osteocyte (see also Fig. 9.4). The periosteum is a membrane that lines the outer surface of all bones, except at joints, and provides nourishment for the bone via its capillaries.

collectively confers on bone both its unusual mechanical properties (to support the weight of the soft tissues of the body) and its ability to serve as a dynamic reservoir for the calcium and phosphate ions needed for mineral homeostasis in the whole organism.

Bone matrix is biosynthesized, secreted, organized, mineralized, and finally destroyed by reabsorption, all in accordance with the local physiological and hormonal signals operative at any particular time. The production of organic matrix by osteoblast cells first involves the intracellular synthesis of protocollagen molecules by the ribosomal system through conventional protein biosynthetic pathways. A schematic summary of collagen biosynthesis and conversion to functional three-stranded cross-linked collagen fibrils is summarized in Fig. 9.3. Each collagen strand of the fibrils has its separate unique amino acid sequence and is glycosylated. The collagen fibrils then become impregnated with the calcium hydroxyapatite that is the precipitated form of $[Ca^{2+}{}_{10}(PO_4{}^{3-})_6(OH^-)_2]$.

In the normal process of human bone formation, there is usually a delay of 5–10 days between the synthesis of the extracellular organic matrix and its ultimate mineralization. After the final secretion of protocollagen by the osteoblasts and the ultimate formation of mature collagen, the osteoblasts differentiate and are incorporated into the bone matrix as osteocytes (see Fig. 9.4). On average, the skeletal system receives 13%–18% of the total cardiovascular output. Given that every day approximately 300 mg of calcium and 160 mg of phosphorus are resorbed from the skeleton and replaced from the circulatory system, then approximately every 10 years all the skeletal calcium and phosphate have been completely replaced.

2 Bone cell types

There are three primary types of bone cells, each with its own sphere of influence on bone activities.

Osteoblasts are mononucleate cells that are responsible for bone formation. The osteoblast's major activity is to biosynthesize the organic matrix that constitutes 40% of the weight of bone. The bone matrix principally comprises collagen that is the most abundant protein in our body (~30% of total body protein). Osteocalcin, which is the fifth most prevalent protein in our body, is also a significant structural component of bone matrix. Surprisingly, it was discovered by Karsenty (2017) that osteocalcin is also a protein hormone secreted by bone that stimulates the secretion of insulin by the pancreas *beta* cells. Osteocalcin also binds to receptors on fat and testis cells, suggesting additional complexity to the endocrine role of osteocalcin (see Fig. 9.5). The osteoblast has receptors for both PTH and for the steroid hormone $1\alpha,25(OH)_2D_3$.

Osteocytes are mononuclear fully mature osteoblasts (see Fig. 9.4) and represent ~90% of all bone cells. The osteocytes in their active osteoblast phase become embedded in bone lacunae (compartments) as a consequence of their active role in bone formation. They send bone canaliculi (slender cytoplasmic processes) that make contact with the canaliculi of other osteocytes. These canaliculi provide the osteoblasts with access to nutrients and oxygen from the bone marrow and bone periosteum (see Fig. 9.4). Osteocytes are a primary source of the hormone FGF23 that is involved in phosphate homeostasis and regulation of the production of $1\alpha,25(OH)_2D_3$ and PTH (see Fig. 9.6).

Osteoclasts are the mediators of bone resorption. They produce small pits or trenches that are generated by the solubilization and export of calcium hydroxyapatite. Osteoclasts are multinucleate and are derived from bone marrow macrophage cells in a complex cell differentiation process that includes direct physical interaction with osteoblasts. The details of the origin of osteoclasts will be discussed later in this chapter (see Fig. 9.28 in Section IV.F).

3 Bone remodeling

In understanding the hormonal regulation of calcium and phosphorus homeostasis, it is important to appreciate that

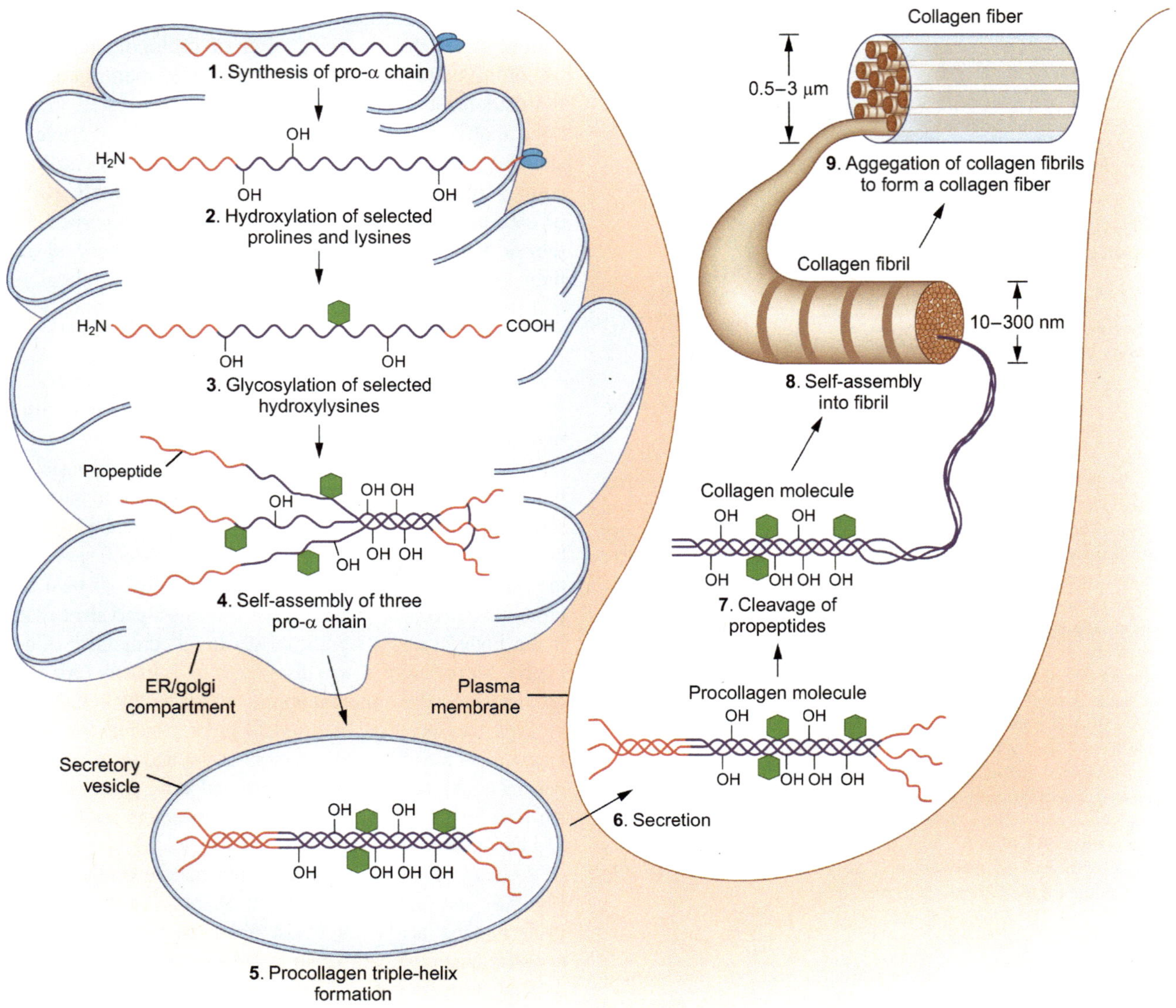

FIGURE 9.3 A schematic summary of the integrated intracellular and extracellular 10 steps involved in the biosynthesis of a collagen fiber. The first five steps occur inside the cell starting with the biosynthesis of a collagen pro-α-chain. This is followed by the hydroxylation of selected lysines and prolines and then glycosylation of selected prolines and lysines. Then the pro-α-chains self-assemble into a procollagen triple helix which is secreted from the osteoblast. The procollagen molecule is next subject to selective protease cleavage to yield a mature collagen molecule which then self-assembles into a mature collagen fiber. These collagen fibers along with other bone proteins (osteocalcin, osteopontin) are the organic matrix that then becomes mineralized with precipitated calcium hydroxyapatite, yielding a solid bone.

bone tissue is quite dynamic, in terms of both the metabolic activity of the various cell populations present and with regard to the kinetics of mineral flux into and out of the several bone compartments. As summarized in Fig. 9.1, there are three functional, but not anatomical, compartments of bone mineral; they are comprised from the total ~1 million milligrams of calcium and ~500,000 mg of phosphate that are incorporated into bone mineral of adult males. The Ca^{2+} and $H_2PO_4^-/HPO_4^{2-}$ precipitate on the surface of the extracellular organic bone matrix as the mineral calcium hydroxyapatite, which has the empirical formula of $[Ca^{2+}{}_{10}(PO_4{}^{3-})_6(OH^-)_2]$. The functional Ca^{2+} compartments present in bone can be identified on the basis of radioactive tracer studies (utilizing either radioactive $^{45}Ca^{2+}$ or $^{47}Ca^{2+}$). These compartments include the following: (1) the rapid-exchange bone mineral compartment (~20,000 mg/day); (2) a slow-exchange compartment (~300 mg/day of Ca^{2+}); and (3) a hormonally regulated compartment (~300 mg/day of Ca^{2+} via bone resorption and ~300 mg/day of Ca^{2+} via bone accretion).

The term *rapid-exchange* implies the capability of a rapid (on a minute-by-minute basis) loss and gain of calcium via

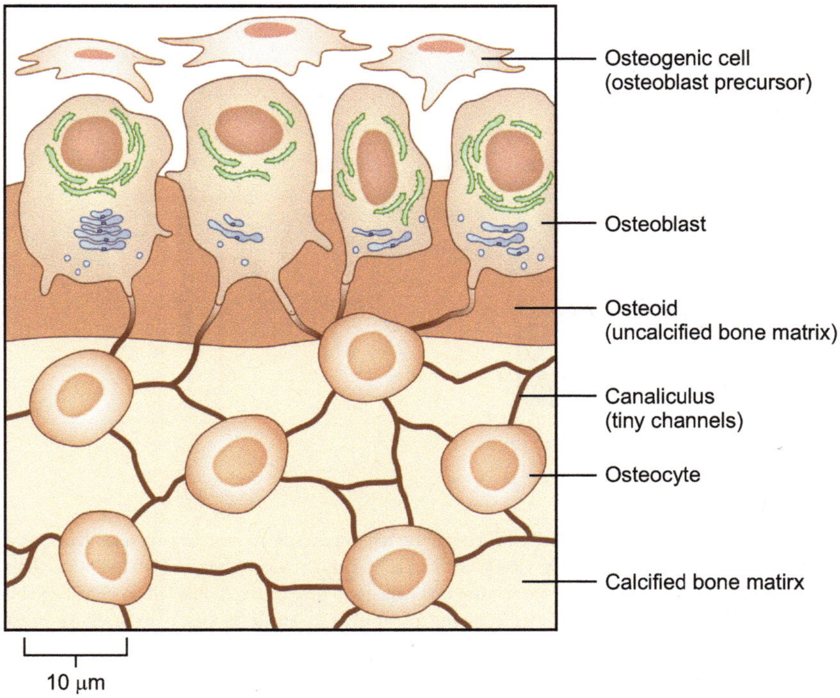

FIGURE 9.4 Role of osteoblasts and osteocytes. Osteoblasts secrete the organic matrix of bone (termed osteoid). The osteoid is principally collagen fibers (see Fig. 9.3) and two other bone proteins, osteonectin and osteopontin. As a consequence of mineralization of the osteoid, each osteoblast becomes surrounded with calcified bone matrix and then becomes converted to an osteocyte. The isolated osteocytes communicate with one another via their cell process, the tiny canaliculus.

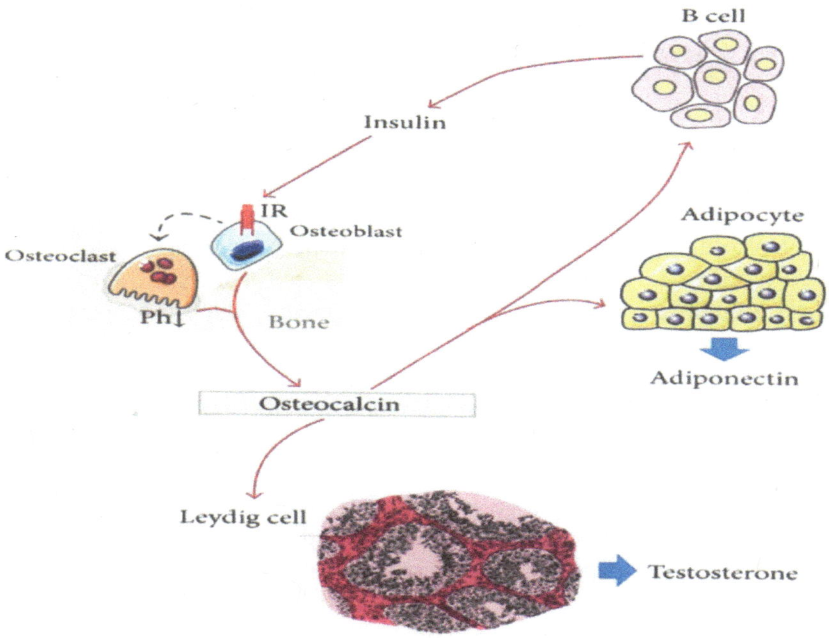

FIGURE 9.5 Endocrine actions of osteocalcin. Circulating osteocalcin and particularly its undercarboxylated fraction (released during active bone resorption) exert a direct effect on β-cells, stimulating insulin production as well as on adipocytes enhancing adiponectin production. Adiponectin itself is able to promote insulin sensitivity. In turn, insulin also acts directly on osteoblast and indirectly on osteoclast. Osteoclast stimulates bone resorption with subsequent release of undercarboxylated osteocalcin in blood circulation. Finally, osteocalcin has a role also on Leydig cells, increasing their activity and testosterone production. *Reproduced form Fig. 2 of A. Patti et al., Endocrine actions of osteocalcin, International Journal of Endocrinology 2013 (2013), https://doi.org/10.1155/2013/846480.*

chemical dissolution from the calcium hydroxyapatite followed immediately by reprecipitation. Thus in 1 day, ~20,000 mg of calcium leaves the bone and another ~20,000 mg returns to the bone in this rapid-exchange "compartment." In the *slow-exchange* compartment, ~300 mg of calcium leave via a slower chemical dissolution and then ~300 mg of calcium return via a reprecipitation process into the bone's calcium hydroxyapatite. Similarly, the third compartment comprises the stable bone mineral (calcium hydroxyapatite) that in the adult male is composed of approximately 1 kg (1 million milligrams) of calcium. However, the "stable bone mineral" compartment is not biologically inert. Both the accretion and resorption processes of the third compartment are those that are mediated on a minute-by-minute basis by the hormonal regulators of calcium metabolism [PTH, CT and $1\alpha,25(OH)_2D_3$], whereas the rapid and slow exchange pools largely represent chemical equilibria between the bone and the bathing extracellular fluids.

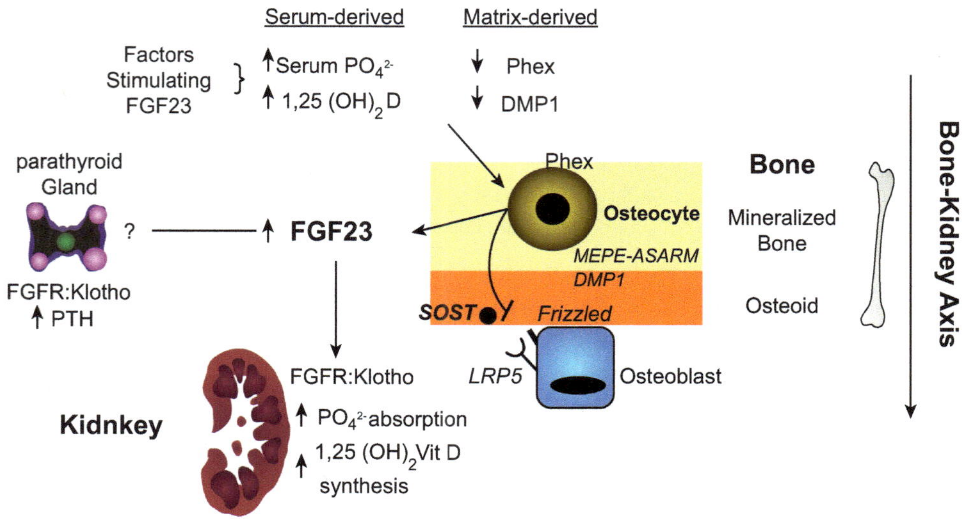

FIGURE 9.6 The bone–kidney axis. In this model, osteocytes in bone are the principal cells that coordinate the mineralization process with renal handling of phosphate. Osteocytes secrete FGF23 and SOST, MEPE, and DMP1 and express Phex. FGF23 targets organs, particularly the kidney, via an FGF receptor: Klotho complex that inhibits sodium-dependent phosphate uptake and 1α-hydroxylase activity by a mechanism that remains to be elucidated. In addition, SOST inhibits osteoblastic formation. Also, the extracellular matrix proteins DMP1 and MEPE, respectively, positively and negatively regulate the process of extracellular matrix mineralization. The inhibition of bone formation would be associated with decreased phosphate buffering capacity of bone and require increased FGF23-mediated phosphaturia, whereas accelerated bone formation and increased bone phosphate utilization would be associated with suppressed FGF23 and renal phosphate retention to meet the need for bone mineralization. *1,25(OH)₂D*, 1,25-Dihydroxyvitamin D; *ASARM*, amino acid sequence Ala–Ser–Ala–Arg–Met; *DMP1*, dentin matrix protein 1; *FGF*, fibroblast growth factor; *FGF23*, fibroblast growth factor 23; *FGFR*, fibroblast growth factor receptor; *Klotho*, type 1 membrane protein related to β-glucuronidases; *LRP5*, cell membrane coreceptor that mediates canonical Wnt signaling and facilitates bone formation; *MEPE*, matrix extracellular phosphoglycoprotein; *Phex*, phosphate-regulating neutral endopeptidase); *PTH*, parathyroid hormone; *SOST*, sclerostin. *Reproduced from Fig. 2 of S. Liu, L.D. Quarles, How fibroblast growth factor23 works, Journal of the American Society of Nephrology 18 (2007) 1637–1647.*

In a normal adult a remodeling of the stable bone mineral is occurring continuously throughout life, such that, on any given day, approximately 300 mg of calcium is resorbed and another 300 mg of calcium is replaced by accretion. Thus in the normal state, there is a balance in the extent of daily bone resorption and bone accretion, leading to no net change in total skeletal calcium content.

C Kidney

The kidney has indispensable homeostatic actions with regard to all the electrolytes of the body and also for the filtration and removal of nitrogenous wastes (primarily as urea). Importantly, the kidney also serves as an endocrine gland for the secretion of several classes of hormones. These include the following hormones: renin and erythropoietin (Chapter 15: Hormones Related to the Kidney and Cardiovascular System), several prostaglandins (Chapter 8: Eicosanoids), and in this chapter the steroid hormone $1\alpha,25(OH)_2D_3$. The anatomical organization of the kidney is presented in detail in Chapter 15, Hormones Related to the Kidney and Cardiovascular System; see Figs. 15.1 and 15.2.

D Parathyroid gland

Parathyroid glands have been identified in all vertebrate species higher than fish. In humans, there are normally four parathyroid glands. Anatomically they are localized on the surface of each side of the thyroid (Fig. 9.7) and total ~120 mg of tissue. The parathyroid glands are derived from the endoderm of the third and fourth pharyngeal pouches.

The two main classes of epithelial cells of the parathyroid gland are the more abundant chief cells and the less abundant oxyphil cells, that appear at puberty, and, as some theorize, could be derived from chief cells that are present at birth. "Oxyphil" means having affinity for oxygen (and acid dyes). Oxyphils express parathyroid-relevant genes, like the chief cells, and produce autocrine/paracrine factors, such as PTH-related protein (PTHrP) and calcitriol. Oxyphils can secrete PTH during secondary parathyroid hyperplasia that contributes to hyperparathyroidism. In this regard, it may be that chronic usage of certain diuretics, that drain the plasma of metal ions (i.e., Ca^{2+}, Mg^{2+}, and others), could cause secondary hyperparathyroidism. In this situation, one or more parathyroids could develop, from hyperplasia, a nonmalignant parathyroid adenoma (owing to the enhanced activity of the calcium-sensing receptor that signals the release of PTH when serum calcium levels decline). The chief cell is responsible for the biosynthesis of PTH. This cell undergoes a series of cyclical changes, which can be observed histologically and are associated with the synthesis, packaging, and secretion of PTH, followed by cellular involution.

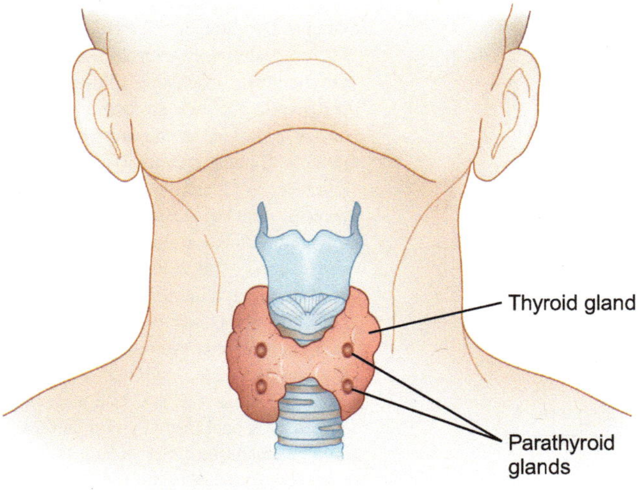

FIGURE 9.7 Human parathyroid glands are embedded in the anterior (frontal) surface of the thyroid gland. There are normally four parathyroid glands. The two higher glands, one on each side, are known as the superior parathyroid glands, while the two lower glands are known as the inferior parathyroid glands. Occasionally, there are only three or five to six glands. The parathyroid glands are normally each the size of a pea and weigh between 30 and 50 milligrams. In instances of hyperparathyroidism, an adenoma in one gland may develop and grow to the size of a walnut.

E Calcitonin-secreting cells

CT is secreted by the parafollicular or C cells, which are located in the thyroid glands of higher animals. The secretion of CT is governed by the serum calcium concentration. The anatomical relationship of the C cells to the thyroid follicles varies with the species. In lower vertebrates, including the teleost fish, elasmobranchs, anurans, urodeles, and aves, the CT-secreting C cells are localized in the anatomically distinct ultimobranchial body.

III Chemistry and biochemistry

A Vitamin D and $1\alpha,25(OH)_2D_3$

The molecular structure of vitamin D_3 is closely allied to that of classical steroids such as cholesterol (see Fig. 9.8). Technically, vitamin D is a secosteroid. Secosteroids are those in which one of the rings of the cyclopentanoperhydrophenanthrene ring structure of the classic steroids (see Fig. 2.2) has undergone fission by breakage of a carbon–carbon bond. In the instance of the vitamins D (both vitamin D_3 and vitamin D_2), this is the 9,10-carbon bond of ring B.

There are several families of vitamin D steroids based upon differences in the structure of the side chain attached to carbon-17. The vitamin D_3 family is derived from cholesterol which has an eight carbon–saturated side chain; see the green side chain in Fig. 9.8. The vitamin D_2 family is derived from ergosterol, which has a 9-carbon side chain attached to carbon-17, with a 22–23 double bond and an additional methyl group on C-24; see the red side chain in Fig. 9.8. Collectively, vitamin D_3 (or cholecalciferol) and vitamin D_2 (or ergocalciferol) can be termed the calciferols or simply vitamin D without a subscript 2 or 3.

Vitamin D_3 is normally produced by exposure to sunlight of the precursor, 7-dehydrocholesterol, present in the skin. In contrast, vitamin D_2 is only produced synthetically via ultraviolet irradiation of the sterol ergosterol. Ergosterol, while present in yeast, is not present in mammals and thus vitamin D_2 cannot be produced via the exposure of skin to sunlight.

A formal definition of a vitamin is that it is a trace dietary constituent required to affect the normal function of a physiological process. The emphasis here is on *trace* and the fact that the vitamin *must* be supplied regularly in the diet; this implies that the body is unable to synthesize the vitamin in question. Thus vitamin D_3 becomes a vitamin only when the animal or human does not have regular access to sunlight or ultraviolet light. Under normal physiological circumstances, all mammals, including humans, can generate via ultraviolet photolysis adequate quantities of vitamin D to meet their nutritionally defined requirements. It is largely through a historical accident that vitamin D_3 has been classified as a vitamin rather than as a steroid hormone. Chemists had certainly appreciated the strong structural similarity between the vitamins D and other steroids (see Fig. 2.3), but this correlation was never widely acknowledged in the biological, clinical, or nutritional sciences until 1965–70.

The chief structural prerequisite of a sterol to be classified as a provitamin D is its ability to be converted, upon ultraviolet irradiation, to a vitamin D; thus it is mandatory that it has in its B ring a $\Delta^{5,7}$-conjugated double-bond system. A summary of the photochemical pathway involved in the production of vitamin D in man and animals is presented in Fig. 9.9. In the skin the principal ultraviolet irradiation product is previtamin D_3. The conversion of previtamin D_3 to vitamin D_3 involves an intramolecular hydrogen transfer from C-19 to C-9 (see the legend for Fig. 9.9 for details); these chemical transformations can occur in the absence of further ultraviolet exposure. The resulting vitamin D_3 is then transported in the general circulatory system by the 50-kDa vitamin D-binding protein (DBP).

Interest has focused on the effects of latitude [diminished ultraviolet (UV)-B intensity], skin pigmentation (concentration of melanin), and photochemical regulation upon the efficiency of conversion of 7-dehydrocholesterol present in the skin to epidermal previtamin D_3. In order of importance, the significant determinants limiting the rate of cutaneous production of previtamin D_3 are (1) photochemical regulation, (2) pigmentation, and (3) latitude.

Although the chemical structure of vitamin D was determined in the 1930s, it was not until the era 1965–95 that the truly unique structural aspects of the molecule became appreciated. In contrast to other steroid hormones

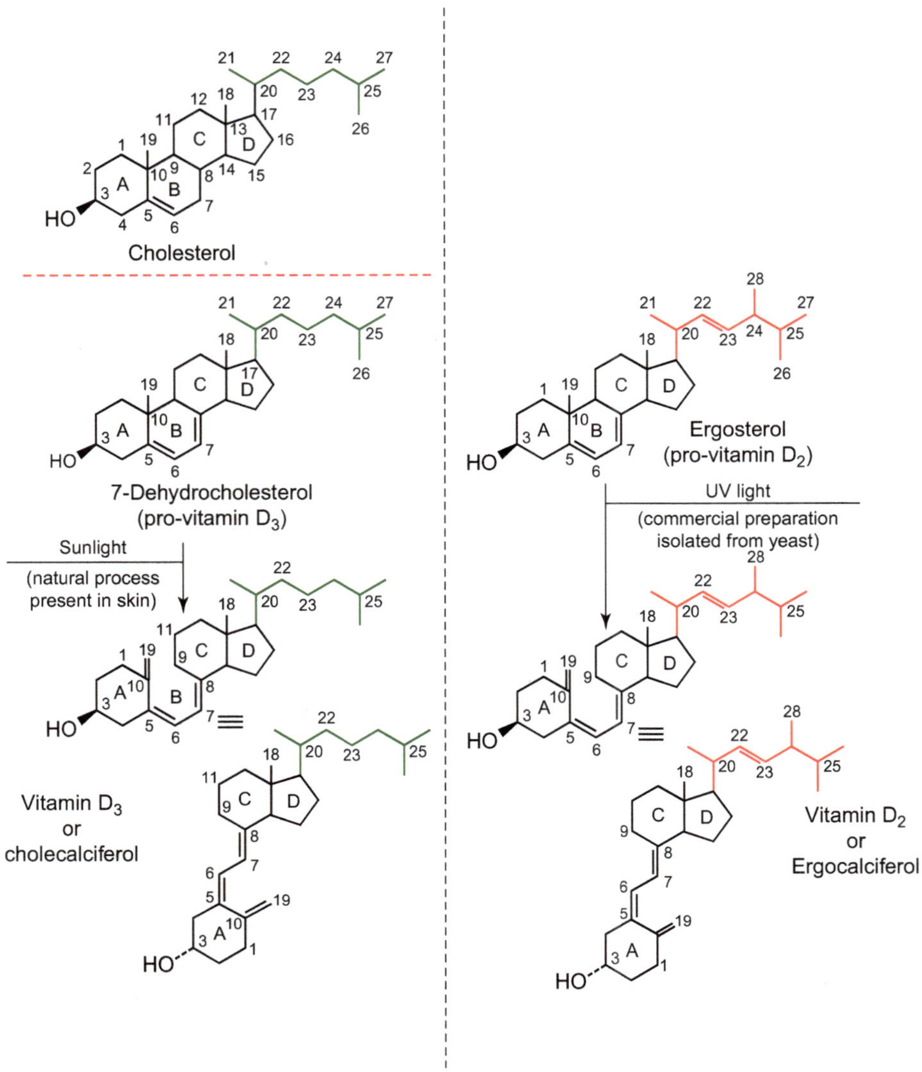

FIGURE 9.8 Structural relationship of vitamin D_3 (cholecalciferol) and vitamin D_2 (ergocalciferol) with their respective provitamins, 7-dehydrocholesterol and ergosterol. The two structural representations presented at the bottom for both vitamin D_3 and vitamin D_2 are equivalent (see text); these are simply different ways of drawing the same molecule and reflect the rotation around the C6–C7 single bond. The only structural difference between vitamin D_3 and vitamin D_2 is the side chain. Vitamin D_3 has the side chain of cholesterol, while vitamin D_2 has the side chain of ergosterol. Vitamin D_2 has a C22=C23 double bond and an additional methyl group on C24. It is to be emphasized that vitamin D_3 is the naturally occurring form of the vitamin; it is produced from 7-dehydrocholesterol, which is present in the skin, by the action of sunlight (see Fig. 9.7). Vitamin D_2 is produced commercially by the irradiation of the plant sterol ergosterol with ultraviolet light. According to Dr. R. P. Heaney vitamin D_2 has only ~30% of the biological activity of vitamin D_3 in humans and only 10% in birds.

(see Fig. 2.18), the vitamin D molecule has three structural features that contribute to the extreme conformational flexibility of this secosteroid molecule; these include the presence of (1) an eight-carbon side chain; (2) a broken B ring, which "unlocks" the A ring; and (3) the ability to undergo chair–chair interchange many times per second. (See Fig. 2.18 for a detailed consideration of the conformational flexibility of vitamin D molecules.)

A totally new era in the field of vitamin D_3 opened in 1967 with the discovery of the metabolism of vitamin D_3 into a steroid hormone. The biologically active form of vitamin D_3 is the steroid 1α25-dihydroxyvitamin D_3 [1α,25$(OH)_2D_3$]. It is now recognized that there is an endocrine system for processing the prohormone, vitamin D_3, into its hormonally active daughter metabolite(s) (see Fig. 9.10). The endocrine gland producing the biologically active form of vitamin D_3 is the kidney. After metabolic conversion of vitamin D_3 into 25-hydroxy vitamin D_3 by a liver microsomal enzyme, this circulating form of the secosteroid serves as a substrate for either the renal 25(OH)D–1α-hydroxylase or the 25(OH)D–24-hydroxylase. Both enzymes are located in the mitochondrial fraction of the kidney proximal tubule. The 1α-hydroxylase is localized in the kidneys of members of every vertebrate class from teleosts, through amphibians, reptiles, and aves, to mammals, including primates. It has been shown that the 25$(OH)D_3$ – 1α-hydroxylase enzyme system is a classical mixed-function steroid hydroxylase similar to the steroid hormone hydroxylases found in the adrenal cortex mitochondria (see Fig. 2.27A). The 1α-hydroxylase is a cytochrome P450-containing enzyme that involves an adrenodoxin component incorporating molecular oxygen into the 1α-hydroxyl functionality of 25(OH)D_3. 1α,25$(OH)_2D_3$ acting as a steroid hormone is also known to genomically induce the renal 24-hydroxylase to permit the coproduction of 24,25$(OH)_2D_3$. Emerging evidence suggests a role for 24,25$(OH)_2D_3$ in mediating normal bone development. Thus 1α,25$(OH)_2D_3$ and 24,25

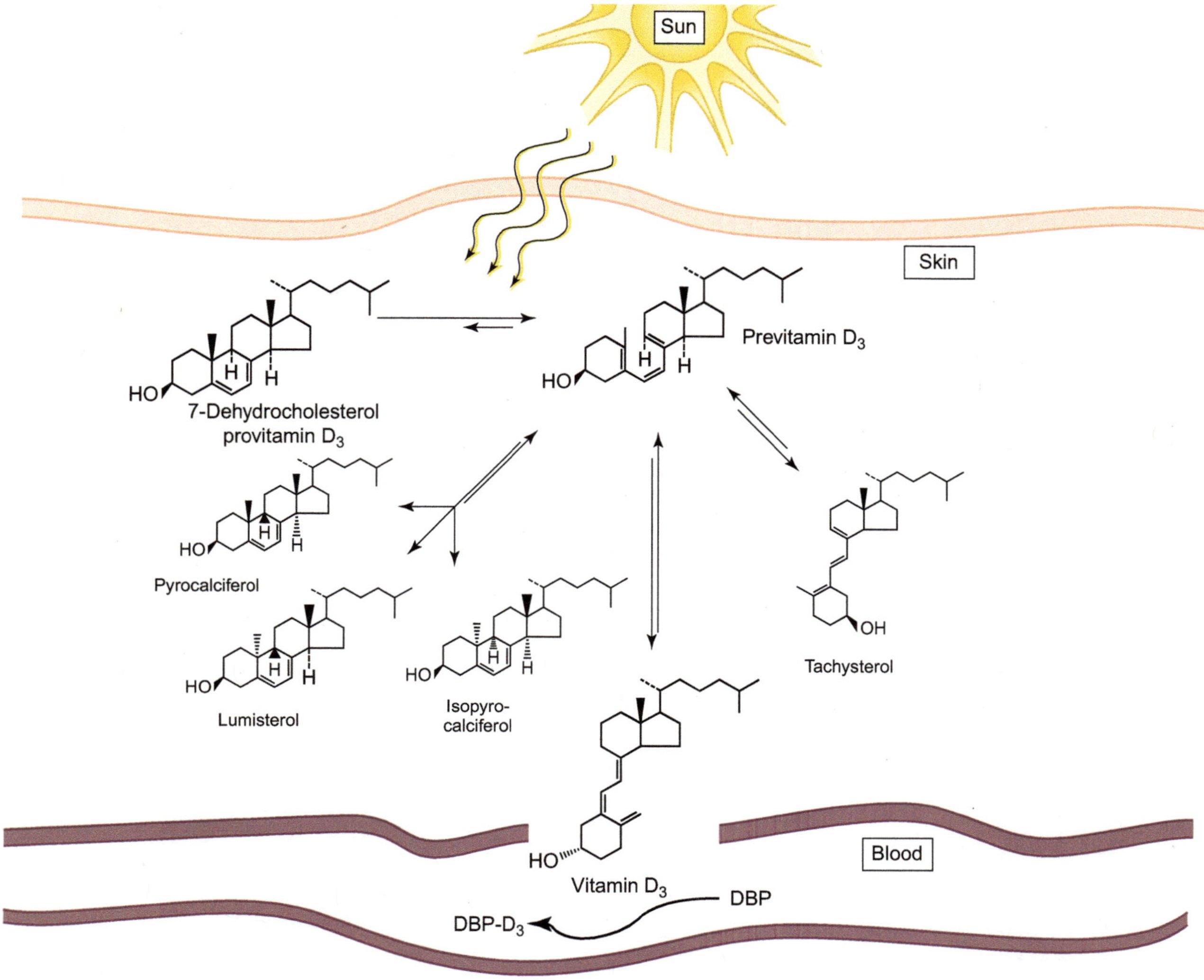

FIGURE 9.9 Photochemical pathway of the production of vitamin D_3 (cholecalciferol) from 7-dehydrocholesterol. The starting point is the irradiation of a provitamin D, which contains the mandatory $\Delta^{5,7}$-conjugated double bonds; in the skin, this is 7-dehydrocholesterol. After the absorption of a quantum of light from sunlight (UV-B), the activated molecule can return to the ground state and generate at least six distinct products. The four steroids that do not have a broken 9,10-carbon bond (provitamin D, lumisterol, pyrocalciferol, and isopyrocalciferol) represent the four diastereomers with either an α- or a β-orientation of the methyl group on carbon-10 and the hydrogen on carbon-9. The three secosteroid products, vitamin D_3, previtamin D_3, and tachysterol$_3$, each have differing positions of the three conjugated double bonds. In the skin the principal product is previtamin D_3, which then undergoes a 1,7-sigmatropic hydrogen transfer from C-19 to C-9, yielding the final vitamin D_3. Vitamin D_3 can be drawn as either a 6-*s*-*trans* representation (this figure) or a 6-*s*-*cis* representation (see Fig. 9.10), depending upon the state of rotation about the 6,7-single bond. The resulting vitamin D_3, which is formed in the skin, is removed by binding to the plasma transport protein, the DBP, present in the capillary bed of the dermis. The DBP-D_3 then enters the general circulatory system. The same overall mechanism applies to the commercial irradiation of ergosterol to yield vitamin D_2. *DBP*, Vitamin D–binding protein; *UV-B*, ultraviolet B.

$(OH)_2D_3$ are the two principal forms of the parent vitamin D_3 that mediate the biological responses characteristic of this vitamin.

B Parathyroid hormone

PTH is an 84-amino acid–containing protein that is secreted by the chief cells of the parathyroid gland. PTH is a single-chain polypeptide of 84 amino acids (molecular weight=9300) with no cysteine residues and hence no disulfide bridges. A fragment of the intact hormone consisting of the first 34 amino acids at the N-terminal region of the molecule is sufficient for the peptide to exert its entire spectrum of characteristic biological effects; see Fig. 9.11.

Because PTH is a secreted protein, it is biosynthesized as a larger precursor by the parathyroid gland. Two precursor species have been identified, including a prepro-PTH with 31 additional amino acids added onto the N-terminal region of PTH, representing the primary translation product of the messenger RNA (mRNA) in the

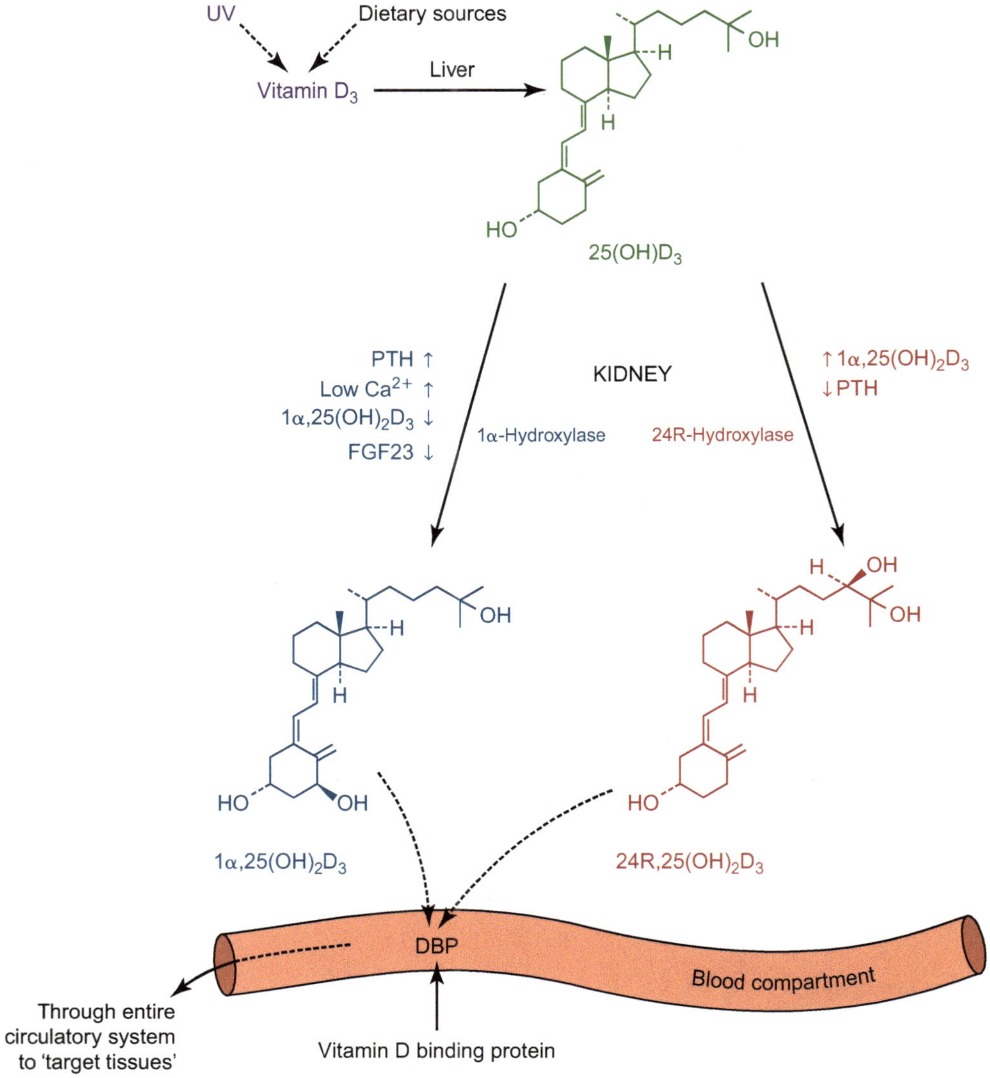

FIGURE 9.10 A summary of the key aspects of vitamin D metabolism. The secosteroid vitamin D_3 itself is biologically inert and does not stimulate or mediate any biological responses. Vitamin D_3 produced photochemically in the skin (see Fig. 9.9) or obtained dietarily is 25-hydroxylated in the liver to generate $25(OH)D_3$ and then further metabolized in the kidney. Thus vitamin D_3 is a precursor to three key daughter metabolites. Accordingly, there are three key enzymes involved in conversion of vitamin D_3 into $25(OH)D_3$, $1\alpha,25(OH)_2D_3$, or $24R,25(OH)_2D_3$. They include the following: (1) the vitamin D_3 − 25-hydroxylase (a liver mitochondrial CYP27A1); (2) a $25(OH)D_3$ − 1α-hydroxylase (the proximal kidney tubule mitochondrial CYP27B1); and (3) a $25(OH)D_3$ − 24R-hydroxylase (the proximal kidney tubule mitochondrial CYP24). The liver 25-hydoyxlase is not subject to physiological regulation. Thus the amount of $25(OH)D_3$ produced is dependent upon the substrate concentration of vitamin D_3 present. In contrast, both the 1α-hydroxylase and the 24R-hydroxylase localized in kidney are highly regulated. As shown in the figure, the activity of the 1α-hydroxylase is increased by PTH, and low serum Ca^{2+} and decreased by FGF-23 and the circulating concentration of $1\alpha,25(OH)_2D_3$. Both the kidney-produced $1\alpha,25(OH)_2D_3$ and $24R,25(OH)_2D_3$ as well as the liver-produced $25(OH)D_3$ move to the circulatory system where they bind to the DBP for transport throughout the circulatory system. Target tissues for $1\alpha,25(OH)_2D_3$ are defined by the presence of the VDR. *DBP*, Vitamin D−binding protein; *FGF*, fibroblast growth factor; *PTH*, parathyroid hormone; VDR, vitamin D receptor.

ribosomal fraction of the parathyroid gland. This prepro-PTH then is secreted into the cisterna of the rough endoplasmic reticulum, where it is processed within seconds into a form known as pro-PTH. This occurs by the removal of the NH_2-terminal methionyl residue and the next 24 amino acids (i.e., residues −30 through −7) after biosynthesis. By 20 minutes after synthesis, pro-PTH reaches the Golgi region where it is stored in vesicles and converted into PTH by the removal of the N-terminal hexapeptide. PTH is stored in the secretory granule until it is released into circulation in response to a fall in the blood concentration of calcium.

The introduction of immunologic assay techniques has permitted the detection of circulating levels of PTH and its changing concentration in disease states. The half-life of the intact PTH molecule in normal human plasma is

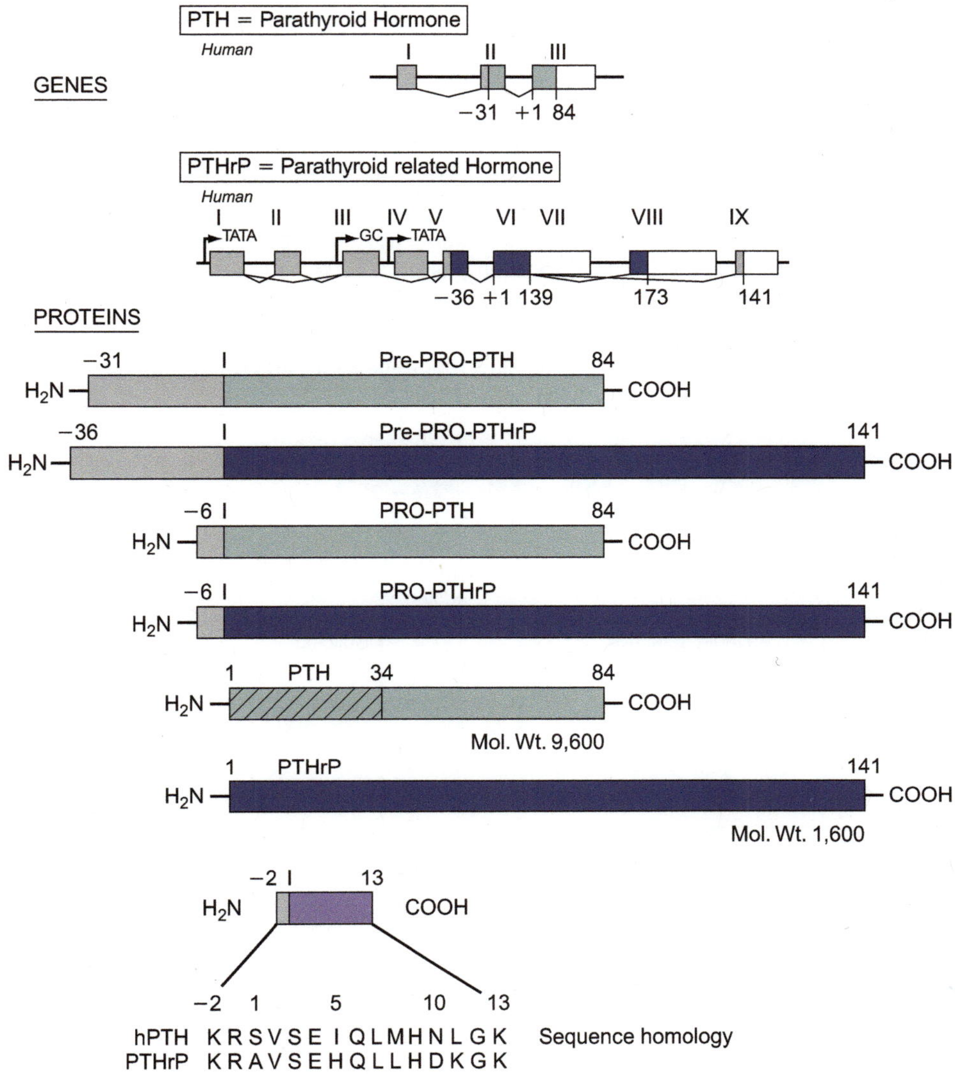

FIGURE 9.11 Comparison of the gene intron/exon organization and the protein size and amino acid length of Prepro-, Pro- for the secreted PTH and PTHrP. The genes for PTH and PTHrP are, respectively, found on chromosome 11p and 12p. In comparison to the gene for PTHrP which has nine exons, the PTH gene has only 3 exons. The secreted and mature forms of PTH and PTHrP each have their own color. The secreted PTH has 84 amino acids, but, as indicated by cross-hatching, all the biological activity resides in the first 34 amino acids. Both the PTH and PTHrP bind to the same G protein−coupled receptor (see Fig. 9.17) to generate their diverse biological responses. The secreted PTHrP has 141 amino acids; after residue 13 there is no sequence homology with PTH. However, the first 13 amino acids of PTH and PTHrP are virtually identical (see bottom of figure). *PTH*, Parathyroid hormone; *PTHrP*, parathyroid hormone−related protein.

only 20 minutes. Several predominant C-terminal peptide species have been identified in human plasma. Conflicting views are held on the importance of these peripheral forms of circulating PTH. One view is that only the 9500 molecular weight material is biologically active, with the smaller species representing degradation products. The other view is that one of the smaller species may have biological importance, which is not yet understood.

C Parathyroid hormone−related protein

PTHrP was originally purified and an N-terminal amino sequence was obtained from human tumors associated with humoral hypercalcemia of malignancy (HHM). This information permitted its molecular cloning and study of the total amino acid sequence. PTHrP is a 141-amino-acid protein; a comparison of the structures of PTH and PTHrP is presented in Fig. 9.11. Of the first 13 amino acids, 8 are identical and 3 more represent conservative amino acid substitutions. The close structural homology of PTHrP with PTH over the first 13 amino acids allows PTHrP to mimic many of the biological actions of PTH (see later discussion).

D Calcitonin

CT is a small polypeptide hormone secreted by the ultimobranchial gland in fish, amphibians, reptiles, and birds; in mammals it is secreted by the specialized C cells that are found primarily in the thyroid gland. The amino acid sequences of nine CTs have been determined; these include five mammalian species—porcine, bovine, ovine, human, and rat—as well as four nonmammalian species—salmon I, II, and III, and eel. All have a similar structure consisting of a straight chain peptide of 32 amino acids, with a seven-membered disulfide ring at the N-terminus and a prolinamide residue at the C-terminus.

Compared to the mammalian CTs, the nonmammalian hormones are much more stable and have a potency 10–50 times greater.

Studies of the molecular biology of the biosynthesis of CT mRNA have demonstrated that the CT gene can support the production of multiple protein products from a single transcription unit (see Fig. 9.12). Thus the same gene produces, as a consequence of alternative processing of RNA transcripts, the hormone CT in the thyroid and a protein termed the CT gene–related peptide (CGRP) in the hypothalamus. CGRP, thus, represents a hormone that was discovered through the application of recombinant DNA methodology without prior knowledge of its existence or its structure or biological function. CGRP is involved with nociception (the ability to perceive noxious stimuli), ingestive behavior, and modulation of the autonomic and nervous systems. Thus it can be argued that CGRP functions as a neurotransmitter and vasodilator more than as a hormone. CGRP has only modest effects on calcium homeostasis but has a wide range of actions as a neuropeptide. They include suppression of appetite and associated gastric acid secretion, modulation of body temperature, potent dilation of vascular beds, and other smooth muscle, chronotropic actions in the heart (increase heart rate) and functions as a paracrine regulator of pituitary hormone release.

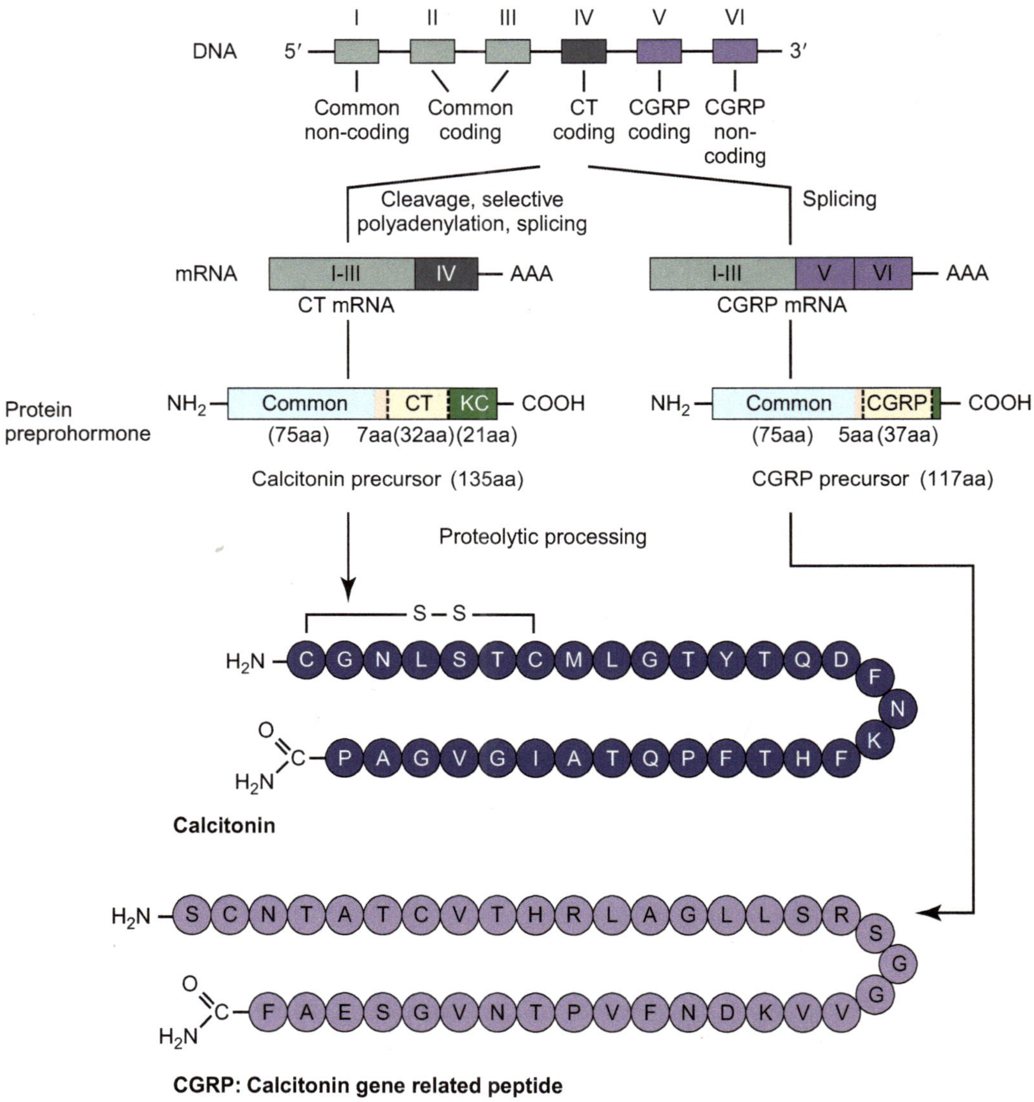

FIGURE 9.12 Model for the production of CT and CGRP via the alternative RNA processing pathways utilized in the expression of the calcitonin gene. The calcitonin gene supports the production of CT in the thyroid and of CGRP in the hypothalamus. Mature CT, which has 32-amino-acid residues, and mature CGRP, which has 37-amino-acid residues, are derived from different precursor proteins. However, these two precursor proteins have an identical "common region" of 75-amino-acid residues, which is derived from the C_1 and C_2 exons of the gene. *CGRP*, Calcitonin gene–related peptide; *CT*, calcitonin. *Reproduced by permission of M. G. Rosenfeld. Nature (London)304 (1983) 129–135. Copyright ©1983. Macmillan Journals Limited.*

The discovery that alternative mRNA processing can produce multiple protein products provides insight into the mechanisms by which endocrine genes increase the diversity and specificity of their biological responses. CT may be measured by immunoassay. The immunoassay for human CT is important in the detection and treatment of medullary carcinoma of the thyroid.

E Fibroblast growth factor 23

FGF23 was proven in 2004 to be an essential regulator of phosphate homeostasis. FGF23 is one of at least 22 known proteins that comprise the family of fibroblast growth factors; there is no FGF19. This family of proteins displays amino acid sequence and structural similarities. The molecular weight of FGF23 is ~31 kDa and it has 251 amino acids; its amino acid sequence of residues 25–176 is very similar to the other members of the FGF family of proteins. The mode of action of FGF23 will be described under the heading of biology and molecular actions.

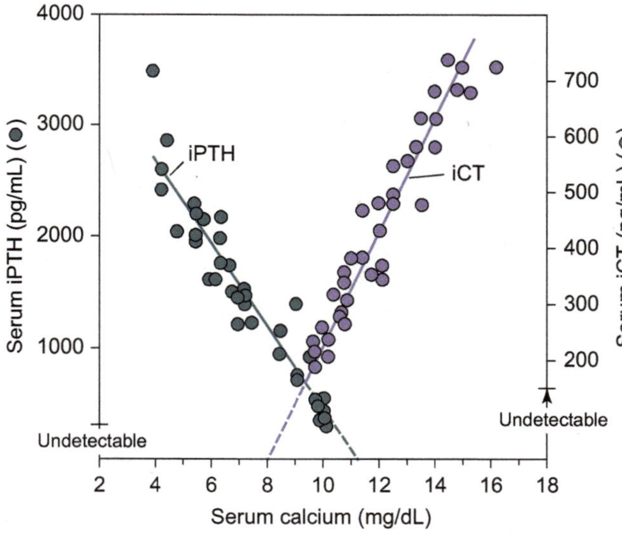

FIGURE 9.13 Changes in plasma levels of iPTH and iCT as a function of plasma total calcium. The data were obtained in pigs given EDTA to decrease plasma calcium or given calcium infusions to increase plasma calcium. Note that, as serum calcium increases, iPTH falls and serum iCT increases; as serum calcium decreases the reverse occurs. *EDTA*, Ethylenediaminetetraacetic acid; *iCT*, immunoreactive calcitonin; *iPTH*, immunoreactive parathyroid hormone. *Reproduced with permission of C. D. Arnaud et al., in: S. Taylor (Ed.), Calcitonin: Proceedings of the Second International Symposium, Heinemann, London, 1970, p. 236.*

IV Biology and molecular actions

A Parathyroid hormone receptor and biological actions

The major target organs for PTH actions are kidney and bone. PTH may also act upon the intestine, but its effects here are indirect and are achieved by virtue of the tropic actions of PTH in stimulating the renal biosynthesis of $1\alpha,25(OH)_2D_3$.

The secretion of PTH is stimulated in response to a lowered blood concentration of calcium (see Fig. 9.13). In terms of the several forms of blood calcium summarized in Table 9.2, it is known that the biosynthesis and secretion of PTH are only responsive to the ionic and not to the protein-bound forms of calcium. The secretion of PTH can be stimulated when hypocalcemia is induced by the infusion of CT or the divalent metal-chelating agent, ethylenediaminetetraacetic acid, or decreased when hypercalcemia is induced by the infusion of calcium. Thus there is an inverse correlation between serum calcium concentration and PTH concentration in the range of 4–10 mg of calcium/100 mL (Fig. 9.13). The most stringent control of serum calcium concentration is achieved in the range of 9–10.5 mg of Ca^{2+}/100 mL of serum, which is considered to be the normal physiological range of this divalent cation. Serum concentrations of calcium that fall above or below this range are likely to be due to the presence of disease states that perturb the integrated calcium–phosphorus homeostatic mechanisms.

A key signal transduction question is related to the identification of the mechanism(s) by which the plasma membrane of a parathyroid cell can sense a fall or rise in the ambient extracellular ionized calcium concentrations and signal either the secretion of PTH or inhibition of PTH secretion. A G protein–coupled Ca^{2+}-sensing receptor (CaR) from bovine parathyroid tissue and then a human adenoma were cloned and sequenced in 1993 and 1995, respectively. The human CaR is a 120-kDa protein (1078 amino acids) and features a large extracellular domain (612 amino acids) with clusters of acidic amino acid residues likely involved in Ca^{2+} binding. The seven membrane–spanning domain encompasses 249 amino acids followed by the 216 intracellular amino acid sequence that translates the activation of the signal transduction response into the release of a second messenger. The CaR is present as a homodimer. Figs. 9.14 and 9.15 show details of the structure of homodimeric CaR in inactive and active conformations. CaR is expressed in the parathyroid cell (and possibly in parathyroid oxyphils), in thyroid gland C cells and in the kidney. It controls the synthesis and secretion of PTH, and CT which reduces the blood level of Ca^{2+}. In addition to the main activation of CaR by Ca^{2+}, it also can be activated by Mg^{2+}, by amino acids and by polyamines. The CaR senses the ionized extracellular Ca^{2+} concentration and generates an intracellular signal transduction activation inside the cell of several phospholipases (C, A_2 and D), or mitogen-activated protein kinase or inhibition of adenylate kinase; despite major efforts, the details of the second messenger generation remain elusive.

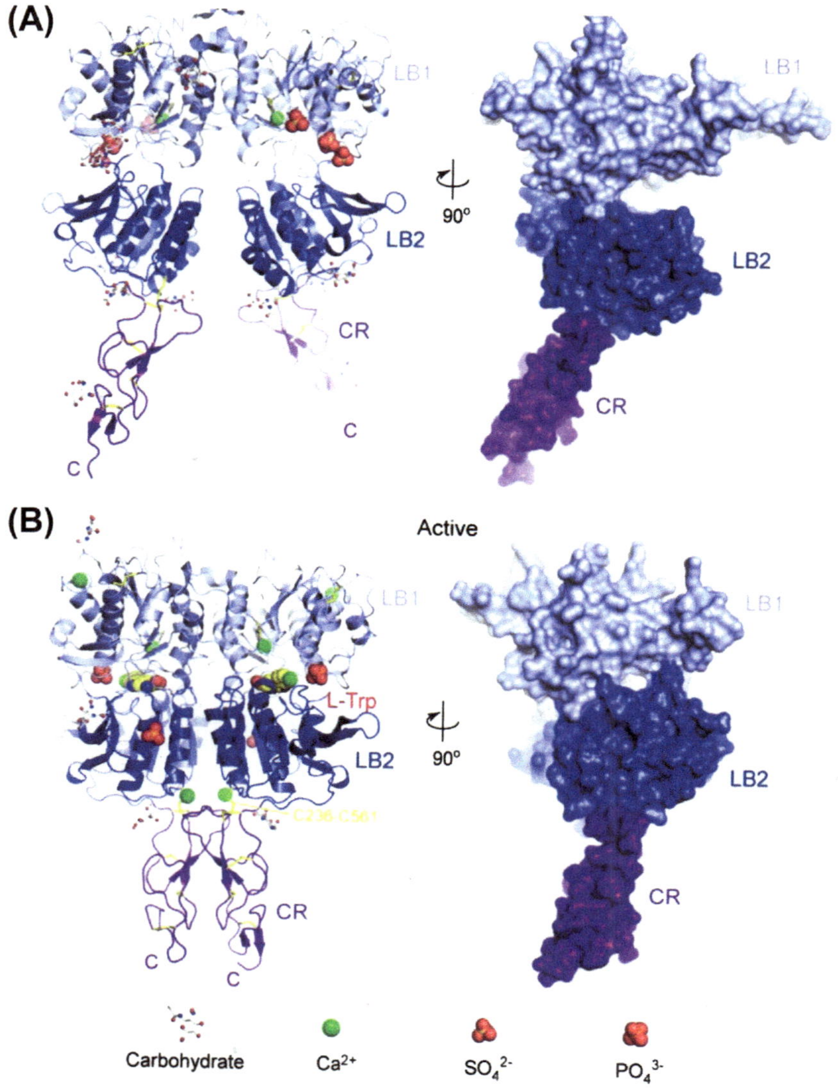

FIGURE 9.14 Crystal structures of human CaSR (CaR) ECD. (A) Inactive-state structure of CaSR ECD homodimer in the presence of 2 mM Ca^{2+}. (B) Active-state structure of CaSR ECD homodimer in the presence of 10-mM Ca^{2+} and 10-mM L-Trp. Each structure is shown in cartoon (front view) representation that are related by a 90-degree rotation about the vertical axis. Each protomer is colored according to its individual domains (LB1, *light blue*; LB2, *blue*; and CR, *purple*). The various ligands (L-Trp, Ca^{2+}, PO_4^{3-}, and SO_4^{2-}) are displayed as space-filling models. Observed carbohydrates are shown as ball-and-stick models in gray. Disulfide bridges are in yellow. *CaR*, Ca^{2+}-sensing receptor; *ECD*, extracellular domain; *LB*, ligand-binding. *Reproduced from Fig. 1 of Y. Geng et al., Structural mechanism of ligand activation in human calcium-sensing receptor, eLife, 5 (2016) e13662; https://doi.org/10.7554/eLife.13662*

The secretion of PTH by the parathyroid gland cells is regulated by changes in the stability of the PTH mRNA. The CaR present in the plasma membrane of a PTH secretory cell can differentiate between hypocalcemia and hypercalcemia and send an appropriate second messenger to increase the stability of the PTH mRNA (hypocalcemia) or decrease the stability of the PTH mRNA (hypercalcemia); see Fig. 9.16. The mediator of changes in the stability of the PTH mRNA is Pin1, a peptidyl-*cis-trans* isomerase. Pin1 responds to the incoming second messenger from the CaR and generates one of two possible changes in the properties of the regulatory K homology-type splicing regulatory protein (KSRP) protein, which is a PTH mRNA-binding protein: (1) in the instance of elevated serum Ca^{2+}, KSRP becomes dephosphorylated which enables the Pin1 isomerase to mediate a conformational change in the KSRP protein, allowing it to bind to the PTH mRNA nucleotide adenine- and uridine-rich elements motif and thereby reducing the stability of the PTH mRNA; this lowers the concentration of PTH mRNA and, therefore, reduces the secretion of PTH; (2) in the instance that KSRP becomes phosphorylated, which blocks its ability to bind to the PTH mRNA thereby increasing the stability of the PTH mRNA, an increase in the concentration of PTH mRNA is caused which increases the secretion of PTH. Further details are provided in the legend to Fig. 9.16.

All of the main biological actions of PTH are mediated by its interaction with the PTH receptor (see Figs. 9.17 and 9.18). In this figure the cryo-electron microscopic structure of the PTH receptor (PTHR) bound with long-acting PTH and in complex with the G protein, G_s, is shown. The lipid layer around the PTHR also is shown, which collectively increases the Ca^{2+} concentration of the blood compartment. The most important biological actions of PTH are the following: (1) to increase

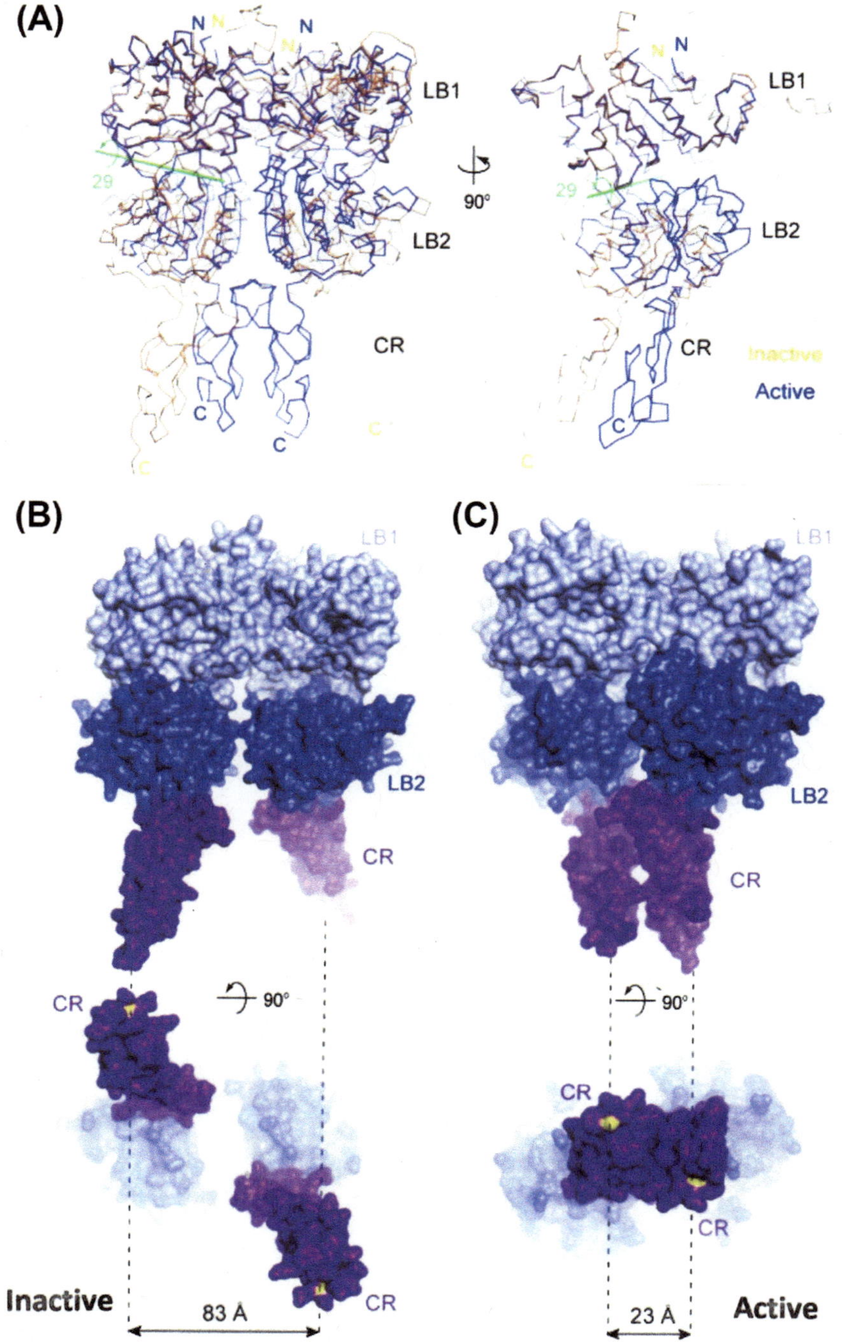

FIGURE 9.15 Agonist-induced conformational changes. (A) Superposition of the inactive (*orange*) and active (*blue*) CaSR ECD structures based on the LB1 (ligand-binding 1) domain of one protomer (smallest structural unit of an oligomeric protein) (front view, left; side view, right). Green line is the axis of rotation that relates the LB23 domains of superimposed protomers (rotation x=29 degrees, screw translation τ_x=2.2 Å). (B and C) Surface representation of the inactive (B) and active (C) structures in front (top) and bottom (bottom) views. Distance between C-termini of the two subunits (*yellow*) is marked by a dashed line for each homodimer. *ECD*, Extracellular domain; *LB*, ligand-binding. *Reproduced from Fig. 2 of Y. Geng et al., Structural mechanism of ligand activation in human calcium-sensing receptor, eLife, 5 (2016) e13662; doi: 10.554/eLife.13662.*

the rate of conversion of 25(OH)D_3 to 1α,25$(OH)_2D_3$ in kidney tissue (see Fig. 9.10) and, thus, increase the serum concentration of 1α,25$(OH)_2D_3$ which will increase intestinal Ca^{2+} absorption; (2) to increase the plasma Ca^{2+} concentration by the action of the proximal convoluted tubule via stimulating the reabsorption of Ca^{2+} from the kidney glomerular filtrate; (3) to increase the number of osteoclasts (see later Fig. 9.28) and, thus, the extent of osteoclastic and osteocytic osteolysis in bone (bone resorption and remodeling); and (4) to increase the urinary excretion of phosphate by inhibiting the renal tubular reabsorption of phosphate. Further details are provided in the legend of Fig. 9.16.

The opossum kidney cell PTH receptor has been cloned, sequenced, and found to be a member of a distinct family of G protein–coupled receptors with seven transmembrane–spanning domains. The mature receptor protein contains 585 amino acids. The PTH receptor (PTH) and the PTHrP have been found to be the same protein (Fig. 9.17). The amino acid sequence of PTH and PTHrP

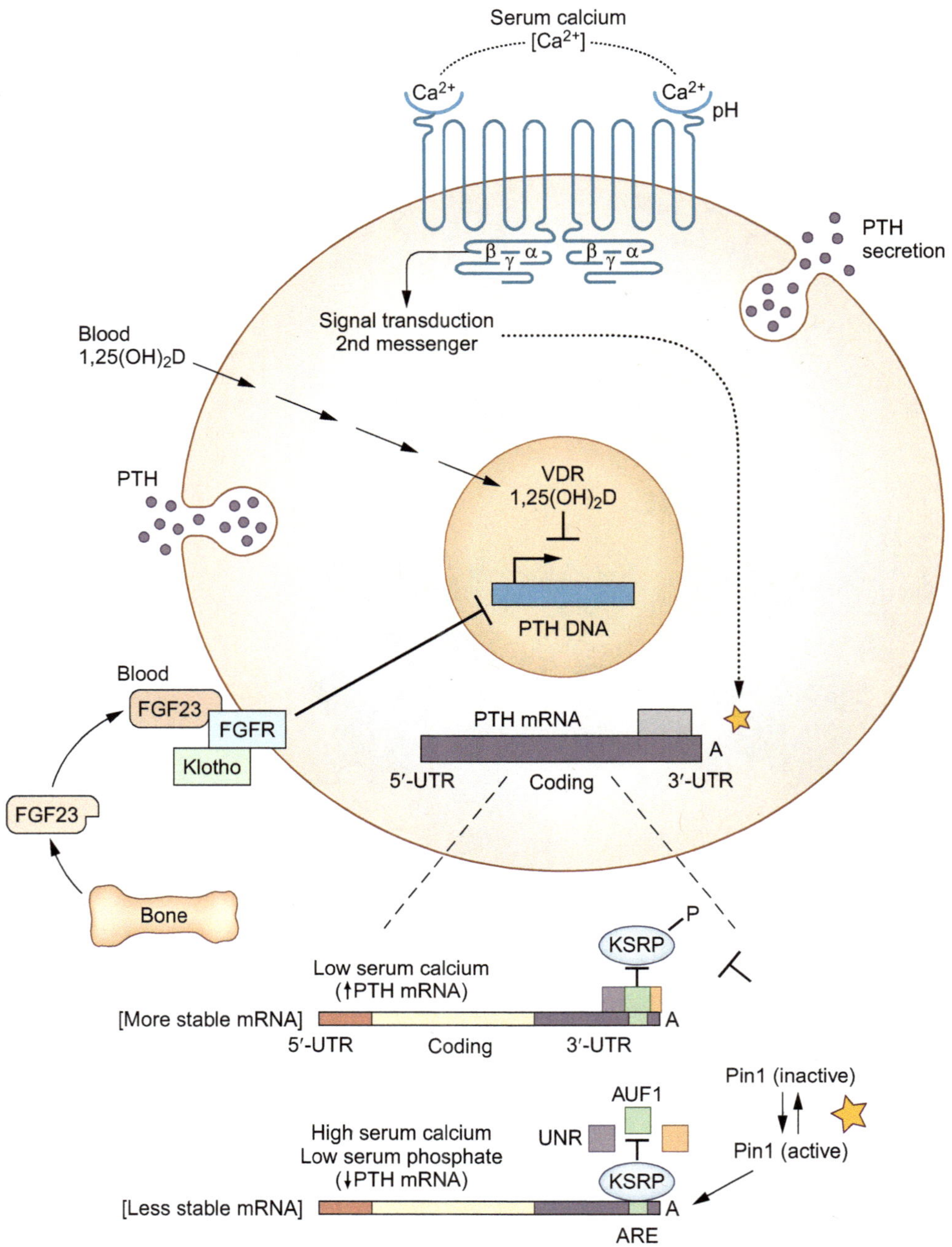

FIGURE 9.16 Regulation of PTH secretion via changes in stability of the PTH mRNA. The plasma membrane Ca^{2+} receptor of the PTH secreting cell senses changes in the serum Ca^{2+} level through a seven-transmembrane G protein linked to phospholipases that sends a second messenger to the cellular site of the regulation of the PTH mRNA concentration (see the two *gold stars*). Changes in the rate of secretion of PTH, PTh secreting cells are mediated by changing the stability of the PTH mRNA. In the inset below the cell, there are two schematic diagrams of the PTH mRNA from 5′-UTR to 3′-UTR. The *More stable* mRNA schematic illustrates the circumstance of hypocalcemia, resulting in an increase in the stability of PTH mRNA and ultimately greater PTH secretion. The *Less stable* mRNA schematic illustrates the circumstance of hypercalcemia (with associated low serum phosphate) resulting in the reduction of PTH mRNA stability and ultimately lower PTH secretion. Two key regulatory proteins (UNR and AUF1) bind to the 3′UTR of the PTH mRNA stabilizing PTH mRNA levels necessary to increase PTH secretion. In contrast the K homology-type splicing regulatory protein (KSRP) also binds to the PTH mRNA 3′-UTR, specifically to the ARE which is a conserved 26-nucleotide sequence that decreases PTH mRNA stability and as a consequence reduces PTH secretion. When KSRP is phosphorylated on serine-181, it cannot bind to the 3′-UTR ARE region (*More stable* mRNA) and there is no reduction of PTH mRNA stability and accordingly PTH secretion is increased. But when KSRP is not phosphorylated, it can bind (*Less stable* mRNA) to the PTH mRNA 3′-UTR ARE region, thus decreasing PTH mRNA stability and thereby reducing the secretion of PTH. Pin1 is a peptidyl-*cis-trans* isomerase that specifically binds to the unphosphorylated Ser/Thr-proprotein motif of KSRP. This catalyzes the *cis/trans* isomerization of the KSRP proline peptide bonds, thus causing a conformational change in KSRP and increasing the biological activity of KSRP so that it can bind to the ARE nucleotide sequence of the PTH mRNA, which then results in a decrease in both the stability of the PTH mRNA and the secretion of PTH. The PTH secreting cell also has receptors for both 1α,25$(OH)_2D_3$ (produced by the kidney) and FGF-23 (produced by bone). Both hormones downregulate PTH gene transcription, thus lowering PTH production and secretion. *3′UTR*, 3′ Untranslated region; *ARE*, adenine- and uridine-rich elements; *KSRP*, K homology-type splicing regulatory protein; *mRNA*, messenger RNA; *PTH*, parathyroid hormone.

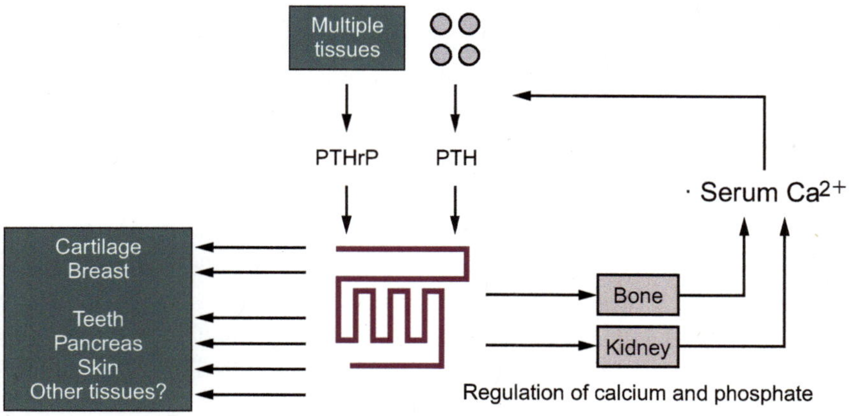

FIGURE 9.17 Both the PTH and PTHrP are ligands for the same receptor yet produce diverse biological responses. PTH (but not PTHrP) acts as a classic hormone and binds to its receptor (lavender boxes), in bone osteoblasts and kidney cells, while PTHrP (but not PTH) acting as an autocrine/paracrine messenger binds to its receptor (*blue box*) in cartilage, teeth, breast, skin, and pancreas cells. The PTH/PTHrP receptor is a G protein–coupled receptor. When either PTH or PTHrP are bound to their receptor (shown in magenta as a seven-membrane receptor), there are two main pathways of response: (1) activation of adenylyl cyclase via $G_s\alpha$ leading to cAMP and activation of PKA; or (2) activation of PLC to produce IP3+DAG that leads to an increase in intracellular Ca^{2+} concentrations and activation of PKC, respectively. *cAMP*, Cyclic antimicrobial peptide; *DAG*, diacylglycerol; *PKA*, protein kinase A; *PKC*, protein kinase C; *PLC*, phospholipase C; *PTH*, parathyroid hormone; *PTHrP*, parathyroid hormone–related protein.

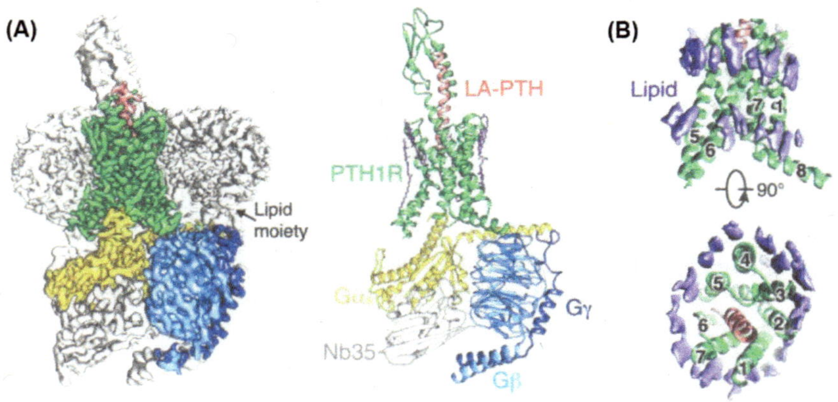

FIGURE 9.18 Cryo-EM structure of the LA-PTH-bound PTHR in complex with G_s. (A) (left) Cut-through view of the cryo-EM density map that illustrates the LA-PTH–LAPTHR–Gs complex and the disk-shaped micelle. The unsharpened cryo-EM density map at the 0.01 threshold shown as light gray surface indicates a micelle diameter of 12 nm. The colored cryo-EM density map at the 0.01 threshold shown as light gray surface indicates a micelle diameter of 12 nm. The colored cryo-EM density map is shown at 0.026 threshold. (Right) Cartoon representation of the LA-PTH–PTHR–G_s complex is shown with annular lipids in purple stick representation. Green, PTHR; orange, LA-PTH; gold, Gs Ras–like domain; light blue, Gβ; medium blue, Gγ; gray, Nb35. (B) Cryo-EM density of the ordered annular lipid layer around the receptor TMD shown in purple; numbers 1 through 7 represent TM1–TM7, receptor ECD and G protein are omitted. *ECD*, Extracellular domain; *LA-PTH*, long-acting parathyroid hormone; *PTHR*, parathyroid hormone receptor. *Reproduced from Fig. 1 of L.-H. Zhao et al. Structure and dynamics of the active human parathyroid hormone receptor-1, Science 364 (2019) 148–153.*

are identical over the first 13 residues (see Fig. 9.11), which is expressed in both kidney and osteoblast cells, which are prime targets for PTH action, and also in aorta, brain, heart, ileum, liver, placenta, skin, uterus, and testes.

Both PTH and PTHrP bind with equal affinity to the cloned and expressed receptor, and both ligands equivalently stimulate adenylyl cyclase (see Fig. 9.17). Intriguingly, there is close structural homology between the PTH receptor, the PTHrP receptor, the CT receptor, and the "secretin" subfamily of closely related receptors. Other family members include secretin, vasoactive intestinal polypeptide, and glucagon receptors. For all of these proteins, receptor occupancy by its cognate ligand results in the activation of adenylyl cyclase and, in some instances, increases in the concentration of intracellular Ca^{2+}.

The actions of PTH on bone are complex and continue to be an area of intense investigation. The response of bone to PTH is biphasic; the immediate action is largely that of bone mineral mobilization (i.e., an elevation of the blood levels of both calcium and phosphate). These effects may be detected within minutes following hormone administration. A second and slower action of PTH is its effect upon bone cell activity.

PTH has been shown to increase the number and size of the bone-resorbing osteoclasts (again, see later Fig. 9.28). Although PTH is a potent bone-resorbing agent, receptors for PTH are not found on osteoclasts and are only present on osteoblasts. Also associated with prolonged bone resorption is an increased release of lysosomal enzymes by the osteoclasts, so that there is a breakdown and solubilization of the bone matrix. This generates two consequences: (1) removal of the proteolyzed bone matrix prepares the bone pit/cavity for replacement with both new bone matrix and calcium hydroxyapatite; and (2) the proteolyzed matrix is further broken down to small peptides that eventually increase the blood concentration of hydroxyproline. Elevated levels of blood and urine hydroxyproline concentrations are used as a marker for excessive bone resorption.

B Parathyroid hormone–related protein receptor and biological actions

The existence of PTHrP was first encountered as a consequence of the disease of HHM. In HHM, patients have significantly elevated serum Ca^{2+} levels and a tumor that secretes PTHrP. Surprisingly, the same gene and receptor protein services both PTH and PTHrP. Thus as emphasized in Fig. 9.17, PTHrP actually has a wider sphere of actions than does PTH. Although PTH and PTHrP are quite different proteins (see Fig. 9.11), because of their high structural homology over the first 13 amino acids, both hormones, utilizing the second messenger, cyclic antimicrobial peptide (cAMP), can generate PTH-like biological responses in bone and kidney.

PTHrP is known to be synthesized in several tissues (growth plate of bone, placenta, breast, skin, hair, and teeth) with most of its actions and biological responses being of a paracrine nature. PTHrP is involved with Ca^{2+} transport associated with pregnancy and lactation. Peptide fragments of PTHrP are able to stimulate Ca^{2+} transfer across the placenta to the fetus. In lactating mice the secretion of PTHrP from the breast results in an increase of bone resorption which increases the serum Ca^{2+} concentration. This Ca^{2+} increase signals to the breast's calcium receptor (CaR) which then sends a second messenger to breast cells that results in an increase in the movement of Ca^{2+} into milk and at the same time a downregulation of PTHrP secretion.

C Calcitonin receptor and biological actions

The dominant biological action of CT is to mediate a lowering of serum calcium levels. CT is secreted in response to elevated blood levels of ionized calcium. CT is secreted by mammals, and fish and birds. The rate of CT secretion is a direct function of the plasma calcium concentration (Fig. 9.13). There is some evidence in rats and pigs that gastrin can function as a secretagogue for CT secretion; however, there is no evidence of this response in humans.

In addition to the onset of hypocalcemia, there is also normally an accompanying hypophosphatemia after the administration of CT. Also, in experimental animals, the blood level of CT is elevated in pregnancy and lactation.

Three important biological functions for CT have been proposed: (1) protection of the young animal or newborn against postprandial hypercalcemia; (2) blocking of the actions of PTH in mobilizing bone calcium and phosphorus; and (3) stimulation of the urinary excretion of both calcium and phosphate in the kidney. The net effect of these three actions is to mediate a reduction in serum calcium levels.

The biological effects of CT are mediated as a consequence of the interaction of CT with its receptor present in the outer membrane of target cells of both skeletal and renal tissues and other target organs. The human CT receptor has been cloned and sequenced and found to belong to the class II (family B) subclass of G protein receptors. This receptor subfamily also includes PTH, PTHrP, secretin, and glucagon G protein–coupled receptors with seven transmembrane–spanning domains. The mature CT receptor comprises 490 amino acids. The activated CT receptor stimulates adenylate cyclase and production of cAMP in many target organs, including the kidney. However, in human osteoclasts, CT stimulates the activation of protein kinase C that mediates the known action of CT to inhibit osteoclast-mediated bone resorption. The CT receptor is known to be expressed in kidney cells, osteoclasts, placenta, testis and spermatozoa, lung, and pituitary. The functional significance of these various isoforms of the CT receptor remains to be established.

D Vitamin D receptor and biological actions

The vitamin D molecule itself is biologically inert; it has no intrinsic biological activity. All biological responses attributed to vitamin D are now known to arise only as a consequence of the metabolism of this secosteroid into its biologically active daughter metabolites, namely, the steroid hormone $1\alpha,25(OH)_2D_3$ and the metabolite 24,25 $(OH)_2D_3$ (see Fig. 9.10).

1 The vitamin D receptor endocrine system for calcium homeostasis, vitamin D metabolism, and the vitamin D–binding protein (DBP)

Fig. 9.19 summarizes that portion of the vitamin D endocrine system concerned with calcium homeostasis. The steroid hormone $1\alpha,25(OH)_2D_3$ is produced only in accord with strict physiological signals dictated by the calcium "demand" of the organism; a bimodal mode of regulation has been suggested. On a time scale of

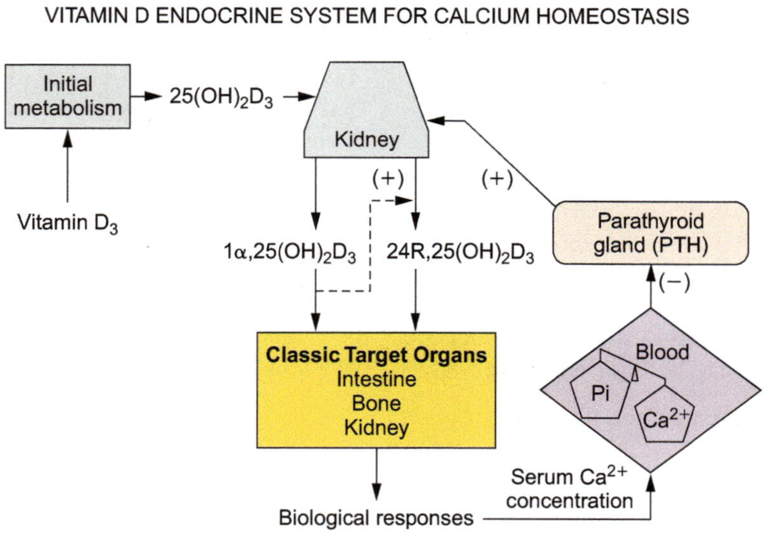

FIGURE 9.19 A schematic of the vitamin D endocrine system governing calcium homeostasis. The process of calcium homeostasis describes the mechanisms by which the serum Ca^{2+} concentration is maintained ~constant at 9.5–10.5 mg/100 mL of serum via the integrated actions of the intestine, bone, and kidney; all three of these organs can contribute Ca^{2+} to the serum compartment based on the actions of three hormones, PTH, 1α,25$(OH)_2D_3$, and FGF23. 1α,25$(OH)_2D_3$ is responsible for stimulating intestinal Ca^{2+} absorption in accordance with the circulating level of 1α,25$(OH)_2D_3$. If the dietary intake of Ca^{2+} is inappropriately low, then Ca^{2+} contribution from the intestine is not adequate and PTH will, therefore, stimulate the bone to solubilize Ca^{2+} and HPO_4^-. At the same time, PTH will interact with its kidney receptor and increase the renal tubular reabsorption of Ca^{2+} and at the same time diminish the renal tubular reabsorption of phosphate (i.e., increase the urinary concentration of phosphate). Also the kidney functions as an endocrine gland in that it enzymatically produces the steroid hormone 1α,25$(OH)_2D_3$ according to the stimulatory actions of PTH on the kidney proximal tubule's 25$(OH)_2D_3$ − 1α-hydroxylase in accordance with the magnitude of the need to elevate the prevailing level of serum Ca^{2+}. The secretion of PTH is governed by the Ca^{2+}receptor of the parathyroid gland which monitors the serum Ca^{2+} concentration (see Fig. 9.16). Finally bone osteocytes secrete the hormone FGF23 when serum Ca^{2+} is inappropriately low. The FGF23 binds to its receptor in the kidney and inhibits the renal tubular reabsorption of phosphate (i.e., increases the urinary concentration of phosphate). *FGF23*, Fibroblast growth factor 23; *PTH*, parathyroid hormone.

minutes, changes in the ionic environment of the kidney mitochondria resulting from the accumulation and release of calcium and/or inorganic phosphate may alter the enzymatic activity of the 25$(OH)D_3$ − 1α-hydroxylase. In addition, PTH has been shown, on a time scale of hours, to be capable of stimulating the production of 1α,25 $(OH)_2D_3$, possibly by stimulating the biosynthesis of the 1α-hydroxylase. It is also intriguing that 1α,25$(OH)_2D_3$ is a stimulant for the renal mitochondrial production of 24,25$(OH)D_3$ (see following paragraph). Thus under normal physiological circumstances, both renal dihydroxylated metabolites are secreted and are circulating in the plasma. There is also evidence of a "short feedback loop" for 1α,25$(OH)_2D_3$ to modulate and/or reduce the secretion of PTH (see Fig. 9.16). The half-life of both 1α,25 $(OH)_2D_3$ and 24 R,25$(OH)_2D_3$ is 6–8 hours. Thus the kidney is clearly an endocrine gland, in the classic sense, in that it has the responsibility, in a physiologically regulated manner, to produce appropriate amounts of the steroid hormone, 1α,25$(OH)_2D_3$.

The biological role of 24 R,25$(OH)_2D_3$ is still under discussion. There is evidence that 24 R,25$(OH)_2D_3$ can play a role in fracture healing, and there is preliminary evidence of a receptor for 24 R,25$(OH)_2D_3$. The enzyme that converts 25$(OH)D_3$ to 24 R,25$(OH)_2D_3$, namely, the cytochrome P450–24A1 hydroxylase (CYP24A1) also converts 1α,25$(OH)_2D_3$ to 1α,24 R,25$(OH)_3D_3$ which represents the first step of catabolism of 1α,25$(OH)_2D_3$. Most target cells for 1α,25$(OH)_2D_3$ which possess the vitamin D receptor (VDR) also have low concentrations of CYP24A1; this ensures a short half-life for 1α,25$(OH)_2D_3$.

The plasma compartment contains the DBP that is utilized to transport vitamin D secosterols. DBP is similar in function to the corticosteroid-binding globulin, which carries glucocorticoids (see Chapter 10: Adrenal Corticoids), and the steroid hormone–binding globulin, which transports estrogens or androgens (see Chapter 12: Androgens). DBP is a slightly acidic (pH=5.2) monomeric glycoprotein of 53,000 Da, which is synthesized and secreted by the liver as a major plasma constituent.

DBP is a multifunctional protein in that it binds both vitamin D and its metabolites and also monomers of the protein actin, thereby preventing their polymerization in the blood compartment. One molecule of DBP has only one ligand-binding domain-binding site for secosterols of the vitamin D family. Thus while a single DBP can carry only one ligand, the relatively high concentration of DBP molecules present in the blood compartment permits the DBP population as a whole to bind the hydrophobic parent, vitamin D_3, as well as the hydrophobic daughter metabolites, 25 $(OH)D_3$, 1α,25$(OH)_2D_3$, and 24,25$(OH)D_3$. Since the total plasma concentration of vitamin D sterols is only ~0.2 μM, while DBP circulates at 9–13 μM, under normal circumstances only a very small proportion of the sterol-binding sites on DBP are occupied.

2 Expansion of the vitamin D endocrine system

The principal mode of action of 1α,25$(OH)_2D_3$ occurs by a steroid hormone receptor-mediated mechanism (sees Fig. 1.19 and 1.21). Definitive biochemical evidence supports the existence of a receptor for 1α,25$(OH)_2D_3$ (VDR) in at least 30 different tissues. While it is not surprising to find 1α,25$(OH)_2D_3$ VDR in the classic target organs of intestine, kidney, and bone linked to calcium homeostasis (Fig. 9.19), the presence of the VDR in such diverse tissues as the parathyroid gland, pancreas, heart and cardiovascular system (arteries+veins), cells of the immune system (B and T lymphocytes), muscle, spermatozoa, and brain (Fig. 9.20) emphasizes the wide physiological responsibility of the vitamin D endocrine system and the pleiotropic actions of 1α,25$(OH)_2D_3$. While the kidney functions as the major source of plasma 1α,25 $(OH)_2D_3$, approximately 12 cell types (e.g., macrophages, pancreas, colon, keratinocytes, prostate, spermatazoa, placenta, and brain) have low levels of the 25$(OH)D_3$ – 1α-

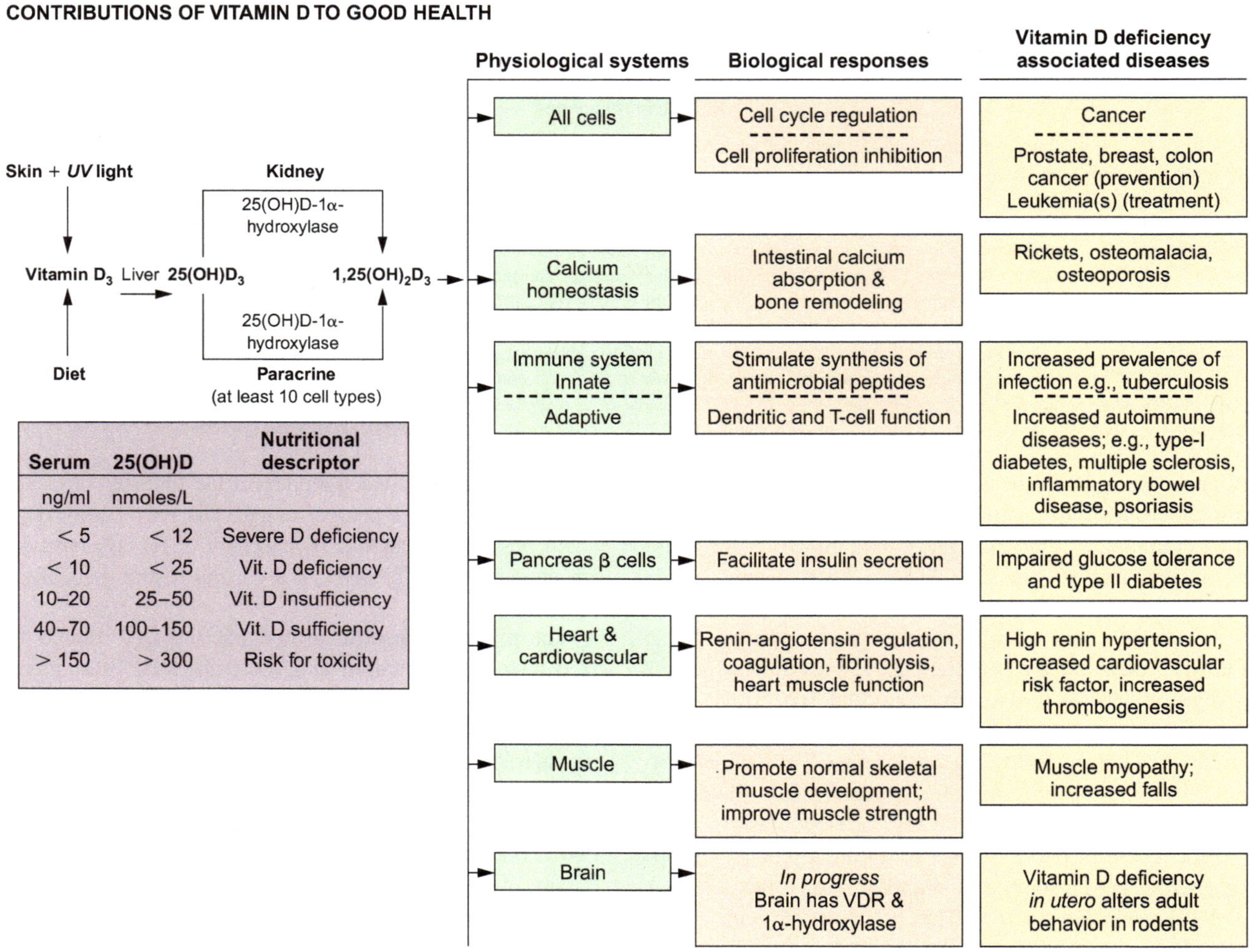

FIGURE 9.20 A summary of the contributions of the vitamin D endocrine system to good health. 1α,25$(OH)_2D_3$, enzymatically produced by either the kidney or paracrine 25$(OH)D_3$ – 1α-hydroxylase, is carried throughout the blood compartment by the DBP. The 1α,25$(OH)_2D_3$ enters any of the some 30 cell types that are known to contain the VDR and stimulates a variety of genomic and rapid responses that can contribute to good health. These responder cell types are organized, in the left-hand column (*light green*), into five primary vitamin D–dependent physiological systems. The 1α,25$(OH)_2D_3$ brain system is at the present time only studied in rodent systems. The center column tabulates for each of the five major vitamin physiological systems the nature of their biological responses. The right-hand column lists known examples of vitamin D deficiency-related diseases that are "associated" with each physiological system. In general, only for the calcium homeostasis physiological system has randomized clinical trials (with humans) been carried out showing that correction of the vitamin D deficiency by increased vitamin D_3 intake can minimize the disease studied. The blood levels of 25(OH)D have been shown to be correlated with varying extents of vitamin D deficiency. The inset table (bottom left) summarizes, for humans, the changes in the circulating levels of 25(OH)D that are linked to varying degrees of vitamin D deficiency (from severe to insufficient to sufficient, etc.). From a clinical perspective, this information can be utilized to determine the recommended daily intake of vitamin D_3 to maintain good vitamin D health. *DBP*, Vitamin D–binding protein; *VDR*, vitamin D receptor.

hydroxylase (CYP2B1) and thus function as a paracrine source of 1α,$25(OH)_2D_3$.

The steroid hormone 1α,$25(OH)_2D_3$ generates biological responses via binding to either the VDR located in the cell nucleus to regulate gene transcription, or to the VDR present in the plasma membrane caveolae (VDR), which is coupled to the rapid opening of voltage-gated channels. These topics will be discussed separately.

The 1α,$25(OH)_2D_3$ receptor is a protein with a molecular weight of 50,000 Da. The cellular distribution of the VDR is 75% in the nucleus, 20% in the cell cytosol, and 3%–5% in the plasma membrane. [Fig. 1.22 shows that VDR can initiate both genomic effects when in the nucleus of the cell and rapid responses with VDR localized in caveolae at the plasma membrane. Fig. 1.21 illustrates the shapes that 1α,$25(OH)_2D_3$ can assume when binding to the VDR with a bowl shape (genomic response) or a planar shape (rapid responses). Fig. 2.8 illustrates how the VDR can "capture" the three hydroxyls of 1α,25 $(OH)_2D_3$ and form a stable ligand/receptor complex.] The VDR binds 1α,$25(OH)_2D_3$ tightly (K_d=$1-5\times10^{-10}$ M) and with great ligand specificity. The absence of the 1α- or 3β-hydroxyl group dramatically reduces interaction with the VDR by >99.9%. Also, when the eight-carbon side chain is lengthened by two carbons or shortened by one carbon, the interaction with the VDR is reduced by 75%. The parent vitamin D_3 blood concentration (~100–500 nM in vitamin D replete individuals) is present at vastly higher concentrations (~1000×) than the plasma concentration of 1α,$25(OH)_2D_3$, (120 pM). However, both vitamin D_3 and vitamin D_2 are not able to bind to the VDR.

3 1α,$25(OH)_2D_3$+VDR–mediated genomic and nongenomic biological responses

The binding of 1α,$25(OH)_2D_3$ to the VDR in the cell nucleus initiates a series of events that ultimately can result in changes in the selective transcription of ~3% of the human genome of ~20,000 genes (i.e., ~600 genes). The overall process of regulating gene transcription is complex and can result in either stimulation or repression of gene transcription. Extensive detail of the VDR structure has emerged as a consequence of X-ray crystallographic as well as mutagenesis studies. The liganded VDR interacts with the unliganded retinoid-X receptor (RXR) to form a heterodimer (VDR-RXR). The heterodimer then recruits either coactivator or corepressor proteins and other transcription factors so as to be able to recognize and interact specifically with vitamin D response elements in the promoter of the selected gene to either upregulate or downregulate its transcription into mRNA (see Figs. 1.19B and 1.22 for additional details).

Besides the VDR-mediated genomic effects, there are many examples of 1α,$25(OH)_2D_3$-mediated rapid responses that can occur in minutes to under 1 hour. These are known to occur in osteoblasts, intestinal cells, endothelial cells, sertoli cells, human spermatazoa, and pancreatic β-cells. Comparable rapid responses are known for all other steroid hormone receptors. The cellular site of initiation of rapid responses is the 1α,$25(OH)_2D_3$-liganded VDR in the caveolae-containing plasma membrane. Depending upon the cell type and circumstances, a wide variety of second messengers are known to be stimulated by the VDR-1α,25 $(OH)_2D_3$; these include opening of chloride or calcium channels, stimulation of exocytosis, and activation of second-messenger signaling pathways (e.g., activation of protein kinase C or phosphoinositide turnover).

4 1α,$25(OH)_2D_3$-dependent calcium-binding proteins

Three distinct 1α,$25(OH)_2D_3$-induced calcium-binding proteins have been isolated and biochemically characterized. These proteins are all produced de novo as a consequence of the $1,25(OH)_2D_3$ VDR receptor complex that has stimulated selective gene activation. The two proteins that tightly bind Ca^{2+} with a K_d of 10^{-7} M are termed calbindins [the 9 kDa CaBP (mammals) and the 28-kDa (birds) CaBP] and belong to a family of tight calcium-binding proteins; other members of this family include the muscle proteins troponin C and parvalbumin and the widely distributed calmodulin. All of these proteins are known to have a high degree of structural similarity in their calcium-binding sites. The third vitamin D–induced calcium-binding protein is osteocalcin; it is an osteoblast secreted extracellular protein associated with bone matrix. It is sometimes referred to as the bone Gla protein, since it has many γ-carboxyglutamic acid residues that bind Ca^{2+} with a K_d of only $\sim1\times10^{-3}$ M. Osteocalcin is the fifth most abundant protein in the body.

5 1α,$25(OH)_2D_3$ actions on intestine and kidney

The principal biological effects of vitamin D on calcium metabolism, mediated by 1α,$25(OH)_2D_3$ are as follows: in the intestine to stimulate the absorption of dietary calcium and to some extent phosphate; in the skeleton to promote both the mineralization of bone matrix and to stimulate osteoclast cell bone resorption (both calcium and phosphate); and in the kidney to reduce the urinary excretion of phosphate and calcium (i.e., to stimulate the renal tubular reabsorption of these two ions). Of these three systems the most thoroughly studied biochemically is the intestinal Ca^{2+} absorption process.

Fig. 9.21 presents a schematic model describing the process of 1α,$25(OH)_2D_3$-mediated intestinal calcium absorption. Both the 9-kDa CaBP (mammals) and the 28-kDa (birds) CaBP species are soluble proteins found exclusively in the cytosol of intestinal and kidney cells, where they constitute 1%–3% of the soluble cellular

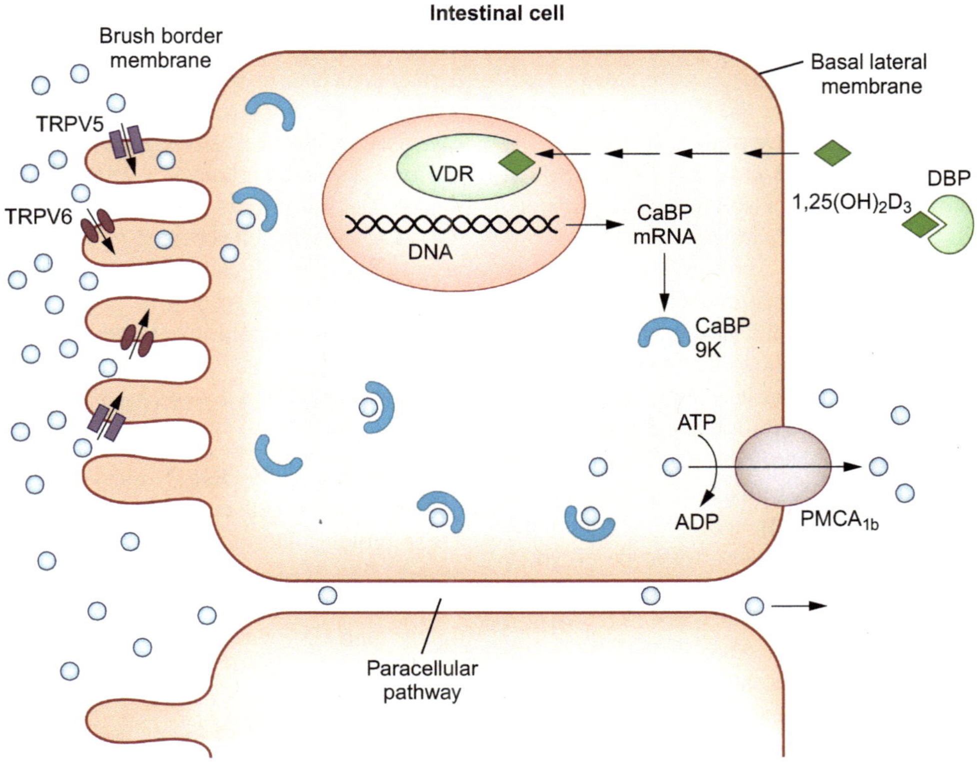

FIGURE 9.21 Schematic model describing the actions of $1\alpha,25(OH)_2D_3$ in the intestine in stimulating intestinal calcium transport. In this model the blood transport protein, DBP, delivers $1\alpha,25(OH)_2D_3$ to the basal lateral membrane of the intestinal cell, where the secosteroid crosses by diffusion and then moves to the cell nucleus and binds to the VDR. VDR with bound $1\alpha,25(OH)_2D_3$ is responsible for upregulating the genes for cytosolic calbindin D9K (mammals) or calbindin 28 K (birds) and TRPV6 (a Ca^{2+} channel located in the brush border membrane). In the intestine, TRPV5 is constitutively expressed and is functional in the absence of $1\alpha,25(OH)_2D_3$. Thus Ca^{2+} present in the lumen of the intestine (*light blue dots*) enters across the intestinal brush border via the Ca^{2+} channels. The "free" Ca^{2+} then binds tightly to the calbindin D9K or D28K (K_d is 0.1 μM) and crosses the interior of the cell to the basal lateral membrane where a calcium pump (plasma membrane Ca^{2+} ATPase or PMCA) moves the Ca^{2+} released from the CaBP across the basal lateral membrane where it enters the cytosol and then blood compartment. Thus the $1\alpha,25(OH)_2D_3$-mediated intestinal Ca^{2+} transport process contributes Ca^{2+} to elevate the serum calcium concentration. An alternate Ca^{2+} absorption pathway, the paracellular pathway, involves the movement of Ca^{2+} between adjacent intestinal cells. The paracellular pathway may provide 10%–15% of the intestinal absorption of Ca^{2+}. *DBP*, Vitamin D–binding protein; *VDR*, vitamin D receptor.

proteins. The intracellular concentration of Ca^{2+} in all cells should be maintained no higher than $\approx 10^{-7}$ M (or 0.1 μM) so that Ca^{2+} at these low concentrations can function as a second messenger. Thus when dietary Ca^{2+} is moved across the brush border through the TRPV5 and TRPV6 calcium channels and enters the cytoplasm of the cell, in the absence of the calcium-binding protein(s), the intracellular Ca^{2+} concentration could rise to >500 μM. Thus the function of the calbindin is to bind all the entering Ca^{2+} very tightly and to move by diffusion across the cell to the basal lateral membrane where the Ca^{2+} is offloaded from the calbindin and pumped by the plasma membrane Ca^{2+} ATPase (PMCA1b) into the extracellular fluid and then the blood compartment. The levels of PMCA1b are regulated by $1\alpha,25(OH)_2D_3$ acting through the VDR.

The effect of $1\alpha,25(OH)_2D_3$ on stimulating the intestinal absorption of phosphate is known, but the mechanisms are not understood. Similarly, the detailed effects of $1\alpha,25(OH)_2D_3$ on increasing the kidney's proximal tubule reabsorption of phosphate and the distal tubule reabsorption of Ca^{2+} are also known but without a clear mechanistic insight. $1\alpha,25(OH)_2D_3$ also has an important effect on increasing the secretion of the phosphaturic hormone, FGF23 (see Fig. 9.22).

6 $1\alpha,25(OH)_2D_3$ and VDR functions in the hematopoietic system

A major new action of $1\alpha,25(OH)_2D_3$ and its receptor, the VDR, has been described within the hematopoietic system (see Fig. 9.23). There are two important involvements of $1\alpha,25(OH)_2D_3$ along with its VDR. Activated lymphocytes express the VDR which then modulates in a complex fashion the mounting of an immune response by B and T lymphocytes. Second, an important biological effect of $1\alpha,25(OH)_2D_3$ bound to its VDR in the bone marrow is to promote cell differentiation along the lineage from the committed stem cell to promonocyte,

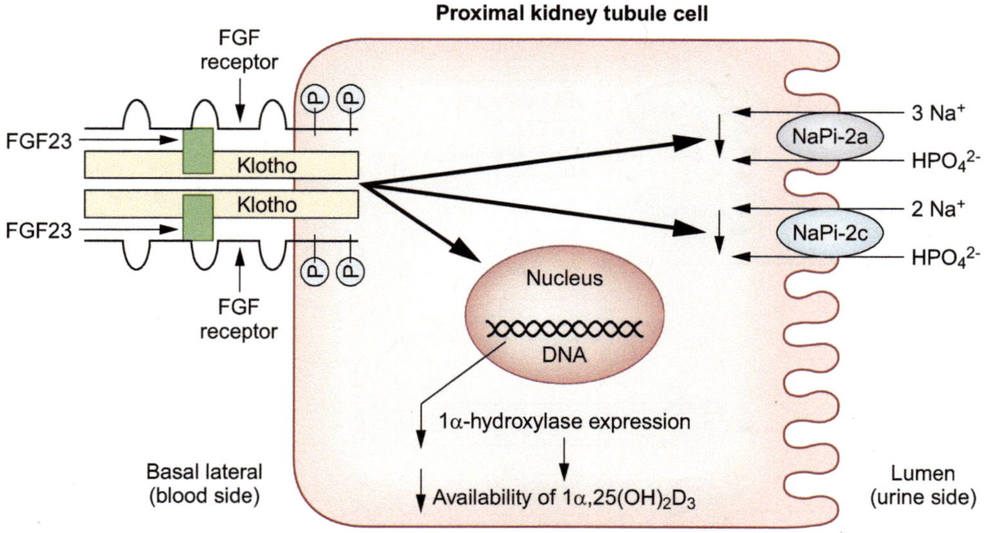

FIGURE 9.22 Schematic of binding of FGF23 to its receptor in a proximal kidney cell, resulting in the reduction of gene expression of two sodium and phosphate transporters and also the $25(OH)D_3 - 1\alpha$-hydroxylase. The FGF23 receptor requires the presence of a protein cofactor known as klotho to generate its signal transduction signals. Generation of biological responses by FGF23 requires the interaction on the surface of the proximal (basal lateral) side of the kidney cell of FGF23 binding to a Klotho–FGF receptor dimeric complex creating the formation of a trimeric complex. This trimeric complex then activates phosphorylation of the FGF receptor and activation of the intracellular signal activating the ERK kinase pathway. This then leads to the reduction of gene expression of the NaPi-2a and NaPi-2c electrogenic phosphate transporters and also the expression of the $25(OH)D_3 - 1\alpha$-hydroxylase. *ERK*, Extracellular signal-related kinase; *FGF23*, fibroblast growth factor 23.

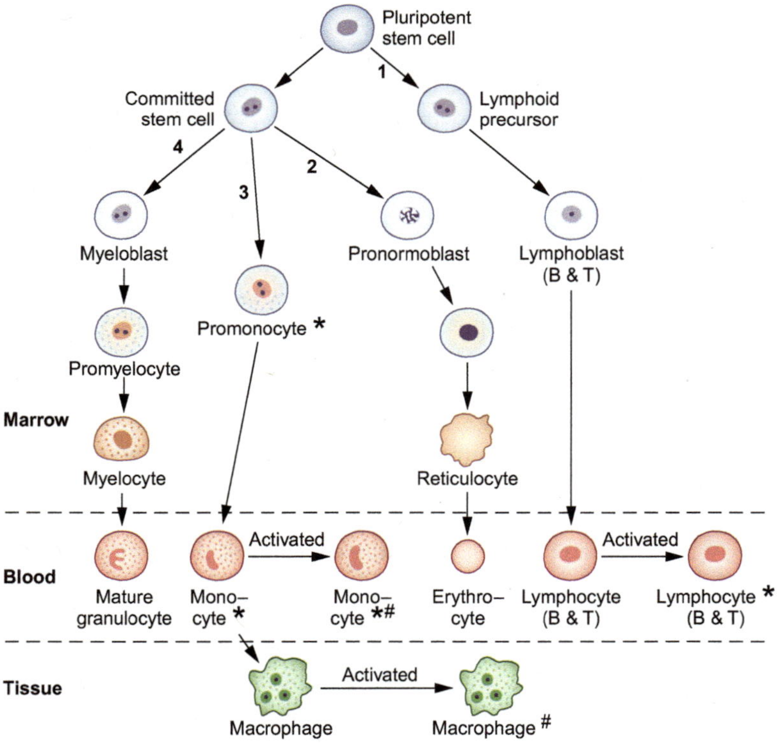

FIGURE 9.23 Flowchart of stem cell differentiation. A pluripotent stem cell can differentiate along one of four lineages; (A) B and T lymphocytes, (B) erythrocytes, (C) monocytes and macrophages, and (D) granulocytes. An asterisk * indicates cells that possess the VDR for $1\alpha,25(OH)_2D_3$ and a pound # indicates cells that can produce small amounts of $1\alpha,25(OH)_2D_3$ that function in paracrine fashion. *VDR*, Vitamin D receptor.

monocyte, to macrophage. Macrophages can be precursors of osteoclasts. Further, γ-interferon derived from activated T lymphocytes will induce bone marrow macrophages to biosynthesize small amounts of $1\alpha,25(OH)_2D_3$. Thus there is a paracrine system operative for $1\alpha,25(OH)_2D_3$, such that "locally" produced $1\alpha,25(OH)_2D_3$ in the bone marrow acts on neighboring cells to promote differentiation toward macrophages and osteoclasts, thereby increasing the bone calcium resorption capability.

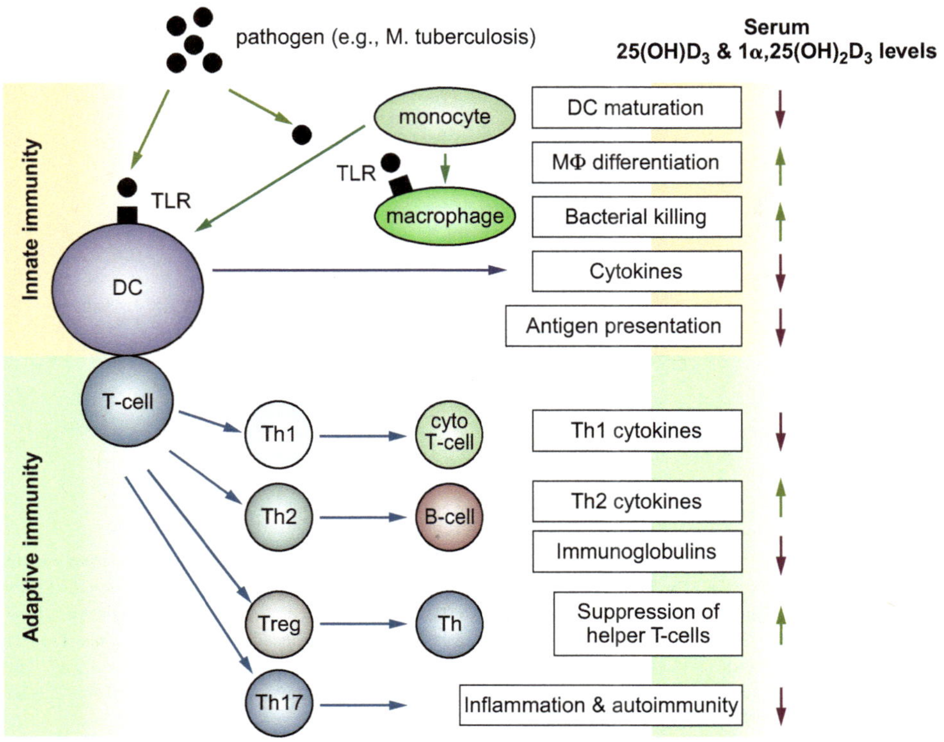

FIGURE 9.24 Schematic organization of the innate and adaptive immune systems and summary of the consequences of the presence of $1\alpha,25(OH)_2D_3$. The top portion lays out the components and responses of the innate immune system in a DC, and the bottom portion illustrates the complexity of the adaptive immune system in a T-cell. The right-hand column (*white boxes*) illustrates the biological responses that are stimulated or repressed (*green and red arrows*) by the actions of $1\alpha,25(OH)_2D_3$ that is produced from $25(OH)D_3$. The presence of infective pathogens is shown by brown circles. The actions of the Toll-like receptors are summarized in Fig. 9.25. *cyto T-cell*, Cytotoxic T-cell; *DC*, dendritic cell; *Mϕ*, macrophage; *T-cell*, T lymphocytes; *Th1*, T helper cells; *TLR*, Toll-like receptor (*black box*); *Treg*, regulatory T cell. *Modified from M. Hewison, Rheumatic Diseases Clinics of North America 38 (2012) 125.*

7 $1\alpha,25(OH)_2D_3$ and VDR function in the innate and adaptive immune systems

The objective of the immune system is to protect the parent organism from infection. The immune system comprises both the innate component and the adaptive component; see Fig. 9.24. The innate immune system provides an immediate, but nonspecific, response against perceived invaders. Innate immune systems are found in all animals and plants. In vertebrates, if pathogens successfully evade the innate response, the adaptive immune system, which is the second line of protection, becomes activated by the innate response; see the bottom portion of Fig. 9.24. Here, the immune system adapts its response during an infection to improve its recognition of the pathogen. This improved response is then biochemically "remembered" after the pathogen has been eliminated. In the event of a second invasion by the same pathogen, the existence of biochemical memory allows the initiation of a more rapid response.

It has become clear that $1\alpha,25(OH)_2D_3$ can function as an immune modulator by binding to the VDR that is expressed by several immune cells, including monocytes, macrophages, dendritic cells (DC), as well as B-lymphocytes and T-lymphocytes. Thus $1\alpha,25(OH)_2D_3$ has been shown to be functionally active in both the innate and adaptive immune responses. Some immune cells express vitamin D−metabolizing enzymes, allowing local conversion of inactive vitamin D_3 into inactive $25(OH)D_3$ and then into functionally active $1\alpha,25(OH)_2D_3$. Taken together, these data indicate that $1\alpha,25(OH)_2D_3$ plays a selective role, depending on the presence or absence of the VDR, in the maintenance of immune homeostasis.

8 Vitamin D and innate immunity

An example of the involvement of vitamin D metabolites in monocytes of the innate system is shown in Fig. 9.25. In this example of subjects with tuberculosis (TB), the pathogen *Mycobacterium tuberculosis* has invaded monocytes and macrophages and activated their plasma membrane toll-like receptors. This results in the gene transcription production of both the VDR and the mitochondrial 25(OH)D−1α-hydroxylase. Next the 25(OH)D in the serum that has migrated into the monocyte's cytosol and then into the mitochondria of the monocytes will be converted to the steroid hormone $1\alpha,25(OH)_2D_3$. Next, the VDR+bound $1\alpha,25(OH)_2D_3$, acting through genomic responses, activates the synthesis of the immune surveillance protein molecule, cathelicidin, which then engages in the expression and activation of autophagy (cell breakdown) related to the TB infection. It is the internalization of the pathogen by the phagosomes and its subsequent breakdown that results in the pathogen's death.

If the individual with TB is vitamin D deficient to the extent that the plasma 25(OH)D level is low (e.g., ~10 ng/mL; see Fig. 9.20), there will be insufficient levels of 25(OH)D available to empower the 25(OH)

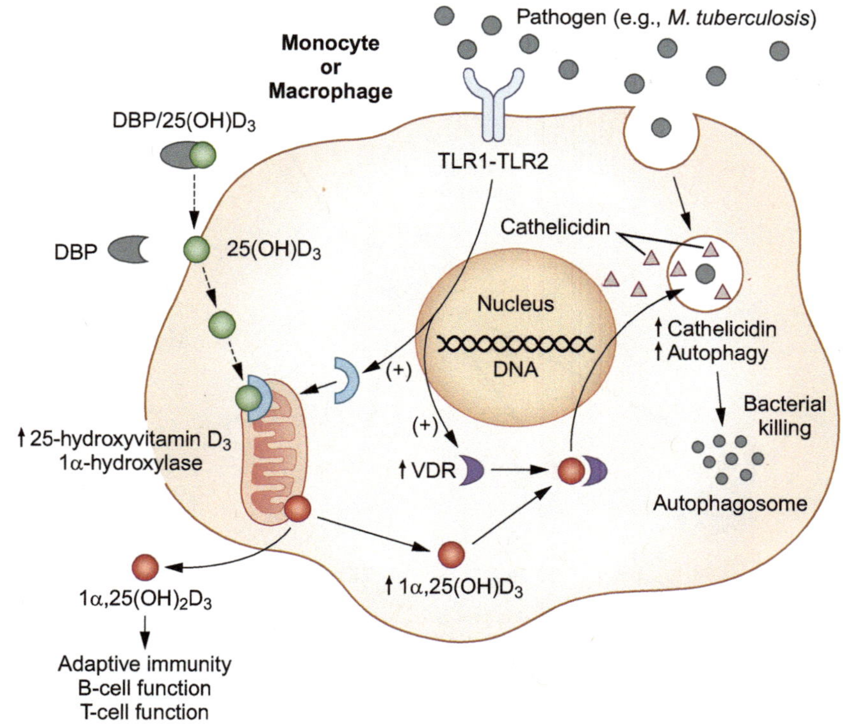

FIGURE 9.25 Activation of a monocyte's or macrophage's Toll-like signaling receptors by infection by a pathogen like *Mycobacterium tuberculosis*; involvement of circulating $25(OH)D_3$, and local $1\alpha,25(OH)_2D_3$ and local VDR. For the innate immune system's dendritic cells, the key players are Toll-like receptors that are single membrane–spanning receptors that recognize and bind as a ligand structurally conserved molecules only derived from invading microbes. This initiates signal transduction pathways that lead to the suppression or induction of genes that orchestrate the inflammatory response against the invading pathogen. In this figure a genomic increase in the production of the mitochondrial $25(OH)D_3-1\alpha$-hydroxylase results in an increased production of $1\alpha,25(OH)_2D_3$. This newly synthesized $1\alpha,25(OH)_2D_3$ binds to the VDR and stimulates the production of cathelicidins that are essential for killing the *M. tuberculosis* pathogen cells. Gray circles, individual invading *M. tuberculosis* microbes; small gray circles, killed *M. tuberculosis* microbes; pink triangles, newly synthesized cathelicidin molecules; green circles, $25(OH)D_3$; red circles, $1\alpha,25(OH)_2D_3$; blue half-moon, 25-hydroxyvitamin $D_3-1\alpha$-hydroxylase; purple icon, the VDR for $1\alpha,25(OH)_2D_3$. *VDR*, Vitamin D receptor. *Modified from M. Hewison, Rheumatic Diseases Clinics of North America 38 (2012) 125.*

$D_3-1\alpha$-hydroxylase to produce adequate levels of $1\alpha,25(OH)_2D_3$ required for the genomic production of cathelicidins. In the absence of adequate levels of cathelicidin, the autophagy innate response will be weak and ineffective.

9 Vitamin D and adaptive immunity

There is emerging evidence that vitamin D status is also linked to the adaptive immune system; see the bottom half of Fig. 9.24. Both the mature DC and macrophages, as well as the activated B- and T-lymphocytes have the $25(OH)D-1\alpha$-hydroxylase and can produce functional amounts of $1\alpha,25(OH)_2D_3$ depending upon the prevailing concentration of the 25(OH)D. Mature DC and macrophages are both able to express an adaptive T-lymphocyte- and B-lymphocyte-mediated immunity via the process of internalizing and processing the invasive pathogens using cathelicidins. As a consequence, presentation of the resulting antigens to both resting T-cells and B-cells results in activation of these cells simultaneously by the adaptive immune responses.

10 Antimicrobial peptides derived from cathelicidins can kill microbes

Mature evolutionarily conserved cathelicidin is a biologically inactive protein of ~19 Da. It is processed by serine proteases into a 94-amino-acid cathelin prosequence and an AMP of 37 amino acids (see Fig. 9.26). This peptide is further processed by serine proteases to generate a family of four unrelated peptides (19, 27, 39, or 31 amino acids), each of which has a selective AMP activity against viruses, bacteria, or fungi.

The structural basis for the mechanism of the antimicrobial activity of a small cathelicidin-derived peptide is illustrated by the wheel plot of the peptide's amino acid residues (see Fig. 9.27). Each peptide is organized into α-helical structure such that the amino acids found on one side are positively charged, while the amino acids on the opposite side are hydrophobic. This evolutionarily created structure enables the positively charged surface of the peptide to interact with negatively charged cell membranes followed by the entry of the peptide into the hydrophobic phase of the cell membrane to create a

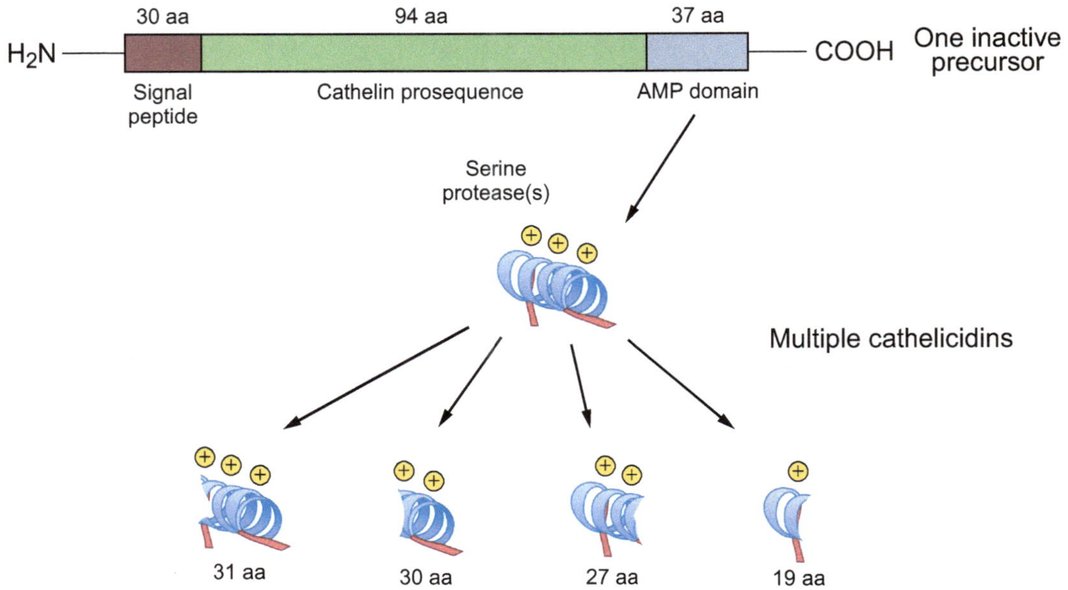

FIGURE 9.26 Illustration of the gene product cathelicidin that is processed by proteinase-3 (a serine protease) to yield a signal peptide ($\sim$30 amino acids, *brown*), a cathelin prosequence (of 94 amino acids, *green*), and a 37-amino-acid antimicrobial protein (LL-37, *blue*). Members of the cathelicidin family of antimicrobial polypeptides are characterized by a highly conserved region (cathelin domain) and a highly variable cathelicidin peptide domain. LL-37 begins its amino acid chain with two leucines. Exposure of the 37-amino-acid AMP to other serine proteases (e.g., from skin at sites of inflammation) results (in this example) in the generation of four unique peptides, with 31, 30, 27, and 19 amino acids, respectively. Each of these peptides will have a different profile of biological actions. AMPs have the ability to participate in diverse functions such as wound healing, chemotaxis, and angiogenesis. The basic mode of action of an AMP is described in Fig. 9.27. AMP, Antimicrobial peptide. *Modified from R.L. Gallo et al. Journal of the American Academy of Dermatology 52 (2005) 381.*

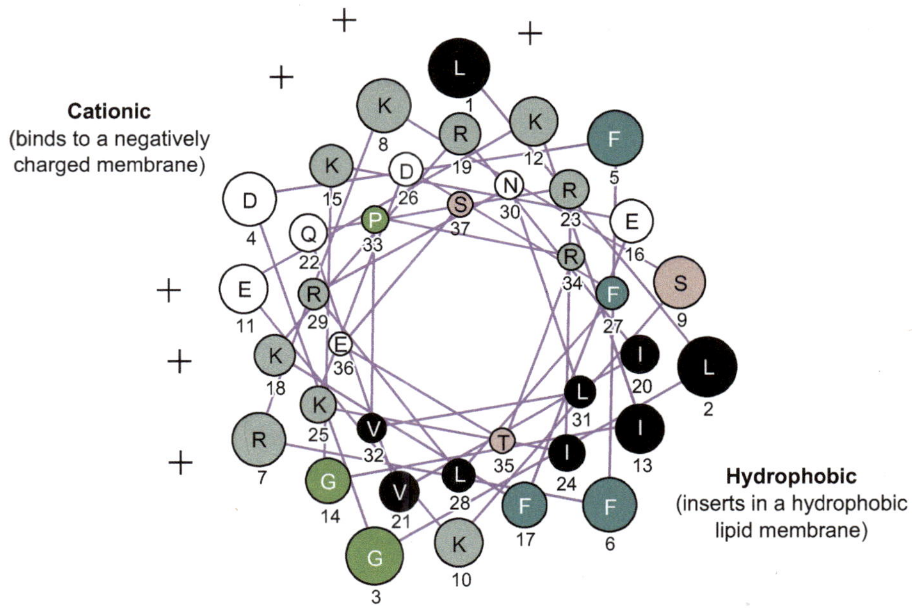

FIGURE 9.27 A schematic wheel plot of a 37-amino-acid cathelicidin's distribution of positively charged and hydrophobic residues. The cathelicidin family in humans is restricted to a single gene, cAMP, with the resulting AMP protein product often referred to as hCAP18/LL-37. The α-helix of the AMP has a hydrophobic surface on one side and a hydrophilic positively charged surface on the opposite side. The schematic wheel of the figure traces the amino acid sequence of the LL-37-amino-acid peptide. The black circles map the hydrophobic amino acids (leucine, isoleucine, phenylalanine, and valine). The white circles denote positively charged (arginine, histidine, and lysine) and blue spheres denote negatively charged (aspartic and glutamic) or amino acids. In the host individual that is experiencing a microbial infection, as a consequence of the activation of the innate immune response, the LL-37 peptide/cathelicidin becomes inserted into the plasma membrane of the invading microbe, which changes the membrane permeability and causes a decrease in protein biosynthesis that ultimately leads to the death of that microbe. *cAMP*, Cyclic antimicrobial peptide. *Modified from R.L. Gallo et al. Journal of the American Academy of Dermatology 52 (2005) 381.*

voltage-dependent ion channel. This results in a change of the cell membrane's permeability, leading to the interference of the cytoplasmic process that ultimately causes lysis of the target cell, leading to a lethal outcome.

E Fibroblast growth factor 23

FGF23 is a relatively recently discovered (2001) key player in maintaining appropriate levels of serum phosphate. There are human genetic diseases that result in either decreased or increased plasma levels of FGF23. Using mouse models where the FGF23 gene was knocked out resulted in an impaired renal phosphate excretion as well as changes in the regulation of the renal 25(OH) $D_3 - 1\alpha$-hydroxylase that led to an overproduction of $1\alpha,25(OH)_2D_3$ resulting in hypercalcemia.

FGF23 is principally secreted by bone osteocytes/osteoblasts but smaller amounts are present in the brain, muscle, heart, thymus, and spleen. FGF23 generates its biological responses through binding to its cognate plasma membrane spanning receptor (see Fig. 9.22). FGF23 also requires the presence of both a protein cofactor known as klotho (1024 amino acids; 122 kDa) and the FGF23 receptor to create a functional trimeric complex. Both the C-termini of klotho and the phosphorylated FGF23 receptor span the plasma membrane as a heterodimer which generates an as yet unknown signal transduction second messenger(s).

In the proximal renal tubule cells (Fig. 9.22), the FGF23 receptor–generated messengers result in two separate responses. One is an impairment of two classes of Na^+-dependent phosphate transporters (NaPi-2a and NaPi-2c) which results in a reduction in renal tubular phosphate reabsorption. The second response works in the cell nucleus to reduce the expression of $25(OH)D_3 - 1\alpha$-hydroxylase leading to a reduction in the production of $1\alpha,25(OH)_2D_3$ and lowering of $1\alpha,25(OH)_2D_3$ plasma levels.

A separate important action of FGF23 occurs in parathyroid secreting cells where FGF23 with its receptor and klotho and also $1\alpha,25(OH)_2D_3$ with its VDR separately work in the nucleus to reduce the production of PTH mRNA (see Fig. 9.16).

F Integrated actions of $1\alpha,25(OH)_2D_3$, PTH, calcitonin, and FGF23 on bone remodeling and calcium homeostasis

As described earlier in this chapter, bone is a metabolically active organ, undergoing throughout life a continual turnover and remodeling process involving bone resorption followed by accretion. The balance between the rates of bone resorption by osteoclasts and bone formation by osteoblasts will determine, both at a local level or globally (the entire skeleton), whether there is a negative, neutral, or positive calcium balance. The biological activities of bone cells are subject to the actions of a multitude of hormones, cytokines, and other physiological regulators (see Table 9.3). A summary of the physiological effects

TABLE 9.3 Modulators of bone cell resorptive activity.

Stimulators of bone resorption	Inhibitors of bone resorption
PTH	Calcitonin
PTHrP	
$1\alpha,25(OH)_2D_3$	Glucocorticoids
Prostaglandins	Estrogens
PGE_2	Androgens
IL-1	Insulin-like growth factor
Tumor necrosis factors	TGF-1
TNF-α	Fluoride (F^-)
TNF-β	
Thyroxine	
Retinol	
Growth factors	
EGF	

EGF, Epidermal growth factor; *IL-1*, Interleukin-1; *PTH*, parathyroid hormone; *PTHrP*, parathyroid hormone–related protein; *TGF-1*, Transforming growth factor 1; *TNF*, tumor necrosis factor.

of CT, PTH, and vitamin D metabolites related to mineral metabolism is presented in Table 9.4.

Considering that PTH and $1\alpha,25(OH)_2D_3$ both stimulate bone resorption, it is surprising that the bone-resorbing cells, the osteoclasts, do not have either a PTH receptor or the $1\alpha,25(OH)_2D_3$ receptor (VDR). The process of osteoclastogenesis is complex (see Fig. 9.28) and is supported by a collaboration between osteoblasts with osteoclast progenitor cells derived from a monocytes/macrophage lineage. The osteoclast progenitor cells are stimulated by macrophage colony-stimulating factor to initiate the production of the cell surface receptor RANK (receptor activator for nuclear factor κB). This is followed by PTH binding to its osteoblast plasma membrane receptor in collaboration with genomic actions of $1\alpha,25(OH)_2D_3$ in the osteoblast to stimulate the production of RANKL, the ligand for the maturing osteoclast's RANK receptor. Note that the ligand for RANK does not diffuse from one cell to the other. Instead the ligand (RANKL) for the osteoclasts' RANK is tethered on the extracellular surface of the osteoblast's plasma membrane. Thus the "message" of the hormone RANKL is basically delivered when the two cells (osteoblast and osteoclast) are in extremely close proximity to one another. The binding of RANKL to its RANK can be inhibited by osteoprotegerin (OPG). OPG, also known as osteoclastogenesis inhibitory factor, is a decoy-soluble receptor for RANKL which can inhibit the production of osteoclasts. The net outcome of this overall process is that the osteoclast precursors (prefusion osteoclasts and multinucleated osteoclasts) are stimulated to fuse and form mature, activated, fully functional osteoclasts.

The process of bone formation and bone resorption can be mediated by a bone remodeling unit (BRU) that comprises the integrated and sequential actions of first osteoclasts followed by osteoblasts. The detailed functioning of this model through the phases of activation, resorption, reversal, formation, and mineralization is described in Fig. 9.29. A series of sequential steps that constitute the operation of the BRU are initiated. They include the following: (1) osteoclasts accumulate on the surface of bone and begin to carry out bone resorption, which results in the creation of a pit or trench; (2) integrated with step (1) is the phagocytosis of the uncovered bone matrix debris; (3) osteoblasts begin to appear at the trench and slowly put in place in the trench osteoid, which is followed by (4) laying down new calcium hydroxyapatite bone mineral so as to completely fill the trench.

Depending on physiological and mechanical needs, the osteoclasts create resorption trenches in pockets of existing cortical or trabecular bone (Fig. 9.30). In any given time interval, it has been calculated that the adult human skeleton has more than one million functional trenches or BRU in operation. For any given BRU the bone resorption removal phase requires several weeks to achieve completion. In healthy bone with active BRU, after the completion of the resorption process, the lost calcium hydroxyapatite is replaced via a process requiring 3–4 months to fill the BRU.

TABLE 9.4 Physiological effects of calcitonin, parathyroid hormone, and vitamin d (metabolites) related to mineral metabolism.

	Calcitonin	Parathyroid hormone	$1\alpha,25(OH)_2D_3$
Intestinal			
Calcium absorption	–	↑ (indirect)	↑
Phosphate absorption		–	↑
Renal			
Phosphate excretion	↑	↑	↓
Calcium excretion	↑	↓	
Adenylyl cyclase activity	↑	↑	–
Skeletal			
Calcium mobilization	↓	↑	↑
Mineralization of bone matrix	–	–	↑
Plasma levels			
Calcium	↓	↑	↑
Phosphate	↓	↓	↑

The important role of bone as a central organ in calcium and phosphorus metabolism, acting both as a source of and a reservoir for these two ions, is discussed in the text. It is apparent that bone remodeling processes may contribute to both short- and long-term events necessary for calcium and phosphorus homeostasis. The relative actions of bone formation and resorption are known to be modulated by various endocrine regulators during times of skeletal growth and lactation and in birds during the process of egg laying. Also, it is not surprising that bone is involved in a wide variety of disease states that reflect perturbations in calcium and phosphorus homeostasis.

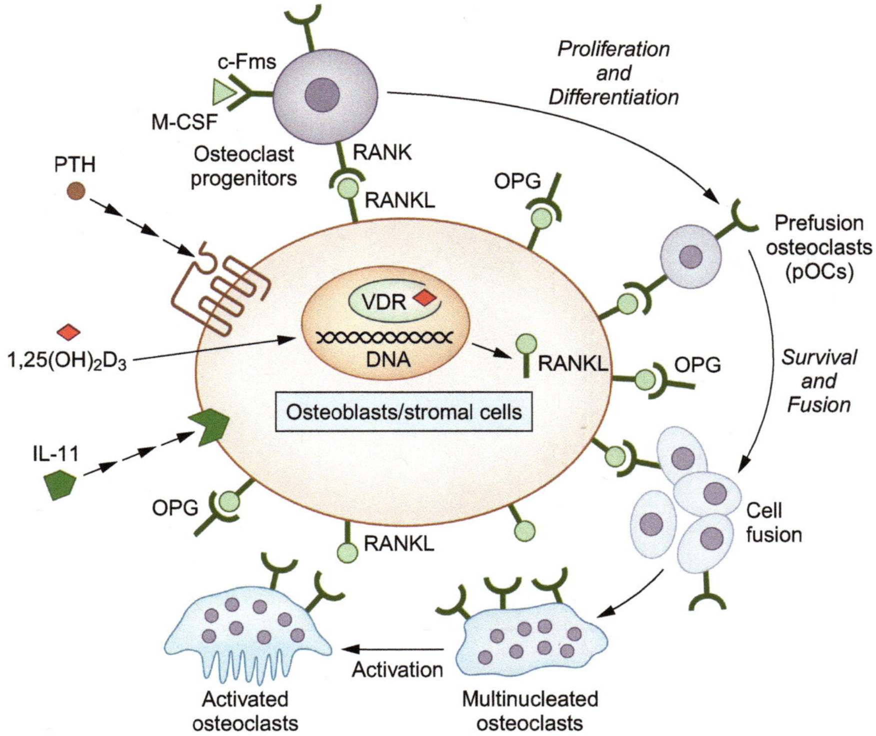

FIGURE 9.28 A schematic pathway of osteoclastogenesis supported by osteoblast/stromal cells. Osteoclast progenitor cells are derived from a monocytes/macrophage lineage (see Fig. 9.23). They enter into a differentiation pathway starting with stimulation by M-CSF that binds to its cell surface receptor, c-Fms. This results in the appearance of the cell surface receptor, RANK (the receptor activator of NF-κB ligand) on several cell types, including osteoclast progenitors, prefusion osteoclasts, multinucleated osteoclasts, and fully activated and functional osteoclasts. RANKL, which is present on the cell surface of osteoblasts/stromal cells, is the ligand for RANK. The binding of RANKL to RANK results in communication between the osteoblast [which possesses PTH, IL-11 and 1α,25 $(OH)_2D_3$ receptors] with the progenitor, prefusion, multinucleated, and functional activated osteoclasts. OPG is a decoy-soluble receptor for RANKL produced by osteoblasts that acts as a decoy receptor for RANKL and thereby inhibits osteoclastogenesis and osteoclast activation by binding to RANKL. IL-11 is a 23-kDa protein that participates in osteoclast progenitor proliferation and differentiation into prefusion osteoclasts. VDR is the receptor for $1\alpha,25(OH)_2D_3$ that is present in osteoblasts but not osteoclasts. *IL-11*, Interleukin-11; *M-CSF*, monocyte/macrophage-colony stimulating factor; *OPG*, osteoprotegerin; *RANK*, receptor activator for nuclear factor κB; *VDR*, vitamin D receptor.

V Clinical aspects

A Vitamin D nutrition

One international unit (IU) of vitamin D_3 is equivalent to 25 ng or 65 pmoles. In the United States the principal form of vitamin D supplementation for food was vitamin D_2 in the interval 1930–65; since 1965 the predominant form of vitamin D supplementation has been vitamin D_3. A joint US/Canadian committee of the Institute of Medicine (IOM) recommended in 2010 that the RDA of vitamin D (either D_2 or D_3) for a person of age 9–70 is 600 IU/day and for persons over 70 years is 800 IU/day.

Fig. 9.20 illustrates the multiple contributions of vitamin D to good health. Besides the classical contributions of vitamin D, via $1\alpha,25(OH)_2D_3$, and its endocrine system responsible for maintaining calcium homeostasis, because of the wide tissue distribution of the VDR, the scope of the vitamin D endocrine system is now known to be much more comprehensive. Thus the vitamin D endocrine system also includes the immune system (both innate and adaptive), the pancreas β-cell secretion of insulin, the heart and cardiovascular system, and muscle.

As emphasized by the IOM committee, an assessment of a person's vitamin D status can be carried out by determining their serum level of 25(OH)D. The inset table in Fig. 9.20 shows for humans the correlation of the serum level of 25(OH)D with the severity of vitamin D deficiency.

The classic deficiency state resulting from a dietary absence of vitamin D or lack of ultraviolet (sunlight) exposure is the bone disease called rickets in children and/or osteomalacia in adults. Historically, the identification of rickets as a disease state allowed G. Mellanby in 1920 to induce rickets in puppies experimentally by nutritionally withholding fat extracts. This work resulted in the discovery of vitamin D_3 and appreciation that in the absence of sunlight, that vitamin D_3 was an essential trace dietary constituent (e.g., a vitamin) that was important for bone mineralization.

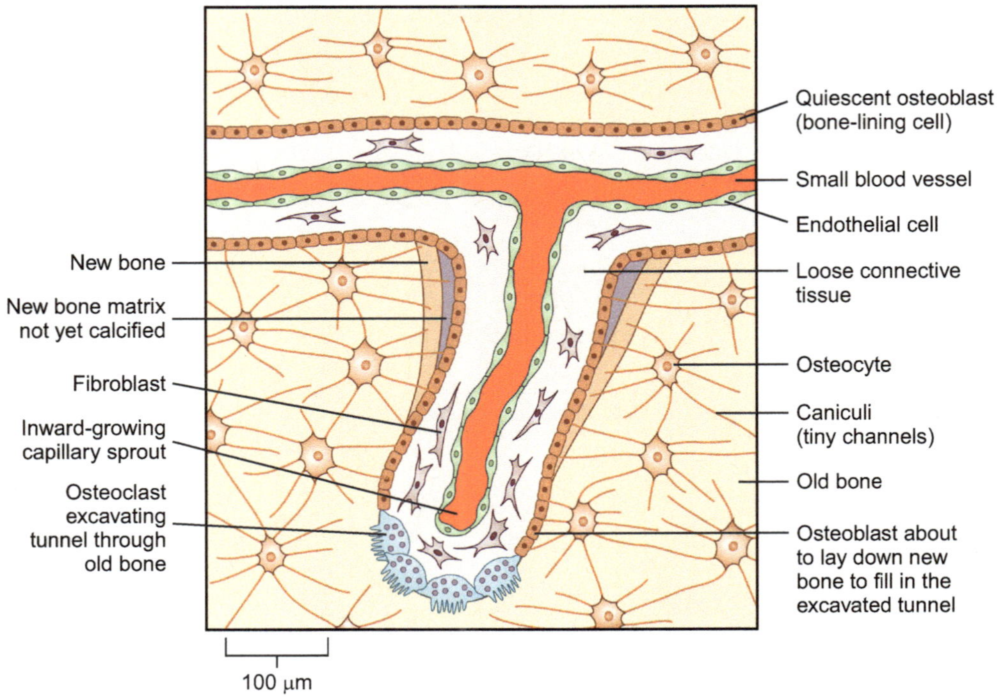

FIGURE 9.29 Schematic illustration of the activities of a BRU. A cluster of osteoclasts has hollowed out in old bone a downward tunnel at a rate of ~50 μm/day. (The biological properties of the working osteoclast cells are summarized in Fig. 9.30.) The osteoclasts are followed by osteoblasts which will lay down a new bone matrix (see Fig. 9.3) and implement a new importation of calcium and phosphate to create the calcium hydroxyapatite precipitate, which will ultimately close the tunnel by concentric layers of new bone at a rate of 1–3 μm/day (a 3–4 month interval). Some of the osteoblast cells will become trapped in the new bone mineral and will become osteocytes (see Fig. 9.4). Associated with the tunneling activity is the generation of a small blood vessel or capillary sprouts that will provide the cells with nutrients and oxygen and remove CO_2 as hydrogen carbonate HCO_3^-. *BRU*, Bone remodeling unit.

The clinical features of rickets and osteomalacia depend upon the age of onset. The classical skeletal disorders of rickets in very young children include deformity of the bones, especially in the knees, wrists, and ankles (see Fig. 9.31) as well as associated changes in the costochondral joint junctions, which have been termed by some as the rachitic rosary. If rickets develops in the first 6 months of life, infants may suffer from convulsions or develop tetany due to a low blood calcium level (usually <7 mg/100 mL) but may have only minor skeletal changes. After 6 months, bone pain as well as tetany is likely to be present. Since osteomalacia occurs after growth and development of the skeleton are complete (i.e., the adult stage of life), its main symptoms are muscular weakness and bone pain, with little bone deformity.

B Osteoporosis

Osteoporosis is the most common generalized disorder of bone. It is estimated that in 2010 in the United States, over 52 million people, 80% of whom are female, had some form of osteoporosis. It is most simply characterized as a state of insufficiently calcified bone and is the end result of a number of metabolic abnormalities that affect the skeleton. Some of the causes of osteoporosis that have been recognized or proposed include the following: (1) adrenal cortical hyperfunction, (2) prolonged dietary calcium deficiency, (3) reduced levels of estrogen, as in postmenopausal women, (4) prolonged skeletal immobilization, and (5) chronic vitamin D deficiency. As a consequence of these "changes," an imbalance results in the remodeling rate of bone, such that there is either an increase in the relative rate of bone resorption or a decrease in the rate of bone formation.

In osteoporosis, there can be an imbalance in the daily hormone-mediated bone remodeling rates so that there is a small net daily loss of calcium from the skeleton. For example, if bone hormone–mediated Ca^{2+} resorption exceeds hormone-mediated bone accretion by only 10 mg/day, when extended for 20 years (e.g., from menopause at age ~50 to age ~70), there will be 10 mg/day × 365 days × 20 years=73,000 mg of calcium lost from the skeleton or about 7.3% of the total skeletal calcium content.

The chief clinical sign of osteoporosis is thinning of the bones with a concomitant increase in the number of fractures (see Fig. 9.32). Each year, in the United States this disease leads to more than 1.5 million fractures,

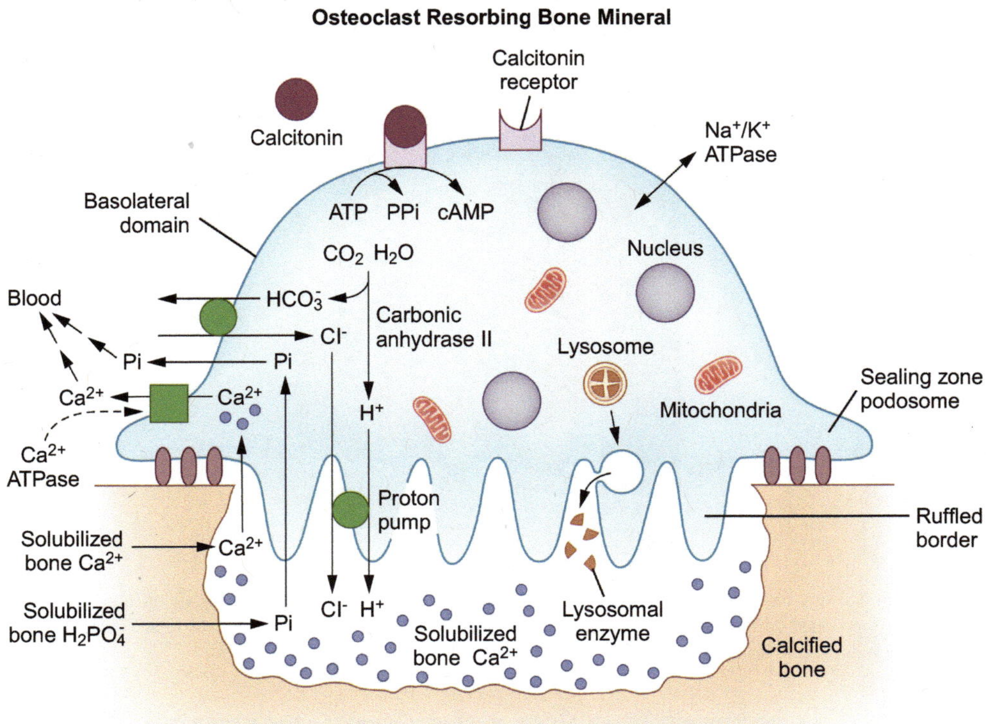

FIGURE 9.30 Schematic illustration of the activities of an osteoclast vigorously engaged in bone resorption. The osteoclast basolateral membrane has ATP-dependent ion transport systems to extrude calcium (which has been solubilized from the bone's precipitated calcium hydroxyapatite). The osteoclast is enriched in mitochondria to provide the large amount of ATP necessary for these pumps as well as for an electrogenic proton pump, which raises the hydrogen ion concentration outside the ruffled border area. Lysosomal enzymes are synthesized and taken up in Golgi vesicles that contain mannose 6-phosphate receptors. These vesicles exocytose across the ruffled border membrane, so as to discharge their enzyme contents outside the osteoclast where Ca^{2+} resorption and local bone matrix breakdown are actively occurring. The sealing zone, with its actin filaments and specialized attachment structures (podosomes), surrounds the ruffled border area and separates it from the extracellular fluid. Also it is postulated that the bone matrix protein, osteopontin, plays a role in anchoring actively resorbing osteoclasts to the bone. Carbonic acid is generated from the CO_2 produced by mitochondrial metabolism and provides the source for hydrogen ions to be pumped across the ruffled border. Any excess HCO_3^- is removed by exchange with chloride. While receptors for calcitonin are expressed in the plasma membrane, no receptors for either PTH or $1\alpha,25(OH)_2D_3$ are present in the osteoclast even though both hormones contribute to bone resorption. *PTH*, Parathyroid hormone.

mostly of the hip, spine, and wrist. More than 350,000 osteoporosis-related life-threatening hip fractures occur annually and they all require hospitalization. The cost to the health-care system associated with osteoporotic fractures is currently approximately $17 billion annually (2010). This converts to more than $45 million a day! Each hip fracture represents an estimated $40,000 in total medical costs.

C Tuberculosis

One-third of the world's population lives with TB, and new infections occur at a rate of about one per second. Also, the number of people who become ill with TB each year continues to grow. The World Health Organization estimates that there are currently 13.7 million chronic active cases of TB, 9.3 million new cases each year, and 1.8 million annual deaths. In 2012 TB is second only to HIV/AIDS as the greatest killer worldwide due to a single infectious agent. The symptoms of TB include a chronic cough, fever, night sweats, and weight loss. Diagnosis relies on a tuberculin skin test, blood tests, X-rays of the lungs, as well as microscopic examination and microbiological culture of bodily fluids.

About 10%–20% of all diagnosed cases of TB infect the skeleton. Bone TB can occur at any age and can infect virtually any bone within the body. Evidence of human TB has been obtained from a 9000-year-old skeleton. Thus TB is an ancient disease and there has not been evolutionary change of the TB genome.

Prof. Robert Koch (1843–1910) received the Nobel Prize in Physiology or Medicine in 1905 for his discovery in 1882 of *M. tuberculosis*, the bacterium that causes TB. The concept of use of a sanatorium for the treatment of patients with TB was included by Prof. Koch in his 1905 Nobel lecture.

The standard TB medical therapy for decades in the preantibiotic era was based on rest in a sanatorium at an

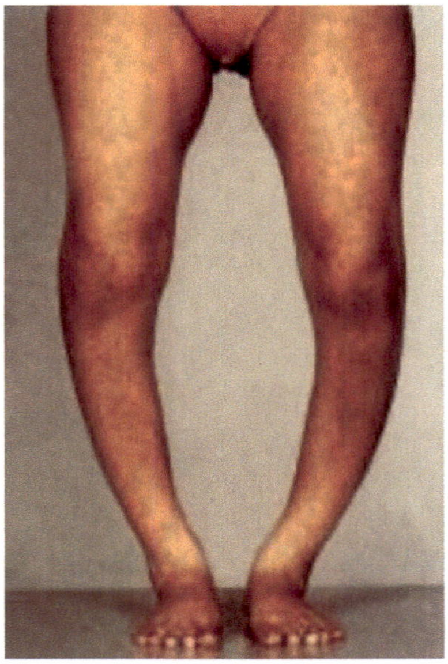

FIGURE 9.31 Photograph of a young child with classic rickets. The bowing of the long bones (femur and tibia/fibula) is diagnostic of a vitamin D deficiency. The D deficiency then results in an inadequate absorption of dietary calcium and a concomitant reduction in the mineral content of the long bones, which results in a reduction in bone strength.

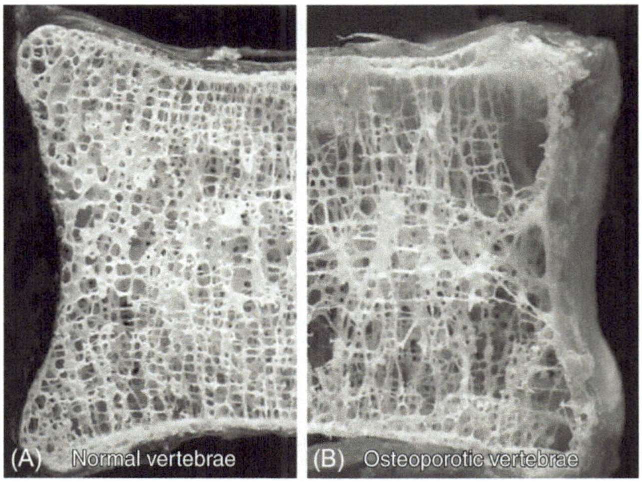

FIGURE 9.32 Effect of long-standing osteoporosis on the mineral content of lumbar vertebrae. Shown are the microradiographs of a lumbar vertebra from a young (A) and an elderly (B) female.

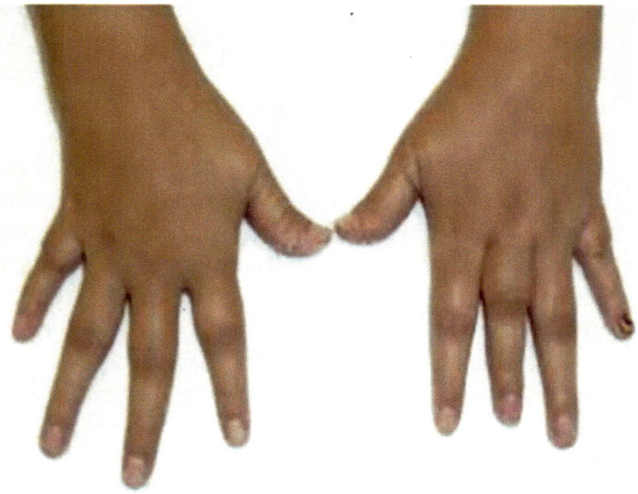

FIGURE 9.33 A shortened left third metacarpal bone is evident in the hands of a 26-year-old woman with PHP. This is diagnostic in patients with PHP. Note that all digits, except the thumbs and index fingers, are shorter than normal. *PHP*, Pseudohypoparathyroidism.

elevation where UV light was amply prevalent. In retrospect, this suggested a linkage between vitamin D status and the disease of TB. In the 1940s a high-dose vitamin D supplementation was introduced for patients with TB. In the past decade, epidemiological studies have linked inadequate vitamin D levels [low serum concentrations of 25(OH)D] to a higher susceptibility of immune-mediated disorders, including chronic infections and autoimmune diseases. In 2006 new data supported a link between TB and vitamin D deficiency as determined by the blood levels of 25(OH)D and by the activation of the innate immunity system via the production of AMPs which can kill the *M. tuberculosis*. The details are summarized in this chapter in Figs. 9.30–9.33.

D Parathyroid hormone

The common disorders of the parathyroid gland fall into two main categories: (1) those associated with poor secretion of PTH or hypoparathyroidism and (2) those associated with excess secretion of PTH or hyperparathyroidism.

Hyperparathyroidism is the most common disorder involving the parathyroid glands. Primary hyperparathyroidism is a disorder of mineral metabolism characterized by a defect in the normal feedback control of PTH secretion by the plasma calcium concentration. Secondary hyperparathyroidism is a disorder characterized by primary disruption of mineral homeostasis, such as impaired production of $1\alpha,25(OH)_2D_3$ or low dietary intake of calcium, which leads to a low blood calcium level and a compensatory increase in parathyroid gland function, size, and secretion of PTH. Hyperparathyroidism can result from a single adenoma, Fig. 9.34 pictures one of the four parathyroid glands that has formed an adenoma (usually nonmalignant), multiple adenomas, hyperplasia, usually of the chief cells, and carcinoma.

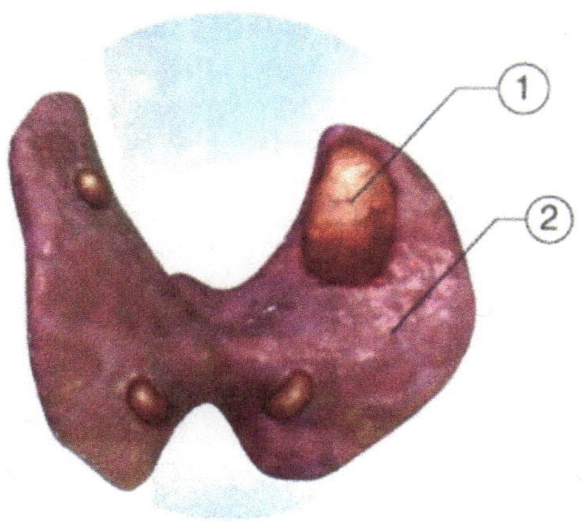

FIGURE 9.34 Appearance of a hyperparathyroid adenoma. 1 (encircled)=parathyroid gland; 2 (encircled)=thyroid gland. *Reproduced in part from https://byjus.com/biology/hyperparathyroidism/.*

The principal clinical features of hyperparathyroidism are a markedly elevated plasma level of serum calcium (11–13 mg/100 mL) and, when present for prolonged intervals of time, extensive resorption of the skeleton. At the kidney level the presence of renal kidney stones or calculi can be a consequence. Deposition of calcium in the collecting tubules of the kidney produces nephrocalcinosis. The hyperactive parathyroid gland may be removed surgically, resulting in a fall in serum calcium levels and often the minimization of the urolithiasis. The principal deleterious consequence of prolonged hyperparathyroidism is the greatly decreased mineral content of the skeleton.

The most common cause of hypoparathyroidism is damage to or removal of the parathyroid glands in the course of an operation on the thyroid gland. Most cases are found in children; approximately twice as many females as males are affected. The most prominent clinical feature of hypoparathyroidism is hypocalcemia, and in extreme instances tetany can be observed.

Pseudohypoparathyroidism is a rare genetic disorder involving bone and mineral metabolism. It is probably inherited as an X-linked dominant trait with variable penetrance and is characterized by signs and symptoms similar to those of hypoparathyroidism. However, in this disease state there is a peripheral resistance to the biological actions of PTH and an elevated secretion of PTH. Characteristic clinical features include a round face, short stature, brachydactylia (abnormal shortness of fingers or toes), especially of the metacarpal and metatarsal bones as a result of early epiphyseal closure (see Fig. 9.33).

E Parathyroid hormone–related protein

Disorders of mineral metabolism are frequently found in patients with cancer. While primary bone tumors rarely lead to systemic disorders of calcium homeostasis, there can be a malignancy that does not alter the skeleton directly but does alter the balance of one or more of the calcium-regulating hormones. Such an example is HHM. In this instance a tumor (e.g., renal or squamous carcinoma or breast cancer) secretes PTHrP that, because of its structural homology to PTH (see Fig. 9.11), acts as an agonist on the PTH receptor in osteoblast cells and leads to an inappropriate activation of bone resorption and the onset of hypercalcemia.

F Calcitonin

The only disease state known to be related to CT is that resulting from a disorder of C-cell function; this is described as a medullary carcinoma of the thyroid, which results in the hypersecretion of CT. The etiology of this condition is unknown. Over 50% of the cases exhibit a familial incidence, and the pattern of inheritance indicates an autosomal dominant mode of transmission. The disease is rare, only occurring in one of every several 1000 patients seeking medical attention. It is one of the few forms of cancer that can be detected by assay of the plasma level of CT. This is carried out by measurement of its concentration before and after a standard provocative infusion with either calcium or pentagastrin.

G Fibroblast growth factor 23

Familial tumoral calcinosis is a heritable autosomal recessive disease that is linked to mutations in FGF23. The disease is characterized by ectopic calcified tumoral masses and soft tissue calcifications that occur because of hyperphosphatemia. Any one of eight known single-point mutations in FGF23 can lead to the disease. This recent new information supporting the role of FGF23 in disorders of phosphate homeostasis has led to it being a new frontier for drug development.

Reference

[1] G. Karsenty, Update on the biology of osteocalcin, Endocrine Practice 10 (2017) 1270–1274. Available from: https://doi.org/10.4158/EP171966.RA. Epub 2017 Jul 13. PMID:28704102.

Further reading

Parathyroid hormone

L.J. Clark, et al., Allosteric interactions in the parathyroid hormone GPCR-arrestin complex formation, Nature Chemical Biology 16 (2020) 1096–1104.

S. Li, F. Liu, Y. Pei, Y. Dong, Y. Shang, Parathyroid type 1 receptor regulates osteosarcoma K7M2 cell growth by interacting with angiotensinogen, Journal of Cellular and Molecular Medicine 25 (2021) 2841–2850.

M. Minezaki, et al., Reduction in parathyroid adenomas by cinacalcet therapy in patients with primary hyperparathyroidism, Journal of Bone and Mineral Metabolism 39 (2021). Available from: https://doi.org/10.1007/s00774-020-01190-2.

L. Oinonen, et al., Parathyroid hormone may play a role in the pathophysiology of primary hypertension, Endocrine Connections 10 (2021) 54–65.

Vitamin D and 1α,25(OH)2D3

S.E. Bollen, P.J. Atherton, Myogenic, genomic and non-genomic influences of the vitamin D axis in skeletal muscle, Cell Biochemistry and Function 39 (2020). Available from: https://doi.org/10.1002/cbf.3595.

R. De Luccia, D. Clegg, D. Sukumar, The implications of vitamin D deficiency on Covid-19 for at-risk populations, Nutrition Reviews 79 (2021) 227–234.

M.R. Haussler, S. Livingston, Z.L. Sabir, C.A. Haussler, P.W. Jurutka, Vitamin D receptor mediates a myriad of biological actions dependent on its 1,25-dihydroxy vitamin D ligand: distinct regulatory themes revealed by induction of Klotho and fibroblast growth factor-23, Journal of Bone and Mineral Research Plus 5 (2021). Available from: https://doi.org/10.1002/jbm4.10432.

A. Pramono, et al., The association between vitamin D receptor polymorphisms and tissue-specific insulin resistance in human obesity, International Journal of Obesity 45 (2021) 818–827.

Calcitonin

R.A. Davey, et al., The calcitonin receptor regulates osteocyte lacunae activity during lactation in mice, Journal of Endocrinology 249 (2021) 31–41.

S.A. Kang, R. Govindarajan, Anti-calcitonin gene-related peptide monoclonal antibodies for neuropathic pain in patients with migraine headache, Muscle and Nerve 63 (2021) 563–567.

D.S. Mathiesen, et al., Amylin and calcitonin: potential therapeutic strategies to reduce body weight and liver fat, Frontiers in Endocrinology 11 (2021). Available from: https://doi.org/10.3389/fendo.2020.617400.

S. Marx, G. Aurbach, J. Gavin, D. Buell, Calcitonin receptors on cultured human lymphocytes, Journal of Biological Chemistry 249 (2021). Available from: https://doi.org/10.1016/s0021-9258(19)42130-3.

Calcium-sensing receptor

X. Chen, et al., Structural insights into the activation of human calcium-sensing receptor, bioRxiv (2021). Available from: https://doi.org/10.1101/22021.03.30.437720.

J. Diao, et al., Therapeutic opportunities of targeting allosteric binding sites on the calcium-sensing receptor, ACS Pharmacology and Translational Science 4 (2021) 666–679.

S. Ling, et al., Structural mechanism of cooperative activation of the human calcium-sensing receptor by Ca^{2+} ions and L-tryptophan, Cell Research 31 (2021) 383–394.

Fibroblast growth factor23 (FGF23)

A. Agrawal, R. Agoro, K.E. White, R.D. Di Marchi, Identification of a second Klotho interaction site in the C-terminus of FGF23, Cell Reports 34 (2021) 108665. Available from: https://doi.org/10.1016/j.celrep.2020.108665.

B.B. Ho, C. Bergwitz, FGF23 signalling and physiology, Journal of Molecular Endocrinology 66 (2021). Available from: https://doi.org/10.1530/JME-20-0178.

Chapter 10

Adrenal corticoids

I Introduction

A Background

The adult adrenal glands function to activate a fairly complex set of endocrine systems. The hormones produced by the cells of the human adrenal cortex in response to appropriate stimuli are selected steroid hormone families and the catecholamines. The key steroid hormones are cortisol and aldosterone. Cortisol is the major steroid secreted in response to stress. Aldosterone is the primary steroid hormone acting on the kidney to conserve sodium ion from being excreted in the urine. Aldosterone also plays an important role in water balance and blood pressure regulation (see Fig. 15.4A/B/C).

The catecholamines' active forms are epinephrine and norepinephrine. They are chemical messengers, that is, they are hormones, but structurally not steroids. They are described in detail in Chapter 11, Hormones of the Adrenal Medulla.

B Glucocorticoids

Glucocorticoids (GC) are a family of steroid hormones that are essential to life. The primary glucocorticoid is cortisol. It acts on different cells in a variety of different ways. Without the secretion of cortisol during stress, a human could not survive. Even in patients who lack the ability to secrete cortisol from the adrenals in the face of stress, the oral administration of a glucocorticoid can be beneficial.

An important result of glucocorticoid action is to stimulate the production of glycogen, especially in the liver, and to maintain blood glucose concentration. The process of gluconeogenesis is a classic biochemical transformation of noncarbohydrate "carbon," for example, glycerol, pyruvate, lactose, glucogenic amino acids (e.g., glutamine, alanine, and all the amino acids of the tricarboxylic acid cycle), and fatty acids with an odd number of carbons, into glucose for storage as glycogen.

The primary focus of this chapter is to describe the biological functions of six steroids important to or related to the GC and three important steroid drug forms. However, these nine biologically important steroids are a small number compared to the ~ 100 steroids that are covered in Chapter 2, Steroid Hormones: Chemistry, Biosynthesis, and Metabolism, which also provides an extensive review of the fundamental chemistry of steroids and their biological production by a large family of steroid enzymes which are required to convert cholesterol into each of the steroids that have functions in human cells.

C Mineralocorticoids

Mineralocorticoids are a class of steroid hormones that regulate salt and water balances. Aldosterone is the primary mineralocorticoid. Mineralocorticoids promote sodium and potassium transport, usually followed by changes in water balance. This function is essential to life. See Chapter 15, Hormones Related to the Kidney and Cardiovascular System, concerning the biological actions of aldosterone. The secretion of aldosterone is stimulated by four factors acting sequentially. The starting point is the plasma α2-globulin, also known as a kininogen (57 kDa); it serves as a substrate for the enzyme renin (42 kDa) which is secreted by the kidney's glomerulosa. The renin clips off a 12-amino-acid peptide (a decapeptide) named kininase II. Next the kininase II (also known as the angiotensin converting enzyme) removes two amino acids from the substrate angiotensin I, thus generating an octapeptide known as angiotensin II. The hormone angiotensin II binds to its receptor in the adrenal cortex's *zona glomerulosa* and initiates the biosynthesis of aldosterone (see Fig. 15.4A/B/C).

II Anatomy

A Adrenal cortex

The adrenal glands are well supplied with blood vessels, including a branch from the aorta and a branch from the renal artery. The blood entering the adrenal cortex is drained out through the veins of the adrenal medulla; see Fig. 10.1.

Hormones. DOI: https://doi.org/10.1016/B978-0-323-90262-5.00009-3

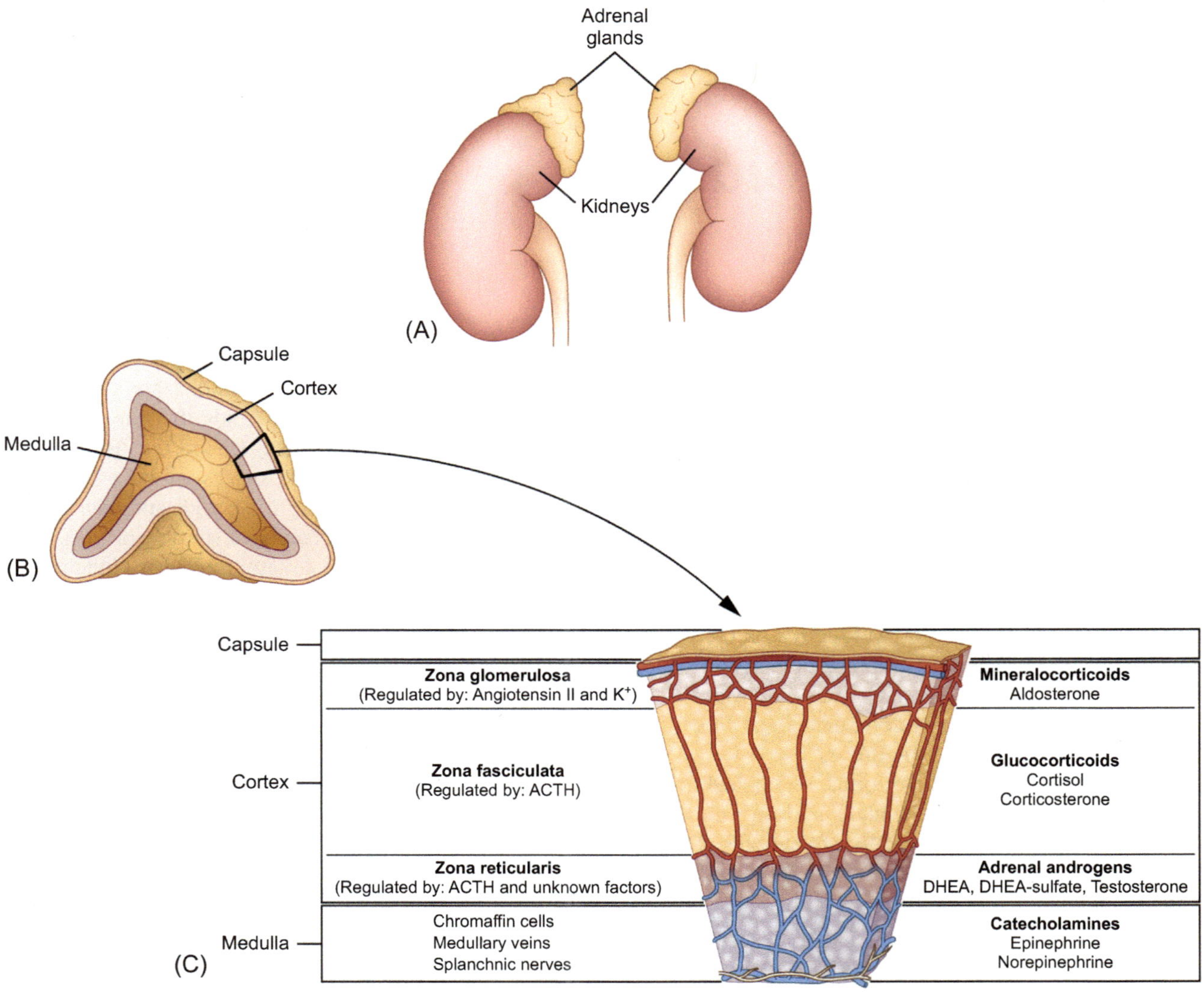

FIGURE 10.1 Structural details of the human adrenal glands. (A) Anatomical relationship between kidneys and their partner adrenal glands. The two adrenal glands have different shapes depending on their anatomical location. The right gland is shaped like a pyramid. In contrast the left gland is crescent shaped. Each adrenal gland sits on top of its partner kidney. (B) The adrenal gland has three distinct layers. The capsule is the outer protective covering of the adrenal gland. Immediately below the capsule is the cortex, which is the location of three *zona* layers. Each of the *zona* layers produces a different steroid hormone family. They are the mineralocorticoids (aldosterone) from the *zona glomerulosa*, the glucocorticoids (cortisol) from the *zona fasciculata*, and the adrenal androgens (dehydroepiandrosterone and testosterone) from the *zona reticularis*. (C) *Function of the medulla layer*. Below the cortex is the medulla layer. The medulla is the site of production of catecholamines, for example, epinephrine and norepinephrine. Readers who want in-depth information on epinephrine and norepinephrine should go to Chapter 11, Hormones of the Adrenal Medulla.

The adrenal gland consists of two major segments, an outer cortex and the central medulla. The three consecutive layers of the cortex are each responsible for the biochemical formation of a different family of steroid hormones. Starting with (1) the outer layer, *zona glomerulosa* produces aldosterone; (2) the middle *zona fasciculata* produces cortisol and corticosterone; and (3) the third innermost *zona reticularis* produces testosterone and dehydroepiandrosterone (DHEA), which are both androgens. Each of these steroid hormone families requires its own steroid metabolic pathway that has the appropriate steroid enzymes. The production of all steroid families starts with cholesterol, followed by three to five successive enzyme alterations that produce a specific steroid hormone. More detail on this process is provided in Section III, "Biochemistry and Biosynthesis of Adrenal Steroids and Testosterone Steroids."

The bottom or most interior layer of the adrenal is the medulla. The medulla produces catecholamines, particularly epinephrine and norepinephrine; see Chapter 11, Hormones of the Adrenal Medulla, for more details.

In fetal life the human adrenal is relatively large in size. The cortex present at birth is often referred to as the "fetal cortex" because of its transient life. At birth and for most of the first postnatal year, the primary cortex begins to atrophy, while the cells of the secondary cortex actively divide. As the fetal cortex disappears, the mature *zona glomerulosa* appears immediately adjacent to the capsule. With continued proliferation, the next inward zone, the *zona fasciculata*, arises and within a few months after birth the most inward layer of cells, the *zona reticularis*, is finally formed.

Also low amounts of corticosterone are secreted. The *zona fasciculata* is the largest of the three zones in the cortex. The *zona reticularis*, the innermost layer, is adjacent to the medulla.

B Liver

The human liver is positioned in the abdominal cavity on the upper right-hand portion just under the diaphragm; see Fig. 10.2A/B. The liver is also positioned on top of the right kidney, the stomach, and the intestines. The liver is a dark reddish-brown organ that is structured like a cone. It weighs approximately 6.6 kg or ~3 lb. Approximately 75% of cells in the liver are hepatocytes, with bile duct cells and Kupffer (scavenging) cells being in the minority.

The liver is divided into the right lobe and left lobe. Each of the lobes is made up of thousands of lobules which are, in turn, connected to small and then larger ducts that collectively form the hepatic duct. The hepatic duct is responsible for transporting the liver-synthesized bile to the gallbladder. When food is taken in and processed by the stomach, the gallbladder releases on demand bile to be delivered to the duodenum (see Chapter 7: Gastrointestinal Hormones concerning gastrointestinal hormones).

The liver contains ~12% or ~600 mL of an adult's total volume of blood, which is 4.7–5 L. The two lobes of the liver are each composed of thousands of lobules. The lobules are hexagonal and create a liver-specific circulatory system. See Fig. 10.2, panel (B). Blood flows through the sinusoids of the liver and empties into the central vein of each lobule. Each lobule is serviced by arterial and venous blood at the same time. The sinusoids are constructed from branches of both the portal vein in the liver and from hepatic arterioles.

The liver is responsible for a wide array of specific responsibilities. Box 10.1 lists some 12 activities carried out by the liver from a longer list of 500 responsibilities; see the box title.

Inspection of Fig. 10.2B will confirm that incoming blood to the lobule is delivered by both the portal vein and the hepatic artery. The two blood compositions then become mixed together via the hepatic sinusoid cells. Sinusoid cells are fenestrated (have windows), which inevitably facilitates mixing of the two blood sources. The Kupffer cells are also associated with the sinusoids and function as macrophages which kill any bacteria that are present in the blood coming from the intestine.

The bottom of the liver is positioned immediately over the gallbladder. The hepatic duct is responsible for transporting the liver-synthesized bile to the gallbladder. The bile is a yellowish brown to dark green fluid produced by the liver. The bile is stored in the gallbladder until there has been food intake, whereupon the bile is transported to the duodenum to assist with the digestion and absorption process of the food taken in. The composition of bile is ~85% water, ~10% bile salts, 1% fat, and 0.6% inorganic salts.

III Biochemistry and biosynthesis of adrenal steroids and testosterone steroids

The structures of the six principal steroid hormones (actually seven structures, including the two forms of aldosterone) enzymatically produced by the adrenal cortex, and three related drug forms, dexamethasone, prednisolone, and RU-486 (mifepristone), are shown in Fig. 10.3. Each background color (*green, peach, and light blue*) defines the appropriate *zona* layer.

The chemical structure and biological relationships of the naturally occurring GC (cortisol and corticosterone) and the drug form, dexamethasone, are members of the ~100 steroid hormones that are reviewed in Chapter 2, Steroid Hormones: Chemistry, Biosynthesis, and Metabolism. Also the detailed metabolic pathway for the production of GC is presented in Chapter 2, Steroid Hormones: Chemistry, Biosynthesis, and Metabolism, in Fig. 2.11 in the green panel (by names) and Fig. 2.19 in the peach panel (by chemical structures).

The naturally occurring cortisol is the most prevalent member of the family of GC and binds tightly to the glucocorticoid nuclear receptor. Whenever the blood concentration of cortisol falls below the normal circulating concentration of 6–20 μg/100 mL, additional cortisol will be produced. The daily secretory rate of cortisol is 10–20 mg/day. Additional cortisol will be secreted in response to a low blood concentration of cortisol; see Table 10.1. Also, when an individual experiences significant levels of stress, there will be an increased production of cortisol.

Cortisol's most important action is to increase and maintain blood glucose levels via the biochemical process of gluconeogenesis in the combined actions of muscle cells, fat cells, and liver cells; see Fig. 10.4. Cortisol, acting through its nuclear receptor in muscle cells inhibits protein synthesis and stimulates proteolysis, which produces free amino acids that will be transmitted via the circulatory system to liver cells and converted to glucose and then stored as glycogen. Similarly, cortisol and its nuclear receptor in fat cells inhibit lipogenesis (conversion of glucose carbons to fatty acids and stored as triglycerides) but stimulate lipolysis (breakdown of triglycerides to fatty acids). The fatty acids will subsequently

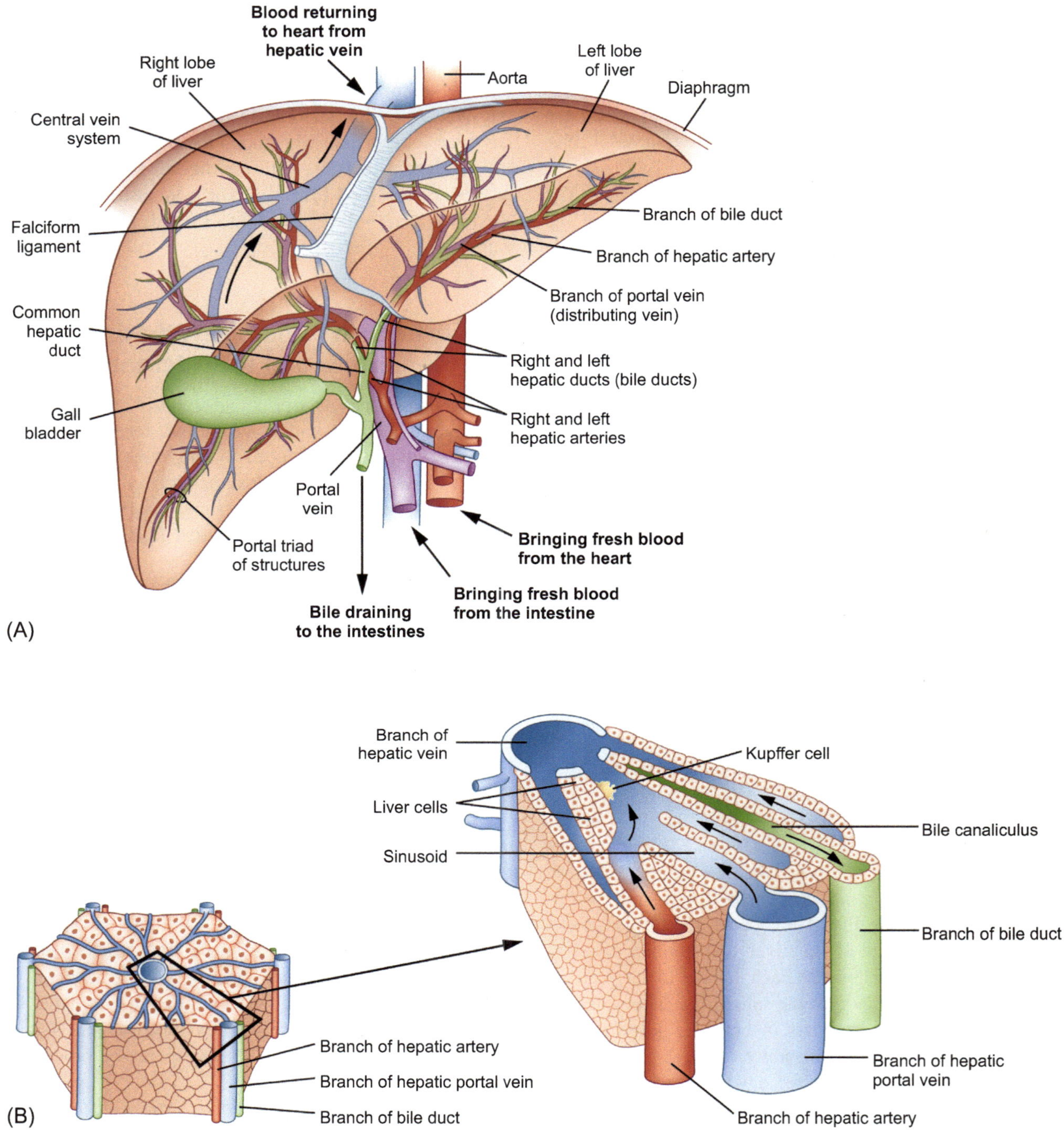

FIGURE 10.2 Structural organization of the human liver. (A) Structure and circulatory organization. The human liver is positioned in the upper right-hand portion of the abdominal cavity just under the diaphragm. The liver is also positioned on top of the right kidney+the right adrenal gland, the stomach, and the intestines. The liver is a soft reddish-brown organ that is structured like a cone. It weighs approximately 6.6 kg or ~3 lb. It is the largest organ inside the body. Approximately 75% of the cells in the liver are hepatocytes, with bile duct cells and Kupffer (scavenging) cells being in the minority. Each of the liver's lobes is made up of thousands of lobules that are, in turn, connected to small and then larger ducts, which collectively form the hepatic duct. (B) The liver circulatory system is constructed of lobules. The liver as a whole comprises lobules. Each of the two liver lobes is composed of thousands of lobules. A single lobule comprises ~1 million individual liver cells that are specifically organized into a single hexagonal polygonal lobule. As shown in panel (B), the lobules are complex constructs organized into polygonal prisms. Each lobule is organized around two branches of the hepatic vein and at the same time to one branch of the hepatic artery, which delivers oxygen to the lobules. It has been stated via a wide number of sources that the liver has over 50 responsibilities. W. W. Li, Faculty of Medicine, Univ. of British Columbia and the Bad-Gut Society (http://www.badgut.org/information-centre/the-liver-2013-an-amazing-organ-1.html). Chapter 2, Steroid Hormones: Chemistry, Biosynthesis, and Metabolism, is dedicated to describing in detail the metabolic pathways that convert cholesterol into a wide number of biologically active steroids. Chapter 2, Steroid Hormones: Chemistry, Biosynthesis, and Metabolism, which has the structures of ~100 steroids, does include the 10 steroid structures that are included here in this figure. The reader may wish to visit Chapter 2, Steroid Hormones: Chemistry, Biosynthesis, and Metabolism, for more detailed information on steroid hormones.

BOX 10.1 Biological functions of the liver.

- Storage of *glycogen,* which acts as a buffer to maintain blood glucose levels.
- Conversion of poisonous *ammonia* into *urea* for urinary excretion.
- *Cleansing* of bacteria from blood.
- Breakdown of *hemoglobin* from worn-out red blood cells to conserve the Fe^{2+}.
- Metabolic conversion of several *steroid hormones* and *vitamin D* metabolism to 25(OH)-vitamin D_3.
- *Detoxification* of many drugs and other substances.
- *Metabolism* of cholesterol and fat.
- Synthesis and endocrine secretion of many *plasma proteins,* including clotting factors.
- *Clearing* the blood of drugs and other poisons.
- *Volume reservoir* for blood.
- *Production of bile* that participates in the catabolism of fats in the small intestine during digestion.
- *Regulation* of blood nutrients.

be transferred through the circulatory system to liver cells, converted to glucose, and stored as glycogen.

Special suggestion for the reader: Chapter 2, Steroid Hormones: Chemistry, Biosynthesis, and Metabolism, is the foundation from a steroid hormone perspective for Chapter 10, Adrenal Corticoids, (cortisol), Chapter 12, Androgens (androgens), Chapter 13, Estrogens and Progestins (estrogens+progesterone), Chapter 14, Hormones of Pregnancy, Parturition and Lactation (pregnancy and lactation), and Chapter 15, Hormones Related to the Kidney and Cardiovascular System (aldosterone). Chapter 2, Steroid Hormones: Chemistry, Biosynthesis, and Metabolism, covers only chemistry, biosynthesis, and metabolism of all biologically produced steroids. This chapter focuses mainly on the anatomical complexity of the adrenal gland and its three *zonas* as well as the production by the *zona fasciculata* of cortisol and corticosterone. Those readers who wish further information on androgens, estrogens, progestins, and mineralocorticoids should visit Chapter 2, Steroid Hormones: Chemistry, Biosynthesis, and Metabolism.

IV Biological and molecular actions of glucocorticoids

A Hypothalamic–pituitary–adrenal axis

The concept and general functionality of the hypothalamic–pituitary–adrenal axis is presented in detail in Chapter 3, The Hypothalamus and Anterior Pituitary, particularly in Figs. 3.1 and 3.2. Signals to the hypothalamus organ can come from the brain cortex, the limbic system, the hippocampus, and also other brain regions. The hippocampus is a clearinghouse for the circumstances of stress. The secretion of cortisol can be regulated by the absence or presence of stress, either acutely or in long-term circumstances.

Shown in Chapter 3, The Hypothalamus and Anterior Pituitary, in Fig. 3.2, is the widespread importance of the hypothalamus in six hormonal systems in the body. Of these, four are steroid hormones: testosterone, estradiol, progesterone, and cortisol. The hypothalamus, in response to incoming messages, will secrete either a hypothalamic release-inhibiting factor or a positive releasing hormone that will travel through the circulatory system to its target. In the present case of stress, the hypothalamus sends the corticotropin-releasing hormone (CRH), a peptide of 41 amino acids, to the pituitary where the hormone ACTH (adrenocorticotropin hormone) is released into the circulatory system and delivered to the adrenal cortex's *zona fasciculata.* This then results in the activation of steroidogenesis which produces cortisol.

B Corticotropin-releasing hormone

The 41-amino-acid primary sequence of CRH peptide is illustrated in Fig. 3.2. Fig. 10.8 in Section IV.F. illustrates the (+) and (−) initiating response of the limbic system with the hippocampus to regulate the production of CRH by the hypothalamus. Following signals from the limbic system to the hypothalamus (Fig. 10.5), CRH (*white box*) is released into a closed portal circulation intimately connected with the anterior pituitary. The pituitary then is stimulated to produce and release ACTH. The ACTH next travels through the circulatory system to its receptors in the *zona fasciculata* where it stimulates the production and release of cortisol.

The catecholamines are considered in-depth in Chapter 11, Hormones of the Adrenal Medulla, but are presented in outline form here because stress reactions involve both GC and catecholamines.

C Transport of glucocorticoids in the blood (corticosteroid-binding globulin)

As touched on in Chapter 1, Hormone Introduction, each of the steroid hormones is quite hydrophobic and is not capable of moving freely through the blood compartment. Thus each steroid hormone has developed its own personalized blood transport protein.

A corticosteroid-binding globulin (CBG) is present in blood. This protein is also referred to as transcortin. It is

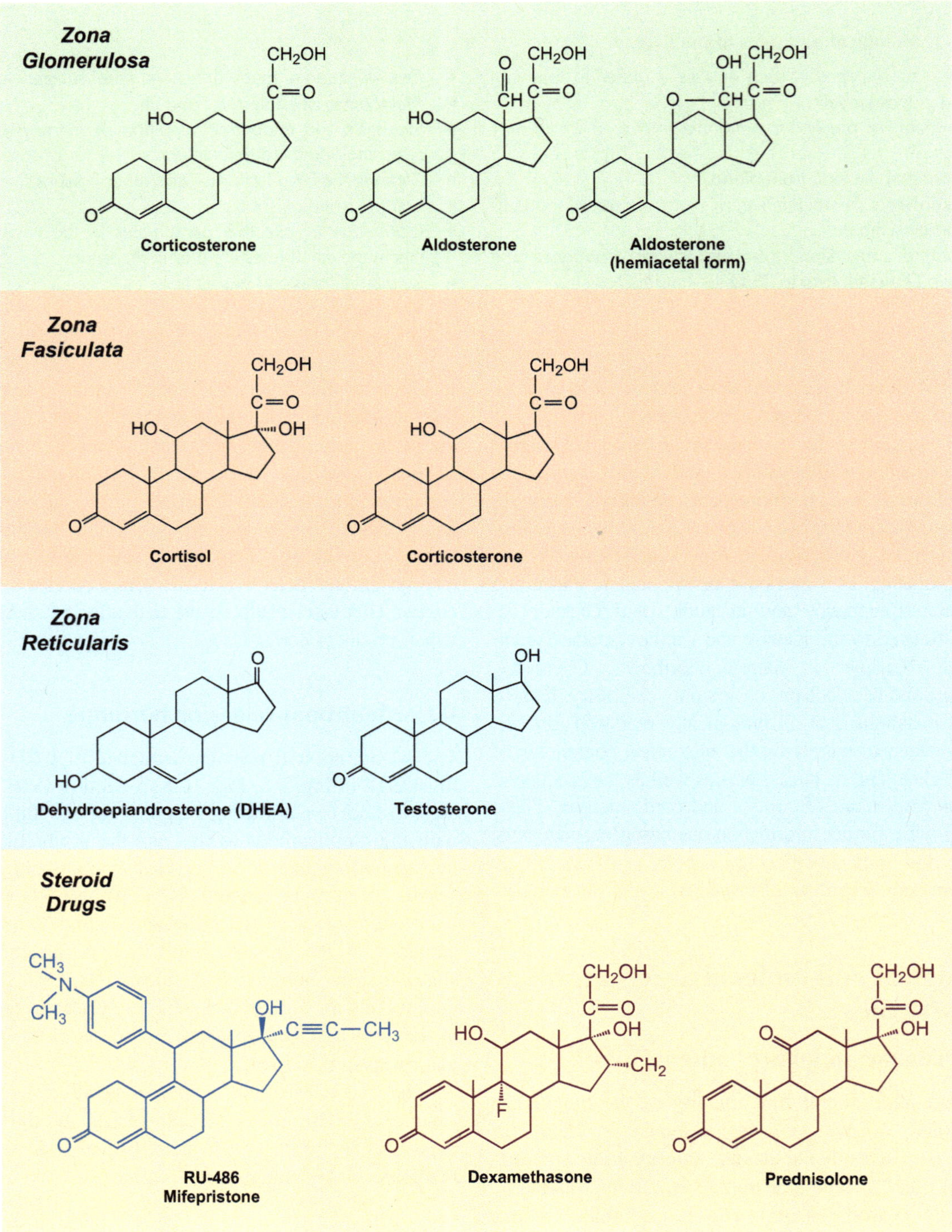

FIGURE 10.3 Presentation of the chemical structure of seven key biosynthesized steroids and three chemically synthesized steroid drugs. The chemical structure of the key steroids produced by each *zona* of the adrenal cortex is as follows. The outermost *zona glomerulosa* produces aldosterone (*green background*). The biological properties of aldosterone are discussed in Chapter 15, Hormones Related to the Kidney and Cardiovascular System. The second layer, the *zona fasciculata*, produces cortisol and corticosterone (*pink background*). The innermost *zona reticularis* is the source of two androgens, namely, DHEA and testosterone (*blue background*). Also shown are the chemical structures of three steroid drugs, namely, dexamethasone, prednisolone, and mifepristone (RU-486). Dexamethasone has both immunosuppressant and antiinflammatory effects. It is ~20× more potent than cortisol and is devoid of aldosterone effects. RU-486 is used to block pregnancies. *DHEA*, Dehydroepiandrosterone.

TABLE 10.1 Normal values for steroids from the adrenal cortex.

Steroid hormones	Normal range	Steroid hormones	Normal range
Production rates of major hormones		Urine excretion	
Cortisol (mg/day)	10–20	Free cortisol	150 μg/day
Aldosterone (μg/day)	60–200	Cortisol in certain disease states	~100,000 μg/day
Testosterone (mg/day)			
Males	5–8	Estrogens (μg/day)	4–25
Females	0.3–0.8	Males	4–60
Dehydroepiandrosterone (mg/day)	25–30	Females	
Plasma levels (μg/100 mL)		Pregnanetriol (mg/day)	0–4
Cortisol	5–20 (at 8:00 a.m.)	Pregnanediol (mg/day)	
Aldosterone	0.002–0.01	Males	0–1.5
Testosterone		Females	
Males	0.5–1.0	Follicular	0.5–1.5
Females	0.03–0.06	Postmenopausal	0.2–1.0
Protein binding (%)		Aldosterone (μg/day)	4–17
Cortisol	80	Free cortisol (μg/day)	0–200
Aldosterone	70		
Testosterone			
Males	90		
Females	95		

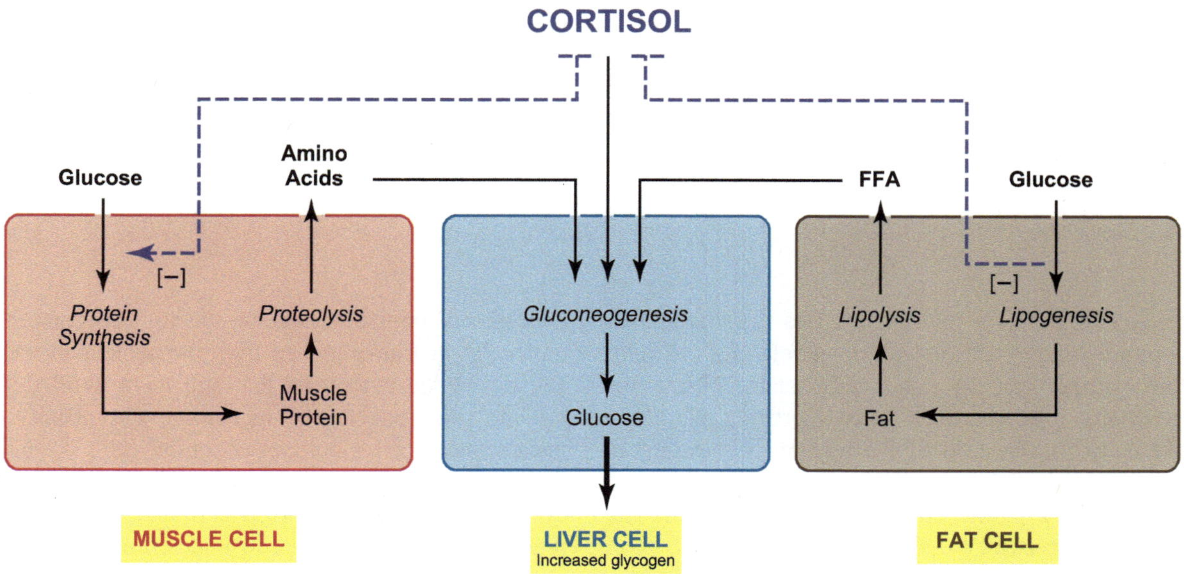

FIGURE 10.4 Cell-specific actions of cortisol on muscle, liver, and fat cells. Cortisol in muscle cells inhibits protein synthesis and in fat cells inhibits lipogenesis. Cortisol, acting through its nuclear receptor in muscle cells, inhibits protein synthesis and stimulates proteolysis which produces free amino acids that will be transmitted via the circulatory system to liver cells and converted to glycogen. Similarly, cortisol and its receptor in fat cells inhibit lipogenesis which is the process of acquiring blood glucose and storing it as fat. Cortisol also stimulates lipolysis, the breakdown of the triglycerides to free fatty acids, and glycerol. The fatty acids will subsequently be transferred through the circulatory system to liver cells and stored as glycogen. Both the fatty acids and the amino acids will subsequently be transferred through the circulatory system to liver cells and stored as glycogen.

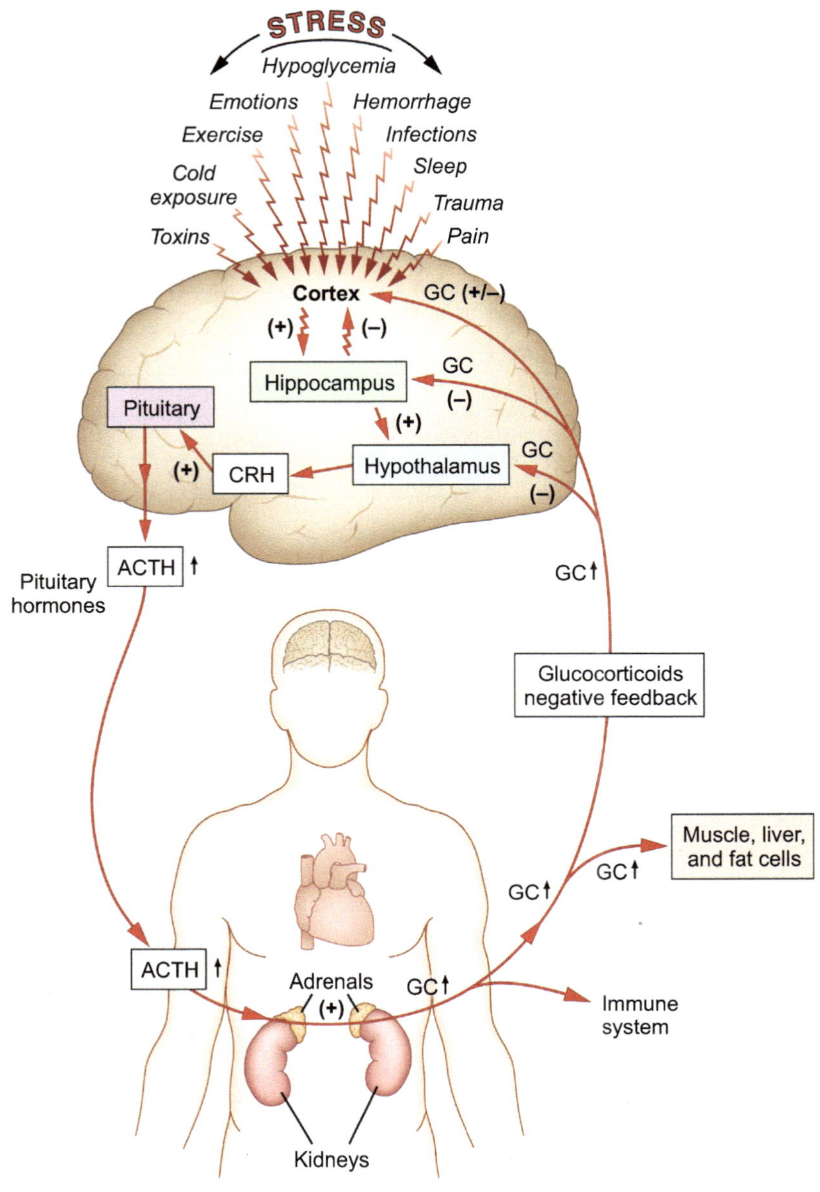

FIGURE 10.5 Communication of the brain cortex, via the hippocampus, hypothalamus, and the pituitary, with the *zona fasciculata* of the adrenal glands to produce ACTH, which, in turn, stimulates the production of glucocorticoids. The glucocorticoids (cortisol and corticosterone) that move through the circulatory system of the body bind to their nuclear receptor protein in target cells, which produces a wide variety of biological responses. One example is in the muscle and fat cells. See Fig. 10.4 where cortisol binds to its nuclear receptor and exports glucose to the liver cell and maximizes glycogen storage there. Here GC generate a collective negative feedback in the brain's cortex, hippocampus, hypothalamus, and pituitary. Also the consequences of an individual's exposure to stress (hemorrhage, pain, infections, emotions, or cold, etc.) stimulate the production of ACTH. ACTH, also known as CRH, is a polypeptide tropic hormone that is secreted by the anterior pituitary gland, which travels through the circulatory system to the adrenal gland's cortex *zona fasciculata* region. *ACTH*, Adrenocorticotropin hormone; *CRH*, corticotropin-releasing hormone; *GC*, glucocorticoids.

synthesized in the liver and exported to the circulation. This protein binds cortisol with relatively high affinity (binding constant $\cong 10^8\ M^{-1}$; the dissociation constant for the reaction, CBG+cortisol ↔ CBG-cortisol, is about 10^{-8} M cortisol). Because of the affinity of the protein for cortisol, most of the hormone circulates in the bound form as reflected in the equilibrium, which favors the complex: CBG+cortisol ↔ CBG-cortisol.

There is only a small amount of the free steroid hormone in a target cell. Nevertheless, at the target cell it is the free steroid that enters across the cell plasma membrane, probably by a free diffusion process. The driving force behind the movement of free hormone in the target cell appears, in part, to be proportionate to the number of unoccupied cortisol hormone-specific nuclear receptor molecules that have moved from the nucleus of the target cell out to the cytoplasm that still have unfilled ligand-binding sites (unoccupied receptors). The affinity of the nuclear receptor for cortisol is similar (20 nM) to that of the circulating CBG–cortisol complex. Steroid hormones cycle into and out of a target cell, and the proportion of retained molecules in the cell is determined by the number of unoccupied nuclear receptors.

D ACTH modes of action

Adrenocorticotropic hormone is the major peptide hormone secreted from the anterior pituitary in response to

CRH. ACTH is a 39-amino-acid peptide. An important cousin peptide hormone is α-melanocyte hormone (αMSH). The 13 amino acids of αMSH are cleaved out by a protease from the first 13 amino acids of ACTH.

The major corticotropic cells of the anterior pituitary are derived from a single gene that produces a key precursor protein, proopiomelanocortin (235 amino acids; pro-POMC). It is cleaved by proteases that produce a family of N-POMC (49 aa), Lys-γ3-MSH (24 aa), ACTH (13 aa), γ-LPH, β-MSH, and β-endorfin (31 aa). See Fig. 3.15 for more detailed information.

ACTH stimulates specific genes that are involved with the biosynthesis of selective steroidogenic enzymes, for example, P450 side-chain cleavage (scc) and the 11β-hydroxylase. Both are crucial to the production of cortisol; see Fig. 2.19. In the intermediary pituitary cells, ACTH is processed extensively into αMSH and CLIP (corticotropin-like intermediary peptide). In humans the peptide CLIP is known to bind to a major histocompatibility complex; however, a detailed understanding is not yet achieved. This occurs by both endopeptidase and exopeptidase cleavage together with acetylation and amidation. The major products of these cells are αMSH, CLIP, and *N*-acetyl-β-endorphin. In diseases (Cushing's disease) characterized by increased concentration of ACTH, usually due to a pituitary tumor, skin pigmentation can occur. This has been proposed to be due to MSH action.

Fig. 10.5 illustrates how stress can be detected in the brain cortex. The brain cortex communicates with the hippocampus, which then communicates with the hypothalamus, both via specific neurons. The hypothalamus then releases, as a messenger, the CRH to the hypothalamic-pituitary portal system. The pituitary then releases ACTH into the blood compartment. The ACTH, acting as a messenger, binds to its specific ACTH receptor in the adrenal cortex's *zona fasciculata.* The ACTH receptor is a G protein–coupled receptor. In Fig. 10.6 is shown the primary structure of the ACTH receptor (ACTHR), actually the melanocortin 2 receptor. Included are the sites of mutations in this receptor. In a second part of this figure, the cellular pathway is illustrated to show the general features of the intracellular movements of the ACTHR until it becomes finally located in the plasma membrane as a homodimer. Increased concentration of ACTH, usually due to a pituitary tumor, skin pigmentation can occur. This then results in the biosynthesis of cortisol and other GC.

E Glucocorticoids and stress

Fig. 10.7 shows two pathways describing the consequences of exposure to stress. The left-hand "stress" column is titled "Loss of control" which can lead to a long-standing circumstance (months and/or years) that includes extensive secretion of ACTH, which will elevate the circulating concentrations of corticosterone in the rat, cortisol in the human. The right-hand *Stress* column is titled *Threat to control.* Depending on the outcome of the *Active (fight–flight)*, there are two outcome paths (*brown*) titled *Control* or *Striving*. It is important to emphasize that these two pathways sometimes operate simultaneously, but they can also be dissociated from one another, especially when the individual is confronted with an emotionally stressful situation and copes on different levels.

The sympathetic adrenomedullary pathway (right-hand side) is activated when a fight or flight response is issued to a specific challenge. On the other hand, the humoral pathway (left-hand side), ending in cortisol release from the adrenal gland, is also operative when the individual becomes immobile, passive, and depressed as loss of control is perceived. A chronic emotional reaction of passivity and defeat to a stressful situation can produce dire consequences, as the adrenal hypertrophy and levels of cortisol continue to increase. This can generate a Cushingoid-like bodily reaction in which visceral fat accumulates and blood pressure becomes elevated, and arteriosclerosis and type II diabetes eventually develop. Sequential episodes of elevated GC cause sufficient repression of glucose uptake in peripheral cells to involve insulin release from the β-cells of the pancreas.

Eventually, chronic repetitions of this scenario can lead to the exhaustion of the β-cells' ability to produce and secrete insulin. The metabolic changes in the organism induced by elevated levels of GC, such as cortisol, are mediated by the amount of available glucocorticoid receptor (GR) proteins located in the cytoplasm of target tissues. Very important organs are liver, lymphocytes (including thymus cells), adipose cells, kidney, anterior pituitary, and various parts of the brain. Some of the effects of cortisol on tissues of normal or adrenalectomized rats are shown in Table 10.2.

GC are secreted in large amounts in humans, up to 25 mg or more per day, which constitutes a major chemical response of the body to stress. A person undergoing long-term stress will have higher amounts of cortisol circulating in the bloodstream than the unstressed person. The ability of cortisol to act on many of the tissues of the body is determined by the distribution and number of glucocorticoid nuclear receptors present in the cell of a particular tissue. The adult rat liver contains about 65,000 receptor molecules per cell which bind corticosterone, which is the principal glucocorticoid in the rat. Other important targets are the lymphoid cells, thymus gland, and kidney. Many other tissues appear to have enough receptor molecules to provide a response to stress, especially if it is long-term. In fact, most tissues of humans and animals or of cells in tissue culture seem to contain measurable amounts of the GR, making it theoretically

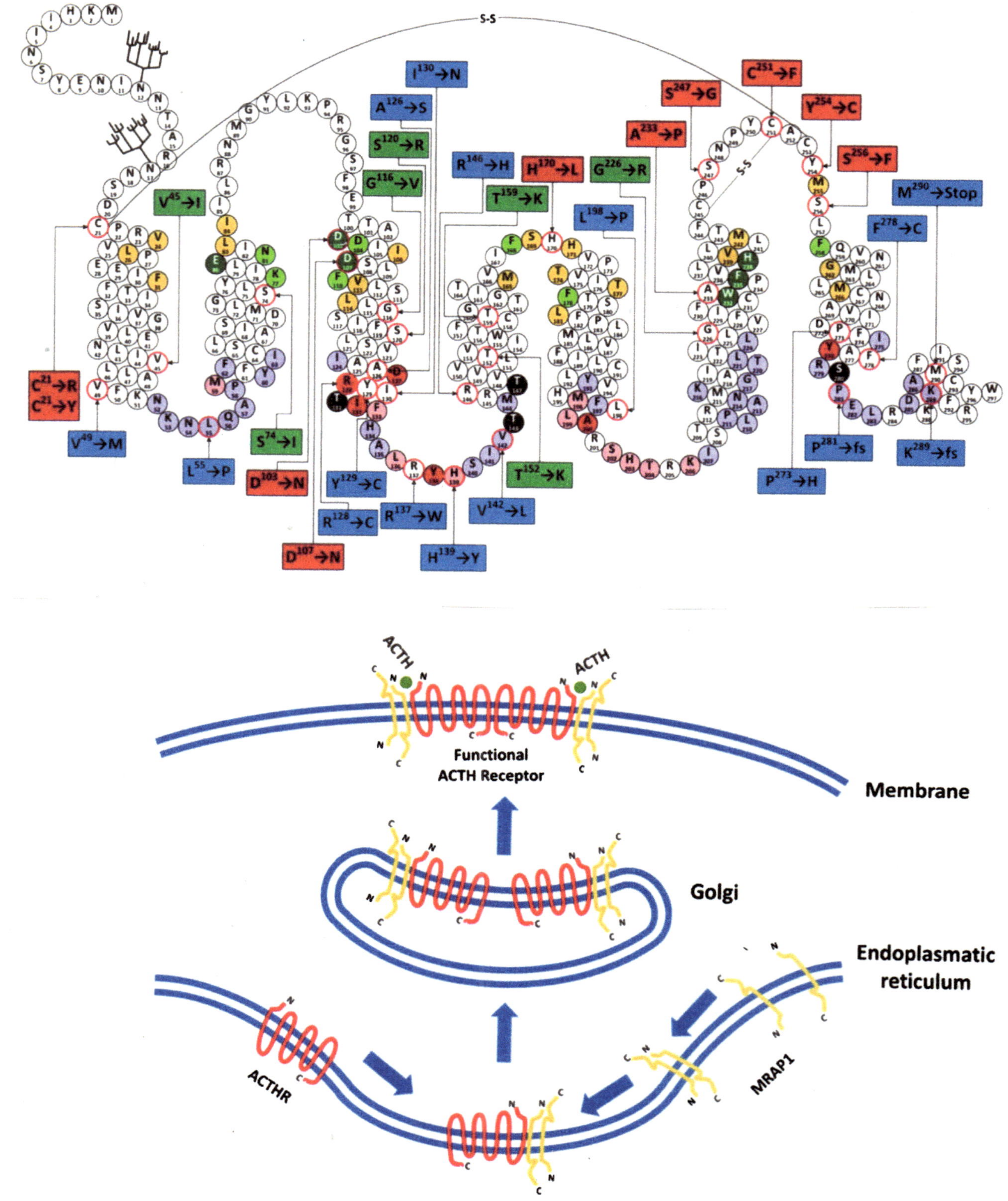

FIGURE 10.6 (Top) Snake-like plot of ACTHR. Forked structures represent glycosylation; -S-S- represents disulfide bonds. Circles with a red outline represent naturally occurring missense mutations (labels enclose information on the resulting residue replacement). Red labels indicate residues located on the extracellular part of the receptor. Green labels indicate residues located within the central part of transmembrane domains. Blue labels indicate residues located within the intracellular part of receptor. Dark green circles represent residues that form the backbone of the MCR-binding pocket. Orange circles represent residues equivalent to those forming the ligand-binding cavity of MC4R based on a previous model. Circles shaded *(Continued)*

possible for nearly all tissues of the body to be affected by stress.

Long-term stress may be distinguished from short-term stress, often referred to as "alarm" or "fright." In short-term stress the minute-to-minute changes in metabolism are under the control of catecholamine hormones, primarily epinephrine, secreted by the adrenal medulla. The secretion of epinephrine is, in turn, controlled by the autonomic nervous system. Both stresses, long- or short term, lead to the release of GC.

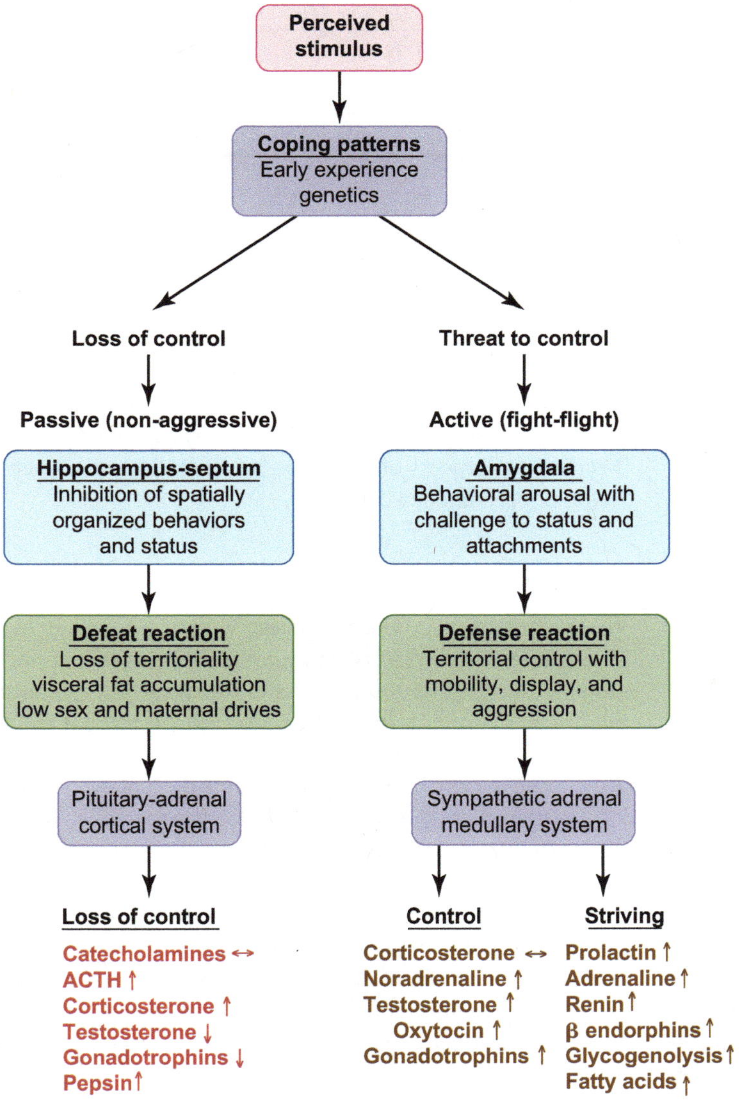

FIGURE 10.7 A defense reaction is activated when the organism is exposed to stress but remains in control (see left column). With the loss of control, there is activation of the hypothalamus–pituitary–adrenal axis, and the gonadotrophic species preservative system shuts down. Visceral fat accumulates in a Cushingoid distribution and there is a shift from active physical defense to a passive nonaggressive coping style. *Reproduced by permission from J.P. Henry, Biological basis of the stress response, Integrative Psychological and Behavioral Science 8 (1993) 69–73.*

in red, pink, and light purple represent residues equivalent to those that directly interact with the Gα, are involved in Gα coupling-related intramolecular interactions, and that form the surface of the Gα-binding cavity, respectively. Circles shaded in black represent S/T residues that are critical for ACTHR expression and functional regulation. Circles with angled cross-hatching represent residues that are unique to ACTHR (yet conserved in other MCRs). (Bottom) Formation of functionally active ACTHR. The functional form of ACTHR has a heterohexameric structure formed from two molecules of ACTHR and four molecules of MRAP1. The antiparallel dimer of MRAP1 is formed with the endoplasmic reticulum, where it couples to one molecule of ACTHR. Afterward, this complex is transported to the Golgi where it "dimerizes," thus forming the heterohexameric functional ACTHR. Symbols representing ACTHR and MRAP1 were selected to highlight the protein secondary structure. The depicted intermolecular interactions are, therefore, general and do not reflect the actual situation in detail. *ACTHR*, Adrenocorticotropic hormone receptor; *MCR*, melanocortin receptor; *MRAP1*, melanocortin-2 receptor accessory protein-1. *(Top) Reproduced form Fig. 6 of D. Fridmanis, A. Roga, J. Klovins, ACTH receptor (MC2R) specificity: what do we know about underlying molecular mechanisms? Frontiers in Endocrinology 8 (2017) 13, https://doi.org/10.3389/fendo.2017.00013. (Bottom) Reproduced from Fig. 3 of D.F. Fridmanis, A. Roga, J. Klevins, ACTH receptor (MC2R) specificity: what do we know about underlying molecular mechanisms? Frontiers in Endocrinology 8 (2017) 13, https://doi.org/10.3389/fendo.2017.00013.*

TABLE 10.2 Biological effects of cortisol.

Focus point	Outcome
Hepatic cells	**Stimulate gluconeogenesis storage of glycogen**
Muscle cells	Protein breakdown to amino acids for export to the liver
Adipose (fat) cells	Lipolysis ≫ free fatty acids for export to the liver
Immune system	Suppressive effects (apoptosis)
Antiinflammatory actions	IL-6 reduction
Response to stress	Adrenal hypertrophy

GC are so named because the steroid hormone influences the polymerization of glucose into the form of glucose macromolecules termed glycogen. This glycogen is an insurance policy that can be utilized on a minute-to-minute basis when the breakdown of glycogen is necessary to enable escape or survive the "fight or flight" or provide "nervous energy." Epinephrine is the primary hormonal signal that instantly activates the release of glucose molecules from the stored glycogen. Many other hormones are involved in stress, including glucagon, growth hormone, prolactin, β-endorphin, vasopressin, angiotensin II, and prostaglandins.

Thus it is possible to classify two types of stressors. One class includes responses that are in effect for a long period of time. These include intensive cold, prolonged loud noise, serious injury, burns, surgery, and significant changes in the environment. These stressful circumstances necessitate the "rapid" response adaptation. If adaptation does not occur, the second type of stressors lasts only for a very short interval. The alarm emergency or shorter term stressor would be an event of a surprise nature, such as fright induced by a specific happening. This type of stressor would primarily evoke the secretion of epinephrine (and norepinephrine) from the adrenal medulla, as well as cortisol secretion from the adrenal cortex. Both types usually occur simultaneously; one change in operation does not preclude the utilization of the other pathway. There are, however, certain conditions that can cause the pathways to operate separately. The sympathetic system generating epinephrine and norepinephrine is activated when the organism attempts to escape from or deal with the environmental challenge or the flight or flight response (see the right arm of Fig. 10.13 in Section V.B). When the organism becomes immobile or passive, the adrenocortical axis is preferentially activated.

F Immunosuppression and apoptosis induced by glucocorticoids

Cortisol can also be a potent inhibitor of the immune system. In some settings, cortisol activates a protein peptidase which, when activated, has selective actions on target proteins/enzymes. See an example in Fig. 10.8. A protease activates an inert protein/endonuclease enzyme into a functional DNAase, which breaks the genomic DNA into 50–300 base pair fragments. This creates a DNA ladder that experimentally can be used to quantitate the proportion of large, medium, or small DNA sections/fragments. Overall, this describes the process of apoptosis, which is the orderly activation of programmed cell death.

G Feedback effects of glucocorticoids

When cortisol is produced in response to ACTH, it has negative feedback effects on various elements of the hormonal cascade system that starts in the brain's cortex; see Fig. 10.5. There is a long feedback loop of cortisol (GC) that is initiated by cortisol release from the adrenal's *zona fasciculata* which then travels through the entire circulatory system to partner with the GR in target cells. Some cortisol will cross the blood–brain barrier and send a (−) signal to both the brain cortex and the hippocampus, so that the hypothalamus diminishes the CRH signal sent from the hypothalamus to the pituitary. This then lowers the rate of secretion of ACTH by the pituitary. The overall feedback inhibition of ACTH is implemented very rapidly. The feedback on the hippocampus may shut off further electrical activity responsible for the release of CRH and the subsequent release of ACTH. These actions are mediated by glucocorticoid nuclear receptors located in these cells that operate transcriptionally.

H The glucocorticoid receptor

The GR is a member of the nuclear steroid–thyroid, vitamin D receptor gene family (see Fig. 1.10). Like the other receptors in this superfamily, the GR can be divided into functional domains (see Fig. 1.10). The major functional regions consist of the ligand-binding domain (C-terminal) and the DNA-binding domain (central). The GR was the first transactivation protein discovered. Its structure has

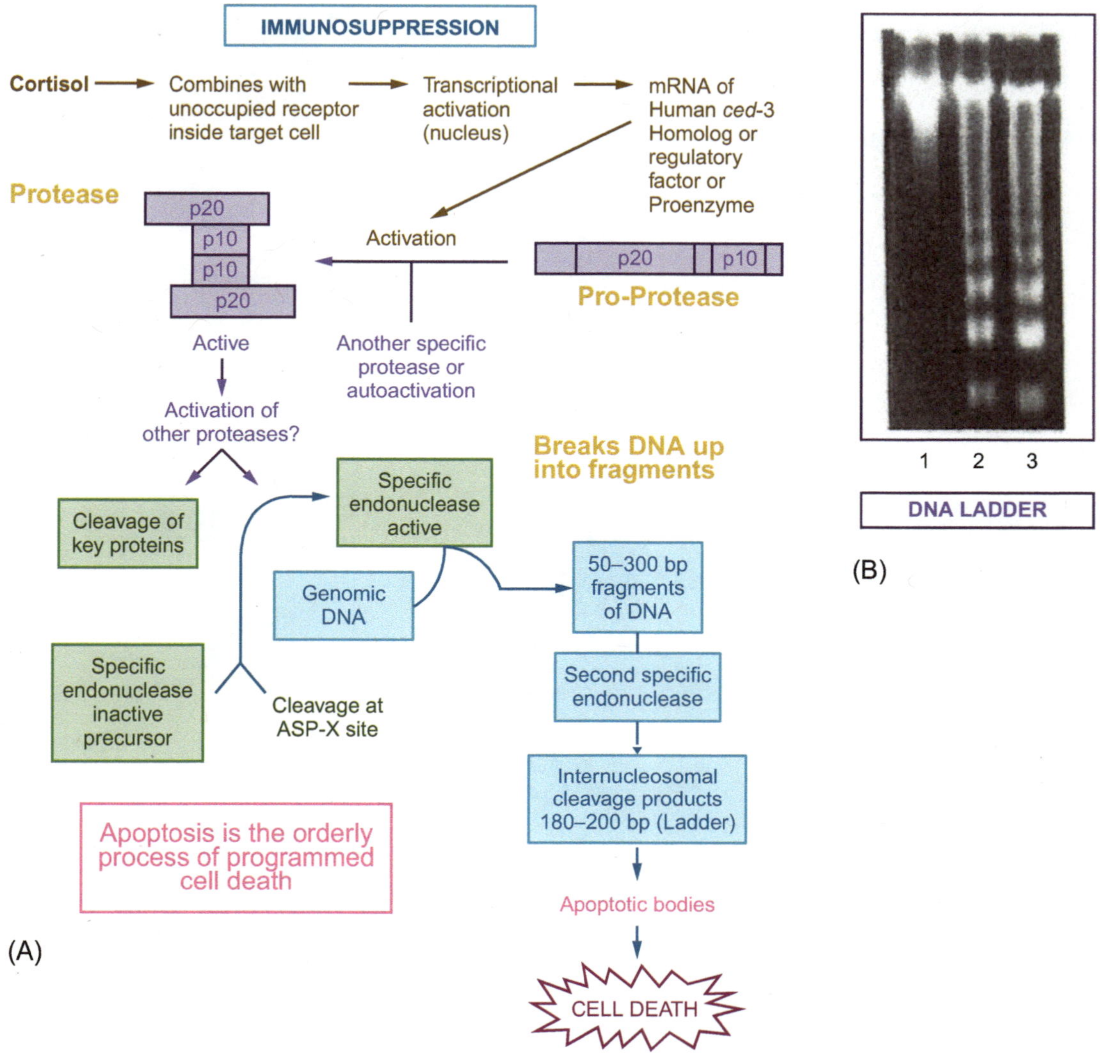

FIGURE 10.8 Scheme for the induction of apoptosis in B or T immune cells by glucocorticoids binding to nuclear receptors and inducing gene transcription. The nuclease becomes selectively activated by a specific protease cleavage, producing a functional protease that can translocate to the nucleus and attack genomic DNA. This generates a family of DNA fragments of 50–300 base pairs. The DNA library of fragments based on size will indicate the extent of destruction of the cells' DNA, which results in the cell's death, known as apoptosis. The smaller the DNA fragments are, the greater is the onset of death. The dead apoptotic cells are quickly recognized by macrophages and removed so that an inflammatory process is not generated. The presence of the DNA fragments can be evaluated by carrying out a chromatographic separation of the DNA fragments based on their physical size. Bigger fragments will be near the top of the column, while smaller fragments will move toward the bottom of a DNA ladder.

been investigated thoroughly and more recently by the method of cryo-electron microscopy that enhances resolution over X-ray crystallography. GR exists in the cell cytoplasm in an inactive form because other proteins form a complex with it so that the DNA-binding site on the receptor is masked. This inactive form consists of the complex, GR-$(HSP90)_2$-p23, as shown in Fig. 10.9 (top) described as the "maturation complex." Also shown in the figure (bottom) is the role of HSP90 in the generation of the maturation complex. Upon activation [binding of the ligand (e.g., cortisol or the artificial ligand, dexamethasone)], the associated molecules detach from the complex and the liberated monomeric GR is dimerized to the homodimer, GR–GR, which is transported into the nucleus where it can interact (through its unmasked DNA-binding domains) with cognate sites on DNA that generate gene expression or gene suppression, as the case may be.

I Dehydroepiandrosterone and zona reticularis

Up to 30 mg of DHEA is secreted daily by the innermost *zona* of the adrenal cortex, and the blood levels of this hormone and its sulfate derivative are fairly high; see Table 10.1. DHEA is considered to be a weak androgen. It can be converted to testosterone and androstenedione,

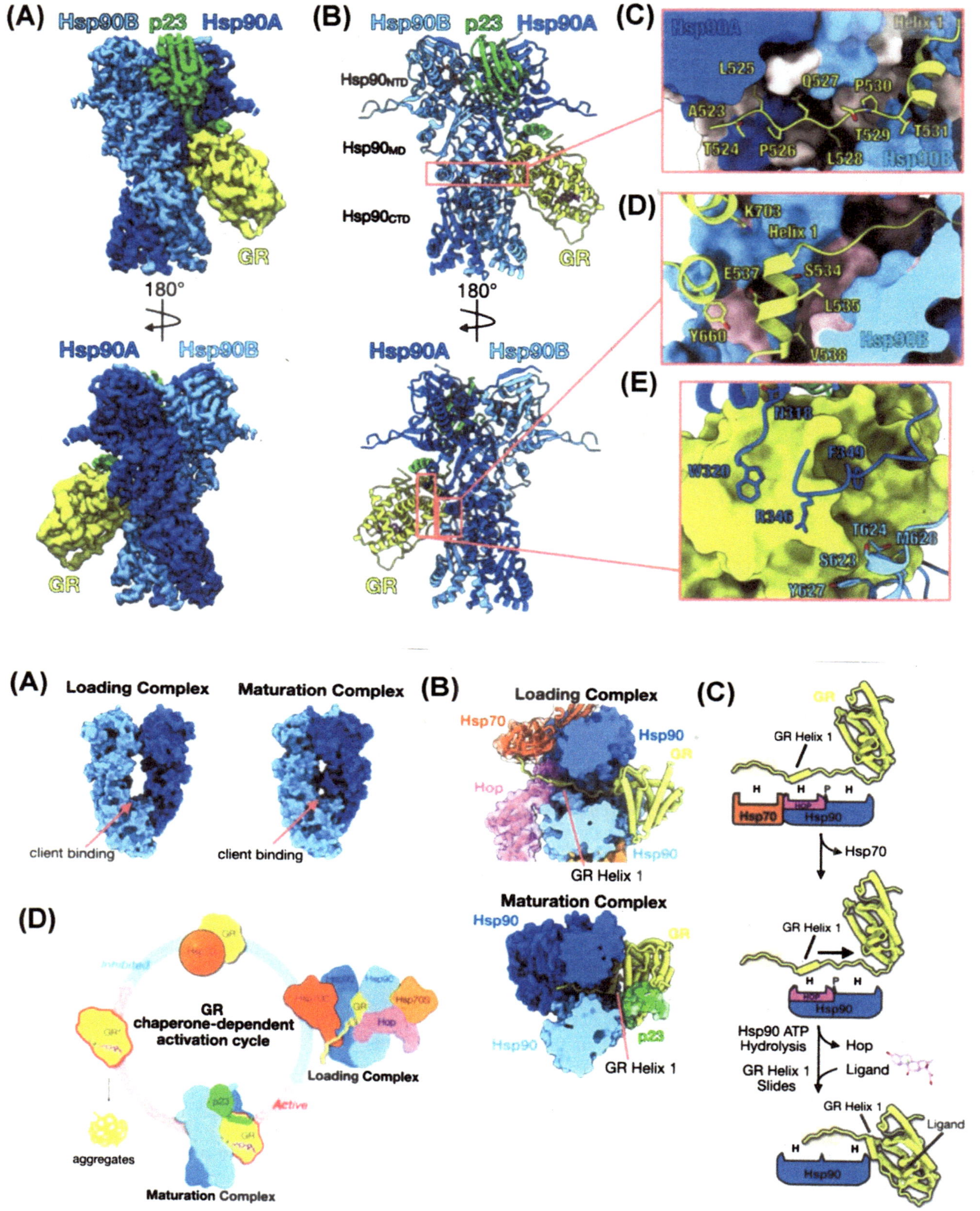

FIGURE 10.9 (Top) Architecture of the GR-maturation complex. (A) Composite cryo-EM map of the GR-maturation complex. Hsp90A (*dark blue*), Hsp90B (*light blue*), GR (*yellow*), and p23 (*green*). This color scheme is maintained in all figures that show the structure. (B) Atomic model built into the cryo-EM map. (*Continued*)

as shown in Fig. 2.20. When the aromatase system of enzymes is present, DHEA can be converted to estrogens, as summarized in Fig. 2.21. The conversion of DHEA to other steroidal compounds that serve as agonists for hormone receptors is reviewed in Fig. 10.10. Although its derivative, DHEA sulfate, is very highly concentrated in the plasma, its level declines with age, a fact that some believe renders it an antiaging factor. For many years, researchers searched for a receptor, specific for DHEA, like the classical hormone receptors for other steroids. Only fairly recently has it been found to be an agonist or antagonist for several neurotransmitter receptors. It is known now to be a positive allosteric modulator of the *N*-methyl-D-aspartate receptor. Also, it is a negative allosteric modulator of the gamma-aminobutyric acid receptor and it is an agonist of the σ1(sigma1) receptor. These actions of DHEA are capitulated in Fig. 10.10 (bottom).

V Clinical aspects

Major clinical problems associated with the adrenal cortex center around hyper- or hypoproduction of adrenal steroids. The causes of these conditions can be quite complicated since there are several organs involved in the hormonal cascade. In the case of the GC, the limbic system of the brain, the hypothalamus, the blood portal system connecting the hypothalamus with the anterior pituitary, and the adrenal gland are all components of the system. See Fig. 10.5. The normal concentrations of steroidal hormones produced by the adrenal are summarized in Table 10.1. When cortisol is overproduced, often by a pituitary tumor causing high levels of circulating ACTH, the resulting clinical problem is known as Cushing's disease. When cortisol is underproduced, the resulting disease is known as Addison's disease, which is most frequently the result of adrenal destruction.

A Cushing's disease

It is estimated that 10–15 million people in the world have Cushing's disease. Cushing's disease is readily diagnosed in adults within the ages of 20–50 years. Women are more prone (70%) to have Cushing's disease than men (30%). Cushing's disease generally occurs after the end of puberty with equal frequency in girls and boys. Cushing's disease is potentially fatal if untreated. The symptoms of Cushing's syndrome result uniformly from an excess of cortisol. At the present time, virtually all individuals with Cushing's syndrome can be treated effectively, and a significant proportion can be cured.

The many effects of Cushing's disease are summarized in Fig. 10.11 and Table 10.3. The figure illustrates the ~10 bodily locations where some aspect of Cushing's disease creates a visible body marker, which can be helpful in making a clinical diagnosis. Hyperproduction of cortisol from the adrenal gland would be expected to occur if there is a tumor of the adrenal producing abnormal amounts of the steroid or if there is a pituitary tumor producing high levels of ACTH. Adrenalectomy is often a treatment of the primary disease, with the concomitant maintenance levels of GC taken orally. Adrenal tumors are more common in the female adult. Fig. 10.12 shows a case of adrenocortical hyperplasia in a 5-year-old girl with congenital adrenogenital syndrome.

Congenital adrenal hyperplasia (CAH) is the terminology to describe several autosomal recessive diseases that result from mutations of enzymes or genes in the several pathways of steroid biosynthesis, for example, conversion of cholesterol to cortisol.

B Addison's disease

Addison's disease, or hypoproduction of cortisol, is usually the result of atrophy of the adrenal cortex by

(C) Interface I of the Hsp90GR interaction depicting the $GR_{preHelix\ I}$ region ($GR^{523-531}$) threading through the Hsp90 lumen. Hsp90A/B are in surface representation. Hydrophobic residues on Hsp90A/B are colored in pink. (D) Interface 2 of the Hsp90GR interaction depicting $GR_{Helix\ I}$ ($GR^{522-539}$) packing against the entrance to the Hsp90 lumen. Hsp90A/B are in surface representation. Hydrophobic residues on Hsp90A/B are colored in pink. (E) Interface 3 of the Hsp90GR interaction depicting residues on the $Hsp90A_{MD}$ loops ($Hsp90A^{N318,W320.P346,F349}$) and $Hsp90B_{amphi\alpha}$ ($Hsp90B^{S623,T624,Y627,M628}$), packing against GR, which is in surface representation. (Bottom) Surface representation of GR activation by Hso90. (A) Surface representation of GR activation by Hsp90 in the GR-loading complex versus the GR-maturation complex showing the change in Hsp90 conformation. Hsp90A (*dark blue*), Hsp90B (*light blue*). (B) Top view of the loading complex (top) where GR_{Helix1} is extended through the Hsp90 lumen versus a top view of the maturation complex (bottom), where GR_{Helix1} is docked back onto GR. Hsp70 (*orange*, partially transparent surface), Hop (*pink*, partially transparent surface), Hsp90A (*dark blue*, surface representation), Hsp90B (*light blue*, surface representation), GR (*yellow*), p23 (*green*, partially transparent surface). (C) Schematic showing the conformational change of GR_{Helix1} region from the loading complex to the maturation complex. Boxes represent chaperone and cochaperone-binding sites along the GR_{Helix1} region (H=hydrophobic interface, P=polar interface). The $GR_{pre\text{-}Helix1}$ strand is highlighted (*tan*). Color scheme is maintained from (B). (D) Schematic of the GR chaperone–dependent activation cycle. Ligand-bound, active GR (left) is aggregation prone under physiological conditions. GR ligand binding is inhibited upon binding to Hsp70. Inactive GR is then loaded onto Hsp90, with Hop (Hsp70/Hsp90 organizing protein), to form the loading complex, where GR_{Helix1} is extended through the Hsp90 lumen. Hsp70 and Hop leave and p23 is incorporated to form the maturation complex, where GR_{Helix1} slides back onto the body of GR and ligand binding is restored. *cryo-EM*, Cryo-electron microscopy; *GR*, glucocorticoid receptor. *(Top) Reproduced from Fig.1 of C.M. Noddings, R.Y.-R. Wang D.A. Agard, GR chaperone cycle mechanism revealed by cryo-EM: reactivation of GR by the GR:Hsp90:p23 client–maturation complex, bioRxiv (2021), https://doi.org/10.1101/2020.09.12294975. (Bottom) Reproduced from Fig. 3 of C.M. Noddings, R.Y.-R. Wang D.A. Agard, GR chaperone cycle mechanism revealed by cryo-EM: reactivation of GR by the GR:Hsp90:p23 client–maturation complex, bioRxiv (2021), https://doi.org/10.1101/2020,09.12294975.*

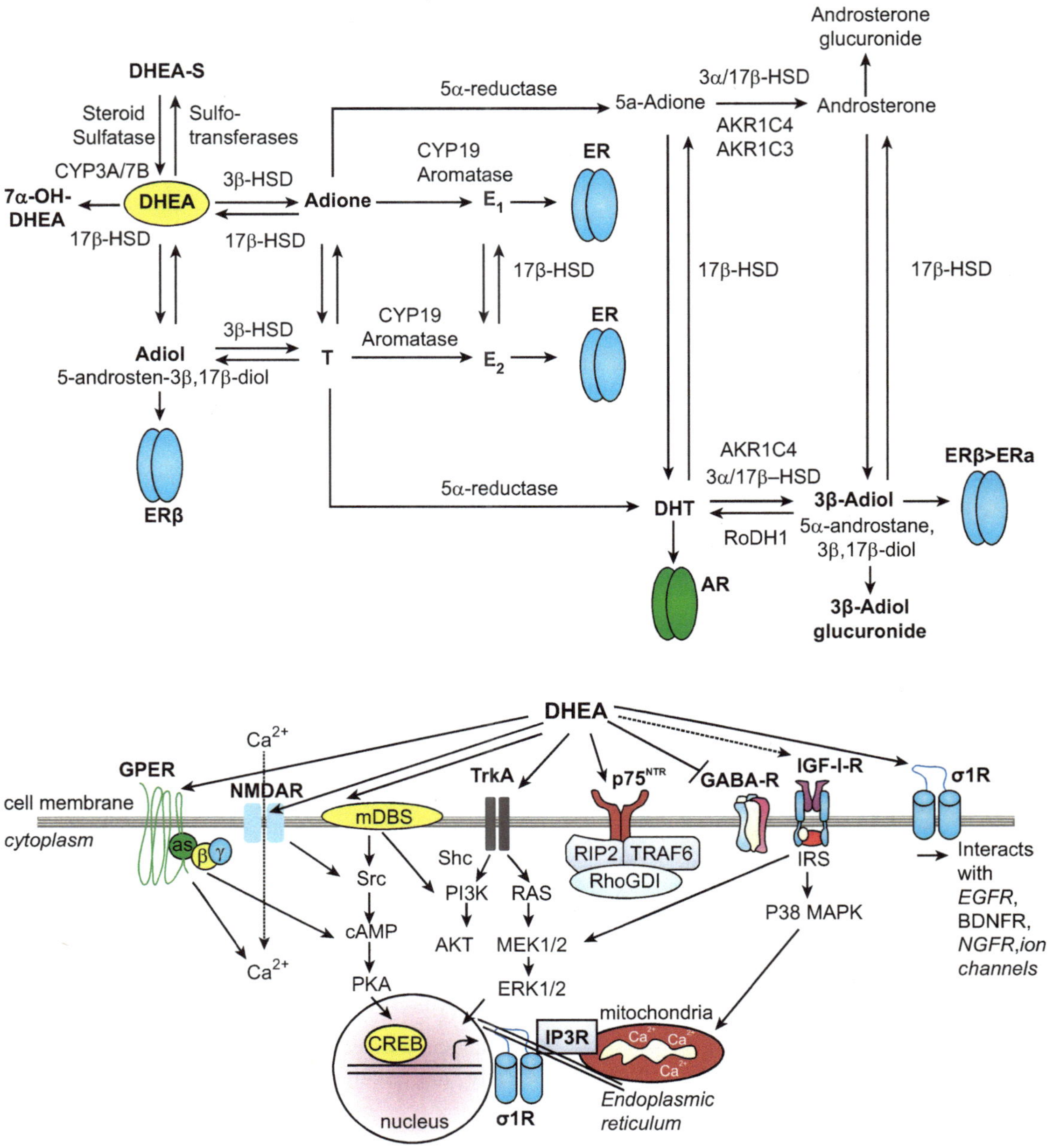

FIGURE 10.10 (Top) Selected aspects of DHEA metabolism. This model shows some of the pathways and enzymes by which DHEA is metabolized to steroids that bind the ER and the AR. AKR1C4 (aldo-keto reductase that reduces ketosteroids to hydroxysteroids with NADH or NADPH) is liver-specific, whereas AKR1C3 is expressed in many tissues, including the adrenal, brain, kidney, liver, lung, mammary gland, placenta, small intestine, colon, spleen, prostate, and testis. *AR*, Androgen receptor; *DHEA*, dehydroepiandrosterone; ER, estrogen receptor. *(Top) Reproduced from Fig. 1 of R.A. Prough, B.J. Clark, C.M. Klinge, Novel mechanisms for DHEA action, Journal of Molecular Endocrinology, 56 (2016) R139–R155. (Bottom) DHEA activation or inhibition of plasma membrane (PM) receptors. This model summarizes the interactions of DHEA with PM receptors. The sigma 1 receptor plays a key role interface between the estrogen receptor (ER) and the mitochondria [referred to as mitochondrial-associated endoplasmic reticulum membrane (MAM)]. IP3R modulates Ca^{2+} homeostasis. Reproduced from Fig. 3 of R.A. Prough, B.J. Clark, C.M. Klinge, Novel mechanisms for DHEA action, Journal of Molecular Endocrinology, 56 (2016) R139–R155.*

unknown causes (idiopathic). This can be the result of failure at other levels of the cascade from the limbic system downward. There is a requirement for life-long treatment with daily steroid replacement. The results of Addison's disease are sodium loss (since aldosterone usually is also deficient), excess potassium in the blood, low blood pressure, hypoglycemia, high levels of ACTH and MSH (loss of negative feedback), and pigmentation of the

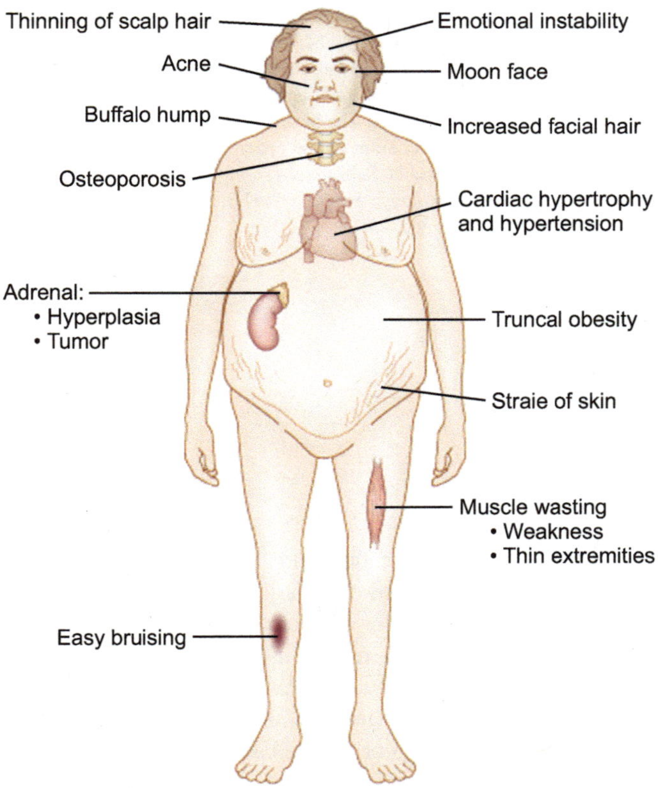

FIGURE 10.11 Consequences of long-standing Cushing's syndrome in a female person. Cushing's syndrome will frequently arise because of having a disease that generates excess cortisol. Also, excess CRH and/or ACTH as well as taking glucocorticoid drugs can sustain Cushing's syndrome. See also Table 10.3 which summarizes effects of overproduction of cortisol (Cushing's syndrome). *ACTH*, Adrenocorticotropin hormone; *CRH*, corticotropin-releasing hormone.

TABLE 10.3 Effects of overproduction of cortisol (Cushing's syndrome).

Protein metabolism	**Overt diabetes mellitus**
Reddish striae	**Centripetal fat distribution**
Thinning of skin	Moon face
Loss of bone matrix	Electrolyte balance
Poor wound healing	Hypertension
Capillary fragility and bruising	Hypervolemia
Muscle weakness and wasting	Sodium retention, potassium loss
Impaired growth (children)	Hematopoietic effects
Carbohydrate metabolism	Erythrocytosis
Abnormal glucose tolerance curve	Eosinopenia, lymphopenia
Overt diabetes mellitus	Polymorphonuclear leukocytosis
General effects	
Gastric ulceration	Hypercalciuria and renal calculi
Psychosis	Impaired immunological tolerance

Cushing's syndrome is most often caused iatrogenically by administration of large amounts of glucocorticoids.
Source: Reproduced from C. Ezrin, J.O. Godden, R. Volpe, R. Wilson (Eds.). Systematic Endocrinology. Harper & Row, New York, 1973.

skin due to the high levels of MSH and/or ACTH (Fig. 10.13). The acute onset of Addison's disease, referred to as Addisonian shock, is shock associated with adrenal insufficiency and most commonly occurs when an individual with long-standing adrenal insufficiency is exposed to a stressful stimulus. It is treated with large intravenous doses of cortisol, fluid replacement, and antibiotics. With the correct daily replacement of medication, the Addisonians are mostly able to continue life as they had before diagnosis of their illness.

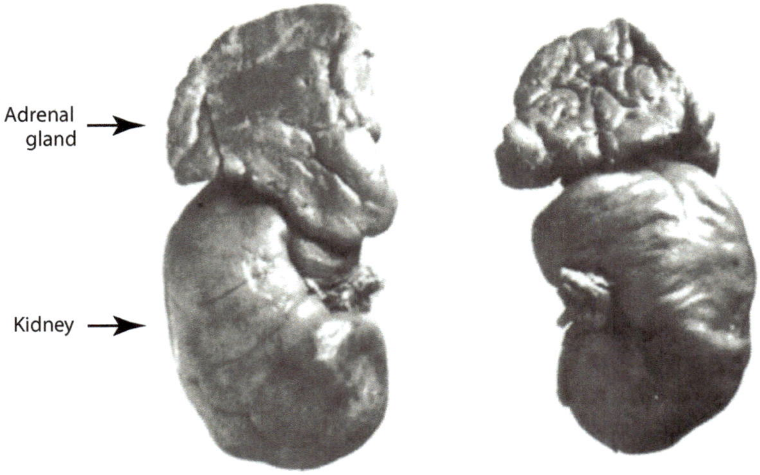

FIGURE 10.12 Bilateral diffuse adrenocortical hyperplasia of the adrenal glands in a 5-year-old girl with congenital adrenogenital syndrome. This photograph illustrates the close anatomical relationship between the adrenal glands and the kidneys. The bean-shaped organ below the adrenal gland is the kidney. The upper arrow points to the hyperplastic glands. *Reproduced from A. Labbart (Ed.), Clinical Endocrinology. Springer-Verlag, Berlin and New York, 1976, p. 362.*

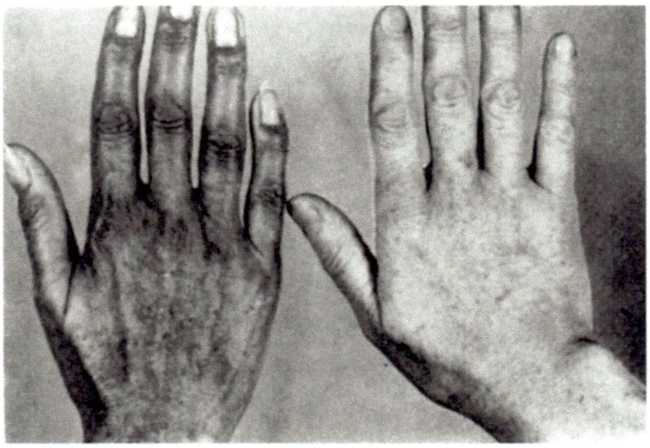

FIGURE 10.13 Example of pigmentation disorders associated with idiopathic Addison's disease. The hand on the left highlights the skin pigmentation problems experienced by a patient with Addison's disease. The hand on the right side is from a healthy person. *Reproduced from C. Ezrin, J.O. Godden, R. Volpe, R. Wilson (Eds.). Systematic Endocrinology. Harper & Row, New York, 1973, p. 178.*

Tuberculosis (TB) is the most common cause of Addison's disease worldwide. TB is a bacterial infection that mostly affects the lungs but can also spread to other parts of the body such as the adrenal gland. There is not a high death rate explicitly attributable to Addison's disease. The death rate of the entire population as a whole varies from 1:25,000 persons to 1:100,000 persons from country to country. With respect to the death rate of persons with Addison's disease, it displays on a country-by-country basis the same proportion of deaths around the world.

Mineralocorticoid excess can be a feature of Cushing's syndrome, as mentioned earlier. It can be the cause of reduced potassium levels in blood, resulting in weakness, sodium retention, and hypertension.

An inborn error of metabolism can result in a deficiency of the adrenal cortical enzyme, C-21 hydroxylase, function of which is required for the synthesis of cortisol. This deficiency results in lowered cortisol levels and enlarged output of ACTH by the anterior pituitary due to reduced negative feedback. High ACTH results in adrenal hyperplasia and hypersecretion, especially in the products of the *zona reticularis* (DHEA), which, together with adrenal testosterone, can lead to masculinization of female babies. Precocious puberty in males can also result from this condition. Treatment is centered on cortisol replacement therapy, which restores the negative feedback mechanism.

C Congenital adrenal hyperplasia and others

CAH refers to any of several autosomal recessive diseases resulting from mutations of genes for enzymes mediating the 3β-hydroxysteroid dehydrogenase, the 11β-hydroxylase deficiency, the 21-hydroxylase, and virilization. CAH can affect both girls and boys. About 1 in 10,000–18,000 children are born with some form of CAH. Under this heading is a family of diseases caused by a single enzymatic defect that results in the reduced synthesis of cortisol. Prenatal diagnosis is especially important in this disease. Table 10.4 lists the properties of the five various forms of adrenal hyperplasia.

Glucocorticoid resistance also is rare and may be based on point mutations in the GR that reduce the effectiveness of negative feedback by cortisol. As a consequence, there will be elevated levels of circulating cortisol. Clinical problems can be mild to excessive due to high levels of circulating cortisol or other steroids. Other variations generating high or low levels of GC or mineralocorticoids can be found in the clinical literature.

D Mifepristone (RU-486)

RU-486 is used widely around the world as an abortifacient in less than 49 days of pregnancy. As discussed in Fig. 13.8,

TABLE 10.4 Forms of congenital adrenal hyperplasia.

Enzyme deficiency	Syndrome	Ambiguous genitalia	Postnatal virilization	Salt metabolism	Steroids increased	Steroids decreased	Enzyme	Frequency
3β-Hydroxysteroid dehydrogenase	**Lipoid hyperplasia**	**Males**	**No**	**Salt wasting**	**None**	**All**	**P450α**	**Rare**
	Classic	Males	Yes		DHEA, 17-OH-pregnenolone	Aldo. Testosterone cortisol	3β-OH-dehydrogenase, Type II	Rare
		Females: mildly virilized				–		
	Nonclassic	No	Yes	Normal	DHEA, 17-OH-pregnenolone	–	3β-OH-dehydrogenase	Frequent
17α-Hydroxylase	–	Males	No	Hypertension	DOC, corticosterone	Cortisol testosterone	$P450_{17}$	Rare
17,20-Lyase	–	Males	No	Normal	–	DHEA, test. Δ^4-A	$P450_{17}$	Rare
21-Hydroxylase	Salt wasting	Females	Yes	Salt wasting	17-OHP, Δ^4-A	Aldo, cortisol	$P450C_{21}$	1/14,000
	Simple	Females virilized	Yes	Normal	17-OHP, Δ^4-A	Cortisol	$P450C_{21}$	–
	Nonclassic	No	Yes	Normal	17-OHP, Δ^4-A	–	$P450C_{21}$	See footnote a
11-Hydroxylase	Classic	Females	Yes	Hypertension	DOC, 11-deoxycortisol	Cortisol ± Aldo	$P450C_{11}$	1/100,000
	Nonclassic	No	Yes	Normal	11-Deoxycortisol	–	$P450C_{11}$	?Frequent
Corticosterone methyl oxidase type II	Salt wasting	No	No	Salt wasting in infancy	18-Hydroxy-corticosterone	Aldo		See footnote b

17-OHP, 17α-hydroxyprogesterone; *Aldo*, Aldosterone; *DHEA*, dehydroepiandrosterone; *DOC*, 11-deoxycorticosterone; *T*, testosterone; *Δ4-A*, Δ^4-androstenedione.
[a] *0.1%–1% (3% in European Jews).*
[b] *Rare (except in Iranian Jews).*
Source: Reproduced from M.I. New, Congenital adrenal hyperplasia, in: L.J. DeGroot (Ed.), Endocrinology, third ed., W.B. Saunders, Philadelphia, PA, 1995, p. 1814.

progesterone must be present in normal concentrations to maintain the pregnancy. RU-486 is an antagonist of the progesterone receptor and its presence can initiate the beginning of an abortion. It works by interfering with the supply of hormones that maintains the interior of the uterus. Millions of women worldwide have safely used mifepristone (RU-486). The use of RU-486 has been officially approved in 29 countries, including the United States.

Mifepristone is also used clinically to control high blood sugar in patients with Cushing's syndrome and who also have type 2 diabetes and have failed surgery or are not candidates for surgery.

Further reading

Cushing's disease

S. Athimulam, et al., Steroid profiling in the diagnosis of mild and overt Cushing's syndrome, Best Practice and Research Clinical Endocrinology and Metabolism 2021 (2021) 101488. Available from: https://doi.org/10.1016/j.beem.2021.201488.

E. Heyer, et al., Chemosensory dysfunction in Cushing's syndrome, Endocrine 2021 (2021). Available from: https://doi.org/10.1007/s12020-021-02707-z.

A. Juszczak, D. Morris, A. Grossman, Cushing's syndrome, in: K.R. Feingold, B. Anawalt, et al. (Eds.), Endotext [Internet] Feingold, South Dartmouth, MA, MD Text.com, Inc.; 2000-, 2021.

S. Vetrivel, et al., Circulating microRNA expression in Cushing's syndrome, Frontiers in Endocrinology (2021). Available from: https://doi.org/10.3389/fendo.2021.620012.

Gene expression, cortisol, corticosteroid-binding protein globulin, receptor and cortisol receptor

J. Gulfo, et al., Corticosteroid-binding globulin (CBG)-deficient mice show high pY216-GSK3β and phosphorylated Tau levels in the hippocampus, PLoS One 16 (2021). Available from: https://doi.org/10.1371/journal.pone.0246930.

J. Kroon, et al., The development of novel glucocorticoid receptor antagonists: from rational chemical design to therapeutic efficacy in metabolic disease models, Pharmacological Research 168 (2021) 105588. Available from: https://doi.org/10.1016/j.phrs.2021.105588.

N. Yoshikawa, et al., The influence of glucocorticoid receptor on sex differences of gene expression profile in skeletal muscle, Endocrine Research 360 (2021). Available from: https://doi.org/10.1080/07435800.2021-1884874.

Clinical aspects and involvement or cortisol

N.A.S.M. Azmi, et al., Cortisol on circadian rhythm and its effect on cardiovascular system, International Journal of Environment and Research in Public Health 18 (2021) 676. Available from: https://doi.org/10.3390/ijerph18020676.

L. Drougat, et al., Genomic and sequence variants of protein kinase A regulatory subunit type 1β (PRKAR1B) in patients with adrenocortical disease and Cushing syndrome, Genetics in Medicine 23 (2021) 174–182.

L. Liu, et al., Glucocorticoid receptor gene (*NR3C1*) is hypermethylated in adult males with aggressive behavior, International Journal of Legal Medicine 135 (2021) 43–51.

J.P. Vago, et al., Exploiting the pro-resolving actions of glucocorticoid-induced proteins annexin A1 and GILZ in infectious diseases, Biomedicine and Pharmacology 133 (2021) 11033. Available from: https://doi.org/10.1016/j.bopha.2020.111033.

Chapter 11

Hormones of the adrenal medulla

I Introduction

In the central nervous system, adrenergic neurons synthesize and release the catecholamine norepinephrine, while cholinergic neurons release acetylcholine. In addition to these two abundant neurotransmitters, other neurotransmitters are secreted by serotonergic neurons, dopaminergic neurons, melatonergic neurons, and enkephalinergic neurons, to name a few. Thus some polypeptides that have clear roles as endocrine hormones also act as neurotransmitters in some contexts, showing that the demarcation between the central nervous system and the endocrine system is not rigid; the two overlap and closely interact with each other. One example of this close relationship between the two communication systems is the hypothalamus, which transduces incoming neural signals from various areas of the brain to hormonal ones that act on the pituitary gland (Chapter 3: The Hypothalamus and Anterior Pituitary). The adrenal medulla, responding to neural input with the secretion of a catecholamine hormone, is another example of this intimate relationship between the two systems.

The adrenal medulla functions in partnership with the sympathetic nervous system in the sympathoadrenal system. The chromaffin cells that make up the adrenal medulla receive neuronal input from the sympathetic nervous system in the form of acetylcholine, which can signal an emergency requiring an immediate response. In response the chromaffin cell secretes a catecholamine, usually epinephrine, transducing a neural signal into a hormonal one. Circulating epinephrine alerts many peripheral tissues and organs to alter their function or metabolism to meet the emergency. The anatomical arrangements that underlie this system and the cellular processes involved in the adrenal medulla's response to stress are the subject of this chapter.

II Anatomical and physiological relationships

A The adrenal gland

The gross anatomy of the adrenal glands, shown in Fig. 11.1A, is described in detail in Chapter 10, Adrenal Corticoids, as are the structure and functions of the layers of the adrenal cortex, which makes up 80%–90% of the gland. The adrenal medulla is the innermost layer of the gland and weighs about half a gram. Fig. 11.1B shows the important circulatory relationship between the adrenal cortex and medulla. Blood flows from the cortex into the medulla, so that the cortical secretions, particularly cortisol, are available to the medulla. The vessels that carry this blood from the cortex to the medulla fuse into the medullary plexus, which drains into the suprarenal vein. During embryonic development the differentiation of the medullary chromaffin cells, which have migrated to their position within the adrenal from the neural crest, is prevented by cortisol from maturing into neurons with dendrites and axons, giving them their hormone-secretory character. In addition, throughout life, cortisol induces the enzyme that converts norepinephrine to epinephrine. Thus cortisol from the adrenal cortex dramatically influences the medulla, linking the two parts of the gland in the body's response to stress.

B The adrenal medulla and the sympathetic nervous system

The mature adrenal medulla is made up of modified neurons that possess neither axons nor dendrites and are essentially cell bodies that have been adapted to secretory function in response to a neural signal. During fetal development, it is cortisol from the fetal adrenal cortex that inhibits the growth of axons and dendrites on the medullary cells. These cells are referred to as the chromaffin cells of the adrenal medulla, because the presence of epinephrine and norepinephrine causes them to stain darkly with dichromate salts. These cells will be discussed in more detail following a discussion of their role in the sympathoadrenomedullary nervous system.

Fig. 11.2A depicts the autonomic nervous system that governs the involuntary workings of many glands and organs in the body. The autonomic centers in the central nervous system that initiate sympathetic responses lie mainly in the hypothalamus and brainstem, with input from the cerebral cortex and limbic system.

Hormones. DOI: https://doi.org/10.1016/B978-0-323-90262-5.00029-9

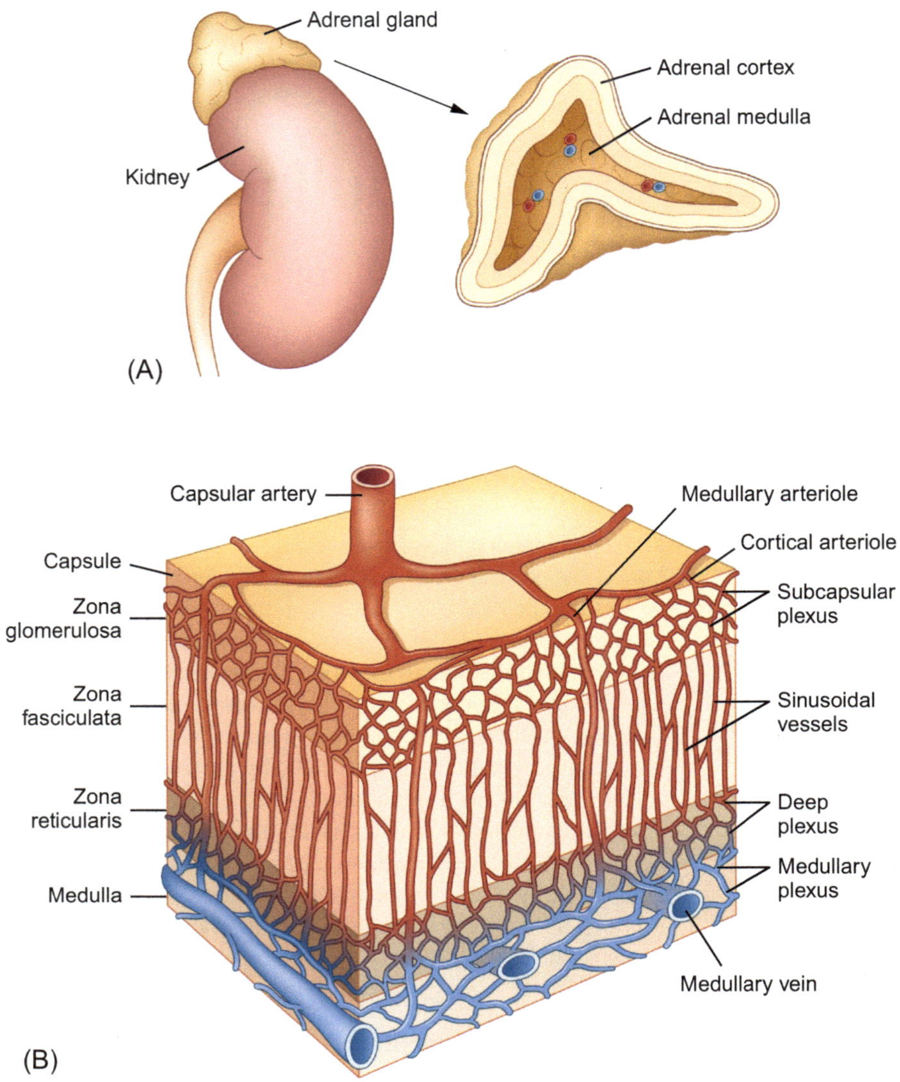

FIGURE 11.1 Anatomical relationships of the adrenal medulla. (A) The adrenal gland, encapsulated in fat, is shown in its position on top of the kidney. The layers of the cortex, of mesodermic origin, are indicated and are shown in more detail in panel (B). The functions of the adrenal cortex are the subject of Chapter 10, Adrenal Corticoids. The medulla, derived from the neural crest, is the innermost layer of the adrenal gland. (B) *The flow of blood through the adrenal gland is depicted.* Sinusoidal vessels branch off from the capsular arteries and carry blood through the cortex to the center of the gland, the medulla. This blood drains into the medullary plexus and the medullary veins to leave the adrenal gland.

The adrenal medulla is part of the sympathetic nervous system and in fact can be thought of as a large and specialized ganglion of it. As shown in Fig. 11.2B, the axons of the greater and lesser splanchnic nerves pass through, without synapsing, the celiac ganglion, one of the collateral ganglia of the sympathetic nervous system, to the adrenal medulla. At the surface of the adrenal medullary chromaffin cell, the neurotransmitter acetylcholine is released. In response to the activation of the acetylcholine receptor on the chromaffin cell and as described in more detail in Section III.A, epinephrine and norepinephrine are secreted into the bloodstream.

C Chromaffin cells of the adrenal

The adrenal medulla chromaffin cell itself is a modified postganglionic sympathetic catecholamine—releasing neuron that is analogous to the second neuron of the sympathetic nervous system. Whereas the postganglionic neurons of the sympathetic nervous system release their neurotransmitters (norepinephrine, acetylcholine, or dopamine) at a particular effector organ or tissue (e.g., heart, vasculature smooth muscle cells, and sweat glands), the chromaffin cells release their catecholamines into the circulatory system to be carried to numerous target organs.

In the medulla, chromaffin cells appear in clusters, with sparse connective tissue between the clusters. In addition to input from the sympathetic preganglionic nerve termini, the cells of these clusters are in communication with each other through gap junctions, so that each cluster may be thought of as a functional unit. Most of the chromaffin cells secrete epinephrine but a small proportion secretes norepinephrine. The two types of cells share general morphological features such as their

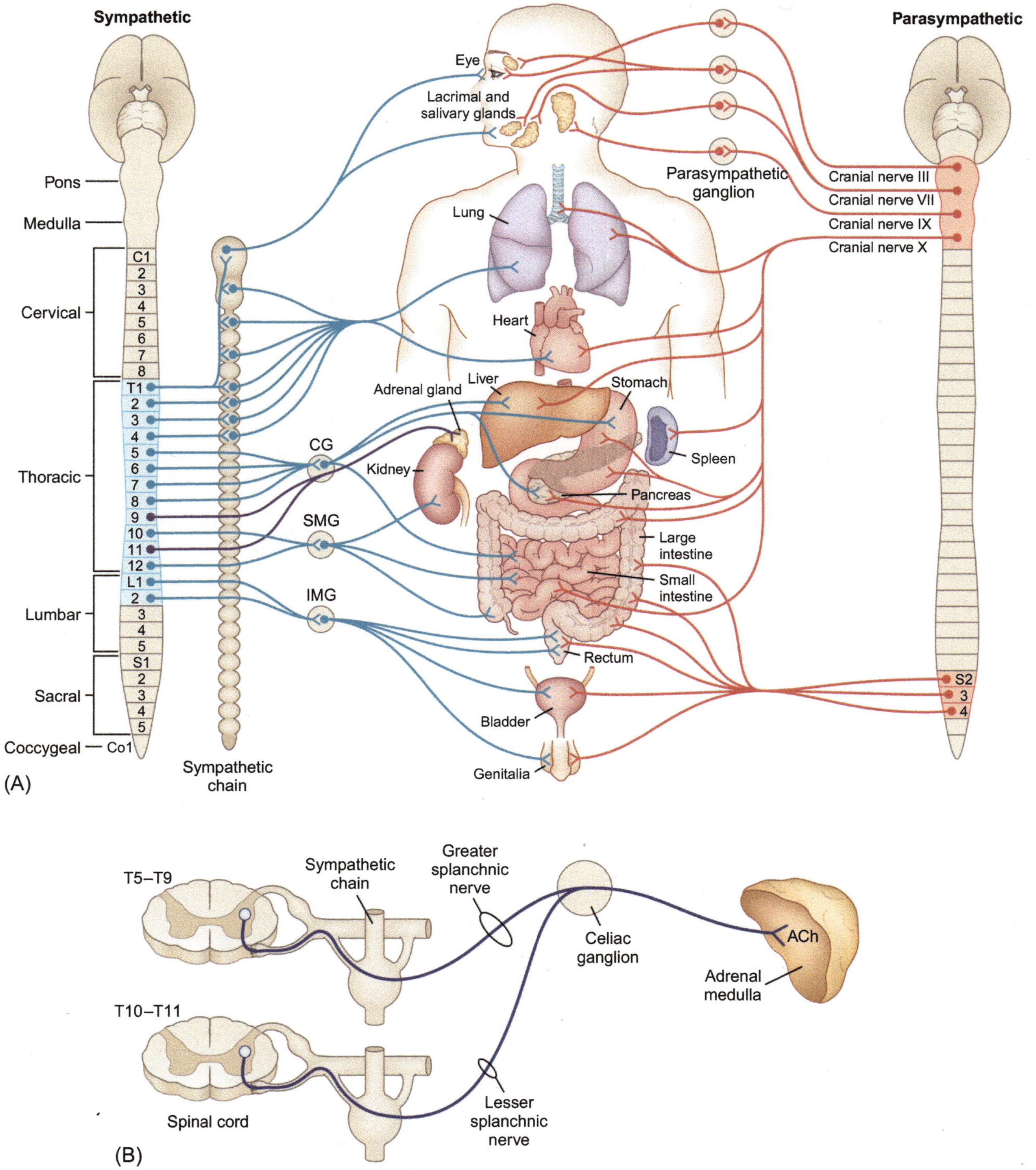

FIGURE 11.2 Relationship of the adrenal medulla to the sympathetic nervous system. (A) The autonomic nervous system is depicted with the sympathetic system on the left and the parasympathetic system on the right. The adrenal medulla is part of the sympathetic nervous system. The spinal nerves of this system, Thoracic nerve 1 through Lumbar nerve 2 (*blue*, T1–L2), receive input from the central nervous system and transmit the information to a synapse with a nerve in the sympathetic chain near the spinal cord or in a collateral ganglion, the CG, SMG, or the IMG. Acetylcholine is the neurotransmitter at these synapses. A postganglionic nerve then carries the signal to the appropriate organ, as shown, to generate the response to the signal from the CNS, with norepinephrine as the neurotransmitter. In contrast the greater and lesser splanchnic nerves (*purple*) pass through the celiac ganglion but do not make a synapse until they reach the adrenal medulla, as shown in part (B). Here the nerve terminal releases ACh at the surface of the chromaffin cells of the adrenal medulla. These cells (see text) serve as postganglionic nerve cells, but instead of making a neural connection of their own, they secrete epinephrine into the blood. *ACh*, Acetylcholine; *CG*, celiac ganglion; *CNS*, central nervous system; *IMG*, inferior mesenteric ganglion; *SMG*, superior mesenteric ganglion.

polyhedral shape and a large round nucleus containing one or more nucleoli, although some differences, some of which depend on the method of preparation, between the two types are seen in electron microscopic studies. Chromaffin cells contain 10,000–30,000 secretory vesicles (also called granules) 150–300 nm in diameter, a small fraction of which are located near or bound to the membrane in preparation for exocytosis.

The large electron-dense core vesicles, as distinguished from the small clear vesicles characteristic of most classical neurotransmitters, make up about 14% of the cytoplasmic volume of the chromaffin cell. These vesicles are formed at the *trans*-Golgi network where intravesicular protein aggregation and condensation take place in the slightly acidic environment. The mature vesicles contain epinephrine or norepinephrine, nucleotides, including ATP, and ascorbic acid. The main structural proteins of the vesicles are the granins, chromagranin A (CGA), chromagranin B (CGB) and in smaller amounts, secretogranin II, also known as chromagranin C. These three proteins constitute at least 80% of the soluble (as opposed to membrane) proteins of the vesicle. They share the property of binding calcium with low affinity (although the affinity of CGA for calcium is greater than that of CGB). Under the low pH and high calcium conditions of the trans-Golgi network, the granins associate with each other as well as with other vesicle constituents such as ATP and other proteins. Thus they are important players in the biogenesis of the secretory granules, although the details of their precise individual roles in this process have not yet been elucidated.

In addition to the chromogranins and other structural proteins, the enzymes and transport proteins required in the vesicle for catecholamine synthesis, and the proteins needed for exocytosis of the vesicle's contents, the chromaffin vesicles contain dozens to hundreds of other peptides and proteins. Some (but by no means all) of these are listed in Table 11.1. They include opioids and an array of autocrine and paracrine peptides that have the potential to modulate blood pressure and the immune system, to name a couple of examples. Several bioactive peptides are derived from one of the chromogranins and these are listed in Table 11.2. The vesicles also contain the protein processing enzymes to produce these peptides as well as the opioids and other peptides from their larger precursors or prohormones. The physiological roles that the secretion of these bioactive peptides from the adrenal medulla might play are not understood and will not be discussed further in this chapter.

TABLE 11.1 Bioactive peptides in adrenal medulla chromaffin vesicles.

General function	Peptide
Regulation of blood pressure	Adrenomedullin
	Natriuretic peptides: atrial (ANF), brain (BNF), and CNF
	Renin–angiotensin system
	Neuropeptide Y
	VIP
	CGA derivatives[a]
Regulation of immune system	Enkephalin
	Substance P
	Bombesin
	TGFβ
Analgesic	Endorphin
	Dynorphin
Other paracrine/autocrine effects	CGRP
	Neurotensin
	Galanin
	VEGF

ANF, Atrial natriuretic factor; *BNF*, brain natriuretic factor; *CGA*, chromagranin A; *CGRP*, calcitonin gene–related peptide; *CNF*, c-type natriuretic factor; *TGFβ*, transforming growth factor β; *VEGF*, vascular endothelial growth factor; *VIP*, vasoactive intestinal peptide.
[a]*Chromogranin A derivatives (see Table 11.2).*

TABLE 11.2 Bioactive peptide derivatives of chromogranin A.

Peptide	Function(s)
Chromagranin A	
Vasostatins I[a] (1–76), II (1–113)	↓ Vascular smooth muscle contraction
	↓ Parathyroid hormone secretion
	↑ Innate immunity processes
Catestatin (332–364)	↓ Catecholamine secretion
	↓ Blood pressure
Cateslytin (344–358)	↑ Innate immunity processes
Chromacin (173–194)	Antimicrobial
Secretolytin (614–626)	Antimicrobial

[a]*Numbers in parentheses are the chromogranin (bovine) amino acids that make up the peptide.*

III Chemistry and biochemistry of the catecholamines

A Biosynthesis and secretion of catecholamines

The pathway for the biosynthesis of epinephrine in the chromaffin cell is shown in Fig. 11.3. Tyrosine enters the cell from the blood and, in the cytoplasm, is converted by tyrosine hydroxylase to dihydroxyphenylalanine (L-DOPA). Tyrosine hydroxylase is the rate-limiting enzyme in the synthesis of the catecholamines in all adrenergic neurons as well as in the adrenal medulla. The enzyme requires the cofactor tetrahydrobiopterin for its hydroxylation activity. The availability of this cofactor can be the rate-limiting factor for this enzymatic step since intracellular concentrations are often subsaturating. Next, L-DOPA is converted to dopamine by an l-amino acid decarboxylase also located in the cytosol. Dopamine must then enter the neurosecretory granule where dopamine β-hydroxylase is located. Ascorbic acid is the cofactor for this mixed function oxidase reaction. In other adrenergic neurons the norepinephrine formed within the granule by dopamine β-hydroxylase remains there until secretion. In the chromaffin cell of the adrenal medulla, the norepinephrine from this reaction perfuses back out to the cytoplasm to be converted to epinephrine by the addition of a methyl group from *S*-adenosyl methionine. The enzyme that catalyzes this final step of this pathway is phenylethanolamine *N*-methyltransferase or PNMT. The final secretory product, epinephrine, must move back into the neurosecretory granule where it remains until the cell receives a signal to release the contents of the vesicles (see Fig. 11.4).

The transport of epinephrine back into the secretory vesicle is carried out by a specific ATP-dependent transporter, the vesicular monoamine transporter (VMAT) that is responsible for transporting epinephrine, norepinephrine, and dopamine in neurons throughout the nervous system. There are two isoforms, VMAT1 and VMAT2, that are encoded by separate genes. Among other characteristics in which the two isoforms vary, VMAT2 has a higher turnover number than VMAT1, making it suitable for rapidly recycling vesicles. Under basal conditions, all chromaffin cells express VMAT1, whereas VMAT2 is found in cells making norepinephrine, but not those synthesizing epinephrine. Under experimental stress conditions, however, VMAT2 in epinephrine-containing cells increases. This observation suggests that one adaptive response to stress is the ability of the secretory granule to accelerate the movement of epinephrine made in the cytoplasm back into the secretory vesicle for rapid secretion.

B Regulation of catecholamine synthesis and secretion

Tyrosine hydroxylase is the rate-limiting enzyme in the overall biosynthesis of epinephrine and as such is subject to several regulatory inputs that operate over both short (minutes) and long (hours) time frames. The enzyme is allosterically inhibited by the penultimate product of the pathway in the adrenal medulla, norepinephrine. Also, since tyrosine hydroxylase rarely functions at maximal velocity, its activity is controlled, to a certain extent, by the level of substrate (tyrosine) and cofactor (tetrahydrobiopterin). A number of external signals stimulate tyrosine hydroxylase activity through signaling pathways that result in the phosphorylation of specific serine residues near the N-terminal domain of the enzyme. This occurs primarily, but not exclusively, at serine-40 and counters the feedback inhibition by norepinephrine. In the longer

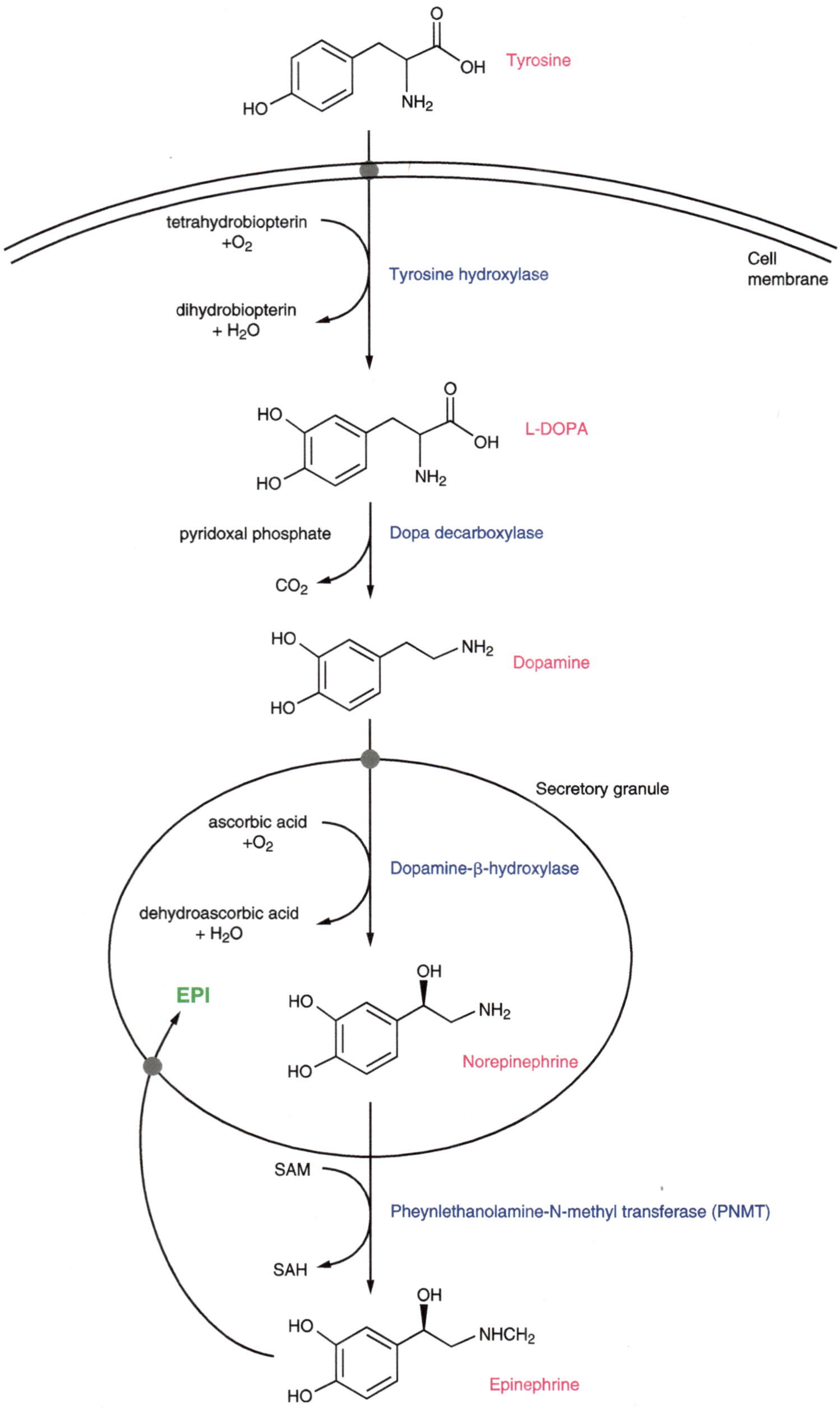

FIGURE 11.3 Epinephrine synthesis in the adrenal medulla. The intermediates, enzymes, and cofactors involved in epinephrine synthesis by the chromaffin cells of the adrenal medulla are shown. The pathway begins with the active uptake of tyrosine from the blood (*closed circle*, top of figure) followed by its hydroxylation by tyrosine hydroxylase. This is the first and rate-limiting step of the pathway. L-DOPA is decarboxylated by DOPA decarboxylase (also known as aromatic L-amino acid decarboxylase) to form dopamine, which is then transported into the secretory granule (*closed circle*, middle of figure) for conversion to norepinephrine by dopamine-β-hydroxylase. In the final step, which in the sympathomedullary system is particular to the adrenal medulla, norepinephrine returns to the cytoplasm for methylation by PNMT using the methyl group from SAM and generating SAH. The resulting final product, epinephrine, is taken back up into the secretory granule through the vesicular monoamine transporter (*closed circle*, bottom of figure). *L-DOPA*, Dihydroxyphenylalanine; *PNMT*, phenylethanolamine-*N*-methyl transferase; *SAH*, *S*-adenosylhomocysteine; *SAM*, *S*-adenosyl methionine.

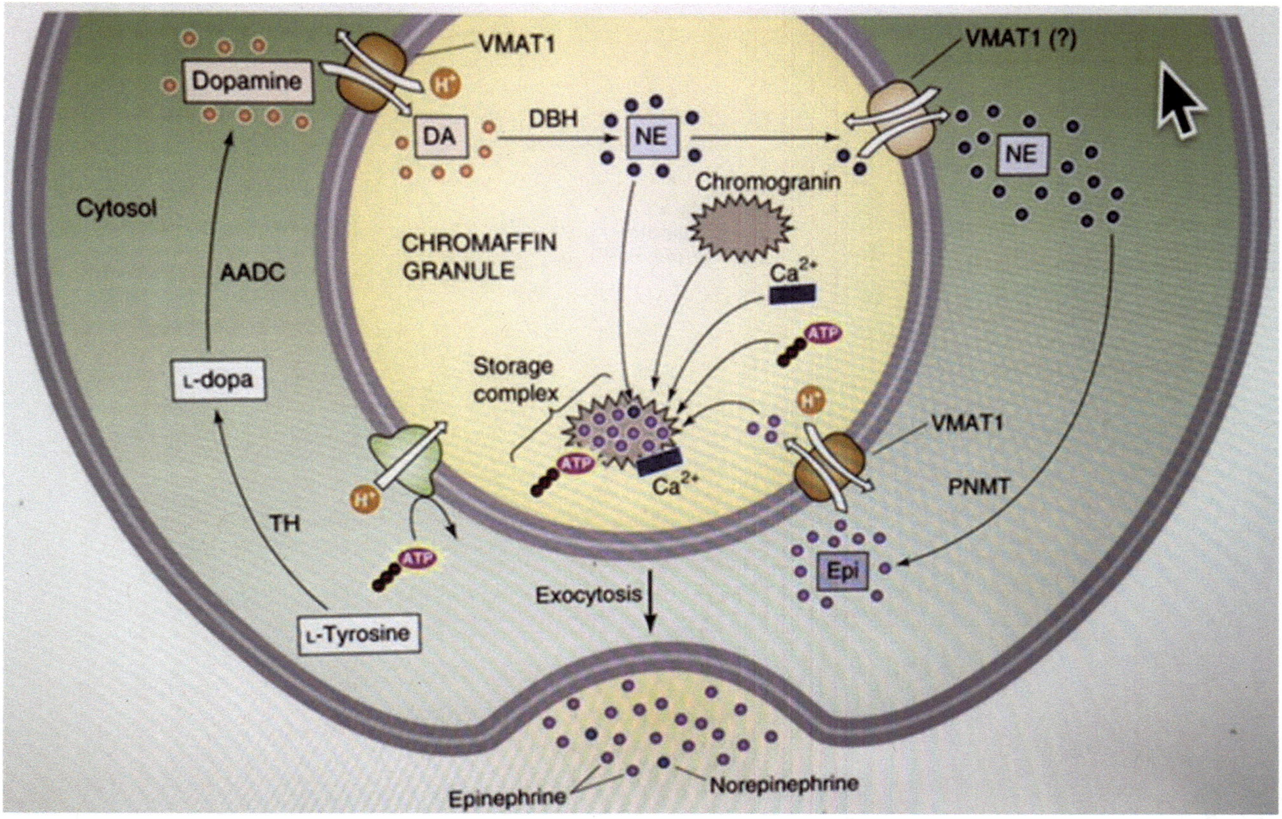

FIGURE 11.4 Cellular view of catecholamine synthesis. The chromaffin cell synthesizes and stores epinephrine in a sequence of four enzymatic and three transport steps. *AADC*, Amino acid decarboxylase; *DA*, dopamine; *DBH*, dopamine β-hydroxylase; *Epi*, epinephrine; *NE*, norepinephrine; *TH*, tyrosine hydroxylase. *Reproduced from Fig. 50.9 of https://doctorlib.info/physiology/medical/278.html.*

term, there is regulation of tyrosine hydroxylase gene transcription, a major way in which the adrenal medulla responds to sustained stress. Increased tyrosine hydroxylase gene transcription is a major part of adaptation to various types of stress in animal studies, including physical stressors, such as immobilization, exercise, or hypoglycemia, and social stress such as crowding. Agents that increase cyclic AMP (cAMP) levels as well as glucocorticoids are mediators of the effects of these stressors on tyrosine hydroxylase gene transcription.

The expression of the final enzyme in the pathway of epinephrine synthesis PNMT in the adrenal medulla is dependent on glucocorticoids. Glucocorticoids also mediate the increase in PNMT transcription in response to stress, which requires the presence of an intact hypothalamic–pituitary–adrenal (HPA) axis. This enzyme, along with tyrosine hydroxylase, is also transcriptionally increased by the 27-amino-acid neuropeptide pituitary adenyl cyclase-activating peptide that is cosecreted with acetylcholine and stimulates catecholamine secretion from the chromaffin cell.

The stimulus for the release of epinephrine from the secretory vesicles into the bloodstream is acetylcholine, secreted by the preganglionic splanchnic nerve terminus. As shown in Fig. 11.5, when acetylcholine binds to α subunits of the nicotinic acetylcholine receptor in the membrane of the chromaffin cell, the ligand gated ion channel opens, allowing the influx of sodium. This causes the depolarization of the membrane, which brings about the activation of one or more voltage-dependent calcium channels. The influx of calcium is the signal for exocytosis of the contents of the chromaffin vesicles into the nearby capillaries.

Although the precise mechanism of calcium-mediated exocytosis in these cells is not yet clear, some features are beginning to emerge from studies with chromaffin cells and neurons. The two exocytotic systems share the properties of Ca^{2+} dependency and secretion from a readily releasable pool of vesicles that have undergone several steps of maturation prior to and following docking at the membrane. Neuronal secretion occurs from active zones in which there is tight coupling between Ca^{2+} entry and the secretion release sites, whereas chromaffin cells do not have such active areas. Nevertheless, the features of the docking process, depicted in Fig. 11.5B, appear to be similar in both of

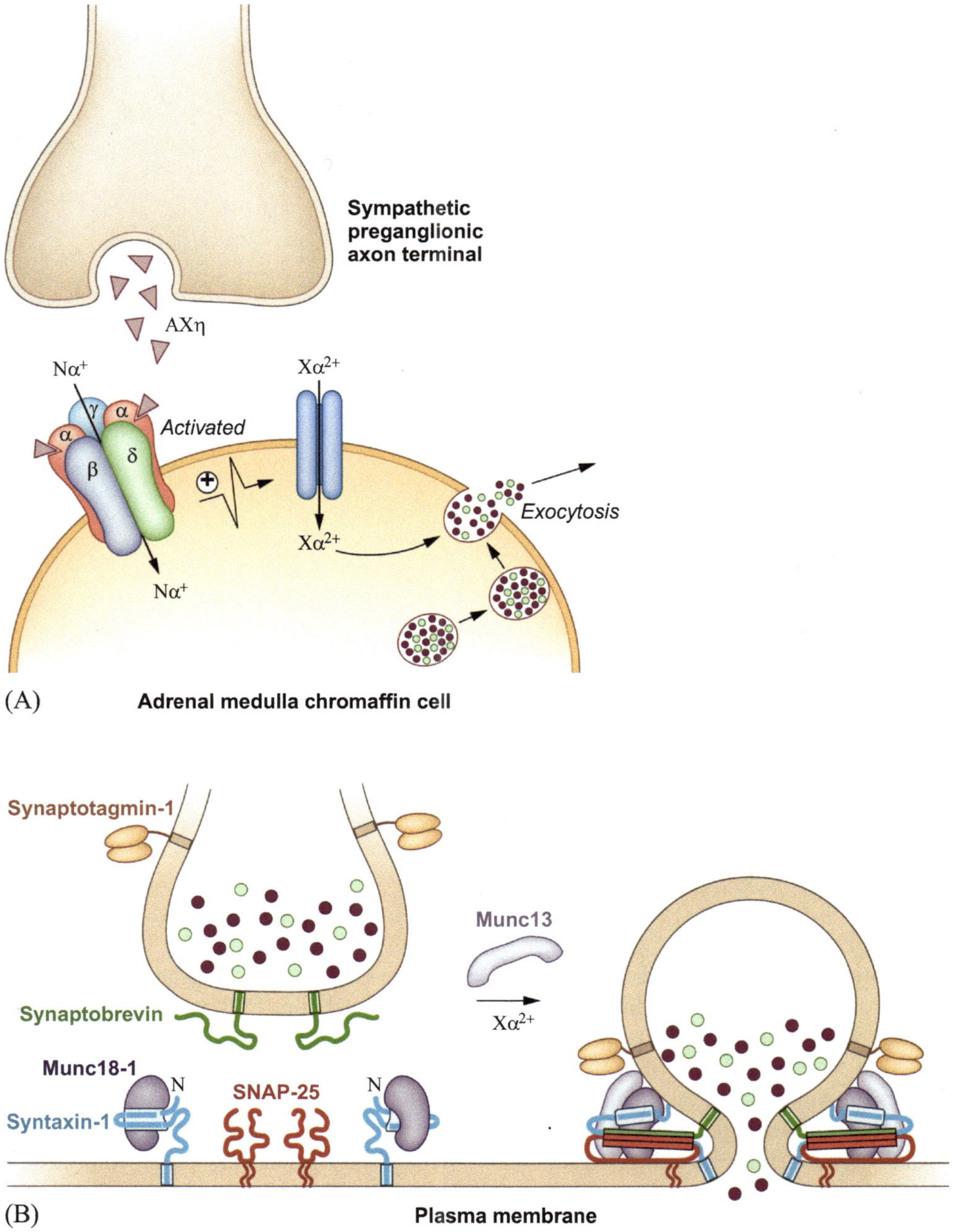

FIGURE 11.5 Epinephrine secretion by the chromaffin cell. (A) Acetylcholine-stimulated secretion of epinephrine by the chromaffin cell. The axon of the preganglionic splanchnic nerve terminates on the chromaffin cell of the adrenal medulla where it releases acetylcholine (*triangles*) which binds to the α subunits of the nicotinic cholinergic receptor leading to an influx of Na^+. The resulting membrane depolarization causes the opening of voltage-dependent calcium channels (*blue*); the rise in intracellular calcium in the vicinity of the mature vesicles brings about their exocytosis. The contents of the vesicle, epinephrine (*red circles*), peptides, and proteins (*blue circles*), and other small molecules such as ATP (not shown) are released. (B) *Proteins involved in the fusion and exocytosis of chromaffin granules*. The SNARE proteins, named receptors for SNAP proteins, shown are synaptobrevin-2 (*green*; also known as VAMP2) in the vesicular membrane; SNAP-25 (*red*); and syntaxin-1 (*blue*) in the plasma membrane. In the unprimed vesicle, syntaxin-1 forms a "closed" complex with Munc18−1 (*purple*), a member of a family of proteins first described in secretion deficient yeast mutants (Munc=mammalian uncoordinated). Munc13 (*light purple*) opens the closed complex, allowing Munc18−1 to bind to the SNAP-25 and synaptobrevin, leading to the formation of a four-helical bundle, tethering the vesicle to the plasma membrane. The increase in intracellular calcium (see panel A), which interacts with synaptotagmin-1 (tan), leads to the opening of the vesicle with the extrusion of its contents. *SNAP protein*, Soluble *N*-ethylmaleimide-sensitive attachment protein.

these cell types and in platelets. The membrane SNARE proteins (see legend, Fig. 11.5) syntaxin-1 and SNAP-25 are present in the docking site along with Munc18, a member of an ancient highly conserved family of secretory proteins also found in yeast, plants, and nematodes. Upon stimulation of secretion, Munc18 participates in the formation of a four-helical bundle made up of the vesicular SNARE protein synaptobrevin and the two membrane SNARE proteins, syntaxin-1 and SNAP-25. This complex, along with Munc13, tethers the vesicle to the plasma membrane. Synaptotagmin-1, in the vesicle membrane, participates as the calcium sensor and, along with other proteins such as calcium-activated proteins of secretion, mediates the stimulatory effect of Ca^{2+} on exocytosis.

C Catabolism of catecholamines

The inactivation and metabolism of circulating epinephrine and norepinephrine occur primarily in the liver and the pathway are shown in Fig. 11.6. Monoamine oxidase, a mitochondrial enzyme in liver, kidney, brain, and adrenergic nerve endings, catalyzes the oxidation and removal of the amino group of the catecholamines. The resulting aldehyde is then oxidized to an acid by an aldehyde dehydrogenase. Catechol-*O*-methyltransferase, which is widely distributed, is especially abundant in liver and kidneys, but is absent from nerve endings, and results in the methylation of the hydroxyl group on carbon 3 of the phenolic ring. The 3-methoxy derivatives, metanephrine and normetanephrine, shown on the right side of Fig. 11.6 make up about 15% of the excreted catecholamines and the remaining 85% are vanillylmandelic acid. The total urinary output of catecholamines includes those derived from both the adrenal medulla and those that diffuse into the circulation from neuronal synapses. The majority of these excretion products are conjugated as glucuronides or sulfates, the latter being the more abundant form.

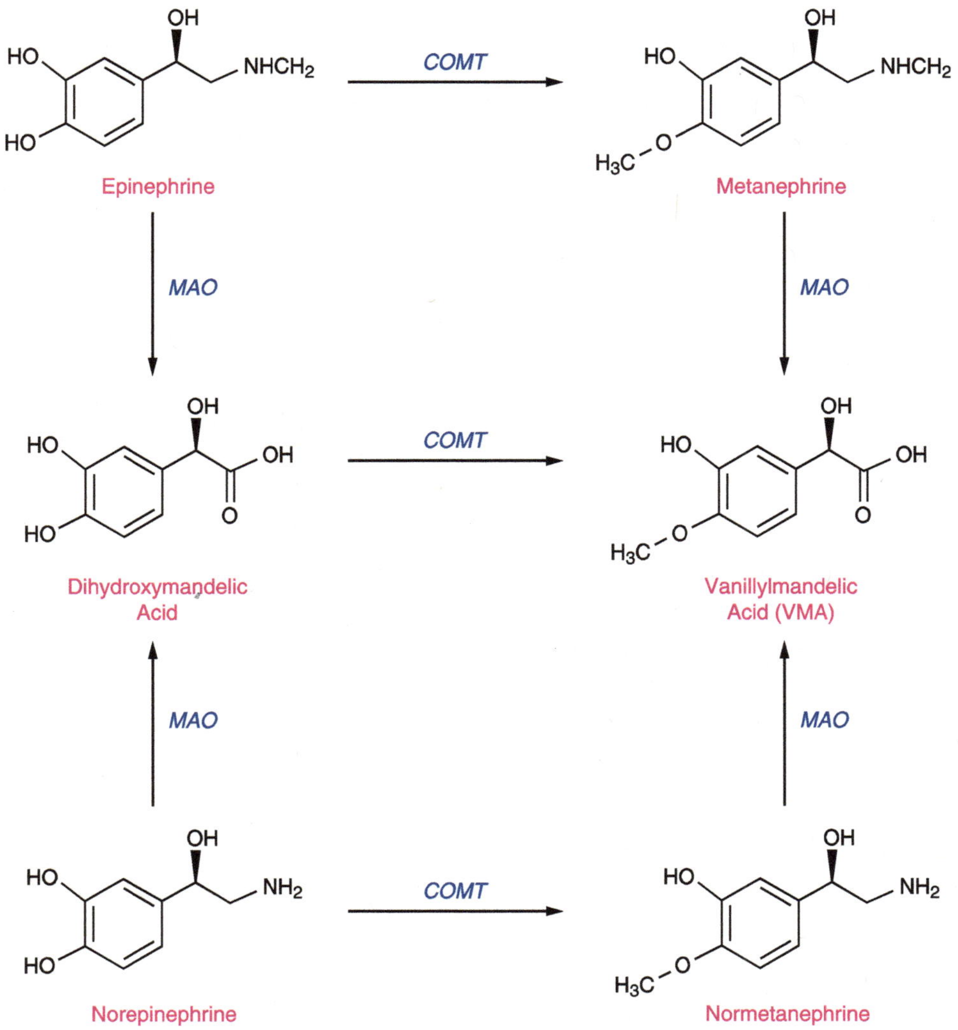

FIGURE 11.6 Catabolism of circulating epinephrine and norepinephrine. Epinephrine and norepinephrine are inactivated in the liver, through the action of one or both of two enzymes, MAO and COMT. The latter uses *S*-adenosyl methionine as the methyl group donor. The metabolites shown on the right are excreted in the urine as glucuronide or sulfate conjugates. *COMT*, Catechol-*O*-methyltransferase; *MAO*, monoamine oxidase.

D Pharmacology of catecholamines

Some of the many pharmacological derivatives of epinephrine and norepinephrine are shown in Fig. 11.7 with an indication of the adrenergic receptors with which they interact (also see Fig. 11.8). A brief description of these examples will serve to indicate the wide-ranging uses of the drugs belonging to this group. Phenylephrine is currently the most commonly marketed decongestant in over-the-counter cold and cough medications. It is not as effective as pseudoephedrine, which has become subject to controls to prevent its use in the production of methamphetamines. Ephedrine and pseudoephedrine, both found in the Chinese herb members of the *Ephedra* family, *ma huang*, act indirectly to increase the biological activity of endogenous catecholamines. Isoproterenol is used to treat bradycardia (a resting heartbeat of less than 50 bpm in a clinical context). In the past, it has also been used as a bronchodilator but albuterol, with greater receptor specificity, has largely replaced isoproterenol for this use.

Propranolol is a blocker of β-adrenergic receptors and is widely used to counter hypertension and anxiety.

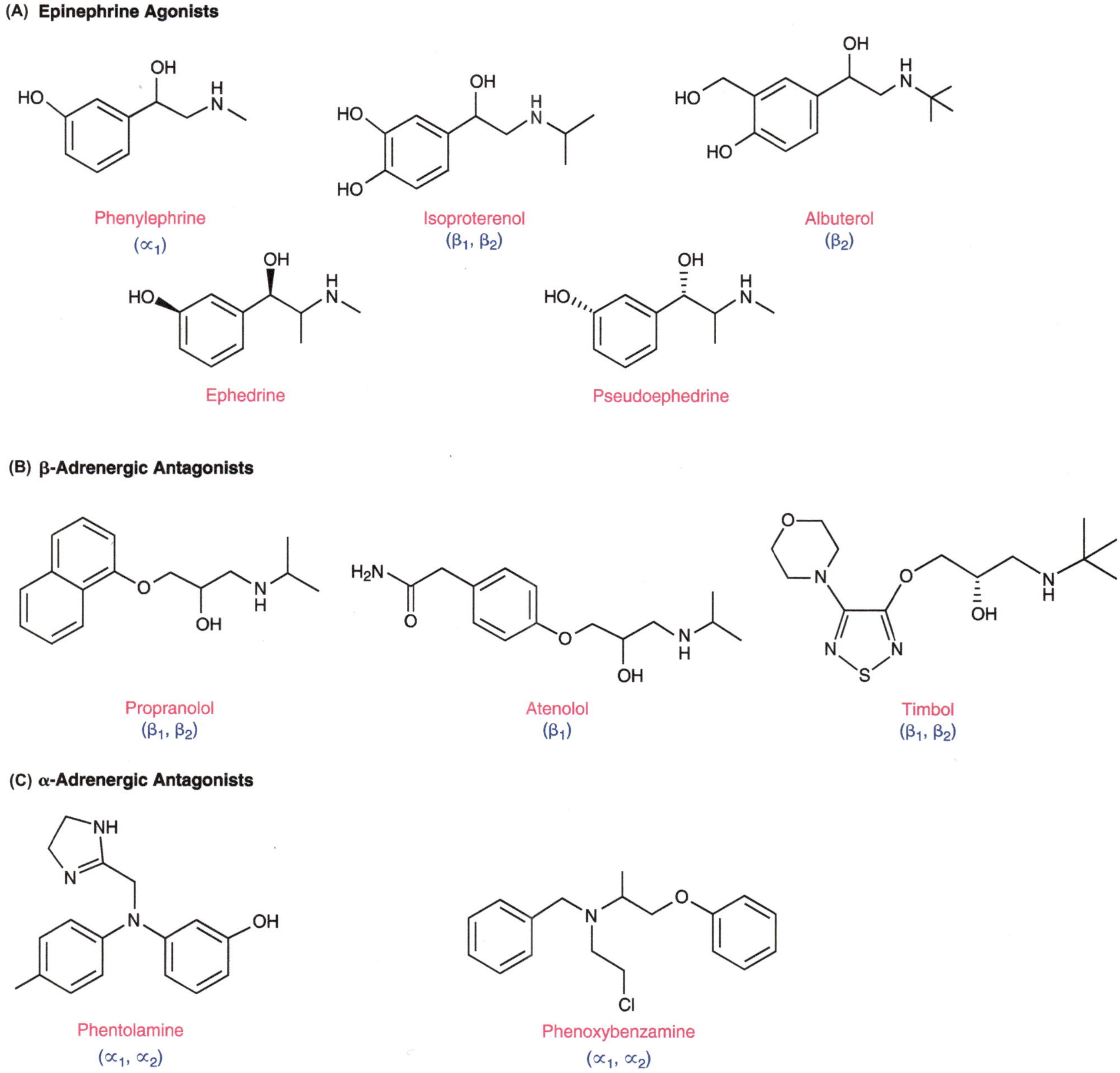

FIGURE 11.7 Some catecholamine pharmacological agents. Three classes of drugs relating to epinephrine and norepinephrine are shown: (A) epinephrine agonists; (B) antagonists interacting with β-adrenergic receptors; (C) antagonists of α-adrenergic receptors. The receptor type(s) with which the molecule interacts most effectively is shown in parenthesis. See Fig. 11.8 for the classification of the adrenergic receptors.

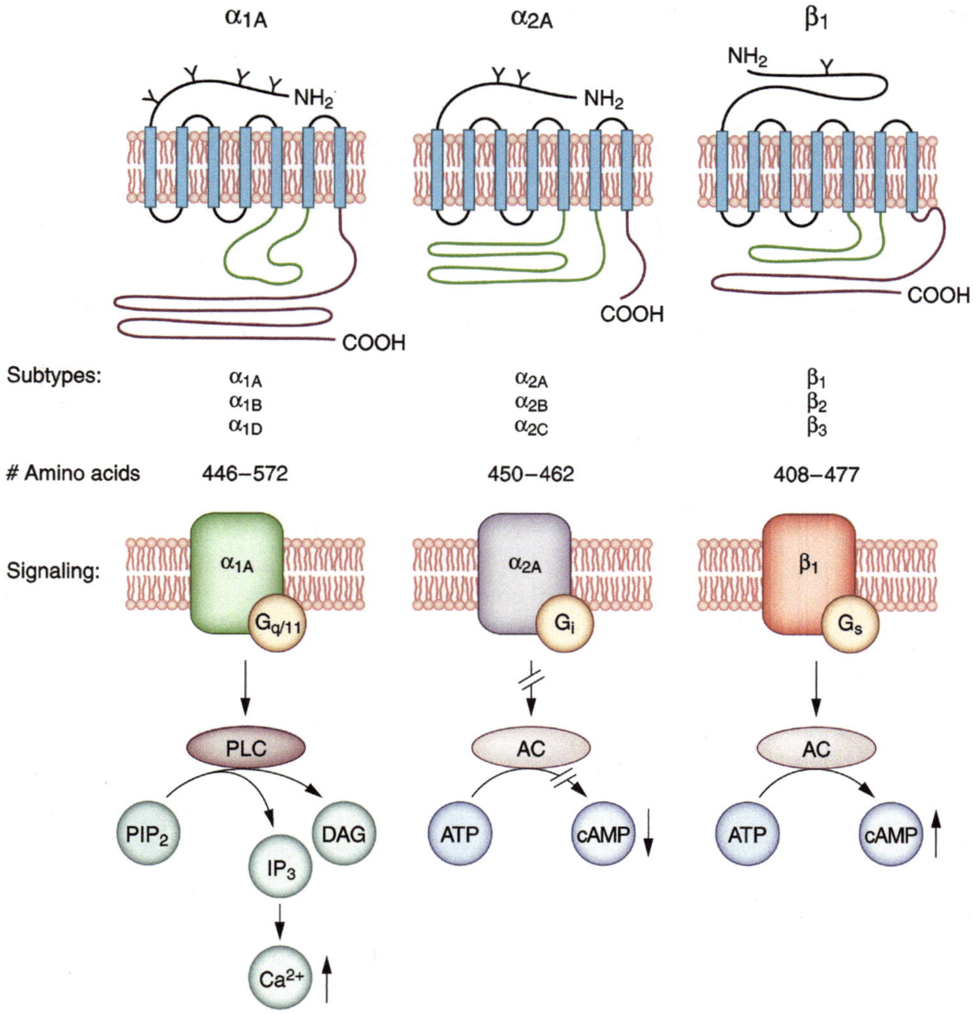

FIGURE 11.8 Adrenergic receptors. The adrenergic receptors, which bind epinephrine and norepinephrine (as well as synthetic compounds such as those in Fig. 11.7), are classified into three major groups, α_1, α_2, and β, based on pharmacological properties and type of signaling system. The features of these three groups are shown in this figure. The subtypes within each group are, listed below, the major example shown. All of the adrenergic receptors are GPCRs containing seven membrane-spanning regions. In the structural depictions of the receptors at the top, the approximate locations of extracellular *N*-glycosylation sites are shown (Y), as are the intracellular loop (*green*) and the carboxy terminal tail (*magenta*). The signaling systems are shown at the bottom of the figure, where the major Gα protein to which the receptor is coupled is indicated (*tan circle*) along with the primary enzymatic activity affected (PLC, AC) and the effect on the second messenger (increased ionized calcium, decreased or increased cyclic AMP). *AC*, Adenylyl cyclase; *GPCRs*, G protein–coupled receptors; *PLC*, phospholipase C.

Atenolol is a more cardio-specific drug used for acute myocardial infarction and tachycardia (more than 100 bpm at rest) due, for example, to hyperthyroidism. Timbol is most often found in opthalmic solutions to treat glaucoma. Phentolamine is used to reverse dental anesthesia and phenoxybenzamine is an antihypertensive agent, particularly used in the treatment of pheochromocytoma (see Section V).

IV Biological actions of catecholamines

A Adrenergic receptors

There are three major types of adrenergic receptors, that is, those receptors that bind the catecholamines epinephrine and norepinephrine. These are known as α_1, α_2, and β and were originally distinguished by their ligand (including agonists and antagonists)-binding characteristics. More recent molecular studies have revealed that each of these groups is composed of three subtypes, each encoded by a separate gene. All of the adrenergic receptors are single polypeptide proteins with seven hydrophobic membrane–spanning regions and are members of the large class of G protein–coupled receptors (GPCRs). Some structural characteristics are shown in the top part of Fig. 11.8 (see also Fig. 1.15A). The receptor classes differ in sites of glycosylation on the external portion and in the sizes of the intracellular loop and carboxyl terminus. These structural differences undoubtedly are the basis for the differences in the ligand specificities of the receptors and in intracellular signaling mechanisms.

Also shown in Fig. 11.8 is the primary G-protein with which each subtype interacts, leading to the functional differences between the receptor subtypes. Thus the intracellular signaling pathway of α_1 adrenergic receptors is generally mediated by an increase in intracellular Ca^{2+} levels; that of α_2 receptors by decreased cAMP; and that of the β receptors by increased cAMP. Although these signaling pathways are characteristic of each of the receptor types, other pathways have been shown to be activated

TABLE 11.3 Relative ligand specificity of adrenergic receptors.

Receptor type	Ligand preference
α_1	E=NE
α_2	E=NE
β_1	E=NE
β_2	E≫NE
β_3	NE≫E

by each of them in certain cells and under certain conditions of, for example, ligand concentration. These other pathways include the activation of K^+ channels, activation or inhibition of Ca^{2+} channels, activation of phospholipase A_2 for prostaglandin synthesis, and mitogen-activation protein kinase.

Table 11.3 lists the approximate relative potencies of the adrenergic receptor subtypes for epinephrine and norepinephrine. The β_3 receptor responds better to norepinephrine than to epinephrine, while the β_2 receptor has a strong, about 10-fold, preference for epinephrine. The α_1, α_2, and β_1 types are fairly nonselective between these two ligands. It should be noted that the receptor preferences indicated are apparent at low (subsaturating) concentrations of the ligands and may be different at higher concentrations of one or both of them.

The β-adrenergic and most of the α-adrenergic receptors undergo desensitization through phosphorylation of the carboxyl terminus and removal of the receptor from the membrane. This process, which is common to many GPCRs, involves GPCR kinase–mediated phosphorylation of the receptor, rendering it capable of binding to β-arrestin followed by endocytosis. This process is one of several ways in which the signal initiated by epinephrine or norepinephrine is terminated in the target cell.

B Biological responses to epinephrine

The response of a given cell to circulating epinephrine and/or norepinephrine will depend on the type(s) of adrenergic receptors present, the relative number of each if there is more than one type, which is often the case, and the prevailing concentrations of the two catecholamines. All of the circulating epinephrine comes from the adrenal medulla and makes up 80% of the catecholamines released from the chromaffin cells, the remaining 20% being norepinephrine. Of the circulating norepinephrine, about one-third comes from the adrenal medulla and the remainder diffuses into the circulatory system from postganglionic sympathetic nerve terminals in which this catecholamine is the neurotransmitter. Thus at any one time target cells are exposed to ligand derived from both glandular and neural sources.

Since the adrenal medulla receives signals directly from the autonomic nervous system, its response to change is very rapid and epinephrine levels, relative to norepinephrine, rise very quickly. This neural signal to the adrenal medulla conveys an emergency situation; the effects of the epinephrine secreted in response to this message are directed toward the dual goals of increasing fuel availability to the cardiac and skeletal muscles and to preserve the supply of glucose and oxygen for the brain. The many organs and pathways involved in this effort, and the receptors that mediate the responses, are summarized in Fig. 11.9.

1 Blood flow

Increased epinephrine from the adrenal medulla binds to the β_1 receptor in the heart to increase the heartbeat rate (chronotropic effect) and strength (inotropic effect). In addition, the rapid relaxation of the ventricle during diastole is facilitated (lusitropic effect). These actions contribute to an overall increase in cardiac output. At the same time, α_2 receptors in the veins are constricted, increasing the rate of return of blood to the heart. Through the vasoconstriction of splanchnic arterioles leading to the gastrointestinal (GI) tract as well as the skin and kidney, blood flow to these parts of the body is decreased. Vasodilation of skeletal and cardiac muscle arterioles, as well as those of the liver and lungs, is mediated by the β_2 receptor and the increased blood flow to these organs completes the diversion of blood to the sites where it is most needed.

2 Gas exchange

Interaction of epinephrine with the β_2 receptors on the smooth muscle cells of the bronchioles (the smallest branches of the airway) increases the flow of gases to and from the alveoli of the lungs. This provides for a faster rate of removal of carbon dioxide from the blood and its replacement with oxygen.

3 Metabolic changes

The many metabolic effects of epinephrine are summarized on the right side of Fig. 11.9. Adipose tissue contributes to the increased availability of glucose in the blood through the increase of lipid breakdown and decreased glucose uptake. In the liver, glucose output is increased through elevated glycogen breakdown and gluconeogenesis.

The liver responds to epinephrine by increasing glycogen breakdown; increasing gluconeogenesis; and increasing its breakdown of free fatty acids (released from adipose tissue; see next). The pathway of epinephrine stimulation of glycogen breakdown in the liver and

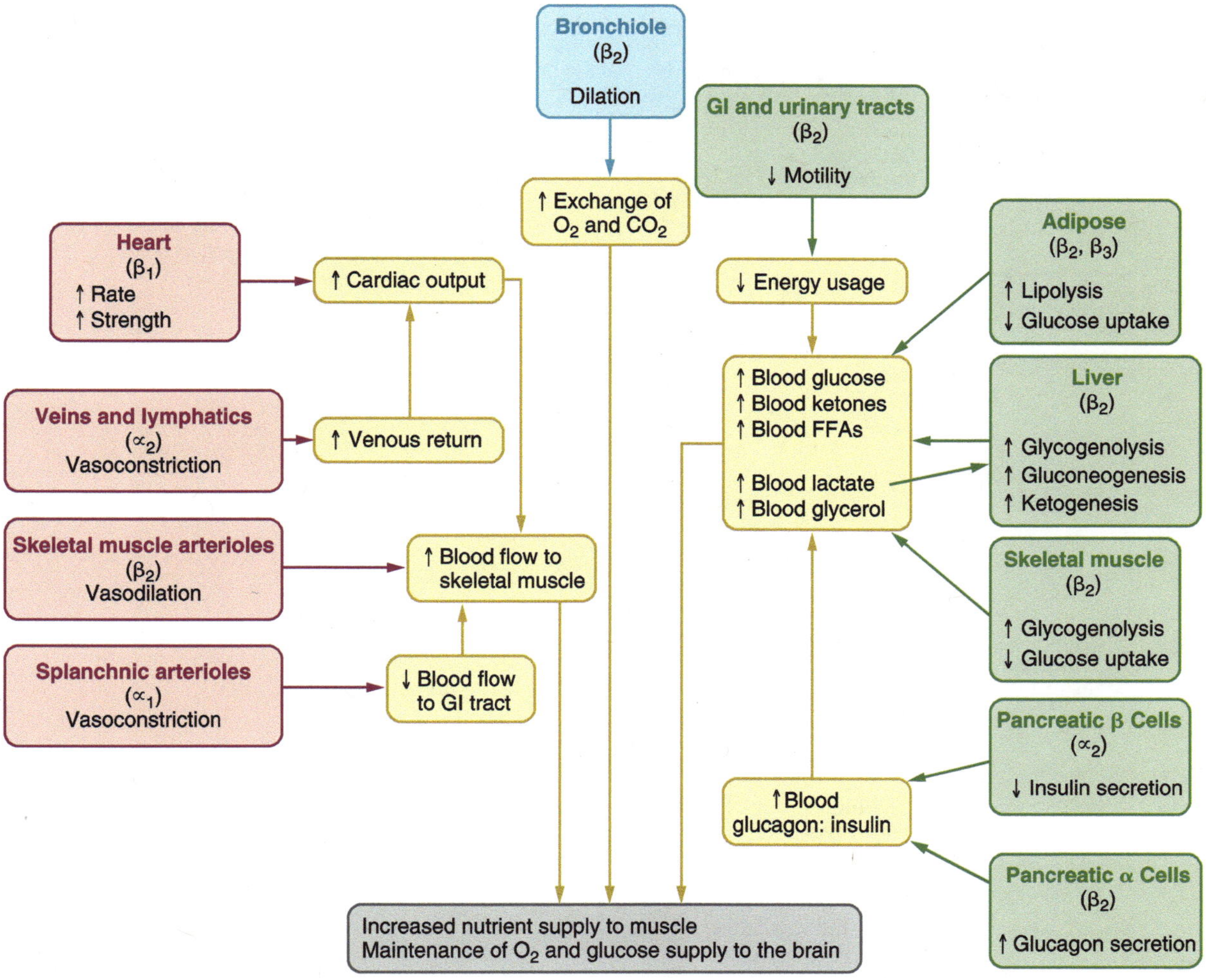

FIGURE 11.9 Catecholamine-mediated responses to acute stress. Three types of responses are shown: on the left (*pink*) are effects of epinephrine on the heart and vasculature which allow increased blood flow to skeletal muscles and decreased flow to the GI tract. In the center (*blue*) is bronchodilation to allow increased gas exchange, thus maintaining the flow of oxygen to the muscles and brain. On the right (*green*) are the metabolic responses that increase the supply of fuel to the muscles and brain. In each box the primary adrenergic receptor responsible for the actions is indicated. *GI*, Gastrointestinal.

skeletal muscle is well known and in fact is the first instance of cAMP as a second messenger to be elucidated and is presented in Chapter 1, Hormone Introduction. Interaction of the occupied β_2 receptor with G_s is followed by the activation of adenylyl cyclase and of cAMP-dependent protein kinase (PKA). Phosphorylation and activation of phosphorylase b by PKA and phosphorylase a by phosphorylase b ensue. Phosphorylase a then phosphorylates glycogen, the first step in the removal of one glucose (as glucose-1-phosphate) molecule from the chain. The same pathway is used in the response of the skeletal muscle to epinephrine shown in Fig. 11.9. The resulting rise in plasma glucose from this burst of hepatic glycogenolysis in the liver is followed by a sustained twofold rise in gluconeogenesis to augment and maintain the output of glucose from the liver. These effects of epinephrine on glucose metabolism in the liver are summarized in Fig. 11.10. Also shown in Fig. 11.10 is the output from the liver of ketone bodies, derived from increased fatty acid breakdown (see Fig. 11.11). These fuels, acetoacetate and D-β-hydroxybutyrate, are important ones for the brain as well as kidney and lung (Fig. 11.9), to maintain their function during a time of stress.

The details of the action of epinephrine to mobilize free fatty acids in adipose tissue are shown in Fig. 11.11. Interaction of epinephrine with a β-adrenergic receptor increases intracellular cAMP (Fig. 11.8), which activates cAMP-dependent protein kinase. PKA phosphorylates hormone-sensitive lipase (HSL), one of three enzymes that removes fatty acids from storage as triglycerides. Epinephrine also results in the cAMP-mediated phosphorylation and activation of perilipin, a protein that coats lipid droplets. In its phosphorylated state, perilipin allows HSL greater access to the glycerides within

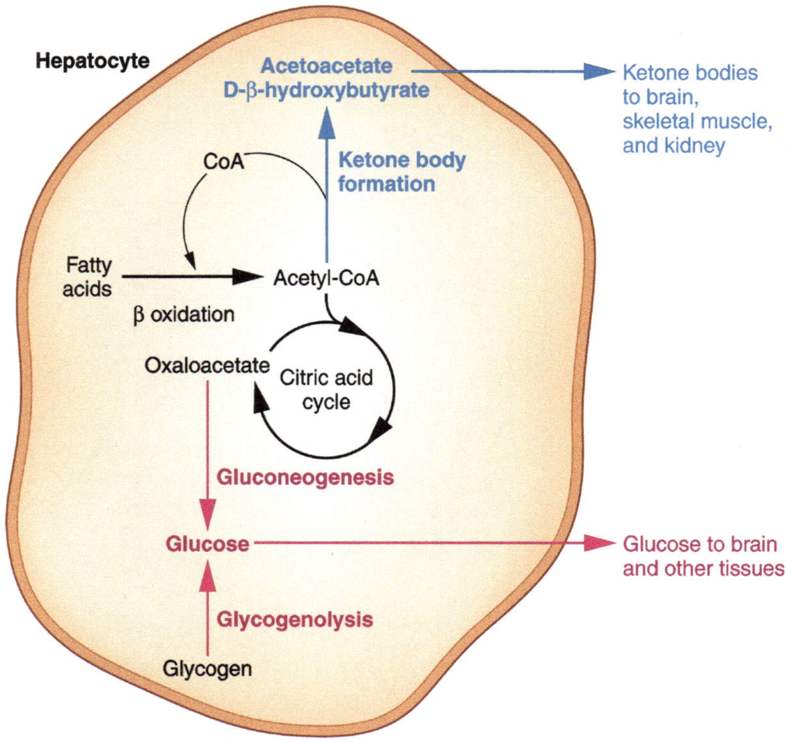

FIGURE 11.10 Epinephrine and liver metabolism. When the hepatocyte is stimulated by epinephrine, the output of glucose is increased by increased glycogen breakdown and gluconeogenesis (*pink pathways*). Increased free fatty acids, from lipolysis adipose tissue, are available for β-oxidation, resulting in increased ketone bodies, acetoacetate and D-β-hydroxybutyrate (*blue pathway*). Both the increased glucose and ketone bodies are released into the circulation to maintain the fuel supply to the brain and other tissues.

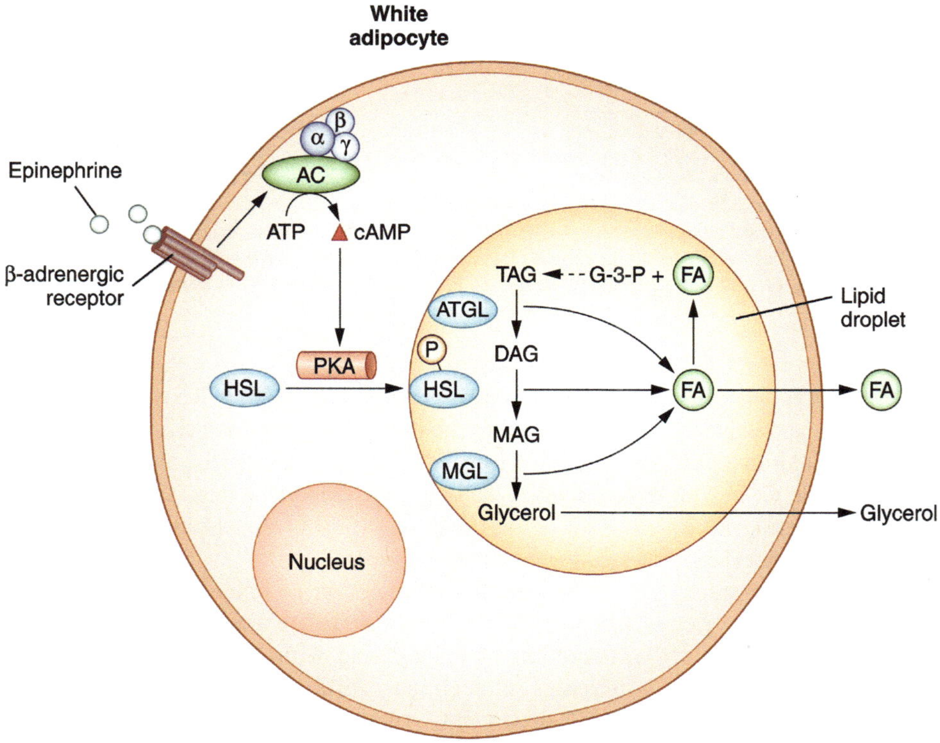

FIGURE 11.11 Effect of epinephrine on adipose cell fatty acid mobilization. When adipose cells are stimulated by epinephrine, the fatty acids stored as triglycerides are released by the action of three lipases (*blue ovals*). TAG are the substrates for ATGL (also known as desnitrin); DAG are the substrates for HSL; and MAG are the substrates for MGL. Activation of the β-adrenergic receptor leads to increased intracellular cyclic AMP and activation of PKA which phosphorylates hormone-sensitive lipase, increasing its activity. The result is the release of fatty acids into the circulation. *ATGL*, Acetyl triglyceride lipase; *DAG*, diacylglycerides; *HSL*, hormone-sensitive lipase; *MAG*, monoacylglycerides; *MGL*, monoglyceride lipase; *PKA*, protein kinase A; *TAG*, triacylglycerides.

the droplet. HSL can remove fatty acids from triacylglycerides, but its most important regulatory role is as the rate-limiting step in their removal from diacylglycerides. As its name suggests, HSL is also sensitive to regulation by other hormones, including insulin (inhibitory) and glucagon (stimulatory). In Fig. 11.12, details of the release of free fatty acids from the adipocyte are shown.

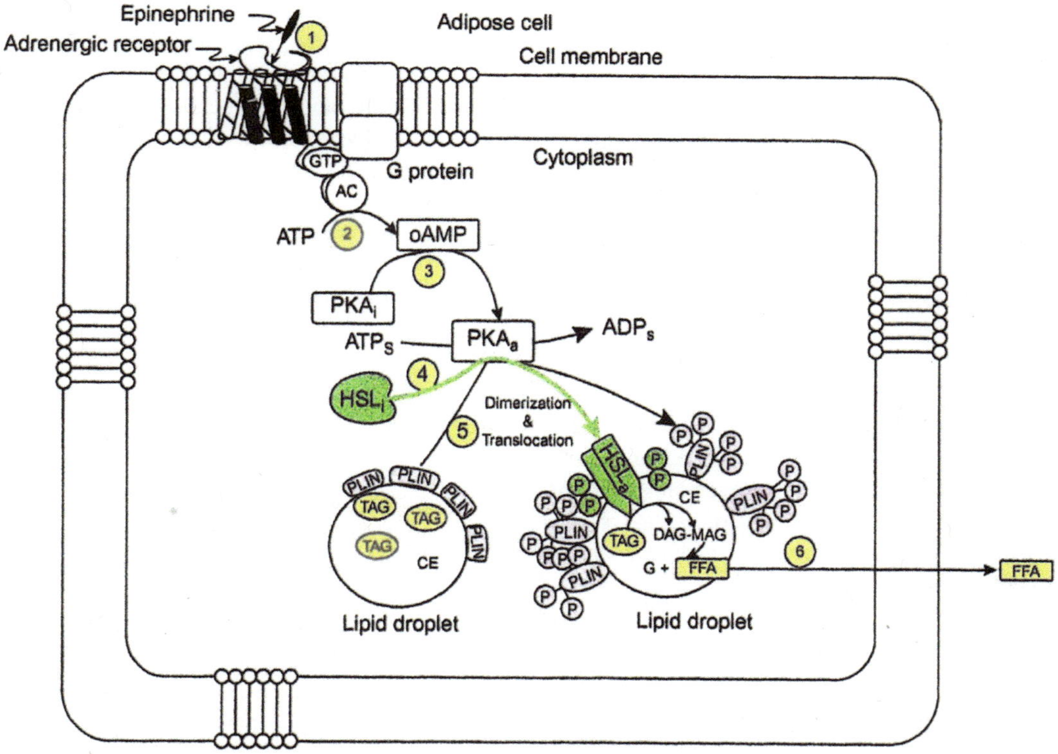

FIGURE 11.12 Cartoon showing the adipocyte with an adrenergic receptor in the cell membrane to which epinephrine binds with high affinity (*step 1*). This is a G protein–coupled receptor and epinephrine binding activates the receptor and its associated G protein that then activates AC. AC catalyzes the conversion of ATP to cAMP (*step 2*). cAMP activates PKA (*step 3*). Active PKA phosphorylates inactive HSL at four sites on the protein (*step 4*). As a result, the phosphorylated HSL dimerizes and translocates to the LD. Simultaneously PKA multiply phosphorylates PLINs on the surface of the LD (*step 5*). The phosphorylated PLINs undergo conformational changes and open gaps on the surface of the LD wide enough to allow entry of HSL. HSL then proceeds to hydrolyze the resident TAGs to DAG, MAG and eventually to glycerol and FFA. Most believe that the enzyme can carry out hydrolysis to the monoglyceride stage and some believe that HSL can also hydrolyze the monoglyceride to free glycerol and fatty acid. Presumably, if HSL is unable to accomplish the last hydrolysis step, other resident lipases (MAG lipase) are able to do so. Free fatty acids then translocate across the cell membrane into the extracellular space (*step 6*) and gain access to the bloodstream. Not shown in this schematic are some details: perilipin is originally complexed to CGI-58 (an enzyme that facilitates hydrolysis of the fat in LDs but is not directly involved in the hydrolytic process). CGI-58 has the activity of 1-acylglycerol-3-phosphate *O*-acyltransferase. Also, the HSL dimer that catalyzes the hydrolysis of triacylglycerol is complexed with FABP. *AC*, Adenylate cyclase; *cAMP*, cyclic AMP; *CE*, cholesteryl ester; *DAG*, diacylglycerol; *FABP*, fatty acid binding protein; *FFA*, free fatty acid; *GTP*, guanosine triphosphate; *HSL*, hormone-stimulated lipase; HSL_a, activated hormone-stimulated lipase that dimerizes; HSL_i, inactive hormone-stimulated lipase; *LD*, lipid droplet; *MAG*, monoacylglycerol; *P*, phosphate; *PKA*, protein kinase A; PKA_a, active protein kinase A; PKA_i, inactive protein kinase A; *PLINs*, perilipins; *TAGs*, triglycerides. *Reproduced from Fig. 9.42 of G. Litwack, Human Biochemistry, Academic Press, London, San Diego, CA, Oxford, 2018, p. 230.*

In addition to the previous direct effect on liver, adipose, and skeletal muscle cells, epinephrine also increases glucagon secretion and decreases insulin secretion from the α and β cells of the pancreas, respectively (Fig. 11.9). The resulting change in the ratio of glucagon to insulin in the blood reinforces the effects of epinephrine to increase blood glucose and mobilize fatty acids. Taken together, the effects of epinephrine on intermediary metabolism, in combination with those of the flow of blood and gases, prepare the organism to meet an emergency with increased muscular capacity for action while maintaining the fuel supply to the brain so that these actions can be coordinated and directed to a desirable outcome.

V Clinical aspects

A Pheochromocytoma

The term *pheochromocytoma* refers to a tumor, usually but not always located in the adrenal medulla, which secretes excessive amounts of catecholamines. The word combines the color ("dusky" from the Greek "phaios") that catecholamines render the tissues in which they occur when exposed to dichromium salts (chromo) with the tissue type (cytoma=tumor). Pheochromocytoma is relatively rare, with an estimated occurrence of 1 in 1700–4500 individuals and 3–8 new cases per million people per year. It is found in 0.1%–1% of patients with hypertension and 5% of patients in whom adrenal masses are discovered incidentally. The average age of diagnosis is 25 years in hereditary cases (approximately 20% of the total) and 44 years in sporadic (nonhereditary) cases. The genders are equally represented in the patient population. Although rare, the tumor is a dangerous one as the result of lack of detection and treatment is often fatal.

The signs and symptoms of the presence of a pheochromocytoma are listed in Table 11.4 and reflect fairly closely the known actions of the catecholamines summarized in Fig. 11.9. The signs and symptoms are presented as characteristic of either the episodic spells of their appearance or their underlying constant (chronic) presence. The episodic symptoms occur as spells that may be spontaneous or brought about by a change in posture, anxiety, or certain medications. Spells may occur several times a day or as infrequently as once every few months; they are usually 15–20 minutes in duration, although they can be much longer or much shorter. It should be noted that most people who have such spells do not have a pheochromocytoma, as there are many other causes for these relatively nonspecific manifestations.

The most sensitive and specific biochemical test for pheochromocytoma is the measurement of plasma or urinary free (nonconjugated) metanephrine and normetanephrine (Fig. 11.6). These are produced continually by the pheochromocytoma itself, independent of catecholamine release, and, therefore, are a good marker of the presence of these cells. Once biochemical confirmation of a catecholamine-secreting tumor has been indicated, imaging studies can determine its localization so that it can be surgically resected. Most (85%) of the tumors will be found in the adrenal glands, although some might lie elsewhere in the abdominal cavity, as will the catecholamine-secreting paragangliomas, which are also detected by metanephrine and normetanephrine measurements.

TABLE 11.4 Signs and symptoms of pheochromocytoma.

Episodic spell	Chronic
Headache	Hypertension
Palpitations	Tachycardia
Sweating	Postural hypotension
Anxiety/nervousness	Pallor
Tremor	Tremor
Nausea/vomiting	Weight loss
Chest/abdominal pain	Fasting hyperglycemia
Hypertension	Increased respiratory rate
	Decreased GI motility

GI, Gastrointestinal.

Surgical removal is the sole treatment for pheochromocytoma and requires great care prior to and during the operation. Preparation involves the control of blood pressure for a period of time (about 2 weeks) prior to surgery and the prevention of a hypertensive crisis due to sudden catecholamine release from the tumor during surgery. Drugs used in this context are α- and β-blockers as well as inhibitors of tyrosine hydroxylase. In addition, anesthetics must be chosen carefully for their safety since some can affect catecholamine release; clearly, cardiovascular and hemodynamic parameters must be closely monitored.

B Chronic stress

The classical concept of the "stress axis" is the HPA axis (Chapter 10: Adrenal Corticoids) but more recently it is being expanded to include the descending neurons from the brainstem which stimulate the release of epinephrine from the adrenal medulla. Thus while epinephrine has long been associated with responses to short-term stress, its role in chronic stress or chronic stress–induced illnesses is emerging. For example, while at low concentrations epinephrine lowers diastolic blood pressure through its interaction with β_2 receptors, at higher concentrations of epinephrine blood pressure is increased due to the activation of α receptors, which are more abundant than β receptors. This provides circumstantial evidence for a role of epinephrine in the hypertension associated with chronic stress.

Recently a number of lines of evidence have converged to implicate activation of epinephrine-activated β_2 receptors in the modulation of cancer progression. In animal studies stresses of various kinds increase the progression of cancers of a number of different types through a decrease in apoptosis and, probably, other mechanisms. A number of correlation studies in humans suggest that beta blockers such as propranolol decrease progression and mortality rates for cancers such as those of breast, prostate, skin, and certain neuroblastomas.

These two examples indicate that while there is still much to learn, there is likely a role for the adrenal medulla in the response of the body to chronic stress and disease states related to it.

Further reading

Chromaffin cells

E. Carbone, Fast inactivation of Nav 1.3 channels by FGF14 proteins: unconventional way to regulate the slow firing of adrenal chromaffin cells, Journal of General Physiology 153 (2021). Available from: https://doi.org/10.1085/jgp.202112879. e2021112879.

A. Gridina, et al., Gaseous transmitter regulation of catecholamine secretion by hypoxia in murine adrenal chromaffin cells, Journal of Neurophysiology 125 (2021) 1533–1542. Available from: https://doi.org/10.1152/jn.00669.2020.

K. Harrada, et al., Mechanisms for establishment of GABA signaling in adrenal medullary chromaffin cells, Journal of the International Society for Neurochemistry 158 (2021) 153–168. Available from: https://doi.org/10.1111/jnc.15345.

P. Tuluc, Exciting link: Na^+ background current makes chromaffin cells burst, The Journal of Physiology 599 (2021) 1727–1728.

Catecholamine secretion

M. Liu, et al., Metastatic pheochromocytoma and paraganglioma: signs and symptoms related to catecholamine secretion, Discover Oncology 12.9 (2021). Available from: https://doi.org/10.1007/s12672-021-00404-x.

W. Shi, et al., The receptor tyrosine kinase EPHB6 regulates catecholamine exocytosis in adrenal gland chromaffin cells, Journal of Biological Chemistry 295 (2020) 7653–7668.

Clinical aspects

E.S. Carlson, et al., Catecholaminergic innervation of the lateral nucleus of the cerebellum modulates cognitive behaviors, The Journal of Neuroscience 141 (2021) 3512–3530.

F. Fornai, S. Puglisi-Allegra, Autophagy status as a gateway for stress-induced catecholamine interplay in neurodegeneration, Neuroscience and Behavioral Reviews 123 (2021) 238–256.

A.C. Mejis, M. Snel, E.P.M. Corssmit, Pheochromocytoma/paraganglioma crisis: case series from a tertiary referral center for pheochromocytomas and paragangliomas, Hormones 20 (2021) 395–403. Available from: https://doi.org/10.1007/s42000-021-00274-6.

Chapter 12

Androgens

I Introduction

The endocrine physiology of the male involves the interplay of the several hormones associated with male sex determination, fetal development, postnatal growth, and sexual maturation as well as the maintenance of the male secondary sex characteristics. Its integrated operation is dependent upon the interaction of signals—both hormonal and neural—between the central nervous system, hypothalamus, pituitary, and testes. The two major functions of the testes are steroid hormone production and spermatogenesis.

The male phenotype can be defined by six characteristics: (1) chromosomal composition; (2) functionally and structurally competent testes; (3) adequate amounts of tonic androgen production; (4) appropriate external and internal genitalia; (5) rearing as a male; and (6) self-acceptance of a male role. Thus the male sexual phenotype is the summation of the physical and biochemical characteristics as well as the psychological characteristics of gender role and identity.

The hormones responsible for the development and maintenance of the male phenotype are the hypothalamic hormones gonadotropin-releasing hormone (GnRH) and kisspeptin; the pituitary gonadotropins, luteinizing hormone (LH) and follicle-stimulating hormone (FSH) (see Chapter 3: The Hypothalamus and Anterior Pituitary), the androgenic steroid hormones, testosterone (produced by the gonads) and 5α-dihydrotestosterone (synthesized by certain target tissues; see Chapter 2: Steroid Hormones: Chemistry, Biosynthesis, and Metabolism); anti-Müllerian hormone (AMH), produced by the fetal testis; and inhibin, a protein hormone produced by the gonads in both sexes. The female steroid hormone, estradiol, also plays important roles in the male.

This chapter discusses the biology and biochemistry of these hormones in the male. Because the male genotype and fetal testicular hormone production play such crucial roles in sexual differentiation, this topic will be covered in this chapter as well, along with its final postnatal developmental step, puberty.

II Anatomy of the male reproductive system

A Testes, ducts, and accessory structures

The male testes have the dual responsibilities for the production and release of the germ cells, the spermatozoa, and for the biosynthesis and secretion of the key androgenic steroid hormone, testosterone. This steroid hormone and others produced from it play a dominant role in the differentiation, growth, and maintenance of the sexual reproductive tissues necessary for the continuation of the species, the growth and maintenance of secondary sex characteristics, and the anabolic effects of androgens on skeletal muscle growth and skeletal growth. The testis and the internal and external genitalia with which it is closely associated provide the venue in which these two important functions, hormonogenesis and spermatogenesis, are carried out and integrated.

The male reproductive system comprises the paired units consisting of the testis, epididymis, and vas deferens, the accessory glands (prostate, seminal vesicles, and bulbourethral glands), and the penis and scrotum. These are illustrated in Fig. 12.1.

In mammals the testes are contained in the scrotum, which is suspended outside the abdominal cavity between the thighs. Spermatogenesis does not take place at the internal body temperature of 35°C–37°C but occurs optimally at the scrotal temperature of 33°C. Descent of the testes from the abdominal cavity into the scrotum occurs just prior to birth in a process that is dependent on androgens and a testis-specific hormone insulin-like hormone 3. Failure of one or both testes to descend into the scrotum is termed cryptorchidism and, if bilateral and not corrected, results in sterility.

The spermatic cord, the wall of which is an extension of the abdominal wall, is the conduit between the testis and the abdominal cavity and contains the vas deferens, blood vessels, and some nerves and lymphatics.

Each adult human testis is a 4- to 5-cm long ellipsoid weighing 35–45 g. As shown in Fig. 12.2, the tunica vaginalis encapsulates the entire surface of the testis except the

Hormones. DOI: https://doi.org/10.1016/B978-0-323-90262-5.00004-4

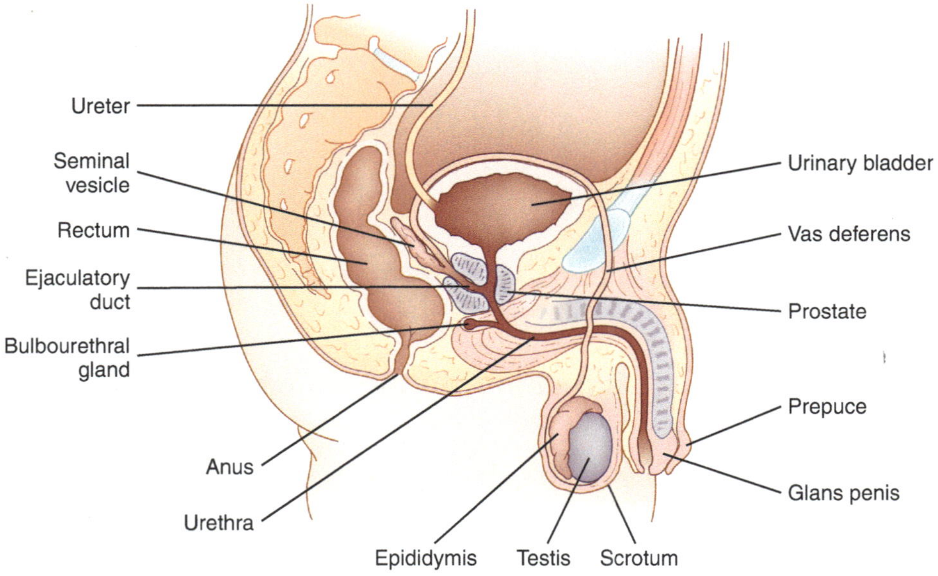

FIGURE 12.1 Anatomy of the male reproductive system. A midsagittal view of the organs, glands, and ducts involved in human male reproduction.

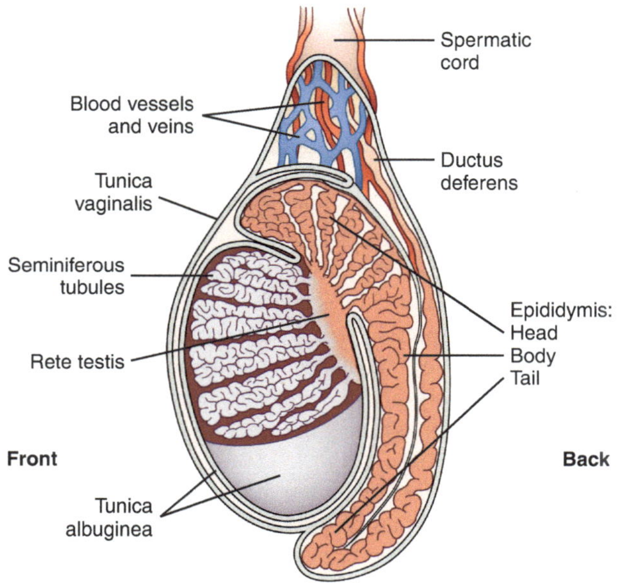

FIGURE 12.2 The testis and epididymis. In the scrotal sac the epididymis lies posterior to the testis, which is covered with the thick tunica albuginea. The spermatic cord is the conduit for blood vessels, nerves, and lymphatics. The sperm produced by the seminiferous tubules are collected in the rete testis and travel through the epididymis to the ductus (vas) deferens.

posterior region that lies in close proximity to the epididymis. More interiorly, the testis is surrounded by the tunica albuginea, a thick capsule that also divides the parenchymal seminiferous tubules into discrete lobules. The seminiferous tubules, which are the site of production of spermatozoa, have a total length of ~500 m. The seminiferous tubules are organized in a highly convoluted irregular pattern, with some blind pouches, ultimately leading to the rete testis and the epididymal duct. The duct system, composed of the epididymis, ductus (vas) deferens, and ejaculatory duct, functions to transport the mature spermatozoa during ejaculation.

The epididymis is attached to the posterior surface of the testis. Approximately 12 days are required for the transit of the spermatozoa through the epididymis to the vas deferens; during this interval, they undergo the morphological and functional changes necessary to provide them with the capacity to fertilize an ovum (see Section V.D). The vas deferens is a 7- to 8-cm tubule connecting the epididymis with the ejaculatory duct (see Fig. 12.1). The passage of the mature sperm through the vas deferens is accomplished by peristaltic contractions of smooth muscle in the duct wall. The ejaculatory duct is located at the merger of the vas deferens and the seminal vesicles and passes the prostate to empty into the urethra.

The accessory structures include the seminal vesicles, prostate gland, and bulbourethral glands. In adult males the two seminal vesicles are 3–5 cm in length and lie distal to the ejaculatory duct. Under the actions of androgens, they produce the mucoid secretions, including high concentrations of fructose, that constitute the major fluid volume of the ejaculate. The prostate gland lies at the junction of the ejaculatory duct with the urethra. The prostatic secretions, which are highly responsive to androgens, consist of simple sugars and some proteolytic enzymes. The two bulbourethral glands, sometimes referred to as Cowper's glands, are 2–3 mm in diameter and secrete prostaglandins and an alkaline mucoid solution to neutralize and coat the urethral lining immediately prior to the arrival of the ejaculate.

The penis carries the urethra to the exterior of the body. Its erectile tissue is highly vascularized, so that when the penis is filled with blood it is in an erect and firm state, facilitating deposition of the spermatozoa in the female vaginal canal.

B Seminiferous tubules

A photomicrograph of a cross section of the adult male seminiferous tubule is shown in Fig. 12.3A. The seminiferous tubule walls are composed of germinal epithelial cells and Sertoli cells. The germ cells undergo meiotic divisions, differentiation, and maturation (spermatogenesis, discussed in Section V.D) as they move

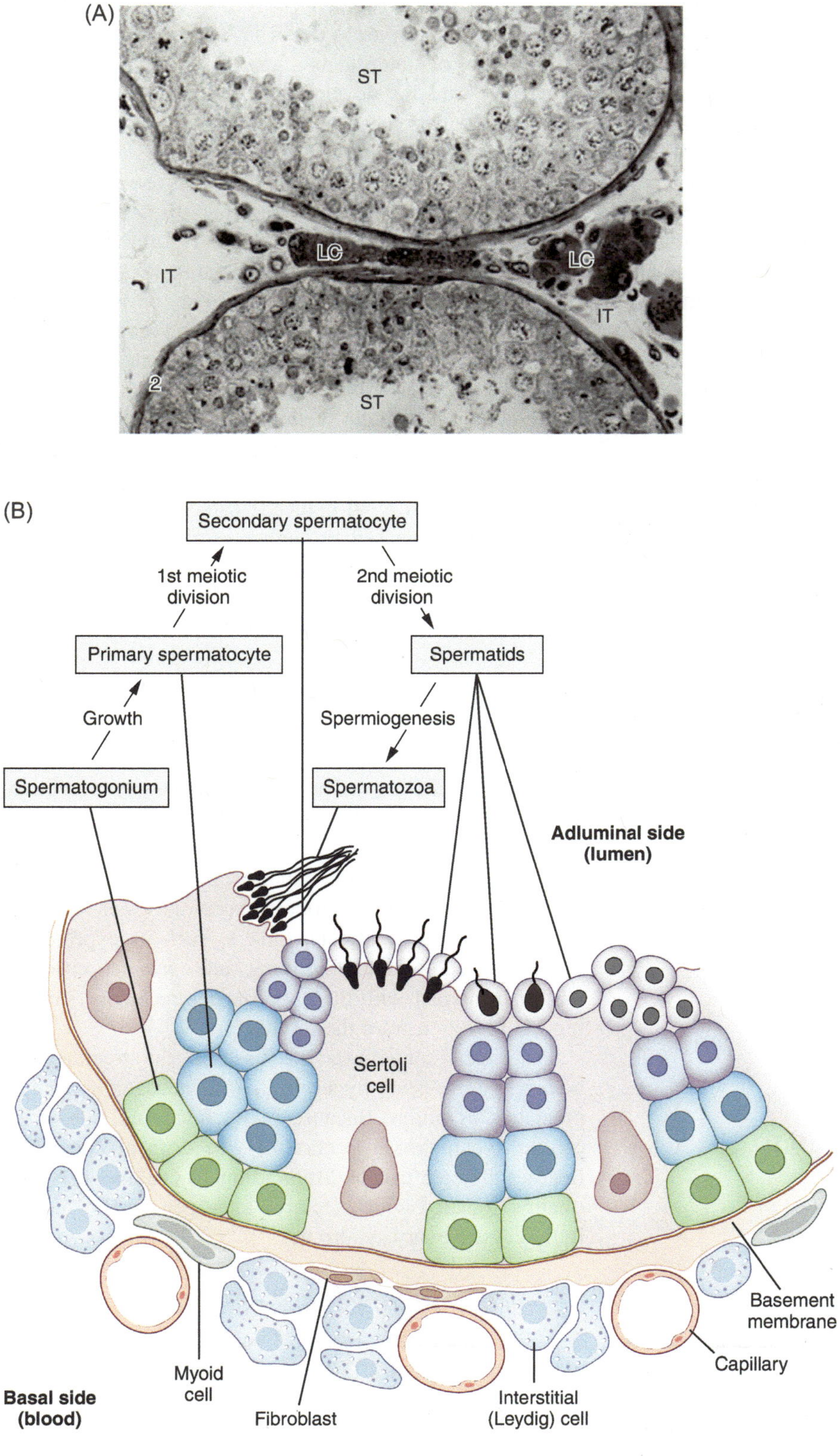

FIGURE 12.3 The seminiferous tubule and surrounding interstitium. (A) The photomicrograph of a testicular section from a normal human male shows portions of two ST and the relationship between them and the LC in the IT. The interstitium also contains myoid cells and capillaries that, along with the Sertoli cells in the tubules, are not clearly visible in this picture. *IT*, Interstitium; *LC*, Leydig cells; *ST*, seminiferous tubules. (B). As depicted in this schematic diagram, the seminiferous tubule is defined by the basement membrane. Outside of the tubules are clusters of Leydig (interstitial) cells, contractile myoid cells, and capillaries. Within the seminiferous tubule the spermatogonia (green), including the subpopulation of spermatogonial stem cells, are in contact with the basement membrane. As the germ cells develop and undergo meiosis to become secondary spermatocytes and spermatids, they move towards the lumen of the tubule. Early to late stages of spermatids are shown from right to left on the luminal side of the tubule. The mature spermatozoa travel from the seminiferous tubule through the duct system as described in Fig. 12.2. The Sertoli cells are in direct contact with the developing germ cells, which are surrounded by thin cytoplasmic extensions of the Sertoli cells. Tight junctions (not shown) between Sertoli cells divide the testis into the basal (containing the spermatogonia) and adluminal compartments. *Reprinted from R.S. Swerdloff, C. Wang, A.P.S. Kikim, Hypothalamic–pituitary–gonadal axis in men, in: D.W. Pfaff, A.P. Arnold, S.E. Farbach, A.M. Etgen, R.T. Rubin (Eds.), Hormones, Brain and Behavior, second ed. Academic Press, 2009, p. 2378.*

toward the center of the lumen, as shown in Fig. 12.3B. Mature spermatozoa are transported through the lumen of the seminiferous tubules to the highly convoluted network of ducts known as the rete testis and then through the efferent ductules to the epididymis, where they are stored.

The Sertoli cells (Fig. 12.3B) line the basement membrane of the seminiferous tubules; they are in close contact with the innermost layer of the basement membrane and, with their thin cytoplasmic extensions, they surround the germ cells as these undergo their maturation process. Tight junctions between Sertoli cells, along with desmosomes and gap junctions, form the blood–testis barrier. One important function of this barrier, which is formed during puberty, is to prevent exposure of more mature germ cells to antibodies from the interstitial compartment. Local disassembly of the tight junctions allows the passage of primary spermatocytes from the basal to the adluminal compartment.

Sertoli cells provide important nurturing for developing sperm and are sometimes referred to as “nurse cells.” Recently it has become clear that Sertoli cells play a crucial role in maintaining the stem cell niche, or microenvironment, required for the self-renewal of progenitor spermatogonia at the basement membrane of the seminiferous tubule. In rodents, Sertoli cells also secrete androgen-binding protein (ABP; same as sex hormone–binding globulin, SHBG, Section III.D) into the seminiferous lumen, where it helps maintain the high testosterone levels necessary for normal spermatogenesis. In both rodents and humans, Sertoli cells secrete inhibin as part of the control of FSH secretion (see Section IV.B).

C Leydig/interstitial cells

Between the seminiferous tubules lies the interstitial tissue (see Fig. 12.3A), where the primary endocrine cells of the testes, the Leydig, or interstitial, cells are located in clusters. The Leydig cells, under the action of LH, are the major site of the production of testosterone, as described in detail in Chapter 2, Steroid Hormones: Chemistry, Biosynthesis, and Metabolism, and following in Section III.A.

III Chemistry and metabolism of androgens

A Androgens and related molecules

1 Natural and synthetic androgens

As reviewed in Chapter 2, Steroid Hormones: Chemistry, Biosynthesis, and Metabolism (see Figs. 2.2 and 2.20), the naturally occurring androgens are 19-carbon steroids. Testosterone is the principal male androgen produced and secreted by the testes. An important biologically active metabolite of testosterone, produced in certain target tissues, is 5α-dihydrotestosterone. The structures of these androgens are presented in Fig. 12.4. Table 12.1 summarizes some of the major steroids and androgenic intermediates found in significant concentrations in the blood of normal adult males.

As discussed in more detail in Section V, the biological actions of androgens can be divided into those directed toward the development and maintenance of the male reproductive system and those that have anabolic effects on skeletal muscle. Anabolic steroids are synthetic analogs of testosterone, actions of which are selective for the anabolic effects on muscle rather than, for example, growth effects on the prostate gland. The reasons for this selectivity have to do with the resistance of the molecule to metabolism into dihydrotestosterone or estradiol, as well as its interaction with the androgen receptor (AR) in specific target tissues. Three examples of the many anabolic testosterone derivatives that have been synthesized are shown in the middle section of Fig. 12.4.

Anabolic androgens are used as adjuvant treatments in diseases with muscle wasting, such as cancer and AIDS, and in glucocorticoid-induced osteoporosis. Anabolic androgens are misused through prescription for nonindicated purposes and abused by athletes who use them to build muscle mass and strength. This topic will be discussed in more detail in Section VI.C.

2 Androgen antagonists

Some synthetic analogs of testosterone compete for binding to the AR but do not generate androgenic effects; such analogs are termed androgen antagonists or antiandrogens. Examples, including one steroidal and two nonsteroidal antiandrogens, are shown in the third row of Fig. 12.4. Some steroidal androgen antagonists tend to interact with the glucocorticoid receptor, which can lead to undesirable side effects. Others, such as cyproterone acetate, in addition to inhibiting androgen action at the AR, also interact with the progesterone receptor, decreasing LH secretion from the pituitary.

Androgen antagonists are used in the treatment of both benign prostate hyperplasia and androgen-dependent prostate carcinoma. Because of its favorable side effect profile, bicalutamide is currently the drug of choice for treating prostate cancer. Antiandrogens are also useful in treating females with excessive androgen levels, such as in polycystic ovary syndrome.

3 Selective androgen receptor modifiers

Selective androgen receptor modifiers (SARMs) are defined as molecules that interact with the AR, eliciting biological responses in the bone and muscle, for example, but not in the prostate; thus the term “selective.”

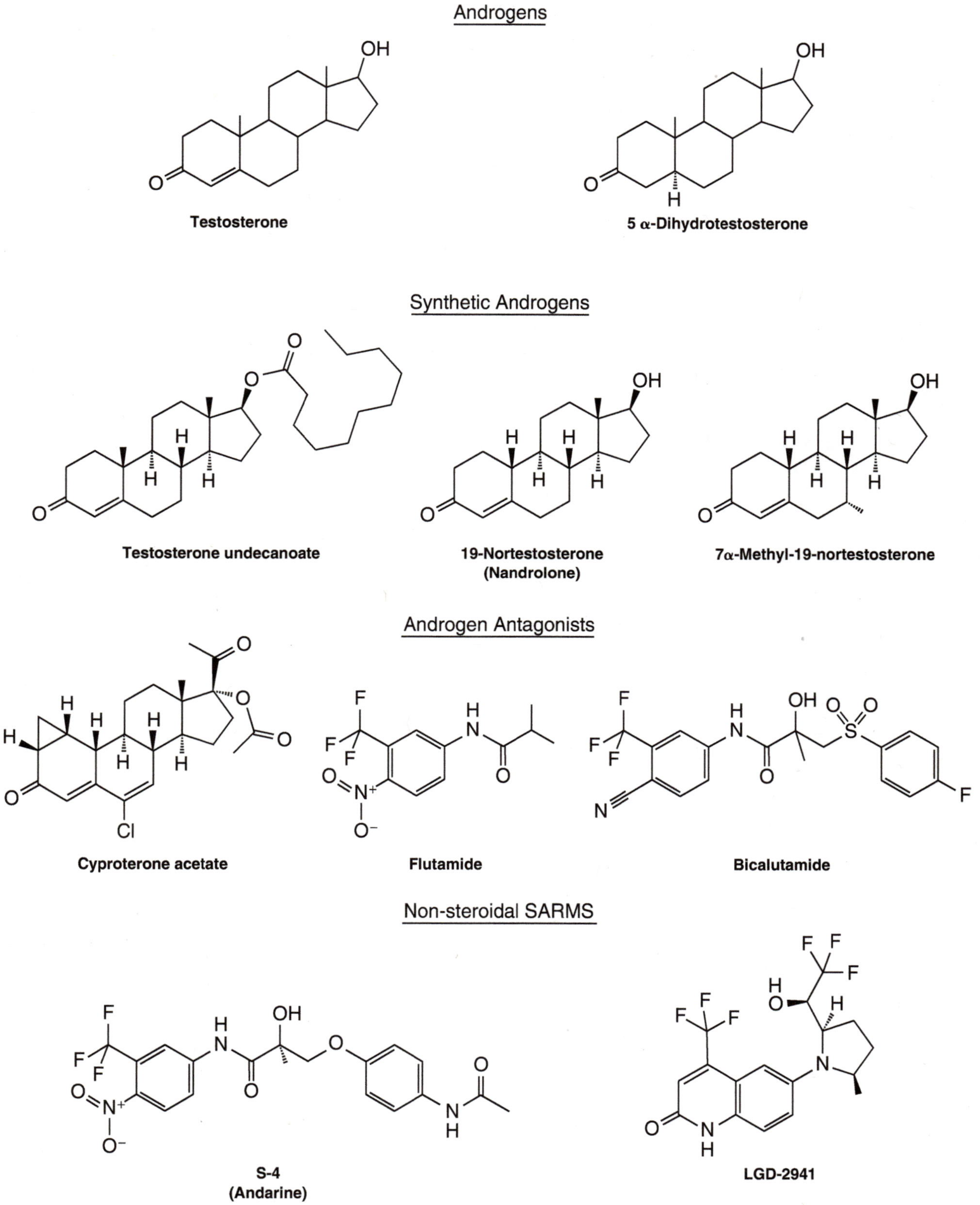

FIGURE 12.4 Androgen-related molecules. The structures of the naturally occurring androgens, testosterone, and 5α-dihydrotestosterone are shown. Later are shown chemically synthesized compounds that mimic some of the effects of the natural androgens (synthetic androgens and SARMs) as well as antiandrogens that oppose the effects of the natural androgens. *SARMs*, Selective androgen receptor modifiers.

According to this definition, they include some of the synthetic androgens described earlier, but there are also a number of nonsteroidal SARMs. The effectiveness of an SARM will depend on its affinity for the AR and the degree of its selectivity. Two examples of nonsteroidal SARMs are shown in Fig. 12.4. S-4 (Andarine) has gone through extensive preclinical testing and LGD-2941 has entered a Phase I clinical trial. Many other nonsteroidal SARMs are currently being synthesized and tested.

B Androgen biosynthesis

The biochemical pathway for the production of androgens is discussed in Chapter 2, Steroid Hormones: Chemistry, Biosynthesis, and Metabolism (see Fig. 2.20). As in other steroidogenic cells, the cleavage of the side chain of cholesterol is confined to the mitochondria of the Leydig, or interstitial, cells of the testes. The role of the steroid acute regulatory protein, StAR (Fig. 2.17), is the same in the Leydig cells as in other cells that carry out this side chain cleavage step, and as in other steroidogenic cells, this is the first and rate-limiting step in the production of testosterone.

Fig. 12.5 summarizes the modifications of the steroid nucleus required to convert the 21-carbon result of cholesterol side chain cleavage, pregnenolone, to the 19-carbon androgens. The two main changes are the removal of carbons 20 and 21 and the oxidation of the 3β-hydroxyl group accompanied by the migration of the double bond from the Δ^5 to the Δ^4 position. Depending on which of these two steps is carried out first, the pathway from pregnenolone to testosterone is called either Δ^5 or Δ^4, as outlined in Fig. 12.5. The relative activities of the Δ^5 and Δ^4 pathway vary among various mammalian species; in the rodent the Δ^4 pathway is predominant, while the human testis uses primarily the Δ^5 pathway. All of these reactions occur in the microsomal fraction of the cell.

Fig. 12.6 depicts the changes in testicular testosterone synthesis, as reflected in plasma hormone levels, throughout the life of a normal human male. The peak during the

TABLE 12.1 Production and circulation of sex steroids in the adult human male.

	Plasma concentration (ng/mL)	Secretion rate (μg/day)
Testosterone	300–1100	5000
5α-DHT	25–75	500[a]
Androstenedione	50–200	2500
Androstane-3α,17β-diol	130	200
Estradiol	2–3	10–15[b]

DHT, Dihydrotestosterone.
[a]*Most circulating DHT comes from nongenital skin and liver.*
[b]*Most circulating estradiol comes from peripheral aromatization.*

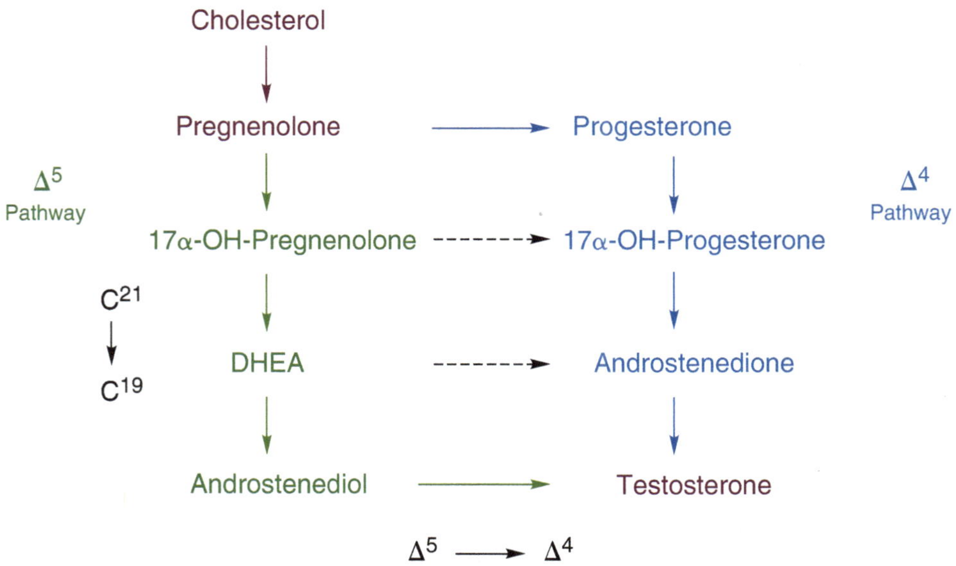

FIGURE 12.5 Pathways of androgen synthesis. Androgens are synthesized in the Leydig cells of the testis by one of two pathways, depending on the species. In the Δ5 pathway from pregnenolone to testosterone (*green*), the cleavage of carbons 20 and 21 takes place before the oxidation of the 3β hydroxyl group and migration of the double bond from the Δ5 to the Δ4 position. In the Δ4 pathway (*blue*), the order of the two steps is reversed. The Δ5 pathway predominates in humans. Fig. 2.20 should be consulted for the structures of the molecules in these pathways and information about the enzymes that catalyze the conversions among them.

second trimester in utero is followed by another during the 6 months after birth. The relative quiescence of the androgen synthetic pathway persists throughout childhood until the beginning of the pubertal period. Plasma testosterone rises to adult levels by the end of puberty and begins to decline in early middle age.

C Modification of testosterone in target tissues

Fig. 12.7 shows the three principal ways in which testosterone can interact at target sites. In many target tissues, typified by the skeletal muscle, unmodified testosterone interacts with the AR to bring about the appropriate biological responses. In others, characterized by (but not limited to) the prostate and hair follicles, testosterone is reduced at the 5α position to form 5α- dihydrotestosterone or DHT (see Fig. 2.20), and it is this steroid that interacts with the AR to cause biological change. Because DHT has a greater affinity for the AR than does testosterone, this is sometimes referred to as the amplification pathway. In some tissues, exemplified by bone and brain, the active derivative of testosterone is estradiol, produced locally by aromatase (see Fig. 2.21), which then interacts with the estrogen receptor (ER). This has been referred to as the diversification pathway.

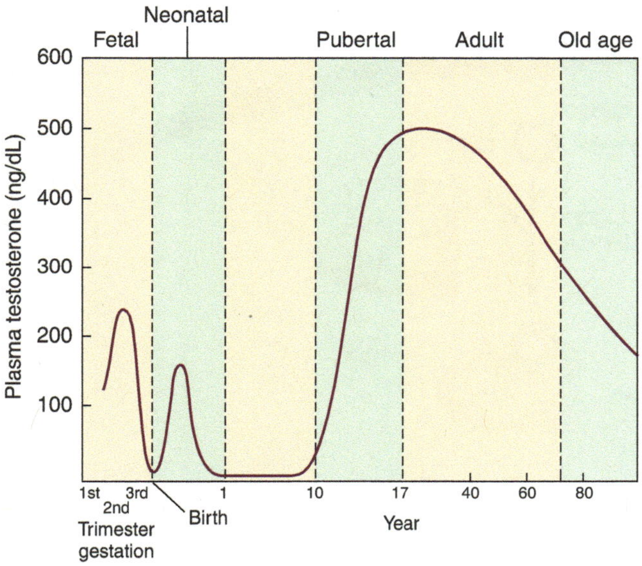

FIGURE 12.6 Changes in human male plasma testosterone levels with age. During the second trimester of fetal development, levels of testosterone rise to nearly adult levels and a second peak is seen during the first half of the first year of postnatal life. Levels remain low during childhood until the pubertal rise to adult levels. This is followed by a decline with age for the remainder of life.

1 Testosterone 5α-reductase

The enzyme responsible for the conversion of T into DHT is a Δ^4-3-ketosteroid-5α-oxidoreductase that requires NADPH as a cofactor. In rodents and humans, there are two forms of 5α-reductase. Type 1 5α-reductase is expressed in skin, liver, and brain and is thought to be responsible for the inactivating catabolism of testosterone. Type 2 5α-reductase is found in the prostate and other male accessory sex glands, hair follicles, liver, and brain. This is the enzyme that catalyzes the activity amplification step of DHT production. The two reductases are encoded by separate genes and share about 50% amino acid homology.

DHT has both a higher affinity for the AR than testosterone and a greater capacity for stimulating changes in gene expression once bound to the AR. In tissues that have Type 2 5α-reductase, testosterone itself is relatively inactive. Evidence for this in humans comes from the study of males with inactivating mutations in the gene for Type 2 5α-reductase. At birth, these individuals display markedly undermasculinized genitalia (see Section VI.A), demonstrating the necessity for the conversion of T to DHT in these tissues during embryogenesis.

A second area of the clinical importance of the 5α-reduction of testosterone involves the potential to control the effects of endogenous testosterone on, for example, the prostate, through inhibitors of the Type 2 5α-reductase. For example, the orally active steroids finasteride and dutasteride are used

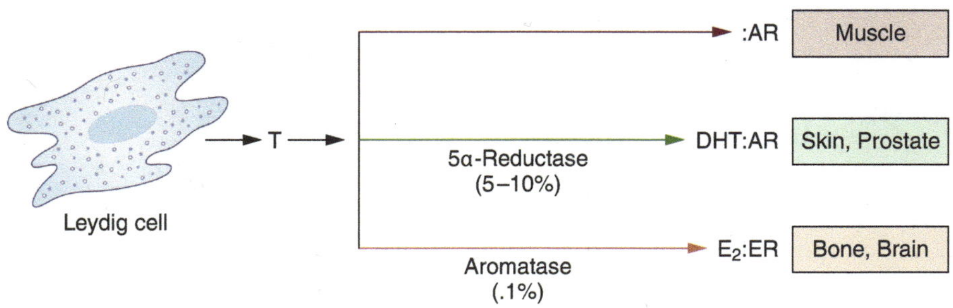

FIGURE 12.7 Testosterone action pathways. In many target tissues of androgens, typified by the skeletal muscle, testosterone (T) is the molecule that interacts with the AR to generate biological responses. In other tissues, of which the hair follicles of the skin and the prostate gland are examples, testosterone undergoes 5α-reduction to form 5α-DHT, which interacts with the AR with greater affinity than does T; 5%–10% of T secreted by the testis undergoes this transformation in target tissues. Finally, a small, but important, amount of T is aromatized in target tissues to form estradiol (E_2) which interacts with the ER. Cells in the bone and the brain provide examples of this pathway. *AR*, Androgen receptor; *DHT*, dihydrotestosterone; *ER*, estrogen receptor.

to treat complications related to benign prostate hyperplasia. Finally, as indicated in Section III.A previously, considerable effort has been directed toward synthesizing androgens that cannot undergo 5α-reduction and will, therefore, have selective actions on tissues sensitive to testosterone such as skeletal muscle, sparing the prostate from growth stimulation.

2 Aromatase

Aromatization of testosterone to estradiol occurs in several tissues of the adult male, including adipose, testis (Sertoli cells and Leydig cells), brain, bone, breast, liver, and blood vessels. In these tissues, androstenedione can also be aromatized, yielding the weak estrogen estrone, which can then be metabolized to estradiol through the reduction of the 17-keto group by 17β-hydroxysteroid dehydrogenase. These pathways are shown in Fig. 2.21. The biological functions of estrogens in the male will be considered in the following Section V.E.

D Sex hormone–binding globulin

Following secretion from the tissue of origin, all steroid hormones are bound to one or more plasma proteins (see

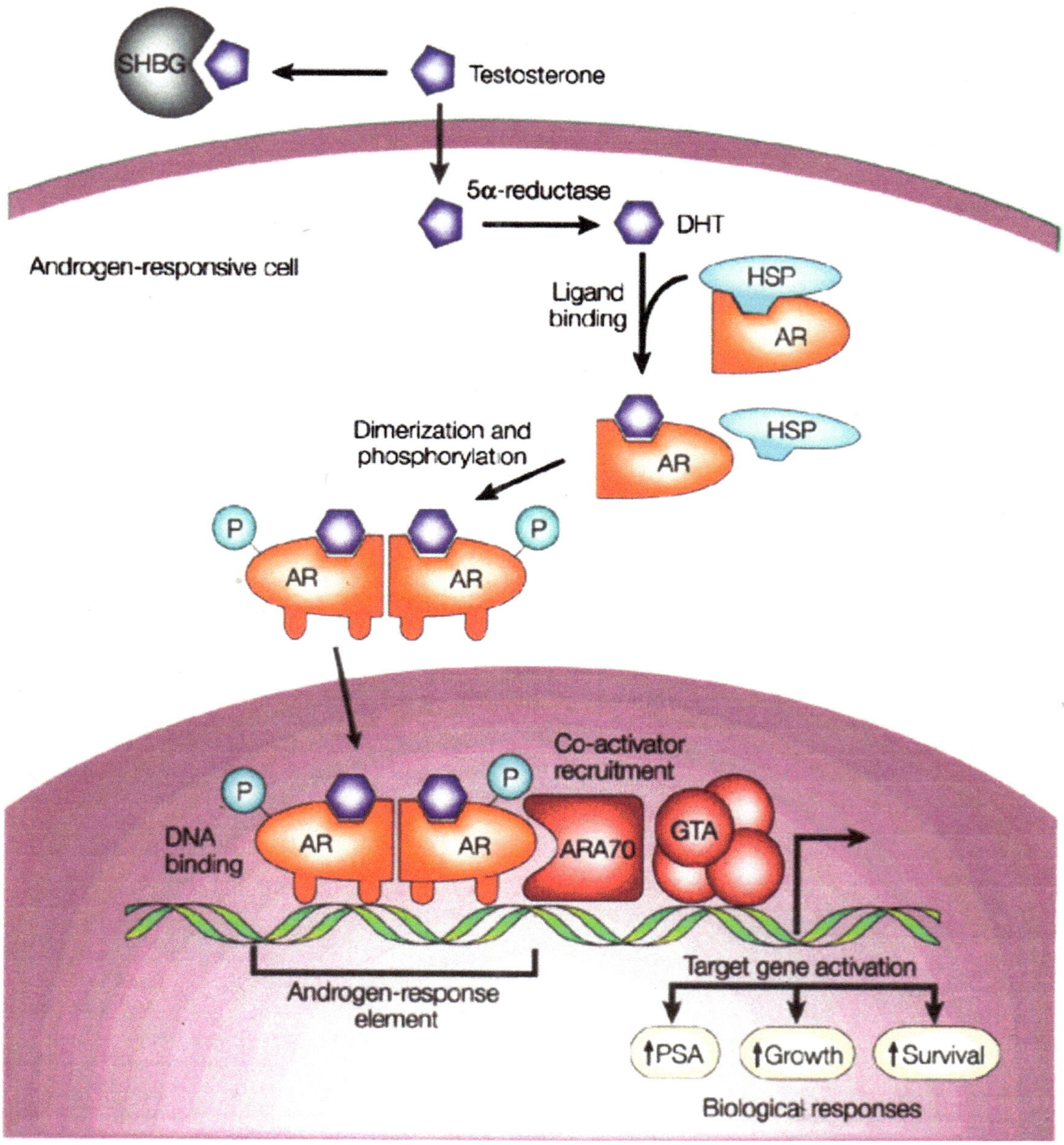

FIGURE 12.8 Summary of the action of testosterone on an androgen-responsive cell. *AR*, Androgen receptor; *ARA70 and GTA*, coactivator proteins; *DHT*, dihydrotestosterone; *HSP*, heat-shock protein; *P*, phosphate; *SHBG*, sex hormone–binding globulin. *Reproduced from https://toddleemd.files.wordpress.com/2015/11/shbgcarrier.gif.*

Table 2.6). Of the total testosterone in the circulation, 0.5%–3% is free (not bound to protein), 54%–68% is bound to albumin with relatively low affinity, and the remainder is bound to SHBG with high affinity. SHBG, synthesized in the liver, is a dimeric glycosylated protein with a molecular mass of 84 kDa. SHBG has a preference for steroids with a 17β-hydroxyl [$K_d \cong (1-5) \times 10^{-10}$ M] so it binds T, DHT, and estradiol, but not, for example, progesterone or cortisol. Thus it serves as the specific transport protein for both testosterone and estradiol. Plasma levels of SHBG are twofold greater in nonpregnant women than in men. The synthesis, and, therefore, the plasma concentration, of SHBG is increased in pregnancy and hyperthyroidism and is decreased by androgens, glucocorticoids, insulin, and growth hormone. Fig. 12.8 summarizes the activity of testosterone on an androgen-responsive cell. Testosterone is bound to circulating proteins, the SHBG having the highest affinity for testosterone. Only a small amount of free, unbound testosterone diffuses into a target cell and is converted to dihydrotestosterone that binds to the inactive receptor bound to proteins, such as the heat-shock protein (HSP). HSP detaches from the receptor after it binds to testosterone converting the receptor to the active form. The activated AR then dimerizes and is phosphorylated and is transported into the nucleus where it forms a complex with an androgen-responsive element of DNA plus coactivator proteins and transcription is initiated.

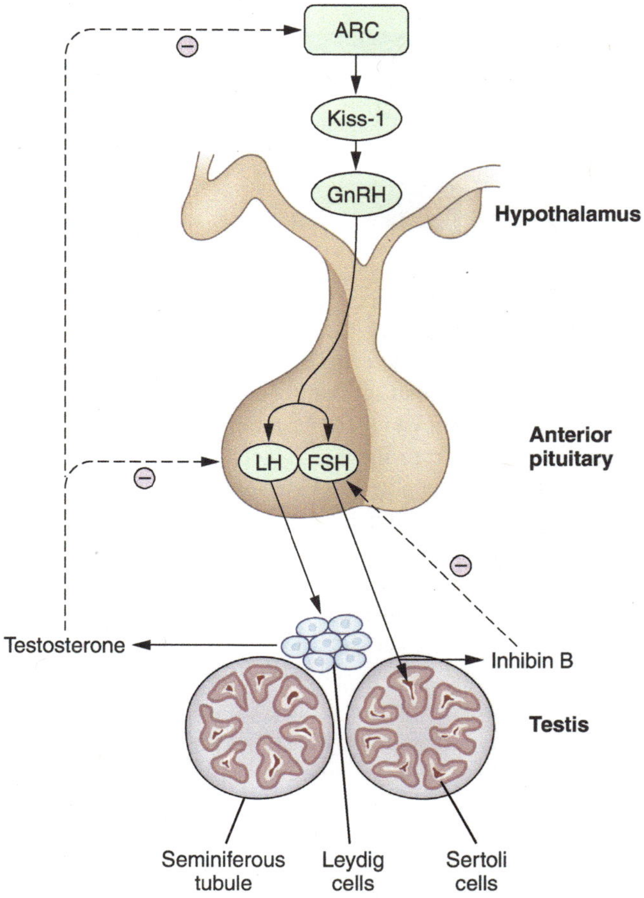

FIGURE 12.9 The hypothalamic–pituitary–testis axis. GnRH is released from hypothalamic neurons in the median eminence and stimulates gonadotrophs to release LH and FSH. GnRH secretion is stimulated by Kiss-1 from neurons of the ARC. Leydig cells in the testis respond to LH stimulation by secreting testosterone. Sertoli cells, the target of FSH, secrete inhibin B. Testosterone exerts negative feedback at the pituitary, the hypothalamus, and on kisspeptide neurons. Inhibin B inhibits the secretion of FSH at the pituitary. *ARC*, Arcuate nucleus; *FSH*, follicle-stimulating hormone; *GnRH*, gonadotropin-releasing hormone; *Kiss-1*, kisspeptin; *LH*, luteinizing hormone.

E Catabolism of androgens

In the liver, testosterone undergoes 5β-reduction, followed by 3α-reduction and reduction of the 17-keto group. DHT undergoes 3α- and 17β-reduction. Both reduced catabolites are then conjugated with glucuronic acid or, to a lesser extent, sulfate, released back into the circulation and removed from the body in the urine or bile.

IV Regulation of androgen production

A Hypothalamic–pituitary–testes axis: hormonal control of the testis

The hormones engaged in the regulation of the hypothalamic–pituitary–testis axis are depicted in Fig. 12.9. The chemistry of GnRH, LH, and FSH was covered in detail in Chapter 3, The Hypothalamus and Anterior Pituitary. The secretion of FSH and LH from the adenohypophysis is governed primarily by hypothalamic GnRH and the blood level of steroid hormones. A relative newcomer to our understanding of this system is kisspeptin, which plays a crucial role in the pulsatile secretion of GnRH. Inhibin B, secreted by the Sertoli cells, also plays an important role in FSH secretion.

1 Kisspeptin

The gene for the peptide now known as kisspeptin and its shorter derivatives was originally identified in 1996 as a tumor suppressor gene and its product was also known as metastin. Peptides that were isolated from human placenta as ligands for the orphan receptor GPR54 were found to be products of the human *KiSS1* gene. Study of the distribution of these peptides soon revealed their presence in what are now known as kisspeptin neurons. In humans, these neurons are located in the arcuate nucleus and the anteroventral periventricular nucleus, both in the hypothalamic area. The axons of these neurons terminate in the GnRH secreting neurons, axons of which release GnRH into the median eminence. From there, GnRH stimulates the release of LH and FSH from pituitary gonadotrophs.

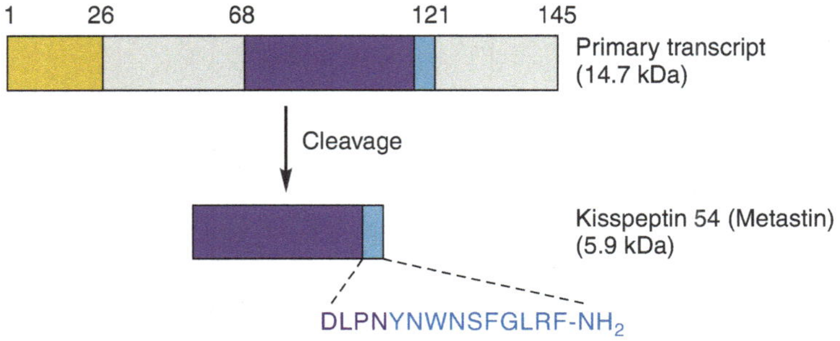

FIGURE 12.10 Kisspeptin. The primary transcript of kisspeptin is shown, with a signal peptide (*yellow*) and the active 5.9-kDa mature peptide, kisspeptin 54 (*purple and blue*). This peptide, originally isolated from human placenta and known for its metastasis suppression properties, has also been referred to as metastin. The amidated C-terminal decapeptide (*blue*) expresses full biological activity. These 10 amino acids, along with the preceding four residues, are completely conserved across mammals and very highly conserved among vertebrates thus far studied.

Thus kisspeptin is a crucial integrator of central and peripheral signals that ultimately control GnRH secretion.

Fig. 12.10 shows the 145-amino-acid primary transcript produced from the *KiSS1* gene and its active cleavage product, kisspeptin 54. C-terminal cleavage products of kisspeptin containing 13 and 14 amino acids have been identified in tissue extracts. It has been determined that only the C-terminal 10 amino acids are required for full kisspeptin activity; these amino acids are completely conserved among primates and only slight variations occur in other mammals and in nonmammalian vertebrates. In Fig. 12.11 are shown a model of the structure of the human kisspeptin primary translated protein (part A), the details of the cleavage products of the primary gene-translated product (part B), and the signaling pathway of the activated kisspeptin receptor (part C).

Kisspeptin signals GnRH neurons through its receptor, KISS1R/GPR54, activating phospholipase C, and ultimately bringing about GnRH release. In both mice and humans, inactivating mutations of either GPR54 or kisspeptin leads to reproductive failure such as hypogonadotrophic hypogonadism and delayed or absent puberty. It is becoming clear that kisspeptin, neurokinin B, and dynorphin are components of the GnRH pulse generator. The exact mechanism for pulse generation of GnRH is under current study.

2 Gonadotropin-releasing hormone

The chemistry, biosynthesis, and control of the secretion of GnRH are covered in Chapter 3, The Hypothalamus and Anterior Pituitary. Briefly, this decapeptide is secreted in a pulsatile fashion under the influence of input from other neurons in the central nervous system, most notably those that secrete kisspeptin, neurokinin B, and dynorphin. The pulsatility of GnRH release is critical to its ability to stimulate gonadotropin secretion. The amplitude and frequency of GnRH pulses are restrained from the age of 4–6 months until the onset of puberty, at which time the increase in both amplitude and frequency of GnRH (and, therefore, gonadotropin) secretion is the hallmark of the onset of reproductive maturation (Section V.C).

3 Luteinizing hormone

The testicular target of LH, the Leydig cells, serves two principal functions: (1) they are the site of production of testosterone, producing, in adult males, approximately 7 mg daily for systemic transport to distal target tissues; and (2) they have paracrine interactions with the immediately adjacent seminiferous tubules to support spermatogenesis. LH-mediated stimulation of testosterone synthesis and secretion is initiated by the binding of LH to specific receptors on the outer membranes of the Leydig cell. Fig. 12.12 (*top*) shows the structure of the LH/hCG (human chorionic gonadotropin) receptor. hCG takes over the function of LH during gestation in the female. Also shown (*bottom*) are the structures of LH (*left*) and hCG (*right*). An increased level of cyclic AMP (cAMP) within the Leydig cell activates protein kinase A (cAMP-dependent protein kinase) which, through phosphorylation of specific transcription factors, induces the synthesis of the steroidogenic acute regulatory protein, StAR (see Chapter 2: Steroid Hormones: Chemistry, Biosynthesis, and Metabolism, Fig. 2.17). This protein is essential for the movement of cholesterol into the mitochondrion where cleavage of its side chain initiates steroidogenesis. Under prolonged stimulation the other enzymes in the pathway from pregnenolone to testosterone are increased by LH, as is the growth of the Leydig cell. The structure of the StAR protein is visualized in Fig. 12.13 (*top*). The mitochondrial function in cholesterol trafficking across the membranes of the mitochondrion and the role of StAR in this process is shown in Fig. 12.13 (*bottom*).

4 Follicle-stimulating hormone

In the male, FSH in conjunction with testosterone acts on the Sertoli cells of the seminiferous tubule at the time of puberty to initiate sperm production. In humans, FSH is required for normal spermatogenesis throughout adult life. FSH binds to its specific G protein–coupled receptor (see Fig. 12.14) on the Sertoli cell to increase, through a cAMP-dependent mechanism, the synthesis of specific proteins, among which are ABP and inhibin, which are discussed just following and in Section IV.B, respectively.

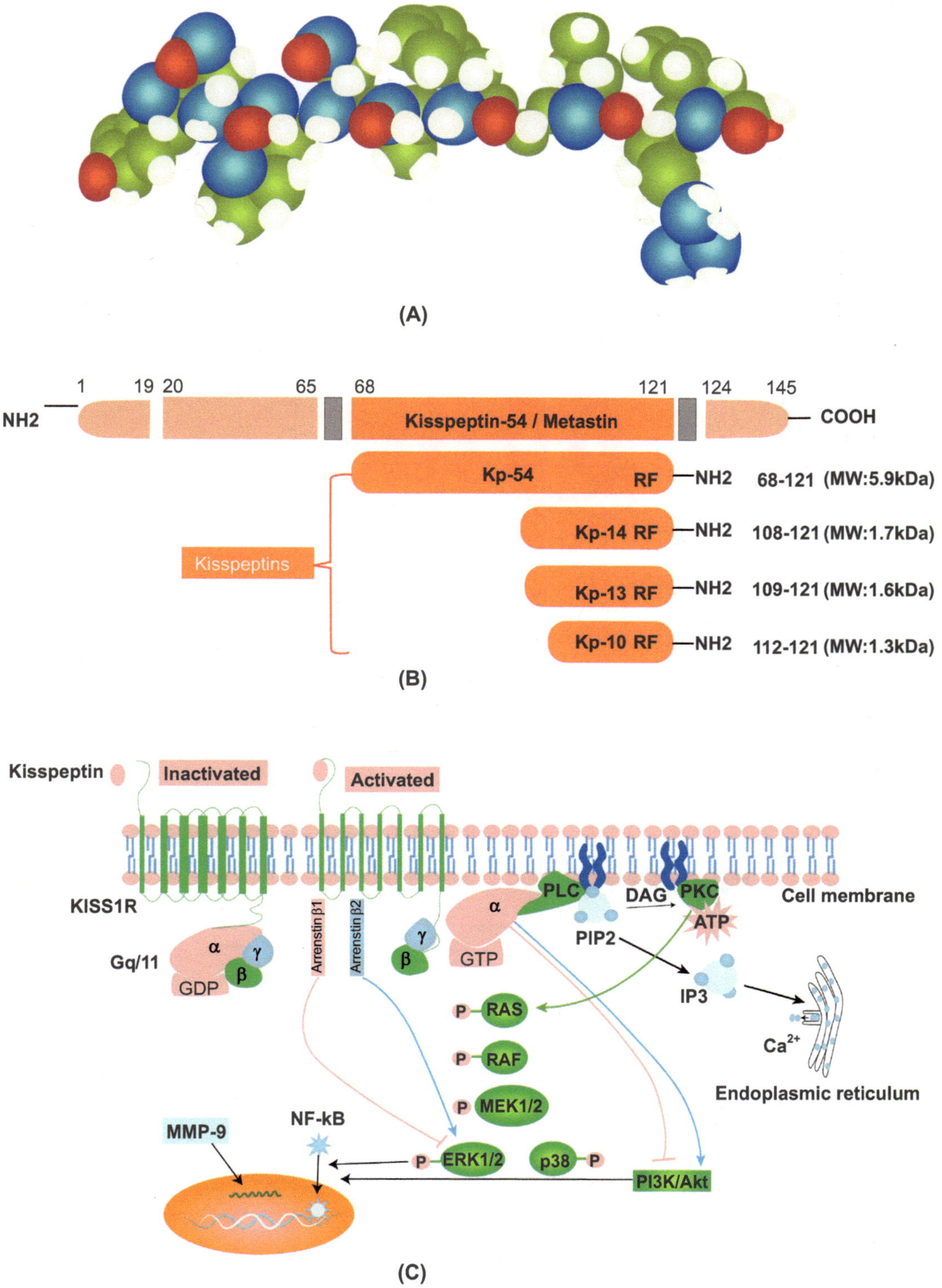

FIGURE 12.11 (A) The primary product, human kisspeptin, consisting with a polyproline-rich domain. (B) Major structural features of human kisspeptins, the products of the *Kiss1* gene. Different kisspeptins are generated by the cleavage from a common precursor, the prepro-kisspeptin. The prepro-kisspeptin contains 145 amino acids with a 19-amino acid signal peptide and a central 54-amino acid region, Kp-54 (formerly termed as metastin). Further cleavage of metastin generates kisspeptins of lower molecular weight: Kp-14, Kp-13 and Kp-10. All kisspeptins contain the RF-amide (Arg-Phe-amide) motif that is able to bind and activate the kisspeptin receptor. (C) KISS1R signaling at a glance. KISS1R is a seven-transmembrane domain G_q/11 coupled receptor. Upon binding of kisspeptin, the intracellular portion of KISS1R phosphorylates G_q/11. The α-subunit of G_q/11 activates PLC, which cleaves PIP2 into IP3 and DAG. IP3 promotes intracellular Ca^{2+} release from the endoplasmic reticulum, while DAG activates a signaling cascade by phosphorylating PKC. PKC activation induces the phosphorylation of MAP kinases, such as ERL1/2 and p38. In addition, activation of KISS1R recruits arrestin-1 and -2, which downregulated and upregulated phosphorylated ERK1/2 levels, respectively. The addition of KISS1R can stimulate or inhibit the phosphorylation of PI3K/Alt, depending on the cell types, but the intermediator is not investigated. Activated KISS1R also enhances the expression of MMP-9 via PI3K/Alt/NF-κB or ERK/NF-κB signaling. *DAG*, Diacylglycerol; *ERK1/2*, extracellular signal-regulated kinase; *IP3*, inositol 1,4,5-triphosphate; *Kiss1*, kisspeptin 1; *KISS1R*, Kisspeptin/kisspeptin receptor; *Kp-14*, kisspeptin-14; *Kp-54*, kisspeptin-54; *MAP*, mitogen-activated protein; *MMP-9*, matrix metalloproteinase-9; *NF-κB*, nuclear factor-κB; *PI3K*, phosphatidylinositol-3-kinase; *PIP2*, phosphatidylinositol 4,5-bisphosphate; *PKC*, protein kinase C; *PLC*, phospholipase C. *(A) Reproduced from http.http://www.biosyn.com/tew/kisspeptin-and-kisspeptide-synthesis.aspx. (B) Reproduced from Fig. 1 of K.-L. Hu, H. Zhao, H.-M. Chang, Y. Yu, J. Qiao, Kisspeptin/kisspeptin receptor system in the ovary, Frontiers in Endocrinology 8 (2018) 365, https://doi.org/10.3389/fendo.2017.00365. (C) Reproduced from Fig. 2 of K.L. Hu, H. Zhao, H.-M. Chang, Y. Yu, J. Qiao, Kisspeptin/kisspeptin receptor system in the ovary, Frontiers in Endocrinology 8 (2018) 365, https://doi.org/10.3389/fendo.2017.00365.*

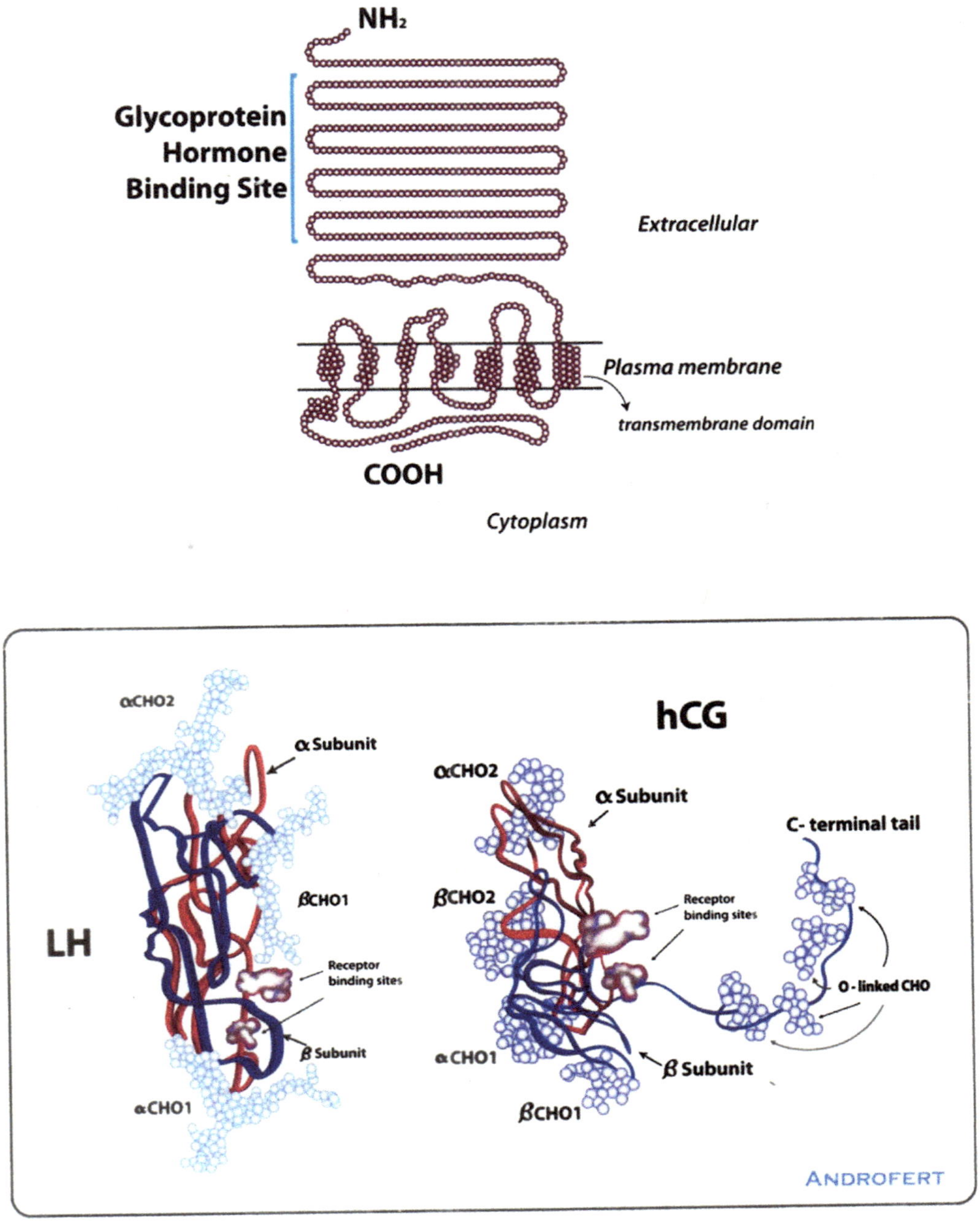

FIGURE 12.12 (Top): LH/hCG receptor. LH receptors are included in the subgroup of G protein–coupled receptors. Such receptors are characterized by having 7-transmembrane domains and a large N-terminal extracellular region, which is the predominant site of hormone binding. (Bottom): Luteinizing hormone and human chorionic gonadotropin molecules. (Left): LH is a glycoprotein with two subunits, the alpha subunit (*red*) similar to that of FSH and hCG with two carbohydrate attachment sites, and the beta subunit (*blue*), with only one carbohydrate attachment site. The light blue balls represent the carbohydrate chains. (Right): hCG is similar in its structural attribute to LH. A notable exception is the presence of a long carboxyl terminal segment that is O-glycosylated (O-linked CHO), conferring longer half-life to hCG. The alpha and beta subunits are represented in red and blue strands, respectively, whereas the light blue balls represent the carbohydrate chains. *FSH*, Follicle-stimulating hormone; *hCG*, human chorionic gonadotropin; *LH*, luteinizing hormone. *(Top) Reproduced from Fig. 16.3 of M. Galano et al., Role of constitutive STAR in Leydig cells, International Journal of Medical Science, 22 (2021): https://doi.org/10.3390/ijms22042021. (Bottom) Reproduced from Fig. 16.1 of M. Galano et al., Role of constitutive STAR in Leydig cells, International Journal of Medical Science, 22 (2021): https://dooi.org/10.3390/ijms22042021.*

ABP was initially characterized as a Sertoli cell product in rodents and other small mammals. In these species, it is thought to function to concentrate androgens in the seminiferous tubules and deliver the steroid hormone to developing spermatocytes and spermatids. ABP is now known to be a homolog of SHBG, the serum-binding protein for androgens and estrogens (Section III.D, previously). In humans, there is little evidence for the presence

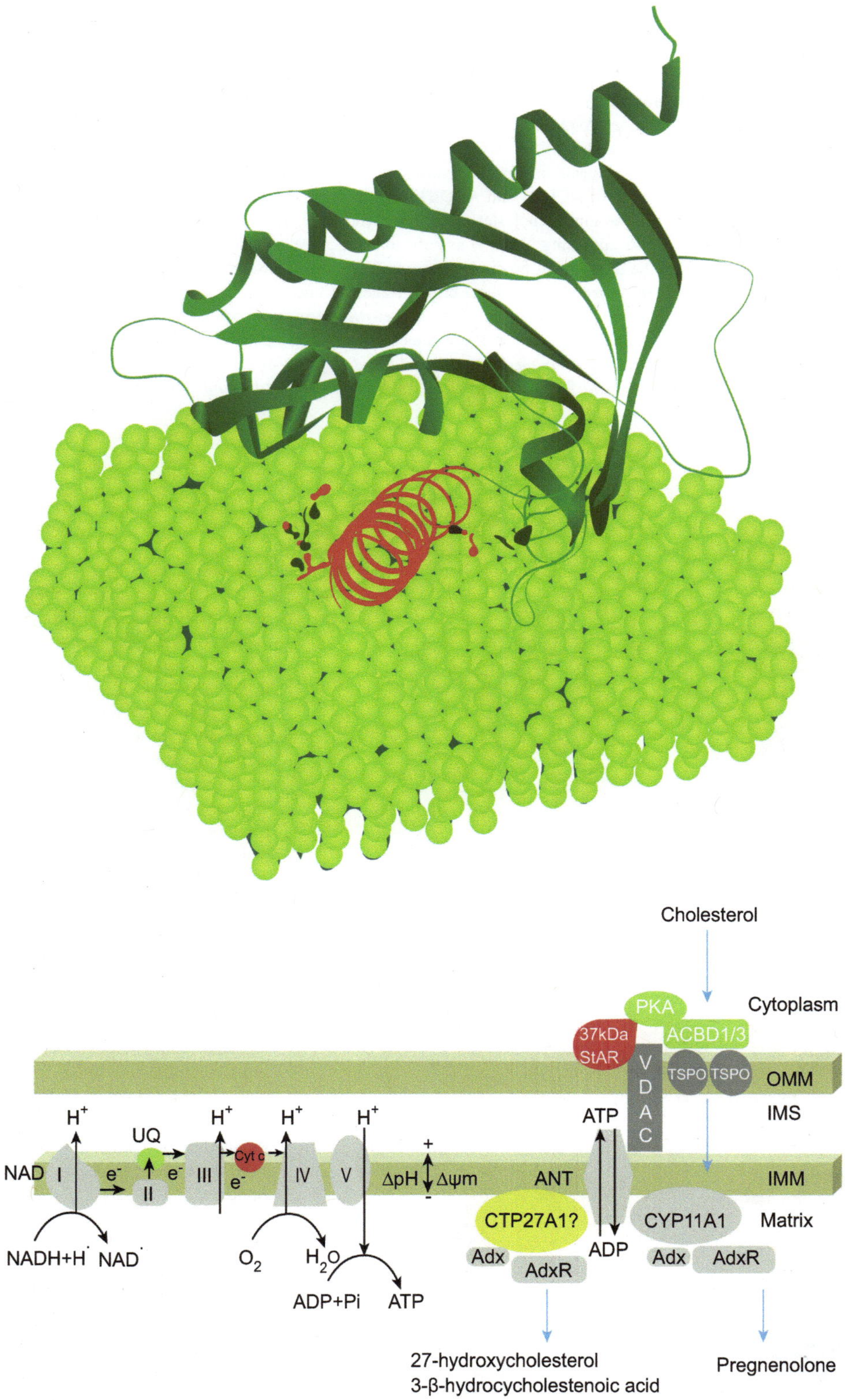

FIGURE 12.13 (Top): Model of N-62 StAR, shown as a *Ribbon Diagram*, sitting on a phospholipid membrane representing the OMM. The N-terminus is in the *upper right-hand corner*, and the C-helix (*in red*) is in the *lower center*. Liposome protection experiments show that only the C-helix, and possibly small portions of the adjacent $\Omega-$ loops, interact with the OMM. Arg272 in the C-helix forms a hydrogen bond with Asp106 in the $\Omega 1-$ loop each shown as *ball-and-stick* figures. This bond limits the movement of the C-helix, closing the cholesterol-binding pocket. Interaction of StAR with protonated phospholipids disrupts this (and other) hydrogen bonds, permitting the C-helix to swing to the right, opening the

(*Continued*)

sterol-binding jacket. (Bottom): Role of mitochondrial function in cholesterol trafficking across mitochondrial membranes. Electrons (e-) from NADH and $FADH_2$ pass sequentially through complex I and complex II, and then to complex III via UQ: Cyt c transfers electrons from complex III to complex IV, reducing oxygen to water. The flow of electrons is accompanied by proton (H^+) transfer across the inner mitochondrial membrane (complexes I, III and IV), creating the electrochemical gradient $\Delta\psi$m. Re-entry of protons to the mitochondrial matrix through complex V utilizes the protomotive force to generate ATP from ADP and P; the same protomotive force also drives the exchange of ATP for ADP via ANT. ANT also forms part of the mitochondrial cholesterol trafficking complex at contact sites between the outer (OMM) and IMM mitochondrial membranes, facilitated by interactions with VDAC and the 18 kDa TSPO. In steroidogenic tissues, after stimulation with gonadotrophin hormones and/or elevation of cAMP, a transduceome complex is generated, with the association of ACDB3 and ACDB1, together with the regulatory subunit of PKA (PKA-R1α) and StAR. This complex trafficks cholesterol from the outer to the inner (cholesterol poor) mitochondrial membrane for conversion to pregnenolone by the action of CYP11A1. In combination with Adx and AdxR, and using NADPH as cofactor. The same complex may traffick cholesterol to CYP27A1/Adx/AdxR in macrophages and endothelial cells within the artery wall, producing oxysterol ligands for LXRα, capable of inducing genes involved in the cholesterol efflux pathway. *Adx*, Adrenodoxin; *AdxR*, adrenodoxin reductase; *ANT*, adenine nucleotide transporter; *cAMP*, cyclic AMP; *Cyt c*, cytochrome C; *IMM*, inner mitochondrial membrane; *OMM*, outer mitochondrial membrane; *PKA*, protein kinase A; *StAR*, steroid acute regulatory protein; *TSPO*, translocator protein; *UQ*, ubiquinol; *VDAC*, voltage-dependent anion channel. *(Top) Reproduced from Fig. 2 of W.L. Miller, StAR search—what we know about how the steroidogenic acute regulatory protein mediates mitochondrial cholesterol import, Molecular Endocrinology 21 (2007) 589–601. (Bottom) Reproduced from Fig. 1 of A.M. Allen, J.M.W. Taylor, A. Graham, Mitochondrial (dys)function and regulation of macrophage cholesterol efflux, Clinical Science 124 (2013) 509–515.*

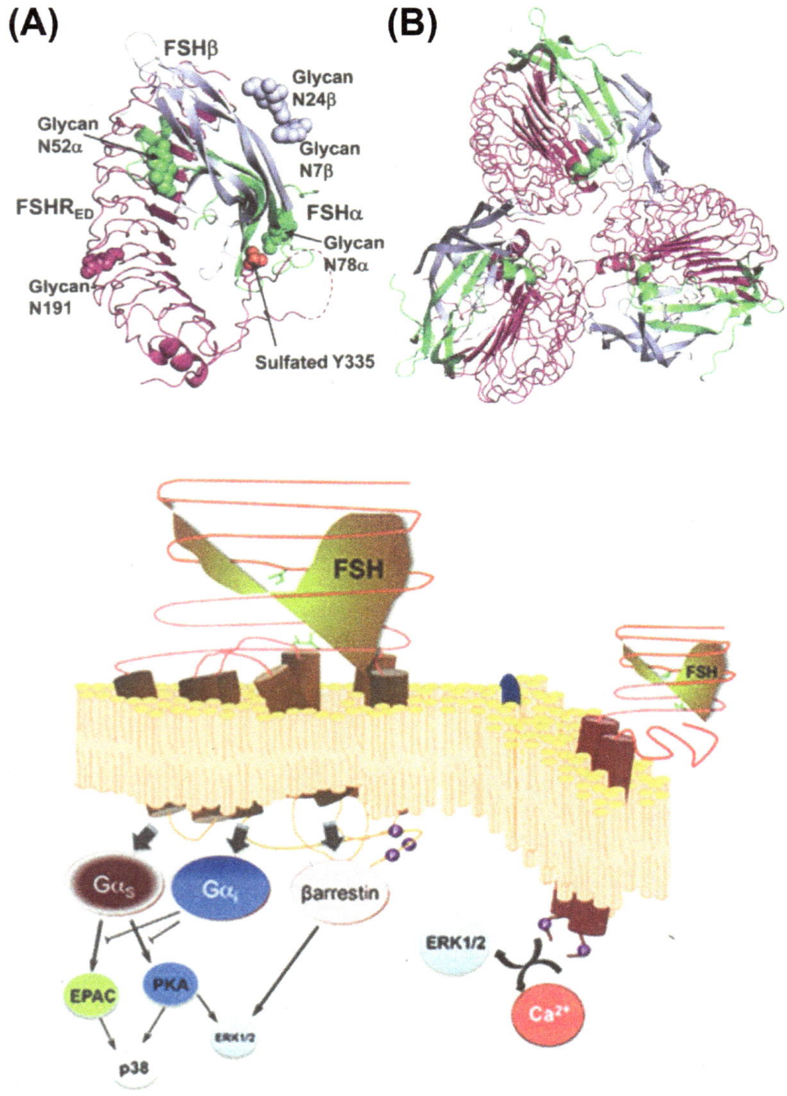

FIGURE 12.14 (Top) Structure of the human FSH in complex with the human $FSHR_{ED}$. (A) Ribbon model of human FSH bound to $FSHR_{ED}$: FSH α-subunit is shown in green, FSH β-subunit is shown in purple, and $FSHR_{ED}$ is shown in magenta. The side chain of sulfated Y335 is depicted as sticks for the tyrosine stem and colored balls for the sulfate group. The carbohydrates are depicted as balls. The disordered residues in the receptors are marked as dashed lines. (B) Top view of the trimer observed in the asymmetrical unit. (Bottom) Follitropin, or FSH, is a G protein-coupled receptor belonging to the glycoprotein hormone receptor family that plays an essential role in reproduction. Although its primary location is in the gonad, the FSHR has been reported in extragonadal tissues including bone, placenta, endometrium, liver and blood vessels from a number of malignant tumors. The recently resolved crystal structure of FSH bound to the entire FSHR ectodomain has been instrumental in more closely defining the role of this domain in ligand binding and receptor activation. Biochemical, biophysical and structural data also indicate that the FSHR exists as a higher order structure and that it may heterodimerize with its closely related receptor. *FSH*, Follicle-stimulating hormone; *FSHR*, follicle-stimulating hormone receptor. *(Top) Reproduced from Fig. 1 of X. Jiang et al., Structure of follicle-stimulating hormone in complex with the entire ectodomain of its receptor, Proceedings of the National Academy of Sciences of the United States of America 109 (2012) 12491–12496. (Bottom) Reproduced from Fig. 1 of A. Ulloa-Aguirre, T. Zarinan, The follitropin receptor: matching structure and function, Molecular Pharmacology 90 (2016) 596–608.*

in FSH-dependent synthesis and secretion of SHBG/ABP by Sertoli cells, although it is abundant in sperm and its levels are positively correlated with sperm motility.

FSH has direct effects on sperm development: it impacts meiosis and the maturation of spermatogonia. In addition, FSH stimulates both the proliferation and differentiation of Sertoli cells in the fetal testis, so that the complement of functional Sertoli cells in the adult human testes is dependent upon FSH-mediated processes in the late phases of fetal development. Since each Sertoli cell can support a limited number of developing sperm, this effect of FSH on Sertoli cell development influences germ cell number.

B Hypothalamic–pituitary–testis axis: feedback control

1 Testosterone

There are three possible levels at which testosterone can exert negative feedback on gonadotropin secretion: the kisspeptin neurons of the arcuate nucleus which regulate the output of GnRH neurons; the GnRH neurons, themselves; and the pituitary gonadotrophs. The relative contribution of each of these components to overall LH and FSH secretion varies with species, but each probably contributes to the negative feedback effect of testosterone in humans.

Kisspeptin neurons contain both androgen and ERs and testosterone represses *KiSS1* gene expression in these neurons in the arcuate nucleus. This leads to decreased GnRH secretion by the hypothalamic neurons that receive kisspeptin signals. Studies in humans in whom aromatase activity is blocked suggest that both testosterone and estradiol suppress kisspeptin expression and activity and that testosterone does not require aromatization for this activity. The relative contribution of the negative feedback effect of testosterone at the kisspeptin neuron and directly at the GnRH neuron in the human is not yet settled, although it is clear that the former is a major component of this action of testosterone.

In the pituitary, testosterone decreases the transcription of the β-subunits of LH and FSH. In contrast to the situation in kisspeptin neurons, in pituitary gonadotrophs, aromatization of testosterone to estrogen plays an important role in the feedback effects of testosterone.

2 Inhibins and activins

The secretion of both LH and FSH from gonadotrophs is stimulated by the same releasing hormone, GnRH. At the same time, it is clear from physiological observations that the two gonadotropins are differentially regulated. For example, in experimental animals in which the testes are removed, FSH levels rise and are not suppressed by the administration of testosterone or estradiol. There are several mechanisms that contribute to this differential regulation. One is that FSH and LH respond differentially to the pulses of GnRH which differ in amplitude and frequency. Additionally, the gonads produce, in response to stimulation by FSH, the peptide hormone inhibin which preferentially inhibits FSH secretion from gonadotrophs. In the testis, inhibin is produced and secreted by the Sertoli cells.

As depicted in Fig. 12.15, the two heterodimeric isoforms of inhibins share a common 18-kDa α subunit and have one of two closely related 14-kDa β subunits. The α and β subunits are joined by disulfide bonds. The two forms are referred to as inhibin A ($\alpha\beta_A$) or inhibin B ($\alpha\beta_B$). Along with the activins, the inhibins belong structurally to the transforming growth factor (TGFβ) multigene family (see Chapter 17: Growth Factors), which also includes the bone morphogenetic proteins, and AMH.

Activins are either heterodimers or homodimers of the two β subunits of inhibin, for example, either $\beta_A\beta_B$, $\beta_A\beta_B$, or $\beta_B\beta_B$ (see Fig. 12.15). Locally produced activins stimulate FSH release from the pituitary, an action that can be blocked by inhibin. Activins play other roles in reproduction in males and females and are also important in such cellular processes as differentiation, proliferation, migration, adhesion, and apoptosis.

Activins stimulate the tonic secretion of FSH by the pituitary. Inhibin (specifically Inhibin B in rodents and humans) binds to the activin type II receptor with low affinity, but this affinity is increased when the coreceptor, TGFβ receptor 3, also known as betaglycan, is present.

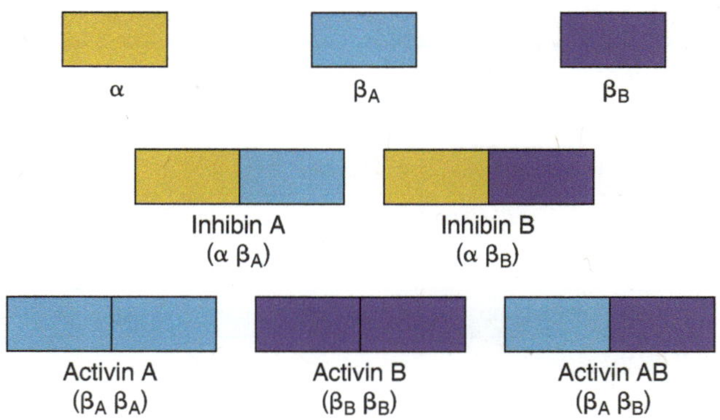

FIGURE 12.15 The inhibin/activin peptides. The inhibins and activins are a group of peptides that are part of the transforming growth factor β superfamily of pleiotropic growth factors. They consist of two subunits joined by a disulfide bond. Inhibins consist of one 18-kDa α subunit and one of the two closely related 14-kDa β subunits, A and B. Activins are composed of two β subunits. Each dimeric activin or inhibin is named according to the β subunit(s) it contains.

This binding blocks activin binding to its receptor, thereby decreasing FSH secretion. In addition to their role in the regulation of FSH secretion, inhibins are important in other aspects of reproduction, for example, in testis development and spermatogenesis.

C Prolactin

The blood levels of the adenohypophyseal hormone prolactin (PRL; see Chapter 3: The Hypothalamus and Anterior Pituitary) in the adult male are approximately one-half those of fertile females. Although the secretion of PRL is reduced in androgen deficiency, the precise biological functions of PRL in the male are not yet known.

Hyperprolactinemia in the male can result from a pituitary tumor and consequences include testicular atrophy, a reduction in plasma testosterone levels, and impotence. All of these symptoms can be reversed by the removal of the tumor. Replacement of testosterone without lowering serum PRL levels does not restore the libido, suggesting a behavioral role for this hormone in male reproduction.

V Biological responses to androgens

A The androgen receptor

Table 12.2 summarizes many of the biological actions of testosterone or one of its biologically active derivatives, 5α-dihydrotestosterone or estradiol (see Fig. 12.7) in target tissues. The biological effects of the androgenic steroids, testosterone or DHT, in the male reproductive system as well as in those tissues associated with the secondary sex characteristics are all believed to be due to the association of the hormone with the AR (see Fig. 12.16). The human AR gene lies on the X chromosome and encodes a protein of 919 amino acids. AR is ubiquitously expressed, although its levels and, therefore, androgen sensitivity varies substantially among nonreproductive tissues.

The AR has a structure typical of its class of nuclear receptors (see Chapter 1: Hormone Introduction) with an N-terminal transactivating domain; a middle DNA-binding domain with two zinc fingers; and, following a short hinge region, a C-terminal ligand-binding domain. There are some homopolymeric repeat sequences that show considerable polymorphism. Among these are the highly variable polyglutamine tract and the less variable glycine variant. The occupied AR acts as a transcription factor that either increases or decreases the expression of specific genes in target tissues to bring about biological responses or phenotypic changes. The length of the polyglutamine tract, which varies between 5 and 35 residues in human populations, is inversely related to the transcriptional efficiency of the AR.

The AR is among the nuclear receptors that, in the absence of ligand, are located in the nucleus or cytoplasm as complexes with chaperone proteins. Upon ligand binding the AR changes conformation, the chaperone proteins dissociate, and the liganded receptor is available to bind to specific AR elements on the DNA.

Several hundred mutations of the AR have been identified and these lead to various degrees of androgen insensitivity. Some are in the ligand-binding domain, affecting the ability of the AR to maintain a stable complex with its ligand; others are in the DNA-binding domain, so that even when ligand is bound, the hormone receptor complex is unable to carry out its function of binding to DNA. Collectively these defects lead to clinical disorders

TABLE 12.2 Biological responses to testosterone and its derivatives.

Response	Active hormone
Differentiation of male internal genitalia	T
Differentiation of male external genitalia	DHT
Differentiation and maintenance of male secondary sex characteristics: for example, hair texture and distribution; voice	T
Skeletal and muscle growth	T
Bone mineralization; epiphyseal closure	E
Prostatic growth	DHT
Libido	T
Sexual differentiation of brain	E
Vascular system: protection from atherosclerosis	E

As shown in Fig. 12.8, in some target tissues testosterone (T) undergoes 5α-reduction to form 5α-dihydrotestosterone (DHT). Both of these androgens interact with the androgen receptor in their respective target tissues. In other androgen targets, testosterone is aromatized to 17β-estradiol (E) which then generates the appropriate biological response through interaction with the estrogen receptor (see Chapter 13: Estrogens and Progestins).

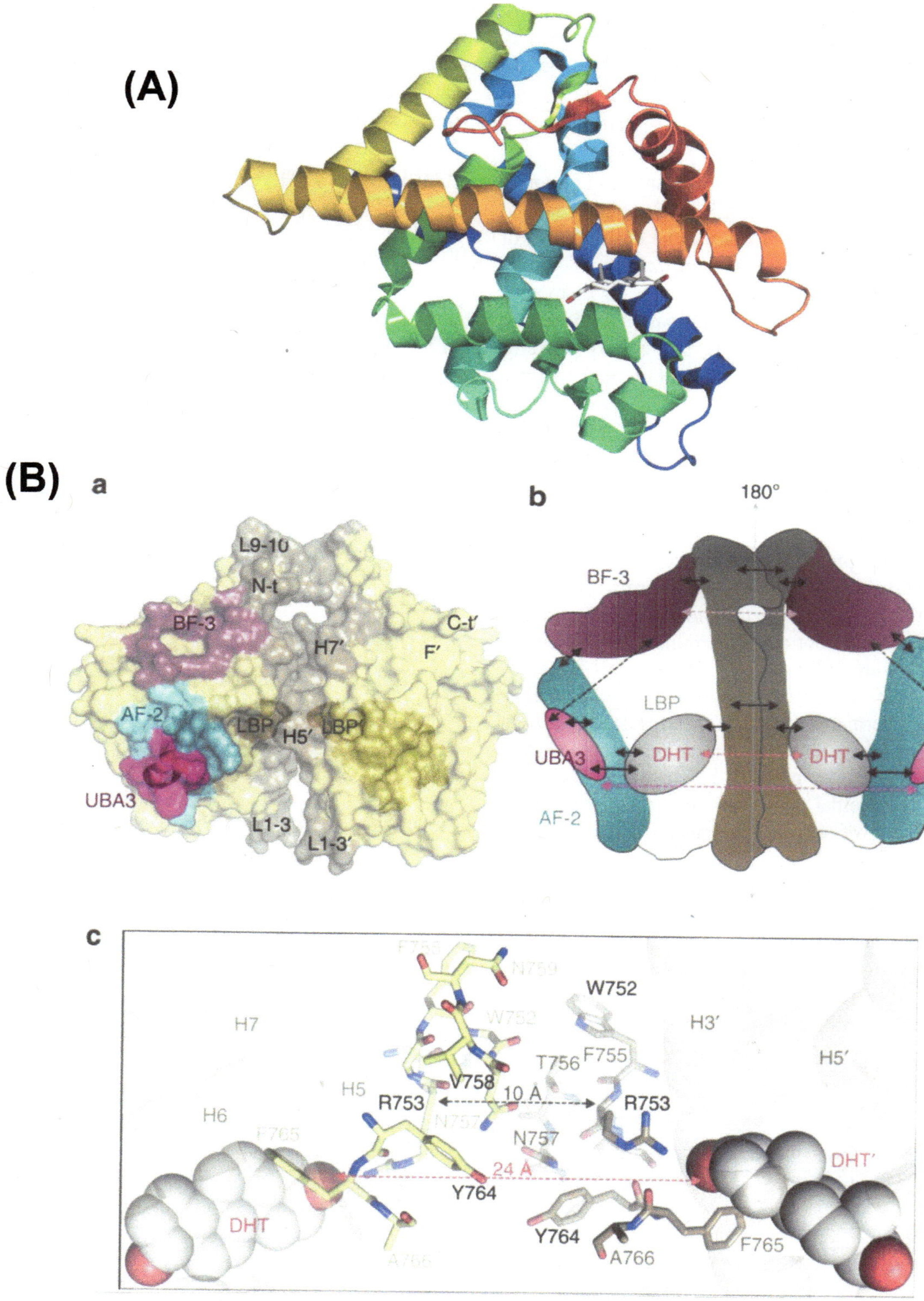

FIGURE 12.16 (A) Ribbon structure of the androgen receptor LBD. The molecule of DHT is in ball-and-stick representation (right lower center). (B) (a) Proposed allosteric communication pathways across the AR-LBD dimer interface. Surface representation of the AR homodimer. The dimer interface (*brown*), the AF-2 groove (*blue*) and the BF-3 (rasberry) are highlighted. Residues that form or line the LBP are shown with a Connolly dot surface and the BA3 peptide as a pink surface. (b) Schematic representation of the proposed intra- and inter-domain allosteric pathways in AR-LBD. Solid arrows indicate short-range communication networks, while dashed arrows point to long-range interactions. (c) Close-up of the dimer interface highlighting allosteric communication between the LBPs across the dimer interface. The distances between the two R753 residues and the two DHT moieties are given. *AF-2*, Activation function-2; *AR*, androgen receptor; *AR-LBD*, androgen receptor-ligand-binding domain; *BF-3*, binding function-3; *DHT*, dihydrotestosterone; *LBD*, ligand-binding domain; *LBP*, ligand-binding pocket. *(A) Reproduced from https://line.17qq.com/articles/ijofbbpbz.html. (B) Reproduced from Fig. 7 of M. Nadal et al., Structure of the homodimeric androgen receptor ligand-binding domain, Nature Communications, 8 (2017) 14388; https://doi.org/10.1038/ncomms14388.*

categorized as the androgen insensitivity syndrome (AIS), which is the most common of the hormone insensitivity conditions and is discussed in Section VI.B following.

B Sexual differentiation

All species have evolved highly coordinated and effective reproductive strategies. These are variable with regard to the complement of sex chromosomes, the development of gonads, the production of gametes, fertilization, and behaviors. We will focus on the process as it is understood in humans, although much of this information is based on studies of animal models. This is particularly true for more recent insights into molecular details gained from genetically altered mice.

1 Sex determination

Sex determination is the process by which the bipotential gonad develops into a testis or ovary which, in turn, is determined by the presence or absence of a Y chromosome. Upon fertilization, union of the spermatozoon with the ovum not only restores the diploid number of chromosomes (46 in humans) but also determines the genetic sex of the new individual. Since the oocyte always has an X chromosome, the sex of the offspring is determined by the fertilizing spermatozoon; an X-bearing spermatozoon will yield a female (XX) zygote and a Y-bearing spermatozoon will result in a male (XY). In some instances, there are chromosomal abnormalities that result in an incorrect number of either autosomal or sex chromosomes. Some of these abnormalities and their effects of sex differentiation are summarized in Table 12.3.

The search for the component of the Y chromosome that determines male sexual development began more than five decades ago. Attention became focused on a portion of the short arm of the Y chromosome, which was termed the sex-determining region of the Y chromosome, *SRY*, and the past two decades have seen elegant studies in mice and humans confirming its importance and identifying its gene product, SRY, as the transcription factor that initiates the male developmental pattern. Confirmation of the crucial role of *SRY* came from the observation that transgenic XX mice also carrying only 14 kb of this region of Y chromosome DNA exhibited a male phenotype. Reports of mutations in the *SRY* region leading to gonadal dysgenesis in 46 XY individuals confirmed the importance of this region of the Y chromosome in sex differentiation in humans.

2 Definition of gender

There are several other parameters besides the genetic sex that contribute to the total definition of the sex of an individual. Gonadal sex encompasses the morphology of the gonads and the hormones that they produce and is of dominant practical importance in defining the sex of an individual. Phenotypic sex refers to the appearance of the external genitalia (i.e., the penis and scrotum or vulva) as well as the secondary sex characteristics, such as facial hair or breasts, while somatic sex refers to the difference in the structures of the internal sex organs. Finally, there is what can be referred to as psychological sex. Thus each individual has a particular gender identity, the identification of oneself as male or female, and gender role, the expression of sexually dimorphic behavior such as aggression and parenting rehearsal behaviors.

3 Gonadal differentiation

Fig. 12.17 outlines the major events of sexual differentiation in human male and female fetuses as a function of time after conception. Early in embryogenesis, the gonads in XX and XY fetuses are morphologically identical; that is, they are bipotential in that they can become either testes or ovaries. The gonadal anlage cells can be identified in the 4-week-old embryo and undergo differentiation into

TABLE 12.3 Some chromosomal disorders of sex differentiation.

Name	Karyotype	Clinical features
Klinefelter syndrome	47,XXY	Small testis with few sperm; tall stature; obesity; and cognitive disability
Turner syndrome	45,X	Streak gonad, immature ovary; lymphedema; short stature; web neck; cardiac defects; and hearing loss
Mixed gonadal	45,X/46XY	Testis or dysgenetic testis; increased dysgenesis risk of gonadal tumors; and some features of Turner's syndrome
Ovotesticular DSD	46,XX/46, XY	Testis, ovary, or ovotestis; increased risk of gonadal tumors

The chromosomal disorders listed earlier are a subset of the broad spectrum of disorders of sex differentiation (DSD). For each the predominant karyotype (or in the case of mosaicism, two karyotypes) is given, although there are variants in most cases.

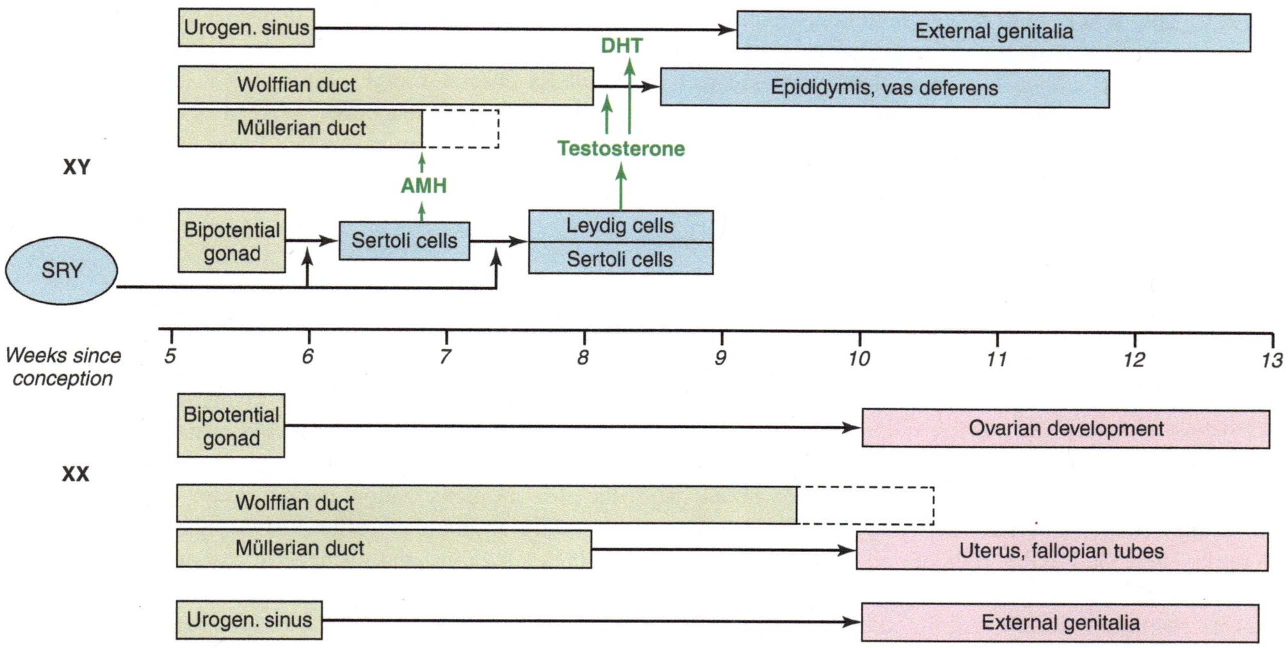

FIGURE 12.17 Major events of sexual differentiation. The processes of sexual differentiation are shown as a function of the number of weeks since conception (time line in the middle of the figure). Moving away from the time line are shown: the differentiation of the bipotential gonad into the testis (upper half) or, later, the ovary (lower half); the regression of the Müllerian ducts and differentiation of the Wolffian ducts into male internal genitalia or regression of the Wolffian ducts and differentiation of the Müllerian ducts into the female internal genitalia; and the differentiation of the male or female external genitalia from the urogenital sinus. *AMH*, Anti-Müllerian hormone (also known as Müllerian inhibiting substance, hormone, or factor, MIS, MIH, or MIF).

the bipotential gonad until the sixth to seventh week post-conception. In the presence of a Y chromosome, the bipotential gonad will develop into a testis and in the absence of a Y and the presence of an X chromosome, it will undergo ovarian differentiation. Gonadal sex then determines internal genitalia and phenotypic sex.

4 SRY or testis-determining factor

The *SRY* gene product, SRY, is detected in the bipotential gonad of XY individuals at about 42 days. The protein contains an high-mobility group box of 79 amino acids that binds to specific regions of DNA, causing the DNA molecule to bend. This bending is thought to make these regions more accessible to other transcription-regulating proteins that ultimately bring about the differentiation of the gonadal cells, some to Sertoli cells, some to Leydig cells. SRY also contains nuclear localization signals, phosphorylation sites, and additional protein–protein interaction sites.

As depicted in Fig. 12.18, a major transcription factor in testicular development is SOX9 (sex-determining region-box 9) an autosomal gene regulated by the SRY protein. In XY fetus, SOX9 is also positively regulated by FGF9 and the product of prostaglandin D synthase, PGD_2 (see Chapter 8: Eicosanoids). SOX9 also has its own positive feedback loop. Sustained elevated levels of SOX9 lead to the normal development of the testes. Since the

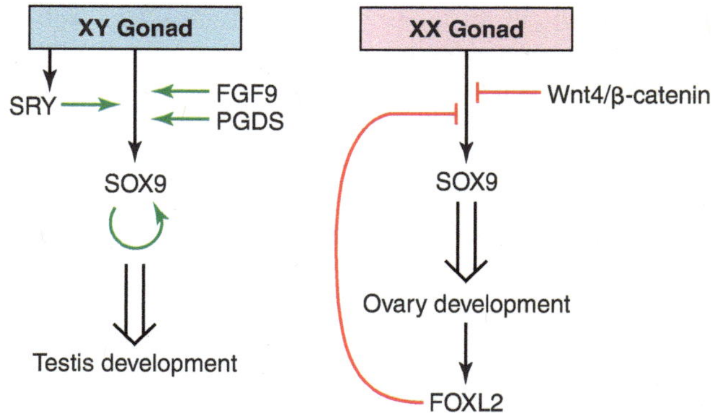

FIGURE 12.18 Signaling pathways in the control of gonad differentiation. In the XY fetus, SRY activates the expression of SOX9 (HMG), a potent transcription factor that is required for the expression of the proteins required for fetal testis differentiation. FGF9 and prostaglandin D_2 (produced by PGDS) also contribute to the maintenance of SOX9 expression in the male. In the embryonic female SOX9 expression is suppressed by the WNT4/β-catenin pathway and its stimulators. In the adult female, normal ovarian function requires the continued suppression of SOX9, which is mediated by the transcription factor FOXL2. *FGF9*, Fibroblast growth factor 9; *FOXL2*, forkhead box L2; *HMG*, high-mobility group; *PGDS*, prostaglandin D synthase; *SOX9*, SRY-related HMG-box-9; *SRY*, sex-determining region of the Y chromosome.

gene for SOX9 is autosomal, it can also be expressed in XX fetuses and adults. Its expression and consequent inappropriate development of testicular cells in the XX fetus are restrained by the Wnt4/β-catenin signaling pathway. This brake on SOX9 expression is required for the maintenance of normal ovarian function in the female throughout reproductive life, and this is provided by Forkhead box L2, among other factors.

By the end of the sixth week after conception, the Sertoli cells of the fetal testis have begun to differentiate and secrete AMH (see later). Within the next 2 weeks the Leydig cells appear and begin to secrete testosterone. Ovarian differentiation follows that of the testis by 4–5 weeks.

5 Differentiation of the internal and external genitalia

During the seventh and eighth weeks, the newly differentiated Sertoli cells secrete AMH which brings about the regression of the Müllerian ducts (Fig. 12.17). A couple of weeks later, the Leydig cells of the fetal testis begin to secrete testosterone leading to Wolffian duct stabilization and differentiation into the internal male genitalia, including the epididymis and the vas deferens. In an XX fetus the absence of AMH leads to the persistence of the Müllerian ducts and their differentiation into the female internal genitalia: the fallopian tubes, uterus, and upper portion of the vagina.

AMH (also known as Müllerian inhibiting substance, factor, or hormone, MIS, MIF, MIH) is a member of the TGFβ family. It is a 24-kDa dimer joined by disulfide bonds that shows some homology to the β subunits of the inhibins and activins. In addition to its crucial role in male sex differentiation, low amounts are made in the ovary and testis of adults, suggesting that it plays additional roles in reproduction.

As with the internal genitalia, the fetal gonad directs the development of the external genitalia. In the male, much of this development is dependent on the conversion of testosterone to DHT, as evidenced by the ambiguous genitalia observed at the birth of XY individuals with a defect in the 5α-reductase. Dozens of transcription and growth factors, supplied in large part by the fetal gonad, are required for these differentiation processes in both XY and XX fetuses.

C Puberty

1 Timing of puberty

Table 12.4 outlines the visible changes in males during puberty and the approximate time span over which they occur. In females the most recorded marker of puberty is the onset of menstruation, the menarche. Thus even though it usually occurs after earlier visible signs of puberty, such as breast budding and changes in axillary and pubic hair, the menarche has been used as a measure of the age of puberty in girls over the decades. The average age of menarche in Caucasian females has been declining for the past two centuries in developed countries. The observations that this trend is now leveling off and that the standard deviation of the average of menarche has also become smaller suggest that earlier effects on the age of menarche, such as those of poor nutrition and disease, are no longer in play in developed countries. The earlier age of menarche in African American girls has been attributed to a higher BMI (body mass index, weight in kg/height in meters squared) as well as genetic influences.

Less definitive data are available for pubertal events in boys and little evidence is available regarding a systematic change in the age of these events over time. In general, obesity in boys is associated with earlier signs of puberty, although some obese boys show delayed puberty.

TABLE 12.4 External events of puberty in the male.

Growth spurt
Pubic hair
Increase in testis size
Increase in penis size
Voice change
Acne
Facial hair

The external events of puberty in the male are indicated in the approximate order in which they occur. Tanner Stage 2 (based on penis size and pubic hair distribution) is the one in which the first changes of pubertal development are apparent and spans 10–14.5 years of age. Stage 5 is characterized by the adult phenotype and is achieved between the ages of 13.5 and 17. The changes listed earlier, with the exception of the onset of the growth spurt, occur within the time frame covered by stages 2–5.

For diagnostic purposes the earliest normal age of pubertal signs for girls is 8 years and for boys is 9 years. If puberty begins prior to these ages, it is referred to as precocious puberty. The upper limits of age are 13 years for girls and 14 for boys; if no signs of puberty are seen by these ages, the individual is said to have constitutional delay of puberty.

2 Signal(s) for the onset of puberty

The hallmark and common proximal event of the onset of puberty is the increase in pulsatile GnRH secretion. The GnRH pulse generator is suppressed during early childhood and is gradually reactivated, usually as the second decade of life begins. The adult pattern of GnRH and, therefore, LH secretion is then achieved during puberty. In contrast to other species such as mice and rats, in primates the suppression of the pulse generator during the juvenile period and its reactivation appear to be independent of gonadal steroid production. In the postpubertal reproductive period, gonadal feedback inhibition is the primary mechanism of GnRH regulation.

Much attention has been directed toward the hypothalamic neurons that produce GnRH in order to identify a "trigger" for the onset of puberty in primates. As the role of kisspeptin in the regulation of GnRH secretion (see Section IV.A) was becoming elucidated, it became clear that it also plays a crucial role in the onset of puberty. Fig. 12.19 depicts a current model of this role of kisspeptin. According to this model, prior to puberty GnRH neurons, even in the presence of stimulatory input such as glutamate, are subject to the restraining influence of inhibitory effect of GABA (γ-aminobutyric acid), one of the most abundant and widespread inhibitory neurotransmitters in the hypothalamus and the rest of the brain. The resulting low levels of pulsatile GnRH secretion result in the juvenile pause in gonadotropin secretion, as compared to the high levels seen during the prenatal and perinatal periods. At the onset of puberty, KNDy neurons express kisspeptin as well as neurokinin B and dynorphin and the GnRH neurons express the kisspeptin receptor, GPR54. The stimulatory effects of kisspeptin (and possibly neurokinin B and dynorphin) overcome the inhibitory effects of GABA so that GnRH is secreted in pulses of increasing amplitude. This leads to the increased gonadotropin secretion which is characteristic of and required for the continuation of pubertal development. Many other neuropeptides and growth factors play a role in establishing the adult pattern of hormone secretion through effects on GnRH neurons and gonadotrophs, but a complete discussion of them is beyond the scope of this chapter.

Because of the association of BMI with the timing of puberty discussed previously and, more specifically, the evidence that the proportion of body mass that is adipose tissue is related to the onset of puberty, a role for the adipocyte hormone leptin (see Chapter 3: The Hypothalamus and Anterior Pituitary and Chapter 6: Pancreatic Hormones: Insulin and Glucagon) has been considered. Although individuals with a severe leptin deficiency display hypogonadotrophic hypogonadism, this hormone most likely plays a permissive role rather than an initiating one in the onset of puberty.

D Spermatogenesis

1 Sperm and semen

A mature human spermatozoon, which contains the haploid number of chromosomes, including either the X or Y

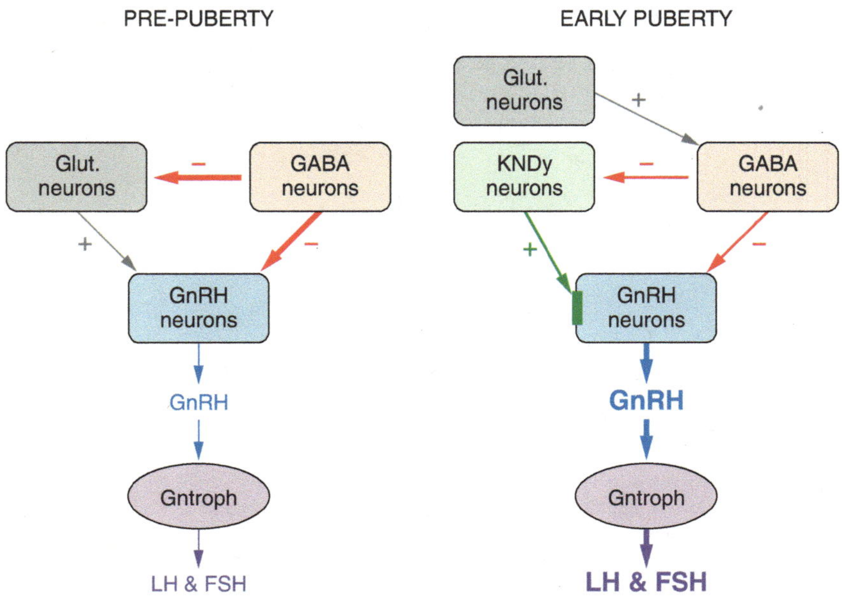

FIGURE 12.19 Onset of Puberty. Prior to puberty (left) the inhibitory influence (*red*) of GABA secreting neurons (*beige*) on GnRH neurons (*blue*) exceeds that of excitatory amino acid transmitters such as glutamate (*gray*), maintaining the juvenile pause in gonadotropin (LH and FSH) secretion. At the onset of puberty KNDy neurons, which can secrete kisspeptin, neurokinin B, and dynorphin (*green*) begin to secrete Kiss1, the ligand for the kisspeptin receptor (*green rectangle*; also known as GPR54) on the surface of the GnRH neurons. This interaction leads to increased secretion of GnRH and the gonadotropins from pituitary gonadotrophs (*magenta*; Gntroph) characteristic of the beginning of puberty. *FSH*, Follicle-stimulating hormone; *GABA*, γ-aminobutyric acid; *GnRH*, gonadotrophin-releasing hormone; *Kiss1*, kisspeptin; *LH*, luteinizing hormone.

chromosome, is ~50- to 60-μm long. Anatomically, the spermatozoon is composed of the head, midpiece, and tail. Under normal circumstances the adult human male produces about 120 million mature sperm per day. Beginning with the commitment of a spermatogonium to maturation, the cycle requires approximately 74 days in humans.

The ejaculate or seminal fluid, normally 2.5–3.5 mL, contains an average of 100 million spermatozoa and a mixture of secretions from the various accessory genital glands. It is rich in fructose, ascorbic acid, prostaglandins, carnitine, and a variety of enzymes, including acid phosphatase, aminotransferases, muramidase, and dehydrogenases. Even though the fertilization of an ovum requires only one sperm, 50% of men with sperm counts of 12.5–25 million/mL and nearly all men with sperm counts under 12.5 million/mL experience severe impairment of fertility.

Gametogenesis in the male is termed spermatogenesis. In contrast to the comparable process in the female (oogenesis), which occurs exclusively during fetal development, spermatogenesis begins in puberty and continues throughout most of the male adult life. In addition to the developing germ cells, at least four cell types are involved in the overall process of spermatogenesis: Sertoli cells, Leydig cells, myoepithelial cells, and the epithelial cells of the duct system.

The three main phases of spermatogenesis, which take place while the developing germ cells are embedded in the wall of the seminiferous tubule, are (1) mitotic phase involving spermatogonia proliferation; (2) the meiotic phase to yield primary spermatocytes, secondary spermatocytes, and spermatids with a haploid chromosomal complement; and (3) the transformation of spermatids into mature spermatozoa. These stages are illustrated in Fig. 12.20.

Subtypes of spermatogonia can be distinguished on the basis of chromosomal staining. Those with dark staining nuclei (A_d) constitute the stem cell renewal pool that provides the continuous supply of germ cells for the reproductive life of the male. Loss of these cells to, for example, high-dose irradiation or vascular compromise leads to lifetime sterility.

A few A_d spermatogonia will emerge from mitosis with pale staining nuclei (A_p) and will proceed to the next stage of development through a type B spermatogonium to a primary spermatocyte. The efficiency of spermatogenesis in humans is limited by the formation and entry of type B spermatogonia into meiosis as primary spermatocytes. The completion of the first round of meiosis results in secondary spermatocytes with a haploid chromosomal complement and another round of division yields spermatids. Meiotic nondisjunction, the improper segregation of chromosomes during meiosis that occurs during secondary spermatocyte and spermatid formation, is responsible for 50% of Klinefelters syndrome (47, XXY) cases and a substantial proportion of trisomy 21 (Down's syndrome) cases.

The maturation of the spermatid into the mature spermatozoan is depicted in Fig. 12.21. The most dramatic features are the loss of cytoplasm, the development of the acrosome in the head, the organization of the mitochondria in the midpiece, and the elaboration of the flagellar tail. The acrosome contains digestive enzymes that will break down the outer membrane of the ovum, the zona pellucida, so that the sperm can penetrate and fertilize the ovum (see Fig. 14.2). The cytoplasm is shed, with only a small amount containing mainly mitochondria remaining in the mature sperm.

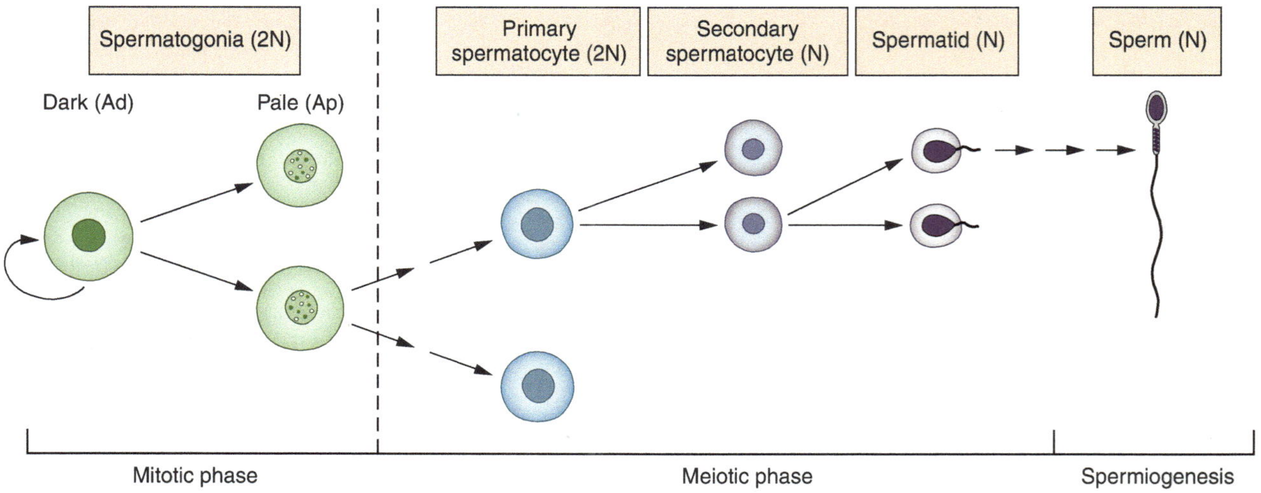

FIGURE 12.20 Spermatogenesis. The three stages of spermatogenesis in the human male are depicted: (1) the mitotic phase in which the stem cell pool is regenerated from Type A_d (dark-staining) spermatogonia, a few of which become Type A_p (pale-staining) spermatogonia; (2) the meiotic phase from which the haploid spermatid emerges; and (3) spermiogenesis during which the spermatid undergoes maturation into a mature spermatozoon (see Fig. 12.21).

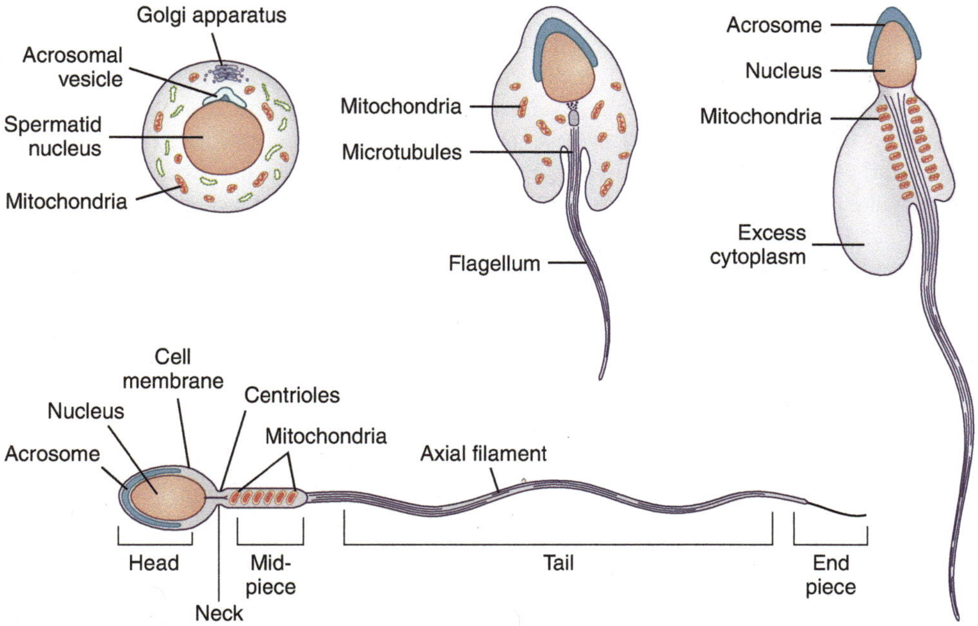

FIGURE 12.21 Spermiogenesis. The haploid spermatid is shown in cross section on the upper left. The remaining diagrams are sagittal sections of the maturing spermatid and, at the bottom of the figure, mature spermatozoon. During maturation the cytoplasmic volume is reduced, the mitochondria are concentrated around what will be the midpiece, and the flagellar tail develops.

E Estrogens in males

The adult human male produces approximately 45 μg of estradiol per day, mostly from aromatization of testosterone (see Fig. 2.22) in adipose tissue, bone, brain, breast, blood vessels, liver, and both the Sertoli and Leydig cells of the testes. As indicated in Fig. 12.7, the aromatization of testosterone is a critical step in its action in bone, where it mediates the closure of the epiphyseal plate during puberty and decreases bone mineral resorption and the brain where it participates in the negative feedback inhibition of testosterone on GnRH secretion. Estradiol produced from testosterone probably affects other areas of brain function, including mood and cognitive function. In the small number of cases of inactivating mutations of aromatase in men, observations have included tall stature, low bone mineral density, and changes in carbohydrate and lipid metabolism.

F Androgens in females

Androgens play important roles in the reproductive functions of females. As indicated in Chapter 2, Steroid Hormones: Chemistry, Biosynthesis, and Metabolism, and discussed again in Chapter 13, Estrogens and Progestins, testosterone and androstenedione are the substrates for aromatase and obligatory intermediates in the production of estradiol and estrone, respectively. Adrenal androgens play a critical role in female puberty bringing about the changes in pubic and axillary hair; the onset of these effects is termed the adrenarche, the adrenal counterpart of the menarche, which is driven by the ovary. Excessive exposure to androgens during uterine life can bring about masculinization of a female fetus, depending on the timing and extent of the exposure. Excess androgens at any point in adult life can have masculinizing effects on the female, manifested as excess hair growth, voice changes, and changes in body composition.

VI Clinical aspects

A 5α-Reductase deficiency

Dozens of mutations in the gene encoding the Type 2 5α-reductase have been identified. In this disorder the production of DHT is diminished or abolished, depending on the severity of the effect of the mutation. The observations from severe cases provided the basis for understanding the role of DHT in the development of the external genitalia. Total 46 XY males with 1 of these mutations are born with an ambiguous external genitalia but with normal development of the testes and male internal ducts. Often these individuals are raised, during childhood, as girls. As testosterone levels rise during puberty, there is usually marked virilization of the external genitalia and in about half of the individuals originally raised as females, gender identity changes to male during this period. Most individuals have decreased spermatogenesis with compromised fertility throughout adulthood.

B Androgen insensitivity syndrome

More than 300 mutations in the human gene for the AR have been identified, leading to a range of phenotypes of AIS. Because the gene for AR is on the X-chromosome, all males carrying the mutation will be affected, whereas heterozygous females are most often phenotypically normal carriers of the mutation on one of two X-chromosomes.

If the result of the mutation is an AR that is lacking all function, leading to complete AIS, the result is an XY individual with an unambiguous female phenotype, including breast development and female fat deposition. Such cases are often identified only when puberty is delayed and the onset of menses fails to occur; upon examination, it is found that the individual has high levels of circulating testosterone from androgen-producing abdominal or inguinal testes. Due to the effects of fetal testicular AMH, there is no fallopian tube or uterine development and the vagina is short with a blind end. Management includes the removal of the testes, estrogen (and some androgen) replacement, and vaginoplasty.

At the other end of the phenotypic variability spectrum is the group of individuals with minimal AIS who present with infertility, gynecomastia (growth of mammary glands), or small penis. In between these two phenotypic extremes lies a broad spectrum of those with partial AIS. In this group, variation in phenotype may occur among those with the same AR mutation, suggesting that genes for other proteins, such as nuclear coregulators, may be involved.

C Androgen abuse

Androgens can be misused—that is, used for medical purposes outside of a recognized medical indication—or abused—that is, used for nonmedical purposes. The most common form of abuse, taking advantage of the anabolic effects of androgens on muscle growth, emerged in the mid-20th century in the arena of international athletic competition. The use of androgens for this purpose has now permeated athletic endeavor at all levels, particularly for power sports and bodybuilding.

Placebo controlled trials have demonstrated the efficacy of androgenic steroids in building muscle size and strength throughout and above the physiological dose range. However, the risks of sustained use of supraphysiological amounts of these steroids include the consumption of compounds of unknown composition and side effects, especially as suppliers produce ever more variations to evade detection in drug screenings. Other negative outcomes associated with illicit androgen use are overall increased mortality, cardiac pathology, abnormal plasma lipoprotein levels (dramatic increase in low-density lipoprotein/high-density lipoprotein), liver pathology, and behavioral changes. Most of these effects of inappropriately high doses of androgens are reversible, with varying time courses following discontinuance of steroid use.

D Prostate cancer

As has been mentioned several times throughout this chapter, androgens promote the growth of components of the male reproductive system, including the prostate gland. Prostate cancer, in its early stages, is androgen dependent. Paradoxically, this disease occurs with increasing frequency as males age, the time that testosterone levels are falling. However, the aromatization of testosterone to estradiol continues, particularly in adipose tissue, so that there is a change in the ratio of the two sex steroids. A considerable amount of evidence suggests that this shift in E/T plays a role in prostate cancer development. Thus during its early stages, testosterone is necessary but not sufficient for prostate cancer development.

During this hormone-dependent stage of prostate cancer, a major mode of treatment depends on establishing an androgen-deprived environment. This can be done through surgical or chemical castration. Most often used for this purpose are GnRH agonists that undermine the normal cyclical pattern of GnRH secretion and after an initial burst of testosterone secretion result in chronic androgen depletion. Although this is not a permanently effective treatment, because prostate cancer cells will eventually become refractory to the androgen-depleted environment, it is a useful treatment if given at the appropriate time in preparation for radiation or chemotherapeutic treatments.

Further reading

Testis anatomy

C. Fan, et al., ACE2 expression in kidney and testis infection in Covid-19 patients, Frontiers in Medicine Nephrology 7 (2021). Available from: https://doi.org/10.3389/fmed.2020.563893.

A.T. Major, M.A. Estermann, C.A. Smith, Anatomy, endocrine regulation and embryonic development and the rete testis, Endocrinology 162 (2021) bqab046. Available from: https://doi.org/10.1210/endocr/bqab046.

Chemistry and metabolism of androgens

N. Appanna, et al., Differential action and expression of human 5β-reductase (AKR1D1) splice variants, Journal of Molecular Endocrinology 66 (2021) 181–194.

I. Asangani, et al., Using biochemistry and biophysics to extinguish androgen receptor signaling in prostate cancer, Journal of Biological Chemistry 296 (2021). Available from: https://doi.org/10.1074/jbcREV120.012411. 2021,100240.

C.A. Frye, J.K. Torgerson, V.F. Lembo, K. Drew, Androgen's effect across the lifespan in men and animal models, Intechopen 9 (2021). Available from: https://doi.org/10.5772/intechopen.96707.

Regulation of androgen production

A.E. Braun, et al., Examining sex differences in the human placental transcriptome during the first fetal androgen peak, Reproductive Sciences 28 (2021) 801–818.

I. Cellai, et al., Insight on the intracrinology of menopause: androgen production within the human vagina, Endocrinology 162 (2021). Available from: https://doi.org/10.1210/endocr/bqaa219. bqaa219.

T. Yazawa, et al., 11-Ketotestosterone is a major androgen produced in porcine adrenal glands and testes, The Journal of Steroid Biochemistry and Molecular Biology 210 (2021) (2021) 105847. Available from: https://doi.org/10.1016/j.jsbmb.2021.108847.

Sex differentiation

F. Han, et al., Identification of SRY-box30 as an age-related essential gatekeeper for male germ-cell meiosis in differentiation, Ageing Cell 20 (2021). Available from: https://doi.org/10.1111/acel.13343.

C. Mayere, et al., Single-cell transcriptome reveal temporal dynamics of critical regulators of germ cell fate during mouse sex determination, The FASEB Journal 35 (2021). Available from: https://doi.org/10.1096/fj.20200242-120R.

Y.-R. Li, et al., Trajectory mapping of the early *Drosophila* germline reveals controls of zygotic action and sex differentiation, Genome Research 31 (2021). Available from: https://doi.org/10.1101/gr.271148.120.

Puberty

L. Malone, C. Bouvattier, U.B. Kaiser, Central precocious puberty: recent advances in understanding the aetiology and in the clinical approach, Clinical Endocrinology (2021). https://doi.org/1-1111/cen.14475.

L. Naule, L. Maione, U.B. Kaiser, Puberty, a sensitive window of hypothalamic development and plasticity, Endocrinology 162 (2021). Available from: https://doi.org/10.1210/endocr/bqaa209. bqaa209.

R.L. Rosenfield, D.W. Cooke, S. Radovicks, Chapter 16—Puberty in the female and its disorders, in: Sperling Pediatric Endocrinology (fifth ed.), Elsevier, 2021, pp. 528–626.

Spermatogenesis

D. Li, et al., Novel gene regulation in normal and abnormal spermatogenesis, Cells 10 (3) (2021) 666. Available from: https://doi.org/10.3390/cells10030666.

E. Pohl, J. Gromoll, J. Wistuba, S. Laurentino, Healthy ageing and spermatogenesis, Reproduction 161 (2021) R89–R101.

X. Tang, et al., Heat shock protein-90α (Hsp90α) stabilizes hypoxia-inducible factor (HIF-1α) in support of spermatogenesis, Cancer Gene Therapy (2021). Available from: https://doi.org/10.1038/s41417-021-00316-6.

Chapter 13

Estrogens and progestins

I Introduction

The integrated operation of the events and processes of female reproduction is dependent upon the interaction of signals—both hormonal and neural—between the central nervous system, the pituitary, and the ovaries. The hormones responsible for the development and maintenance of the female phenotype comprise the gonadotropin-releasing hormone (GnRH); the gonadotropins, luteinizing hormone (LH) and follicle-stimulating hormone (FSH), both produced by the anterior pituitary; and the female steroid hormones, the estrogens and progestins.

Six characteristics contribute to the female genotype: (1) chromosomal composition and structure; (2) gonads that are functionally and structurally ovaries; (3) female sex hormone production (cyclical in the adult); (4) external and internal genitalia that are morphologically appropriate for a female; (5) rearing as a female; and (6) self-acceptance of a female role. The focus of this chapter will be on the structure and function of the ovary; the secretion of the female sex hormones and its regulation; and the actions of hormones, both ovarian and from other sources, in regulating and integrating the processes necessary for successful reproduction.

II Anatomy of the female reproductive system

A Components of the female reproductive system

The components of the human female reproductive system are shown in Fig. 13.1 and 13.2. They include the paired ovaries, which will be discussed in more detail following the fallopian tubes, the uterus, and vagina. The mammary glands will be discussed in Chapter 14, Hormones of Pregnancy, Parturition, and Lactation. The key steps of differentiation of the chromosomally determined but sexually indifferent female gonad are discussed in Chapter 12, Androgens, along with the events of puberty in both the male and the female. The focus in this chapter will be on the functional adult female nonpregnant reproductive system.

1 Fallopian tubes

The fallopian tubes (oviducts) are 11- to 12-cm long and extend from the trumpet-shaped infundibulum, which is in contact with the ovary, to the proximal end, which opens into the uterus. During ovulation the small finger-like projections or *fimbriae* of the fallopian tubes undulate to draw the released ovum into the oviduct where it can move down to the uterus. This is accomplished by the peristaltic-like contractions of the smooth muscle cells lining the fallopian tube, which also act to propel the sperm upward toward the ovum. The structure and function of the cells of the fallopian tube vary, largely under the influence of estrogen, during the reproductive cycle so that activities to capture and move the ovum are maximal at the time of ovulation. During the tubular passage of the oocyte it will, if not fertilized, eventually disintegrate and disappear. If the ovum is fertilized, it promptly completes its final meiotic division and becomes a zygote (see Fig. 13.5 in Section II.B). After 4–5 days the blastocyst that has resulted from cell division of the zygote will reach the uterus, where it will implant and continue to develop during pregnancy (see Chapter 14: Hormones of Pregnancy, Parturition and Lactation).

2 Uterus

In the nonpregnant adult woman, the uterus (womb) is a muscular, thick-walled, pear-shaped organ about 5-cm wide × 7.5-cm long × 2.5-cm thick. The cavity of the uterus communicates below with the vagina and above with the fallopian tubes. The uterine wall comprises two main functional compartments, the endometrium and the myometrium (Fig. 13.2A). The thin outer layer, equivalent to the peritoneum, is the perimetrium. The endometrium is composed of two layers: the thin basal layer and the outer functional layer (Fig. 13.2B). The functional layer changes dramatically under hormonal influence during the menstrual cycle and is almost completely lost during the process of menstruation. In the secretory phase of the menstrual cycle (14–28 days), the cells are quite tall and columnar with numerous microvilli (see Section IV. D). The myometrium consists of smooth muscle tissue

Hormones. DOI: https://doi.org/10.1016/B978-0-323-90262-5.00001-9

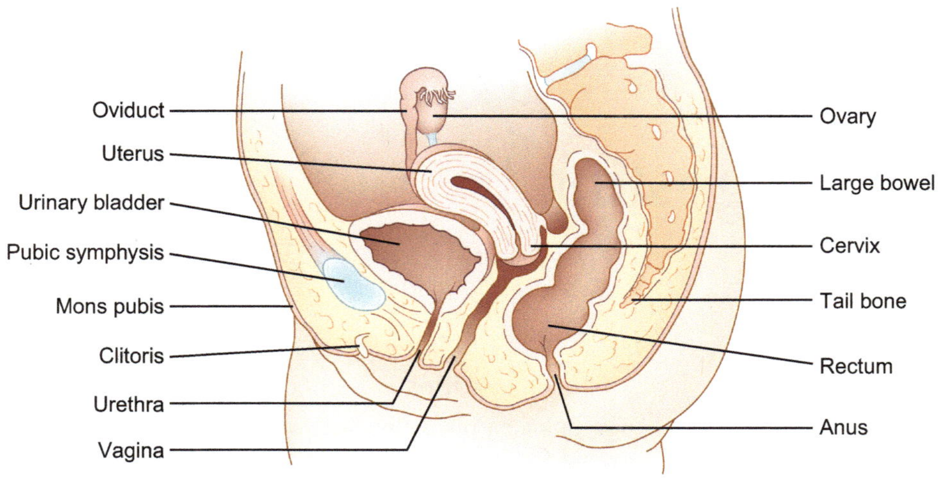

FIGURE 13.1 Anatomy of the female reproductive system. A mid-sagittal view is shown of the organs and structures involved in human female reproduction.

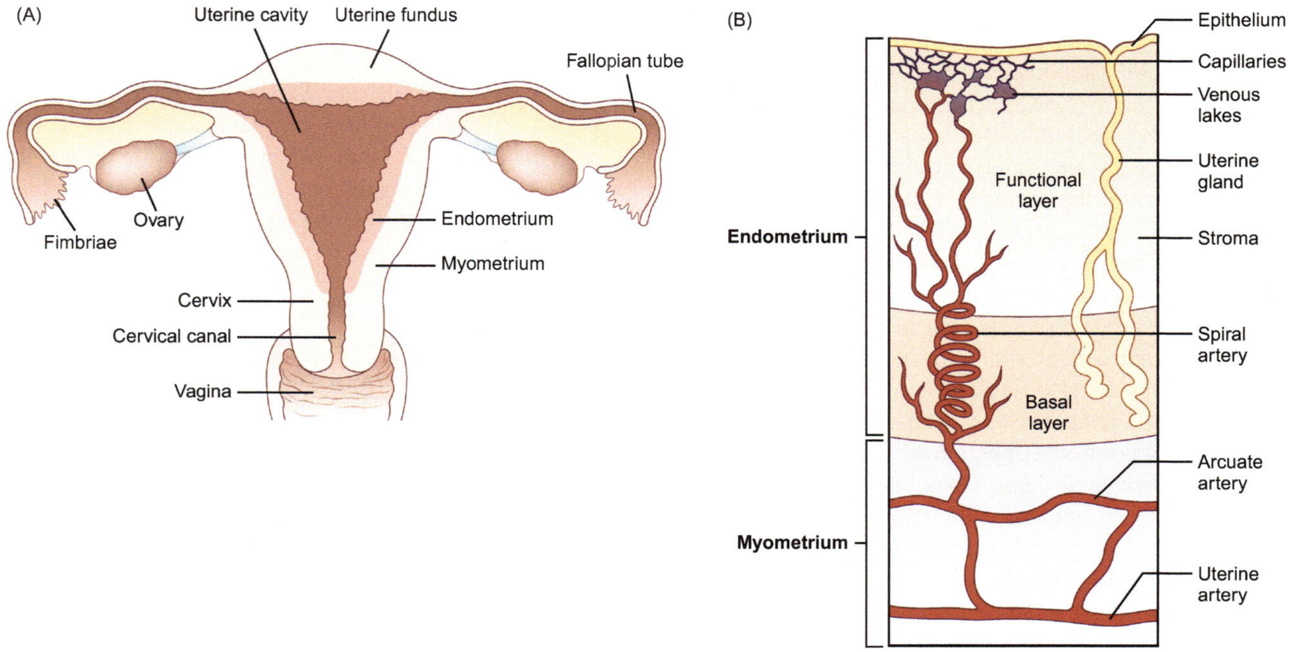

FIGURE 13.2 Internal female reproductive organs. (A) The arrangement of the ovary, fallopian tubes (oviducts), uterus, and vagina is illustrated in a mid-frontal section. (B) Detail of the uterine wall showing the outer muscular wall, the myometrium, and the two main layers, basal (basalis) and functional (functionalis), of the endometrium. The illustration shows the endometrium during the second half of the menstrual cycle, known, relative to the uterus, as the secretory phase. The uterine glands are lined with a single layer of epithelial cells. The rich vasculature is characterized by the arterial supply in the myometrium, the spiral arteries in the basal layer, and venous lakes near the surface.

which functions to evacuate the contents of the uterus at the end of an ovulatory cycle if no fertilization has occurred or at the end of a pregnancy, either aborted or full term.

The volume capacity of the uterine cavity changes from 2–5 mL before pregnancy to 5000–7000 mL at term; the increase in uterine mass is from 60 g at puberty to 1000 g at the end of pregnancy, a 16-fold increase. The primary stimulus for the growth of the myometrium during pregnancy is estrogen produced by the placenta (see Chapter 14: Hormones of Pregnancy, Parturition, and Lactation).

The function of the cervix of the uterus is to prevent uterine contents from emptying during pregnancy. Its rigidity is maintained by progesterone and its estrogen-induced softening and opening (dilation) is an important step in childbirth, as described in Chapter 14, Hormones of Pregnancy, Parturition, and Lactation. The Papanicolaou test, or Pap smear, which consists of a light microscopic examination of cells obtained from the cervix of the uterus, is of considerable clinical importance in the early diagnosis of cancer of the cervix. The cancerous cells can easily be distinguished from the regular uniform appearance of the normal epithelial cells.

3 *Vagina*

The vagina is a 9-cm membranous tube that extends from the lower portion or cervix of the uterus downward and forward to the external opening in the vestibule (see Fig. 13.1). The wall of the vagina is composed of fibroelastic tissue and smooth muscle, which is lined with a mucous membrane formed of squamous epithelial cells. These cells are quite responsive to estrogen, and their physical appearance under the light microscope changes during the ovulatory cycle.

B The ovarian follicle and corpus luteum

The ovary is responsible for the production and release of the ovum as well as the biosynthesis and secretion of the key steroid hormones, progesterone and estrogen. These steroid hormones play a dominant role in the maintenance of and changes in the sexual reproductive tissues during both the ovulatory cycle and pregnancy.

The adult human ovary is 2–5 cm in length and is almond-shaped. As shown in Fig. 13.3A, each ovary consists of the cortex, or germinal epithelium, where the follicles are located and the medullary (inner) zone, consisting largely of stromal scar-like tissue of atretic follicles and the corpus albicans formed in each cycle. The progression of the development of the ovarian follicle from the primordial state to the preovulatory Graafian follicle is depicted. During this process, which several follicles begin each cycle, most follicles will undergo atresia (absence or disappearance). Fig. 13.3B shows some of these stages in the human ovary. Part of the development of the follicle includes the elaboration of the granulosa layer of cells surrounding the ovum as well as that of the thecal cells forming the outside layer of the follicle. Both these cell layers play important roles in governing the reproductive cycle as discussed later in this chapter. Following ovulation, the remaining cells, from both granulosa and theca layers, undergo differentiation into steroid hormone producing cells. The resulting structure, known as the corpus luteum, is filled with lipids, including carotenoids, particularly lutein, which give the structure its color and name, "yellow body." In the human the corpus luteum is large relative to the size of the ovary, ranging in size from 1.5 to 4 cm. If the released ovum is not fertilized, the corpus luteum regresses in a process known as luteolysis after about 14 days and the fibrous remnant remains as the corpus albicans.

The development of the follicle and ovum are shown in more detail in Fig. 13.4. The primordial follicle, the stage of the entire reserve of ova of the human female at birth, consists of an ovum arrested in meiotic prophase I surrounded by a single layer of squamous (flattened) granulosa cells and has no access to the circulatory system. During puberty, in a complex process that is not thoroughly understood, locally produced signaling factors bring about the recruitment of a cohort of these follicles to begin development to the stage of the primary follicle, in which the egg has increased in size and the single layer of granulosa cells has become cuboidal. Proliferation of these granulosa cells and accumulation of stromal-like cells to become the theca characterize the secondary follicle, which contains a fully grown ovum. This first phase, from primordial to secondary follicle, is termed initiation and requires several months for each cohort throughout the woman's reproductive years.

During the next phase of development, the granulosa and thecal layers continue to increase in size and a small cavity or antrum appears. The antrum is filled with follicular fluid from the surrounding cells. At this point, further growth and development are dependent on pituitary gonadotrophins. The size range of antral follicles is large, 0.4–20 mm in diameter. During the first half of the cycle, one follicle (usually) is selected for the final maturation prior to ovulation. This selection process will be discussed in more detail later in this chapter.

The meiotic maturation of the ovum is depicted in Fig. 13.5. During the transition from oogonium to oocyte, the DNA of the chromatids is replicated and the meiotic process is arrested in prophase I. This status remains unchanged until the follicle containing the ovum becomes the dominant one destined for ovulation. During the preovulatory LH surge (Fig. 13.11 in Section IV.D), meiosis is resumed, and the first meiotic division is completed followed by a rest. If fertilization occurs, the second round of meiosis is completed to achieve the haploid state. Completion of fertilization with the haploid sperm results in the diploid zygote.

In contrast to gametogenesis in the male (spermatogenesis), which is initiated in puberty and continues throughout most of adult life, oogenesis in the female occurs exclusively in embryonic life. Thus all the ova, in primordial follicles, that a woman will have for the approximately 400 ovulations of her reproductive life are present at birth. As shown in Fig. 13.6, by the sixth month of gestation there are ~7 million germ cells (oocytes) in each ovary. This peak number of oogonia in the pool has been achieved by the combined processes of oogonia mitosis which increases and atresia which decreases the pool. In addition, the entry of some oogonia into meiosis decreases the pool of cells that can undergo mitosis. As described previously, these cells, now termed oocytes, arrest in prophase and are later surrounded by a single layer of granulosa cells, forming the primordial follicle. Oogonia that have not entered meiosis by the end of the seventh month of gestation undergo atresia. The loss of these cells, along with that of many primordial follicles, reduces the total pool of germ cells present at birth to about 20% of its peak, which is further reduced to the 4%–5% remaining at puberty. Except for the few hundred oocytes that will participate in ovulation, all of these will be lost to atresia until the entire pool is depleted approximately 50 years after birth.

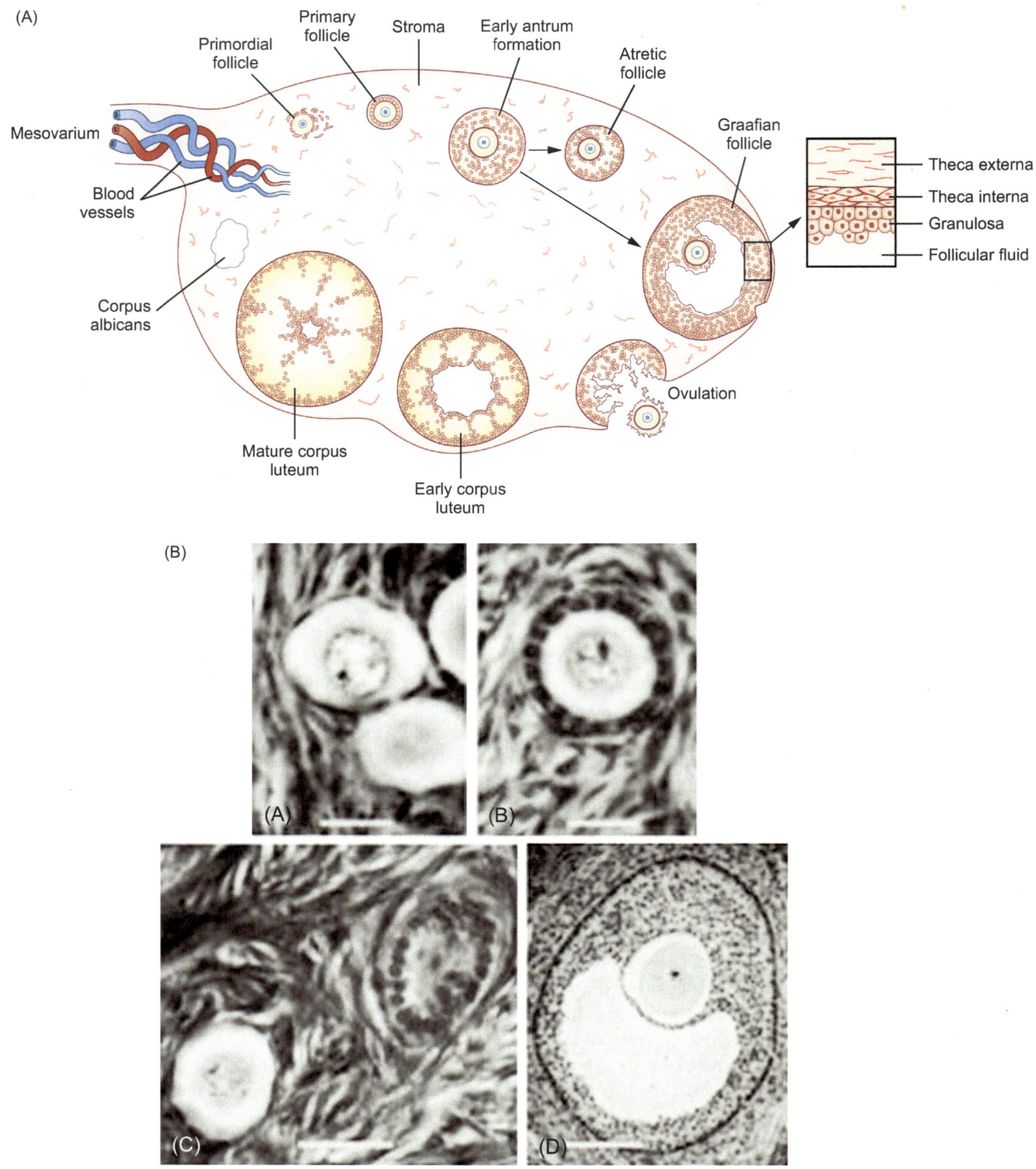

FIGURE 13.3 The ovary in cross section. (A) The stages of follicle development are depicted clockwise beginning with the pool of primordial follicles up to the Graafian, preovulatory stage. See Fig. 13.4 for more detailed description. Of the many follicles that begin development, the majority become atretic, while, usually, one is selected for final maturation and ovulation. At this point the follicle has a highly differentiated wall consisting of the theca externa, theca interna, and granulosa, as seen in the inset. Following ovulation, the remaining granulosa and thecal cells of the follicle differentiate into the lutein cells of the corpus luteum which, in the absence of pregnancy, degenerates after about 2 weeks into the corpus albicans. (B) Micrographs of human ovary: (a) primordial follicle; (b) primary follicle; (c) atretic primary follicle, top right; and (d) early antral follicle. *Adapted with permission from A, Gougeon, Human ovarian folliculat development: from activation of resting follicles to preovulatory maturation. Annales d'Endocrinologie 71 (2010) 132–143.*

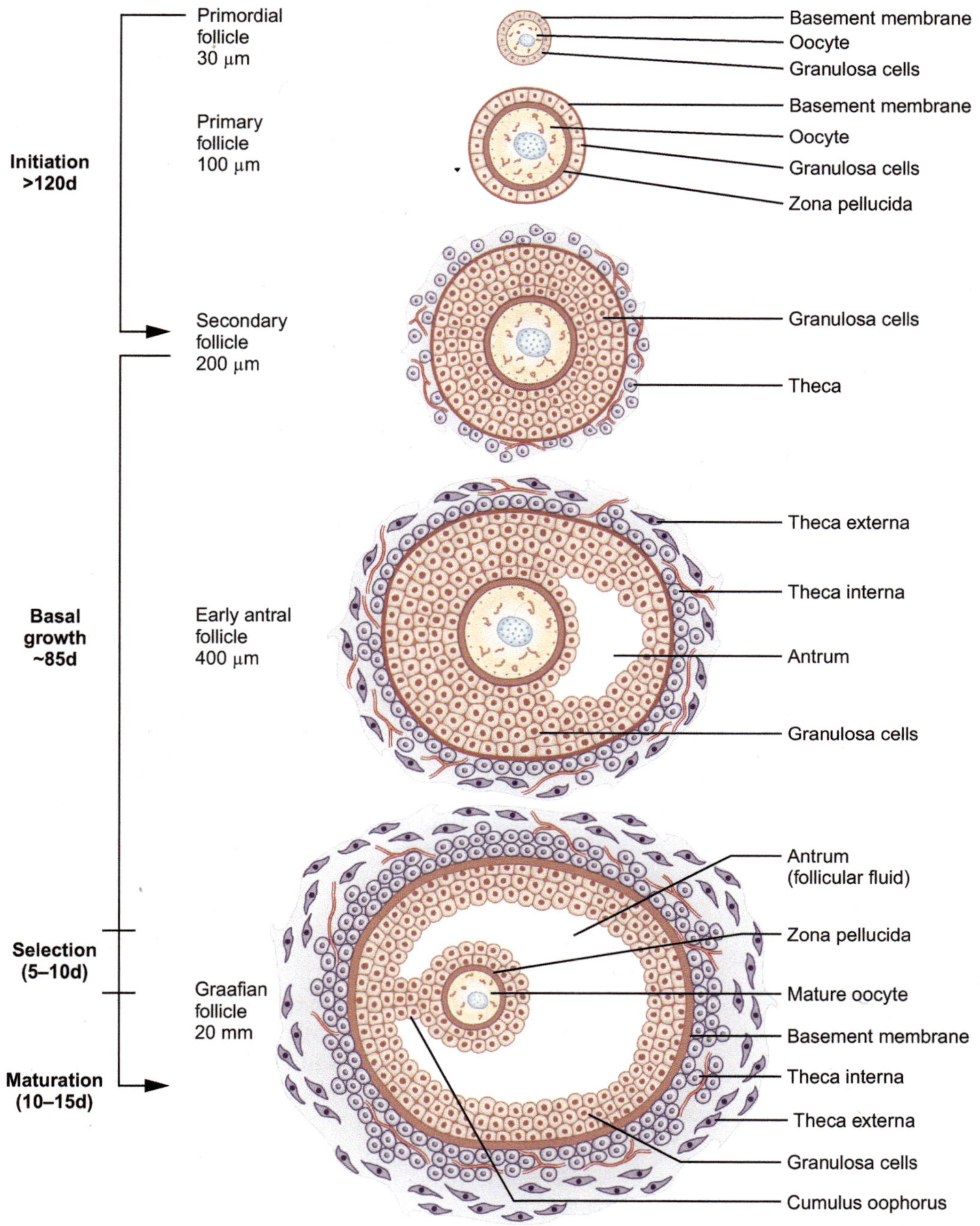

FIGURE 13.4 Development of the ovarian follicle. The stages in the development of the follicle are depicted. The primordial follicle consists of an immature ovum surrounded by a single layer of flattened granulosa cells and a basal lamina. The primary follicle is characterized by a single layer of cuboidal granulosa cells and a fully grown oocyte. The secondary follicle, in which the ovum has reached its full size, has a greater abundance of granulosa cells and the beginning of the thecal layer. The next step is the development of the antrum and selection of, usually, a single follicle for final maturation and ovulation.

III Chemistry and metabolism of female steroid hormones

A Female steroid hormones

1 Structures

The two most important steroid hormones of the adult female are estradiol and progesterone. In addition, estrone and estriol, both weak estrogens, are present in pregnancy (Chapter 14: Hormones of Pregnancy, Parturition, and Lactation). The structures of these compounds are presented in Figs. 13.7 and 13.8.

As reviewed in Chapter 2, Steroid Hormones: Chemistry, Biosynthesis, and Metabolism, the naturally occurring estrogens are typically 18-carbon steroids that have an aromatic A ring with a phenolic hydroxyl; the

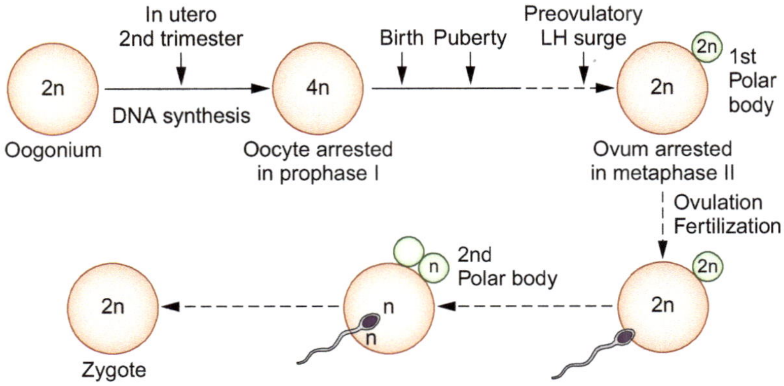

FIGURE 13.5 Meiotic maturation of the ovum. The pathway of the conversion of the chromosomal complement of the oogonium of the fetal ovary to that of the fertilized ovum, the zygote, is shown. The key steps involve the beginning of the first meiotic division to form the oocyte, which is arrested in prophase I after synthesis of the new chromatids and crossing over. At the onset of reproductive maturation, the first round of meiosis is completed and the second round arrests at metaphase II until ovulation. The second round of meiosis is completed during fertilization and the haploid ovum becomes the diploid zygote.

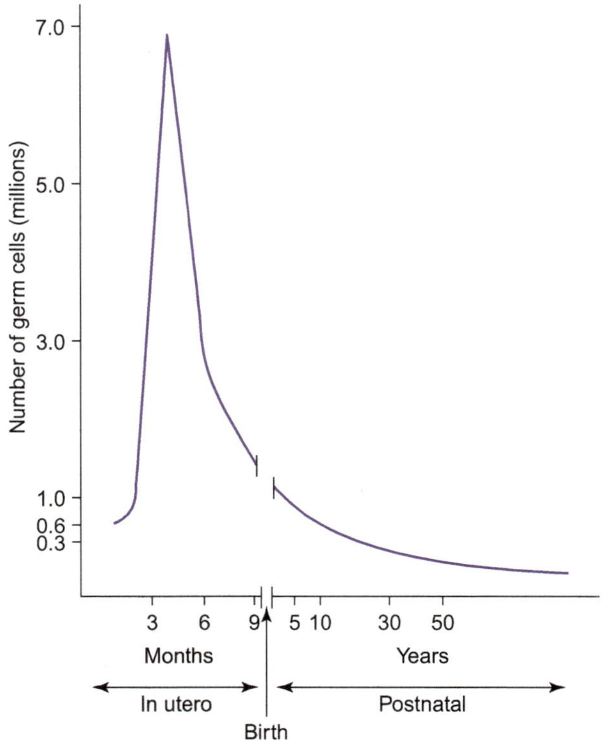

FIGURE 13.6 Oocyte number as a function of age in the human ovary. The number of oocytes peaks at about 6 months of gestation then decreases sharply during the third trimester. This decline continues after birth until the store is depleted at menopause. Of the 6–7 million oocytes present at the peak, all undergo atresia except the 400 or so that are ovulated. *From T.G. Baker, A quantitative and cytological study of germ cells in the human ovaries. Proceedings of the Royal Society B: Biological Sciences 158 (1963) 417–433.*

naturally occurring progestin, progesterone, has 21 carbons, with an oxo functionality on both C-3 and C-20. The pathways of synthesis of estradiol and progesterone are covered in Chapter 2, Steroid Hormones: Chemistry, Biosynthesis, and Metabolism, and in Section III.B following. The variations in secretion of these hormones during the menstrual cycle are covered in Section IV.D.

2 Agonists and antagonists

The emergence of a detailed understanding of how steroid hormones interact with receptors to regulate gene transcription led to a new field of research centering on the design of steroid hormone agonists and antagonists, which either mimic or inhibit the natural hormone, respectively. Often the synthesized molecule has turned out to act as an agonist or antagonist in one tissue or group of tissues and have the opposite or no effect in other tissues. In the case of estrogens, these molecules are called selective estrogen receptor modulators, SERMs; the analogous molecules for progesterone are termed selective progesterone receptor (PR) molecules.

Shown in Fig. 13.7 are an estrogen antagonist (antiestrogen), clomiphene, and three SERMs. Clomiphene is a very effective fertility drug for women who fail to ovulate. Because it blocks the feedback inhibition of estrogen on the pituitary and hypothalamus (Section IV.C), FSH levels are increased, stimulating the growth of the follicle and increasing the chance of ovulation. Tamoxifen acts as an antagonist in breast tissue (and an agonist in the uterine endometrium) and is used in the treatment of hormone-receptor positive breast cancer (Chapter 14: Hormones of Pregnancy, Parturition, and Lactation). Raloxifene behaves as an agonist in bone tissue (and an antagonist in the uterus and breast) and is used to ameliorate the bone loss that occurs and leads to osteoporosis in postmenopausal women. Bazedoxifene is a newer SERM, similar in its actions and use (in the EU and Japan) to raloxifene.

Fig. 13.8 shows the structures of two progesterone antagonists. Because progesterone from the corpus luteum is absolutely essential for the maintenance of pregnancy in its early weeks, the antagonistic activity of RU-486 has been used for inducing abortions shortly after conception. Both RU-486 and ulipristal acetate are under investigation for their ability to reduce uterine fibroids in women during their reproductive years, thus obviating the need for surgery for this condition.

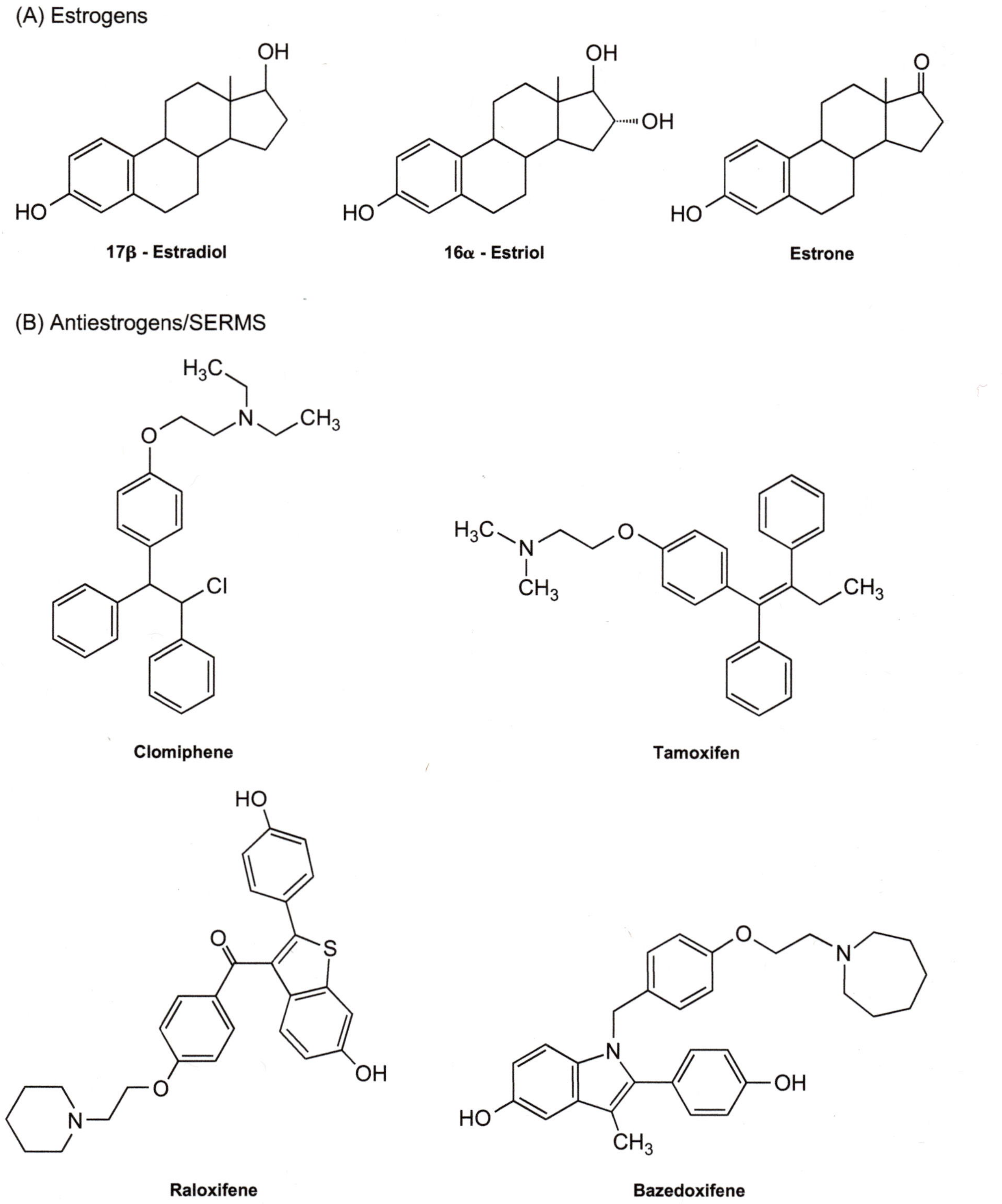

FIGURE 13.7 Estrogens and related molecules. (A) The naturally occurring estrogens, of which 17β-estradiol is by far the most active, are shown. (B) An estrogen antagonist (clomiphene) and three SERMs with mixed agonist/antagonist activities are shown. *SERMs*, Selective estrogen receptor modulators.

B Synthesis of progesterone and estrogens

The pathways of the biosynthesis of progesterone and estradiol from cholesterol are shown in Fig. 2.21. The amounts of estradiol, estrone, and progesterone produced by the ovary and circulating during different phases of the menstrual cycle are shown in Table 13.1.

The specific manner in which estradiol synthesis is accomplished in the preovulatory follicle is shown in Fig. 13.9 and involves the cooperation of two cell types and the activity of two gonadotrophins. The cells of the theca interna contain cholesterol side chain cleavage activity (CYP11A1/P450scc). These are the ovarian cells that express receptors for LH, the response to which is an increase in steroid acute regulatory protein and the cleavage of the side chain of cholesterol. In the endoplasmic reticulum of the thecal cells, pregnenolone is converted to androstenedione by the removal of carbons 20 and 21 and

(A) Progestin

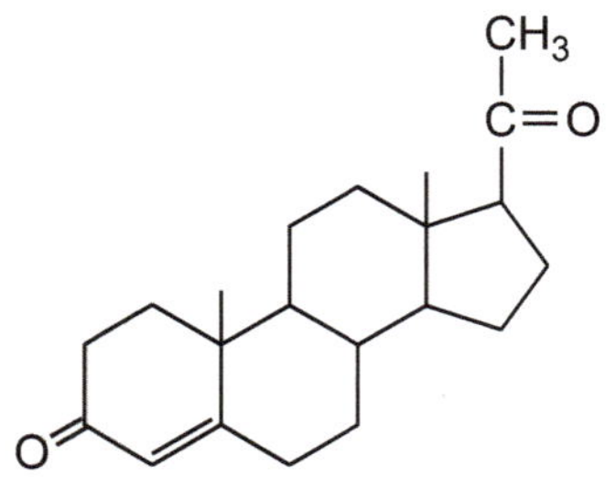

Progesterone

(B) Progesterone Antagonists/SPRMS

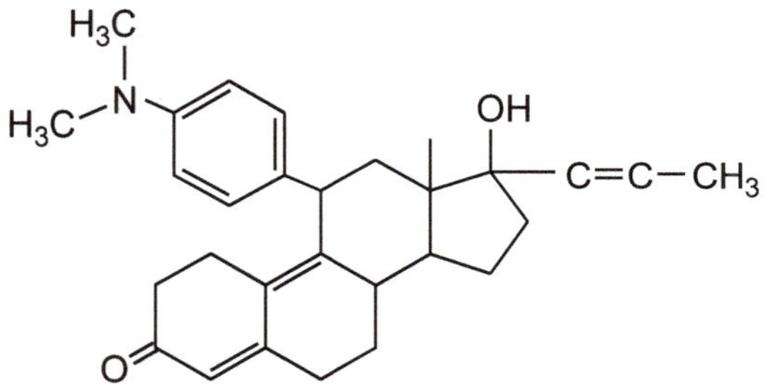

CDB-2477
(RU-486; mifepristone)

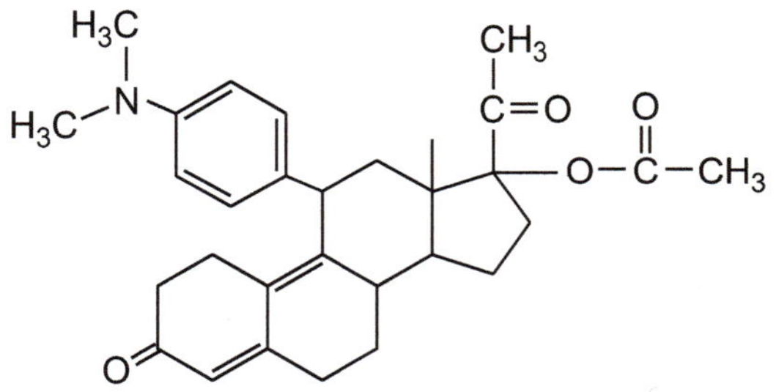

CBD-2914
(ulipristal acetate)

FIGURE 13.8 Progesterone and related compounds. (A) Progesterone is the only naturally occurring ligand for the progesterone receptor. (B) Many antagonistic ligands for the progesterone receptor have been synthesized of which two are illustrated.

TABLE 13.1 Production rates and plasma levels of ovarian steroid hormones.

	Early follicular	Late follicular	Mid-luteal
17β-Estradiol μg/day	70	400–800	300
ng/100 mL	6	300	20
Estrone μg/day	80	300–600	200
ng/100 mL	4	15–30	10
Progesterone[a] mg/day	1	4	24
mg/100 mL	0.2	0.7	1–1.5

[a]*Note the change in units from the earlier 17β-estradiol and estrone.*
Source: Abstracted from G.E. Tagatz, E. Gurpide, Hormone secretion by the normal human ovary, in: R.O. Greep, E.B. Astwood (Eds.), Handbook of Physiology, vol. II, Part I, Section 7, American Physiological Society, Bethesda, MD, 1973, pp. 603–613.

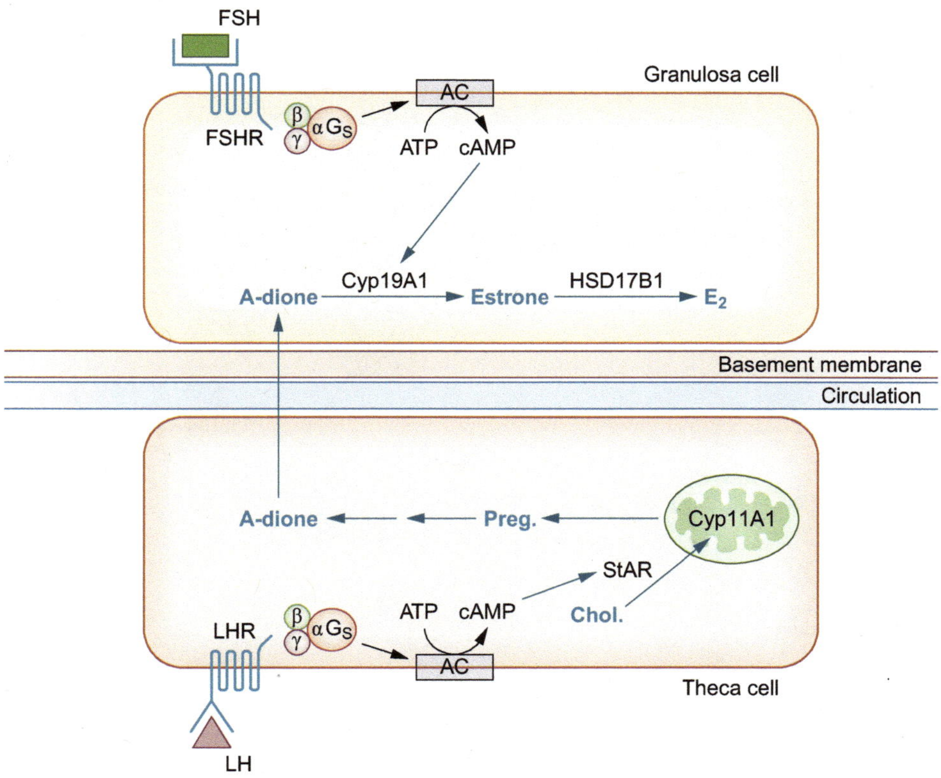

FIGURE 13.9 Estrogen synthesis in the ovary. The two-cell model for estrogen synthesis in the preovulatory follicle is depicted. Under the stimulation of LH, mediated by AC activation leading to increased StAR, Chol. is converted by Cyp 11A1 (cholesterol side chain cleavage; also known as P450scc) to Preg. in the mitochondria of the thecal cell. Pregnenolone is converted to androstenedione (A-dione) which crosses the basement membrane surrounding the follicle and enters the granulosa cells. Cyp 19A1 (aromatase), elevated by FSH stimulation of the granulosa cells, converts A-dione to estrone (see Fig. 2.21). Finally, estrone is converted to estradiol by 17β-hydroxysteroid dehydrogenase (HSD17B1). *AC*, Adenylyl cyclase; *Chol.*, cholesterol; *FSH*, follicle-stimulating hormone; *LH*, luteinizing hormone; *Preg.*, pregnenolone; *StAR*, steroid acute regulatory protein.

the isomerization of the double bond from the B to the A ring of the steroid nucleus (Fig. 2.21). Androstenedione diffuses across the basement membrane of the follicle into the follicular fluid from which it is taken up by granulosa cells. These cells express the FSH receptor (see Fig. 12.9D) which, when activated by FSH, brings about, through adenylyl cyclase activation, the synthesis of Cyp19A1 (P450arom), the aromatase that converts androstenedione to estrone. Action of 17β-hydroxysteroid dehydrogenase (17-ketosteroid reductase; HSD17B1) converts estrone to estradiol (Fig. 2.21). Thus in order for sufficient amounts of estrogen to be synthesized for maturation and ovulation of the follicle, both granulosa and thecal cells must be functional and both gonadotrophins must be secreted in appropriate amounts.

Following ovulation the major steroid produced by the luteinized cells of the corpus luteum is progesterone, although estrogen continues to be synthesized and secreted as well. Steroidogenesis in the corpus luteum is dependent on stimulation of the cholesterol side chain cleavage step by LH and the conversion of the resulting pregnenolone to progesterone (Fig. 2.21).

C Catabolism of progesterone and estrogens

Progesterone is inactivated by the reduction of the ketone group at C-20 and, in the A ring, of the double bond and the ketone group at C-3. Estradiol is inactivated by hydroxylation at any one of a number of positions, including C-2, C-16, C-14. As with androgens (Chapter 12: Androgens), excretion in the bile or urine is preceded by the formation of either sulfate or glucuronide conjugates. The majority of this catabolism takes place in the liver.

IV Hypothalamic–pituitary–ovary axis

A Introduction

Reproductive function in females is characterized by cycles of follicle development, ovulation, and preparation of the uterine endometrium for the implantation of the blastocyst resulting from a fertilized ovum. These events are orchestrated and synchronized by the changes in hormones that constitute the hypothalamic–pituitary–ovarian axis, depicted in Fig. 13.10. There are two differences that distinguish this system from its counterpart in the male (Fig. 12.8): (1) the target cells of the pituitary gonadotrophins undergo a marked structural and functional differentiation so that there are two distinct phases of the cycle, follicular and luteal; (2) at one brief specific point in the cycle, the pituitary and hypothalamic centers respond to circulating estrogen levels with a positive rather than a negative response. In this section, we will examine the functioning of the hypothalamic–pituitary–ovarian axis as a control system and in the next section we will see how its functioning coordinates the events of the human menstrual cycle.

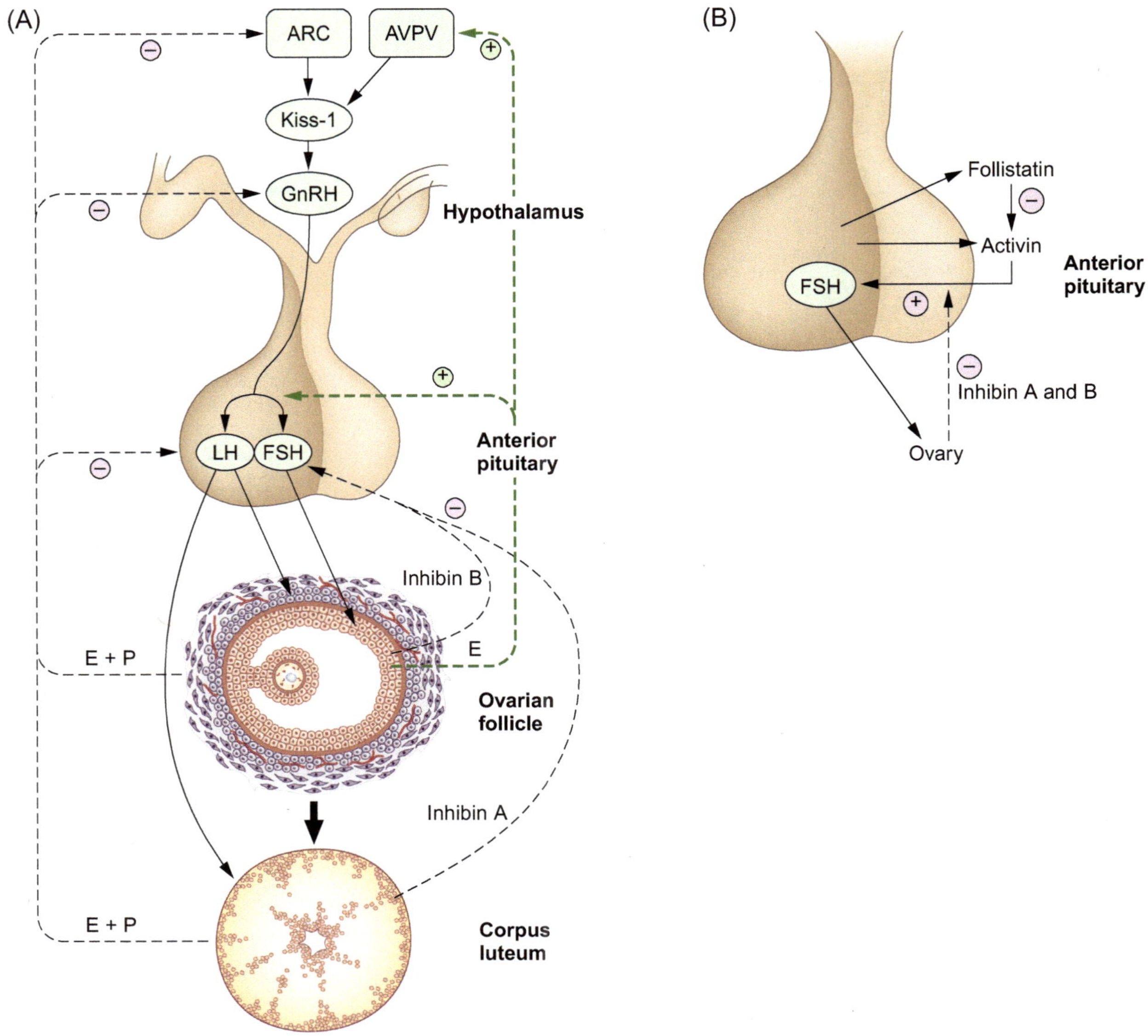

FIGURE 13.10 The hypothalamic–pituitary–ovarian axis. (A) Under the influence of Kiss-1 peptide from the ARC or the AVPV, hypothalamic neurons in the median eminence release GnRH in a pulsatile fashion. GnRH stimulates pituitary gonadotrophs to release LH and FSH. Prior to ovulation, the target cells for LH are the thecal cells of the follicle and those of FSH are the granulosa cells. As described in Fig. 13.9, these two components of the follicle collaborate to synthesize estrogen. After ovulation (the luteal phase of the cycle), LH stimulates the production of progesterone and estrogen by the corpus luteum. These steroids exert negative feedback inhibition on LH and FSH release at the level of the pituitary and on GnRH secretion, primarily at the arcuate nucleus by inhibiting Kiss-1 secretion. Just prior to ovulation, estrogen has a positive (*green-dashed line*) feedback effect on GnRH secretion, through Kiss-1 secretion by the AVPV and at the pituitary. Inhibin B, secreted by the follicle and inhibin A, secreted by the corpus luteum, exerts negative feedback on FSH secretion by the pituitary. (B) FSH secretion is under local control by activin which stimulates its secretion and follistatin which binds to and blocks the effect of activin. As shown in part (A), ovarian inhibins inhibit FSH secretion by blocking activin binding to its receptor. *ARC*, Arcuate nucleus; *AVPV*, anteroventral periventricular nucleus; *FSH*, follicle-stimulating hormone; *GnRH*, gonadotropin-releasing hormone; *Kiss-1*, kisspeptin-1; *LH*, luteinizing hormone.

B Control of the ovary by the hypothalamus and pituitary

Fig. 13.10A depicts the major features of the regulation of the hypothalamic–pituitary–ovarian axis. The pulsatile secretion of GnRH (see Chapter 3: The Hypothalamus and Anterior Pituitary) from the median eminence is under the control of several inputs from various areas of the brain. These neurotransmitters may include dopamine, norepinephrine, and opioids. Circadian rhythms, which do influence GnRH and gonadotropin secretion in other mammalian species, do not have a strong influence in humans, but sleep itself appears to have direct effects on the nature of LH secretion, which vary with the reproductive status of the individual. As discussed in Chapter 12, Androgens, it is now believed that by far the most crucial input to GnRH neurons is the peptide kisspeptin-1 (Kiss-1) from the arcuate nucleus; in the female, there are also kisspeptide neurons in the anteroventral periventricular nucleus (AVPV). Stimulation of GnRH secretion is manifested in both the amplitude and especially the frequency of its pulses.

As described in Chapter 3, The Hypothalamus and Anterior Pituitary, GnRH is transported to the gonadotrophs in the anterior pituitary where the secretion of FSH and LH is increased. However, the responses of FSH and LH secretion to GnRH are quantitatively and qualitatively different. For example, blocking the GnRH receptor inhibits LH secretion almost completely but reduces FSH secretion by only 50%. In another example of differential control of the gonadotrophins, increased GnRH pulse frequency favors LH secretion and decreased pulse frequency leads to increased FSH and decreased LH secretion.

During the majority of the follicular phase (first half) of the cycle the target cells of LH are the thecal cells and those of FSH are the granulosa cells. This responsiveness is determined by the type of receptor each cell expresses. Both gonadotrophins play a role in stimulating steroidogenesis by the ovarian follicle, as detailed in Fig. 13.9. FSH is necessary for the continued growth and maturation of the follicle until ovulation. During the last stages of follicle development, the granulosa cells also express the LH receptor in preparation for the direct participation of the midcycle surge of this gonadotrophin in ovulation. During the last stages of follicle development, the granulosa cells also express the LH receptor in preparation for ovulation in which the midcycle surge of the gonadotrophin participates directly.

Following ovulation, when thecal and granulosa cells of the follicle have differentiated into the corpus luteum, LH from the pituitary is required for the production of progesterone, which is necessary for the growth of the uterine endometrium.

C Feedback effects on the hypothalamus and pituitary

1 Steroid hormones

Fig. 13.10A shows the role of the steroid hormones, estrogens and progesterone, in controlling gonadotrophin secretion. These effects were clearly established by experiments in many mammalian species that demonstrated that the removal of the gonads leads to increased GnRH, LH, and FSH. In humans failure to synthesize estrogens, due to aromatase deficiency, for example, has the same consequences. Both estradiol and progesterone exert negative feedback inhibition on GnRH secretion by the hypothalamus as well as by direct inhibition of gonadotrophin secretion at the pituitary. The latter involves alterations in the expression of the genes for all three of the subunits required for LH and FSH synthesis as well as modulation of the sensitivity of pituitary gonadotrophs to GnRH. The negative feedback effect of estrogen predominates during the follicular phase of the reproductive cycle and that of progesterone, synthesized in large amounts by the corpus luteum, predominates during the luteal phase.

It is currently thought that the primary feedback effects of the steroid hormones on GnRH secretion are exerted at the level of Kiss-1 secretion, although direct effects on the GnRH neurons have not been definitively ruled out. The current model is depicted in Fig. 13.10A with the negative feedback effects of estradiol and progesterone being mediated by the Kiss-1 neurons in the arcuate nucleus.

The midcycle positive effect of estradiol on GnRH and LH secretion is mediated by a group of Kiss-1 neurons in the AVPV. These cells respond to the rapid rise in estrogens produced by the maturing follicle with increased Kiss-1 secretion and stimulation of GnRH secretion that drives the midcycle LH-surge. The positive feedback of estrogen on LH (and FSH) secretion is also exerted at the pituitary gland and involves enhancement of the sensitivity to GnRH. The relative role of the pituitary, as opposed to changes in GnRH secretion, in mediating the midcycle LH surge varies with species. In primates, including humans, the pituitary appears to be the predominant site of this regulatory event. The molecular mechanism of the switch from negative to positive feedback by estrogen is not understood.

2 Activins, inhibins, and follistatin

Although all three of these peptides are produced throughout the body, the focus in this section will be on their presence in the pituitary gland and their role in the regulation of FSH secretion. The inhibins and activins were introduced in Fig. 12.11. All are involved in the GnRH-independent regulation of FSH secretion from the gonadotrophs in the pituitary, as depicted in Fig. 13.10B. The synthesis of the FSHβ subunit is very sensitive to locally produced activin. Activins also participate in maintaining and increasing the number of GnRH receptors, and thus pituitary sensitivity to GnRH. Follistatin is a monomer of around 300 amino acids (several isoforms exist) which binds to and inhibits the function of activins in the pituitary. Inhibins A and B, which show little difference in their activities, bind to activin receptors, blocking their ability to increase FSH secretion. Thus whereas LH secretion from gonadotrophs is largely responsive to GnRH, FSH secretion is also modulated by the interactions of these three peptide hormones, the locally produced activin and follistatin and the inhibins from the gonads. Similar interactions between activins, inhibins, and follistatin occur in many tissues outside the pituitary in the control of cell growth and function.

D The human menstrual cycle

The events of the human menstrual cycle are the result of myriad hormonal (both endocrine and paracrine/autocrine) and neuroendocrine interactions, many of which were outlined in Fig. 13.10. The coordination of events in the ovary and uterus with and by the circulating hormone levels are shown in Fig. 13.11. By convention the cycle begins with the onset of menses, the shedding of the endometrium from the previous cycle. The cycle is typically shown to have a length of 28 days, the median length in studies of large numbers of women, with a range of 25–35 days.

1 Follicular phase

The first 14 days of the cycle constitute the follicular phase during which the dominant follicle is selected and undergoes the final steps of maturation shown in Figs. 13.4 and 13.5. This growth is dependent on the slightly elevated FSH levels seen at the end of the luteal phase of the previous cycle and still present at the beginning of the current cycle (Fig. 13.11, panel B). As shown in Fig. 13.12, several follicles are recruited to begin growing in the early follicular phase. By 5–7 days the leading

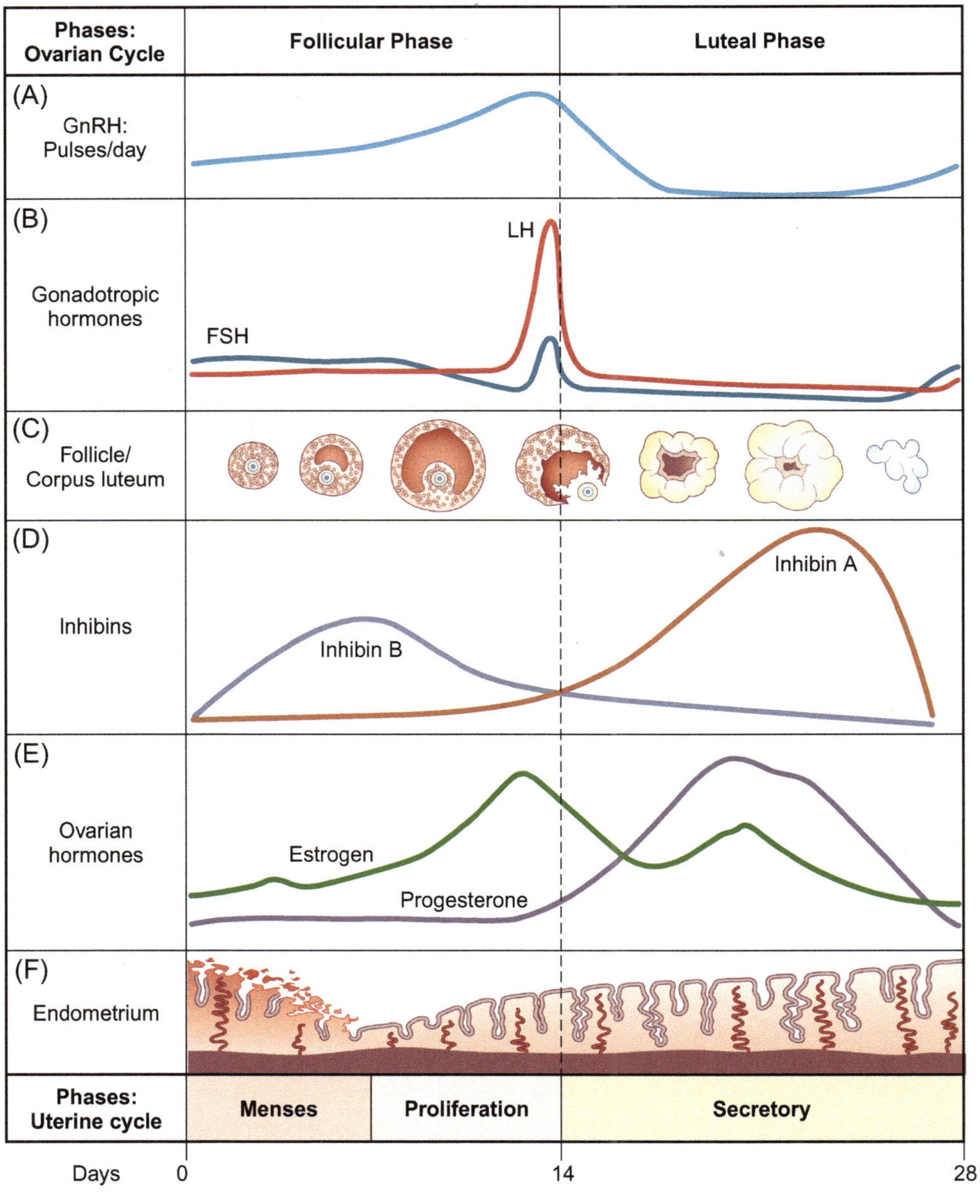

FIGURE 13.11 Events of the human menstrual cycle. The ovarian cycle is divided into the follicular and luteal phases, separated by ovulation as depicted in panel (C). The uterine or endometrial cycle is divided into three phases (panel F), beginning with the onset of menses, the proliferative phase and the secretory phase. Panels (A and B) show, respectively, the increased frequency of GnRH pulses during the follicular phase and the secretion of gonadotrophins. The steroid hormone patterns, with estrogen predominating during the follicular phase and progesterone during the luteal phase, are shown in panel (E). It is the positive feedback effect (see Fig. 13.10) of E_2 that brings about the midcycle surge of LH (and to a lesser degree, FSH), which is, in turn, directly responsible for ovulation. As the corpus luteum regresses (in the absence of fertilization; panel C), progesterone and estrogen levels fall and the endometrial lining degenerates and is shed, during days 1–5 of the next cycle. *FSH*, Follicle-stimulating hormone; *GnRH*, gonadotropin-releasing hormone; *LH*, luteinizing hormone.

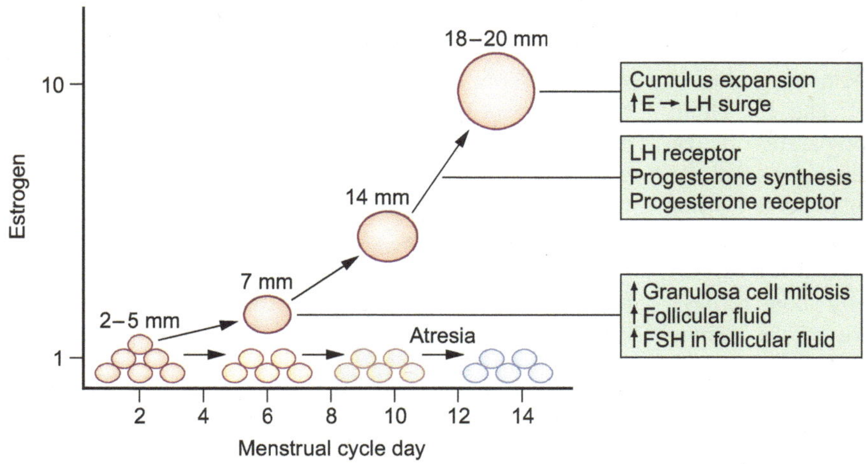

FIGURE 13.12 The development of the dominant follicle. A cohort of follicles is recruited late in the luteal phase of the previous cycle and these follicles grow during the first 5–7 days of the new cycle. During this time, one becomes larger and displays characteristics, shown in the box on the right, that distinguish it from the others. This becomes the dominant follicle which continues to grow and change as seen in the box until it produces enough estradiol to bring about the midcycle surge in LH. The remaining follicles undergo atresia in which the cells die and are replaced by fibrous tissue. *LH*, Luteinizing hormone.

follicle can be distinguished by its increase in size and concomitant increased mitotic activity of its granulosa cells, as well as the amount and FSH content of its follicular fluid. The remaining follicles of the cohort, which have undetectable FSH levels, begin to undergo atresia: the granulosa cells and oocyte die and are replaced by fibrous tissue and the thecal cells dedifferentiate into stromal cells.

The dominant follicle is the source of circulating estrogen, which rises (Fig. 13.11, panel E) as the follicle becomes larger and more mature (panel C). Note also that during the late follicular phase, GnRH pulses (panel A) are increasing in frequency. Late in the follicular phase, the preovulatory follicle undergoes qualitative changes in preparation for ovulation. These include the appearance of LH receptors, the increased capacity to synthesize progesterone and the expression of the receptor for this steroid hormone (Fig. 13.12).

When estrogen secreted by the ovulatory follicle reaches a sufficient threshold and is sustained for a sufficient time, the Kisspeptide neurons in the AVPV and the pituitary gonadotrophs (see Fig. 13.10) react positively resulting, approximately 24 hours following the peak estrogen levels, in the midcycle surge of LH. The small midcycle increase in progesterone contributes to the timing of the surge, which is delayed when PRs are blocked by mifepristone (RU-486; Fig. 13.8) administration. As indicated in Section IV.C, the predominant positive feedback effect of estradiol in primates, including humans, is at the pituitary, with that at the hypothalamus being required in a permissive role.

It is the action of LH on the mature Graafian follicle that is directly responsible for ovulation. The peak in FSH at this time is likely due to both increased GnRH and the fall in inhibin B from the follicle's granulosa cells.

The two phases of the menstrual cycle are generally named in reference to changes in the ovary, but changes also occur in the endometrium which is in its proliferative phase prior to ovulation. As seen in Fig. 13.2B, the endometrium has two components, the functional layer, which is shed in each menses, and the basal layer, which is permanent. Under the influence of estrogen, the cells of the functional layer proliferate to regenerate the tissue lost during menses, as shown in Fig. 13.11, panel F. In addition, estradiol stimulates the expression of the PR in the cells of both the basal and functional layers in preparation for their response to this steroid during the second half of the cycle.

2 Ovulation

Ovulation, the release of the ovum from the follicle (Fig. 13.11, panel C) in close proximity to the fimbriae of the fallopian tube, occurs about 36 hours following the onset of the LH surge and 10–12 hours following its peak. Many changes take place during this final stage of follicle maturation. Under the influence of LH, meiosis resumes temporarily then is rearrested in metaphase II (Fig. 13.5). The cumulus continues to expand. LH, FSH, and progesterone all play a role in the activation of proteolytic enzymes and the synthesis of prostaglandin E_2 synthase (Chapter 8: Eicosanoids), both of which contribute to the breakdown of the follicle wall through which the egg and cumulus cells will be extruded.

3 Luteal phase

Following the release of the ovum, the granulosa and inner layers of the thecal cells differentiate into the corpus luteum as the cells become filled with lipids. New capillaries are formed as the corpus luteum becomes highly vascularized, reaching its maximum vascularization in 8–9 days. The extent of cellular reorganization involved in the conversion of the follicle into the corpus luteum is evident from the changes in the expression of at least 100 genes. LH, in addition to inducing ovulation and the luteinization of the corpus luteum, plays the central role in maintaining its function. This is primarily to produce the large amounts of progesterone required to support the growth and function of the endometrium during its secretory phase (Fig. 13.11, panels E and F). These processes are, in turn, required for the preparation of the endometrium for the implantation of a blastocyst from the fertilized ovum. If fertilization and implantation do not occur by 8–9 days following ovulation, the corpus luteum begins to regress ($PGF_{2\alpha}$ plays a role in luteal regression) and degenerates into the corpus albicans by about 14 days (day 28 of the cycle). With the demise of the corpus luteum, steroid hormone levels fall as shown in panel E of Fig. 13.11.

Following ovulation, the endometrium enters its secretory phase. This is characterized by (1) increasing thickness of the functional layer; (2) increasing secretion of glycogen and other molecules by the uterine glands; and (3) increasing vascularization and formation of the spiral arteries (see Fig. 13.3B). Toward the end of the luteal phase, the corpus luteum slows then ceases its production of estradiol and progesterone. One consequence of this is the vasoconstriction of the spiral arteries leading to ischemia (oxygen deprivation) of the surrounding endometrial tissue, which is then destroyed, prior to its shedding, by leukocyte infiltration, autolysis, and autophagy. It is interesting to note that whereas all mammalian species experience a loss of steroid hormones at the end of nonfertile cycles, only humans and a handful of other primates shed the uterine lining in each ovulatory cycle.

E Menopause

Menopause is defined as having occurred 12 months after the woman's last menses and is the result of depletion of viable ova and the cessation of ovulation. The average age is about 51 for women of European descent and somewhat younger for African Americans and Hispanic women of Mexican descent. The process usually begins about 5 years prior to the final menses with inconsistent and decreasing ovarian estradiol synthesis. This leads to increasingly irregular menstrual periods. In addition many women experience one or more of a range of symptoms, including hot flashes, vaginal irritation, sleep disruption, fatigue, mood lability, changes in hair texture and distribution, and weight gain. The exact symptoms and the severity of them vary greatly from one individual to another.

Changes in hormone levels in the years just before and after the menopause are shown in Fig. 13.13. As the store of viable follicles is depleted, ovarian estradiol synthesis decreases and eventually ceases, and estrone from the peripheral aromatization of androstenedione becomes the main circulating estrogen. Some of this estrone can be converted to estradiol by 17β-HSD (Fig. 2.21) so that low levels of estradiol remain after menopause. Peripheral aromatization takes place largely in adipose tissue, so that levels of circulating estrone and estradiol in obese women are substantially higher than in women of normal weight. The steep decline of inhibins A and B takes place 2–4 years prior to the last menses. The rise in FSH, due to declining E_2 and inhibin levels, is steeper than that of LH, which is primarily reflective of E_2 levels. Both FSH and LH remain elevated for the remainder of the individual's lifetime, about ninefold and fourfold greater than premenopausal levels, respectively.

As discussed in Section V, estrogens have effects on many tissues and physiological processes outside of the reproductive system, such as bone. The idea of replacing the estrogen lost at menopause to alleviate the perimenopausal symptomology as well as to offset the declines in

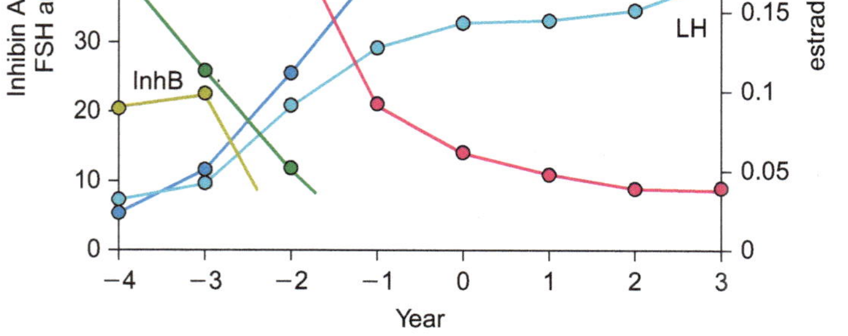

FIGURE 13.13 Perimenopausal hormonal changes. The graph depicts changes in estradiol (*red*), the inhibins (*green*), and the gonadotrophins (*blue*) for the 4 years prior to and the 3 years following the last menstrual period (indicated by 0 on the *x*-axis). Inhibins A and B levels fall to or below the limit of detectability by 2 years prior to the last menstrual period and gonadotrophins remain elevated during the remainder of life.

these extrareproductive systems has been around since the 1950s. Recent assessments of this type of hormone replacement therapy (HRT) approach will be discussed in Section VI.

V Biological responses to female sex steroids

A Estrogen and progesterone receptors

1 Estrogen receptor

The estrogen receptor (ER) was first identified as a specific binding protein in the 1950s by Elwood Jensen, a body of work that laid the foundation for the steroid nuclear receptor field that was to follow. The human ER and its gene and cDNA were characterized in the 1980s. About 10 years later, examination of mice lacking this receptor (ER^-/ER^- mouse) led to the discovery of a second form of the receptor. This new form was named ERβ and the original ER became ERα. The two forms of ER, depicted in the top portion of Fig. 13.14, are products of separate genes, ERα on chromosome 6 and ERβ on 14. ERα and ERβ show 97% amino acid identity in their DNA-binding domains and 55% identity in the ligand-binding domains. Subforms of human ERα, or hERα, are the N-terminal truncated hERα − 46, missing most of the A/B region, and hER-39 which is also missing almost 2/3 of the ligand-binding domain. ERβ also appears in other forms resulting from splice variants resulting in changes in the carboxyl terminus of the protein or deletion of exon 5 leading to the absence of the entire ligand-binding domain.

The tissue distribution of ERα and ERβ messenger RNA (mRNA) expression in the mouse is shown in Table 13.2. Clearly ERα has much broader distribution than ERβ, which is largely restricted to the ovary and hypothalamus. Thus the actions of estrogen in nonreproductive tissues such as adipose tissue and skin are most likely mediated by ERα. Both ERα and ERβ are expressed in human breast cancer cells (Chapter 14: Hormones of Pregnancy, Parturition, and Lactation). Not shown in Table 13.2 are several male reproductive tissues in which ERα mRNA is found. These include vas deferens, epididymis, prostate, seminal vesicles, and testes. The levels of ERα in these tissues vary over the range from that seen in the pituitary (vas deferens,++) to that of the liver (testes,+).

The binding of estradiol to either form of ER leads to homodimerization followed by the binding of the homodimer to response elements on DNA to alter gene expression. In Fig. 13.15, are shown structures of the estradiol receptor and the signaling pathways following estrogen binding to

TABLE 13.2 Tissue distribution of estrogen receptor (ER)α and ERβ mRNA.[a]

Tissue	ERα	ERβ
Uterus	++++	−
Ovary	+++	++++
Pituitary	+++	−
BAT[b]	++	−
WAT[b]	++	−
Skin	++	−
Kidney	++	−
Sk. muscle	++	−
Liver	+	−
Bone	+	−
Hypothalamus	+	+

[a] *1 mRNA levels of ERα and ERβ were measured as arbitrary units in mouse tissues.++++, > 1.0; +++, 0.5–0.99; ++, 0.1–0.49; +, 0.02–0.095.*
[b] BAT, *Brown adipose tissue;* WAT, *white adipose tissue.*
Source: Data abstracted from NURSA (Nuclear Receptor Signaling Atlas): www.nursa.org/10.1621.datasets.02001, A.L. Bookout, Y. Jeong, M. Downes, R. Yu, R.M. Evans, D.J. Mangelsdorf, 2005.

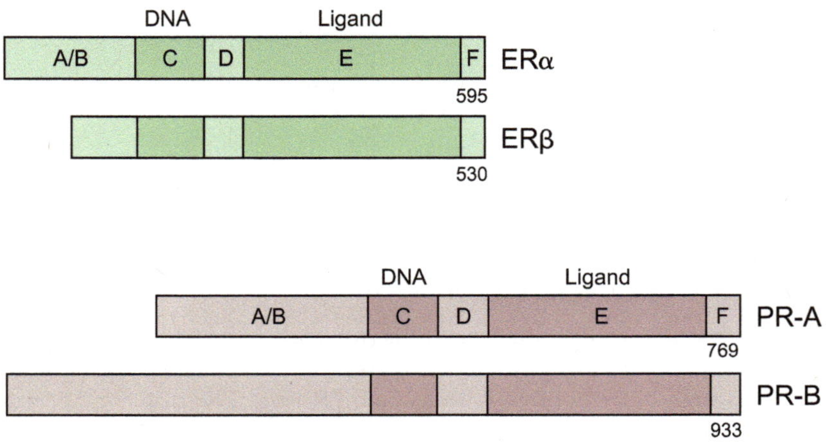

FIGURE 13.14 Receptors for estrogen and progesterone. The two main isoforms of the ERα and ERβ (encoded by two different genes) are shown above and the two forms of the PR-A and PR-B (which arise from the use of alternative promoters in the same gene) are below. As discussed in Chapter 1, Hormone Introduction, the nuclear receptors share common functional domains. These are A/B-transactivation; C-DNA binding; D-hinge; E-ligand binding; and F-transactivation. The numbers below each receptor indicate the number of amino acids in the human protein. *ER*, Estrogen receptor; *PR*, progesterone receptor.

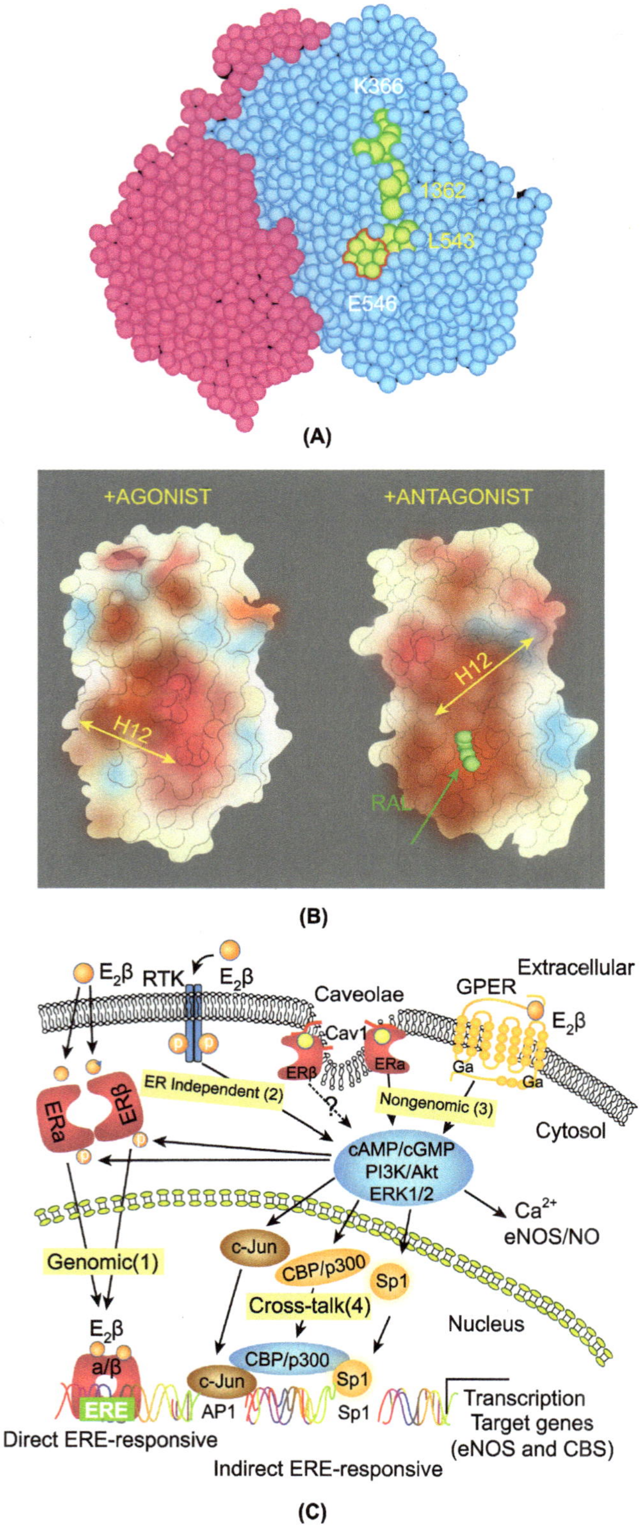

FIGURE 13.15 (A) Crystallographic structure of E2-bound LBD dimer of ERα. The diagram was generated from the human ERα LBD with 17β-E2 (Protein Data Bank ID:1ERE). The dimer form of pink and blue ERα LBD molecules is shown. The residues suggest the position corresponding to the I363, K366, L543, and E546 of mouse ERα. I362 and L543 are hydrophobic amino acids (*yellow*), K366 is a basic amino acid (*blue*), and E546 is an acidic amino acid (*red*). (B) Crystal structure of the ligand binding domain of ERb in the presence of an agonist (genistein) and an antagonist (raloxifene). The picture depicts the two distinct positions of helix 12 which are adopted with the two ligands. (C) Signaling pathways of estrogen actions. There are different signaling mechanisms that mediate estrogen actions, including (1) nuclear actions: ERs bind estrogens; liganded ERs translocate to the nucleus where they form homo- or heterodimer transcription factors to induce the transcription of estrogen-responsive genes with or without EREs, (2) ER-independent estrogen signaling: estrogen transactivates RTKs, that is, EGFR and FGFR to initiate ER-independent signaling such as an increase in second messengers cAMP and cGMP and activation of protein kinase cascades such as ERK1/2 and PI3K/AKT, (3) extranuclear nongenomic actions: estrogens bind to plasma membrane ERs, including ERα and ERβ, and GPER, resulting in increase in second messengers and activation of protein kinase cascades, and (4) interactions between nuclear and extranuclear estrogen signaling pathways: ER-independent estrogen signaling via the transactivation of RTKs and plasma membrane ER-dependent nongenomic estrogen signaling can interact with nuclear estrogen signaling via protein kinase-mediated ER phosphorylation and activation of gene transcription via ER-interacting transcription factors such as AP-1 and SP-1. *AKT*, Protein kinase B; *AP-1*, activator protein-1; *cAMP*, cyclic AMP; *cGMP*, cyclic guanosine monophosphate; *EGFR*, epidermal growth factor receptor; *ER*, estrogen receptor; *EREs*, estrogen-responsive elements; *ERK1/2*, extracellular signal-related kinases 1/2; *FGFR*, fibroblast growth factor receptor; *GPER*, G protein–coupled estrogen receptor; *LBD*, ligand-binding domain; *PI3K*, phosphoinositide 3-kinases; *RTKs*, receptor tyrosine kinases; *SP-1*, specificity protein-1. *(A) Reproduced from Fig. 2 of Y. Arao, K.S. Korach, Transactivation function-1-mediated partial agonist activity of selective estrogen receptor modulator requires homo-dimerization of the estrogen receptor α ligand biding domain, International Journal of Molecular Science 20 (2019) 378–390 (B) Reproduced form Fig. 2 of S. Nilsson et al., Mechanisms of estrogen action, Physiological Reviews 81 (4) (2001) 1535–1565; https://doi.org/10.1152/PHYSREV.2001.81.4.1535. (C) Reproduced from Fig. 2 of J. Bai et al. Estrogen receptors and estrogen-induced uterine vasodilation in pregnancy, International Journal of Molecular Science 21 (2020) 4349; https://doi.org/10.3390/ijms21124349.*

ER-α and ER-β. In part (C), E2β represents the plasma membrane form of the receptor. The figure also shows, in parts (A) and (B) the dimeric structure of the ER ligand-binding domain as well as the appearance of the ligand-binding domains in the presence of an agonist ligand and an antagonist ligand (part B).

In cells in which both ERα and ERβ occur, either naturally or as the result of experimental manipulation,

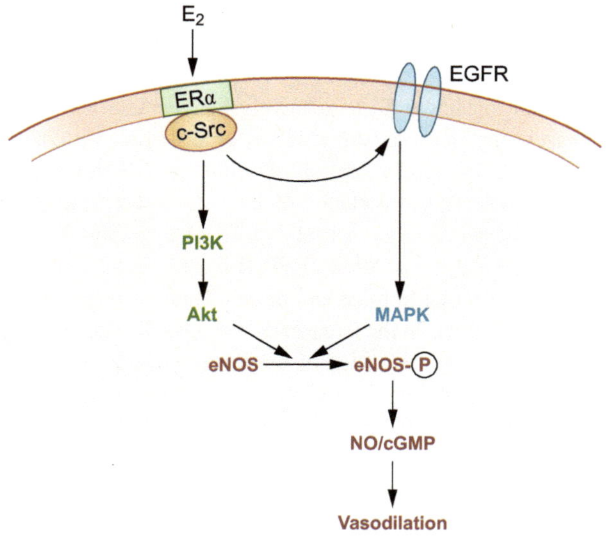

FIGURE 13.16 Mechanism of E2-mediated vasodilation in vascular endothelial cells. The model depicts the interaction of E_2 with a subpopulation of ERα that resides in the plasma membrane. The hormone–ER complex interacts with intracellular proteins, such as c-Src, 111a nonreceptor tyrosine kinase, which can activate the PI3 kinase pathway or, through EGFR, the MAP kinase pathway. The result is the activation of eNOS by the phosphorylation of serine-1177, which through the NO/cGMP pathway results in vasodilation. *cGMP*, Cyclic guanosine monophosphate; *EGFR*, epidermal growth factor receptor; *eNOS*, endothelial NO synthase; *ER*, estrogen receptor; *MAP*, mitogen-activated protein.

heterodimerization can take place but the circumstances under which this happens normally and its biological significance is still under investigation.

In addition to the effects of ER on gene transcription, its classical mechanism of action, it is now understood that it, along with other steroid hormone receptors, exerts membrane-initiated effects in certain cell types (see Chapter 1: Hormone Introduction). These effects are sometimes referred to as "rapid" because they can be seen within seconds or minutes of treatment, "nongenomic" because the proximal events do not involve nuclear localization or DNA binding, or "membrane-initiated," since they are ascribed to a population of the receptor that is localized to the plasma membrane. One example of this type of action mediated by the ER is the modulation of cardiovascular function through the activation of phosphatidylinositol-3-OH kinase (PI3K), leading ultimately to increase in NO and vasodilation, described in more detail following (Fig. 13.16).

2 Progesterone receptor

The structures of the two forms of the PR are shown in the bottom half of Fig. 13.14. Both forms are encoded by the same gene on chromosome 11. Use of an alternative promoter generates a 5′ truncated message from which PR-A is generated by the use of a translation initiation site at codon 165. Thus the two forms of PR are identical except for the missing portion (164 amino acids) of the A/B domain in PR-A.

Table 13.3 shows the tissue distribution of the PR mRNA in the mouse. In contrast to ERα, PR expression is largely limited to the uterus, ovary, and several areas of the brain as well as the pituitary. Very little if any PR mRNA is expressed in male reproductive tissues and male mice lacking the gene for PR display no adverse effects on reproduction. PR expression is stimulated by and dependent on estrogen and the action of the ER in the target tissues.

Both the A and B forms of the PR function, like all steroid receptors, as transcriptional regulatory proteins. Although there is evidence for the formation of PR-A/PR-B dimers under some experimental conditions, the transcriptional activities of the two homodimers have been reported to differ depending on cell and promoter contexts. PR-B is a strong transcriptional activator of target genes that are unresponsive to PR-A. PR-A often acts as a repressor of PR-B action, as well of those of other steroid hormone receptors such as ER and androgen receptor. Clearly there is the likelihood of a wide variety of responses to progesterone depending on the PR-A and PR-B content of the target cell.

In addition to its actions in the uterus and mammary tissue (see the following and Chapter 14: Hormones of Pregnancy, Parturition, and Lactation), progesterone also participates in the acrosome reaction of sperm during fertilization, by inducing a rapid Ca^{2+} influx. The rapidity of this effect (seconds) and the fact that the mature sperm DNA is highly condensed and transcriptionally inactive

TABLE 13.3 Tissue distribution of progesterone receptor (PR) messenger RNA (mRNA).[a]

Tissue	PR mRNA
Uterus	++++
Ovary	+++
Cerebrum	++
Hypothalamus	++
Brain stem	++
Spinal cord	++
Corpus striatum	++
Cerebellum	++
Kidney	+
Pituitary	+

[a]*Levels of PR mRNA were measured as arbitrary units in mouse tissues:++++, > 10.0; +++, 2.0–9.9; ++, 0.7–1.9; +, 0.1–0.6.*
Source: Data abstracted from NURSA (Nuclear Receptor Signaling Atlas): http://www.nursa.org/10.1621.datasets.02001, A.L. Bookout, Y. Jeong, M. Downes, R. Yu, R.M. Evans, D.J. Mangelsdorf, 2005.

suggests that this is an example of a membrane initiated nongenomic action of progesterone.

B Uterine effects of estrogen and progesterone

The functional layer of the endometrium of the uterus (see Fig. 13.2B) undergoes repeated cycles of proliferation and differentiation and, if no implantation occurs, breakdown and sloughing of tissue. These processes are under the control of the ovarian steroid hormones which define three distinct hormonal environments: estrogen dominance during the proliferative phase; progesterone dominance during the secretory phase; and progesterone withdrawal in the breakdown phase. The window of implantation occurs during the midsecretory phase (see Fig. 13.11; Chapter 14: Hormones of Pregnancy, Parturition, and Lactation), 6–10 days following the LH surge.

Estradiol (E_2) is required for the changes in the endometrium during the proliferative phase and for preparing it to respond to progesterone during the secretory phase. E_2 governs uterine responsivity to itself and to progesterone by increasing the transcription of the mRNA for both ERα and PR.

The effects of E_2 during the proliferative phase are mediated primarily by ERα, the expression of which in the endometrium is greatest during the proliferative phase. ERα expression declines following ovulation due to the suppressive effects of progesterone on transcription of its mRNA. During the proliferative phase, E_2 interacts with ERα in stromal cells of the endometrium which respond by producing growth factors and cytokines. Among these are insulin-like growth factor 1, transforming growth factor β, and epidermal growth factor (EGF) (see Chapter 17: Growth Factors, for further discussion of these and other growth factors). These growth factors and cytokines exert paracrine effects on the epithelial cells of the endometrial glands, increasing mitosis, and growth.

The signaling pathways that E_2 stimulates in the stromal cells also operate in the endothelial cells of the blood vessels to promote angiogenesis. Currently the best candidate for a role in the increased blood vessel growth and differentiation seen during each menstrual cycle is vascular endothelial growth factor, VEGF, although many other signaling molecules are also likely to be involved.

Progesterone levels begin to rise just prior to ovulation and peak in the midsecretory phase of the uterine cycle, during the window of implantation. Progesterone is required for ovulation, for fertilization, and for the differentiation of the endocrine cells of the endometrial glands for their secretory functions. These secretions include glycogen, mucus, chemokines, cytokines, and other substances required for successful implantation.

Progesterone limits the effectiveness of E_2 on the growth of the endometrium during the secretory phase of the menstrual cycle through three mechanisms. It blocks the transcription of the mRNA for ERα. It stimulates the transcription of mRNA for and, therefore, the activity of 17βHSD2 (same activity as 17βHSD1, Fig. 2.21), thereby increasing the conversion of active E_2 to relatively inactive estrone. And it increases the levels of estradiol sulfotransferase, an enzyme that catalyzes the incorporation of sulfate at C-3 to inactivate the molecule. The importance of the role of progesterone in limiting the stimulatory activity of E_2 is apparent from the observation that the absence of PR-B is associated with endometrial hyperplasia in mice and in the condition of endometriosis in the human. In the latter case, endometrial tissue outside the uterus undergoes estradiol stimulated hyperplasia unopposed by progesterone. This tissue contains PR-A but not PR-B, suggesting that PR-B is responsible for the progesterone restraint on estrogen stimulation of endometrial growth.

C Effects of estrogen on bone

Bone structure and formation, the types of bone, and the cells that participate in its metabolism are all discussed in Chapter 9, Calcium-Regulating Hormones: Vitamin D, Parathyroid Hormone, Calcitonin; Fibroblast Growth Factor-23 (Figs. 9.2–9.4). Estrogens and androgens are involved in several aspects of bone physiology in all vertebrate species, including (1) skeletal sexual dimorphism; (2) peak bone mass; (3) closure of the epiphyseal growth plate; (4) mineral homeostasis during reproduction, when calcium is removed from the maternal skeleton for that of the offspring; and (5) maintenance of bone mass and architecture in adults. Both androgens and estrogens contribute to bone physiology. The focus here will be on estrogens and their effects as they are understood in humans. While androgens do have a role in skeletal physiology, particularly in promoting long bone growth prior to closure of the epiphyseal plate in both males and females, in other actions the role of androgen action independent of that of estrogens has been difficult to establish due to potential conversion to estrogens in bone target cells that contain aromatase.

Table 13.4 summarizes the current understanding of the effects of the estrogen environment (deficient or replete) in vivo on bone during growth and in the adult skeleton. Effects vary by bone type, in keeping with the different functions of the two types of bone. Cortical bone is largely responsible for the mechanical strength of the skeleton, whereas cancellous (endocortical, trabecular, and spongy) bone contributes less to strength but is more available for Ca^{2+} exchange and, therefore, maintenance of mineral homeostasis.

In estrogen deficiency the skeleton lengthens due to delayed closure of the epiphyseal growth plate. The role of estrogen in this process in both males and females is substantiated by the skeletal lengthening seen in (extremely rare) males with defects in aromatase or ERα. In females, aromatase deficiency is detected and treated with estrogen replacement early in childhood so that the bone manifestations at and following puberty are averted.

TABLE 13.4 **Effect of estrogen environment on bone.**

	Growing		
	Bone length	**Cortical bone**	**Cancellous bone**
Estrogen deficiency	↑	↑[a]	↓[b]
Estrogen treatment	↓	↓[a]	↑c
	Adult		
	Cortical bone	*Cancellous bone*	
Estrogen deficiency	↓[b]	↓[b]	
Estrogen treatment	→	→	

[a]*Due to change in periosteal expansion (cortical bone thickness).*
[b]*Due to net increased resorption.*
[c]*Due to net decreased resorption.*
Source: Modified from R.T. Turner, D.R. Rickard, U.T. Iwaniec, T.C. Spelsberg, Estrogens and progestins, in: J.P. Bilezikian, L.G. Raisz, T.J. Martin (Eds.), Principles of Bone Biology, vol. 1, 2008, pp. 855–885.

As shown in Table 13.4, in estrogen deficiency or resistance, the mass of cortical bone is increased due to increased width resulting from the unopposed effects of androgens; cancellous bone is decreased due to increased resorption due to estrogen deficiency. The opposite effects are seen prior to puberty in situations where an individual has been treated for estrogen deficiency with 17β-estradiol or a synthetic estrogen.

In the adult, estrogen deficiency brought about by, for example, menopause, ovariectomy, or treatment with aromatase inhibitors, leads to bone loss due to net increased resorption. This is the major underlying cause of postmenopausal osteoporosis in women (see Chapter 9: Calcium-Regulating Hormones: Vitamin D, Parathyroid Hormone, Calcitonin; Fibroblast Growth Factor-23). Estrogen replacement can forestall this loss of bone mineral.

As seen in Table 13.2, bone has very little ER of either type compared to the reproductive tissues. In more detailed studies of specific cell types, both ERα and, in lesser amounts, ERβ are found in osteoblasts of humans and in other species. Both are also present in the cells that line the cancellous bones and in growth plate chondrocytes, particularly the resting and proliferation zones. How the relative amounts of the two receptors play a role in the response of these cells to estrogens is as yet unknown. ERα is also seen in osteoclasts, in which estradiol induces apoptosis, and in osteoclast progenitor cells, speeding their differentiation into osteoclasts. Thus estrogen could exert effects on osteoclast numbers through both osteoclastogenesis and by altering the lifespan of the mature osteoclast.

D Estrogen and the cardiovascular system

A comprehensive discussion of the interaction of estrogens with the components of the cardiovascular system is beyond the scope of this book, so only a few well-known examples will be given here. It should be noted that in all these situations, estradiol is acting in a milieu that contains many other hormones and growth factors and, therefore, is only one of many factors influencing these systems. One physiological importance of the role of E_2 in the cardiovascular system in humans lies in the effects of its withdrawal at menopause and the clinical approaches to replacing it following this withdrawal.

One of the most widely studied and best understood effects of estrogens in the cardiovascular system is vasorelaxation. This is brought about by the stimulation of endothelial NO synthase (eNOS) in the endothelial cells. The effects of E_2 on the endothelial cell involve nuclear actions of ERα leading to the increased transcription and stability of mRNA for eNOS. In addition, as outlined in Fig. 13.16, E_2 acts through rapid membrane-initiated events. Studies of the latter effects have implicated ERα and/or ERβ located in caveolae of the plasma membrane, which activates the PI3K/Akt pathway leading to phosphorylation of eNOS at S1177. E_2-mediated activation of EGF also stimulates eNOS phosphorylation and thus vasodilation. It should be noted that enzymes that affect the local concentrations of active E_2, such as aromatase, are also located in vascular endothelial cells.

Other effects of E_2 on the cardiovascular system include its inhibition of vascular smooth muscle proliferation, particularly in response to hyperglycemia. E_2 also exerts a protective effect on the heart by inhibiting apoptosis and necrosis of myocardial cells. Following a myocardial infarction, E_2 promotes restorative cardiac remodeling and angiogenesis. Many studies suggest that these cardioprotective effects are mediated by ERβ. Finally, E_2, by increasing expression of genes for antioxidants, helps protect the vasculature against the damage done by reactive oxygen species.

TABLE 13.5 Relative effectiveness of contraceptive methods.

Method	Pregnancies/100 women-years
Rhythm[a]	15.5
Withdrawal[b]	6.7
Condom	3.6
Diaphragm	1.9
IUD	1.2
Combined oral contraceptive	0.3
Tubal ligation (female)	0.13
Vasectomy (male)	0.02

IUD, Intrauterine device.
[a]*Abstinence from coitus during presumed fertile periods.*
[b]*Prior to ejaculation.*

VI Clinical aspects

A Hormonal contraception

A major clinical topic related to female endocrinology, for both individual women as well as on a societal level, concerns the ability of a woman to control her periods of fertility. Table 13.5 summarizes the effectiveness of various contraceptive methods. These fall into three main categories: (1) reversible physical methods of prevention of fertilization; (2) reversible chemical methods of inhibition of ovulation; and (3) permanent (normally) prevention of fertilization. Our discussion will focus on (2), the use of oral contraceptives, the most effective of the reversible methods of contraception.

The combined oral contraceptive pill (COCP) has been available for more than 50 years. The forty or so formulations available today consist of combinations of estrogen and progestin, which, like the original pills, are designed to mimic the normal menstrual cycle with 21 days of steroid-containing pills followed by 7 days of placebo pills, followed by menstrual bleeding. In some formulations the amount of progesterone (and less frequently, estrogen) changes during the 21-day period. Structures of commonly used steroids in these formulations are shown in Fig. 13.17.

The primary mechanism of action of the estrogen/progestin combination is to inhibit gonadotrophin secretion at the hypothalamus and the pituitary, thereby preventing final ovarian follicle development and ovulation. The efficacy of the oral contraceptive pills is highly dependent on adherence to the daily taking of the pill. Ovulation may resume during the first cycle following cessation of taking the pill or may take up to several months to become re-established.

There are nonreproductive health benefits associated with oral contraceptive use. Probably of greatest importance is the decrease in epithelial ovarian cancer (40%–80% depending on length of COCP use) and endometrial cancer (50%). Oral contraceptive use also decreases irregular and dysfunctional uterine bleeding and ameliorates the effects of acne and premenstrual dysphoric disorder. Health risks include a small increase in the risk of a thromboembolic event (0.4→1–1.5/10,000 woman-years) and a very small increase in the risk of cervical cancer. Currently used formulations used by nonsmokers pose no increase in myocardial infarctions, atherosclerosis, or breast cancer in those without *BRCA* mutations.

Other modes of delivery of contraceptive hormones have been developed for specific purposes. These include regimens with extended lengths of hormone administration and more widely spaced hormone-free periods that bring about uterine bleeding. For example, a 365-day regimen with only one hormone-free period is currently available. In addition, contraceptive hormones can be delivered by vaginal ring, skin patch, or injection. Each of the methods has its applications and advantages as well as disadvantages and all are equally efficacious if used properly. The safety of these delivery methods, while not studied as thoroughly as for the COCP, is believed to be about the same. Finally, for women for whom estrogen is contraindicated, a progestin-only contraceptive is available.

B Postmenopausal hormone replacement therapy

The use of estrogen to alleviate the symptoms of menopause was approved by the FDA in 1942. This approach, which became known as HRT, involves giving estrogen alone or in combination with a progestin. As more was learned about the actions of estrogen in other tissues such

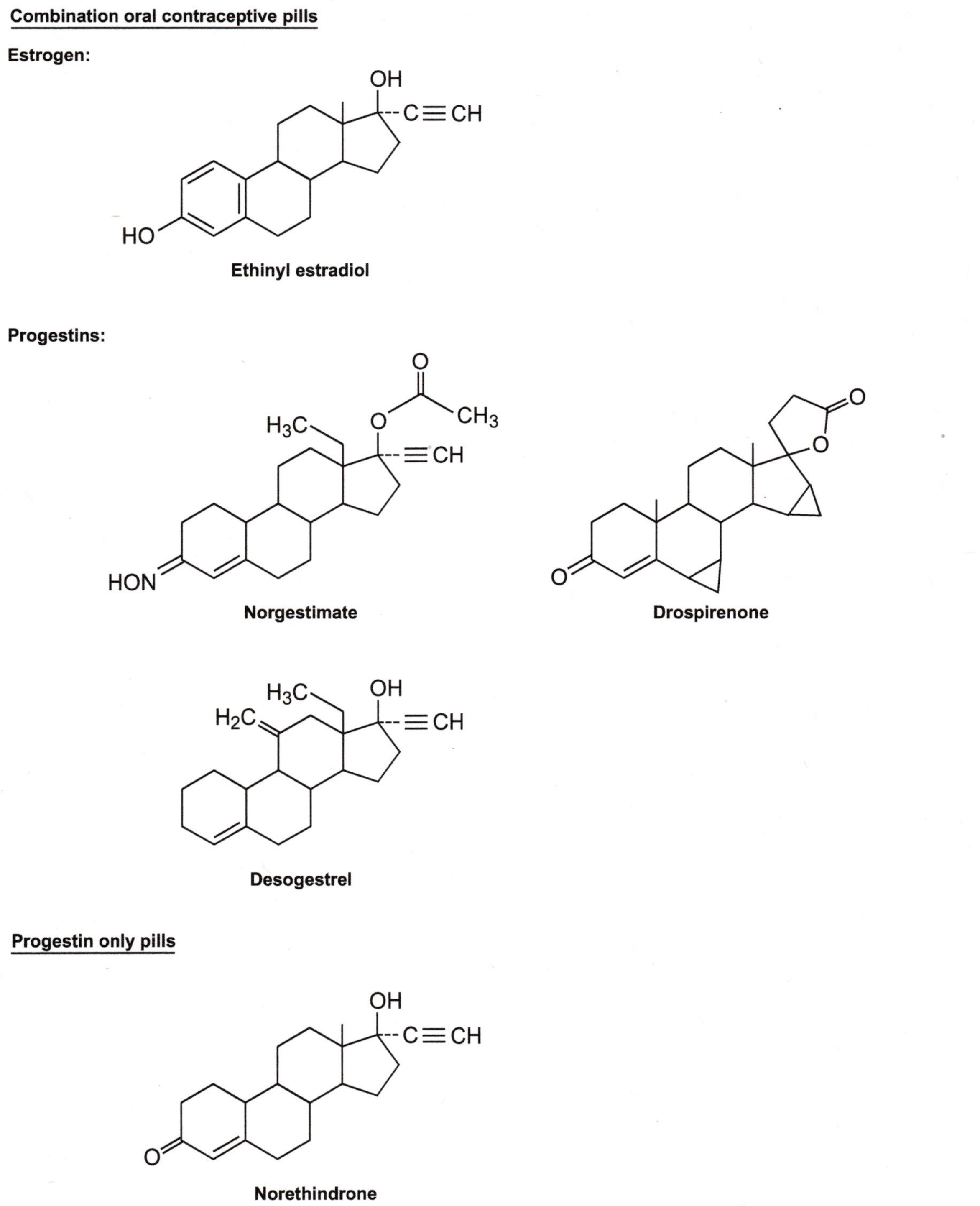

FIGURE 13.17 Oral contraceptives. The structures of the estrogen and progestin components of most modern combination oral contraceptive pills are shown in the top part of the figure. At the bottom is the structure of norethindrone, the only progestin-only oral contraceptive available in the United States.

as bone and the cardiovascular system, it was believed, based on many observational studies, that continued use beyond the completion of menopause contributed to the health of these tissues and others.

In 2002, 11 years after its initiation, the Women's Health Initiative (WHI), the first large randomized controlled trial to test these ideas, was terminated early because HRT was observed to be associated with an increase in rates of cardiovascular disease and breast cancer. Many women discontinued or did not begin HRT due to apprehension about these possible negative outcomes. In the ensuing years, much discussion and analysis of the results of the WHI have led many to believe that the relatively late start of HRT in the women participating in the WHI (on average 12 years after the start of menopause) may not accurately reflect the situation in peri- and early postmenopausal women. For example, heart disease might have already become established at the time that HRT was started and was no longer susceptible to the salutary effects of estrogen. At least one randomized control trial is currently underway to study the use of low doses of estrogen in these younger women closer to the time of menopause. With regard to breast cancer, the negative effect was seen with combination (both estrogen and progestin), while subjects receiving estrogen only showed lowered risk, relative to those receiving no HRT.

As more is learned about the effects of estrogen-replacement therapy in different populations of women, physicians and patients will have better grounds upon which to balance its positive effects such as reducing the unpleasant side effects of menopause and, in later years, the risk of fractures due to osteoporosis, and the possible negative ones such as increased risk for cardiovascular or breast disease.

C Polycystic ovary syndrome

Polycystic ovary syndrome, or PCOS, is among the most common female endocrine disorders in the world. It is typically recognized during puberty and is estimated to affect 5%–10% of reproductive age women. There is a familial disposition for the condition, but no definitive attributions to specific genes have yet been possible. In 1935 Stein and Leventhal formalized the recognition of the association of hyperandrogenism, as evidenced by hirsutism, amenorrhea (irregular or absent menses), and polycystic ovaries, in the syndrome that would carry their name for decades. The ovaries contain on the order of 12 antral follicles in which growth of the granulosa cells is arrested by an unknown mechanism (SMAD3 could be involved). A decreased rate of follicular atresia accounts for the high number of antral follicles.

The thecal cells of the follicles in PCOS, which are hypertrophic and may also have an increased sensitivity to LH, are responsible for the high circulating testosterone and androstenedione levels seen in this condition. Estrogen levels in the follicular fluid in PCOS are low, but circulating levels are in the normal range due to the peripheral aromatization of circulating testosterone and androstenedione. Levels of steroid hormone-binding globulin, the sex steroid–specific serum-binding protein, are decreased, further increasing the bioavailability of both androgens and estrogens. LH, but not FSH, secretion is increased through an increase of both frequency and amplitude of its pulses. Although several hypotheses have been and continue to be studied, the primary defect in reproductive hormone secretion has yet to be elucidated.

The previous changes in reproductive hormone status in PCOS are often accompanied by metabolic dysfunction which includes insulin resistance, obesity, and dyslipidemia. Insulin resistance is the inability of the body to respond to a given dose of insulin in terms of glucose uptake and utilization. The mechanism of the insulin resistance is not clear; both the number and binding effectiveness of insulin receptors appear to be normal, suggestive of a defect in a downstream signaling step. In PCOS, fat deposition tends to be in central and upper body areas with an increased waist to hip ratio compared to obese females without PCOS. This type of fat distribution is termed android obesity as it occurs more typically in males than in females. Total and low-density lipoprotein cholesterol, as well as triglycerides are increased while high-density lipoprotein cholesterol is decreased. Although the pathogenic mechanisms of PCOS, along with the relationship between the reproductive and metabolic symptoms, have not yet been elucidated, it has been found that insulin suppresses the response to GnRH in women with and without PCOS. Thus insulin resistance in PCOS could at least contribute to greater LH, resulting in hypertrophic thecal cells producing increased androgens.

In addition to antidiabetic drugs such as metformin, treatment of PCOS relies on oral contraceptives to reduce LH secretion and, therefore, androgen production, thus addressing one of the most pressing problems for these individuals, excessive and inappropriate hair growth. The use of oral contraceptives also addresses the long-term deleterious consequences of unopposed actions of estrogens on the endometrium and provides contraception for those not wishing to become pregnant. For those who do wish to become pregnant clomiphene citrate (Fig. 13.7), an ER ligand with mixed agonist/antagonist properties, is used to normalize, probably at the hypothalamus, secretion of LH and FSH.

Further reading

SERMs, SPMs, and endocrine disruptors

E.E. Dogan, Computational binding analysis and toxicity evaluation of estrogen receptor with estradiol and the approved SERMs raloxifene, tamoxifen, and toremifene, Medical Science International Medical Journal 10 (2021). Available from: https://doi.org/10.5455/medscience.org.

A. Ferramosca, et al., Herbicides glyphosate and glufosinate ammonium negatively affect human sperm mitochondria respiration efficiency, Reproductive Toxicology 99 (2021) 48–55.

K.-Y. Wong, et al., Water extract of Er-xian decoction selectively exerts estrogenic activities and interacts with SERMs in estrogen-sensitive tissues, Journal of Ethnopharmacology 275 (2021) 114096. Available from: https://doi.org/10.1016/j.jep.2021.114096.

Hypothalamic–pituitary–ovary axis

C.C. Davande, A.R. Gade, A.S. Kanekar, Pregnancy complications and management related to polycystic ovarian syndrome: an overview, International Journal of Research and Review 8 (1) (2021).

X.-L. Xu, S.-L. Deng, Z.-X. Lian, K. Yu, Estrogen receptors in polycystic ovary syndrome, Cells 10 (2021) 4359. Available from: https://doi.org/10.3390/cells10020459.

Estrogen receptor

E.E. Dogan, Computational binding analysis and toxicity evaluation of estrogen receptor with estradiol and the approved SERMs raloxifene, tamoxifen, and toremifene, Medical Science International Medical Journal 10 (2021). Available from: https://doi.org/10.5455/medscience.org.

Estrogens and the cardiovascular system

J.S. De Silva, et al., Estrogen receptors: therapeutic perspectives for the treatment of cardiac dysfunction after myocardial infarction, International Journal of Molecular Sciences 22 (2021) 525. Available from: https://doi.org/10.3390/ijms22020525.

Oral contraception

R. Bonfiglio, M.L. DiPietro, The impact of oral contraceptive use on breast cancer risk, Seminars in Cancer Biology (2021). Available from: https://doi.org/10.1016/semcancer.2020.10.008.

L.R. McCloskey, et al., Contraception for women with psychiatric disorders, The American Journal of Psychiatry 10 (2021) 2020. Available from: https://doi.org/10.1176/appl.ajp.2020.20020154.

A. Shah, et al., Effects of oral contraception and lifestyle modification on incretins and TGF-β superfamily hormones in PCOS, The Journal of Clinical Endocrinology and Metabolism 106 (2021) 108–119.

Hormone replacement therapy

R.D. Langer, H.N. Hodis, R.A. Lobo, M.A. Allison, Hormone replacement therapy–where are we now? Climacteric 24 (2021). Available from: https://doi.org/10.1080/13697137.2020.1851183.

L. Skeith, G. LeGal, M.A. Rodger, Oral contraceptives and hormone replacement therapy: how strong a risk factor for venous thromboembolism? Thrombosis Research 202 (2021) 134–138.

V. Varagiu, et al., Hormone replacement and therapy and cervical cancer: a systematic review of the literature, Climacteric 24 (2021) 2021. Available from: https://doi.org/10.1080/13697137.2020.1826426.

Polycystic ovary syndrome

A.S. Moin, et al., Metabolic consequences of obesity on the hypercoagulable state of polycystic ovary syndrome, Scientific Reports 11 (2021) 537. Available from: https://doi.org/10.1038/s41598-021-84586-y.

A. Subramanian, et al., Increased Covid-19 infections in women with polycystic ovary syndrome: a population-based study, European Journal of Endocrinology 184 (2021) 637–645.

Chapter 14

Hormones of pregnancy, parturition, and lactation

I Introduction

The ultimate purpose of reproduction is to produce a new female or male to maintain the species. This process, known as procreation, requires the meeting of a competent sperm with a fertilizable ovum, the development of the resulting zygote into an individual (pregnancy), and the delivery of that individual from the uterus to the outside world (parturition). There is a window of time in which each of these events must occur for an optimal outcome so that coordination of several biological processes is required throughout each successful procreative undertaking. In mammals, procreation includes the required nourishment of the offspring from the mother for a period of time following birth (lactation). As described in Chapter 12, Androgens, and Chapter 13, Estrogens and Progestins, for the male and female, respectively, there is extensive anatomical and endocrinological specialization that determines the individual's ability to participate successfully in procreation. However, the fundamental unit of reproductive biology is not the individual but the triad of the mother, father, and offspring. This chapter describes the human endocrinology that is unique to the processes of pregnancy, parturition, and lactation. For the purposes of this chapter, the result of conception from zygote until 8–10 weeks of gestation when the major organ systems have been formed will be referred to as the *embryo*; after this time point the term *fetus* will be used.

Human pregnancy is divided into thirds, each referred to as a trimester. As shown in Table 14.1, the perceived length of a pregnancy depends on what the observer defines as the starting point. A common and convenient starting point for the woman and her physician is the beginning of the previous menstrual period, that is, day one of the cycle in which fertilization occurred (Table 14.1, left column). Others may date the pregnancy from the time of the first missed menstrual period or a positive pregnancy test shortly thereafter (Table 14.1, right column). Biologically, fertilization itself marks the beginning of the pregnancy, generally 14 days following the onset of menses (Table 14.1, middle column). This is the timing of events that is most often used in research and many clinical settings and will be used in this chapter. Since the normative time from fertilization to parturition is 38 weeks, utilization of the other two start points will result in lengths of the total pregnancy that differ from those using fertilization as the start point.

II Anatomical relationships and the beginning of pregnancy

The first two steps in the establishment of a pregnancy are fertilization of the ovum and implantation of the resulting blastocyst in the uterus. These events, which take place during the luteal phase of the menstrual cycle (see Fig. 13.11), are depicted in overview in Fig. 14.1 and are discussed in more detail following. The human ovum must be fertilized within 24 hours of ovulation, following which it completes its final meiotic division (see Fig. 13.5). The zygote undergoes its first and second mitotic divisions on days 2 and 3 following fertilization and by the following day, it has become a compact ball of 16–32 cells called the morula. It arrives in the uterus as a blastocyst of 70–100 cells, which continues to increase in cell number, but not overall size. By this time the cells of the blastocyst have differentiated into two types of cells, trophoblast, which will develop into the fetal part of the placenta and an inner cell mass that will become the embryo. Implantation begins with the positioning of the fluid-filled blastocyst on the uterine wall on days 6–7. Note that this corresponds to the time at which preparation of the uterine lining by progesterone and estrogen from the corpus luteum has reached its maximum receptivity, called the window of implantation in the secretory phase (Chapter 13: Estrogens and Progestins, Section V. B). If the implantation is successful, the uterine lining will not be shed and this missed menstruation will be one of the first signals to the woman that she is pregnant.

Hormones. DOI: https://doi.org/10.1016/B978-0-323-90262-5.00026-3

TABLE 14.1 Stages of pregnancy with different start points.

Start point	First day of previous menstrual period	Fertilization	First missed period (preg. test) menstrual
Total apparent length of pregnancy	40 weeks (9.2 months)	38 weeks (8.75 months)	36 weeks (8.3 months)

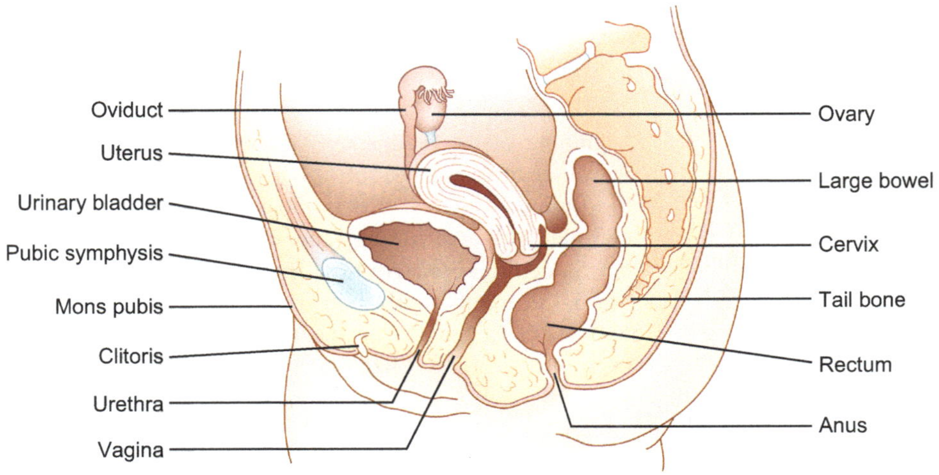

FIGURE 14.1 Overview of ovulation, fertilization, and implantation. Fertilization (see Fig. 14.2), which takes place within 24 h of ovulation (see Fig. 14.2), is designated as Day 0. The first cleavage is accomplished on day 2 and is followed by more rapid divisions, resulting, on day 4, in a compact ball of 16–32 cells called the morula. Continued division without an increase in size leads to the blastocyst that arrives in the uterus consisting of 70–100 cells on day 5 following fertilization. The process of implantation takes place on days 7–9 and is shown in Fig. 14.3.

A Fertilization

In the 1950s, it was demonstrated that the sperm of many mammalian species, including humans, must undergo some kind of modification in the female reproductive tract to be capable of fertilizing an ovum. This process is termed capacitation and the length of time required for it varies with species but is usually several hours. The meaning of the term capacitation has undergone some change over the past decades as more has been learned about the process. It is now generally understood to mean the removal or destabilization of proteins in and around the acrosomal region of the sperm head to enhance its penetration ability, as well as effects on the flagellum to increase motility. The recognition of the existence of this process and the ensuing success in replicating it outside the female reproductive tract have had a profound effect on the development of assisted reproductive technology (ART) in humans (Section VI.A) as well as on artificial insemination in domesticated species.

The steps in mammalian fertilization are shown in Fig. 14.2. The initial binding of the sperm to the zona pellucida surrounding the ovum involves a species-specific interaction between the sperm membrane and one or more of the glycoproteins, ZP1, ZP2, ZP3, and (in humans) ZP4, that make up the zona pellucida. This interaction, through an increase in intracellular Ca^{2+}, cyclic AMP, and pH, triggers the acrosome reaction in which hydrolytic enzymes, including hyaluronidase and a protease known as acrosin, are released from the acrosome. The digestive action of these enzymes forms a tunnel in the zona pellucida around the head of the sperm which, along with the propulsive force of the flagellum, enables the sperm to traverse it and bind to the outer surface of the ovum. The plasma membranes of the sperm and egg then fuse, allowing the sperm's pronucleus to enter the ovum and move toward the ovum's pronucleus, formed from the completion of the ovum's final meiotic division (see Fig. 13.5). The two pronuclei join, forming the zygote nucleus. During this time the fusion of the two plasma membranes initiates the cortical reaction in which the enzymes contained in the cortical granules lying just inside the ovum's plasma membrane are released into the perivitelline space. This renders the zona pellucida refractory to sperm penetration, which, along with the destruction of ZP proteins, blocks the entrance of any other sperm. IP_3 (inositol triphosphate) and diacylglycerol signaling pathways are involved in this important step of blocking polyspermy.

B Implantation

The fertilized ovum undergoes several rounds of cell division as described in Fig. 14.1 and reaches the uterus as a blastocyst, still surrounded by the zona pellucida and little changed in size from the unfertilized ovum. The blastocyst consists of a layer of trophoblasts and an inner cell

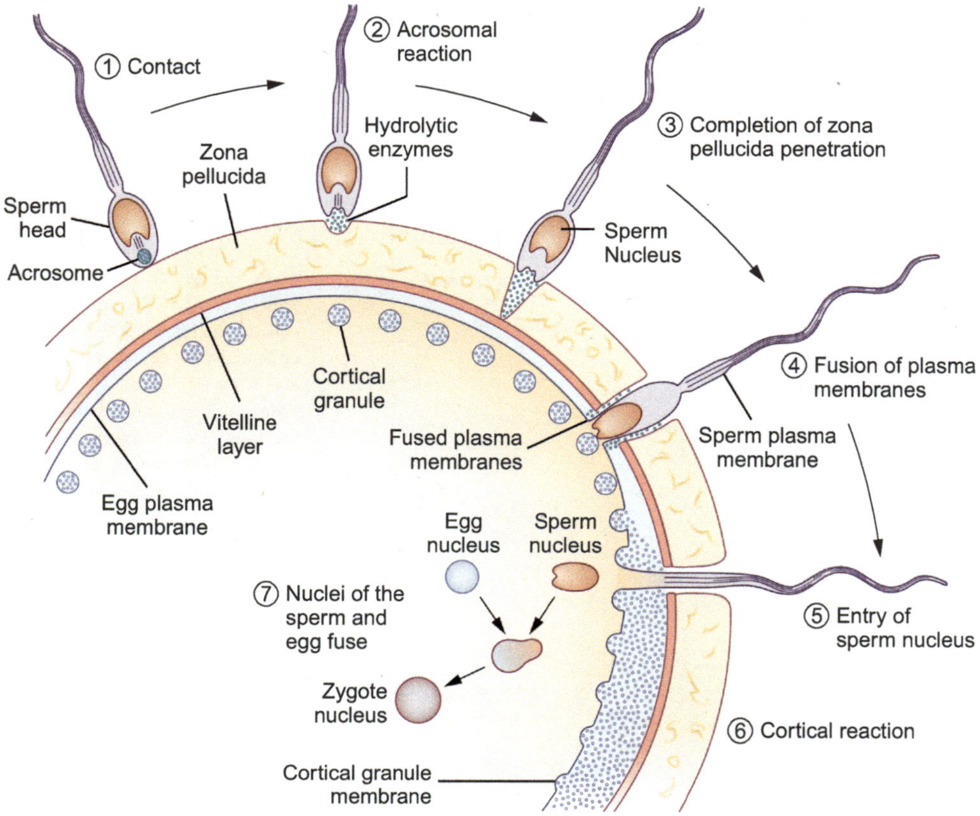

FIGURE 14.2 Fertilization. The outer layer of the ovum in the fallopian tube is the zona pellucida, consisting of the glycoproteins ZP1, ZP2, ZP3, and ZP4. When the sperm binds to one or more of these proteins (1), the proteolytic enzymes in the acrosome are released (2) and the sperm penetrates this coating. The penetration of the zona pellucida is completed (3), leading to the fusion of plasma membranes of the sperm and the egg (4). The sperm pronucleus is released into the egg (5); it and the ovum pronucleus move toward each other and fuse, forming the zygote nucleus (7). Meanwhile, the contents of the cortical granules lining the intracellular face of the ovum's plasma membrane are released to form the cortical granule membrane (6), which prevents the entry of other sperm.

mass, both of which surround a cavity, the blastocoele (Fig. 14.3A). The blastocyst remains near but not attached to the uterine lining for a day or two, being nourished by the secretions of the endometrial glands. Near the end of this period, the endometrial glands secrete an enzyme that dissolves the zona pellucida in a process referred to as "hatching" (not shown). Shortly thereafter, the trophoblastic cells differentiate into an outer layer, the syncytiotrophoblast, and an inner layer, the cytotrophoblastic cells. The development of the embryo to this point and the preparation of the uterine lining for its reception are closely synchronized events. Delay in the embryo's development is likely to result in the failure of the potential pregnancy. For this reason, this preimplantation stage is one of a high rate of fetal loss.

During the implantation window (days 20–24 of the menstrual cycle), implantation sites known as pinopods develop in the endometrial epithelium. The cells at these sites express chemokines (molecules that attract cells) and cell adhesion molecules such as mucins, integrins, and troponins. At one of these sites the syncytiotrophoblastic cells invade the endometrium by secreting their own cell adhesion molecules as well as enzymes that break down stromal cells (Fig. 14.3B). In the deciduoma response the stromal cells begin to grow over the invading conceptus (Fig. 14.3C) until implantation is complete on about day 10 following fertilization (Fig. 14.3D). By this time, sinusoids derived from endometrial spiral arteries (see Fig. 13.2B) and containing maternal blood have appeared in the syncytiotrophoblast, an initial step in the establishment of the maternal–fetal blood circulatory system.

C Placental development

The fetoplacental unit has three important functions: (1) a source of protein and steroid hormones, which are delivered to the maternal circulation; (2) a selective barrier that determines the nature of the communication and interaction between the maternal and fetal endocrinological and physiological systems; and (3) a participant in the control of fetal growth, development, endocrine function, and parturition (birth). In all these functions the fetus and maternal placenta have separate circulatory systems which, as shown in Fig. 14.4, exchange nutrients, gases, and waste products in the blood pools of the maternal portion of the placenta. To accomplish this, blood poor in oxygen and nutrients leaves the fetus through the umbilical arteries and enters the capillaries of the chorionic villus. In these capillaries, blood is oxygenated by, gains nutrients from, and gives up wastes to the maternal blood in the pools before it returns to the fetus in the umbilical vein. Oxygen-depleted blood leaves the maternal blood pool through the maternal veins. Thus from the time blood

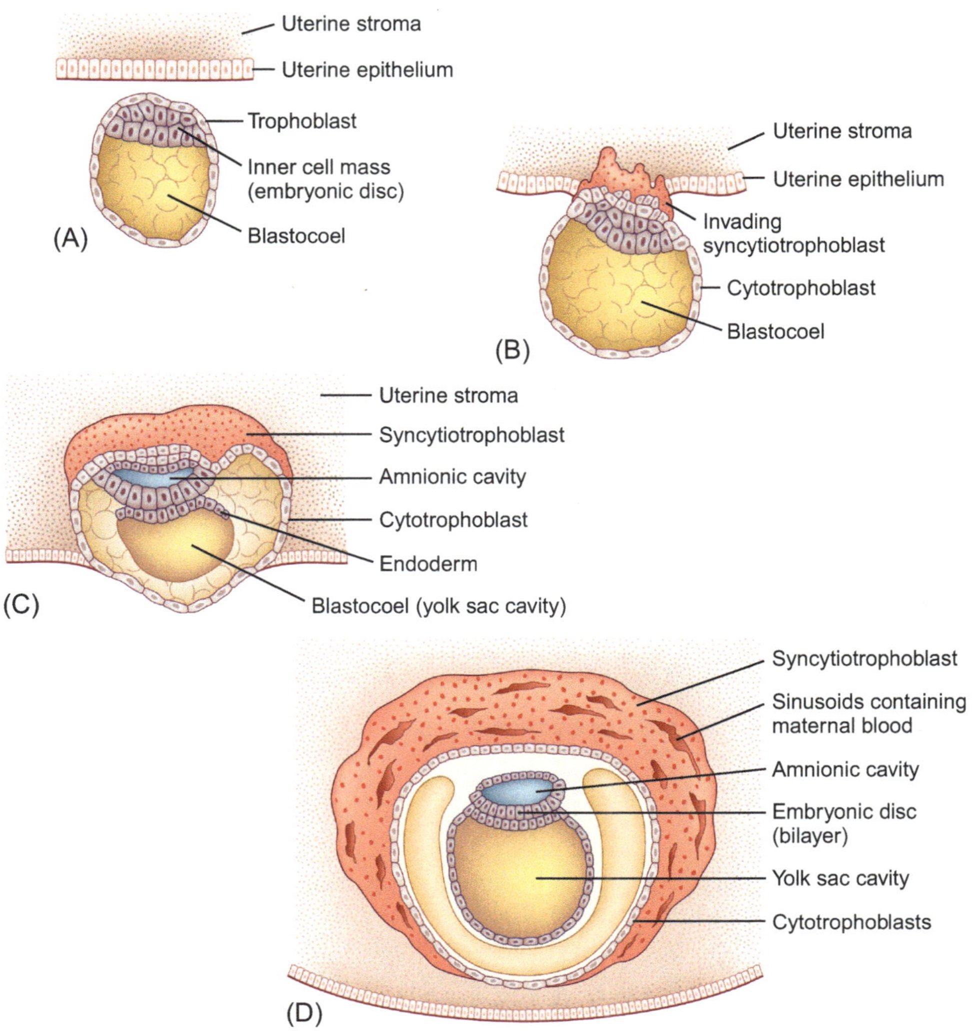

FIGURE 14.3 Implantation. Implantation of the blastocyst takes place on days 7–10 following fertilization of the ovum. (A) The blastocyst (Fig. 14.1, day 5) consists of a layer of trophoblastic cells, which will develop into the fetal portion of the placenta, an inner cell mass which will develop into the embryo, and a cavity, the blastocoel, which will become the yolk sac. (B) The trophoblastic cells of the blastocyst attach to the epithelial cells of the endometrium and some of them form a syncytium (a large multinucleate cell resulting from fusion of many cells). This portion of the structure is the syncytiotrophoblast, which begins to invade the uterine lining. (C) The blastocyst continues to move into the stroma of the endometrium and the amniotic cavity appears. (D) Uterine tissue grows over the blastocyst and irregular blood sinusoids appear in the syncytiotrophoblast.

flow is established at 3 weeks following fertilization, fetal oxygenated blood is carried by veins rather than arteries (and oxygen-poor blood by arteries rather than veins), a reversal from the situation after birth when the neonatal circulatory system becomes functional.

Following implantation, the trophoblast rapidly proliferates new chorionic villi that infiltrate the endometrial decidua and become the chorion frondosum, the fetal portion of the placenta as shown in Fig. 14.5. The portion of the placenta adjacent to the chorion frondosum, the decidua basalis, constitutes the maternal portion of the placenta. The fetus, in the amniotic sac, actually lies within the uterine wall. As the fetus grows, the amnion eventually fuses with the chorion as the fetus in the amniotic sac fills the uterine cavity. The villous surface area continues to grow throughout gestation, but the growth rate falls off at 36–38 weeks. By this time the placenta is a disk 15–20 cm in diameter and weighs 500–600 g; the umbilical cord is about 55-cm long.

D Mammary glands

The breasts or mammary glands are functionally related to the female reproductive system in that they secrete milk for nourishment of the newborn and this function relies on the steroid hormones progesterone and estrogen as well as pituitary hormones prolactin (anterior lobe) and oxytocin (posterior lobe) and others discussed in Section V. The anatomy of the adult human female breast is shown in Fig. 14.6. Each breast, which overlies the second to sixth ribs and the major chest muscles, the pectorals, comprises 15–20 lobes. Each lobe consists of glandular lobules that, in turn, contain the alveoli (or acini), consisting of a single layer of milk-producing epithelial cells surrounding a lumen. The milk drains from the lumen into ductules that empty into larger ducts, which eventually carry the milk to the nipple, where it is ejected during suckling. The epithelial cells of the alveoli and the duct cells have distinct morphologies, functions, and responses to steroid hormones, as will be described in Section V.

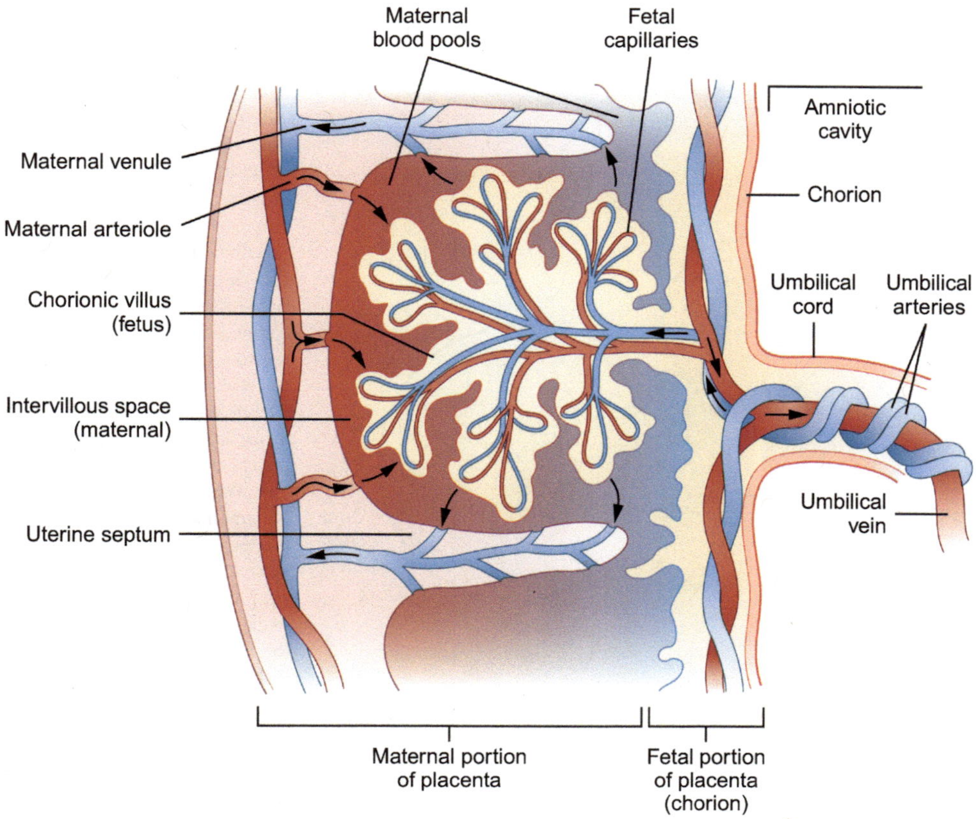

FIGURE 14.4 Fetal–maternal circulation in the placenta. The transfer of gasses, nutrients, waste products, and other molecules between the maternal and fetal circulation takes place in the maternal blood pools that have arisen from the spiral arteries of the endometrium (see Fig. 13.2B). Freshly oxygenated blood from the mother enters the pools through the maternal arteries (left side, *red*). Blood that is poor in oxygen and nutrients leaves the fetus through the umbilical arteries (right side, *blue*), enters the capillaries of the chorionic villus where it is oxygenated and gains nutrients before it returns to the fetus in the umbilical vein (right side, *red*). Oxygen-depleted blood leaves the maternal blood pool through the maternal veins (left side, *blue*). Note that in fetal life, the situation is reversed from that after birth, when oxygenated blood is carried by the arteries.

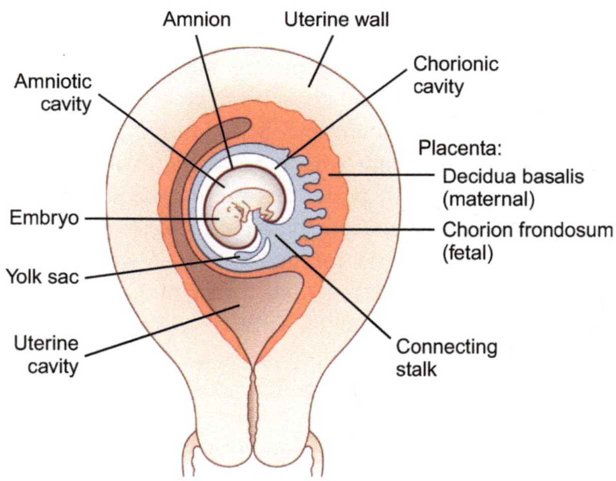

FIGURE 14.5 The development of the placenta. By the end of 4 months, the placenta is anatomically complete. The fetal portion of the placenta, where fetal blood exchanges gases and molecules with maternal blood, is the chorion frondosum (*blue*, right side). The portion of the decidua which lies near the chorion frondosum, the decidua basalis, is the maternal portion of the placenta (*orange*, right side). Here the amniotic membrane is shown separately, but as the fetus and placenta continue to gain in size, the amniotic and chorionic membranes fuse to surround the amniotic cavity. Note that the fetus lies within the uterine wall, not outside it in the uterine cavity (*brown*).

III Chemistry, biochemistry, and activity of the hormones of pregnancy

Table 14.2 lists many (but by no means all) of the hormones produced by the fetoplacental unit. Some of these are produced by the newly formed embryo (e.g., human chorionic gonadotrophin, hCG), and, after 8–10 weeks, the fetus (e.g., intermediates in steroid hormone synthesis). Most are produced by the placenta (e.g., human placental lactogen, hPL). Many of these hormones mediate communication between the fetus and the mother, one primary goal of which is to modify the mother's intermediary metabolism so that sufficient nutritional resources are delivered to the growing, developing fetus.

A Human chorionic gonadotrophin

The establishment and maintenance of a pregnancy is dependent on progesterone, and during the first 6–7 weeks this is produced by the corpus luteum. In humans the continued production of progesterone by the corpus luteum depends on chorionic gonadotrophin, hCG, secreted by the trophoblast cells of the embryo within 1

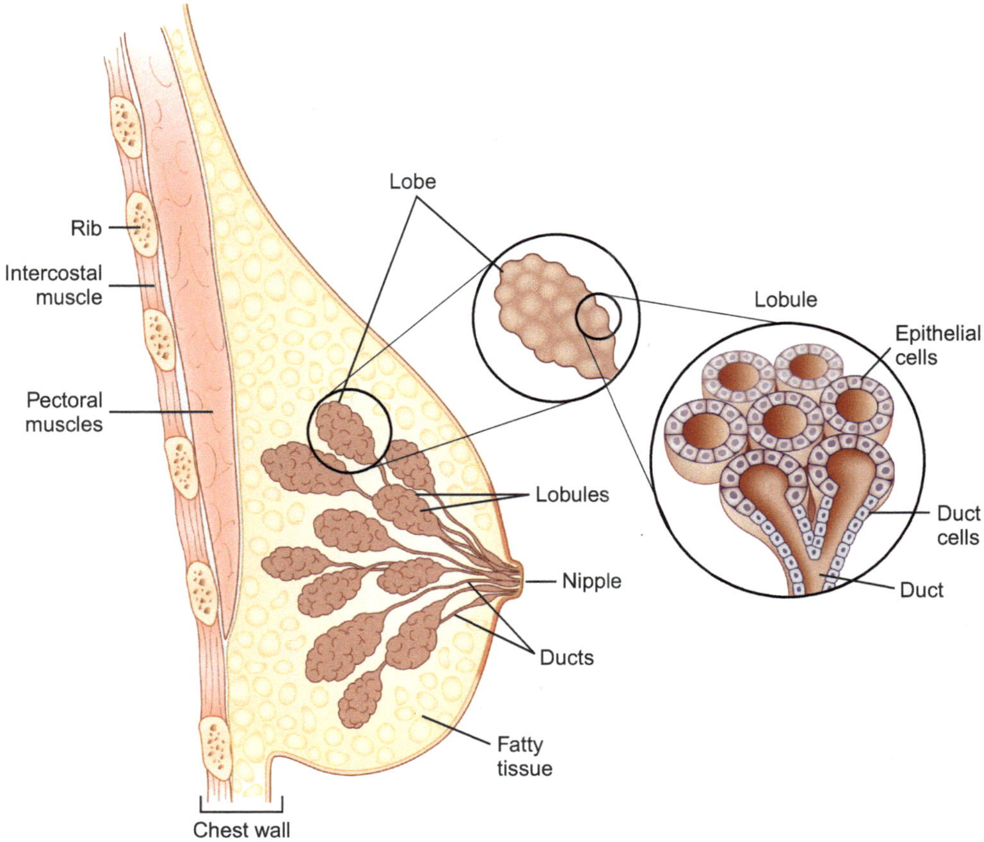

FIGURE 14.6 The adult human female breast. The breast, shown in sagittal section on the left, has three main components: lobes of milk-producing tissue, ducts to carry the milk to the nipple, and adipose tissue. Each lobe (center) consists of lobules, which, in turn (right), are made up of the functional unit of the breast, the alveolae (acini) composed of a single layer of milk-producing epithelial cells surrounding a lumen into which milk is secreted. Each alveolus has a ductule, in which the milk travels to the lobule's duct and is eventually emptied into the major duct system which converges on the nipple from which the milk is ejected. The breast, located between the second and sixth ribs, contains variable amounts of fat tissue and is supported by the musculature of the chest wall. The nipple is surrounded by an areola of variable color.

TABLE 14.2 Hormones produced by the human placenta.

Pituitary analogs	Steroids	Neuropeptides
ACTH	Progesterone	CRH
TSH	Estrogens	TRH
LH	Estradiol	GnRH
FSH	Estrone	GHRH
hCG	Estriol	Somatostatin
hPL	Estetrol	VIP
Prolactin		Neuropeptide Y
Oxytocin		Substance P
Leptin		Urocortin
Activin		
Inhibin		

ACTH, Adrenocorticotrophic hormone; *CRH*, corticotrophin-releasing hormone; *FSH*, follicle-stimulating hormone; *GH*, growth hormone; *GHRH*, growth hormone–releasing hormone; *GnRH*, gonadotrophin-releasing hormone; *hCG*, human chorionic gonadotrophin; *hPL*, human placental lactogen; *LH*, luteinizing hormone; *TRH*, thyrotrophin-releasing hormone; *TSH*, thyroid-stimulating hormone; *VIP*, vasoactive intestinal peptide.

day of the beginning of implantation. This is 9–11 days following the midcycle luteinizing hormone (LH) peak and 8 days following ovulation. If the embryo is unable to secrete sufficient amounts of hCG during this crucial time, the corpus luteum will undergo regression, progesterone production will cease, and the pregnancy will fail. The dependency of the pregnancy on progesterone from the corpus luteum lasts for the first 6–7 weeks of the pregnancy when the placenta takes over progesterone production in the luteal-placental shift (see Section III.D).

Human chorionic gonadotropin is a heterodimeric glycoprotein of 57 kDa. Its 92-amino-acid α subunit is identical in amino acid sequence to those of human LH, follicle stimulating hormone (FSH), and thyroid-stimulating hormone (see Fig. 3.12). The β subunit of hCG is very similar to that of LH (80% homology for the first 115 amino acids) except that the carboxy terminus of hCG has an extra 22 amino acids. Noncovalent hydrophobic bonding holds the two subunits together. The single gene for the α subunit is located on chromosome 6 and is expressed in both the cytotrophoblast and the syncytiotrophoblast. A cluster of six genes on chromosome 19 encode the hCG β subunit, three of which are expressed in the syncytiotrophoblast, which also carries out the glycosylation of both subunits and the assembly and secretion of the mature dimeric hormone. LH and hCG share the same receptor, encoded by the single copy *LHCGR* gene on chromosome 2.

The regulation of hCG synthesis and secretion by the trophoblast are not clearly understood. Many placental molecules, including gonadotrophin-releasing hormone (GnRH), activin, and inhibin, have been implicated, but a clear picture has not yet emerged of the integrated actions of these and other hormones on hCG production.

The levels of hCG rise exponentially for about the first 8 weeks of pregnancy (Fig. 14.7), doubling approximately every 72 hours if there is a single fetus and more rapidly if there are two or more viable embryos. Levels then fall off, as the fetoplacental unit takes over in the production of progesterone. Several functions for hCG, in addition to maintenance of progesterone production by the corpus luteum, have been postulated. One for which there is considerable evidence is its role in the stimulation of the fetal testis to secrete testosterone early in development prior to the ability of the fetal pituitary to secrete sufficient amounts of LH for this function.

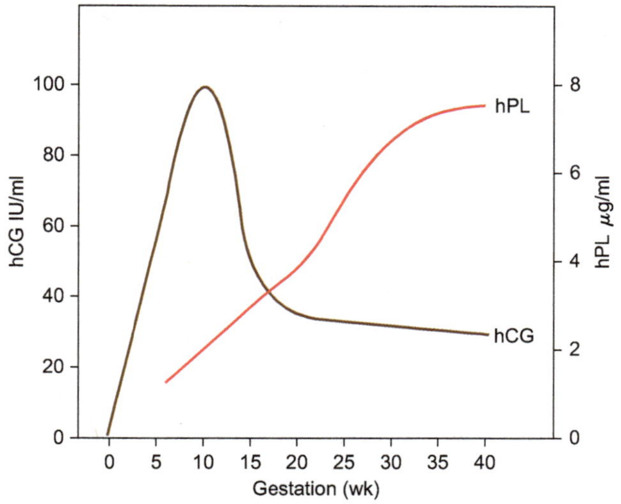

FIGURE 14.7 Chorionic gonadotrophin and placental lactogen levels during pregnancy. The graph shows the levels of hCG and hPL throughout pregnancy. hPL is also known as chorionic somatomammotrophin, CS or hCS. hCG is detectable a few days following fertilization, rises exponentially to a peak at about 8 weeks, then declines to a plateau for the remainder of the pregnancy. hPL is first detectable at about 5 weeks and continues to rise until term. *CS*, Chorionic somatomammotrophin; *hCG*, human chorionic gonadotrophin; *hPL*, human placental lactogen.

Immunoassays can detect the β subunit of hCG in blood at the level of 1 mIU/mL and a 10.5-kDa core fragment of the β subunit in urine at a level 10–100 mIU/mL. This level of sensitivity is sufficient to detect a pregnancy with a blood sample (5 mIU or greater) within a day of implantation and within the next 2–4 days using urine, as in home pregnancy tests.

B Placental lactogen and placental growth hormone

Fig. 14.8 shows the human gene cluster that encodes pituitary growth hormone (Chapter 3: The Hypothalamus and Anterior Pituitary) and three related proteins produced in the placenta (the fifth gene, Pl-L, is a pseudogene with a nonfunctional splice junction). Placental growth hormone, PGH (GH-V, with the V standing for variant), differs from the pituitary protein in only 13 amino acids. The genes for PL-A and PL-B encode the same mature protein product, a glycoprotein containing 191 amino acids, which is 85% identical to pituitary growth hormone.

The growth hormone gene cluster is controlled by a remote upstream locus control region which contains five hypersensitive sites: two of these are specific for pituitary expression of GH-N, another is specific for the placental cluster, and two are active in both tissues. These are the sites at which chromosomal alterations (e.g., histone acetylation and removal, making the sites available to transcription factors) leading to changes in gene expression are initiated. As expected, the patterns of these and consequent alterations differ for the five genes in the cluster.

hPL is first detectable in the placenta at about 18 days after fertilization and in the maternal circulation at 5–6

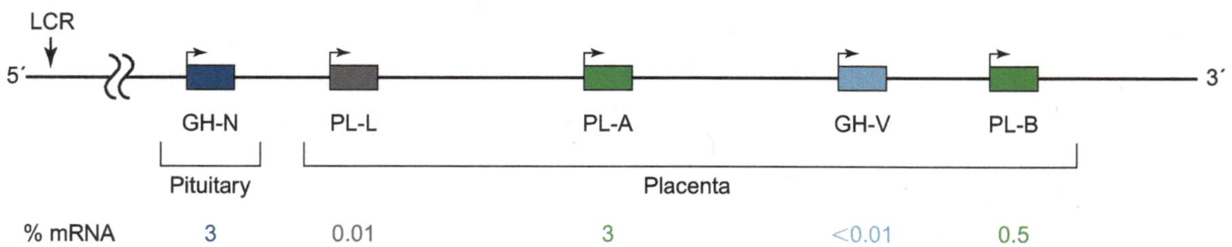

FIGURE 14.8 The human growth hormone/placental lactogen gene cluster. The figure depicts the five genes in the 66-kb cluster, located on the long arm of chromosome 17 and the proteins they encode: pituitary growth hormone (GH-N); placental growth hormone (GH-V); PL-A and PL-B that are identical proteins; and PL-L, a nonfunctional protein encoded by a pseudogene. All five genes are oriented in the same transcriptional direction and each consists of five exons and four introns, having arisen from a common ancestral gene. Placental lactogen is also known as CS and the human gene bears this name. The LCR lies 14–32 kb upstream of the first gene in the cluster. The relative amount of total cellular mRNA expressed from each gene in either the pituitary or the placenta is shown in the bottom line. *GH*, Growth hormone; *LCR*, locus control region; *mRNA*, messenger RNA; *PL*, placental lactogen.

weeks. As shown in Fig. 14.7 the levels of hPL rise throughout pregnancy. In fact it has been shown that the amount of maternal circulating hPL is exactly proportional to placental mass and, as shown in Fig. 14.8, at least 3% of placental messenger RNA (mRNA) encodes this hormone. As term approaches, the secretion rate of hPL approaches 1.0 g/day, greater than that of any other protein hormone; hPL becomes nondetectable in maternal blood within 1 day of the delivery of the placenta. A small amount of hPL is secreted into the fetal circulation, where levels also rise during the pregnancy, but are about 1000-fold lower than those in the maternal circulation.

Maternal serum levels of PGH (GH-V) are three orders of magnitude lower than those of hPL and it cannot be detected in maternal serum until about 22 weeks. PGH increases exponentially during the last trimester of pregnancy; at the same time, maternal pituitary growth hormone falls to undetectable levels. It is likely that this suppression of maternal GH secretion is the result of the negative feedback effect of increased insulin-like growth factor-1 (about twofold from week 25 to term) in response to both hPL and PGH. Maternal pituitary GH levels (GH-N) return to normal within 48 hours of delivery. No PGH is found in the fetus.

Although hPL was originally identified by its lactogenic properties in rodents and in bioassays using cells in tissue culture, a role for the hormone in lactogenesis in humans has not been clearly established. It is, however, known to alter maternal metabolism leading to the diabetogenic state of pregnancy in which fuels, especially glucose, are preferentially made available for the fetus. Features of this state include decreased glucose tolerance, decreased insulin sensitivity, and increased lipolysis in adipocytes with increased availability of free fatty acids. The latter are used by the mother to spare glucose for the fetus and also are made available to the fetus, along with the ketone bodies that result from their oxidation by the mother, providing additional fuels for the fetus.

These effects of hPL are similar to those of growth hormone, including PGH. Since no receptor for specific PL receptors has been identified, it is thought that hPL, PGH, and GH share the same receptor. Although hPL binds to the GH receptor with relatively low affinity, the large amounts of the hormone could be sufficient to activate the receptor. Women with inactivating mutations of hPL deliver normal babies, while those missing both PGH and hPL have babies of very low birth weight. There have been no reported cases of women lacking only PGH. Taken together, these observations suggest that even though hPL is normally produced in very high amounts, its absence is compensated for by the presence of PGH, indicating some redundancy in the important function of providing fuel for the fetus from the maternal metabolic system.

C Other peptide hormones

1 Corticotrophin releasing hormone

A 41-amino-acid peptide identical to hypothalamic corticotrophin releasing hormone, CRH (see Chapter 3: The Hypothalamus and Anterior Pituitary), is produced by both the cytotrophoblast and the syncytiotrophoblast. Placental CRH mRNA can be detected at about 7 weeks of pregnancy and the peptide is detectable in maternal serum at 15 weeks. By the final 5–7 weeks of pregnancy, circulating maternal CRH levels are 1000 times those in a nonpregnant woman. Some CRH is also released into the fetal circulation.

CRH-binding protein is a liver-derived serum protein that binds the peptide with high specificity and affinity, making it biologically unavailable to target tissues. C-reactive protein-binding protein (CRP-BP) levels fall during the last few weeks of pregnancy, while CRH itself is rising, increasing even further the amount of circulating biologically active CRH. The maternal pituitary is not responsive to these changes in CRH levels and maternal pituitary adrenocorticotrophic hormone (ACTH) remains low throughout pregnancy.

Placental CRH production is stimulated by glucocorticoids, in contrast to the situation in the hypothalamus where glucocorticoids exert negative feedback inhibition on CRH expression. Since some placental CRH is released into the fetal circulation, a positive feedback loop is established in which placental CRH stimulates the fetal pituitary to secrete ACTH, increasing fetal adrenal glucocorticoid production. Such a loop is thought to play a role in human parturition (see Section IV). CRH may also be a secretagogue for fetal adrenal androgen production, particularly dehydroepiandrosterone sulfate (DHEAS), which, as will be seen in Section III.D, leads to increased placental estrogen production.

2 Gonadotrophin-releasing hormone

Cells of both the cytotrophoblast and syncytiotrophoblast synthesize and secrete the same decapeptide, GnRH, that is produced in the hypothalamus. Considerable evidence supports the involvement of this GnRH in the stimulation of hCG secretion, particularly through the increased synthesis of its β subunit. Early in the first trimester, this stimulation is not required for the very high levels of hCG that are produced to maintain the corpus luteum, but the stimulatory effect of GnRH (and its inhibition by GnRH antagonists) is clear by the second trimester. In addition to its role in hCG secretion, GnRH, which stimulates a certain cadre of proteinases, contributes to the breakdown of the extracellular matrix of the uterine stromal tissue during the invasion and implantation of the embryo.

3 Proopiomelanocortin

Proopiomelanocortin (POMC) is expressed by syncytiotrophoblasts. Some of it is processed into its constituent peptides as described in Fig. 3.15. However, in contrast to the pituitary, where POMC is completely processed, a significant amount of intact POMC is released by the placenta into the maternal circulation. The physiological significance of this peptide and those derived from it is not yet understood. Placental ACTH, which has no effect on the fetal endocrine system, may partially offset feedback inhibition by glucocorticoids of maternal ACTH secretion, allowing for a greater maternal adrenal capacity for responsiveness to stress.

4 Inhibins, activins, and follistatin

The three inhibin subunits (Fig. 12.10) are expressed in the syncytiotrophoblast and inhibin A is the main form released into the maternal circulation. Increased circulating levels of inhibin in early pregnancy are largely due to ovarian production and the placental contribution maintains relatively unchanging levels from week 12 until week 30, when they rise again. Activin and follistatin are also produced by the placenta and by fetal membranes. Research with human trophoblast cells in culture indicates that these peptides play autocrine and paracrine roles in the regulation of GnRH and hCG synthesis and secretion. In addition, inhibin A and activin A are associated with several gestational disorders such as hypertensive disorders of pregnancy, intrauterine growth restriction, and preterm delivery. In particular, in the prenatal screen of amniotic fluid often used at 16–18 weeks of pregnancy, the presence of high levels of inhibin A along with decreased estriol is suggestive of a fetus with Down's syndrome.

5 Leptin

During pregnancy the placenta is the main source of leptin (see Chapter 6: Pancreatic Hormones: Insulin and Glucagon). At the end of the first trimester, maternal leptin levels are twice those seen in the nonpregnant state; they continue to rise throughout pregnancy then return to the nonpregnant level shortly after delivery. At the same time, the hypothalamus becomes resistant to leptin, a state associated with obesity in the nonpregnant individual, so that appetite and food intake increase throughout pregnancy. It is not surprising, then, that humans and other mammals gain more weight and fat deposition during pregnancy than is required for fetal growth. The precise role of leptin in maternal physiology is not yet clear, but given the presence of leptin receptors in many peripheral tissues, including liver, muscle, and pancreas, it is likely that it contributes to increased lipolysis and decreased insulin sensitivity which are characteristic of pregnancy and which spare glucose for the fetus. In the placenta, which also has leptin receptors, it has been suggested that leptin affects placental growth and hCG production.

D Steroid hormones

As shown in Fig. 14.9, progesterone and the three major estrogens increase progressively throughout pregnancy. During the menstrual cycle, serum estradiol, even at its midcycle peak, is less than 1 ng/mL and progesterone during the luteal phase is in the range of 12–15 ng/mL. Estrone circulates at considerably lower concentrations than estradiol and estriol is not secreted by the ovary in the absence of pregnancy. Because of the role of the fetoplacental unit in the production of the steroid hormones of pregnancy beginning during the 8th week of gestation (see Fig. 14.10), estriol levels can be an indicator of fetal health.

During the first 6–7 weeks the corpus luteum, under the influence of hCG, is the sole source of the progesterone which is absolutely required for the maintenance of the pregnancy. The fetoplacental unit begins to produce significant amounts of both progesterone and estrogens during the 7th–8th week, and by 10–12 weeks it has taken over providing both of these hormones for the remainder of the pregnancy.

Fig. 14.10 depicts the synthesis of progesterone in the placenta and how the fetal and placental compartments collaborate in the production of estrogens. Cholesterol, as a component of low-density lipoprotein in the maternal plasma, is available to the placenta which readily converts it to pregnenolone and progesterone. Some of this progesterone is released into the maternal circulation and some

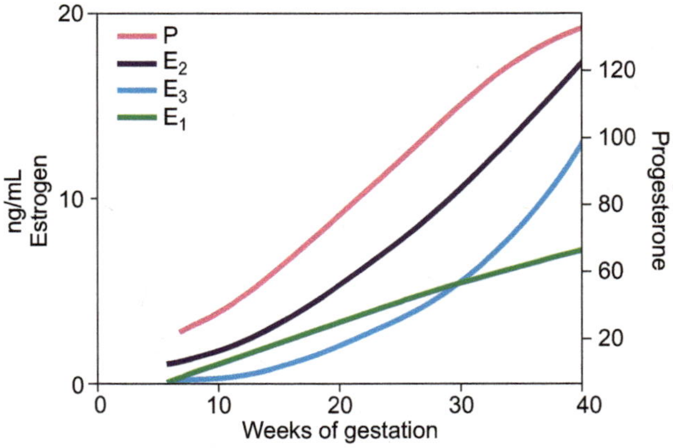

FIGURE 14.9 Circulating steroid hormones during pregnancy. The graphs depict, in ng/mL serum, the relative circulating concentrations of progesterone (P), estrone (E_1), estradiol, (E_2), and estriol (E_3) as a function of time during pregnancy. Serum levels of estradiol and estrone in nonpregnant women usually do not exceed 1 ng/mL and estriol is not produced by the ovary in the absence of pregnancy. Peak progesterone levels during nonpregnant menstrual cycles are in the range of 12–15 ng/mL.

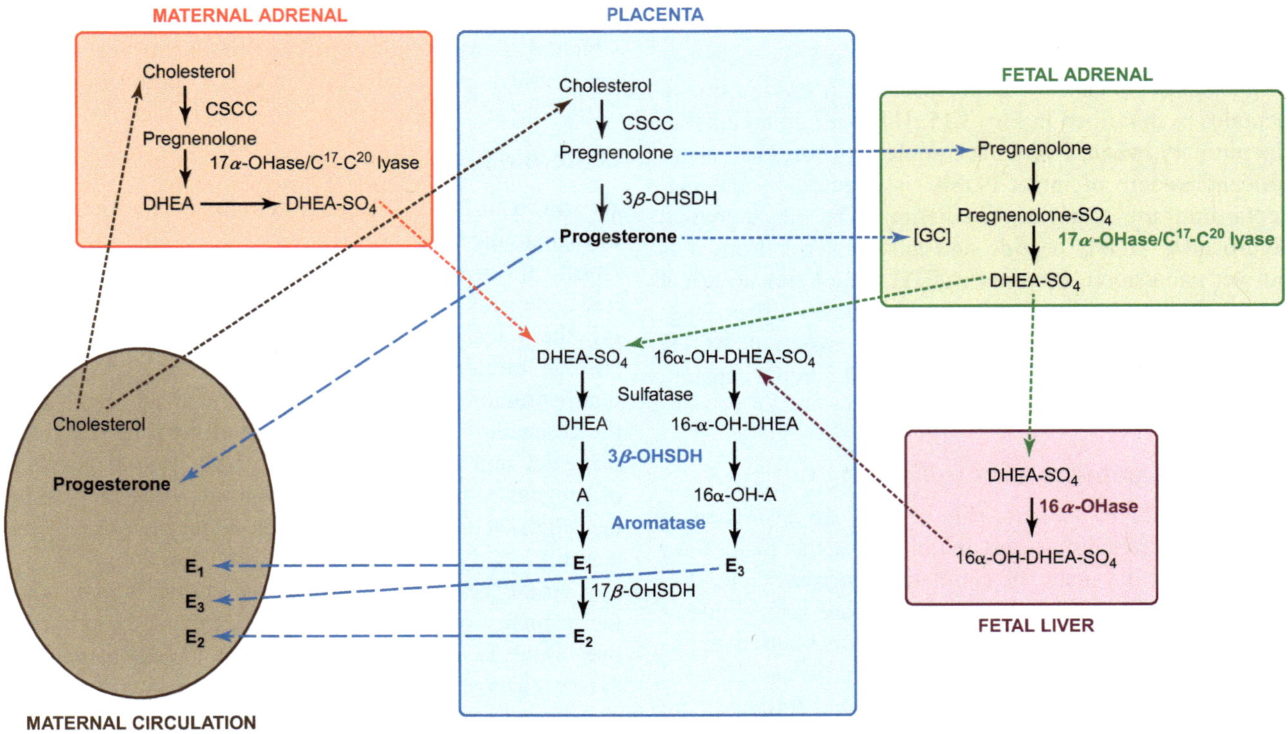

FIGURE 14.10 Production of estrogens and progesterone by the fetoplacental unit. See Figs. 2.20–2.23 for steroid synthetic pathways and structures. The steroid hormones required for the maintenance of pregnancy require the placenta for progesterone synthesis and both the fetal and placental compartments for estrogen synthesis. Cholesterol is the substrate for steroid hormone production and in the placenta (*blue box*) is converted, through the *CSCC* (P450scc), to pregnenolone. This is readily converted to progesterone which is secreted in large amounts into the maternal circulation (*brown oval*). Because the placenta lacks the ability to convert C-21 to C-19 steroids, pregnenolone is also transported to the fetal adrenal gland (*green box*), sulfated, and converted to DHEA-SO_4 by 17αOHase/C^{17}-C^{20} lyase (P450C17). DHEA-SO_4 serves as the substrate for both estradiol (E_2) and estrone (E_1) biosynthesis in the placenta which has abundant 3β-OHSDH (lacking in the fetal adrenal) and aromatase. Some fetal adrenal DHEA-SO_4 is transported to the liver to become further hydroxylated at the C-16 position, which is then aromatized in the placenta leading to estriol (E_3). The maternal adrenal makes a smaller contribution of DHEA-SO_4 to the placenta. Note that steroid intermediates are sulfated while in the fetal compartment; the sulfate groups are removed by sulfatases in the placental compartment (details not shown). A, androstenedione; 16α-OH-A,16α-OH-androstenedione; DHEA, dehydroepiandrosterone. *16α-OH-A*, 16α-OH-androstenedione; *A*, androstenedione; *CSCC*, cleavage of its side chain; *DHEA*, dehydroepiandrosterone.

is transported to the fetal adrenal for conversion to glucocorticoids. Some progesterone is also hydroxylated at either the C-16 or C-17 positions in the fetal compartment, so that circulating levels of these progesterone derivatives are also seen to increase during pregnancy.

In contrast to its ability to produce progesterone, the placenta lacks the ability to convert C-21 to C-19 steroids, the substrates for estrogen synthesis, due to its lack of the 17α-hydroxylase/C^{17}–C^{20} lyase (see Figs. 2.20–2.23 for structures and pathways). This enzymatic activity is supplied by the fetal adrenal, beginning at about 7 weeks of gestation. Thus pregnenolone is transported to the fetal adrenal, where it is conjugated with sulfate; most steroid hormones and their intermediates in the fetal compartment are present as the sulfates, probably to protect fetal tissues from exposure to active hormones. The two carbons are cleaved to form DHEA-SO_4, the C-19 androgen precursor for the estrogens. DHEA-SO_4 returns to the placenta where the sulfate is removed and the C-19 steroid converted in two steps to estrone and then to estradiol through the action of 17β-OHSDH. Both of these estrogens enter the maternal circulation from the placenta. Finally, back in the fetal compartment, some DHEA-SO_4 is hydroxylated at the 16α-position in the liver. The product is transported to the placenta where, like its unhydroxylated counterpart, it is converted to estriol, which is released into the maternal circulation.

E Maternal adaptations to pregnancy

In addition to and in concert with the changes in maternal fuel metabolism already discussed, virtually every aspect of maternal physiology is affected by pregnancy. Table 14.3 summarizes many of these. The fetus, placenta, and amniotic fluid account for just under 40% of the average weight gain during pregnancy, with the rest attributable to uterine and breast growth (11%), blood and extravascular fluid (23%), and maternal fat stores (26%).

TABLE 14.3 Changes in maternal physiology during pregnancy.

System	Change[a]
Physiology	
Weight	↑12.5 kg: fetus, placenta, amniotic fluid=4.8 kg
Blood volume	↑45%
Renal-GFR	↑60%
• Blood flow	↑(Increased ANP)
• Serum creatinine	↓(Increased clearance)
Cardiovascular output	↑30%–50%
• Heart rate	↑10–15 bpm
Pulmonary–respiratory reserve	↓40% (Elevated diaphragm)
GI-stomach emptying rate	↓50%
• Esophageal sphincter tone	↓(↑ GERD)[b]
Endocrinology	
Pituitary-lactotrophs	↑Lactotroph # and size
• Serum PRL	↑
• GH	↓[c]
• FSH, LH	↓
• TSH	↓In first trimester[d]
• ACTH	↑Fourfold
• Oxytocin	↑[e]
Thyroid-size	↑15%–20%
• Total T_4, T_3	↑
• Free T_4, T_3	→[f]
Adrenal gland–total cortisol	↑ Eightfold
• Free cortisol	↑ Fourfold
• Aldosterone	↑ 10-fold (↑ renin and angiotensin II)

ACTH, Adrenocorticotrophic hormone; *ANP*, atrial natriuretic protein; *FSH*, follicle stimulating hormone; *GFR*, growth factor receptor; *GH*, growth hormone; *LH*, luteinizing hormone; *PRL*, prolactin.
[a]*Values given are at term.*
[b]*GERD, Gastroesophageal reflux disease.*
[c]*Increase in placental GH keeps serum GH constant during pregnancy.*
[d]*CG, chorionic gonadotrophin has significant TSH activity and the relationship between it and TSH in early pregnancy is a reciprocal one.*
[e]*See Section V for more detail.*
[f]*Due to elevated liver synthesis of serum thyroxine-binding globulin; see Chapter 5, Thyroid Hormones.*

To maintain the needed rate of delivery of oxygen, nutrients, and other molecules to the fetus and the removal of CO_2 and waste from it, both the volume of blood and the rate at which it flows are dramatically increased; these changes are attributable to the high capacity and low resistance of the uteroplacental vessels (Fig. 14.4).

Pituitary size increases by 35%–40%, due primarily to the increase and size of the prolactin secreting lactotrophs. Changes in other pituitary hormones are seen as outlined in Table 14.3. Increased estrogen and progesterone levels are probably responsible for the suppression of the pituitary gonadotrophins.

In the thyroid gland, both the size of the follicles and the colloid content (see Chapter 5: Thyroid Hormones, Fig. 5.3) contribute to a modest increase in the gland's size. Although increased synthesis of the thyroid hormones leads to their increased total amounts in the blood, the amounts of free T_4 and T_3 are unchanged. This is due to the stimulation by estrogens of the liver synthesis of the thyroid hormone serum-binding protein, thyroid-binding globulin (Table 5.2). The rise in total blood cortisol levels is due to increased ACTH secretion but also in part to estrogen-induced increase in the hepatic synthesis of cortisol-binding globulin, the specific serum-binding protein for this steroid hormone. Free cortisol in the blood also rises, in parallel with the increase in ACTH.

IV Parturition

A Introduction

Parturition is the final step in the ultimate goal of the reproductive apparatus which has been discussed in this chapter as well as in Chapter 12, Androgens, and Chapter 13, Estrogens and Progestins. Fig. 14.11 summarizes the estimated outcome, in the United States, of 100 fertilization events that lead on average to 54 live births of which 1 neonate has a serious defect. These successes depend on the health of the embryo/fetus and the mother as well as a correctly timed and unhampered parturition process.

For all mammals the length of gestation is a precise interval having evolved in concert with the degree of maturity attained by birth, which varies considerably by species. Some newborns, such as rodents, are quite undeveloped at birth (referred to as altricial). Others are highly developed (precocial) species, for example, horses and whales, which are able to undertake at birth some adult behaviors such as locomotion and have fully operable sensory systems. While humans are precocial in some physiological respects, in terms of neurological and behavioral development, they are altricial and highly dependent on parental care for an extended period following birth. The important point is that in every viviparous species, gestation must end and parturition begin at the time that the fetus has achieved the appropriate level of maturation. The average time for delivery of the newborn human is 38 weeks following conception, although a

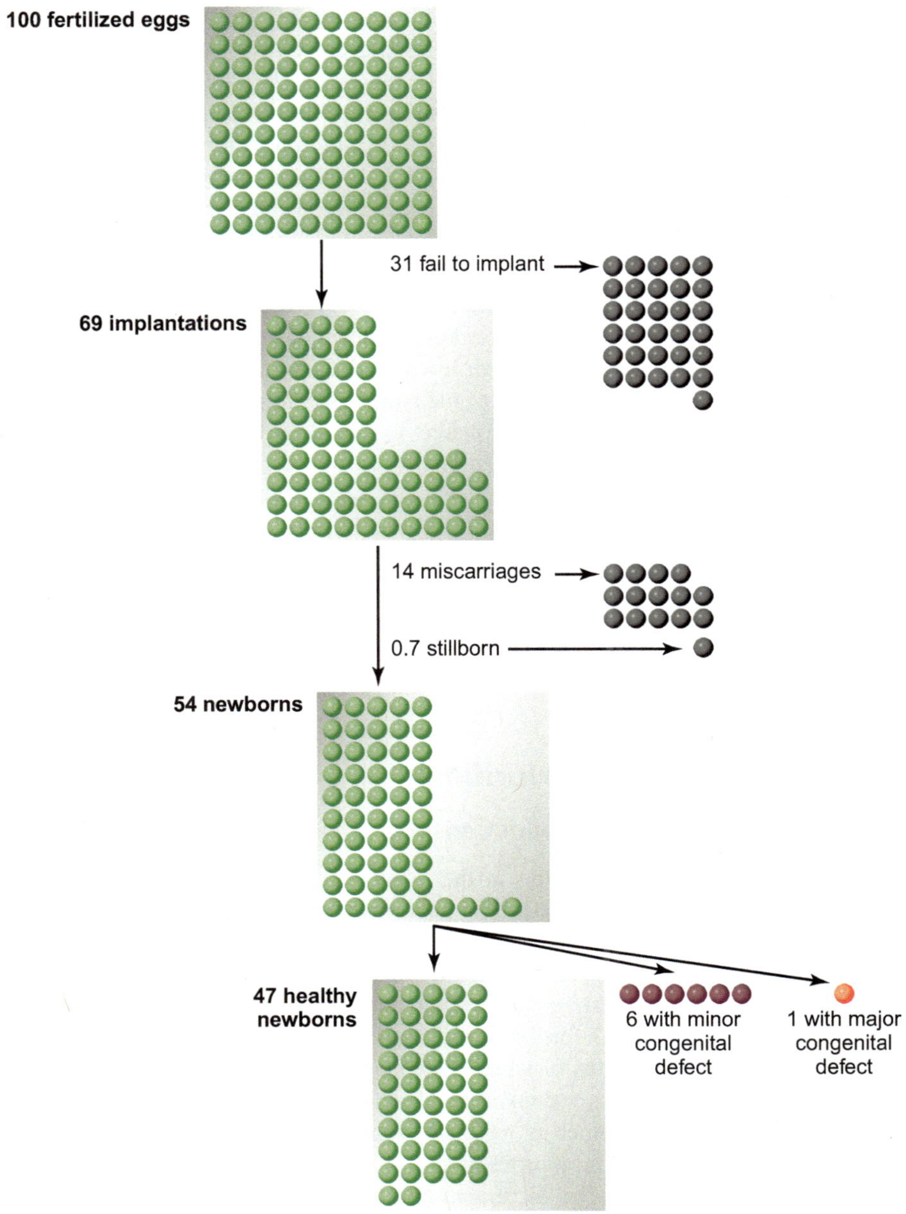

FIGURE 14.11 Outcomes for fertilized ova. The average losses between fertilization and parturition are shown at implantation (31%) and during pregnancy (21% of implantations). These losses are due to either embryonic/fetal problems such as chromosomal abnormalities, other genetic disorders, or to endocrine disorders such as maternal hormonal imbalance. Of the 54 newborns resulting from the 100 fertilizations, 47 have no congenital defects and only one is affected in a major way.

normal birth can occur within 2 weeks on either side of this time.

The signals that govern the onset of parturition have been studied in a variety of experimental animal models, including rodents, sheep, and nonhuman primates and, although in lesser detail, humans. Although variations exist in the strategies used to coordinate fetal maturation with parturition, three features appear to be shared: (1) parturition is initiated when the fetoplacental unit shifts from a progesterone-dominated environment to an estrogen-dominated one; (2) there is a link between fetal organ maturation and the timing of parturition; and (3) specific agents, including but not limited to oxytocin and prostaglandins (PG), are necessary for the strong myometrial contractions required for the expulsion of the fetus and, subsequently, that of the placenta.

B Progesterone and estrogen

As shown in Fig. 14.12 progesterone, directly and indirectly through the inhibition of the effects of estrogen, maintains pregnancy by inhibiting uterine contractility to maintain quiescence and by promoting cervical rigidity. In all mammalian species, these actions of progesterone ensure that the fetus will not be expelled from the uterus prematurely. At the onset of parturition, there is a shift from progesterone dominance to estrogen dominance in the fetoplacental unit. As a result, uterine contractions are

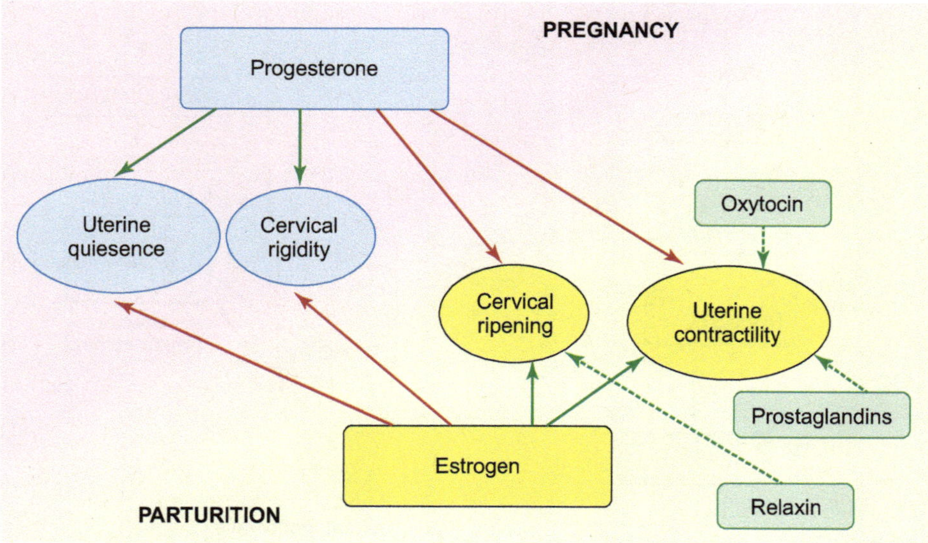

FIGURE 14.12 Progesterone and estrogen in pregnancy and parturition. Green arrows indicate stimulation and red arrows indicate the inhibition of the process. During gestation, progesterone maintains cervical rigidity and uterine quiescence by inhibiting the contractions of the myometrium of the uterus and inhibiting cervical ripening. These actions ensure the retention of the fetus in the uterus. Estrogen opposes the effects of progesterone and stimulates cervical dilation and uterine contractility. The initiation of parturition involves a shift from the progesterone-dominated environment to an estrogen-dominated one. Among other hormones involved in the process, ovarian relaxin contributes to cervical dilation, and oxytocin from the posterior pituitary as well as locally produced prostaglandin-stimulated uterine contractions.

stimulated and the processes required to soften and dilate the cervix are initiated. As labor progresses (see Fig. 14.13), oxytocin and PG increase uterine contractions; relaxin from the ovary promotes cervical softening; and estrogen continues to contribute to this process by antagonizing the effects of progesterone.

In many species, this progesterone to estrogen shift is accomplished by dramatic changes in the circulating concentrations of the two hormones, increasing the E/P ratio. In humans, such a change in circulating progesterone and estrogen is not seen. Rather, it is likely that the shift from progesterone to estrogen dominance, at least in the uterus, is due to changes in the responsiveness of the myometrial tissue to the two steroid hormones, that is, through their receptors. As described in Chapter 13, Estrogens and Progestins, the progesterone receptor exists in two forms, PR-A and PR-B, the latter of which is the most active regulator of gene transcription in progesterone's target tissues. PR-A can block the transcriptional activity of PR-B, so that an increase in the ratio of PR-A/PR-B would, effectively, diminish the activity of progesterone, bringing about functional progesterone withdrawal. Several studies support this hypothesis. In addition, there is evidence that the number of estrogen receptors (ERα), the predominant active form of the estrogen receptor in myometrium, is positively correlated with PR-A/PR-B ratios, suggesting that the same shift in the ratio that contributes to the decreased response to progesterone brings about the increased response to estrogen that is essential to parturition.

The softening of the cervix involves inflammatory infiltration and release of metalloproteinases to break down and reorganize its collagen and other extracellular matrix components. The peptide hormone relaxin from the ovary plays a role in this process along with the steroid hormones. The effects of progesterone and estrogen are, as indicated in Fig. 14.12, to maintain rigidity or promote softening, respectively, but the details of how the shift between the two hormones in the cervix takes place are not understood.

C The fetal adrenal, placenta, and posterior pituitary

The hormonal contributions of and interactions between the fetus and the placenta in parturition are shown in Fig. 14.13. Placental CRH (see Chapter 3: The Hypothalamus and Anterior Pituitary and Section III.C) plays a central role in these events. Maternal serum levels of CRH rise exponentially during the final weeks prior to parturition until they are three orders of magnitude above the nonpregnant level. This rise occurs earlier in preterm deliveries and later in those that occur significantly past the due date. CRH levels return to the nonpregnant state within 24 hours of the delivery of the placenta.

CRH increases uterine contractions through direct effects on the myometrial cells and through the increased production of PG (PGE and PGF; see Chapter 8: Eicosanoids), which play a crucial role in parturition in

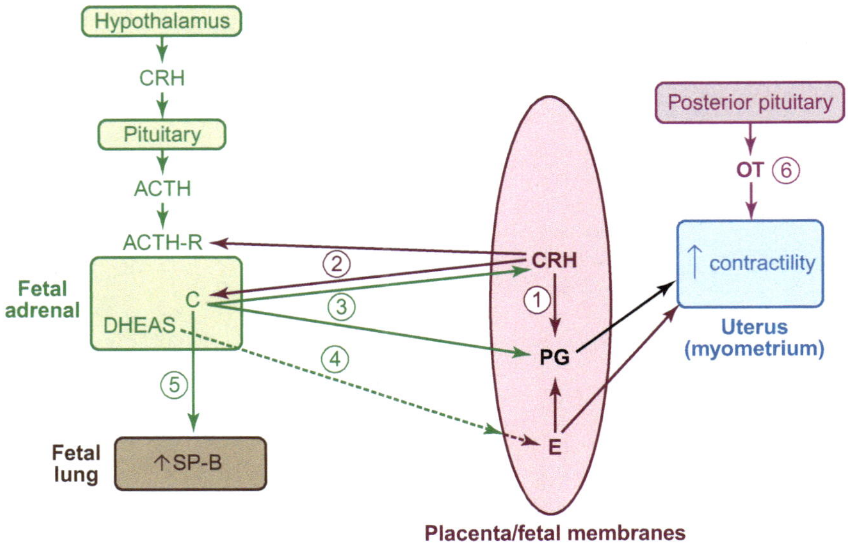

FIGURE 14.13 Hormonal interactions in parturition. Placental CRH plays a central role in parturition. (1) Placental (*pink oval*) CRH stimulates myometrial contractile activity through the increased production of PG, which are directly responsible for increasing uterine contractions (*black arrow*). (2) CRH also stimulates the fetal adrenal (*green box*) to increase steroidogenesis both directly and indirectly by increasing the ACTH receptors and thus the responsiveness of the fetal adrenal to the pituitary hormone (*magenta arrows*). (3) Increased cortisol from the fetal adrenal stimulates increased CRH production by the placenta in a positive feedback loop and also stimulates PG levels (*green arrows*). (4) Increased fetal adrenal steroidogenesis results in greater sulfated DHEAS leading to increased estrogen synthesis in the placenta (*dashed green arrow*). Estrogen stimulates contractions of the uterus both directly and by stimulation of PG production (*magenta arrows*). (5) Cortisol from the fetal adrenal is required for several aspects of the final development of the fetus, including lung (*brown box*) maturation through the increased synthesis of SP-B (*green arrows*). (6) As labor progresses, secretion of OT from the posterior pituitary increases and the myometrium, through an increase in OT receptor number, becomes more sensitive to the hormone, further strengthening and coordinating uterine contractions. *CRH*, Corticotrophin releasing hormone; *PG*, prostaglandins; *ACTH*, adrenocorticotrophic hormone; *DHEAS*, dehydroepiandrosterone sulfate; *SP-B*, surfactant protein B; *OT*, oxytocin.

the direct stimulation of uterine muscle contraction. PG are produced in large amounts in the amniotic portion of the fetal membranes, which has CRH receptors. PGE_2 interacts with the EP1 and EP3 receptors (see Chapter 8: Eicosanoids) in the myometrium to increase intracellular Ca^{2+}, decrease cyclic AMP, and stimulate contractions. Similarly, $PGF_{2}\alpha$ interacts with the FP receptor (see Chapter 8: Eicosanoids) to stimulate contractions through increased intracellular Ca^{2+}.

CRH increases the cortisol output of the fetal hypothalamic–pituitary–adrenal axis through a direct stimulatory effect on the steroidogenic enzymes that produce cortisol in the fetal adrenal zone and by increasing the number of ACTH receptors, increasing the sensitivity of these cells to fetal pituitary ACTH. Cortisol from the fetal adrenal stimulates CRH production by the placenta, forming a positive feedback signaling loop, propelling the parturition process forward. Fetal adrenal cortisol also stimulates myometrial prostaglandin production, thereby directly participating in increased uterine contractility.

An important link between fetal maturation and the onset of parturition is the effect of cortisol on the final steps in the maturation of several organ systems, including the lung. Pulmonary surfactant lines the alveoli of the lung and lowers the surface tension so that upon birth, the lung can make the transition to exposure to air rather than fluid. Failure of this step in lung development (or delivery prior to its achievement) leads to respiratory distress syndrome. One of the proteins that makes up the surfactant, which is 90% lipid, is surfactant protein B, synthesis of which in alveolar epithelial cells is dramatically increased by cortisol and other glucocorticoids. Thus the maturity of the fetal adrenal and its ability to secrete adequate amounts of cortisol for lung maturation occurs at the same time that it contributes to increasing uterine contractions through stimulation of CRH.

The increased steroidogenic activity of the fetal adrenal also results in increased DHEAS synthesis which, as detailed in Fig. 14.10, is converted by the placenta into estrogen. Estrogen, as already discussed, stimulates uterine contractility directly and through increased PG production in the fetal membranes; progesterone has the opposite effect. These actions are mediated in part by ERα stimulation of and PR-B inhibition of cyclooxygenase-2 (the enzyme responsible for prostaglandin synthesis—see Chapter 8: Eicosanoids—in the fetal membranes).

Finally, as the uterine contractions become stronger and more frequent, stretch receptors in the cervix signal the posterior pituitary to release oxytocin (Chapter 4: Posterior Pituitary Hormones) which further strengthens and coordinates uterine contractions. In humans, oxytocin does not initiate parturition; the uterus must become sensitized to its actions by estrogen action in stimulating oxytocin receptor numbers in late pregnancy and early parturition. Clinically, an analog of oxytocin, pitocin, is used to induce labor at term or strengthen contractions when labor appears to have stalled.

Note that in the complex hormonal interactions of hormones in parturition, there are no negative feedback loops. There is also redundancy in that several systems converge on increasing uterine contractions (and cervical dilation). These two features help ensure that the ultimate goal, the birth of a healthy offspring, will happen at the right time and is less likely to be compromised once it gets underway.

V Lactation

A Mammogenesis

The development of the breast tissue, primary function of which in reproduction is supplying nutrition for the newborn, is called mammogenesis and its major phases are shown in Fig. 14.14. In the embryo the ectoderm gives rise to the mammary buds in which a network of tubules is formed, which will eventually become the ducts. Estrogens secreted by the ovary at puberty increase ductal and overall mammary growth, including the accumulation of adipose tissue. Only rudimentary alveoli are present at this time and the breast will remain at this stage of development until pregnancy occurs. The high levels of progesterone, beginning in early pregnancy and lasting until term, bring about the differentiation of the alveoli into the mature structures in which milk synthesis takes place. During pregnancy, high levels of prolactin and other lactogenic hormones such as PL and PGH contribute to the growth of the breast tissue through cell proliferation and elaboration of the alveoli. Increased growth and branching of the duct system takes place, largely under the influence of estrogen. Until parturition, progesterone inhibits the effects of prolactin on milk production. During lactation the myometrial cells surrounding the epithelial cells of the alveoli are the site of action of oxytocin to stimulate milk secretion.

B Lactogenesis

Inhibition of milk synthesis by progesterone is relieved when the placenta is delivered and within hours, the human breast secretes colostrum, which has a high protein content (8%) and is enriched in antimicrobial peptides and immunoglobulins. Over the next few days, the composition and quantity of milk synthesis matures and by 1 week postpartum, the rate of milk production is normally 500 mL/day; after 3 weeks, milk production increases to 800–1000 mL/day. The top portion of Table 14.4 compares human and bovine milk in terms of the macronutrients and in the bottom half of the table some of the micronutrients of human milk are listed. In addition to minerals and vitamins, human milk contains immunoglobulins, antimicrobial peptides, growth factors, and small molecules that play regulatory as well as structural roles. In Table 14.4 the values of the micronutrients are averages and

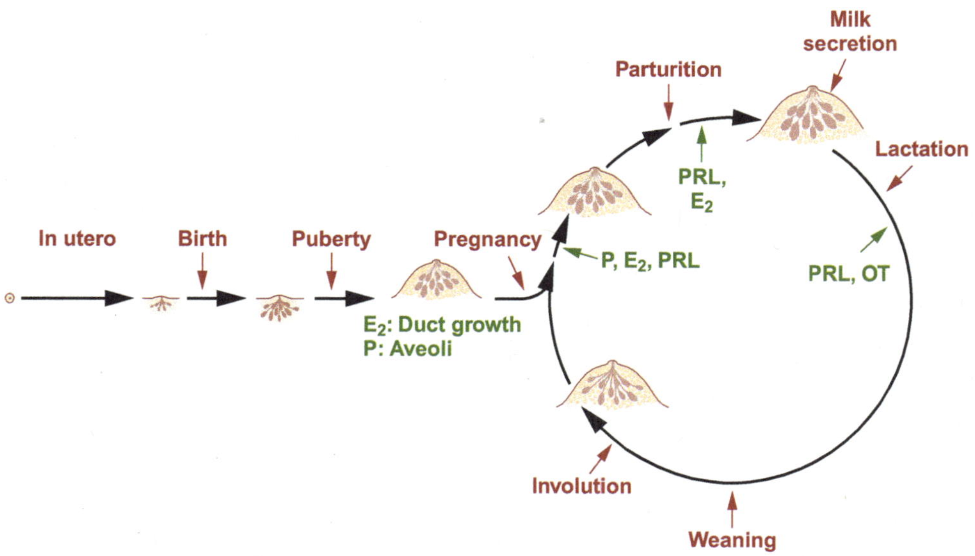

FIGURE 14.14 The development of the mammary gland. At birth, breast tissue consists of a rudimentary ductal system. Prior to puberty breast tissue growth parallels linear growth. At puberty increased ovarian steroid hormones bring about the further growth of the ducts by estrogen (E_2) and initial development of alveoli under the influence of P. During pregnancy and after parturition, PRL contributes to the effects of E_2 and P as the alveoli mature and express the enzymes in the pathways necessary for milk synthesis and secretion. During lactation, prolactin is required for continued milk synthesis and secretion and oxytocin stimulates milk ejection brought about by suckling. *P*, Progesterone; *PRL*, prolactin.

TABLE 14.4 Composition of milk.

Component	Human milk	Bovine milk
Macronutrients (g/100 mL)		
Water	88	87
Carbohydrate (lactose)	7	4.5
Total protein	1.2	3.3
Casein	0.4	2.8
Lactoglobulin	0.2	0.2
Lactoalbumin	0.3	0.4
Lipids (mostly triglycerides)[a]	3.8	3.7
Other components of human milk		
Minerals (mg/ 100 mL)	Vitamins	Proteins
	Water sol (μg)	
Potassium: 55	Ascorbic acid: 4000	Immunoglobulins
Calcium: 33	Thiamine: 17	Lysozyme
Phosphorus: 15	Riboflavin: 40	Lactoferrin
Sodium: 15	Nicotinic acid: 120	Cytokines
Magnesium: 4	Pantothenic acid: 200	IGF-1, EGF
Zinc: 0.53	Folic Acid: 1.3	
Iron: 0.15	Biotin: 0.4	*Other*
Copper: 0.04	Vitamin B_{12}: 0.03	Amino sugars
Iodine: 0.014	*Fat sol (IU)*	Gangliosides
	Vitamin A: 160	Nucleotides
	Vitamin D: 1.4	Oligosaccharides

EGF, Epidermal growth factor; *IGF*, insulin-like growth factor.
[a]*Cholesterol=6–20 mg/100 mL.*

the range for individual components may be quite large. It should be noted also that if a lactating mother is nutritionally deficient in one or more of these components, this will likely be reflected in her milk and, therefore, her infant's nutritional status.

Although the molecular details of milk synthesis in humans are not all fully understood, there has been progress in elucidating the pathways and their regulation of two of the major components, casein, the major milk protein, and lactose, the primary and milk-specific carbohydrate.

Casein is a mixture of at least three different subtypes of the protein. Each is a polypeptide chain of 150–200 amino acids containing distinct hydrophobic and hydrophilic regions. The proteins are synthesized on the rough endoplasmic reticulum and released into the Golgi complex. Casein molecules undergo posttranslational serine phosphorylation, which enables them to form stable micelles which include calcium ions. In the epithelial glandular cells of the breast, these micelles are transported from the Golgi complex to the lumen of the acini where they undergo exocytosis into the lumen. Transcription of the β-casein gene is stimulated by prolactin through the Jak2-Stat-5 signaling pathway. Glucocorticoids act synergistically with prolactin to stimulate casein gene expression, either by enhancing Stat-5 action at the β-casein gene promoter directly or by altering the nuclear retention of Stat-5.

Lactose, the disaccharide composed of one molecule of galactose and one of glucose in a β1–4 linkage, is the main carbohydrate in the milk of most mammalian species and is only found in milk. It is synthesized in the Golgi complex of the epithelial cells of the breast alveoli by the pathway shown in Fig. 14.15. The irreversible formation of lactose is catalyzed by the lactose synthesizing system, composed of a 1:1 complex of β4-galactosyl transferase and lactalbumin. This 14-kDa protein, which is only expressed in the mammary gland, is absolutely required for the enzyme to transfer the galactose moiety to glucose at physiological glucose concentrations. The expression of lactalbumin is elevated prior to parturition, whereas most of the other enzymes in the pathway shown in Fig. 14.15 increase in the first 4 days following the birth of the newborn. Newly synthesized lactose is unable to diffuse out of the Golgi complex so water enters to balance the osmotic pressure, and in this way lactose synthesis plays an important role in milk volume and production. Milk is secreted from the epithelial cells into the duct as secretory granules are formed from the Golgi membrane.

Progesterone inhibits and prolactin stimulates milk synthesis. These hormones undoubtedly act at many points in the various pathways required for the production of milk, but certainly the synthesis of casein and lactalbumin are two critical points of control. In addition many other hormones, including glucocorticoids, growth hormone, thyroxine, and insulin, are required for the maintenance of normal lactogenesis.

C Suckling

Suckling is necessary for the maintenance of lactation both because it stimulates pituitary release of hormones necessary for lactogenesis and lactation and because it empties the breast, a requirement for continued milk synthesis. The pathways of lactational control are shown in Fig. 14.16. At the end of parturition, prolactin levels are high due to the inhibition of dopamine secretion by estrogen and to the release of the lactotrophs from

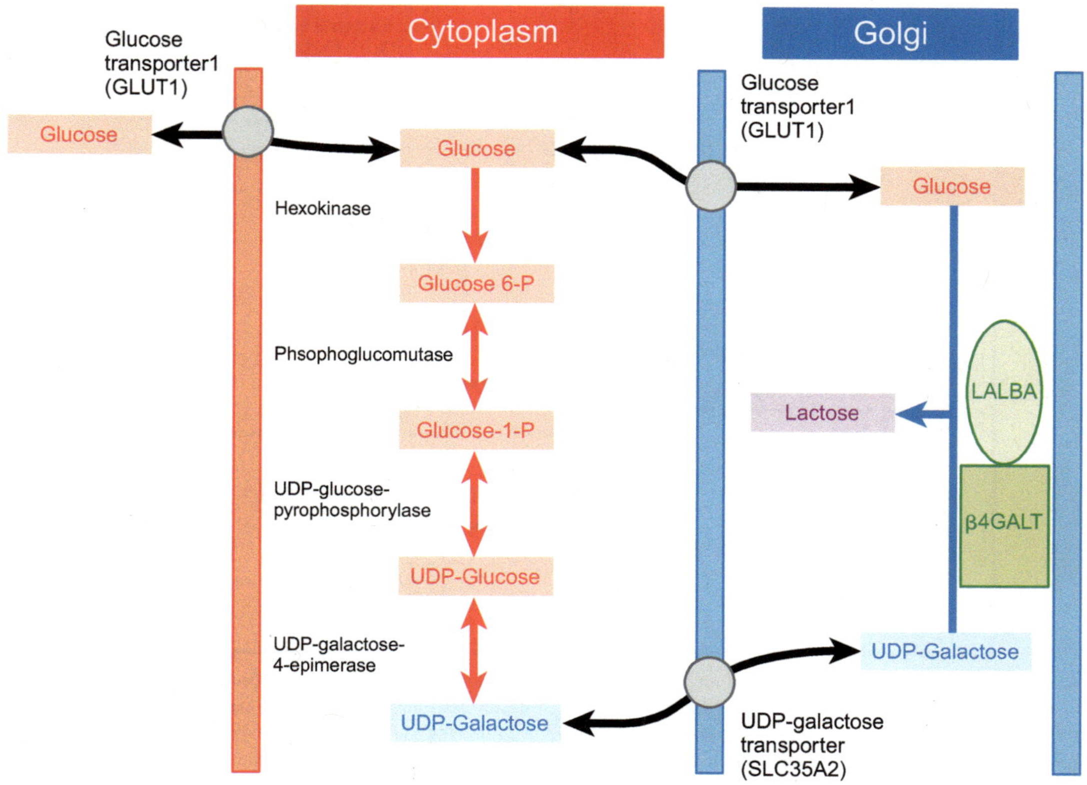

FIGURE 14.15 Lactose synthesis. The pathway of lactose synthesis in the epithelial cells of the breast alveoli is shown. GLUT1 brings glucose into the cell and also transfers it into the Golgi. In the cytoplasm, glucose is converted to uridine diphosphate (UDP)-galactose, a process requiring energy input. UDP-galactose is actively transported into the Golgi by a specific transporter, where the complex of LALBA and β4GALT catalyzes the formation of the disaccharide lactose. *GLUT1*, Glucose transporter 1; *LALBA*, lactalbumin; *β4GALT*, β4-galactosyl transferase.

progesterone inhibition of prolactin secretion. Prolactin stimulates milk synthesis in the glandular cells of the alveoli. Once breast feeding begins, the nerve endings in and around the nipple respond to suckling by sending neural signals directly to the neurons in the hypothalamus that control prolactin secretion as well as those that produce oxytocin. Through the former, milk synthesis is maintained at a level appropriate to the amount of feeding. At the same time, oxytocin stimulates myoepithelial cell contractions, causing milk ejection or let-down. Within several days after birth, a reflex pathway from the nipple through the spinal column to higher brain centers is established such that external signals such as the sight of the baby or the sound of a baby's cry will stimulate the release of oxytocin and milk ejection. Since oxytocin is also subject to input from higher brain centers, other external influences such as stress or anxiety can negatively affect milk secretion.

The high levels of prolactin during the first 6 months or so of lactation are sufficient to block GnRH secretion from the gonadotrophs in the anterior pituitary. As a consequence, ovarian function is suppressed, leading to lactational amenorrhea. Basal prolactin levels fall gradually during the 6 months following birth and normal ovarian function resumes. If maternal nutrition is inadequate or the mother has a low body mass index, suppression of ovarian function during lactation tends to be longer. Thus lactation tends to increase the interval between pregnancies, but due to the wide range of individual variation in the completeness and length of the GnRH blockage and the length of it, the continuance of lactation is an unreliable method of contraception.

VI Clinical aspects

A Assisted reproductive technology

The term *ART* covers a variety of medical approaches to infertility. These include in vitro fertilization (IVF), cryopreservation of gametes and embryos, embryo transfer (ET), gamete intrafallopian tube transfer, zygote intrafallopian tube transfer, and intracytoplasmic sperm injection. In this section, attention will focus on the combination of IVF-ET, which allows thousands of births each year to otherwise infertile couples. The first successful IVF-ET started with the transfer of an eight-cell embryo produced by in vitro fertilization into the uterus of the mother and resulted in the birth of a healthy child in 1978. In the ensuing years, as

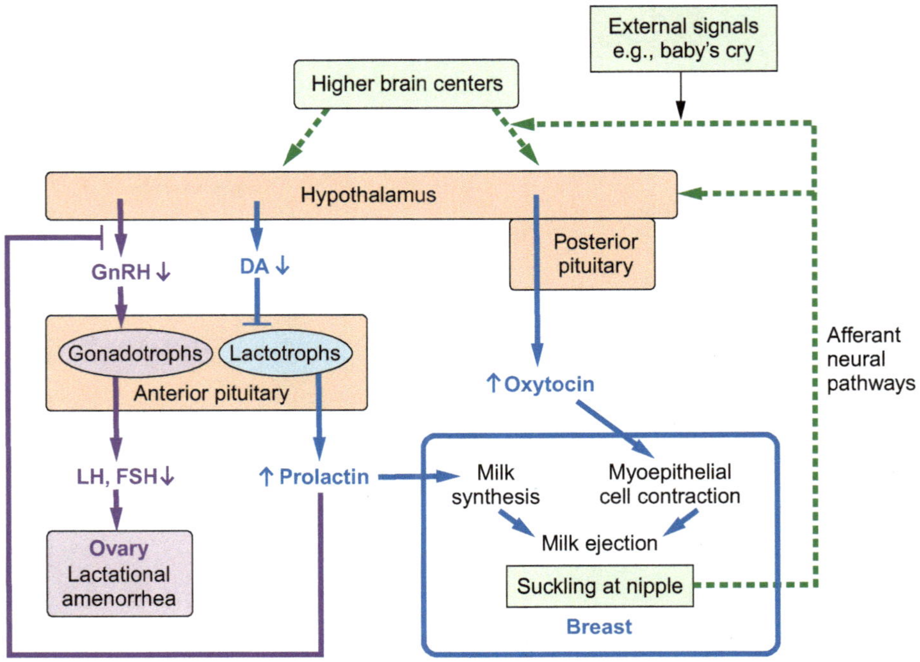

FIGURE 14.16 Neuroendocrine regulation of lactation. Late in pregnancy and following parturition, the decreased hypothalamic secretion of dopamine brings about an increase in PRL secretion by pituitary lactotrophs. Prolactin is required for the development and maintenance of milk synthesis in the epithelial cells of the alveoli (*blue* pathway from hypothalamus, left). Following parturition, OT, synthesized in the magnocellular neurons of the hypothalamus and secreted from the posterior pituitary, stimulates the contraction of the myoepithelial cells in the ducts, leading to milk ejection (*blue* pathway from hypothalamus, right). Suckling, through nerves in and around the nipple, stimulates the hypothalamic neurons that produce oxytocin and inhibits the ones that secrete the DA, thereby stimulating PRL secretion (*green pathways*). In humans a neuroendocrine reflex is soon established such that external signals related to suckling bring about oxytocin release and milk ejection in the absence of the suckling infant. During the first few months of lactation, elevated levels of prolactin inhibit GnRH secretion, thereby suppressing ovarian function during this period (*purple* pathway from hypothalamus). *DA*, Dopamine; *GnRH*, gonadotrophin-releasing hormone; *OT*, oxytocin; *PRL*, prolactin.

TABLE 14.5 Outcomes of in vitro fertilization (IVF) cycles, 2010.[a]

Stage	Number	Percent
Cycles begun[b,c]	100,824	100.0
Egg retrievals	90,241	89.5
Embryo transfers	82,624	81.9
Pregnancies	37,191	36.9
Live-birth deliveries[d]	30,425	30.2

[a]*From the CDC ART 2010 National Summary Report: http://www.cdc.gov/art/ART2010/section2a.htm#f6.*
[b]*Data are for cycles using fresh nondonor eggs or embryos.*
[c]*Some women begin more than one cycle, so this does not represent the number of patients undergoing IVF.*
[d]*Due to multiple births in some deliveries, the number of newborns is higher than the number of deliveries.*

every aspect of this process has been studied and refined, a greater understanding of the cellular processes involved in it has been achieved. Today in the United States approximately 1% of live births result from ART.

The Center for Disease Control has kept detailed statistics on many aspects of ART, examples of which are shown in Table 14.5 and Fig. 14.17. Table 14.5 shows the outcome in 2010 of ART cycles started using fresh nondonor eggs or embryos. Many women undergo more than one cycle, so the 100% number does not reflect the number of patients undergoing the procedure. Many factors influence the prognosis for a successful outcome of IVF-ET. Foremost among these is ovum vitality, which declines with the age of the ovum donor. As shown in Fig. 14.17, maternal age has little effect on outcome until early in the fourth decade. At age 33, both the number of pregnancies and the number of live births begin to fall and by age 40, the chances for IVF-ET success are reduced to 53% and 40%, respectively, of what they were at age 24 for pregnancy and live birth.

About half of the patients using IVF-ET have one of the following widely accepted indications for the procedure, for which the procedure has the most successful outcomes: poor Fallopian tube function, endometriosis, mild-to-moderate problem of the male partner, and unexplained infertility. Other reasons, for which the success of the treatment is more variable, include polycystic ovary syndrome (see Chapter 13: Estrogens and Progestins), serious problem of the male partner, and uterine dysfunction. Overall, however, in the absence of multiple causes,

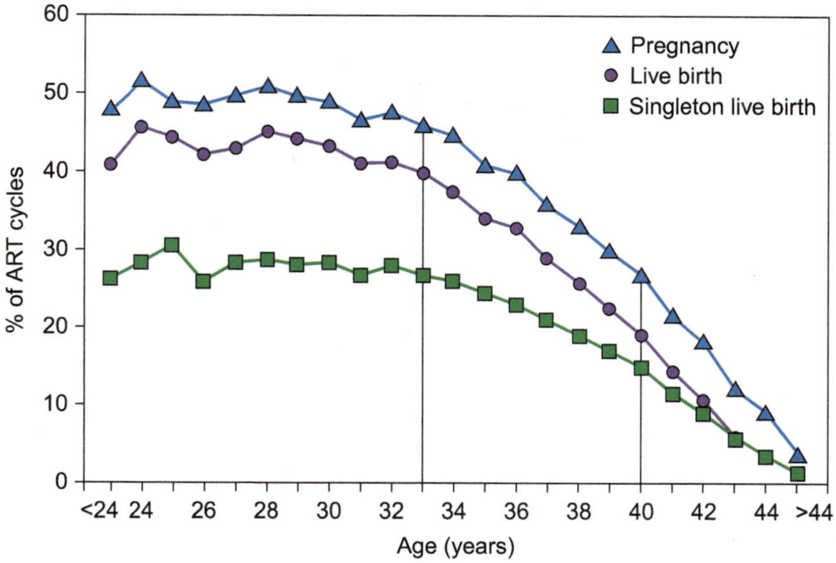

FIGURE 14.17 Effect of age on IVF success rates. The percentages of IVF cycles resulting in pregnancy (*blue*), a live delivery (*purple*) which may be a multiple birth and a singleton live birth (*green*) are shown. The vertical lines show that the age-related decline in success begins at about 33 years of age (vertical line) and at age of 40 the success rate by all three measures is 40%–54% of what it was at age 30. *IVF*, In vitro fertilization. *From the 2010 CDC Assisted Reproductive Technology (ART) Report, http://www.cdc.gov/art/ART2010/sect2_fig6-15.htm#14.*

the single most important determinant of success is ovum age and ovarian reserve.

The many IVF-ET protocols used in the approximately 450 fertility clinics in the United States differ from one another in major ways and in more minor details. However, all protocols share common major steps. In the first phase of hormonal manipulation, controlled ovarian hyperstimulation, the aims are to suppress the natural hormonal cycle, overstimulate the ovary to induce the development of multiple mature follicles, and synchronize the final maturation of the follicles. In the second phase of hormonal treatment, the uterus is prepared for implantation and maintenance of the pregnancy for the first 8–10 weeks of gestation. Following retrieval of the ova prior to ovulation, in vitro fertilization of the ova, and cell division of the zygote, the embryo is transferred to the uterus and maintenance of the pregnancy for the first 8–10 weeks of gestation is provided.

The hormonal treatments given to meet these objectives are depicted in Fig. 14.18. Three typical protocols are shown and in practice there are many variations on these. To interrupt the natural hormonal cycle (Fig. 13.11) either a gonadotrophin agonist (Fig. 14.18A and B) or antagonist (Fig. 14.18C) is given. High continuous levels of GnRH inactivate the normal hypothalamic–pituitary axis, which depends on the pulsatile secretion of the releasing hormone. The "long" version of this protocol, in which GnRH is given during the preceding cycle, also empties the pituitary of endogenous gonadotrophins so that the levels of these hormones can be controlled completely by their exogenous administration. Blocking GnRH secretion or action prevents premature ovulation, which would lead to the cancellation of the current cycle. In the "short" protocol (Fig. 14.18B), the GnRH agonist is given on day 3 of the treatment cycle, so that endogenous gonadotrophins are added to those being given exogenously. GnRH antagonists (Fig. 14.18C) are given somewhat later in the follicular phase of the cycle to block the LH surge at the level of the GnRH receptor in the gonadotrophs and, therefore, prevent premature ovulation. Each approach has advantages and drawbacks and the choice among them depends on many factors such as the reasons behind the infertility, coexisting conditions in the patient, cost, and individual clinic practices.

The structures of agonists and antagonists of GnRH are shown in Fig. 14.19. For the agonists, substitution of a D-amino acid at position 6 increases the half-life of the peptide and derivatization at the C-terminal increases biological activity by one or two orders of magnitude. GnRH antagonists have at least five-amino-acid substitutions, including three at the N-terminus where the GnRH receptor activation function lies. As with the length of treatment, the choice between GnRH analogs depends on a number of considerations, including speed of reversibility, cost, and possible side effects.

Supraphysiological levels of gonadotrophins are given during the follicular phase of the treatment cycle to override the normal hormonal interactions limiting ovulation to a single ovum. Ultrasound monitoring, measurement of blood hormone levels, and individualized dosing help minimize untimely harvesting of pre- or postmature ova. As discussed in Chapter 13, Estrogens and Progestins, levels of both LH and FSH rise in postmenopausal women as estrogen levels fall and cease to inhibit the pituitary output of these hormones (see Fig. 13.13). Human menopause gonadotrophins refers to the mixture of the two gonadotrophins purified from the urine of postmenopausal women and is a common source of the

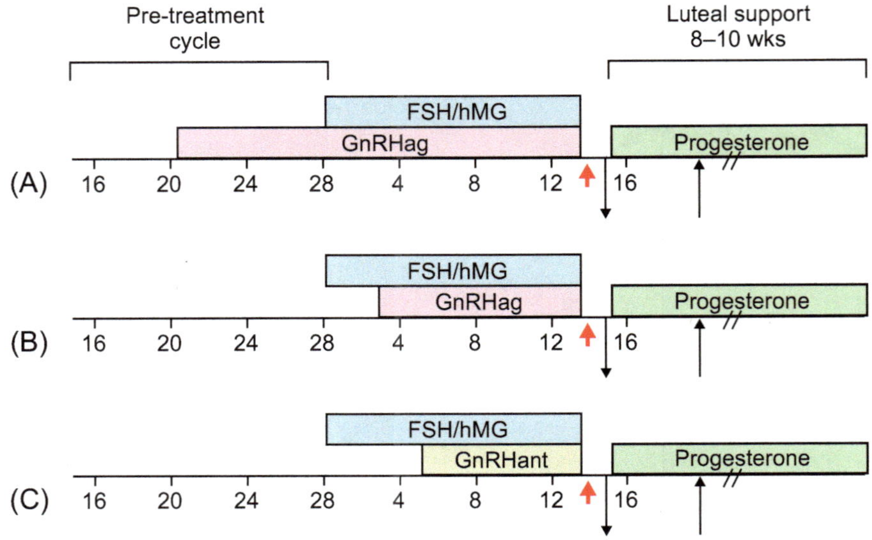

FIGURE 14.18 Typical IVF protocols. Three typical IVF protocols are shown. Although all have the same goals, there are variations in how the hormonal milieu is manipulated. Across the bottom of each protocol is shown the cycle day. Hormonal treatments are indicated by the bars and the red arrow. Ova retrieval and embryo transfer are depicted by the downward and upward black arrows, respectively. (A) In this protocol (termed a GnRH agonist long protocol), a GnRH agonist (*pink bar*) is begun during the cycle prior to treatment and continued during the follicular phase of the treatment cycle to suppress the natural hormonal cycle. Gonadotrophins (*blue bar*) are added at the beginning of the treatment cycle to stimulate the growth of several follicles to maturity. Release of the ova is stimulated by a single large dose of hCG (*red arrow*) and eggs are retrieved 36 h later (↓). Embryo transfer takes place three (eight cells) to 6 (blastocyst) days later (↑). Luteal support (*green bar*) refers to the administration of progesterone to support the growth and maintenance of the endometrium. (B) This protocol (termed the GnRH agonist short protocol) differs from (A) in that the GnRH agonist is started 3 days into the treatment cycle. (C) In this protocol a GnRH antagonist (*tan bar*), rather than an agonist, is used to block the natural hormonal cycling. The structures of GnRH agonists and antagonists are shown in Fig. 14.19. Gonadotrophins used are rhFSH or FSH purified from urine or human pituitary glands and/or hMG, a combination of LH and FSH isolated from the urine of postmenopausal women. *FSH*, Follicle stimulating hormone; *GnRH*, gonadotrophin-releasing hormone; *hCG*, human chorionic gonadotrophin; *hMG*, human menopausal gonadotrophin; *IVF*, in vitro fertilization; *LH*, luteinizing hormone; *rhFSH*, recombinant human FSH.

FIGURE 14.19 GnRH agonists and antagonists. The amino acid sequence of the decapeptide GnRH is shown along with two agonists and two antagonists that are used in IVF treatment cycles. D, dextro; DCit, D-citrulline; DhArg, D-homoarginine; D-Nal, D-naphthylalanine; DPal, D-pyridylalanine; hArg-homoarginine; pGlu, pyroglutamate. *D*, Dextro; *DCit*, D-citrulline; *DhArg*, D-homoarginine; *D-Nal*, D-naphthylalanine; *DPal*, D-pyridylalanine; *GnRH*, gonadotrophin-releasing hormone; *hArg*, homoarginine; *IVF*, in vitro fertilization; *pGlu*, pyroglutamate.

hormones for IVF treatment. Each preparation is standardized to an equal amount (biological activity) of each gonadotrophin, with addition of hCG to raise LH activity if necessary. Thus the actual amount of gonadotrophin given is standardized across protocols and patients. In other protocols, FSH alone is used. It is either highly purified from urine (uFSH) or produced by recombinant technology (rFSH). Patient history and dosing convenience are among the considerations used to determine the gonadotrophin regimen to be used.

Toward the end of the follicular phase of the treatment cycle, when tests have shown a desirable number and maturity of ova, a bolus of hCG is given to synchronize the follicles and mimic the LH surge prior to harvest. The ova are collected by aspiration from the ovary, incubated with sperm and the fertilized eggs allowed to undergo division to either the eight-cell stage or to the blastocyst (Figs. 14.1 and 14.3) stage. If necessary, the embryos undergo cryopreservation until the time (normally 3–5 days following retrieval) for transfer to the uterus. Beginning at the time of egg retrieval through the next 8–10 weeks, progesterone is given, at first to prepare the uterus for implantation, and then to maintain the pregnancy as the corpus luteum stimulated by hCG would have done in an unassisted pregnancy.

This discussion has been based on procedures when nondonor ova are being used for ART. In those cases in which a woman is unable to produce her own eggs and must rely on those of a donor, hormonal preparation of both the donor and recipient is required so that there are sufficient eggs from the donor and a prepared uterus in the recipient. Other variations include cryopreservation of unused embryos for later use; embryos produced from both donor eggs and donor sperm and, therefore, genetically unrelated to either of the prospective parents; and implantation of an in vitro generated embryo into the uterus of a surrogate for gestation in cases where the woman is unable to carry a pregnancy to term.

B Breast cancer

Breast cancer is the most prevalent malignancy in developed countries. Approximately one of nine women can expect to develop invasive breast cancer during her lifetime. About 20% of cases are diagnosed before and 80% after the age of 50. Death rates from breast cancer in the United States have been declining since 1990, particularly in women under 50, due in part to advances in treatment and in part to early detection and awareness.

1 Risk factors

Some of the risk factors for breast cancer are listed in Table 14.6. It is well known that there is a familial component to the disease and this accounts for approximately 15% of cases. For example, if a first-degree relative of a woman has been diagnosed with breast cancer, her risk for developing it is increased two- to threefold. A second relative with the diagnosis increases the risk still further. Some of this genetic risk (estimated at approximately one half) is attributable to the three genes listed in Table 14.6.

TABLE 14.6 Some risk factors for breast cancer.

Genetic risk factors[a]
BRCA1
BRCA2
TP53
PTEN
Reproductive risk factors
Early menarche
Late menopause
No pregnancy
No breastfeeding
Other risk factors
Obesity (postmenopausal)
Alcohol use

[a]BRCA, *Breast cancer susceptibility gene;* BRCA1 *and* BRCA2 *encode DNA repair enzymes;* TP53, PTEN *encode the proteins p53 and PTEN, respectively; both of these proteins are involved in cell cycle regulation.*

BRCA1 and *BRCA2* (breast cancer susceptibility gene) are genes found on chromosomes 17 and 13, respectively, that encode two different DNA repair enzymes. DNA is subject to constant damage from such sources as radiation, chemicals, and inherent instabilities in the nucleotides themselves. A mammalian cell may undergo as many as 1000 mutations due to DNA damage in a 24-hour period. The vast majority of these are repaired by one of the many DNA repair enzyme systems found in cells. If these repairs are not made, mutations accumulate. Since some of these mutations will affect genes involved in the control of cell proliferation, the outcome of their accumulation can be the uncontrolled proliferation of the cell, or cancer. The *BRCA1* and *BRCA2* genes, for reasons that are not fully understood, show tissue specificity in that they are linked to breast and, particularly in the case of *BRCA1*, ovarian cancer. The lifetime risk for breast or ovarian cancer for people (men and women) having an inactivating mutation in *BRCA1* or *BRCA2* is increased about fivefold over those not carrying such a mutation. This estimate may be somewhat high since many of the studies are carried out in large families in which other genetic similarities also occur. Other genes have also been associated with increased risk for breast cancer (see examples in Table 14.6), but these two make the largest known contribution (about half to two-thirds) of the genetic risk that can be attributed to specific genes.

Consideration of the reproductive risks listed in Table 14.6 suggest that the length and extent of exposure to an estrogen-rich environment is an important contributor to the risk for breast cancer. For example, greater incidence is seen in women whose reproductive cycles began early or ended late. As described later, these extensive observations have led to the hormonal therapy that has been a mainstay for a subset of breast cancer patients.

Other risk factors that have been studied in detail are obesity, particularly after menopause. Adipose cells contain aromatase and convert adrenal androgens to estrogen in the absence of ovarian activity, thus extending the length of estrogen exposure of these individuals. Alcohol use, while protective of cardiovascular tissues, appears to contribute to the risk of breast cancer, possibly by decreasing the rate of inactivation of active estrogens.

The use of oral contraceptives does not appear to be a significant factor (either positive or negative) in the risk of breast cancer. The possible role of postmenopausal hormone replacement therapy in breast cancer risk is complex. Long-term (more than 5 years) use of estradiol/progestin or estradiol alone is associated with a small-to-moderate increased risk of breast cancer. Conjugated equine estrogens (CEE), however, especially when taken for fewer than 5 years, do not appear to increase the risk. The complex mixture of the CEE, containing sulfated estrone, 17α-estradiol, and $\Delta^{8,9}$-dehydroestrone, may contain compounds with both agonist and antagonist activities, offering an explanation for the difference between this estrogen source and estradiol alone.

2 Treatment

The goals of the treatment of any solid tumor type of cancer are to remove or at least control the primary tumor site; minimize the possibility or effects of metastasis; and maintain, as much as possible, the patient's quality of life. With regard to the primary site, surgery (either mastectomy or lumpectomy) followed by radiation is the primary approach. To reduce the risk of or treat metastatic disease, chemotherapy, hormonal therapy, and, more recently, immunotherapy can be used. Whether and when to use chemotherapy depends on many factors, including the stage of the cancer, the age of the patient, and the patient's medical history. The details of the chemicals used and the course of chemotherapy treatment are variable and beyond the scope of this chapter.

Since it is clear that an estrogenic environment contributes to the development of breast cancer, blocking estrogen action is an obvious way to combat it. Therefore it is not surprising that attention has been given to both estrogen antagonists and aromatase inhibitors as agents to decrease actions of estrogens on breast cancer cells. Not all breast tumors, however, contain the ERα. In fact, approximately 30% are ER negative and these patients do not, in general, receive hormone therapy.

For the cancers that are ER positive (60%–70%), the main form of treatment is with tamoxifen (Fig. 13.7), which acts as an estrogen antagonist in breast tissue (but is an agonist in liver, bone, and endometrium). Tamoxifen decreases the rate of growth of human breast cancer cells in vitro. Given over 5 years, it reduces the incidence of recurrence of breast cancer. Although there are no life-threatening side effects, tamoxifen does increase the relative risk of endometrial, colon, and stomach cancers as well as blood clots. It also can lead to unpleasant perimenopausal symptoms such as hot flashes and menstrual irregularities. Another selective ER modulators, raloxifene (Fig. 13.3), developed for the prevention and treatment of postmenopausal osteoporosis, was found to decrease the incidence of breast cancer in women in this age group. In a direct trial of tamoxifen and raloxifene in postmenopausal women, the latter was slightly less effective in preventing breast cancer. However, because the agonist profile of raloxifene differs from that of tamoxifen, it does not lead to the increase in uterine cancer or some of the other side effects seen with tamoxifen. Postmenopausal women who are at increased risk for breast cancer may undergo long-term treatment with either of these drugs to reduce the risk of breast cancer with the choice varying with the individual patient.

Recently it has become clear that some (about 30% of cases) breast cancer cells also contain ERβ. In fact, it appears that ERβ may, in contrast to ERα, be antiproliferative in breast cancer cells in vitro and that, in humans, the presence of ERβ is correlated with a good prognosis. If screening of breast cancer patients for ERβ becomes routine as for ERα, a clearer picture of its role in breast cancer will undoubtedly emerge.

Another approach to reducing the estrogen exposure of breast tissue is to reduce its endogenous synthesis by inhibiting aromatase. The structures of two clinically important nonsteroidal aromatase inhibitors are shown in Fig. 14.20. Within the past decade, these compounds have become a first line of treatment for adjuvant therapy, that is, accompanying chemotherapy or radiation following surgery. Both anastrozole and letrozole have been shown to be as effective at reducing the recurrence of disease in the same breast as well as the subsequent development of cancer in the contralateral breast. In addition, both result in fewer of the side effects of tamoxifen. The one disadvantage of the aromatase inhibitors is that they increase the number of osteoporotic fractures. Thus measures, such as bone density measurements and, if warranted, supplemental calcium, vitamin D, and bisphosphonates, are given when aromatase inhibitors are used in breast cancer treatment.

In addition to inactivating mutations in tumor suppressor genes such as *BRCA1* and *BRCA2*, the loss of control of cell proliferation can also be attributed to the overexpression of proto-oncogenes. These are genes that are required for cell growth and division in normal cells but when overexpressed cause uncontrolled proliferation to

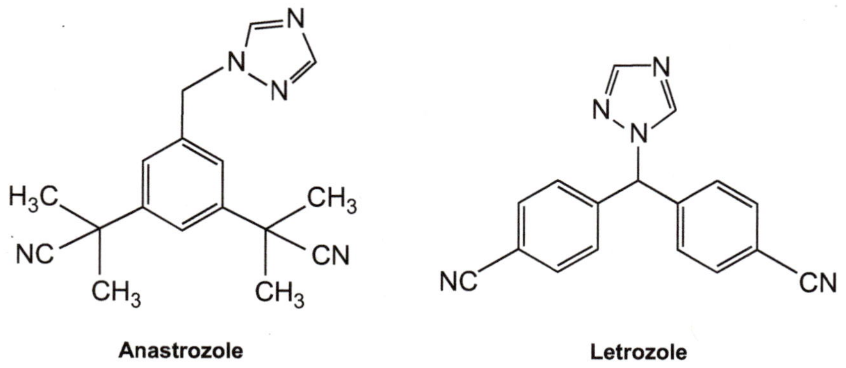

FIGURE 14.20 Aromatase inhibitors. The structures of two nonsteroidal aromatase inhibitors that are used clinically for adjuvant breast cancer treatment are shown.

occur. An example in the case of breast cancer (and several other solid tumors such as those found in ovary, pancreas, stomach, and bladder) is the gene for HER2, a tyrosine kinase membrane receptor which is a member (numbered 2) of the EGFR (epidermal growth factor receptor, see Chapter 17: Growth Factors) receptor family. There is no known natural ligand for HER2 but it is thought to mediate its complex actions on cell growth, survival, and differentiation through heterodimerization with other members of the EGFR family through which ligand binding and activation occurs. All cells contain two copies of the *HER2* gene on chromosome 17 and express small amounts of the HER2 receptor. In some cells the number of *HER2* genes is amplified 10- to 100-fold, leading to greatly increased expression of the HER2 receptor and ultimately to tumorigenesis and metastasis.

The recognition of the role of HER2 in the unrestrained proliferation of breast cancer cells led to the development of a monoclonal antibody against the receptor which blocks its ability to dimerize with other receptors of the EGFR family. Breast cancer patients are screened for the presence of increased HER2 expression and if positive (about 25%–30%) treatment with this antibody, trastuzumab (Herceptin), is an effective adjunct treatment with chemotherapy (before or after surgery) to slow the progression of disease. The choice to use it depends on whether the tumor is HER2 positive and what chemotherapeutic agents are being used, as the drug shows some cardiotoxic effects in the presence of certain anticancer chemicals. The molecular mechanism of action of trastuzumab is not fully understood but may include, among other actions, increased degradation of the receptor.

Further reading

Fertilization

S.J. Gunderson, et al., Machine-learning algorithm incorporating capacitated sperm intracellular pH predicts conventional in vitro fertilization success in normospermic patients, Fertility and Sterility 115 (2021) 930–939.

G.A. Sayed, et al., Mitochondrial DNA in fresh vs frozen embryo culture media of polycystic ovarian syndrome patients undergoing in vitro fertilization: a possible predictive marker of a successful pregnancy, Pharmgenomics and Personalized Medicine 14 (2021) 27–38.

L. Turnova, et al., Ligands and receptors involved in the sperm-zona pellucida interactions in mammals, Cells 10 (2021) 133. Available from: https://doi.org/10.3390/cells10010133.

Implantation

S. Cai, et al., Impact of vitamin D on embryo implantation—a prospective cohort study in women undergoing fresh embryo transfer, Fertility and Sterility 115 (2021) 655–664.

M. Marinic, V.J. Lynch, Evolutionary transcriptomics indicates HAND2 in the origins of implantation and regulation of gestation length, eLife 10 (2021) e61257. Available from: https://doi.org/10.7554/eLife.61257.

A. Mishra, N. Ashary, R. Sharma, D. Modi, Extracellular vesicles in embryo implantation and disorders of the endometrium, America Journal of Reproductive Immunology 85 (2021) e13360. Available from: https://doi.org/10.1111/aji.13360.

Human chorionic gonadotropin

J. Fink, et al., Human chorionic gonadotropin treatment: a viable option for management of secondary hypogonadism and male infertility, Expert Review of Endocrinology and Metabolism 16 (2021). Available from: https://doi.org/10.1080/17446651.2021.1863783.

C. Lazzaretti, et al., Identification of key receptor residues discriminating human chorionic gonadotropin (hCG)- and luteinizing hormone (LH)-specific signaling, International Journal of Molecular Sciences 22 (2021). Available from: https://doi.org/10.3390/ijms22010151.

K. Mofomura, et al., Human chorionic gonadotropin modulates the transcriptome of the myometrium and cervix in late gestation, Reproductive Sciences 28 (2021). Available from: https://doi.org/10.1007/s43032-020-00454-6.

Parturition

O. Guzeloglu-Kayisli, et al., Decidual cell FKBP51-progesterone receptor binding mediates maternal stress-induced preterm birth, Proceedings of the National Academy of Sciences of the United States of America 118 (2021). Available from: https://doi.org/10.1073/pnas.2010282118. e2010282118.

N. Khader, V.M. Shchuka, O. Shylova, J.A. Mitchell, Transcriptional control of parturition: insights from gene regulation studies in the myometrium, Molecular Human Reproduction 27 (2021). Available from: https://doi.org/10.1093/molehr/gaab024.

J.K.H. Li, M.R. Johnson, Transcription factors regulated by cAMP in smooth muscle of the myometrium at human parturition, Biochemical Society Transactions 49 (2021) 997–1011.

Chapter 15

Hormones related to the kidney and cardiovascular system

I Introduction

A Background

The kidney, heart, and circulatory system together play an indispensable role in the maintenance of life in higher organisms. This includes maintaining the constancy of many components of the extracellular fluid (ECF), including the filtering of nitrogenous wastes and maintenance of normal blood pressure, but also as major sources of hormones.

The kidney can function as an endocrine gland and is the site of production of the following five hormones: (1) renin, which is an enzyme that also functions as a hormone; (2) erythropoietin (EPO), which is a peptide hormone essential for the process of erythropoiesis or red blood cell (RBC) formation by the bone marrow; (3) 1α,25-dihydroxyvitamin D_3, the steroid hormone active form of vitamin D_3, which is essential for calcium homeostasis and many other functions (see Chapter 9: Hormones of the Adrenal Medulla, Fig. 9.16); (4) the kallikreins, a group of serine proteases that act on blood proteins to produce bradykinin, a potent vasodilator; and (5) prostaglandins. In addition, the kidney serves as an endocrine target organ for at least 12 hormones. Table 15.1 summarizes the endocrine aspects of the kidney both as a source of hormones and as an endocrine hormone target organ.

An important physiological function of the kidney was observed by J. Peters in 1835; he wrote the following: "The kidneys appear to serve as the ultimate guardians of the constitution of the internal environment." Thus it became clear that the kidney occupies a unique position within the physiological network of the living organism. The kidney functions as the ultimate organ for the monitoring and conservation of body water and all its electrolyte constituents, as well as many small organic molecules. In addition, the kidney plays an important physiological role in maintaining a correct acid–base balance in the circulatory system. It is, therefore, not surprising to see an array of hormone production sites, as well as sites of hormone action, concentrated in the kidney.

Many aspects of these physiological processes are also closely linked to the process of blood pressure regulation. Accordingly, a secondary objective of this chapter is to introduce the hormonal aspects of the cardiovascular system. Table 15.2 provides a brief introduction to 11 hormones that are produced by or have a major action on the cardiovascular system.

II Anatomical, morphological, and physiological relationships

A The kidney

1 Introduction

The two kidneys, along with the ureters, urinary bladder, and male or female urethra, comprise the anatomical units responsible for the multiplicity of endocrine, metabolic, and filtering actions carried out by the kidney. In the normal human, $\sim$65% of the total body weight is water, whereas in an overweight human only $\sim$45% of the body weight is water. The average person has a blood volume total of 5–6 L. A typical kidney filters $\sim$180 L of plasma a day. Thus the total blood volume is filtered $\sim$30 times/day.

Fig. 15.1A presents an anatomical overview of the human urinary system. With the completion of the metabolic and filtration actions of the kidneys, the resulting urine, containing a wide variety of bodily wastes, is conveyed via the two ureters (one per kidney) to the urinary bladder for temporary storage. Then, at the time of micturition or emptying of the urinary bladder, the urine is voided through the urethra. The normal output of urine per day in nonobese adults may range from 600 to 2500 mL. Table 15.3 lists the principal electrolytes of both the intracellular and the extracellular compartments and as well the contents of urine excreted per day.

Hormones. DOI: https://doi.org/10.1016/B978-0-323-90262-5.00017-2

TABLE 15.1 Hormones that are produced or have major actions in the kidney.

Hormones	Major target organ(s)	Function(s)
Produced by the kidney		
Erythropoietin	Bone marrow	Stimulate red blood cell formation
1α,25-Dihydroxyvitamin D_3 (+other vitamin D metabolites)	Intestine, bone, kidney	Maintenance of calcium homeostasis and regulation of gene transcription in many target organs
Renin (an enzyme)[a]	Blood to mediate production of the hormonal angiotensins	Mediate production of aldosterone (by the adrenal cortex)
Prekallikreins	Serum protein α_2-globulins	Produce kinins (e.g., bradykinin) that are potent vasodilators
Prostaglandins	Kidney	Facilitate Na+-deprivation induced the stimulation of renin secretion
Acting on the kidney		
Adrenomedullin		Facilitate renin release
Aldosterone		Stimulate Na^+ reabsorption
ANF		Increase glomerular filtration rate Increase urine sodium excretion
1α,25-Dihydroxyvitamin D_3		Stimulate Ca^{2+} reabsorption
Endothelins		Secondary actions
Vasopressin		Promote H_2O reabsorption
Prostaglandins		
Catecholamines (epinephrine, norepinephrine)		See Chapter 11, Hormones of the Adrenal Medulla
Cortisol		Diverse
Insulin		Diverse
Glucagon		Diverse
Thyroxine		Diverse

ANF, Atrial natriuretic factor.
[a]*Technically renin, an enzyme, does not fit the classical definition of a hormone (see Chapter 1: Hormone Introduction).*

TABLE 15.2 Hormones that are produced by or have a major action on the cardiovascular system.[a,b]

Hormones	Major target organ(s)	Function(s)
Produced by cardiovascular system		
ANF	Kidney glomerulus	Increase filtration rate
ET-1, ET-2, ET-3	Endothelial tissue, brain	Vasoconstriction
Nitric oxide (NO)	Vascular smooth muscle	Vasorelaxation
Acting on the cardiovascular peripheral system		
Endothelins		Vasoconstriction
Angiotensin II		Vasoconstriction
Vasopressin		Vasoconstriction
Prostaglandins (thromboxanes)		Vasoconstriction
Epinephrine, norepinephrine		Vasoconstriction
Bradykinin		Vasorelaxation
Nitric oxide		Vasorelaxation
Atrial natriuretic factor		Vasorelaxation

ANF, Atrial natriuretic factor; *ET*, endothelin.
[a]*The cardiovascular system comprises the heart and the arterial and venous circulatory systems.*
[b]*This table excludes hormones that are produced in the blood (e.g., angiotensin) and hormones that use blood components as substrates (e.g., renin).*

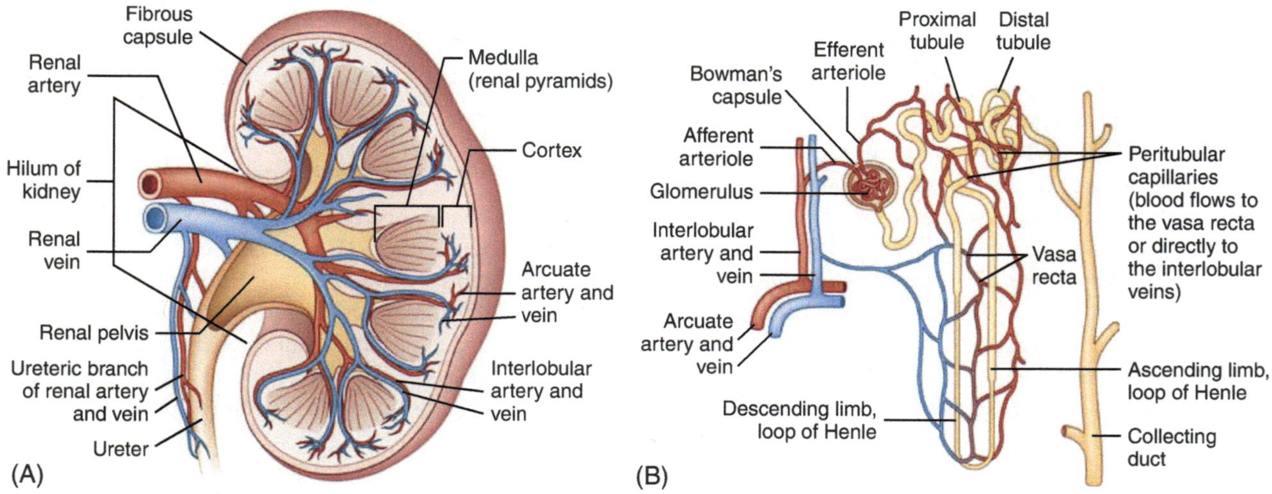

FIGURE 15.1 The human kidney: (A) Cross section of the human kidney. This emphasizes the complex circulatory network of renal arteries and renal veins in the kidney medulla and cortex. (B) View of the kidney's loop of Henle, the nephron's arteries, and venules. Each kidney has ~850,000 to 1.4 million nephrons. The nephron interacts first with the glomerulus then followed by the proximal tubule, the loop of Henle, and the distal tubules. The distal tubules individually connect directly to their own collecting duct which then carries away the urine to the ureters (not illustrated in Fig. 15.2) which, in turn, transfer the urine some distance (~25 cm) to the urinary bladder where the urine is stored until excretion.

TABLE 15.3 Principal constituents of major body fluids in the normal adult.

Constituent	Intracellular compartment (mEq/L)	Extracellular compartment (mEq/L)[a]	Urine[b] (g)[c]	Urine ratio (plasma)
Cations				
Sodium	11	142	2–4	1.0
Potassium	162	4	1.5–2	10–15
Calcium	2	5	0.1–0.3	
Magnesium	28	2	0.1–0.2	
Anions				
Chloride	3	101	17.1	1.5
Bicarbonate	10	27	0.15–2.5	1.5
Phosphate (HPO_4^{2})	102	2	0.6–1.8	25
Sulfate (SO_4^{2})	20	1		50
Other				
Protein	65	16		
Urea			30.0	35
Creatinine			1.0	70
Organic acids	3	6		
Glucose (mg/100 mL)		80–110	6–8	

[a]*The values reported for the extracellular compartment are also applicable to plasma.*
[b]*Calculated by assuming a daily urinary output of 2 L.*
[c]*Amount excreted in a 24-h interval by a normal man consuming an average mixed diet.*

2 Kidney–gross structure

The two kidneys in humans are located in a posterior position in the abdominal cavity; they present a "bean-like" appearance and are enclosed in a fibrous capsule. Diagramed in Fig. 15.1A/B is the cross-sectional appearance of one kidney. In the central portion of the concave border of the kidney is the hilum or entrance portal to the kidney by the blood vessels and nerves for the kidney as well as the ureter. There are three general regions of the kidney: the renal pelvis, the renal cortex, and the renal medulla; they are each called out in Fig. 15.1A. The renal pelvis, which is attached to the upper end of the ureter, collects the urine from all regions of the kidney. The renal medulla, which is the inner portion of the kidney, is composed of cone-shaped renal masses termed renal pyramids. The renal cortex constitutes the outer portion of the kidney.

3 Kidney nephron at the microscopic level

The fundamental operational unit of the kidney is the nephron; see Fig. 15.1B and Fig. 15.2. Each human kidney contains ~850,000 to 1.4 million nephrons. The nephron is the anatomical structure responsible for the filtration of all the ions and smaller organic molecules present in the blood. This is followed by a concomitant selective reabsorption of ions and small organic molecules

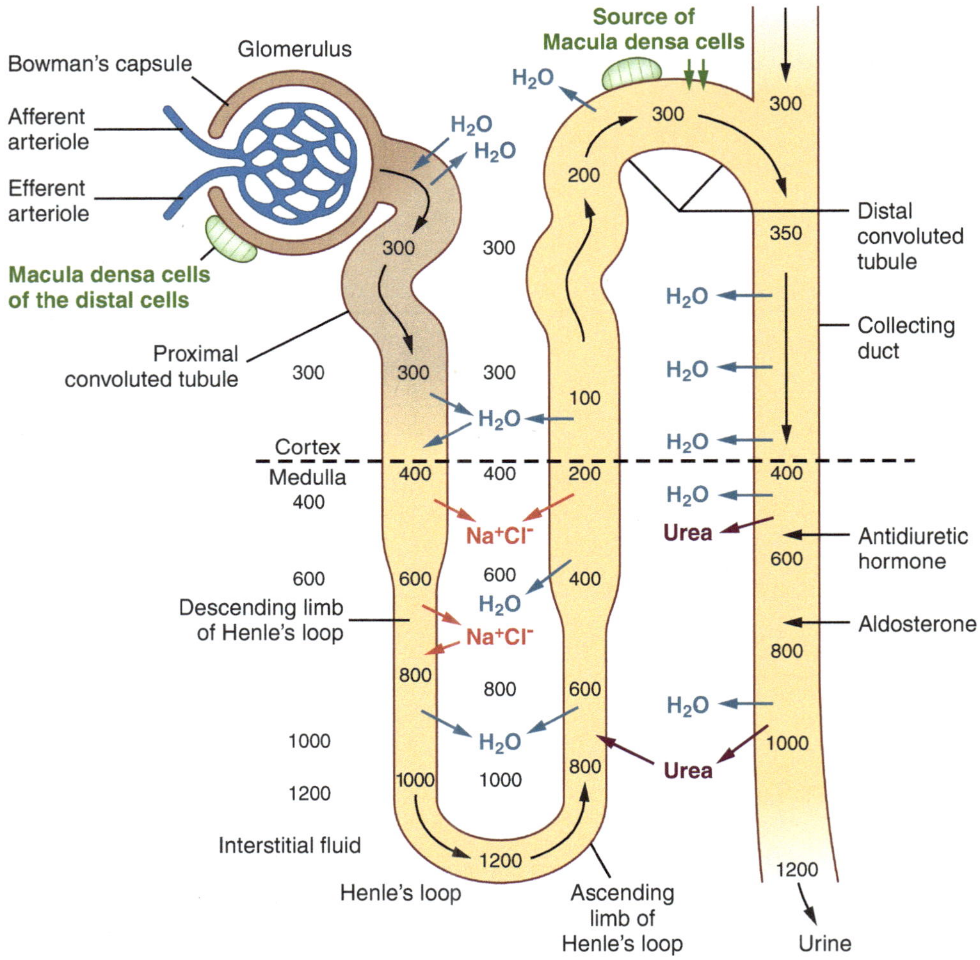

FIGURE 15.2 Schematic diagram of a kidney nephron with its sequential components that filter small molecules from the blood. The nephron comprises eight sequential structural components that collectively facilitate the overall filtering of the blood. The filtering process begins in the nephron's Bowman's capsule (1) that surrounds the glomerulus (2), followed by the proximal convoluted tubule (3), the descending limb of Henle's loop (4), the loop of Henle (5), the ascending limb of Henle's loop (6), followed by the distal convoluted tubule (7) (where the *macula densa* is primarily located; see next paragraph 7A), which merges into the collecting duct (8) that ultimately ends up in the urine bladder (9; not shown). 7A illustrates the green-color *macula densa* cells on the outer surface of the distal convoluted tubule (7), which also contains the glomerulus that comprises the afferent arteriole (blood's entrance) and the efferent arteriole (filtered blood's exit) which are both part of the blood's circulatory system (see *blue*). The distal convoluted tubule empties into a branch of the collecting duct that ultimately leads to a main collecting duct, and then delivery to the urine bladder. In the kidney the *macula densa* (see *green ovals*) is an area of closely packed specialized cells lining the wall of the distal convoluted tubule. Not illustrated in this figure is the fact that in vivo, the distal convoluted tubule with its *macula densa* cells folds back and makes physical contact with the Bowman's capsule/glomerulus/efferent arteriole (see *green oval*) at the point of return of the nephron to its parent glomerulus.

that are brought back into the blood compartment. The remainder of ions and small organic molecules that were not reabsorbed participate in the formation of the urine.

Each nephron consists of a renal tubule, largely located in the medulla region (located in the cortex region; see Fig. 15.1A). The renal tubule, in turn, is composed of a glomerulus enclosed by the Bowman's capsule, which is often described as a corpuscle. The glomerulus is an exquisite anatomical structure containing a group of capillaries interposed between two arterioles. The vessel transporting blood to the glomerulus (see *blue coloring* in Fig. 15.2) is known as the afferent arteriole; it divides into 40–60 capillary loops, which then ultimately reunite to form the exit pathway for the blood known as the efferent arteriole. Associated with the cells of the efferent arteriole is a cluster of cells known as the juxtaglomerular or JG cells. The JG cells are the site of production of the enzyme/hormone renin (see Section III.B.). The crystal structure of glycosylated renin is shown in Fig. 15.3, part A. Renin, as well as being a hormone, has the activity of a peptidase in the bloodstream that cleaves angiotensinogen to angiotensin I, thus, being an angiotensinogenase. In the second part of the figure, part B, preprorenin is cleaved into prorenin in a cleavage that is associated with v-ATPase. The N-terminus of the cleavage product is truncated, but active, prorenin. Fig. 15.3, part C, shows the attachment of prorenin to the extracellular domain protruding into the lumen where it is bound by its substrate, angiotensinogen, and cleaves it to angiotensin I. Angiotensin I is cleaved further by the enzyme angiotensinogen converting enzyme (ACE) in the cellular membrane to angiotensin II.

Bowman's capsule constitutes the beginning of the renal tubule. Specialized cells, termed podocytes, in Bowman's capsule form slits or pores with molecular dimensions that restrict and effectively prevent the passage of large protein macromolecules in the blood into the top of the nephron. Thus only ions and small organic molecules pass into the filtrate and are selectively reabsorbed by the ascending limb of Henle's loop, while any small molecules that are retained in the ascending limb of Henle's loop are delivered to the distal convoluted tubule and are ultimately included in the urine.

The renal tubule is anatomically structured to permit countercurrent distributive processes between its proximal convoluted tubule and the loop of Henle and the distal convoluted tubule (see the *yellow–brown coloration* in Fig. 15.2).

The *macula densa* cells at the top end of the loop of Henle can fold back and make physical contact with the afferent arteriole of its own glomerulosa. The *macula densa* in Fig. 15.2 are illustrated as two separate lime green lumps. One "lump" is located in the distal convoluted tubule just before the junction with the collecting duct and the second "lump" is when the distal convoluted tubule folds back on itself to make a physical interaction with the efferent arteriole.

The function of the macula densa is to act as a salt sensor. It is capable of detecting any increase in the $Na^+ \times Cl^-$ concentration in the distal convoluted tubule. When this occurs, it secretes a paracrine vasopressor that binds to its receptor on a nearby afferent arteriole. This then results in a decrease of the glomerulus filtration rate (GFR). Thus this anatomical specialization allows for physiological, metabolic, and hormonal communication to occur between the distal convoluted tubule (containing the exiting urine) and the afferent arteriole (containing the entering blood, which will be filtered). A fall in blood pressure typically results in a reduction in the rate of glomerular filtration of Na^+ and Cl^- by the renal tubule. This reduction promptly signals the macula densa to release a hormone signal to the efferent arteriole that promptly increases the reabsorption of water, Na^+, and Cl^-, which collectively normalizes the blood pressure.

The nephron's fundamental function is to regulate the concentration in the water of the relatively soluble small molecules (e.g., ions, Na^+, K^+, Cl^-, Ca^{2+}, and Mg^{2+}) and small organic molecules (e.g., glucose, lactose, and peptides). Proteins in the range of ~30–60 kDA (e.g., albumin) present in the blood plasma are readily filtered through the capillaries of the glomerulus and deposited into the Bowman's capsule. About 99% of this fluid, including a surprisingly large proportion of the available proteins, and many of the dissolved solutes are selectively reabsorbed by the tubules of the kidney as they pass through the upper portion of the proximal tubule (see *dark brown color* in Fig. 15.2) that leads to Henle's loop (*yellow*) and finally to the collecting duct where the urine is collected. The reabsorption of protein is strikingly efficient; approximately 99% of the filtered protein is recovered and only 1% of the originally filtered protein is present in the urine.

The numbers in the descending and ascending limbs of the loop of Henle of Fig. 15.2 indicate the ion gradients in osmolarity that result as a consequence of simple diffusion and transport or exchange. The filtrate, as it enters the loop of Henle, becomes further concentrated by diffusion of water from the inside of the descending loop of Henle into the adjacent "space" (interstitium) outside of the descending loop of Henle. In the ascending limb of the loop of Henle, Na^+ is actively transported out of the filtrate (thereby diluting it) into the interstitium, where it is concentrated. The osmolarity of the interstitial fluid can increase from 300 to 1200 mOsm/kg Hg as the papilla is approached. Then, as the distal tubule passes back past the glomerulus (see Fig. 15.2), additional Na^+ may be reabsorbed in exchange for H^+ or K^+ ions. Eventually water is reabsorbed in the distal tubule and collecting

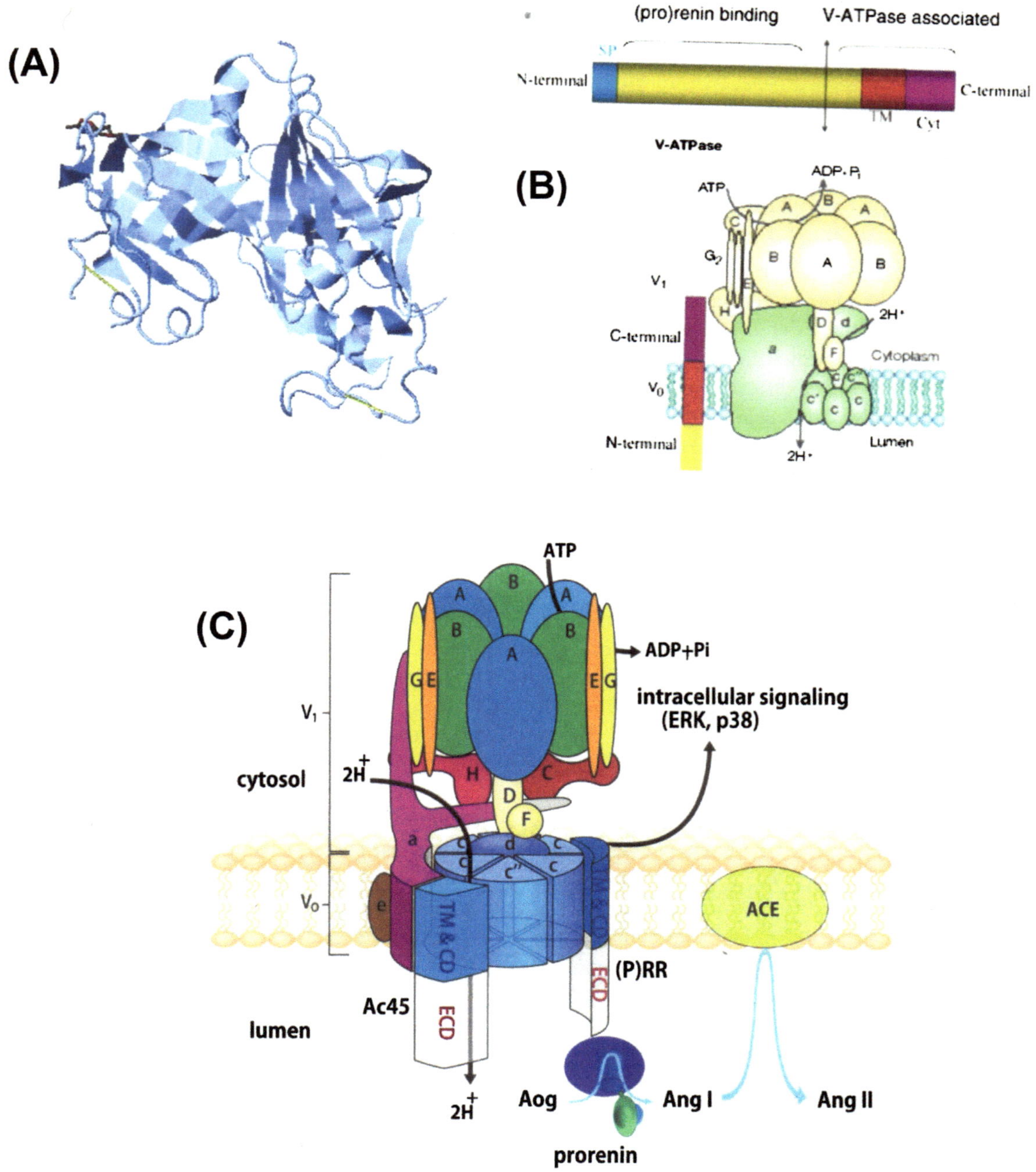

FIGURE 15.3 (A) Structure of glycosylated human renin. (B) Schematic structure of the prorenin receptor (P)RR and its association with vacuolar ATPase (V-ATPase) (bottom). The arrow in the top diagram shows a putative cleavage site generating a truncated form of (P)RR composed of the C-terminal part of the (pro)renin receptor that remains associated with V-ATPase The N-terminal part in the (pro)renin-binding region of the (P)RR (C) Putative structure of the mammalian vacuolar-type H+-ATPase and (pro)renin receptor. (Pro)renin receptor bound prorenin undergoes a nonproteolytic conformational change in prorenin, generating local angiotensin I from angiotensinogen. Both (pro)renin receptor and Ac45 have the furin cleavage sites between transmembrane and extracellular domains. *Ac45*, Accessory subunit45; *ACE*, angiotensin converting enzyme; *Ang 2*, angiotensin 2; *Ang I*, angiotensin I; *Aog*, angiotensinogen; *CD*, cytosolic domain; *cyt*, cytoplasmic domain; *ECD*, extracellular domain; *ERK*, extracellular signal-regulated kinase; *SP*, signal peptide; *TM*, transmembrane domain. *(A) Reproduced from the Protein Data Bank, PDB code 2ren. (B) Reproduced from Fig. 2 of G. Nguyen, A.H.J. Danser, Prorenin sand (pro)renin receptor: a review of available data from in vitro studies and experimental models in rodents, Experimental Physiology, 93 (2008) 557–563. (C) Reproduced from Fig. 1 of A. Ichihara, K. Kinouchi, Current knowledge of (pro)renin receptor as an accessory protein of vacuolar H+-ATPase, Journal of the Renin–Angiotensin–Aldosterone System 12 (4) (2011) 638–640; https://doi.org/10.1177/1470320311428227.*

duct. Aldosterone and antidiuretic hormone also act principally on the collecting duct (see Fig. 15.2).

4 Physiological processes

The kidney is the principal organ responsible for the homeostasis of a wide spectrum of electrolytes, as well as the conservation of body water. The kidney carries out its homeostatic actions sequentially by selective glomerular filtration (mediated by a high blood pressure in the glomerulus), tubular secretion, and tubular reabsorption. These are all processes that collectively regulate the concentration of the metabolic end products, osmotic pressure, ionic composition, and fluid volume of the internal environment. Fig. 15.2 is a schematic diagram of a typical kidney nephron, indicating the sites of reabsorption of the various ionic substances. The key to the achievement of homeostasis is the process of countercurrent distribution. Countercurrent distribution occurs as a consequence of the anatomical organization of the nephron and is supported by the processes of passive diffusion, tubular reabsorption, and tubular secretion. Both of the latter processes utilize energy-dependent active transport mechanisms. The end result of these activities is the formation of the residual urine containing bodily wastes (including ionic, organic, and nitrogenous); however, in the process of forming the urine from the glomerular filtrate, the tubule has returned many essential nutrients and electrolytes to the blood. A comparison of the extracellular concentrations to urine concentrations of major urine constituents is presented in Table 15.3.

B Cardiovascular system

1 Heart

The human heart, although weighing less than 450 g, is a remarkable organ. Its continuous beating provides the driving force to ensure the delivery of oxygen, nutrients, hormones, and other regulatory molecules to every cell in the body, as well as effecting the continuous removal of metabolic waste products. Over an average lifetime, the heart pumps approximately 300 million L ($\sim$80 million gallons) of blood. The dynamic range of the heart pumping capacity is also impressive; cardiac output can range from a low of 5 to an upper limit of 35 L of blood per minute. The heart is also the source of an important hormone, atrial natriuretic factor (ANF), which is involved with the regulation of blood pressure. See the later section on ANFs.

The human heart is divided longitudinally into right and left halves, each of which consists of two chambers, the atrium and the ventricle. The function of the left ventricle of the heart is to pump the oxygenated blood (*red*) into the arterial system (head and arms, internal organs, and legs). The right ventricle of the heart functions to receive deoxygenated blood (*blue*) from the superior *venae cavae* (the arms and head), the inferior *vena cava* (the legs and internal organs), and from the peripheral tissues. This is then pumped (*blue*) into the left and right pulmonary arteries for circulation to the lungs. The left side of the heart receives the oxygen-enriched blood back from the lungs via the pulmonary veins; this blood is then pumped into the aorta for disbursement to all of the peripheral tissues.

The pressure necessary for these tasks is created by the strong contractions of the various musculatures of the heart. However, the force and rate of these heart contractions, as well as the volume of blood that may exit the heart with each heartbeat, are determined via a complex set of physiological and hormonal signals operative in the circulatory system. This will be discussed in the next sections.

2 Cardiovascular system

The circulatory system or cardiovascular system comprises the heart and blood vessels, or the body's "plumbing system"; see Fig. 15.4A. The heart is the pumping organ that ejects blood into the arterial network. The blood is then returned to the heart via the venous system. The capillaries represent microscopic vesicles that interconnect between the small arteries (arterioles) and small veins (venules); see Fig. 15.4B and C. The walls of the capillaries are only one endothelial cell in thickness. Tiny openings, or fenestrations (small windows), facilitate the delivery and exchange between the circulating blood in the capillary and the interstitial fluid that surrounds and bathes the neighboring cells.

3 Regulation of blood pressure

Table 15.4 classifies normal and hypertensive blood pressures. The principal determinants of mean arterial blood pressure are the volume of cardiac output (stroke volume and heart rate), systemic blood flow, and resistance to blood flow in the various perfused organs. These functions are regulated by a variety of hormones, neurotransmitters, and local paracrine factors, as well as by the health and vigor of the heart.

III Homeostasis of fluid, electrolytes, and blood pressure

A Introduction

Hormones are intimately involved in the regulation of both cardiovascular and renal functions. The maintenance of salt homeostasis, circulatory volume, and blood pressure requires the integrated actions of the

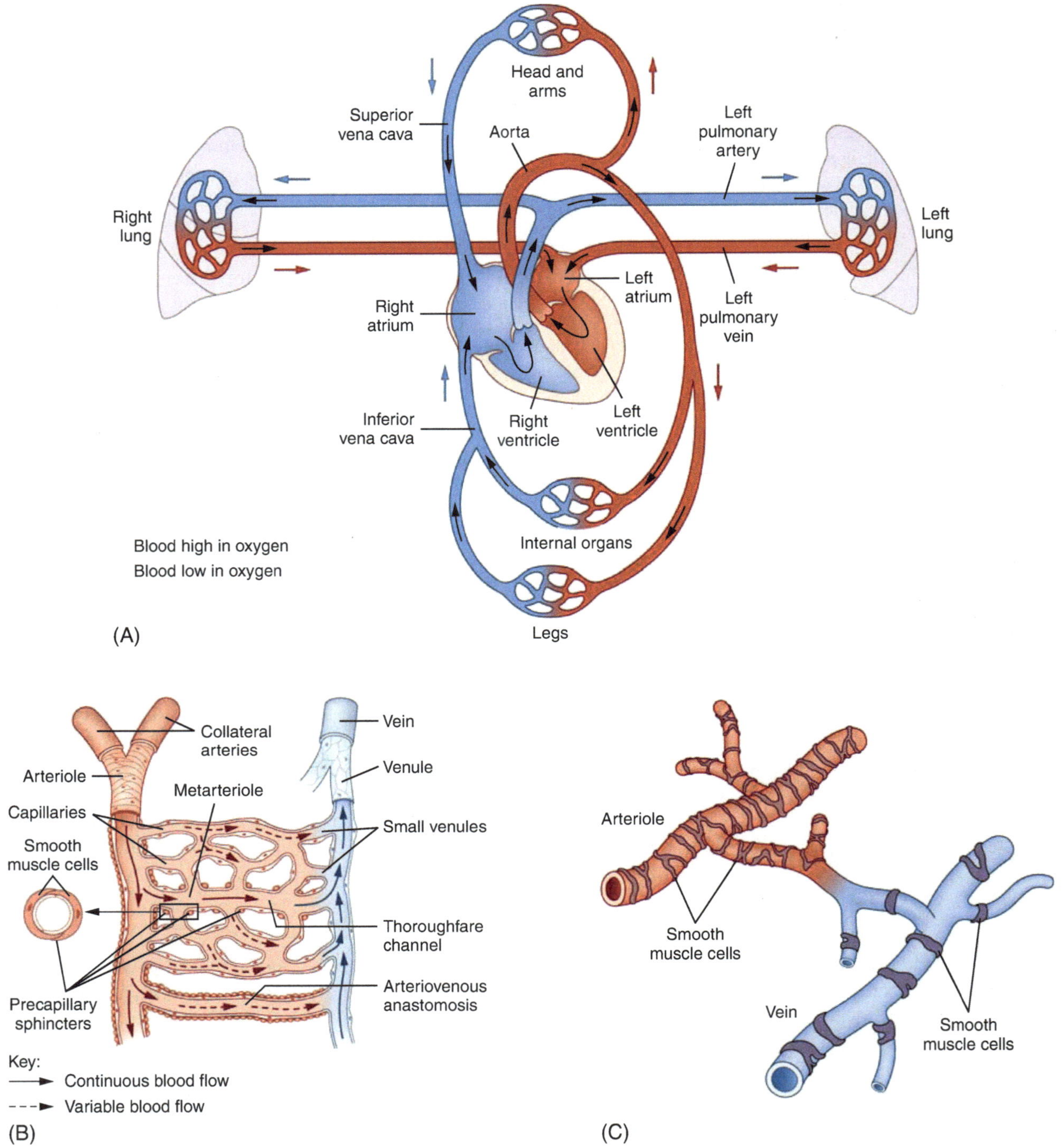

FIGURE 15.4 Diagram of the human circulatory system. (A) *Illustration of the pulmonary (heart) and the body's systemic circulatory systems.* The systemic circulation system includes all of the arteries, veins, and capillaries of the body. In general, the arteries are shown in red and the veins in blue. The exception is the pulmonary circulation system that includes the left and right pulmonary veins that carry the oxygenated blood from the lungs (*red*) to the left antrum. (B) *A greatly magnified portion of the human capillaries circulating system's arteriole and its partner the venule.* The smallest blood vessels are the capillaries which are 5–10 μM in diameter; they connect to arterioles and venules. Precapillary sphincters are relaxed, thus permitting the flow of blood through the capillary network. Also visible in the capillaries are (A) the several cell layers that contribute to the walls of both the arterioles and venules and (B) the fenestration (window) openings. These "windows" may be opened or closed for short intervals to permit the diffusion of solutes (e.g., K^+, Na^+, Cl^-, Mg^{2+}, and Ca^{2+}), carbohydrate molecules (glucose, etc.), small peptides, waste molecules, water, oxygen, and CO_2 between the plasma and the interstitial fluids. (C) *The presence of the precapillary sphincter created by the smooth muscle cells is functionally important.* The wrapping around the surface of both the capillaries and veins with smooth muscle fibers permits the regulation of the volume of blood flow entry into the artery and to exit via the veins.

TABLE 15.4 Classification of blood pressure levels.

Classification[a]	Systolic (mm Hg)	Diastolic (mm Hg)
Normal	<130	<85
High normal	130–139	85–89
Hypertension		
Mild	140–159	90–99
Moderate	161–179	100–109
Severe	180–209	110–119

[a]*The classifications are for age >18 years. The data are obtained from the Joint National Committee on Detection, Evaluation, and Treatment of High Blood Pressure.*

TABLE 15.5 Factors regulating renin release.

Inhibitory factors	Stimulatory factors
Increased renal arterial pressure	Decreased renal arterial pressure
Increased extracellular fluid volume	Decreased extracellular fluid volume
Angiotensin II	
ANF	PGI_2
Vasopressin	ACTH
α-Adrenergic stimulation (dopamine)	β-Adrenergic stimulation (catecholamines)

ACTH, Adrenocorticotropin hormone; *ANF*, atrial natriuretic factor.

renin–angiotensin–aldosterone system, kallikreins and kinins, the adrenergic nervous system, vasopressin, ANF, endothelins (ETs), prostaglandins, the nitric oxide system, and adrenomedullin (AM).

Some of these hormones can affect heart rate and contractility directly or effect vasoconstriction or vasodilation of the arteries and veins. In addition, the growth factor properties of hormones are able to influence cardiovascular development and muscular hyperplasia or hypertrophy, as well as being involved with pathological changes that can be associated with atherosclerosis, cardiac hypertrophy, and heart failure. These hormone systems are also capable of mediating important biological effects in the kidney.

The volume of the ECF is governed by its Na^+ concentration. The Na^+ concentration of the ECF is determined by regulating the extent of excretion of Na^+ in the urine. The principal, although not exclusive, factor governing the excretion of Na^+ is the steroid hormone aldosterone and the glomerular filtration rate. The glomerular filtration rate of the kidney can be markedly increased by the actions of the hormone, atrial natriuretic protein (ANP), thus increasing the extraction of blood Na^+. Also, ANP acts on the smooth muscle present in large arteries and vascular beds to effect relaxation and thus achieve a reduction in blood pressure.

As discussed in Chapter 10, Adrenal Corticoids, bilateral adrenalectomy is fatal. This results from the absence of aldosterone, which, in turn, leads to an increased loss of Na^+ in the urine, and a concomitant retention of K^+ in the ECF, and loss of water from both the extracellular and intracellular compartments. If this process continues, death inevitably follows.

B Renin–angiotensin II

Renin is a zinc-containing enzyme with 340 amino acids of molecular mass ~37 kDa. Renin can be isolated from both human and rat kidneys and mouse submaxillary glands. The original isolation of the pure protein required a 3 million-fold purification; the primary sequence of human renin has been deduced through cloning and sequence analysis of cDNA prepared from human kidney messenger RNA (mRNA). Renin as an enzyme belongs to the class of aspartyl proteinases.

Renin is biosynthesized as a preprorenin consisting of 406 amino acids. The prepro form is converted to prorenin by the removal of the first 20 amino acids of the presegment followed by the removal of a prosegment of 46 amino acids. Approximately 50% of the prorenin in the juxtaglomerular cells is constitutively secreted continuously without proteolytic processing, while the other 50% of the prorenin is sorted to secretory granules, where it is proteolytically processed and stored as renin. The release of this renin is regulated by cyclic AMP (cAMP). Renin in the circulatory system is variably glycosylated and circulates as 4–5 isoenzymes.

In renin the aspartyl residues at positions 38 and 226 are believed to be important for catalytic activity. The X-ray crystallographic structure of human renin has been determined to the 2.5 Å level. The general shape of renin is bilobal with a long deep cleft that contains the aspartyl 38 and 226 residues.

Changes in the secretion of renin by the renal juxtaglomerular cells can occur in response to changes in renal arterial pressure, via sympathetic nervous system signals and also via changes in the status of a number of hormones (see Table 15.5). The secretion of renin is governed at the level of the renal glomerulus by both long and short feedback loops. The dominant negative long feedback loop, which diminishes renin secretion, occurs when there is an increased renal arterial pressure, which is sensed by stretch receptors present in the glomerular afferent arteriolar wall so as to result in a reduction in renin release and, ultimately, a diminution in aldosterone production. A negative short feedback loop is mediated by angiotensin II, which directly inhibits renin release.

C Angiotensins I and II

The natural substrate for renin is the plasma protein α_2-globulin, which is also known as angiotensinogen. Angiotensinogen is a glycoprotein of 57 kDa that is synthesized and secreted into the bloodstream by the liver. The biosynthesis of angiotensinogen is increased by glucocorticoids, estrogens. The general details of the conversion of angiotensinogen into angiotensin I (a decapeptide) and then into angiotensin II (an octapeptide) are summarized in Fig. 15.5A/B.

In the circulatory system, renin hydrolyzes the Leu–Leu bond of angiotensinogen between residues 10 and 11 to generate the biologically inactive decapeptide angiotensin I. Angiotensin I is then converted by the ACE (molecular mass ~200 kDa) that removes the carboxyl-terminal His-Leu dipeptide to yield the octapeptide hormone angiotensin II.

ACE is a zinc-containing protein. Its most important actions are (1) to convert angiotensin I→II and (2) to inactivate bradykinin (a very potent vasodilator).

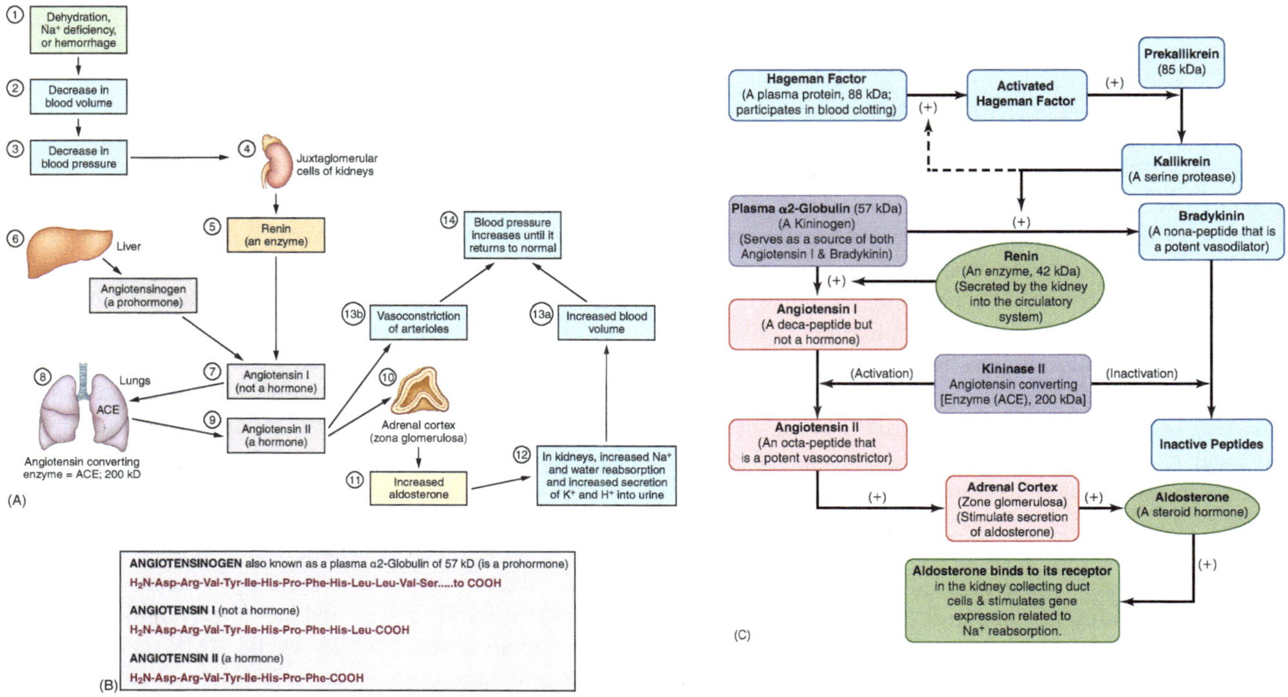

FIGURE 15.5 Fundamentals of blood pressure regulation by integrated actions of the peptide hormones renin, and angiotensin II, and the steroid hormone aldosterone. (A) A flowchart of 14 sequential steps that becomes activated starting with an acute decrease either in blood volume and/or blood pressure that is detected by the kidney, resulting in the release of the enzyme protease, renin, which functions as a hormone. Renin, also known as angiotensinogenase, has 406 amino acids, while the functional mature renin has 340 amino acids (~37 kDa). Renin is secreted by the granular cells associated with the glomerulus (see Figs. 15.1B or 15.2) in response to a drop in blood pressure and/or a decrease in Na^+ concentration. Renin's substrate is the blood protein, α2-globulin (57 kDa), that has been secreted by the liver and then localized within the capillaries of the lungs. This is described in steps 4–7 of panel 4A. The hormone angiotensin II is an octapeptide produced by the ACE (step 8) acting on the precursor, angiotensin I. Angiotensin II is a hormone that acts on the *zona glomerulosa* of the adrenal cortex where it stimulates the production and secretion of the steroid hormone, aldosterone, which binds to its receptor in the kidneys. Then in steps 11–14, angiotensin II stimulates the release of the steroid hormone aldosterone from the adrenal cortex (*zona glomerulosa*) which then binds to its receptor localized in the kidney's collecting duct where it increases the reabsorption of both Na^+ and water and, also, increases secretion of H^+ and K^+ into the urine, thus leading to an increased blood volume, which increases blood pressure until it has returned to normal. (B) *Amino acid sequence of the hormone angiotensin II and two precursor peptides.* Panel 4B presents the following: (1) the 13-amino-acid prohormone portion cleaved off angiotensinogen which has 453 amino acids; (2) the 10-amino-acid peptide sequence of angiotensin I (not a hormone); and (3) the 8-amino-acid peptide sequence of the active hormone, angiotensin II. (C) *A schematic illustration of the sources of production of the peptide hormones renin, angiotensin II, and bradykinin and the steroid hormone, aldosterone.* There are three important protein sources: (1) the plasma α2-globulin (57 kDa) that is the substrate for (2) the enzyme, renin (42 kDa), that is secreted by the kidney's juxtaglomerular cells based on changes in renal arterial pressure; and (3) the prekallikrein (86 kDa; see upper right *blue rectangle*) that is converted by the activated Hageman factor (also known as blood coagulation factor XI; 80 kDa) that cleaves off an 85-kDa portion of prekallikrein. When the enzyme kininase II serine protease becomes available, it inactivates bradykinin by removing one or two amino acids from the active bradykinin (a potent vasodilator). The same kininase II enzyme (ACE) also can cleave a 2-amino-acid peptide from angiotensin I, which then generates angiotensin II, which is a potent vasoconstrictor. The same angiotensin II, besides being a potent vasoconstrictor, also is a stimulating agent for the biosynthesis of aldosterone in the *zona glomerulosa* of the adrenal cortex. *ACE*, Angiotensin-converting enzyme.

TABLE 15.6 Biological actions of the angiotensins.

Activity	Angiotensin peptides
Stimulation of aldosterone biosynthesis and secretion by the adrenals	II=III >>> I
Elevation of blood pressure via vasoconstriction	II > III
Stimulation of release of catecholamines by adrenal medulla	II
Stimulation of thirst by action on central nervous system	II > III

The principal site of conversion of angiotensin I to II is in the vascular epithelium of the lung; however ACE activity is also present in the kidney vascular epithelium, heart, brain, and testis. The converting enzyme is also named kininase II because of its action on bradykinin (see Fig. 15.5C). Angiotensin II, a nonapeptide, is produced by the action of an N-terminal peptidase acting on angiotensin I. Table 15.6 summarizes the biological actions of the angiotensins. Angiotensin II is the most potent vasoconstrictive agent known. All of the components of the renin–angiotensin system that are necessary to produce angiotensin II are also present in the brain, where the angiotensin II may function as a neurotransmitter.

There are two principal biological actions of angiotensin II; see Table 15.6. First it functions as a highly potent vasoconstrictor of the smooth muscle of the blood vessels of the circulatory system (see Fig. 15.4C) where it acts within seconds. This increases the resistance of these arteries for the heart.

D Aldosterone biosynthesis and actions in renal tubular reabsorption

1 Aldosterone biosynthesis

The site of production of the mineralocorticoid aldosterone is the adrenal *zona glomerulosa* cell, which employs a variety of signal transduction mechanisms that include receptors for angiotensin II, adrenocorticotropin hormone, dopamine, and the ANP as well as a K^+-modulated voltage-gated Ca^{2+} channel. The dominant agonist is angiotensin II, but under appropriate physiological circumstances the other agonists are also effective. Within seconds of exposure to angiotensin II, IP_3, diacylglycerol (DAG), and intracellular Ca^{2+} concentrations all increase, followed shortly thereafter by an elevated biosynthesis of aldosterone.

2 Aldosterone actions

In humans the principal biologically active mineralocorticoid is aldosterone; the structure of aldosterone is shown in Fig. 15.6A.

The details of the structure and the biosynthesis of aldosterone and some related steroids are summarized in Figs. 2.10, 2.11, and 2.19. The mineralocorticoids mediate their actions on ion balance principally in the kidney, and also to some limited extent in the salivary glands, gut, and sweat glands.

In the kidney the actions of the aldosterone result in an increased cortical collecting tubule reabsorption of Na^+ with a concomitant secretion of K^+. Only a fraction of the Na^+ filtered by the glomerulus is actually reabsorbed as a consequence of aldosterone's action. However, this fraction can effect significant consequences on electrolyte balance.

The steroid hormone aldosterone by binding to its nuclear receptor produces biological responses via the stimulation of selected gene expression related to kidney tubule Na^+ transport. The focus of aldosterone is on the mucosal cells of the kidney's cortical collecting duct (see Fig. 15.2). In the absence of aldosterone, there is only a minimal basal uptake of Na^+ from the lumen (extracellular) side of the collecting duct cell by the Na^+ permease. Fig. 15.6B illustrates the four steps in the cortical collecting duct cell which are operative in the presence of aldosterone. The presence of aldosterone binding to its nuclear receptor stimulates the gene transcription of several proteins that are essential to supporting the overall entrance of Na^+ into the cell via a Na^+ permease. Since the extracellular concentration of Na^+ is significantly higher ($\sim$140 mM) than the intracellular concentration of Na^+ ($\sim$10 mM; see Table 15.3), the Na^+ in the cortical collecting duct cell's cytosol must utilize the Na^+/K-ATPase (see Fig. 15.6B) to transfer the Na^+ across the peritubular cell membrane, against a concentration gradient, and release it into the ECF.

E Atrial natriuretic protein (ANP) system

1 Introduction

The heart and brain are the source of a family of peptide hormones referred to as the ANP system. Cardiac ANP is secreted by atrial cells of the heart following volume expansion of the heart or in circumstances of elevated blood pressure. Table 15.7 summarizes the physiological actions of ANP.

A chief property of a natriuretic peptide (NP) is that it stimulates the secretion of sodium in the urine; this is the process of natriuresis. The biological functions of the three NPs are briefly described here. ANP functions to significantly reduce central blood pressure and also

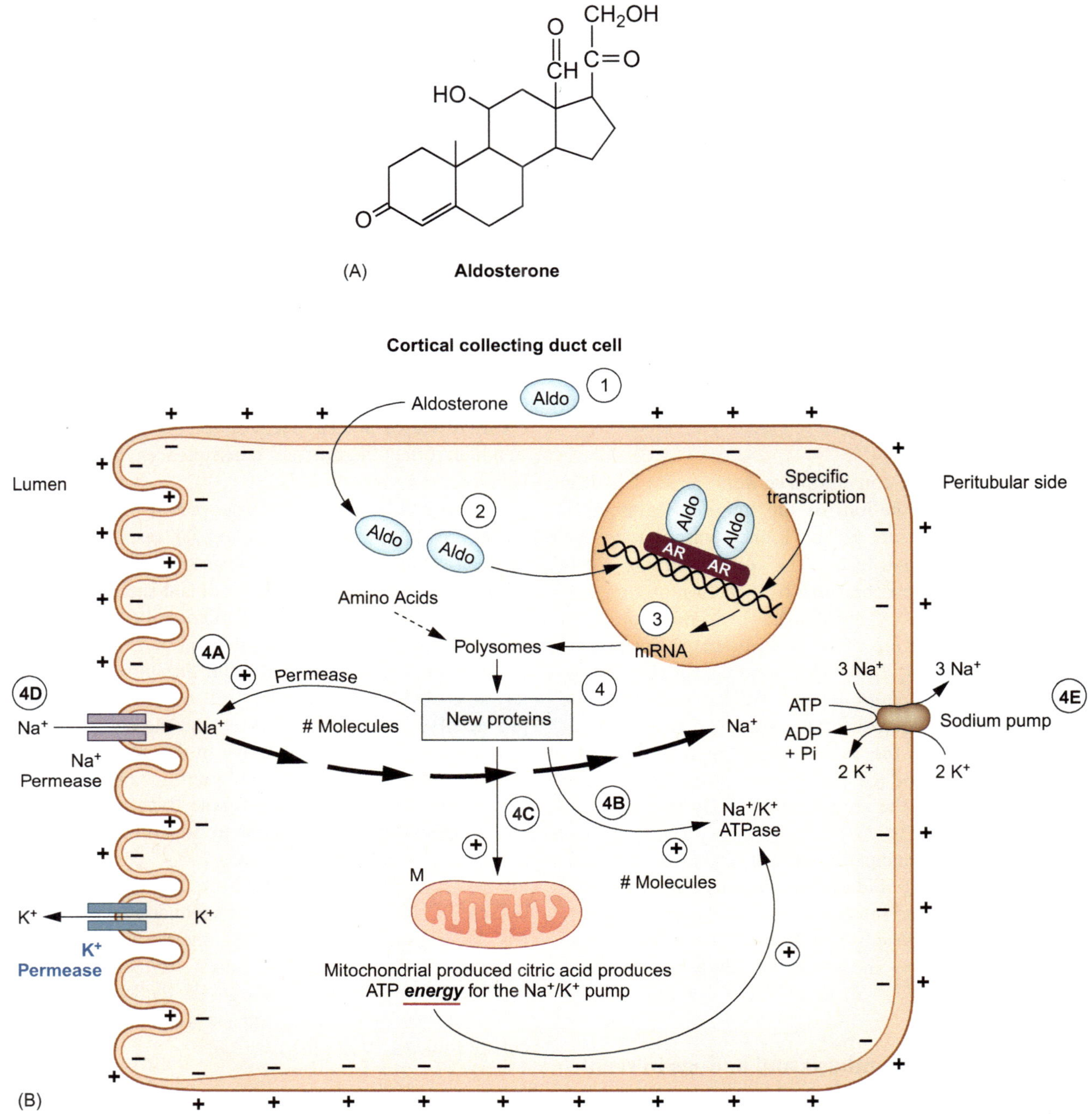

FIGURE 15.6 Structure and production of aldosterone and a cellular model of the actions of aldosterone on Na^+ and K^+ transport processes in the mucosal cells of the cortical-collecting duct of the kidney nephron. (A) The structure of aldosterone is shown. Chapter 2, Steroid Hormones: Chemistry, Biosynthesis, and Metabolism, of *Hormones* focuses on steroid chemistry, biosynthesis, and metabolism of all the steroid hormones, including aldosterone. The chemical structure of aldosterone, a mineralocorticoid, is shown. Chapter 2, Steroid Hormones: Chemistry, Biosynthesis, and Metabolism, Fig. 2.10 shows the relationships of stimulatory peptides like angiotensin II for each of the six steroid hormones. Figs. 2.11 and 2.19 summarize the details of the metabolic pathways that convert cholesterol into aldosterone. (B) Mode of action of aldosterone in a kidney's cortical collecting duct cell in stimulating the reabsorption of Na^+. The steroid hormone aldosterone produces biological responses via the stimulation of gene expression related to kidney tubule Na^+ reabsorption. The focus is on the role of aldosterone in mucosal cells of the kidney's cortical collecting duct (see Fig. 15.2). The figure illustrates the four steps that are operative in the presence of aldosterone. In step ①, aldosterone (Aldo) enters the cell (*blue oval*) and in ② localizes in the nucleus where it binds to its specific high-affinity AR (*magenta*) creating a 1:1 homodimer of the AR-Aldo. The AR-aldosterone dimer then is translocated to the specific hormone DNA response elements in the promoter region of the specified genes ③. This stimulates the rate of transcription of selected mRNAs that generate several aldosterone-induced proteins indicated in the green box as "new proteins" ④. These include the α- and β-subunits of the Lumen Na^+ permease pump 4A and 4D (all step labels are inside a circle), which are believed to be the rate-limiting step in Na^+ reabsorption. The newly synthesized Na^+/K^+ ATPase (step 4B) is responsible for (1) pumping Na^+ from the cortical collecting duct cell's cytoplasm on the peritubular side of the cell, into the cell membrane and (2) release of Na^+ into the peritubular extracellular fluid. This increases the removal of Na^+ from the interior of the cell (step 4E) on its way to the blood. Increasing the concentration of the citrate lyase (step 4C) results in an increase in the rate of biosynthesis of citric acid which, after traversing the mitochondrial oxidative pathways, gives rise to the production of correspondingly more ATP which is required as a substrate for the Na^+/K^+ ATPase (step 4E). *AR*, Aldosterone receptor; *mRNA*, messenger RNA.

TABLE 15.7 Physiological actions of atrial natriuretic peptide.

Site of action	Effect produced
Kidney	(a) Increased GFR leading to an increased Na^+ excretion in urine and thus a correlated reduction in plasma volume
	(b) Suppression of renin secretion
Smooth muscle of arteries, capillaries, and veins	Muscle relaxation with reduction in blood pressure
Adrenal *zona glomerulosa* cells	Antagonism of angiotensin II stimulation of aldosterone biosynthesis
Brain	Complex set of neuroendocrine and neurotransmitter effects

GFR, Glomerular filtration rate.

to decrease cardiac hypertrophy, which is the enlargement of the ventricles (lower chambers) of the heart. The properties of BNP (brain natriuretic protein) are similar to that of ANP. Surprisingly, C-type NP (CNP) was found to have an unanticipated biological response. Overexpression of CNP leads to the overexpression of long bones with bone anomalies, for example, the femur, tibia, and humerus. The basis of this observation has not yet been found.

2 Chemistry, biosynthesis, and secretion of ANP

There are three structural forms of the NP, designated as ANP, BNP, and CNP. The amino acid sequence relationships for ANP (27 amino acids), BNP (32 amino acids), and CNP (22 amino acids) are illustrated in Fig. 15.7A.

All three NPs are characterized by a 17-member amino acid ring formed by a disulfide bridge between 2 cysteine residues; see Fig. 15.7B. Each of the three NPs is the product of a separate gene and mRNA. The genes that code for ANP produce a prepro-ANP of 151 amino acids. Loss of the 25-amino-acid hydrophobic leader sequence generates a 126-amino-acid pro-ANP, which is stored in granules of the atrial myocytes. Mature ANP is generated at the time of secretion by the cleavage of the pro-ANP at Arg-98-Ser-99 to yield the mature peptide of 28-amino-acid residues (residues 98–126 of prepro-ANP) and the N-terminal peptide (residues 1–98 of prepro-ANP). The secretion of ANF is stimulated by any of the following: (1) atrial stretch caused by volume expansion; (2) elevated blood pressure; (3) atrial tachycardia; and (4) high-salt diet. BNP is secreted by the ventricles of the heart.

3 Biological actions of atrial natriuretic peptide

The ANP is produced, stored, and released mainly by the cardiac myocytes of the heart's atrium. A similar system is operative for the release of the "basic" BNP by the ventricles of the heart. In each instance an elevated vascular volume results in a stretching of the heart's atrium and ventricles, which initiates the cleavage and secretion, respectively, of both ANP and BNP from the heart.

The several biological effects of ANP (see Table 15.7) includes the inhibition of angiotensin II stimulation of aldosterone biosynthesis, stimulation of the renal GFR so as to promote Na^+ natriuresis, and vasorelaxation. All of these biological effects are individually mediated by the appropriate peptide hormone (ANP, BNP, or CNP) binding to their partner receptor protein (NP receptor, NPR). As shown in Fig. 15.7C, three distinct receptors (NPR-A, NPR-B, and NPR-C) have been identified through recombinant, DNA techniques.

Fig. 15.7C presents a schematic overview of the structural details of the three families of atrial natriuretic receptors. The NPR-A and NPR-B receptors comprise two identical receptor amino acid sequences (of 1037 amino acids) that are only covalently linked together inside the cell (see *green boxes*). Both the NPR-B and NPR-A receptor proteins, but not the NPR-C receptor, span the cell's plasma membrane and, therefore, have an extracellular portion (450-amino-acid residues), a hydrophobic membrane spanning region (~21-amino-acid residues), and an intracellular domain (566-amino-acid residues). The intracellular portion of the receptor protein is divided into three distinct domains: (1) a 250-amino-acid kinase homology domain (see *blue ovals*) with 3–4 attached phosphates; (2) a 41-amino-acid hinge-dimerization region; and (3) a 250-amino-acid C-terminal guanylyl cyclase enzyme catalytic domain. The two receptor peptide chains interact only on the inside of the cell to form a dimer (see the schematic *green* "box") which is characteristic of catalytic guanylyl cyclase units. In contrast the NPR-C receptors form a dimer (via a disulfide bond) outside the cell. The NPR-C has a highly truncated intracellular peptide of only ~80-amino-acid residues and consequently has no catalytic guanylyl cyclase units.

The NPR-A, NPR-B, and NPR-C receptor-favored ligands are shown by the long black arrows. NPR-C is the most prevalent natriuretic receptor. It largely exists to remove NPs from the circulatory system by internalizing and digesting them. The very short bold black lines with a black circle at the end indicate sites on the receptor amino acid sequence of N-linked glycosylation locations outside

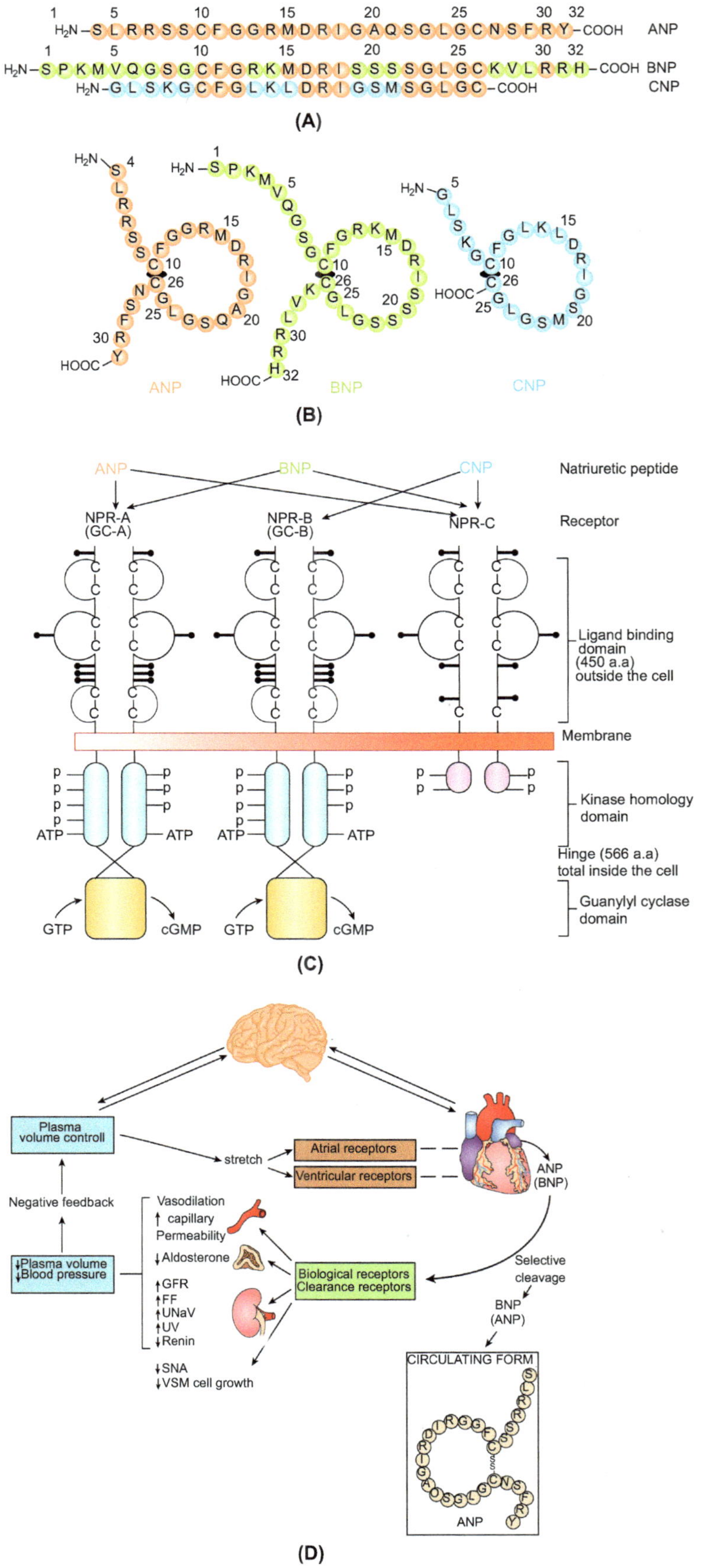
ANP
BNP
CNP
(A)
(B)
Natriuretic peptide
NPR-A (GC-A)
NPR-B (GC-B)
NPR-C
Receptor
Ligand binding domain (450 a.a) outside the cell
Membrane
Kinase homology domain
Hinge (566 a.a) total inside the cell
Guanylyl cyclase domain
ATP
GTP
cGMP
(C)
Plasma volume controll
stretch
Atrial receptors
Ventricular receptors
ANP (BNP)
Negative feedback
Vasodilation
↑ capillary Permeability
↓Aldosterone
↑GFR
↑FF
↑UNaV
↑UV
↓Renin
↓SNA
↓VSM cell growth
↓Plasma volume
↓Blood pressure
Biological receptors
Clearance receptors
Selective cleavage
BNP (ANP)
CIRCULATING FORM
(D)

(Continued)

the cell membrane. Also each partner receptor protein has three disulfide bonds located outside the cell; they are identified by the "R-C-C-R."

The longitudinal blue ovals that are side-by-side inside the cell each have four phosphorylation sites which are adjacent to the cell membrane. The COOH portion of the GC-A and GC-B receptors (inside the cell) is a green square signifying that the two intracellular portions of the two-receptor amino acid strands are covalently linked together. The green box for both receptors is the guanylyl cyclase domain that initiates the production of their biological responses via their second messenger, cyclic guanosine monophosphate (cGMP). Both the ANP and BNP receptors (1016-amino-acid residues) have full enzymatic activity on their COOH domain (inside the cell). But the NPR-C clearance receptor (~475-amino-acid residues) has lost the intracellular amino acid chain of 566 residues and, thus, has no guanylyl enzymatic activity.

The ANP is produced, stored, and released mainly by the cardiac myocytes of the heart's atrium (Fig. 15.7D). A similar system is operative for the release of the "basic" BNP, by the ventricles of the heart. In each instance an elevated vascular volume results in a stretching of the heart's atrium and ventricles, which initiates the cleavage and secretion, respectively, of both ANP and BNP from the heart. BNP was originally thought to be only secreted by the brain into the circulatory system. More study revealed that BNP was more prevalent in the cardiac ventricles.

The released ANP acts on the kidney (glomeruli) to increase the GFR so as to increase renal blood flow which in a short time increases both urine volume and Na^+ excretion. ANP also decreases plasma renin activity. Natriuresis and diuresis are also enhanced by the suppression of aldosterone production and its biological actions and by the release from the posterior pituitary of arginine vasopressin.

The function of the ANPR-C is still emerging. It is widely distributed in the vascular system, but since the ANPR-C has no intracellular domain, it, therefore, has no kinase and guanylyl cyclase actions. However, CNP has been shown to both relax the vascular smooth muscle and also is a much stronger dilator of veins than either BNP or ANP. A gene knockout of CNP was only lethal for one-half of the mice. Intriguingly, the surviving mice had defects in their endochondral bone development and a shortened body length similar to that of human dwarfs.

F Endothelins

1 Introduction

The vascular endothelium, which provides the barrier between the blood and the vascular wall (see Fig. 15.4B), is the site of production of three hormones that are essential to the regulation of blood pressure; these are the ETs (discussed in this section) and prostacyclin and nitric oxide.

A thin layer of endothelial cells constitutes the interior surface of the arteries on the surface facing the blood.

◀ FIGURE 15.7 Structural organization of the atrial natriuretic peptides, their receptors, and their biological responses. (A) *Amino acid sequences of ANP, BNP, and CNP.* The translation of the three color codes for the atrial natriuretic peptides is as follows. The blue color represents the ANP that is secreted by the human myocytes of the cardiac atrial portion of the heart; this is the reference peptide for the other amino acid sequences. The red color is the B-type BNP that is produced by both the human brain and the cardiac ventricles; the relative contributions of the two sources of BNP are not clear. The green color is the CNP. Each of the three natriuretic peptides is the product of a separate human gene transcript. The extent of the homology of the peptide sequences of BNP and CNP with the reference ANP can be judged by the position and number of the blue residues present in the red residues of BNP and in the green residues of CNP. Both BNP and CNP have precisely the same homology of 12- to 13-amino-acid residues with ANP. Some evidence supports the view that BNP reduces ventricular fibrosis which is the abnormal thickening of heart valves. (B) *Conversion of the linear amino acid sequences of ANP, BNP, and CNP into a circular structure created by a disulfide bond.* All three natriuretic peptides are characterized by the presence of a 17-member amino acid ring formed by a disulfide bridge between two cysteine residues. (C) *Ligand specificity of ANP, BNP, and CNP for binding to three classes of receptor.* Schematic diagram of the interaction of atrial natriuretic peptide hormones binding to their receptors to generate initiation of biological responses. Panel C presents a schematic overview of the structural details of the three families of atrial natriuretic receptors. Both the NPR-A and NPR-B receptor proteins, but not the NPR-C receptor, comprise two identical receptor amino acid sequences (of 1037 amino acids) that are only covalently linked together inside the cell (see the *blue and green boxes*). The extracellular domains have 450-amino-acid residues, a hydrophobic membrane spanning region (~21-amino-acid residues), and an intracellular domain (566-amino-acid residues). The intracellular portion of the receptor protein is divided into three distinct domains: (1) a 250-amino-acid kinase homology domain (see *blue ovals* with several attached phosphates); (2) a 41-amino-acid hinge-dimerization region; and (3) a 250-amin-acid C-terminal guanylyl cyclase enzyme catalytic domain (see *green boxes*). In contrast, two NPR-C receptors form a dimer (via a disulfide bond) outside the cell. The NPR-C has a highly truncated intracellular peptide of only ~80-amino-acid residues and consequently has no catalytic guanylyl cyclase units. Definition of abbreviations: ANP, atrial natriuretic peptide; BNP, basic natriuretic peptide and/or brain natriuretic peptide. (D) *Atrial natriuretic hormones secreted by the heart and the wide sphere of action of their biological responses.* ANP and BNP are true circulating hormones. They are released from the heart's atria and ventricles into the circulatory system where they are taken up by the kidney. There are a variety of responses. These include inhibition of renin secretion, and also inhibition of the *zona glomerulosa* stimulating of aldosterone biosynthesis. ANP increases the glomerular filtration rate as well as the amount of Na^+ excretion in the urine. The capillaries experience an increase in permeability when exposed to ANP, thereby increasing the flow of blood. *ANP*, Atrial natriuretic peptide; *BNP*, brain natriuretic peptide; *CNP*, C-type natriuretic peptide; *FF*, filtration fraction; *FF*, fractional flow; *GFR*, glomerular filtration rate; *NPR*, natriuretic peptide receptor; *SNA*, sympathetic nervous activity; *UNaV*, urine sodium excretion; *UV*, urinary volume; *VSM cell growth*, vascular smooth muscle cell growth. *Abstracted from M.T. Rademaker, E. Espiner, The endocrine heart, in: K.L. Becker (Ed.), Principles and Practice of Endocrinology and Metabolism, third ed., Lippincott Williams & Wilkens, Philadelphia, PA, 2001.*

The endothelial cells are present in arteries leading from the heart down to the capillaries.

2 Chemistry, biosynthesis, and secretion

ET is a 21-amino-acid peptide that is a highly potent vasoconstrictor. The three isoforms of ET 21 are termed ETs 1, 2, and 3 (ET-1, ET-2, and ET-3). The structures of the ETs and their corresponding pro-ETs are presented in Fig. 15.8A/B/C. Three separate ET genes have been identified; each codes for a separate and distinct ET. After transcription of the ET-1 gene, the mature mRNA codes for a prepro-ET-1 of 212 amino acids. Removal of the amino-terminal 20-amino-acid leader sequence generates a 192-amino-acid protein that is an intermediate form of ET-1. Processing of this "intermediate" ET-1 by a pair of dibasic-specific endopeptidases (at Lys51-Arg52 and Arg91-Arg92) generates the 38-amino-acid polypeptide pro-ET-1 (amino acid residues 53–73) which is secreted; it is sometimes referred to as "big" ET-1 (Fig. 15.8B). Pro-ET-1 represents the secreted form of ET; further processing occurs outside the ET cell by the action of an endothelin-converting enzyme (ECE). The sites of ECE cleavage of the pro-ETs are between residues 73 and 74 which are shown in Fig. 15.8B; this generates mature ET-1. The same processing occurs for ET-2 and ET-3. Pro-ET-1 has only 1% of the biological activity of mature ET-1.

3 Biological actions of endothelins

In endothelial cells, ET-1 is predominant, while ET-2 and ET-3 are virtually undetectable. The mRNAs for all three ETs are present in human kidney and the intestinal jejunum, while in the nervous system the mRNA for ET-1 is the major species. Radioligand-binding studies with the isoforms of ET have indicated the presence of two classes of receptors. Type I receptors are involved with vasoconstriction, bronchoconstriction, and stimulation of aldosterone biosynthesis. Type II receptors are

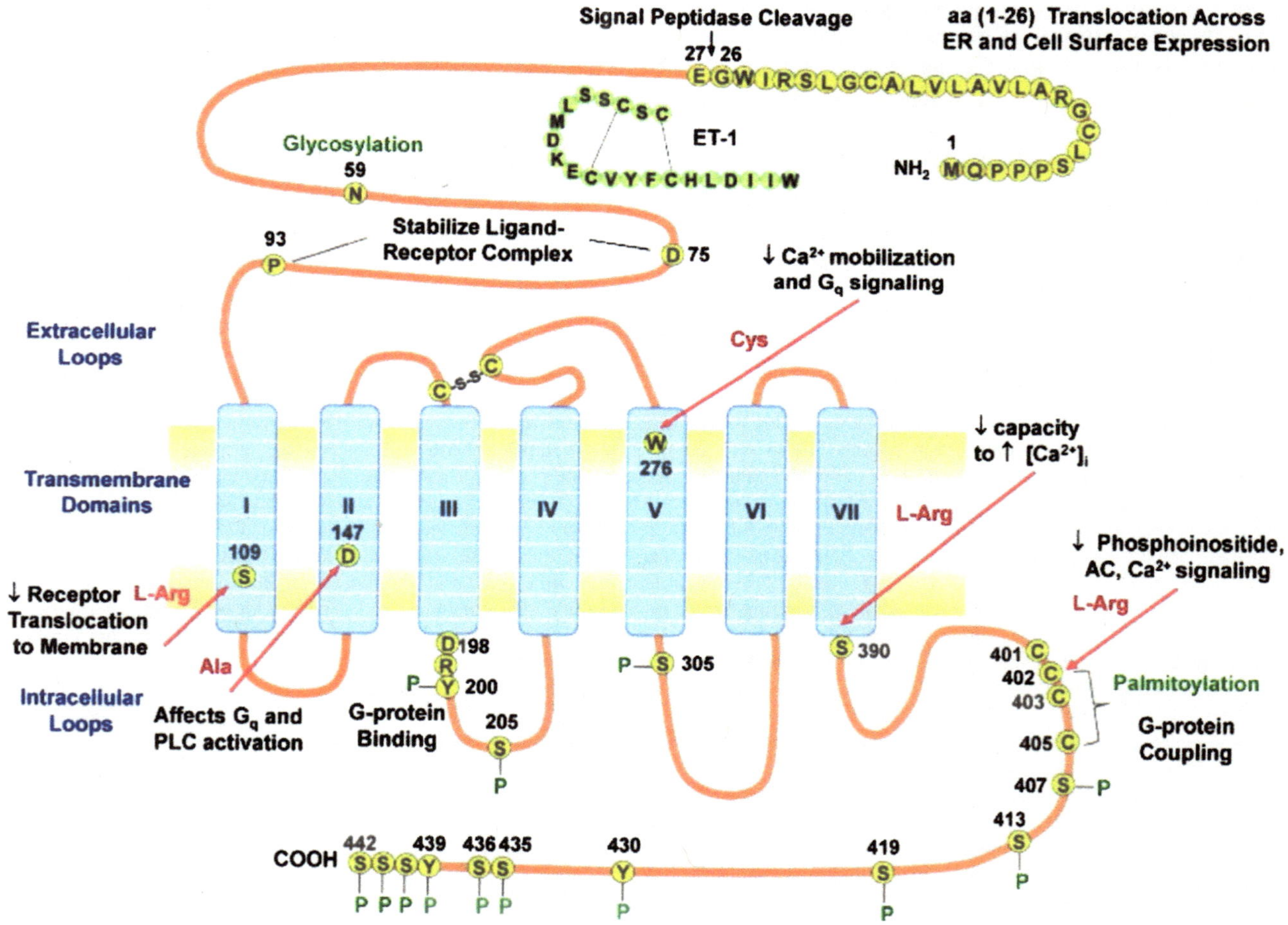

FIGURE 15.8 Human ET_BR structure. ET_BR is composed of a long N-terminus, seven helical TMD extracellular loops, three intracellular loops, and an intracellular C-terminal tail. The receptor has a conserved G protein–coupled receptor DRY sequence and Cys402/403/405. Mutations in some of these sites could affect receptor binding/signaling. Also, palmitoylation, phosphorylation, and glycosylation of ET_BR could affect intracellular signaling. *ANP*, Atrial natriuretic peptide; *BNP*, brain natriuretic peptide; *ET*, Endothelin; *TMD*, transmembrane. *Reproduced from Fig. 1 of M.O. Mazzuca, R.A. Khalil, Vascular endothelin receptor type B: structure, function and dysregulation in vascular disease, Biochemical Pharmacology, 84 (2012) 147–162.*

linked to the inhibition of platelet aggregation and vasorelaxation. Through molecular cloning of ET receptors, the deduced amino acid sequence of seven ET receptors has been elucidated. They can be divided into two classes designated ET_A and ET_B; see Table 15.8. Both classes of ET receptors belong to the seven-transmembrane, G protein–coupled superfamily.

The current view is that ET_A ET receptors mediate the paracrine (vasoconstrictor) actions of ET-1, while the ET_B receptors mediate the more "nonselective" actions of ETs, including autocrine actions related to vasodilation (Fig. 15.8D).

At present, there are at least four known ET receptors; they are designated as ET_A, ET_{B1}, ET_{B2}, etc., all of which are G protein–coupled receptors. Ligand binding of the ETs to their appropriate receptor(s) results in the activation of the receptor and stimulates an increase of the concentration of the intracellular free calcium. Activation of the ET_A receptor leads to vasoconstriction, whereas activation of the ET_{B1} receptor leads to vasodilation. The biological function of the receptor etc. is not yet clear.

Simultaneous biological responses mediated by ETs in adjacent cells are illustrated in Fig. 15.8D. In this model, receptors for ET-1 are present in endothelial cells of the vascular system throughout the body and also the smooth muscle cells that are associated with the endothelial cells. Table 15.8 summarizes the relative ligand affinity of ET-1 and ET-3 for the three receptor isoforms, ET_A, ET_{B1}, and ET_{B2}.

Based on the DNA sequence, it has been determined that the sequence homology is ~88%–90% identical between human and rat receptors for both the ET_A and ET_B receptors. In contrast, surprisingly, the degree of homology between the ET_A and ET_B receptors within a given species (e.g., the human) is only 55%, suggesting that they have quite different responsibilities. Both of these receptors mechanistically are G protein–coupled and when they are ligand activated, the signal transduction for both receptors results in an increase in intracellular free calcium concentration. As shown in Fig. 15.8D for smooth muscle cells, depending upon which signal transduction pathway(s) are used, the end result can be quite different.

TABLE 15.8 Properties of endothelium receptors.

Receptor	Tissue localization	Ligand affinity
ET_A	Vascular smooth muscle	ET-1 > ET-3
ET_{B1}	Vascular endothelium	ET-1=ET-3
ET_{B2}	Vascular smooth muscle	ET-1 ET-3
etc.	Anterior pituitary, renal papilla, vas deferens, myocardium	ET-3 > ET-1

ET, Endothelin.
Source: Abstracted from D.M. Pollock, T.L. Keith, R.F. Highsmith, Endothelin receptors and calcium signaling, The FASEB Journal 9 (1995) 1196–1204.

Receptors for ET-1 are present in endothelial cells of both the vascular system throughout the body and also the associated smooth muscle cells that are intimately associated with the endothelial cells; see Fig. 15.4B/C.

Table 15.8 summarizes the relative ligand affinity of ET-1 and ET-3 for the three receptor isoforms, ET_A, ET_{B1}, and ET_{B2}. In Fig. 15.8D the two endothelial cells are colored green, while the two adjacent smooth muscle cells are colored tan. The endothelial cell on the left has the capability to positively respond to five hormones (see large rectangle with the+on the arrow) so as to stimulate gene transcription and to secrete ET-1. In addition, the green endothelial cell on the left has the capability to partially respond to five different hormones (see rectangle with—on the arrow) which diminishes the transcription of the ET-1 genes and the secretion of ET-1. The positive endothelial cell's ET-1 secretion of ET-1 is sent to the smooth muscle cell on the bottom left where it binds to the ET-1 plasma membrane receptor labeled (A). This results in the activation of phospholipase C which then stimulates IP_3 which, in turn, increases the intracellular Ca^{2+} concentration and also activates the H^+/Na^+ channels that collectively cause the smooth muscle cells to engage in constriction.

The endothelial cell membrane's ET-1 receptor labeled (B), on the upper right endothelial cell, responds to the incoming diminished ET-1 "signal" from the second group of five hormones received by the left endothelial cell. However, there is still a significant increase in the right-hand endothelial cell's intracellular Ca^{2+} concentration. This causes the calmodulin, by binding Ca^{2+}, to become active and activate the endothelial NO synthase (NOS) enzyme to secrete nitric oxide (NO). The NO, which has a half-life in cells of ~5 seconds, can rapidly diffuse through the cell membrane of the right tan cell and bind to a nearby sGC (soluble guanylyl cyclase) which activates the cGMP pathway. This, in turn, causes the smooth muscle cells to engage in "relaxation."

Note the three labels underneath the two smooth muscle cells; they are *constriction, relaxation, and proliferation.* Each label describes the response initiated by the ET-1 signal transduction received. The right-hand smooth muscle cell also can respond directly to ET-1 binding to its membrane receptor labeled *A*, which activates phospholipase C, to produce DAG, which then activates protein kinase C and finally mitogen-activated protein kinase. This leads to the proliferation of this smooth muscle cell. In the left-hand smooth muscle cell, it is described that an increase in intracellular Ca^{2+} can lead to the stimulation of "constriction" of smooth muscle cell membranes. This can contribute to an increase in blood pressure.

The ET_A is known to mediate the process of vasoconstriction (raises blood pressure), while ET_B receptors mediate the process of vasodilation (lowers blood pressure). As for the subtypes of the ET_B receptor, ET_{B1} mediates vasodilation, while ET_{B2} mediates vasoconstriction. In Fig. 15.9 the structure of the endothelinβ receptor ($ET_{\beta}R$) is shown. The receptor has a long N-terminus and seven transmembrane domains. $ET_{\beta}R$ is a member of the rhodopsin-type superfamily of G protein–coupled receptors. Activities of various domains of the receptor are indicated in the figure.

Based on the DNA sequence, it has been determined that the sequence homology is ~88%–90% identical between human and rat receptors for both the ET_A and ET_B receptors. In contrast, and surprisingly, the degree of homology between the ET_A and ET_B receptors within a given species (e.g., the human) is only 55%, suggesting that they have quite different responsibilities.

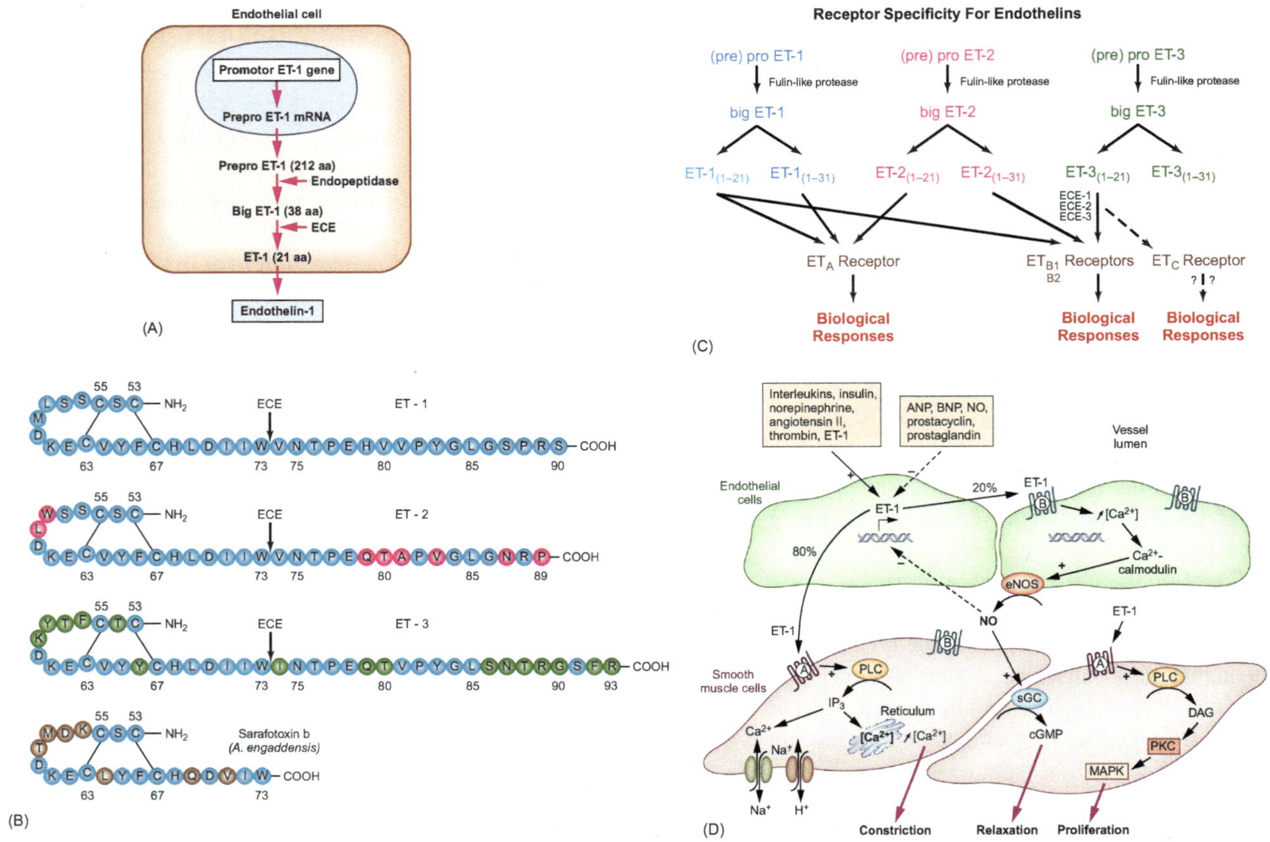

FIGURE 15.9 The endothelin gene, their hormones, and receptor families and receptor signal transduction events in the vasculature resulting in a reduction of blood pressure. (A) *Processing of the endothelin prepro ET-1 mRNA*. The human genome has three separate genes that individually code for three endothelin amino acid sequences; they are ET-1, ET-2, and ET-3. The ET-1 prepro endothelial-1 has 212 amino acids, the "big" ET-1 has 38 amino acids, and the mature ET-1 has 21 amino acids. (B) *Isoforms of endothelins*. There are three separate biologically active isoforms of endothelin which are all about the same prosize; they are ET-1 (90 aa, *blue*), ET-2 (89 aa, *red*), and ET-3 (93 aa, *green*). Panel B shows the amino acid sequence for ET-1, ET-2, and ET-3. The reference amino acid sequence is ET-1; all the residues are colored blue. ET-2 has eight-amino-acid residues different from that of ET-1; each is marked by the red color. ET-3 has 16-aa residues (*green*) that are different from ET-1. The vertical arrow indicates the site of proteolytic cleavage by the ECE, which generates the three mature, biologically active endothelins. The two solid lines at the left end loops indicate the presence of their two disulfide linkages for each of the three endothelins. Also shown in the bottom row is the amino acid sequence of the structurally homologous sarafotoxin b, which is present in the venom of the burrowing Egyptian asp, *Atractaspis engaddensis*. The sarafotoxin b is a potent vasoconstrictor in cerebral arteries of some mammals where it is employed to protect the asp. (C) *Schematic presentation of four families of endothelin receptors*. At present there are at least four known endothelin receptors; they are designated as ET_A, ET_{B1}, ET_{B2}, etc., all of which are *G protein–coupled receptors*. Ligand binding of the endothelins to their appropriate receptor(s) results in the activation of the receptor and stimulates an increase of the concentration of the intracellular-free calcium. Activation of the ET_A receptor leads to vasoconstriction, whereas the activation of the ET_{B1} receptor leads to vasodilation. The biological function of the receptor etc. is not yet clear. (D) *Biological responses mediated by endothelins in the presence of smooth muscle*. In panel (D) the two endothelial producing cells are colored green, while the two adjacent endothelial responding smooth muscle cells are colored tan. The endothelial cell on the left has the capability to positively respond to 10 hormones (see left rectangle+) and to stimulate gene transcription and production of ET-1 for secretion. This panel is described in more detail in Section III.F.3. *aa*, Amino acids; *Ca^{2+}*, cytosolic calcium; *DAG*, diacylglycerol; *ECE*, endothelin-converting enzyme; *ET*, endothelin; *IP_3*; inositol triphosphate; *mRNA*, messenger RNA; *NO*, nitric oxide; *PKC*, protein kinase C; *PLC*, phospholipase C; *PLD*; phospholipase D; *PT*, pertussis toxin; *TK*, tyrosine kinase.

G Nitric oxide system

1 Introduction

A relatively surprising addition to the family of chemical messengers is nitric oxide (NO). NO is a free radical gas of limited solubility in water. Since NO is not charged, it can rapidly diffuse across cell membranes and into cells. It has been shown to act as both an intracellular and an intercellular (paracrine) messenger to elicit a wide spectrum of biological responses. NO is now known to be an integral participant in the signal transduction processes associated with the vascular, immune, and neural systems. Also, the gas, NO, is a hormone that initiates vasorelaxation and lowers blood pressure.

2 Chemistry, biosynthesis, and secretion

The Nobel Prize in Physiology and Medicine was awarded in 1999 for the discovery in the early 1980s that the gas NO was found to be a hormone in the cardiovascular system. The recipients of the award were R.F. Furchgott, L.J. Ignarro, and F. Murad.

Biochemical properties of the nitric acid synthase enzyme. Nitric acid (NO) is enzymatically produced from the amino acid L-arginine by a Ca^{2+}-dependent NOS that is present in many mammalian tissues. Particularly, arteries and their endothelial cells are able to produce NO.

The formation and production of NO is an enzyme-mediated reaction; see Fig. 15.10A and B. NOS are a family of three enzymes catalyzing the production of nitric oxide (NO) from L-arginine. Each of the three NOS enzymes is encoded by different genes. The nitrogen donor is the amino acid L-arginine and the oxygen donor is molecular oxygen. The reaction is catalyzed by an NADPH requiring the NOS. NOS enzymes are structurally related to cytochrome P450 reductase (see Chapter 2: Steroid Hormones: Chemistry, Biosynthesis, and Metabolism) and range in size from 130 to 160 kDa. NOS exists as a constitutive enzyme, which is regulated by both the intracellular Ca^{2+} and the calcium-binding protein, calmodulin.

The hormone NO is a free radical gas that has a limited solubility in water. As previously mentioned, a very helpful property is that, because NO is not charged, it is readily able to diffuse from endothelial cells through the cellular membrane and gain entry into nearby cells, such as smooth muscle cells where biological responses can be initiated. However, NO is inherently unstable. The time interval from its enzymatic formulation to its chemical breakdown is only several seconds. The NO undergoes

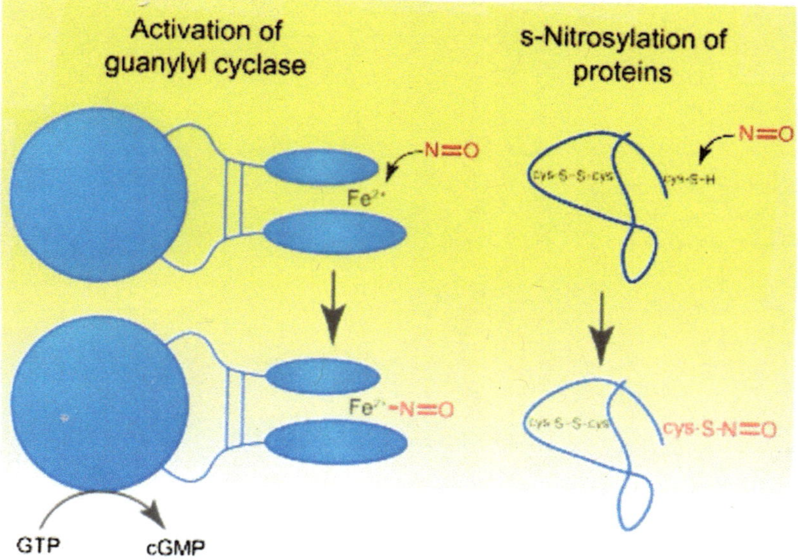

FIGURE 15.10 The gas, nitric oxide, is a hormone that initiates vasorelaxation and lowers blood pressure. (A) *Biochemical properties of the nitric acid synthase enzyme.* Nitric acid (NO) is enzymatically produced from the amino acid L-arginine by a Ca^{2+}-dependent NO synthase that is present in many mammalian tissues. The reaction is catalyzed by an NADPH requiring NOS. The nitrogen donor (*red*) is located on the guanidino group of the side chain of the amino acid substrate L-arginine. The oxygen donor (*red*) is molecular oxygen. The final products of the enzymatic reaction are the release of the gas nitric oxide, NO, and the amino acid L-citrulline. (B) *Nitric oxide signaling pathways that generate biological responses in both endothelial cells and smooth muscle cells.* Endothelial cells have a variety of supporting signal transduction pathways that lead to the activation of nitrogen oxide synthase, which then diffuses to adjacent smooth muscle cells. Here it enters the cell and, via signal activation, causes the smooth muscle cell to engage in lowering the blood pressure because of the reduction of peripheral vascular resistance and at the same time increase. Ultimately the generation of NO leads to the relaxation of both the endothelial cells and its partner smooth muscle cell(s). In these cells, there is guanylyl cyclase that has a heme group that binds the NO hormone, thus activating the production of the second messenger, cGMP, which results in muscle relaxation. *cGMP*, Cyclic guanosine monophosphate; *NOS*, nitric oxide synthase.

spontaneous chemical oxidation to either nitrate (NHO_3^-) or nitrite (NHO_2^-) neither of which has hormonal or biological activities. However, since there is an excess of the NOS enzyme, there is no difficulty in maintaining a continuous availability of NO in the local environment. NO binds to its receptor in the low nanomolar (10^{-9} M) concentration range. A prominent action of nitric oxide (NO) is in its binding to guanylyl (guanylate) cyclase leading to the activation of the enzyme, as illustrated in Fig. 15.11. NO binds to an iron molecule in the enzyme forming protein-Fe^{2+}-N=O. Also illustrated in this figure is the s-nitrosylation of various proteins where NO binds to protein–Cys–S–H to form protein–Cys–S–N=O.

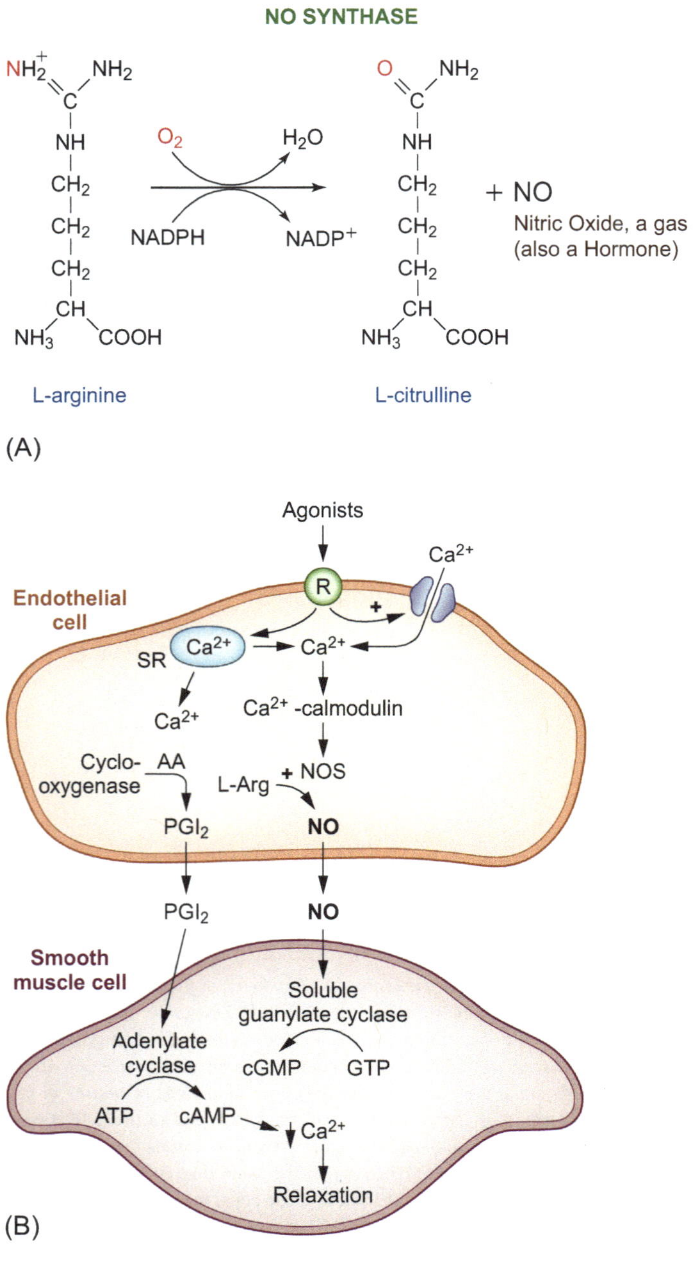

FIGURE 15.11 Common nitric oxide reactions in the nervous system that mediate signaling. (Left) Guanylyl cyclase can be activated by nitric oxide (N=O) when it reacts with an iron molecule that is coordinated between two domains of the guanylyl cyclase enzyme. The reaction activates the catalytic subunit of the enzyme, generating cGMP from GTP. (Right) Proteins can be covalently modified when nitric acid reacts with free cysteine amino acids in a protein backbone. Cysteines within the protein that form a disulfide bond with another cysteine (Cys–S–S–Cys) do not react with nitric oxide. This nitrosylation of cysteines in the protein backbone changes the conformation of the protein, changing its function. *cGMP*, Cyclic guanosine monophosphate; *GTP*, guanosine triphosphate. *Reproduced from Fig. 20.4 of S.D. Meriney, E.E. Fanselow, Gaseous Neurotransmitters in Nitric Oxide, 2019, https://www.sciencedirect.com/topics/neuroscience/nitric-oxide.*

3 Biological actions of NO

NO functions as an important cellular signaling molecule. It modulates insulin secretion, airway, and vascular tone and has been shown to be involved in angiogenesis and neural development. Further examples of biological actions of NO in the cardiovascular, nervous, and host defense systems are summarized in Table 15.9. Only the actions of NO in the cardiovascular system will be discussed in this chapter. After the generation of NO in the endothelial cell by the NOS enzyme, the NO diffuses to an adjacent smooth muscle where it acts as an agonist to initiate biological responses.

TABLE 15.9 Biological actions of nitric oxide.

System	Response
Cardiovascular	
Smooth muscle	Initiate vasorelaxation; control of regional blood flow and blood pressure
Platelets	Limitation of aggregation and adhesion
Nervous	
Peripheral	Neurotransmission (penile erection, gastric emptying)
Central	Neurotransmission; long-term potentiation (appetite control, nociception)
Host defense	
Macrophages	Defense against bacteria, fungi, protozoans, parasites, and viruses
Leukocytes	
Monocytes	

See Fig. 15.10B. In the endothelial cell the arrival of the hormone ET-2 activates its membrane receptor which, in turn, increases the intracellular Ca^{2+} concentration and as well allows the entrance of Ca^{2+} into the cell through a Ca^{2+} permease. The Ca^{2+} then binds to the Ca^{2+}-binding protein, calmodulin, which interacts with the NOS and releases NO. The hormone NO then diffuses through the cell membrane of the endothelial cell and then diffuses across the cell membrane of a nearby smooth muscle cell and binds to the soluble guanylate cyclase and generates cGMP. At the same time, the fatty acid arachidonic acid stimulates the release of prostacyclin (PGI_2), which then diffuses through the endothelial cell plasma membrane followed by an entrance into the smooth muscle cell where it stimulates the adenylate cyclase to produce the second messenger cAMP, which causes the smooth muscle cell to engage in relaxation.

Virtually all of the known biological actions of NO on the cardiovascular system are mediated by the activation of a soluble guanylate cyclase. The guanylate cyclase has heme as a prosthetic group, and NO binds tightly to this moiety in the smooth muscle cells. The resulting activation of the guanylate cyclase results in the production of cGMP, which is then postulated to have actions on protein kinases, nucleotide-sensitive phosphodiesterases, or ion channels, which are linked to the generation of the smooth muscle response of vasodilation and relaxation.

H Kallikreins and kinins

1 Introduction

The kallikreins are a group of serine proteases known as kininogens, which act on plasma α_2-globulins to release kinins such as bradykinin. Bradykinin is the most potent vasodilator substance known.

The kallikrein–kinin system consists of proteins present in blood and other body fluids such as lymph, urine, saliva, pancreatic juices. In the blood, they play a role in coagulation, blood pressure control, and inflammation.

2 Biochemistry and physiology

Two classes of kallikreins have been identified: (1) those present in plasma and (2) those present in organs, including the kidney, salivary glands, and pancreas. The plasma kallikreins have molecular weights of ~70,000 and ~110,000 Da. All of the kallikreins are serine proteases. There are 14 kallikreins.

The plasma kallikrein is normally found as a proenzyme termed prekallikrein. Prekallikrein is activated to kallikrein by one of the blood-clotting factors, factor XII, or Hageman factor. In turn, the Hageman factor is activated by the plasma kallikrein (see Fig. 15.5C). The kallikreins utilize their protease activity to release peptide kinins from their precursor substrate forms.

The receptors for bradykinin are a group of G protein–coupled proteins that, when activated, increase the cytosol Ca^{2+} concentration. The bradykinin receptor is present on the human BDKRB1 gene. The principal kinin is a nonapeptide named bradykinin. It has the following amino acid sequence: Arg–Pro–Pro–Gly–Phe–Ser–Pro–Phe–Arg. The bradykinin receptor, which is principally associated with endothelial cells, has been cloned and found to belong to the seven-transmembrane G protein–coupled receptor superfamily.

It is noteworthy that the principal mode of inactivation of bradykinin is the removal from the COOH-terminal of a dipeptide by an enzyme termed kininase II; see Fig. 15.5C. The enzymatic activities of kininase II and the enzymatic activities of ACE have shown them to be the same protein. Thus ACE, which generates the potent vasoconstrictor, angiotensin II, inactivates the vasodilator bradykinin.

I Adrenomedullin

1 Adrenomedullin peptide structure and related topics

The amino acid sequence of AM is shown in Fig. 15.12A. The AM gene, which is present in humans, generates a prepro 185-amino-acid peptide, followed by a pro-aa164 peptide and a final 52-amino-acid sequence of AM which has a widespread list of biological properties (see later). Panel A lists the final 52-amino-acid sequence of AM. It includes a disulfide bond from residue 16 to 21, thus creating a 6-amino-acid loop.

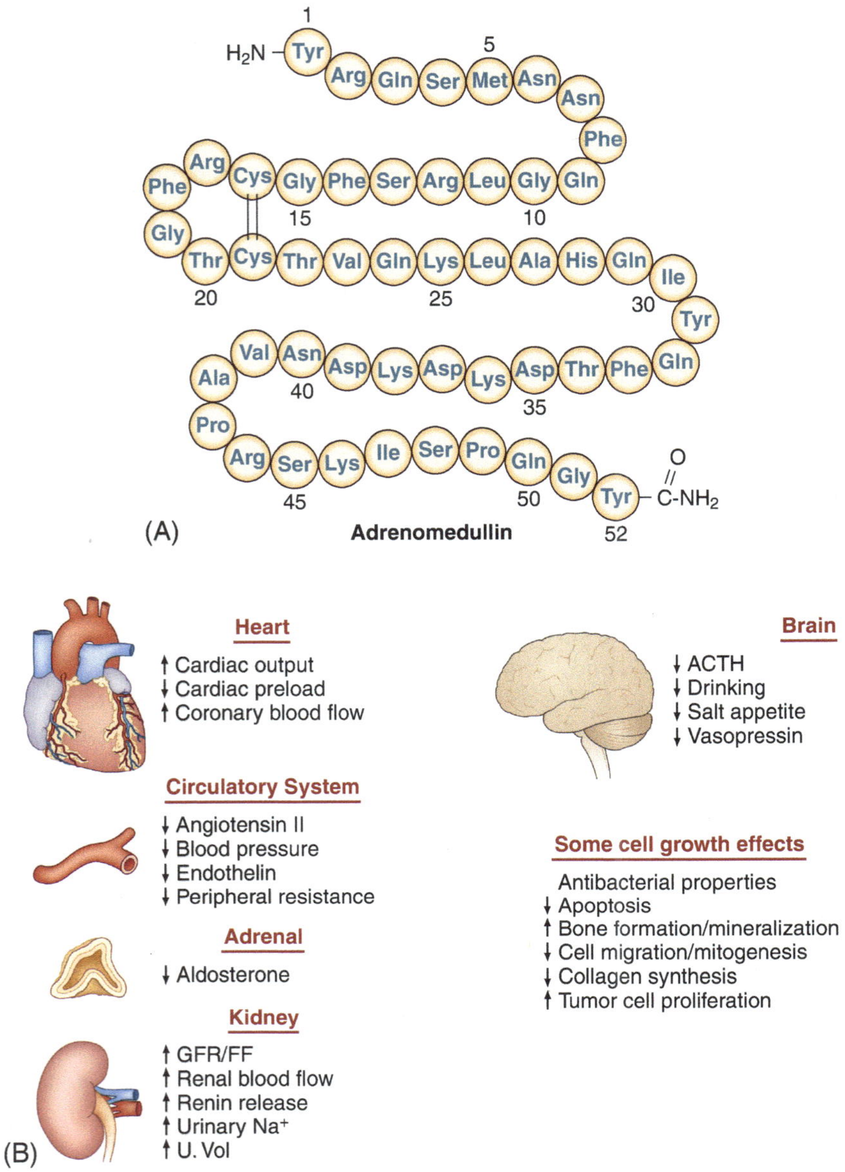

FIGURE 15.12 Biochemical and biological properties for the peptide hormone adrenomedullin (AM). (A) *AM peptide structure*. The AM gene, which is present in humans, generates a prepro 185-amino-acid peptide, followed by a pro-aa164 peptide and a final 52-amino-acid sequence of AM which has a widespread list of biological properties (see the following). Panel (A) lists the final 52-amino-acid sequence of AM. It includes a disulfide bond from residue 16 to 21, thus creating a 6-amino-acid loop. The AM carboxyl terminus has an amide group. In addition, a second biologically active PAMP is produced. (B) *Physiological functions of AM*. The normal circulatory concentration of AM in humans is in the range of 2–8 pmol/L. The acute administration of AM to humans results in a prompt fall in both blood pressure and vascular resistance. *AM*, Adrenomedullin; *PAMP*, pro-AM N-terminal 20-amino-acid peptide.

The AM carboxyl terminus has an amide group. In addition, a second biologically active pro-AM N-terminal 20-amino-acid peptide is produced. The amino acid sequence and properties of AM display a moderate similarity to the calcitonin gene–related peptide superfamily; see Tables 7.1 and 7.2 of Chapter 7, Gastrointestinal Hormones. The biological properties supported by the hormone AM are separately dependent upon one of two separate AM1 and AM2 receptors. Both of the receptors belong to the seven-transmembrane G protein–coupled family which produces the second messenger cAMP. However, there are reports describing other second messengers, including nitric oxide, NO, changes in intracellular Ca^{2+}, the involvement of prostaglandins and MAPK. The T1/2 of AM is ~20 minutes, which is reflective of the presence of a lively production of AM as well as a relatively vigorous breakdown mechanism.

2 Physiological functions of adrenomedullin

The physiological functions of AM are schematically shown in Fig. 15.12B. The normal circulatory concentration of AM in humans is in the range of 2.8 pmol/L. The acute administration of AM to humans results in a prompt fall in both blood pressure and vascular resistance. This occurs because of the stimulation of the second messengers cAMP and cGMP.

In addition to the sharp reduction in the circulatory system's blood pressure, there is also a fall in peripheral resistance, and reductions in the concentration of angiotensin II and ETs. In the kidney, administration of AM to humans results in a rapid increase in renal blood flow, renin release, and an increase in the glomerular filtration rate and an increase in the adversary effects of both urinary Na^+ excretion and water excretion. The presence of AM in the circulatory system and its adverse effects on the kidney lead to a reduction in the production by angiotensin II and the adrenals *zona glomerulosa* of aldosterone. As shown in Fig. 15.12B under the heading of "Some cell growth effects," the presence of AM stimulates bone formation, and under certain conditions it increases tumor cell proliferation. AM can inhibit apoptosis, cell migration/mitogenesis, and collagen synthesis. In animal studies, where there was a complete homozygous embryonic knockout of the gene for AM, there was a 100% lethality. In contrast, in the heterozygous AM mice there was an increase in blood pressure mediated by the in vivo production of some circulating-level AM. The conclusion of the scientists is that AM "plays an important role in the regulation of blood pressure."

Because of combined vascular relaxation, natriuresis, and contraction of plasma volume, the NP system is likely the primary defense against hypervolemia. However, with the discovery of AM and the wide array of AM-mediated responses, it may be possible eventually to understand and define the contribution of AM.

J Summary

In the normal physiological state in humans, the distribution, composition, and volume of body fluids are held within relatively narrow limits, despite wide variations in the intake of water and Na^+. Such homeostasis or stability of the internal environment requires an intricate collaboration of the hormones and their physiological processes that affect electrolyte and water metabolism. Thus as has been reviewed in the preceding sections of this chapter, a diverse array of hormones (aldosterone, angiotensin II, renin, ANP, NO, ET, AM, vasopressin, prostaglandins, and kinins) and related signal transduction systems, each responding to different stimuli, is integrated to provide the kidney and cardiovascular system with coherent messages to effect the appropriate regulation of blood pressure, electrolyte concentration, and water volume.

IV Hormones and blood cell production

A Introduction

It is known that an average human consumes 550 L of oxygen daily, which is 19 $ft.^3$/day or 631 lbs a year. The starting point for this process is the inhalation of air, which is 20% oxygen and 80% nitrogen, and the exhaled air is reduced to 15% oxygen. The explanation for the reduction from 20% to 15% is because 5% of the inhaled oxygen has been absorbed by the hemoglobin present in the erythrocyte cells of the blood as they pass through the lung. After leaving the lung the RBCs travel throughout the total circulatory system of the body so that it can offload the oxygen bound to the hemoglobin. The released O_2 can then diffuse to all the cells throughout the body where it is utilized by metabolic processes. This analysis explains the basis for the statement that an adult human uses ~550 L or 19 $ft.^3$ of oxygen per day. The key to understanding in detail this information is to appreciate the properties of the protein hemoglobin and the related process of hematopoiesis.

The various cellular components of blood are produced by the hematopoietic system. The main components of the hematopoietic system are the bone marrow and all the blood present in the circulatory system. In addition, the liver and spleen are key accessory organs. The spleen, which is located in the left upper quarter of the abdomen, is $\sim 1'' \times 3'' \times 5''$ in size. The spleen is important for many functions in the body, from degrading and filtering out old retired RBCs to creating antibodies. But surprisingly, the spleen is not essential for life. The spleen also stores platelets that normally help in blood clotting and coagulation. The liver stores vitamin B_{12} which is essential for erythrocyte production as well as the biosynthesis of hemoglobin.

Erythrocytes are also described as RBCs. All erythrocytes contain significant amounts of hemoglobin. The main biological property of hemoglobin's function is to capture and specifically bind the oxygen that has been inhaled by the lungs. The oxygen binds specifically to a heme group present in each hemoglobin molecule (this will be discussed in more detail under the topic *erythropoiesis*). The major function of the oxygenated hemoglobin is for the erythrocytes to deliver the hemoglobin/O_2 through the circulatory system (arteries, capillaries, venules, etc.) to *all the body's cells*. In addition, after the hemoglobin has released its bound oxygen, it has the equally important responsibility to carry a significant proportion of the waste carbon dioxide (CO_2) from all metabolizing cells back to the lungs for exhalation (excretion from the body).

Bone marrow is the flexible tissue in the interior of bones and it contributes ~4% of the total body weight. Each erythrocyte is created from a stem cell released in the bone marrow. The term "erythron" describes the total mass of the circulating precursors and the mature RBCs that are present in the blood compartment. Another way to describe the erythron is as a dispersed organ, prime function of which is the transport of oxygen in the outbound direction (from the lungs) to all the cells of the body and the transport of carbon dioxide in the inbound direction (back to the lungs), and as well, the maintenance of the blood pH. In adults the erythron is in a steady state where the loss of cells is precisely balanced by the new production of cells. Every minute of every day an adult human must biosynthesize approximately 150 million erythrocytes and 120 million granulocytes, as well as numerous mononuclear cells and platelets; see Table 15.10. Or stated differently, adult humans' erythrocytes are produced at a rate of 2–4 million every one second!! There are numerous hormones and regulatory processes that control the crucial aspects of the various cell populations of the erythron.

A detailed consideration of the complexities of the overall hematopoietic process is beyond the scope of this presentation. The focus will be on the process of erythropoiesis, the protein EPO, and the function of RBCs.

B Process of erythropoiesis

The process of producing RBCs is labeled erythropoiesis. In humans the erythropoietic process occurs exclusively in the red bone marrow of all the bones from birth to ~ages 5–20 years. Over the interval of 5–20 years, the long bone marrow slowly loses its capability to provide RBCs. By the age of 20 the responsibility of carrying out erythropoiesis is retained only by the bone marrow of the ribs, sternum, and the pelvis.

The process of producing RBCs is completely dependent upon the availability of a protein hormone, EPO, which is secreted by the kidney. When the kidney detects low levels of oxygen in the blood, it promptly releases the protein EPO, which then arrives shortly at the body's bone marrow via traveling through the blood circulatory system; see Fig. 15.13A.

EPO is biosynthesized in the peritubular capillary lining cells of the renal cortex region of the kidney (see Fig. 15.1A). The EPO then moves through the circulatory system and ultimately binds to the EPO receptors associated with proerythroblasts present in the bone marrow of the entire skeleton; see Fig. 15.13, panels A/B/C. This is the site of production of the hemoglobin that is produced for all the RBCs.

C Erythropoietin (the protein)

EPO is the hematopoietic hormone secreted by the kidney that functions as a major stimulator for the production of erythrocytes. EPO is a secreted glycoprotein of 165 amino acids, but with a mature molecular mass of 30–34 kDa due to the presence of a significant amount of covalently linked carbohydrate. Hypoxia (a low-oxygen environment) can stimulate the synthesis of EPO mRNA in the renal capillary tubular cells by unknown mechanism(s). In the adult human, EPO is a classic hormone, secreted by the kidney and transported systemically to its site of action in the erythron.

Each molecule of hemoglobin is composed of two paired amino acid chains of $\alpha_2\beta_2$ with a molecular weight of ~16,000. The four heme prosthetic group is synthesized in the cytosol and mitochondria of the immature RBCs; see the proerythroblasts, basophilic erythroblasts, and erythroblasts in Fig. 15.13, panel C. Each heme group binds one O_2 and the intact hemoglobin molecule $\alpha_2\beta_2$

TABLE 15.10 Cellular components of blood.

Cell type	Normal range (number)	Function
Red blood cells (erythrocytes)	Male 5.6×10^6	Oxygen transport
	Female 4.8×10^6	
Platelets (thrombocytes)	$130–370 \times 10^6$	Essential for clotting
WBC; also known as leukocytes	$5–10 \times 10^3$	Indicator of the presence of disease
Lymphocytes	25%–30% of WBC	
B cells		Antibody production
T cells		Participate in cell's mediated immune responses
Monocytes	6% of WBC	Precursors of macrophages
Macrophages		Phagocytosis
Neutrophils	60% of WBC	Phagocytosis
Eosinophils	1%–3% of WBC	Participate in allergic response
Basophils	1% of WBC	Prevention of inappropriate clotting

WBC, White blood cells.

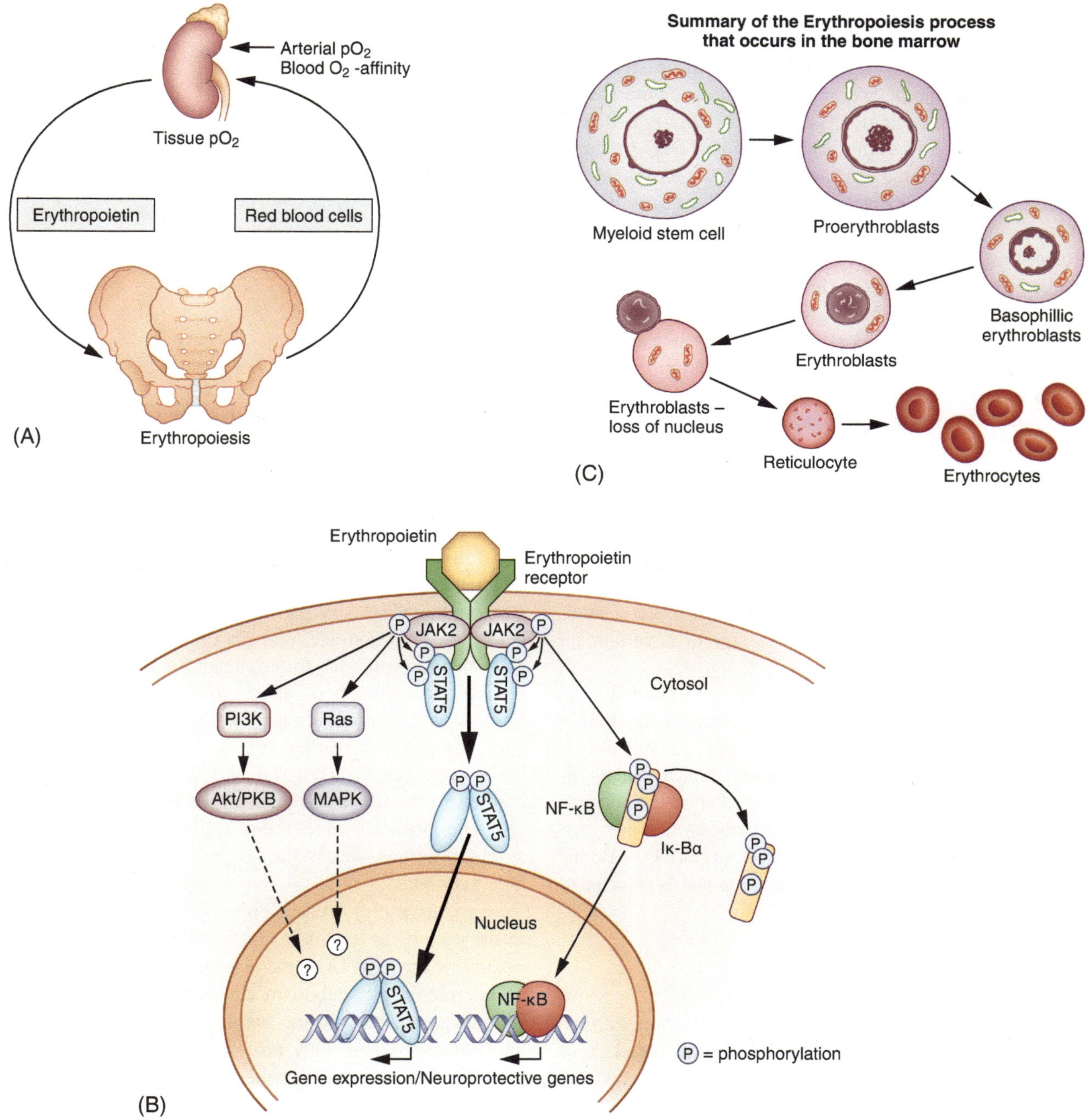

FIGURE 15.13 Role of the kidney in blood erythropoiesis. (A) *Illustration of the collaboration between the kidney as a source of the hormone EPO and the skeletal system as a bone marrow site to biosynthesize hemoglobin for the red blood cells.* The protein hormone EPO is critical for the production of the red blood cells for the whole body. The renal cortex of the kidney (see Fig. 15.1A) is the source of production of EPO. When the kidney senses the lack of oxygen (hypoxia is the condition when oxygen is lower than the normal level), it mobilizes the production and secretion of the glycoprotein hormone, EPO. EPO is biosynthesized in the peritubular capillary lining cells of the renal cortex region of the kidney (see Fig. 15.1A). The EPO then moves through the circulatory system and ultimately binds to the EPO receptors associated with proerythroblasts present in the bone marrow of the entire skeleton. This is the site of production of the hemoglobin that is produced for all the red blood cells. (B) *Mode of action of EPO in the regulation of gene expression.* After the secretion of EPO into the circulatory system, it ultimately is taken up by erythrocytic progenitor myeloid stem cells present in the bone marrow. One EPO hormone molecule binds to two collaborative receptors present in the cell membrane of myeloid stem cells. The receptors activate the JAK2 and STATS pathways of PI3K—Akt/PKB, Ras-MAPK, and the NF-kB pathway. Collectively these pathways stimulate the production of differentiation and growth of the myeloid stem cells into intermediate cell types (basophilic erythroblasts and then more mature erythroblasts that have lost their nucleus). (C) *Summary of the process of erythropoiesis which throughout life is producing new erythrocytes necessary to transport oxygen to all cells in the body.* The starting point is activation of the myeloid stem cell, followed by differentiation into proerythroblasts, basophilic erythroblasts, with their cell nucleus and then without their cell nucleus, followed by differentiation into the final reticulocyte, which can join the large family of erythrocytes. *Akt*, Protein kinase B; *EPO*, erythropoietin; *JAK2*, Janus-2 kinases; *MAPK*, mitogen-activated protein kinase; *NF-kB*, nuclear factor kappa B; *PI3K*, phosphatidylinositol-3-kinase; *PKB*, protein kinase B; *STAT*, signal transducer and activator of transcription.

normally carries four oxygen O_2 molecules linked to four separate heme groups. The intact human erythrocytes survive only ~120 days, which justifies the necessity for a vigorous production of erythrocytes in the bone marrow.

The important biological property of EPO is that it is antiapoptotic. See Fig. 15.13B. Thus when EPO binds to the EPO receptor as a homodimer in the plasma membrane, it results in the activation of the cytoplasmic Janus-2 kinases pathway that catalyzes the phosphorylation of tyrosine residues of the EPO-receptor as well as of intracellular proteins (transcription factors and enzymes). Collectively these actions facilitate the production and maturation of the reticulocytes in the bone marrow compartment which supports the essential differentiation and growth properties of the myeloid stem cells.

Fig. 15.13C illustrates a summary of the process of erythropoiesis which throughout life is producing new erythrocytes necessary to transport oxygen to all cells in the body. This is dependent upon converting the myeloid stem cell successfully into proerythroblasts, then basophilic erythroblasts, erythroblasts (with and then without a nucleus), and finally into reticulocytes. They are then released/secreted and circulate in the body's blood compartment for ~1 day while they complete development into mature erythrocyte RBCs.

The RBCs are functionally active for ~3–4 months and then they begin to deteriorate. This is because the RBC's membranes have become fragile as a result of many passages through narrow arterioles and capillaries, which can physically disrupt and break the RBC. These damaged RBCs are eventually taken out of the general circulation when they pass through the spleen, which breaks them down further. Also the iron is released from the hemoglobin group and is recycled to be incorporated into a new RBC.

D Hemoglobin

Both H^+ and O_2 bind directly to hemoglobin with an inverse affinity that is pH dependent. In the presence of high oxygen, hemoglobin binds oxygen and releases protons. But when oxygen concentration is low, hemoglobin binds H^+ and oxygen is released. As will be discussed next, hemoglobin binds O_2 specifically to iron Fe^{2+} in a heme group. The proton, H^+, binds randomly to several of hemoglobin's amino acids, such as histidine.

Hemoglobin binds oxygen cooperatively and generates two conformations designated as the R state (relaxed) and T state (taut). Oxygen has a much higher affinity for hemoglobin in the R state. When the oxygen concentration is reduced or absent, the T state of hemoglobin is dominant. Some of the consequences of changes in pH are illustrated in panel C of Fig. 15.14.

When the hemoglobin that is present in the circulatory system passes through the lungs (in the presence of high

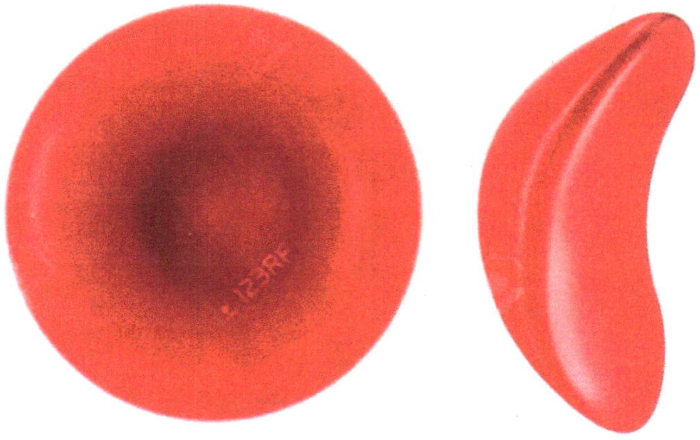

FIGURE 15.14 On the left is shown a normal red blood cell. On the right is shown a sickle cell. *Reproduced from http://www.123rf.com/photo_76762063_stock-vector-sickle-cell-and-normal-red-blood-cell-illustration.html.*

concentrations of oxygen in the blood from inhalation), the pH is ~7.6, and it is relatively easy to saturate the hemoglobin with O_2. See Fig. 15.14, panel C. In contrast, when the blood hemoglobin is passing in the peripheral tissues, the pH is on the low side (~pH 7.2, or more acidic) and the extracellular O_2 concentration is intrinsically much lower (due to the metabolic use of O_2 in intermediary metabolism of all the tissues). Thus the newly O_2-saturated hemoglobin arriving in the peripheral tissues from the lungs is readily able to dissociate O_2 from the heme state (Fe^{2+}). This then allows the hemoglobin to convert to the relaxed state where it is optimized to carry two end products of cellular respiration, CO_2 and H^+, from the tissues back to the lungs and kidneys where they are excreted.

Another useful consequence of hemoglobin's T state in the peripheral tissues is that ~35% of the total H^+ and ~20% of the total CO_2 that is released as a waste product of intermediary metabolism can bind to hemoglobin and be transported back to the lungs, where they are exhaled, and kidneys, where they are excreted. The remainder of the CO_2 is dissolved in water along with the H^+ which generates bicarbonate (HCO3–) and moves through the circulatory system back to the lungs, where it is converted back to CO_2 and is exhaled.

The heart of the hemoglobin structure is shown in Fig. 15.14, panels A/B. The covalent linkage of four pyrrole rings to one another collectively confers upon the hemoglobin protein the ability to transport oxygen bound to the Fe^{2+} in the center of the heme group. The four planar pyrrole rings are each connected, in an "oval," by methylene bridges. Each of the four pyrrole rings in Fig. 15.14B has a peach color. The stoichiometry is one pyrrole ring per one Fe^{2+} ion which has six coordination bonds. Four of the bonds are in the plane of the overall pyrrole ring, and the other two bonds are perpendicular (one above to an oxygen and one below to a histidine residue of hemoglobin). The Fe^{2+} ion has a tight bond to O_2 and ensures that the O_2 will be safely delivered from the lungs through the arteriole system to the peripheral cells.

The four groups of four "circular" pyrrole rings are visible (*green*) in *each* of the hemoglobin's $\alpha_2\beta_2$ subunits (Fig. 15.14A). Thus one molecule of $\alpha_2\beta_2$ hemoglobin binds four molecules of O_2. Also, unfortunately, each pyrrole ring can also bind one carbon monoxide molecule (CO) to the heme group. Since hemoglobin binds CO with a ~250-fold greater affinity than O_2, the CO will supplant the normally present O_2. This describes why the circumstances of extensive exposure to CO gas can be lethal due to the blocking of transfer of O_2 from the lungs to the peripheral tissues.

See Fig. 15.14, panel C; the right panel shows two oxygen-binding curves to hemoglobin, one for a normal healthy individual (*dark green line*) and one for a person with sickle cell anemia (*red line*). Sickle cell anemia is a genetic disorder and the anemia can be passed through family generations. The RBCs that are normally shaped like a disk take on a bold sickling or rigid crescent shape, which makes it difficult for the RBCs to move through large arteries in the lung and small capillaries in many locations of the body. The specific cause is a mutation in the hemoglobin gene. Thus as shown in Fig. 15.14C on the right side, the hemoglobin from an anemic individual is only able to bind ~50% of the required oxygen (*red line*) as compared with the ~100% for a normal individual.

V Clinical aspects

A Anemia

A working definition of an anemia is when a person has fewer RBCs than normal and consequently feels very weak. While there are well over 300 types of anemia, the large proportion of them are rare and difficult to uniquely diagnose. The four most common types are as follows: iron-deficient anemia; pernicious anemia (PA); sickle cell anemia, and trauma-caused anemia.

Trauma-caused anemia frequently occurs as a consequence of an automobile accident or other major accident, such as serious gunshot wound. The rapid onset of the anemia results from a loss of a large proportion of RBCs all at one time, which impairs the ability to transfer oxygen and nutrients through the circulatory system in the absence of RBCs.

Genetic disorders impair hemoglobin synthesis. Iron-deficiency anemia can arise when the body cannot produce enough hemoglobin. Mutations in the coding sequence of globin genes particularly around the heme group have been shown to alter hemoglobin's oxygen affinity. Hemoglobin disorders are currently the best studied and defined of all human genetic diseases. This type of anemia is especially common among women of childbearing age. A woman's need for iron increases during menstruation and pregnancy.

Sickle cell anemia is an inherited form of anemia. Its hallmark is a shortage of enough healthy RBCs to carry adequate oxygen throughout our body. See Fig. 15.14, panel C (right panel) as an example of the inability to properly deliver oxygen to the peripheral tissues. Normal human RBCs are round, flexible, and move easily through the blood vessels. In contrast, sickle cell anemia cells become tacky and are shaped like crescent moons or flattened biconvex ellipsoidal disks. These are shown in Fig. 15.15. These abnormally shaped cells particularly get stuck in small blood vessels. This can block or slow the delivery of oxygen/blood flow to all the precincts of the body. Some 120,000 infants are still born with sickle cell disease (SCD) every year worldwide. For African Americans, 1/375 is diagnosed per year. In 1973 the SCD life expectancy was 14 years. Now the SCD life span is ~50 years.

PA also is known as vitamin B-12 anemia. PA occurs when the intestine does not absorb enough vitamin B-12, which is required for the body to make RBCs. PA was a fatal disorder until 1920 when dietary raw liver was used as a treatment. PA is treated in the same manner as vitamin B-12 deficiency, with vitamin B-12 shots or pills. Approximately 10–20 cases of PA are now diagnosed per 100,000 people/year.

B Cardiovascular events in hypertension

A heart attack, or acute myocardial infarction, can be a frequent occurrence when there is an inadequate flow of nutrient-enriched blood and oxygenated blood; this can result in the death of a portion of the heart muscle which can lead to death of the individual. Having a heart attack is a leading killer of women and men in developed countries; estimated worldwide deaths due to coronary heart disease are 7.1 million/year. Each year in the USA approximately 365,000 women and 555,000 men are diagnosed as having had heart attacks, according to the American Heart Association. In part because women have heart attacks at older ages than men do, women are more likely to die from an attack within a few weeks. Thus ~800,000 Americans experience a heart attack every year. In 2006, 631,636 men and women died of heart disease. The cost for comprehensive diagnosis and treatment of coronary heart disease in the United States exceeded $156 billion in 2008 (estimated by Genetech, Inc). Also of relevance is that individuals who survive their acute heart attack have a chance of forthcoming illness and death that is 1.5–15 times greater than that of the general population, depending on their sex and clinical outcomes.

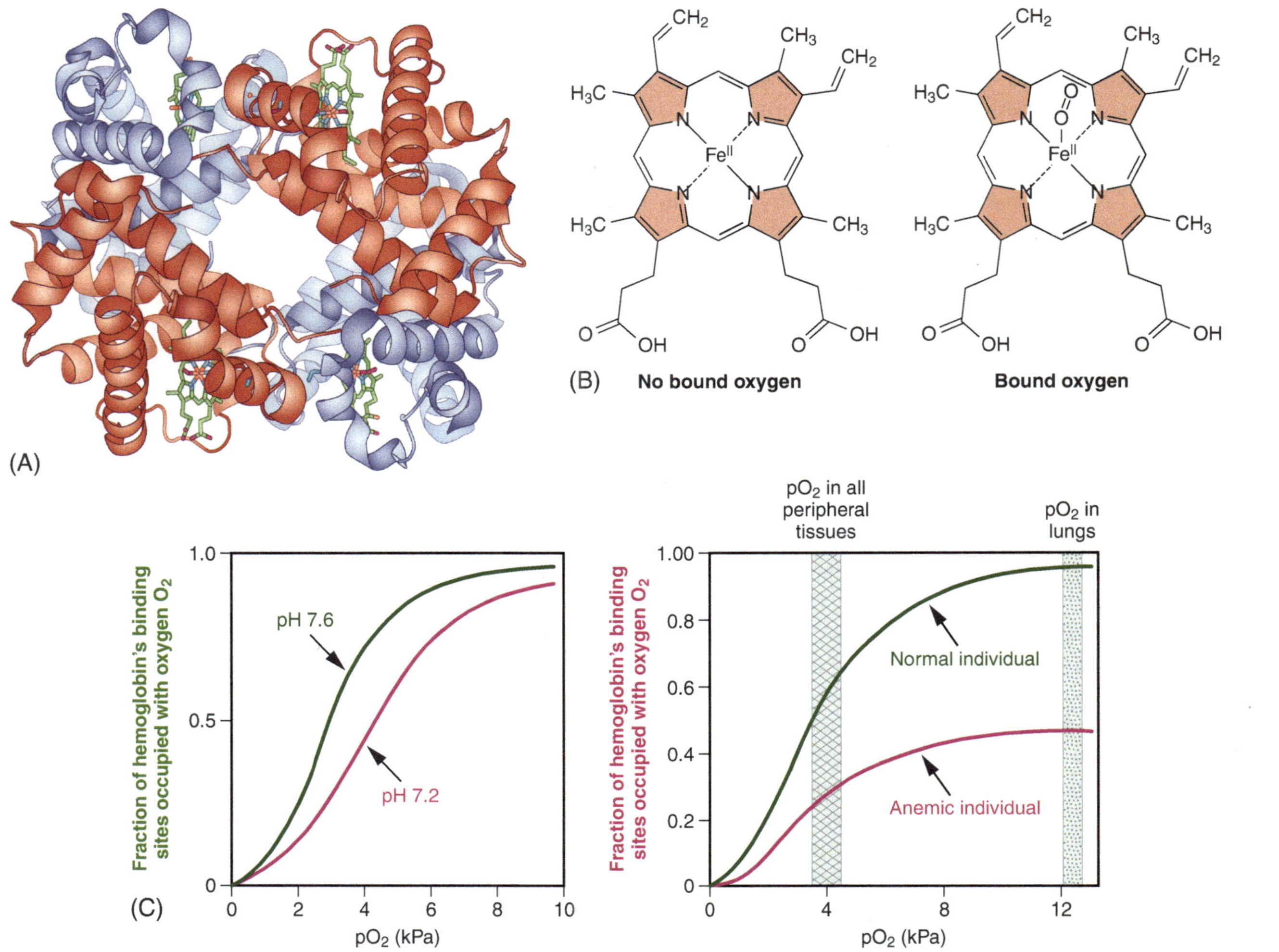

FIGURE 15.15 Hemoglobin as a carrier of oxygen in the circulatory system. (A) *X-ray crystallographic structure of hemoglobin.* Hemoglobin has a molecular weight of 64,500. It comprises three components: (1) two α chains (each 146 amino acids), each folded as a red α helix; (2) two β chains (each 141 amino acids), each folded as a blue α helix; and (3) four heme prosthetic groups that are separately associated with one of the four α-helices. Each molecule of hemoglobin can bind four molecules of oxygen (O_2) with one O_2 per heme group. (B) *Structure of hemoglobins heme group with bound oxygen.* Each of the four heme groups, known as protoporphyrin IX, has a bound iron atom in its ferrous oxidation state (Fe^{2+}). The four protoporphyrin IXs each has four planar pyrrole rings that are connected together by methylene bridges. Each of the four pyrrole rings is colored peach. (C1) *The pH dependency of oxygen binding to hemoglobin.* This figure reports the difference in fractional occupancy of the oxygen-binding sites on the heme groups of hemoglobin as a function of the pH of the local environment. The comparison was made between a pH 7.6 (less acidic) environment (*green line*) and that of a lower pH 7.2 (more acidic) environment (*red line*). At the lower 7.2 pH the fractional binding of 50% was achieved at a pressure of 4.2 kPa. In contrast, at the higher pH of 7.6, a reduction of the pressure to 2.6 was still adequate to achieve a 50% occupancy of the hemoglobin-binding sites. Or to analyze the data differently, at an oxygen pressure of 4.2 kPa, for a pH 7.2, only a 50% occupancy was achieved, while for the pH of 7.6 a higher occupancy of ~75% was achieved. (C2) *Physiological effects of oxygen binding to hemoglobin.* This panel shows two oxygen-binding curves to hemoglobin, one for a normal healthy individual (*dark green line*) and one for a person with sickle cell anemia (*red line*). Sickle cell anemia is a genetic disorder and the anemia can be passed through family generations. The red blood cells that are normally shaped like a disk take on a bold sickling or crescent shape, which makes it difficult for the red blood cells to move through large arteries in the lung and small capillaries in many locations of the body. The specific cause is a mutation in the hemoglobin gene. The classic circular shape of the erythrocytes becomes grossly altered to an abnormal rigid sickle shape (see Fig. 15.13, panel C). Thus as shown here, the hemoglobin from an anemic individual is only able to bind ~50% of the oxygen (*red line*) as compared with the ~100% for a normal individual (*green line*).

Another contributor to a heart attack is hypertension. It is a disorder that may occur throughout life. Hypertension is currently defined as physician office systolic blood pressure ≥140 mm Hg and diastolic pressure of ≥90 mm Hg; see the categories in Table 15.4. Hypertension has a worldwide incidence of 9%–18% in the adult population. As many as 50 million Americans have elevated blood pressure. In the presence of continuing uncontrolled hypertension, there is frequently a high incidence of heart disorders. These include congestive failure, coronary heart disease, stroke, kidney disease, and/or aneurysms of the aorta.

An important component of the diagnosis of the etiology of hypertension is a determination of the level of Na^+ excretion in relation to the plasma levels of renin and also the daily amount of aldosterone excreted in the urine. Such an analysis may provide the formulation of a rational basis of treatment, which can include the use of inhibitors of the renin–angiotensin system such as the antagonist saralasin, use of β-blocking drugs such as propranolol, or dietary restriction of salt intake. It is beyond the scope of this presentation to consider this topic in detail.

Further reading

Angiotensin-converting enzyme and the renin–angiotensin system

A. Bartolome, et al., Angiotensin converting enzyme 2 is a novel target of the γ-secretase complex, Nature Scientific Reports 11 (9803) (2021). Available from: https://doi.org/10.1038/s41598-021-89379-x.

G.E. Cozier, L. Lubbe, E.D. Sturrock, K.R. Acharya, Angiotensin-converting enzyme open for business: structural insights into the subdomain dynamics, The FEBS Journal 288 (2021) 2238–2256.

J.F. Giani, et al., Novel roles of the renal angiotensin-converting enzyme, Molecular and Cellular Endocrinology 529 (2021) 111257. Available from: https://doi.org/10.1016/j.mce.2021.111257.

A. Vitiello, C. Pelliccia, F. Ferrara, Drugs acting on the renin angiotensin system and Sars-CoV-2, Drug Discovery Today 26 (2021) 870–874.

Atrial natriuretic peptide (ANP) actions on the nitric oxide synthase

M.A. Costa, et al., Role of nitric oxide as a key mediator on cardiovascular actions of atrial natriuretic peptide in spontaneously hypertensive rats, American Journal of Physiology Heart Circulation Physiology 298 (2010) H778–H786.

A. Frees, et al., Natriuretic peptides relax human intrarenal arteries through natriuretic peptide receptor type-A recapitulated by soluble guanylyl cyclase agonists, Acta Physiologica 231 (2021). Available from: https://doi.org/10.1111/apha.13565.

Z. Liang, et al., Interleukin-9 deficiency affects lipopolysaccharide-induced macrophage-related oxidative stress and myocardial cell apoptosis via the Nrf2 pathway both in vivo and in vitro, Biofactors (2021). Available from: https://doi.org/10.1002/biof.1754.

L. Yu, B.M. Park, Y.J. Ahn, G.-J. Lee, S.H. Kim, Hydrogen sulfide donor, NaHS, stimulates ANP secretion via the KATP channel and the NOS/sGC pathway in rat atria, Peptides 89 (2019) 97. Available from: https://doi.org/10.1016/j.peptides.2018.04.005.

Endothelins and their receptors

A.B. Mekuria, Z.D. Kifle, M. Abdelwuhab, Endothelin system and therapeutic application of endothelin receptor antagonists, Journal of Clinical and Experimental Pharmacology 10 (2021) 272. Available from: https://doi.org/10.35248/2161-1459.21.10.272.

A. Russignan, et al., Targeting the endothelin-1 receptors curtails tumor growth and angiogenesis in multiple myeloma, Frontiers in Oncology 10 (2021) 600025. Available from: https://doi.org/10.3389/fonc.2020.600025.

A.P. Young, et al., Endothelin B receptor dysfunction mediates elevated myogenic tone in cerebral arteries from aged female Fischer 344 rats, GeroScience (2021). Available from: https://doi.org/10.1007/s11357-020-00399-7.

Erythropoietin and its receptor

D. Fujimoto, et al., Efficacy of continuous erythropoietin receptor activator for end-stage renal disease patients with renal anemia before and after peritoneal dialysis initiation, Clinical and Experimental Nephrology 25 (2021) 191–199.

M. Oshima, et al., Comparison of circulating biomarkers in predicting diabetic kidney disease progression with autoantibodies to erythropoietin receptor, Kidney International Reports 6 (2021) 284–295.

M.A.H. Talukder, et al., Obligatory role of Schwann cell-specific erythropoietin receptors in erythropoietin-induced functional recovery and neurogenic muscle atrophy after nerve injury, Muscle and Nerve (2021). Available from: https://doi.org/10.1002/mus.27121.

Some adrenomedullin actions

N. Cui, et al., Adrenomedullin-RAMP2 and RAMP3 systems regulate cardiac homeostasis during cardiovascular stress, Endocrinology 162 (2021). Available from: https://doi.org/10.1210/endocr/bqab001.bqab001.

T. Koyama, et al., Mid-regional pro-adrenomedullin is a novel biomarker for arterial stiffness as the criterion for vascular failure in a cross-sectional study, Scientific Reports 11 (2021) 305. Available from: https://doi.org/10.1038/s41598-020-79525-2.

M. Ogoshi, Adrenomedullin 2 and 5 activate the calcitonin receptor-like receptor (clr)-receptor activity-modifying protein 3 (ramp3) receptor complex in Xenopus tropicalis, General and Comparative Endocrinology 306 (2021) 113752. Available from: https://doi.org/10.1016/j.ygcen.2021.113752.

Chapter 16

The pineal gland

I Introduction

The pineal gland was identified as a distinct structural component of the vertebrate brain for several centuries prior to its description in the human by Galen of Pergamon (c.130–200). Its distinct features as a highly vascularized unpaired organ in the center of the brain likely provided the basis for the idea of René Descartes (1596–1650) that it is an organizer of psychological and physiological functions or the "seat of the soul." As the study of comparative anatomy grew and improved microscopic techniques became available in the late 19th century, the view developed that the mammalian pineal gland was related evolutionarily to the "third eye," the photoreceptive organ located on the top of the brain in lower vertebrates. The ability of crude extracts of the gland to alter the color of frog skin was discovered in the early 20th century and the isolation of melatonin in 1958 was quickly followed by observations of the light–dark variations in the secretion of this hormone in the 1960s and 1970s.

It is now understood that the main function of the pineal is to receive information about the current state of the light–dark cycle from the environment and convey this information through the secretion of the hormone melatonin to the internal physiological systems of the body. Relative to other endocrine and neuroendocrine systems, the pineal gland has undergone a great deal of change during its evolutionary development. In cold-blooded vertebrates the cells of the pineal gland include photoreceptors that contact neurons to communicate with other organs in the body. The avian pineal gland is also a directly photosensory organ and secretes melatonin in response to this signal. In mammals, pinealocytes producing melatonin have replaced photoreceptors and the gland receives its information about light and darkness from the retina through multiple neuronal connections. Melatonin is secreted during the dark period of the day and, through its elevated blood levels, informs, through specific cell receptors, the peripheral organs and tissues regarding the light/dark cycle.

Many biological processes in all living plant and animal species follow 24-hour cycles. In mammals the central 24-hour cycle generator lies in the bilateral suprachiasmatic nuclei (SCN). Through the rhythmic expression of genes for transcription factors and a network of autoregulatory positive and negative feedback loops, the neurons of the SCN send information about the 24-hour cycle to the periphery of the body through the sympathetic and parasympathetic systems. In addition to its function of sending information to the periphery regarding light and dark, melatonin from the pineal gland also contributes to the entrainment of the central SCN oscillator to the light–dark cycle.

II Anatomical features of the pineal gland

A Anatomical location and cellular composition

The human pineal gland is located at the midline of the brain and is attached by a short stalk to the roof of the third ventricle (Fig. 16.1). In the adult, it is 5–9 mm in its longest dimension and weighs 100–200 mg in the adult. The gland tends to accumulate calcium deposits with age and, while the effect of these deposits on pineal function is not known, they provide an excellent imaging marker for the center of the brain. The pineal gland is highly vascularized; it is one of several circumventricular organs that surround the third ventricle and lie outside the blood–brain barrier and the protections from the general circulation that this barrier provides.

The main type of cell found in the pineal gland is the pinealocyte. These are large cells with round nuclei, comprise 95% of the cells in the pineal gland, and are the site of the synthesis and secretion of the hormone melatonin. In addition to pinealocytes the gland contains scattered glial cells of uncertain function.

B Connection with the visual system

Fig. 16.2 depicts the neural connection between the visual system and the pineal gland. In the top part of the figure is shown a schematic diagram of the layers of the retina. Light passes through the inner layers, which are made up of ganglia and neural cells, to the densely packed rods and cones of the outer retina. These

Hormones. DOI: https://doi.org/10.1016/B978-0-323-90262-5.00023-8

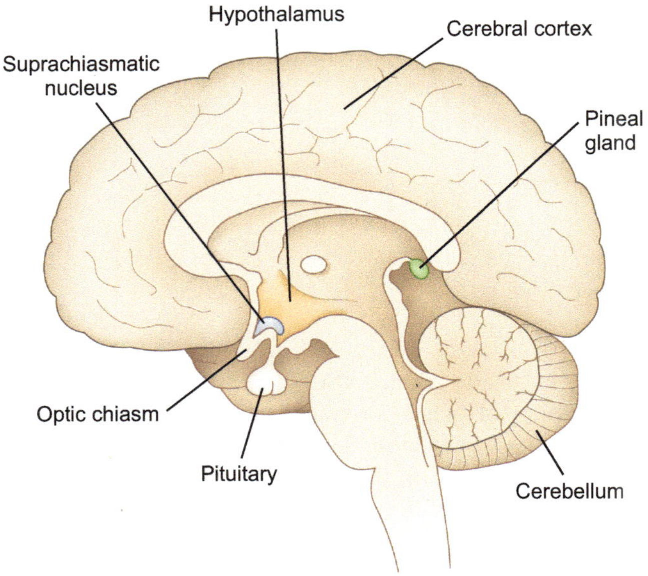

FIGURE 16.1 Anatomical relationships of the pineal gland. The pineal gland (*green*), named for its shape, which is similar to that of a pine cone, is attached to the posterior end of the roof of the third ventricle of the brain, here shown in sagittal section. This places it very near the midline of the brain. It is some distance from the suprachiasmatic nucleus and the hypothalamus, with which it has neuronal connections as shown in Fig. 16.2.

photoreceptors, using rhodopsin, transmit the visual signal through the inner retina to the retinal ganglion cells (RGC) from which emerge the optic nerve fibers. These impulses will go to the visual, or image-forming, centers of the brain.

A few (1%–2%) of the RGCs contain another visual pigment, melanopsin, and are the intrinsically photosensitive, the ipRGCs. Melanopsin-containing ipRGCs carry out several functions that are not related to image formation, that is, those mediated by rods and cones. These NIF (nonimage-forming) activities can be identified in humans by their presence in profoundly blind people (no rod or cone activity). They include the pupillary light reflex, increased alertness in light, and light exacerbation of migraine headaches. In addition, one of the most important NIF effects of light in the context of this chapter is the acute suppression of melatonin secretion. It is these cells that detect the light and send the signal to the suprachiasmatic nucleus, which informs the pineal gland whether it is light or dark. Peak light sensitivity of the ipRGCs is 480 nm, while that for three types of cones is 420, 530, and 560 nm and for rods is 500 nm. Together, the rods, cones, and ipRGCs are responsible for all known mammalian ocular photoresponses.

The lower part of Fig. 16.2 shows the pathway this light takes from the retina along the retinohypothalamic tract to the SCN, where the central circadian clock is located. The SCN sends the signal to the pineal gland through the paraventricular nucleus (PVN) of the hypothalamus. Axons from the PVN descend through the brainstem to the intermedialateral cell column, and synapse with preganglionic sympathetic neurons in the spinal column. The axons of these cells leave the spinal cord along with other nerves carrying sympathetic responses to the superior cervical ganglion (SCG); the postganglionic axon conveys the message to the pineal gland. As will be discussed in more detail following, the axon from the SCG releases norepinephrine (NE) which is the trigger for the pinealocyte to produce and secrete melatonin.

III Synthesis and secretion of melatonin

A Melatonin biosynthetic pathway: catabolism

The pinealocyte takes up tryptophan from the blood and converts it to melatonin through the pathway shown in Fig. 16.3. The hydroxylation of the aromatic ring of tryptophan is followed by the decarboxylation of the side chain to form serotonin (5-hydroxytryptamine). The transfer of an acetyl group from acetyl CoA is the rate-limiting step in the formation of melatonin and is catalyzed by arylalkylamine-*N*-acetyltransferase (AANAT). This is the step at which melatonin production by the pinealocyte is regulated as discussed later. The product of this reaction, *N*-acetyl serotonin, is converted to melatonin by the transfer of a methyl group from *S*-adenosyl methionine to the hydroxyl at carbon 5. Other indoles are produced and secreted by the mammalian pineal gland, but melatonin is by far the most abundant and physiologically important of them. Other cell types from skin, the gastrointestinal tract, and placenta have been reported to synthesize melatonin from tryptophan, likely for functions in the local physiology of these tissues.

The main catabolic pathway of melatonin is its 6-hydroxylation by a microsomal liver enzyme. This is followed by esterification with sulfate and excretion in the urine. As seen later, melatonin exhibits striking periodicity in its secretion. The appearance of sulfated 6-hydroxymelatonin in the urine follows the same pattern with an approximate 2-hour offset and is, therefore, an excellent marker of melatonin synthetic activity by the pineal gland.

B Patterns and regulation of melatonin secretion

1 Melatonin secretion patterns

As shown in Fig. 16.4, melatonin from the pineal gland sends two kinds of information to the rest of the body. The first is whether, during any given 24-hour period, it is dark or light outside. In the dark, synthesis and circulating levels of melatonin are high; when it is light, synthesis ceases and circulating levels are diminished. It is these

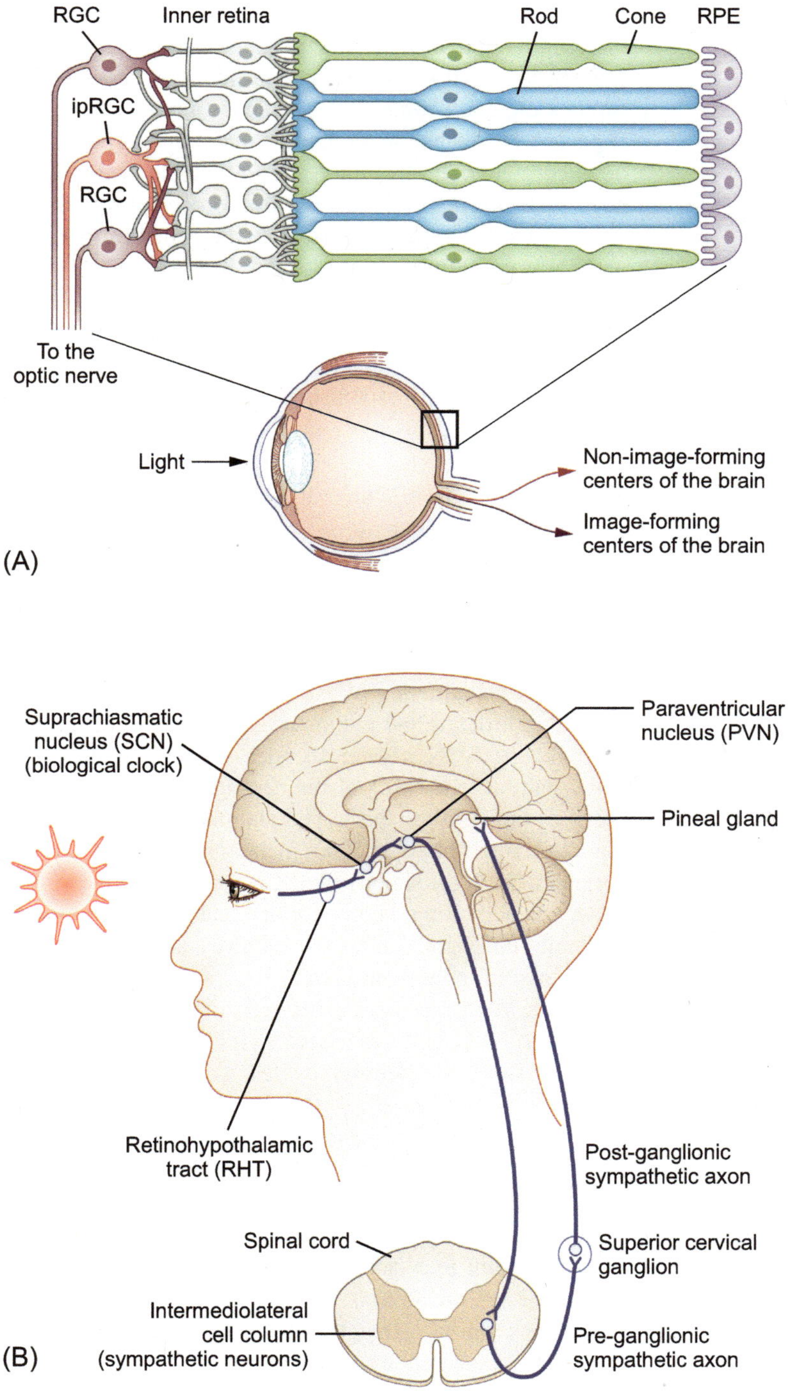

FIGURE 16.2 Retinal–pineal neural connections. (A) Light travels through the layer of RGC and the neural cells in the inner retina to the rods and cones in the photoreceptor layer of the retina. The rods and cones send neural signals back through the inner retina to the ganglia, and through the optic nerve to the visual areas of the brain. A small number of the RGCs contain melanopsin and have intrinsic photoreceptor capability (ipRGC). These cells send neural signals to nonvisual (nonimage-forming) areas of the brain. Among these signals are those that eventually reach the pineal gland. (B) The neural pathway that connects the melanopsin-containing RGC cells to the pineal gland is shown. The photic information from the retina is first sent to the SCN, the location of the central biological clock (see Fig. 16.5). From there the signal travels to the paraventricular nucleus of the hypothalamus, the intermediolateral nucleus, and the superior cervical ganglion which transmits the signal to the pineal gland. The neurotransmitters involved in transmitting the signals in the pathway are detailed in Fig. 16.5. *RGC*, Retinal ganglion cells; *RPE*, retinal pigment epithelium; *SCN*, suprachiasmatic nucleus.

changes in serum melatonin levels through which organs with melatonin receptors sense whether it is light or dark. In humans the melatonin rhythm is absent at birth but is apparent at 6 months of age and the magnitude of the nocturnal secretion of melatonin peaks in early childhood. Throughout adulthood, pineal size and hormone production continue to decrease with each decade of life (compare the different curves in Fig. 16.4). In a given individual the pattern of melatonin secretion is highly reproducible, with much greater variability seen between individuals than within one person over time. This interindividual variability is particularly notable with the amplitude of the melatonin peak during darkness.

For human adults in temperate zones, melatonin levels begin to increase around 10:00 p.m. in the evening (shortly after dark), peak in the middle of the night, and decrease by 7:00 a.m. the next morning (in the morning light). As shown in Fig. 16.4B, the pineal gland sends a

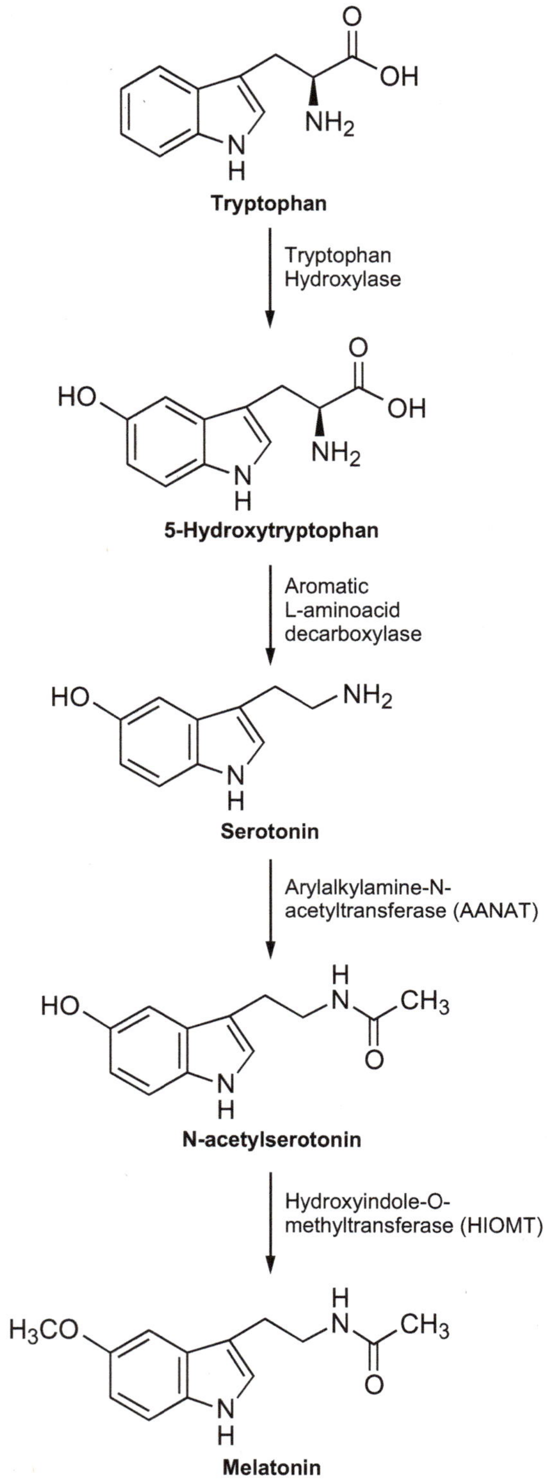

FIGURE 16.3 Pathway of melatonin biosynthesis. The modifications of tryptophan that take place in melatonin biosynthesis in the pinealocyte are shown. The step that is regulated by the dark–light cycle is the conversion of serotonin to *N*-acetyl serotonin, catalyzed by arylalkylamine-*N*-acetyltransferase.

second kind of information to the periphery concerning the length of light and dark periods, that is, day length. Longer days will keep melatonin production and secretion suppressed, and shorter days will lead to longer periods of exposure of peripheral organs to melatonin. These functions of melatonin are highly important in species which need such seasonal information to prepare for hibernation or reproductive functions that must be timed properly for optimally timed birth of the young. Some reports indicate that humans also respond to changes in day length with altered melatonin secretion, but, because these measurements are quite difficult to control for, this point has not yet been definitively established.

2 Regulation of melatonin synthesis

Fig. 16.5A depicts in more detail the regulation of the synthesis of melatonin by the pinealocyte. The neurotransmitters that transmit the information from the melanopsin-containing ipRGCs (see Section II.B.) in the inner layer of the retina do so through the secretion of the neurotransmitters PACAP (pituitary adenylyl cyclase activating protein) and glutamate. When the light signal is positive, the SCN neuron secretes γ-amino butyric acid which inhibits the firing of the neurons with which it synapses in the PVN of the hypothalamus. The signal to the pineal is interrupted and no melatonin is made. During darkness (no light signal) the SCN neuron secretes glutamate, an activating neuropeptide, and the PVN neuron transmits the signal along the pathway seen in Fig. 16.2 and shown in more detail here.

At the final connection between the axon of the neuron of the SCG and the pinealocyte, NE is secreted. NE interacts with its adenylyl cyclase-coupled β-adrenergic receptors leading to the activation of protein kinase A. PKA phosphorylates cyclic AMP (cAMP) response element binding protein; see Chapter 1: Hormone Introduction, activating the transcription of the messenger RNA (mRNA) encoding the enzyme (AANAT) that converts serotonin to *N*-acetylserotonin, which is then methylated to form melatonin.

As shown in Fig. 16.6, the regulation of AANAT activity is also accomplished by phosphorylation of specific serine residues of the enzyme protein molecule by PKA activated by the elevated levels of cAMP in response to NE release during darkness. Phosphorylated AANAT forms a complex with a cytoplasmic protein, 14–3–3, in which the enzyme is more active and, most importantly, is protected from degradation by proteasomes. 14–3–3 proteins comprise a family that is able to bind to and thereby influence the activity and longevity of a variety of intracellular proteins. In the light, as NE and cAMP

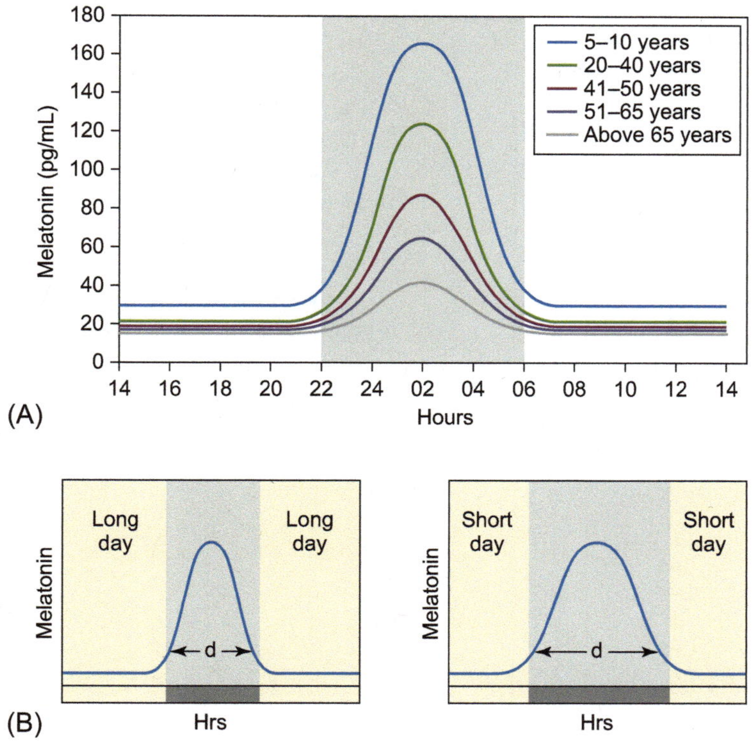

FIGURE 16.4 Patterns of melatonin secretion. (A) The pineal conveys information about the light–dark cycle of the current day because melatonin is secreted only during darkness. Serum levels of melatonin rise several folds to a peak in the midpoint of the dark cycle. Peak melatonin levels are seen during childhood and decrease progressively during adulthood and aging. (B) The pineal gland also conveys seasonal information by varying with the length of days. As the areas under these two curves indicate, the shorter the day, the more melatonin is produced and the longer it is present in the bloodstream.

levels fall, nonphosphorylated and noncomplexed AANAT is rapidly degraded and melatonin synthesis and secretion diminishes and then ceases. The relative contribution of transcriptional and posttranslational control of AANAT's activity varies with species. In ungulates and primates, including humans, enzyme protein phosphorylation is the predominant if not the only mechanism of AANAT regulation.

Melatonin is released into the vascular system from which it reaches the peripheral organs that have melatonin receptors (see Section IV). Some melatonin also circulates to the SCN where, along with glutamate and PACAP from the ipRGC, the central 24-hour clock is entrained to the light/dark cycle. Since melatonin is readily released into the bloodstream and not stored within the gland, the regulation of its secretion and serum levels lies entirely in the regulation of its production at the step of the acetylation of serotonin.

3 Circadian rhythmicity of melatonin secretion

The circadian rhythmicity of the pineal gland is disrupted if the connection with the SCN is severed, emphasizing the SCN's role as the generator of the melatonin rhythm, as well as some other circadian rhythms in the body. In the absence of light the approximate 24-hour cycle of the SCN will keep melatonin synthesis on this schedule, but serum levels of the hormones will not continue to convey information about light and darkness. The basic operating mechanism of the central biological clock in the SCN is an interlinked set of positive and negative feedback loops in the transcription and translation of certain genes. One such loop is shown in Fig. 16.5B. Brain and muscleARN-1-like (BMAL1) and circadian locomotor output cycles kaput (CLOCK) are positive transcription factors that bind to the Enhancer box in the promoter region of the genes for, among other proteins, Per (period) and Cry (cryptochrome). In the presence of BMAL1 and CLOCK, the mRNA for these proteins increases and is translated in the cytoplasm. As the levels of the proteins in the cytoplasm increase, they, along with other cytoplasmic proteins, form a complex that translocates to the nucleus where Cry–Per represses transcription so that further synthesis of the genes is shut down. Per and Cry undergo protein degradation, lowering their levels, the repression is released, and the cycle begins again. Other regulatory loops interact with this one, stabilizing its rhythmicity. In addition, posttranslational modifications such as phosphorylation and acetylation of the positive and negative regulatory proteins also play roles in maintaining the rhythm of the SCN central clock.

IV Biological actions of melatonin

A The melatonin receptors

The first melatonin receptor was cloned from frog skin in 1994 and shortly thereafter two human versions of the receptor, now known as MT1 and MT2, were cloned and characterized. MT1 and MT2 receptors are typical seven

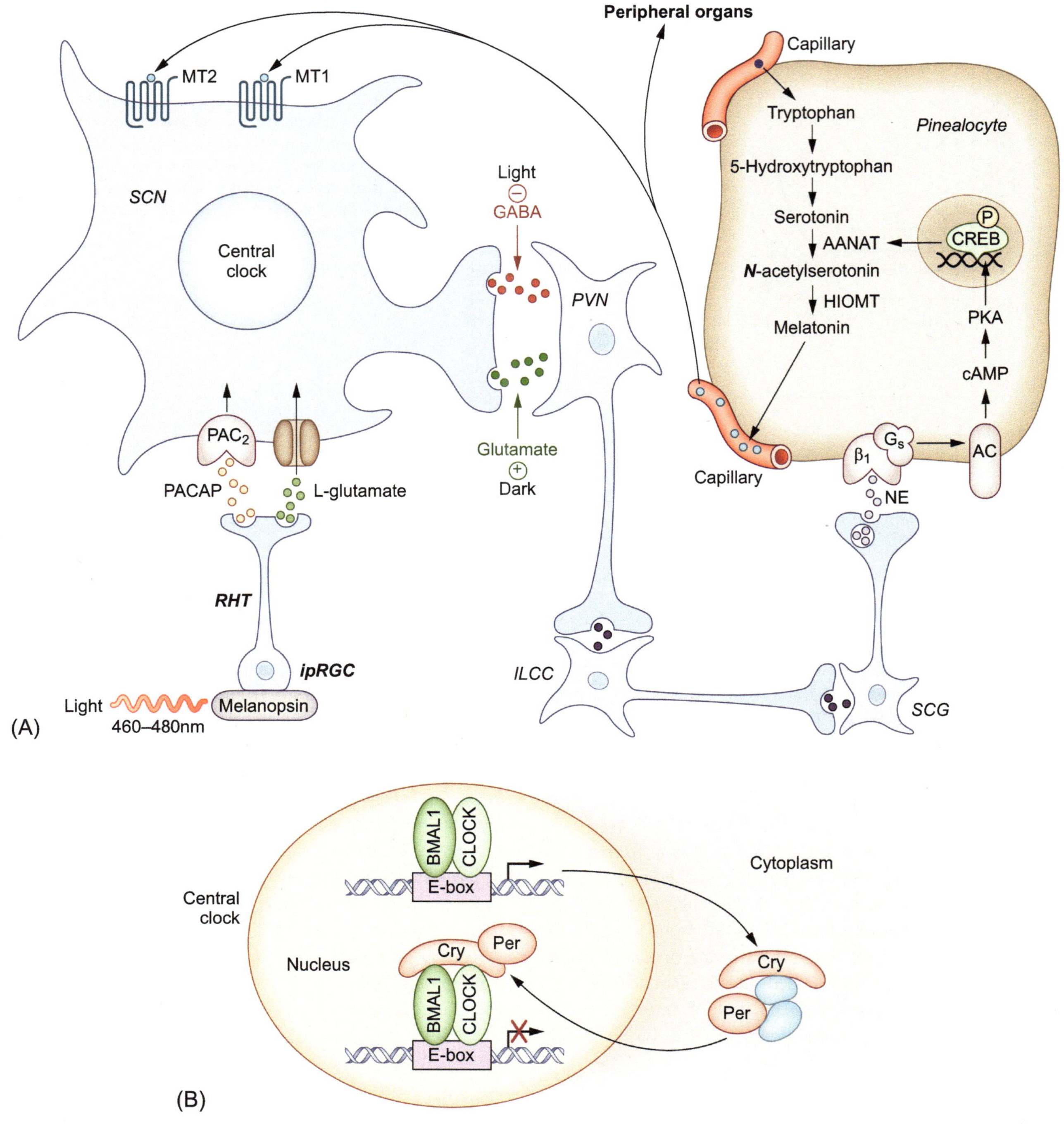

FIGURE 16.5 Regulation of melatonin synthesis. (A) As described in Fig. 16.3, light interacts with an ipRGC, the axon of which passes along the RHT and terminates on a neuron of the SCN. The SCN neuron releases GABA (*red*) which inhibits the firing of the neuron of the PVN of the hypothalamus. In the absence of light, this cell releases glutamate that stimulates the firing of the PVN neuron so that the signal continues through the ILCC neuron to the neurons of the SCG. These neurons release NE that interacts with its β-adrenergic receptor to stimulate intracellular cyclic AMP levels, leading to increased synthesis and translation of mRNA-encoding *N*-acetyltransferase (AANAT) required for the conversion of serotonin to *N*-acetylserotonin. Melatonin is released into capillaries and carried to peripheral organs to transmit information about the light/dark cycle and to the SCN to contribute to the entrainment of the 24-hour central clock to the light–dark cycle. (B) A simplified model of the biological clock. Two transcription factors, BMAL1 and CLOCK form a heterodimer that activates the transcription of the genes for two proteins, Per (Period) and Cry (Cryptochrome). Cry and Per, along with other proteins, form a repressor complex that translocates back into the nucleus, allowing the Per–Cry complex to shut off the transcription of these genes in a negative feedback loop, creating the oscillator. *AANAT*, Arylalkylamine-*N*-acetyltransferase; *BMAL1*, brain and muscleARN-1-like; *CLOCK*, circadian locomotor output cycles kaput; *GABA*, γ-amino-butyric acid; *ILCC*, intermediolateral cell column; *ipRGC*, intrinsically photosensitive retinal ganglion cell; *mRNA*, messenger RNA; *NE*, norepinephrine; *PVN*, paraventricular cell; *RHT*, retinohypothalamic tract; *SCG*, superior cervical ganglion; *SCN*, suprachiasmatic nucleus.

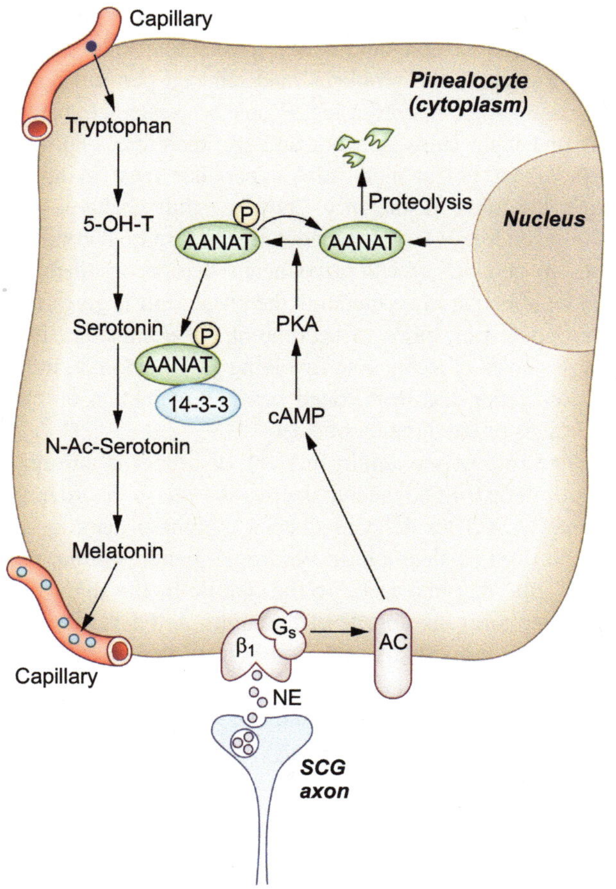

FIGURE 16.6 Posttranslational control of AANAT. Transcriptional control of AANAT, the enzyme that catalyzes the rate-limiting step of melatonin production, was described in Fig. 16.5A. A second, equally important, mechanism of posttranslational control is presented there. cAMP, stimulated by NE release, activates cAMP-dependent PKA. PKA phosphorylates specific sites on the AANAT molecule, increasing its affinity for a regulatory protein, 14–3–3. When bound to 14–3–3, AANAT is protected from degradation through proteolysis and catalyzes serotonin acetylation more efficiently. The activation of AANAT is rapidly reversed at the beginning of the light cycle when NE release from the SCG axon terminal ceases. *AANAT*, Arylalkylamine-*N*-acetyltransferase; *cAMP*, cyclic AMP; *NE*, norepinephrine; *PKA*, protein kinase A; *SCG*, superior cervical ganglion.

membrane–spanning G protein–coupled receptors of 350 and 363 amino acids, respectively. They share about 60% amino acid homology. Functionally, the two types of receptors are distinguished by their affinity for melatonin that of the MT1 receptor being approximately fivefold greater than that of the MT2 receptor.

Table 16.1 lists many of the tissues in which the melatonin receptors have been reported, indicating their wide and often overlapping distribution. It should be noted that within tissues composed of many regions or cell types, the cellular distribution of the two receptors may be different. For example, MT1, MT2, or both have been found in more than 100 locations in the mammalian brain. Locations in the brain where MT1 has been identified include the hypothalamus, including the SCN, cerebellum, hippocampus, ventral tegmental area, and substantia nigra; MT2 has been found in the SCN and other areas of the hypothalamus.

TABLE 16.1 Tissue distribution of melatonin receptors.

Tissue	Receptor type
Liver	M1
Pancreas	M1
Spleen	M1
Adrenal cortex	M1
Placenta	M1
Ovary	M1
Testis	M1, M2
Brain	M1, M2
Retina	M1, M2
CV system[a]	M1, M2
Immune system	M1, M2
Breast	M1, M2
Kidney	M1, M2
Skin	M1, M2
Pituitary	M2
Gastrointestinal tract	M2
Adipose tissue	M2

[a]*CV, cardiovascular system, includes peripheral blood vessels, aorta, and heart.*

There is also overlap in the signaling pathways used by the two receptors. There are two primary G protein–coupled pathways used by MT1 and MT2 and, as shown in Fig. 16.7, either receptor is capable of activating either pathway, depending on the cell type. In some cells the receptor (either MT1 or MT2) is coupled to G_i resulting in the inhibition of adenylyl cyclase and decreased cAMP levels. In other cells the receptor is coupled to G_q, activating phospholipase C (PLC), the cleavage of phosphatidylinositol diphosphate to release IP_3 to raise intracellular Ca^{2+} levels and diacylglycerol to activate protein kinase C (PKC). Several possible downstream events can follow these initial signals and these, too, vary with cell type, although specific details have not been elucidated in all target cells of melatonin.

Although there are some high affinity agonists for MT1 and MT2 receptors that do not distinguish between the two as well as some that are somewhat selective for each of them, no agonists have yet been identified/synthesized that bind one or the other receptor exclusively. This

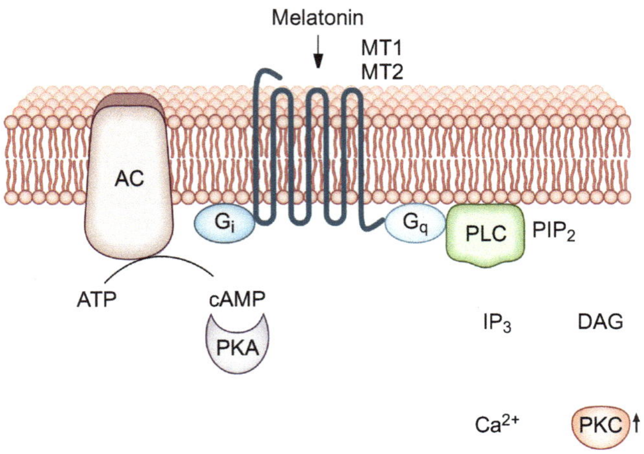

FIGURE 16.7 Membrane signaling by the melatonin receptor. The MT1 and MT2 receptors for melatonin, both G protein–coupled receptors, use one of two signaling pathways, depending on the cell type. On the left is shown the interaction of G_i with adenylyl cyclase to decrease cyclic AMP levels and, therefore, cyclic AMP-dependent PKA. On the right is shown the interaction of G_q with PLC leading to the cleavage of phosphatidylinositol diphosphate (PIP_2) into inositol triphosphate (IP_3) and DAG. These second messengers stimulate increased intracellular Ca^{2+} and PKC, respectively. The downstream effects of these events at the membrane vary with cell type. *DAG*, Diacylglycerol; *PKA*, protein kinase activity; *PKC*, protein kinase C; *PLC*, phospholipase C.

has hampered developing a clear understanding of the role of each of the receptors in the many tissues in which both are found. MT1 or MT2, individually, appear to act in the same way so that the action of either one results in the inhibition of adenylate cyclase leading to a reduction in the concentration of intracellular cAMP. On the other hand, when both receptors are present in the same cell, a heterodimer (MT1–MT2) can be formed. The action of the heterodimer is different from that of a single receptor and leads to a signaling pathway that involves the intracellular concentration of Ca^{2+} and the activities of PKC and PLC. These pathways as well as the primary structures of either receptor are shown in Fig. 16.8.

B Sleep and jet lag

The clearest and most important role of melatonin in the human is the one it plays in influencing, in partnership with the retina and the SCN, circadian rhythms throughout the body. In the SCN, where the central clock in the SCN has an autonomous period of slightly longer than 24 hours, the feedback effect of melatonin on SCN firing helps maintain entrainment of the central clock to the external light/dark cycle. It is now known that many peripheral cells have their own internal oscillators and melatonin is important in helping to synchronize many of these with the light/dark cycle.

In humans, who are characteristically active by day and at rest during the night, manifestations of disruption of the normal circadian rhythm occur when an individual travels through several time zones, a condition known as "jet lag." Fatigue, sleep problems, and reduced performance are common symptoms of jet lag. People who alternate between day and night shifts of work and are, therefore, exposed to light of 480 nm at night, also experience these symptoms. After the initial disturbance (travel or shift change), it can take several days for endogenous rhythms and environmental cues to become synchronized again. Exogenous melatonin can be effective in accelerating the phase shift if given at the appropriate time prior to bedtime at the destination. In the many studies of sleep, both involving jet lag and not, little to no association has been found between melatonin on either the stages or the duration of sleep.

The role of melatonin in sleep disorders is an area of great interest. In advanced sleep-wake phase disorder syndrome (ASPS) or delayed sleep-wake phase disorder syndrome (DSPS) sleep phase syndrome, patients either sleep during the day and wake in the middle of the night or do not sleep until the middle of the night. ASPS is often seen in elderly individuals and DSPS in children with some neurodevelopmental disorders. In both of these settings, restoring the melatonin during the night with controlled release hormone improved sleep patterns.

C Adrenal cortex

The actions of melatonin in peripheral tissues may be direct ones on the transcription, translation, or posttranslational alterations of cellular components such as enzymes and other proteins or may take place through effects on the endogenous oscillator of the cell. Many peripheral tissues have their own oscillators that operate along the same mechanistic lines of the simple model shown in Fig. 16.5B. These oscillators, which have varying degrees of autonomy from the SCN central clock, have been characterized in the adrenal cortex, pancreatic islets and the exocrine pancreas, adipose tissue, liver, heart, blood vessels, and stomach, among others. The adrenal cortex is one example of a tissue in which melatonin appears to act in the context of the cell's oscillator.

The adrenal cortex is also a physiologically significant example of peripheral oscillators and their central control, because of the high amplitude change in the secretion of glucocorticoids and their broad influence on gene expression, including those of the peripheral oscillators, in target cells. Although the adrenal oscillator will lose its synchrony if the SCN is abolished and it is also under strong influence from the hypothalamic–pituitary–adrenal axis (see Chapter 3: The Hypothalamus and Anterior Pituitary and Chapter 10: Adrenal Corticoids) as well as neural innervation, there is clear evidence for additional controls on its rhythm. In diurnal species, including humans and nonhuman primates, glucocorticoid levels are low shortly

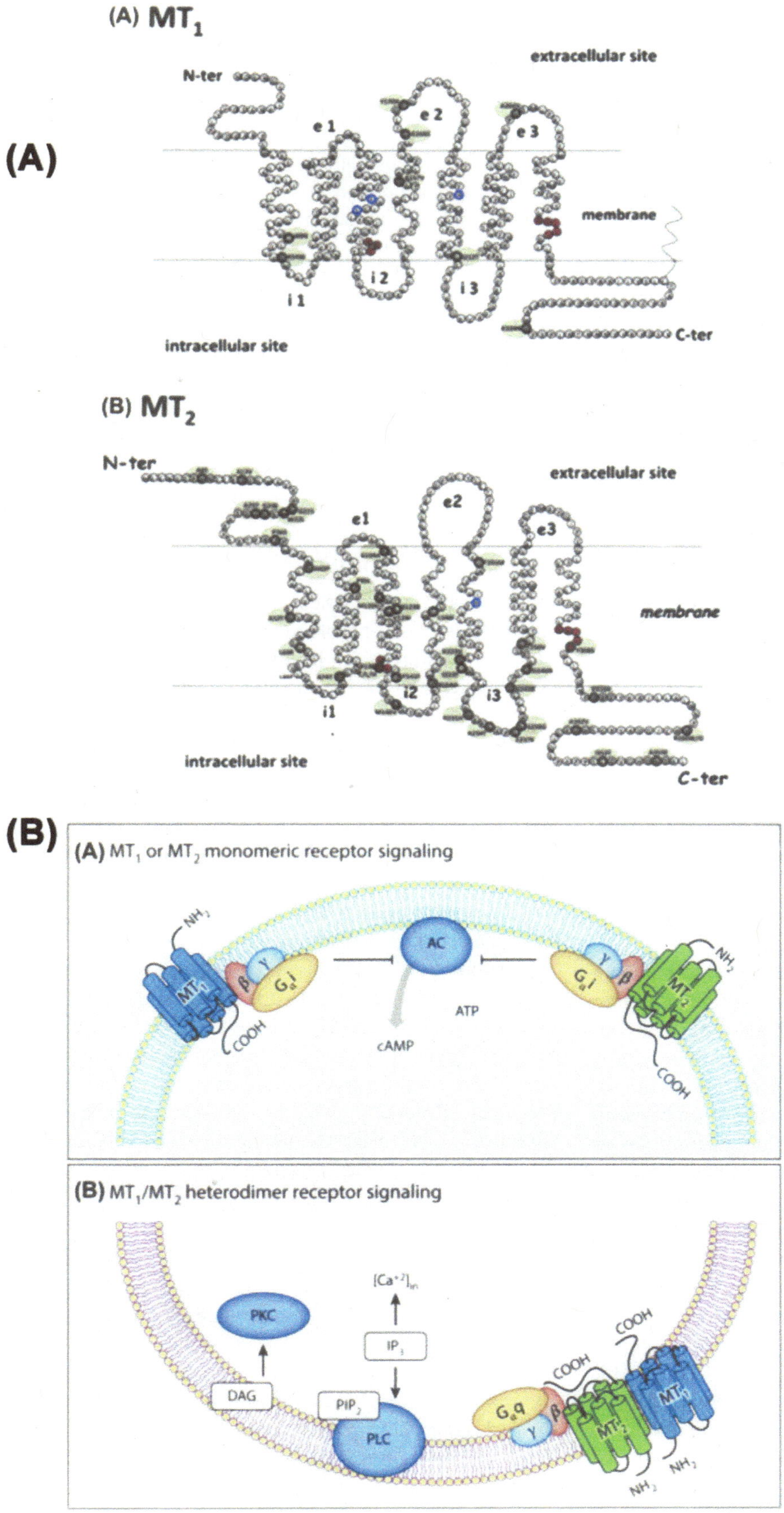

FIGURE 16.8 (A) Distribution of nonsynonymous MT1 (A) and MT2 (B) receptor variants identified in various human populations. Positions of variants are highlighted in *light brown*. Typical signatures of MT receptors, such as the $^{3.49}$NRY$^{3.51}$ motif and the $^{7.49}$NAXXY$^{7.53}$ motif, are highlighted in *red*. Residues suspected to be directly involved in melatonin binding (S3.35 and S3.39 in MT1 and H5,46 in both MT1 and MT2 receptors) are highlighted in a *blue circle*. The positive palmitoylation site at C314 is indicated in MT1 receptors. (B) Melatonin signaling. (A) MT1 or MT2 monomeric receptor signaling. Both MT1 and MT2 melatonin receptors couple to pertussis toxin-sensitive $G\alpha I$, β, γ and insensitive G_q, β, and γ proteins inhibiting forskolin-stimulated cAMP, protein kinase A signaling, and CREB phosphorylation. MT1/MT2 heterodimer receptor signaling. Human MT1 and MT2 receptors form homo- and hetero-dimer receptor signaling. Human MT1 and MT2 receptors form homo- and hetero-oligomers between themselves, altering the pharmacological properties of the individual receptors. Formation densities of the MT1/MT2 heterodimer and the MT1/MT2 homodimer are similar and three- to fourfold lower than the MT1/MT1 homodimer. Native functional MT1/MT2 heterodimers have been characterized in mouse rod photoreceptors, where they mediate the enhancement of scotopic light sensitivity by melatonin through a heterodimer-specific PLC and PKC pathway. *AC*, Adenylate cyclase; *CREB*, cyclic AMP response element binding protein; *DAG*, diacylglycerol; *IP3*, inositol-1,4,5-trisphosphate; *PIP2*, phosphatidylinositol-4,5-bisphosphate; *PKC*, protein kinase C; *PLC*, phospholipase C. *Part (A) was reproduced from Fig. 6.6 and part (B) was reproduced from Fig. 6.7 of R.J. Wurtman, Chapter 6, The pineal as a gland and melatonin as a hormone, in: G. Litwack (Ed.), Hormonal Signaling in Biology and Medicine, Academic Press, 2020.*

after the onset of darkness and begin to rise after the middle of the night when melatonin levels are falling. This timing is consistent with the inhibitory effects of melatonin, acting through the MT1 receptor, on glucocorticoid production and of its response to adrenocorticotropin hormone from the pituitary gland (see Chapter 10: Adrenal Corticoids) observed in monkeys and humans. The observation of direct inhibition of clock genes by melatonin in adrenal explants from monkeys strongly suggests a role for melatonin in the regulation of this important peripheral oscillator and, by extension, the rhythms of other tissues that are influenced by glucocorticoids.

D Reproduction

1 Photoperiodicity

In seasonally breeding animals, information about the time of year is crucial to reproductive success and the role of the pineal gland in obtaining and conveying this information was described in Section III.B, Fig. 16.4. Examples of seasonal functions of melatonin include, in addition to those required for successful reproduction, changes in the growth and color of the coat and appetite changes in preparation for or recovery from hibernation. As a species, humans do not display overt seasonality in reproductive events, although those who live near the Arctic circle have been reported to have decreased pituitary function and conception rates during the winter darkness (when melatonin levels are high) relative to those in the summer.

2 Melatonin and the hypothalamic–pituitary axis

In nonprimates the seasonal and other effects of melatonin on the reproductive axis are exerted through the inhibition of gonadotrophin-releasing hormone (GnRH) at the hypothalamus, leading to decreased gonadotrophin secretion from the pituitary and a dampening effect on the reproductive organs. In humans a negative correlation between nocturnal serum luteinizing hormone and melatonin levels during the night has been reported and clinical observations support this relationship. For example, women who experience exercise-induced amenorrhea have high serum melatonin levels. In both normal and hypogonadal men, melatonin suppresses gonadotrophin secretion. The mechanism of melatonin inhibition of GnRH secretion is not certain, but recently it has been suggested that kisspeptin neurons (see Chapter 12: Androgens and Chapter 13: Estrogens and Progestins) may mediate these effects. It appears that the suppression of gonadotrophin secretion occurs by melatonin's inhibition of the release of gonadotrophin-releasing hormone at the level of the hypothalamus.

3 Melatonin and puberty

The hypothalamic–pituitary–gonadal axis is active in the human fetus but after the first year of life, it becomes quiescent until approximately the age of 10 years, when an increase in the pulses of GnRH secretion leads to increased gonadotrophin levels (see Chapter 12: Androgens). Melatonin levels peak during childhood and begin to fall around the end of the first decade, suggesting that the diminution of inhibition of the hypothalamic–pituitary axis by melatonin may be involved in the timing of the onset of puberty. In children with precocious puberty (prior to the age of 9–10) or delayed puberty (later than 13–14 years old) nocturnal levels of melatonin secretion are, relative to age-matched control children, low or high, respectively. These observations support an association between melatonin secretion and the timing of human puberty but do not provide evidence of a causal relationship between the two processes.

4 Melatonin in pregnancy and parturition

Preeclampsia is a medical condition of late pregnancy characterized by high blood pressure and urinary protein loss. If untreated, it can lead to eclampsia, seizures during labor, which can be life-threatening. Patients with severe preeclampsia have lower nocturnal melatonin secretion than normal pregnant women at the same stage of pregnancy, suggesting a blood pressure–lowering role for melatonin. This is supported by observations that humans have lower blood pressure, heart rate, and cardiac output at night than during the day. Another aspect of melatonin in preeclampsia is its possible role in reducing the oxidative stress (see Section IV.F) which is thought to contribute to this condition in pregnancy.

The onset of labor (parturition) in humans tends to occur more often during the night (midnight to 5:00 a.m.) than at other times during the day. Recent observations that uterine receptors for melatonin and oxytocin (the hormone responsible for uterine contractions, see Chapter 14: Hormones of Pregnancy, Parturition and Lactation) increase in parallel at the onset of labor indicates that melatonin could be the circadian signal that underlies its timing.

E Cancer

Two broad areas of research are suggestive of a role for melatonin in the incidence and progression of cancer. The first includes the many studies of melatonin as an oncostatic agent at the cellular level, one that can slow the progression of cancer. The majority of these studies have focused on breast or prostate cancer and implicate melatonin in one or more of the following processes: decreased cell proliferation; decreased metastasis; increased apoptosis; and increased differentiation. Although MT1 receptors have been identified and activation of signaling pathways (see Fig. 16.7) confirmed in many cancer cell types, the mechanisms by which melatonin exerts these effects have not been elucidated.

The other major type of evidence implicating melatonin in the onset and/or progression of cancer, again particularly that of the breast and prostate, are observations of the association between exposure to disruption in the normal dark/light rhythms, and, therefore, of melatonin production, on cancer incidence in humans and progression in experimental animals. As one example of the former, many studies have been carried out in people who engage in night shift work, such as those in the health care, transportation, communication, and hospitality sectors. In general, an inverse relationship between melatonin production, altered by changes in work schedules, and cancer incidence has been observed. In an experimental setting with animal models, constant light exposure, dim light during darkness, or pinealectomy all stimulate cancer initiation or growth. Taken together, these results provide support for a role for melatonin in cancer incidence and possibly progression in humans, but a firm cause-and-effect relationship remains to be established.

F Melatonin as an antioxidant

Several properties of melatonin make a role for it as an antioxidant and reactive oxygen species scavenger a very attractive possibility. It is small and is amphipathic (both lipo- and hydrophilic) and, therefore, can enter most cellular compartments. Chemically, melatonin is four times as effective as glutathione as an antioxidant. The antioxidant properties and free radical scavenging abilities of melatonin, by itself and through its metabolite N^1-acetyl-N^2-formyl-5-methoxykynuramine, have been amply demonstrated in vitro and in many animal model systems. Much of this work and some studies in humans involve supraphysiological doses of melatonin and these may suggest important uses for pharmacological amounts of melatonin. As yet, however, a clear role of the hormone's antioxidative activities in normal human physiology remains to be established.

V Clinical aspects

From the previous brief descriptions of some of the biological effects of melatonin, it appears that there could be opportunities to use the hormone in clinical settings, at the very least as an adjuvant to other treatments. Clinical presentations with their basis in the pineal, however, primarily involve lesions of the gland itself. Benign cysts occur relatively frequently, being seen on 4% of MRIs and up to 40% of routine autopsies. Such cysts need no follow-up care.

Solid tumors of the pineal are relatively rare, representing less than 1% of space-occupying intracranial tumors. They tend to occur during the first two decades of life, although one type, parenchymal cell tumors, occurs with equal frequency in adults and children. Pineal tumors show considerable histological variation and are now classified into three broad groups, depending on tissue of origin: germ cell, parenchymal cells (pinealocytes), and glial and other cells. All tumors in the pineal region of the brain share common symptoms, some of which are secondary to hydrocephalus, including vision dysfunction, nausea and vomiting, headaches, and difficulty walking. In children, pineal tumors are often associated with either delayed or precocious puberty.

The relative rarity of pineal tumors combined with the variety of types has made it difficult to establish a clear standard of care. In general, treatment depends on the type and aggression of the tumor and may consist only of surgery, radiation, or chemotherapy or a combination of two or three of these. Techniques in all three of these modalities have rapidly evolved recently. This is particularly true for surgical approaches which were hampered by difficulty in access to the gland until recent developments in imaging and microsurgery. For example, mortality and morbidity attributed to pineal surgery dropped from 20%–70% in 1950 to 0%–2% in 2011.

Further reading

The retinal–pineal system

G.A. Lincoln, F.J.P. Ebling, Photoperiodism and circannual timing: introduction and historical perspective, in: Neuroendocrine Clocks and Calendars, 2021. Springer. Available from: https://doi.org/10.1007/978-3-030-55643-3_1.

R. Soni, C. Haldar, C.M. Chaturvedi, Retinal and extra-retinal photoreceptor responses and reproductive performance of Japanese quail (*Coturnix coturnix japonica*) following exposure to different photoperiodic regime, General and Comparative Endocrinology 302 (2021) 113667. Available from: https://doi.org/10.1016/j.ygcen.2020.113667.

Regulation of melatonin synthesis

H.-R. Ahn, et al., Circadian rhythms are associated with various abiotic stresses, Plants 10 (2021) 129. Available from: https://doi.org/10.3390/plants10010129.

R.J. Reiter, et al., Melatonin synthesis in and uptake by mitochondria: Implications for diseased cells with dysfunctional mitochondria, Future Medicinal Chemistry 13 (2021). Available from: https://doi.org/10.4155/fmc-2020-0326.

Y. Wei, et al., Lighting the way: advances in transcriptional regulation and integrative crosstalk of melatonin biosynthetic enzymes in cassava, Journal of Experimental Botany 72 (2021) 161–166.

W. Zhou, et al., Whole genome sequence data of Hypericum perforatum and functional characterization of melatonin biosynthesis by *N*-acetyltransferase, Journal of Pineal Research 70 (2020). Available from: https://doi.org/10.1111/jpi.12709.

The melatonin receptors

Y. Cui, R. Jockers, O. Lahuna, Heterodimerization and subcellular distribution of melatonin and cannabinoid type 1 receptors, Journal of the Endocrine Society 3 (2021) A504. Available from: https://doi.org/10.1210/jendso/bvab048.1029.

A.M. Berbets, et al., Melatonin 1A and 1B receptors' expression decreases in the placenta of women with fetal growth restriction, Reproductive Sciences 28 (2021) 197–206.

R. Gerbier, et al., Pharmacological evidence for transactivation within melatonin MT2 and serotonin 5-HT_{2C} receptor heteromers in mouse brain, The FASEB Journal 35 (2020). Available from: https://doi.org/10.1096/fj.2020x00305R.

G.S. Kinker, et al., MT1 and MT2 melatonin receptors play opposite roles in brain cancer progression, Journal of Molecular Medicine 99 (2021) 289–301.

M. Stazi, et al., Melatonin promotes regeneration of injured motor axons via MT1 receptors, Journal of Pineal Research 70 (2020). Available from: https://doi.org/10.1111/jpi.12695.

Biological actions of melatonin

R. Reiter, R. Sharma, Central and peripheral actions of melatonin on reproduction in seasonal and continuous breeding mammals, General and Comparative Endocrinology 300 (2021) 113620. Available from: https://doi.org/10.1016/j.ygcen.2020.113620.

M. Segovia-Roldan, E.R. Diez, E. Pueyo, Melatonin to rescue the aged heart: antiarrhythmic and antioxidant benefits, Oxidative Medicine and Cellular Longevity 2021 (2021). Available from: https://doi.org/10.1111/fcp.12566.

M. Vlachou, et al., Pineal hormone melatonin as an adjuvant and treatment for COVID-19 (review), International Journal of Molecular Medicine 47 (2021). Available from: https://doi.org/10.3892/ijmm.2021.4880.

Clinical aspects

S. Da Silveira, et al., Disrupted nocturnal melatonin in autism: association with tumor necrosis factor and sleep disturbances, Journal of Pineal Research 70 (2021). Available from: https://doi.org/10.1111/jpi.12715.

L.S. Sartorelli, et al., Blood melatonin level can serve as a potential biomarker for prostate and hepatocellular carcinomas, Melatonin Research 4 (2021). Available from: https://doi.org/10.32794/mr11250094.

D. Shen, et al., The inhibitory effect of melatonin on human prostate cancer, Cell Communication and Signaling 19 (34) (2021) 2021. Available from: https://doi.org/10.1186/s12964-021-00723-0.

A.C. Tonon, et al., Melatonin and depression: a translational perspective from animal models to clinical studies, Frontiers in Psychiatry 12 (2021). Available from: https://doi.org/10.3389/tpsyt.2021.638981.

Chapter 17

Growth factors

I Introduction

A large group of molecules are referred to and named using the term *growth factor*. These molecules have one or more of the following activities that lead to an altered growth rate of a particular cell type: (1) increasing cell mass by stimulating the production and/or inhibiting the degradation of macromolecules; (2) increasing cell number by stimulating the rate of cell division, referred to as mitogenic activity; (3) increasing cell number through increased cell survival due to inhibition of apoptosis.

The normal physiological actions of the growth factors concern cell growth and differentiation in the context of morphogenesis as well as homeostatic mechanisms such as wound healing. Growth factors can have either positive or negative influences on these processes and their activities vary with cell type and the developmental stage of the organism. The activity of any one growth factor in a particular cell will be determined by its own state (active or latent); the presence of a receptor and, in some cases, a coreceptor in the cell membrane; and the necessary signaling machinery inside the cell. In addition, the presence of other extracellular signaling molecules can impact the activity of a growth factor at a target cell.

The goal of this chapter is to introduce five classes of growth factors with the emphasis on the structures of their members and the receptors that mediate their actions. Because of the great diversity of specific biological outcomes of the action of the growth factors on particular cells, no attempt is made to enumerate these. Rather the focus is on understanding the nature of the molecules engaged in these processes so that approaches to moderating their activities, particularly in the clinical setting of cancer, can be understood.

II Epidermal growth factor

A Structure and synthesis

The amino acid sequences of the founding member of this family, epidermal growth factor (EGF), as well as that of a closely related member, transforming growth factor α (TGFα) are shown in Fig. 17.1. All the members contain the conserved amino acid sequence $CX_7CX_{4-5}CX_{10-13}CXCX_8C$ where X is any amino acid. The six cysteine residues in this sequence participate in three intramolecular disulfide bonds, forming three structural loops that are characteristic of these growth factors and necessary for high-affinity binding to their receptor(s).

All members of the EGF family of growth factors, listed in Table 17.1, are synthesized as proproteins with a signal sequence directing it to the plasma membrane. A single membrane-spanning region anchors EGF to the membrane, generating an N-terminal extracellular domain and an intracellular cytoplasmic tail. Cleavage of the extracellular domain to release the soluble active form of the growth factor is carried out by a member of the ADAM (a disintegrin and metalloproteinase) family of metalloproteinases. For example, ADAM17 has been implicated in the cleavage heparin-binding (HB)-EGF, TGFα, amphiregulin, whereas ADAM10 catalyzes the release of soluble betacellulin. This process is highly regulated by other cell signaling systems in ways that are not yet completely understood. Although there is some in vitro evidence that the anchored form of EGF-related growth factors can activate receptors on nearby cells (termed a juxtacrine effect), it is well accepted that the release of the soluble form from the cell membrane is necessary for its physiological role in normal development and in tumorigenesis.

B EGF receptors and signaling

1 The EGF/HER/ErbB receptor family

Fig. 17.2 shows the four distinct receptor protomers for the members of the EGF family of growth factors. The nomenclature is shown beneath the protomers: EGF receptor (EGFR), which binds EGF, TGFα, and amphiregulin; human epidermal growth factor receptor (HER)2, for which no ligand has been identified (HER2 seems not to have a ligand-binding domain but HER2 can form heterodimers with the other HER molecules); HER3 that binds neuregulins 1 and 2; and HER4, which binds neuregulins 3 and 4, as well as 1 and 2. In addition, both EGFR and HER4 bind betacellulin, HB-EGF, and epiregulin. Because of its role in the control of cell growth, the EGFR is a proto-oncogene and in this context it

Hormones. DOI: https://doi.org/10.1016/B978-0-323-90262-5.00025-1

is known as c-ErbB1 or ErbB1. The three related receptors then, in addition to their HER (human EGFR) designations, are also known as ErbB2, ErbB3, or ErbB4.

Each of the four EGFR protomers consists of a cytoplasmic (carboxy-terminal) portion containing a tyrosine kinase domain, a single-pass membrane-spanning region, and an N-terminal extracellular portion consisting of four distinct domains, one of which is the dimerization domain. In EGFR, HER3, and HER4, ligand binding induces a conformational change that brings about dimerization. HER2, with no known ligand, is always in the dimerization conformation and can, therefore, in the absence of ligand, form a heterodimer with any of the other protomers. Although heterodimers can form between any of the receptor protomers, HER2 is the preferred heterodimerization partner for the others. EGFR, HER3, and HER4 can all form homodimers in the presence of ligand, but HER2 does not form homodimers. The tyrosine kinase domain of HER3 is nonfunctional, but the receptor can be phosphorylated by a heterodimerization partner.

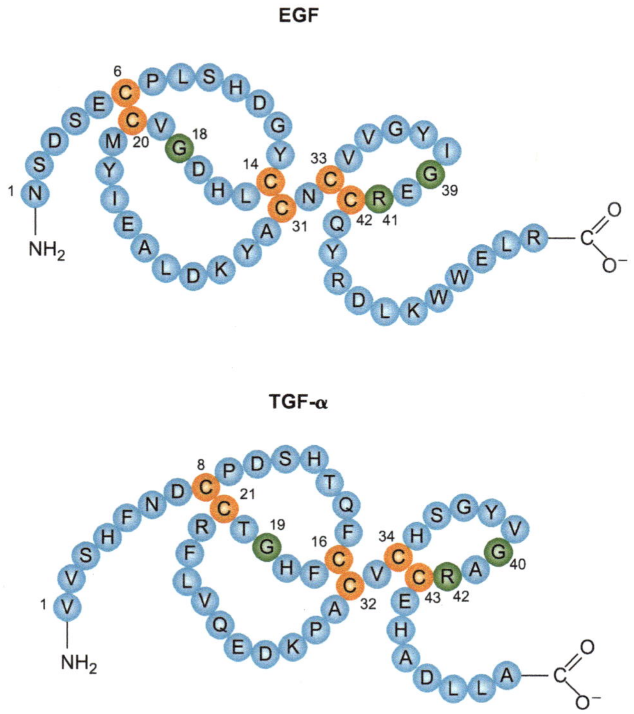

FIGURE 17.1 Structural features of the family of epidermal growth factors. The figure shows the amino acid sequences of two closely related members of the EGF family, EGF itself and TGFα. Highly conserved among all the members of this family are the three disulfide bonds (*orange*) that result in the characteristic loop structure of the peptides. There are also three highly conserved amino acids (*green*) in this portion of all members of this peptide family, leading to the overall motif: $CX_7CX_{4-5}CX_{10-13}CXCX_8C$ where X is any amino acid. *EGF*, Epidermal growth factor; *TGF*, transforming growth factor.

2 EGFR intracellular signaling

Upon dimerization the tyrosine kinase domains of EGFR and related receptors are activated, bringing about phosphorylation of the hydroxyl group of specific tyrosines on the cytoplasmic tail. This creates binding sites for specific signaling molecules which initiate further intracellular signaling events, as shown in Fig. 17.3. Four possible pathways are shown; the one (or more) that is activated depends upon the dimer partners and the activating ligands. Each of the four pathways, generated by binding and activating phospholipase C, p85/phosphatidylinositol-3-kinase, Src homology 2 domain containing, or signal transducer and activator of transcription (STAT), ultimately leads to changes in the expression of genes involved in the fundamental processes of cell growth, division, and survival.

Usually the result of EGF/HER receptor activation is the normal growth and differentiation of specific tissues. For example, EGF and TGFα are critical participants in the normal development of the skin, lung, intestine, eye, and

TABLE 17.1 Members of the epidermal growth factor family.

Name	Abbreviation	No. of amino acids in soluble form
Epidermal growth factor	EGF	53
Transforming growth factor α	TGFα	50
Heparin-binding EGF-like growth factor	HB-EGF	86
Amphiregulin	AR, AREG	84
Epiregulin	ERG, EREG	49
Betacellulin	BTC	80
Neuregulin-1[a]	NRG1	222
Neuregulin-2	NRG2	293
Neuregulin-3	NRG3	359
Neuregulin-4	NRG4	61

[a] *Neuregulins are also known as heregulins.*

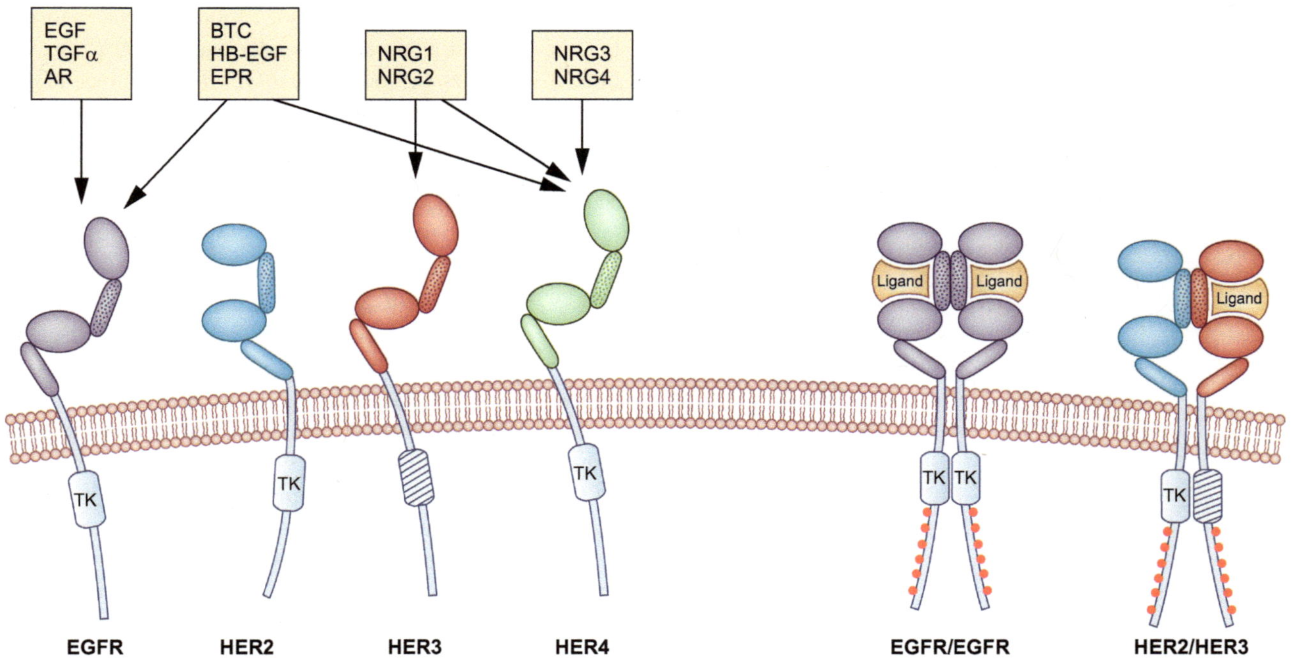

FIGURE 17.2 Structural features of the EGFR and related receptors. The four receptor tyrosine kinases that make up this family are shown in their protomeric forms on the left side of the figure. Above each are the ligands which bind with high affinity. From left to right the protomers are known as EGFR (ErbB-1); HER2 (neu/ErbB-2); HER3 (ErbB-3); and HER4 (ErbB-4). These receptors are single membrane-spanning receptor tyrosine kinases. Both heterodimers and homodimers can form and an example of each is shown on the right side of the figure. Each of the protomers has four extracellular domains which, upon binding ligand, change conformation, exposing a dimerization face. HER2 has no known ligands and is in the conformation exposing the dimerization face in the absence of a ligand. The intracellular portion of each of the four protomers contains a TK. The TK of HER3 is inactive. When heterodimers form, TK is activated, bringing about autophosphorylation and phosphorylation of adjacent proteins to initiate signaling cascades (see Fig. 17.3). HER3 is thought to be phosphorylated by its dimerization partner, HER2, in the figure. See Table 17.1 for definition of ligand abbreviations.

mammary gland. However, if there is overexpression or mutation of the receptor proteins, the expression of genes controlling cell proliferation and differentiation can be altered such that the rate of growth escapes the normal controls. For example, the inappropriate overexpression of one of the receptor proteins is linked to several forms of cancer. This is especially true of the favored dimerization partner, HER2, which can become activated in the absence of a ligand and has become an important target of cancer chemotherapy (see Section VII and Chapter 14: Pancreatic Hormones: Insulin and Glucagon).

The normal physiological functions are known for only some of the EGF/HER receptor ligands listed in Table 17.1. For example, amphiregulin is mitogenic for a range of targets, including astrocytes, Schwann cells, and fibroblasts. Betacellulin is a potent mitogen for retinal pigment epithelial cells and neuregulin 1 stimulates the expression of the acetylcholine receptor in the cells of synaptic nuclei.

III Fibroblast growth factor family

A Members of the FGF family

1 Nomenclature

It has been nearly 75 years since an extract of bovine brain was shown to promote the proliferation of fibroblasts in cell culture and was called fibroblast growth factor or FGF. About 25 years later, the mitogenic activity was attributed to a purified basic (based on isoelectric point) 15-kDa protein. At around the same time, an acidic form of the protein was identified and the two proteins became known as aFGF and bFGF. Upon the discovery of the binding activity of these proteins to heparan sulfate (HS), heparin affinity chromatography led to the discovery of further proteins of this family. The original two were renamed with numbers, FGF1 and FGF2, and the numbering system was applied to the members of the family discovered through protein purification, biological activity, and genetic analysis over the next few decades.

2 FGF subfamilies: structure and receptor binding

The structural features and receptor-binding activity of the FGFs are summarized in Fig. 17.4. The size of the proteins varies between 155 and 257 amino acids and they share a conserved core sequence of about 120 amino acids (Fig. 17.4A). These are arranged in three sets of four antiparallel β-sheets that make up the globular portion of the molecules. The N- and C-terminal portions form loops that distinguish the shapes of the proteins and

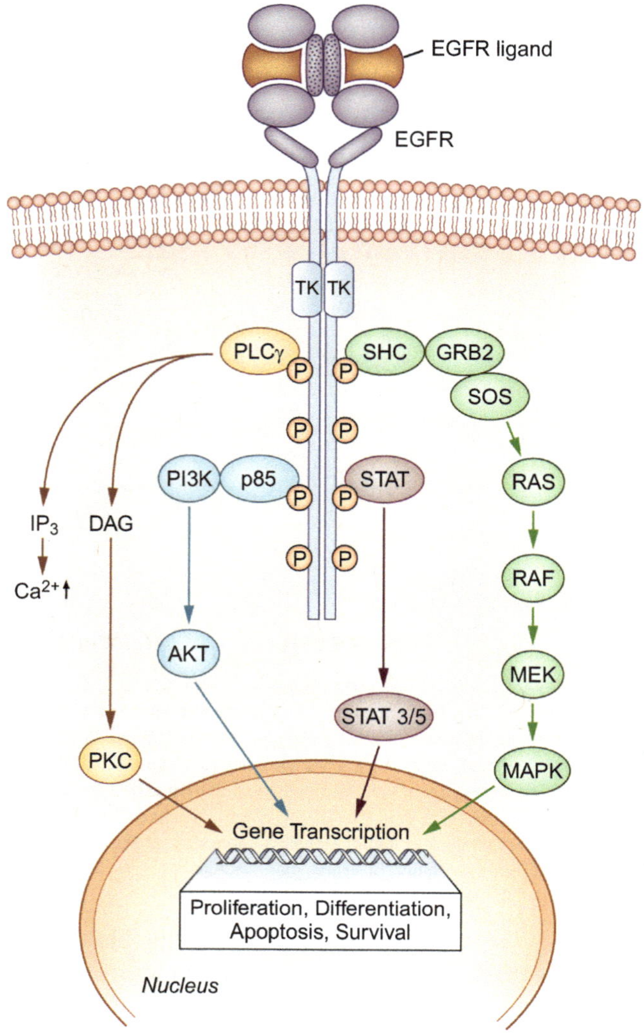

FIGURE 17.3 EGFR family signaling pathways. Four intracellular signaling pathways can be initiated by a homo- or heterodimer of the EGFR/HER receptor family. The first step in each is the tyrosine kinase–mediated phosphorylation of a protein that generates a second intracellular signal. For example, the phosphorylation of PLCγ leads to the release of IP_3 and DAG, leading to the activation of PKC. Phosphorylation of p85 activates PI3K leading to the activation of Akt (also known as PKB). The STAT pathway and the MAP kinase pathway are also activated by the receptor tyrosine kinase activity as shown. Depending on the pathway(s) activated, the cell type, and its environment (e.g., exposure to other signaling inputs) these pathways, alone or in combination, will lead to changes in gene expression, which will influence the ability of the cell to divide, differentiate, undergo, or resist apoptosis and survive. GRB2, growth factor receptor bound-2, an adaptor protein; SOS, son of seven less guanine nucleotide exchange protein. *DAG*, Diacylglycerol; *EGFR*, epidermal growth factor receptor; *GRB2*, growth factor receptor bound-2, an adaptor protein; *HER*, human epidermal growth factor receptor; *IP_3*, inositol triphosphate; *PI3K*, phosphatidylinositol-3-kinase; *PKB*, protein kinase B; *PKC*, protein kinase C; *PLCγ*, phospholipase Cγ; *SOS*, son of seven less guanine nucleotide exchange protein; *STAT*, signal transducer and activator of transcription.

account for their differences in receptor binding and activity. The FGFs are divided into numbered subfamilies, as shown by the grouping by color in 17.4B based on phylogeny and sequence homology.

Based on their mechanism of action, the FGF subfamilies are grouped as paracrine, endocrine, or intracrine. The largest of these, the paracrine FGFs (subfamilies FGF-1, -4, -7, -8, and -9), contain both an N-terminal signal sequence (except for subfamily 1; see legend to Fig. 17.4A) and a HS binding region in the C-terminus. The latter is necessary for high-affinity FGF receptor (FGFR) binding. Thus this group of subfamilies is secreted from the cells that make them to bind nearby target cells, in a 2:2:2 complex of FGF:HS:FGFR (see Fig. 17.5). The members of the endocrine group (subfamily 19) lack the HS-binding site and, therefore, are not retained in the region in which they are secreted. They are, instead, released into the general circulation for action at distant target cells, acting as classical endocrine hormones. As will be discussed following, the binding of these FGFs to the FGFR in their target cells requires the presence in the membrane of a coreceptor protein, klotho. Finally, since the FGFs of subfamily 11 lack the signal sequence, they are not secreted from the cells in which they are made and their biological activity is confined to these cells. Fig. 17.4B shows that the six secreted subfamilies of FGFs have distinct receptor-binding affinities and, therefore, different biological activities in their target cells. The receptor subtypes are discussed in Section III.B.

FGFs 11–14, which are also known as FGF-homologous factors are not always included as *bona fide* members of the FGF family since their mechanism of action differs so substantially from those of the other FGFs. Rather than interacting with the extracellular domain of the FGFR (see Fig. 17.4B), these proteins bind to and regulate the activity of voltage-gated sodium channels and are largely active in neurons and cardiomyocytes. The remainder of our discussion of FGFs in this chapter will focus on their interaction with their cognate FGFRs, consequent intracellular signaling, and biological activities in various cells and tissues and will not, therefore, include the intracrine FGF11 subfamily.

B The FGF receptors and signaling

1 Structure and types of the FGF receptor

There are four forms of the FGFR, encoded by different genes, and designated FGFR1–FGFR4. Each has the common structural elements shown in Fig. 17.5A. The extracellular portion of the receptor is composed of three immunoglobulin-like domains of which the two nearest the membrane (D2 and D3) are required for FGF binding. The purple section in loop D3 is subject to alternative splicing in FGFRs 1–3, giving rise to the subtypes FGFR1b (epithelial cells) and FGFR1c (mesenchymal cells), etc., shown in Fig. 17.4. Table 17.2 gives a partial list of the presence of FGFRs in various cells and tissues. FGFR has

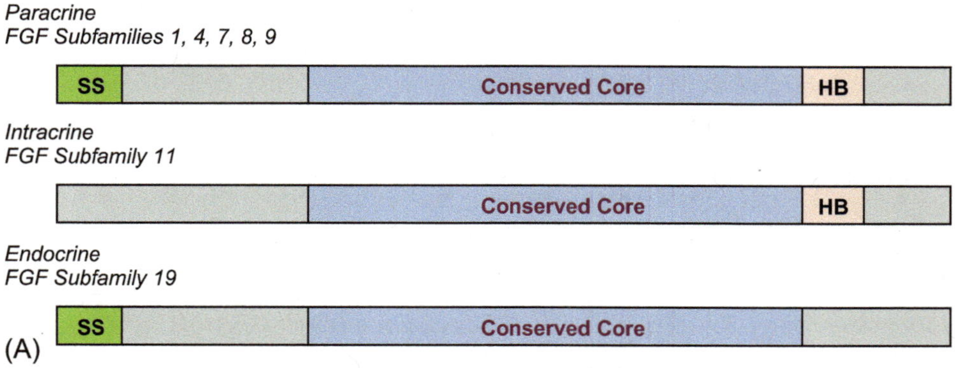

(A)

Subfamily	Ligand	1b	1c	2b	2c	3b	3c	4
		FGF Receptor Binding						
Paracrine								
FGF-1	FGF1	●	●	●	●	●	●	●
	FGF2	●	●		●		●	●
FGF-4	FGF4		●		●		●	●
	FGF5		●		●			
	FGF6		●		●			●
FGF-7	FGF3	●		●				
	FGF7			●				
	FGF10	●		●				
	FGF22	●		●				
FGF-8	FGF8				●		●	●
	FGF17		●		●		●	●
	FGF18				●		●	●
FGF-9	FGF9				●	●	●	●
	FGF16				●		●	●
	FGF20		●		●	●	●	●
Intracrine								
FGF-11	FGF11							
	FGF12							
	FGF13							
	FGF14							
Endocrine								
FGF-19	FGF19		●		●		●	●
	FGF21		●		●		●	●
	FGF23		●		●		●	●

(B)

FIGURE 17.4 FGF family. (A) Structural schematic of FGF structure. The 22-member family of human fibroblast growth factors, each encoded by a separate gene, range in length from 155 to 267 amino acids. All share a common conserved globular core composed of 12 antiparallel β-sheets (*blue*). The degree of conservation of this core as well as some functional considerations have led to the grouping of the 22 FGFs into 7 subfamilies as listed in parts (A) and (B) of the figure. These, in turn, can be placed into three groups based on differences in the N- and C-termini of the proteins which lead to distinct functional differences. Four of the five paracrine subfamilies that act on nearby cells have a SS, which allows them to be secreted from the cell and a heparin-binding sequence (HB) which ensures their accumulation and binding to receptors on nearby cells. FGFs 1 and 2 lack the N-terminal signal sequence and are released by damaged cells or by a non-Golgi exocytotic mechanism. FGFs 11−14, sometimes referred to as FHF, comprise the intracrine family because, having no signal sequence to bring about their secretion, they work in the cells in which they are produced. The members of endocrine subfamily, FGFs-19, −21, and −23, lack a functional heparin-binding site and thus are not concentrated in the vicinity of the cell but are released into the circulation to reach distant target cells. (B) *Binding of FGFs to the FGF receptor.* The binding of each FGF in a mitogenic bioassay to the seven forms of the FGF receptor (see legend to Fig. 17.5) is shown by a closed circle. All measurements are relative to FGF1 and, for the purposes of this table, 15% relative binding was the arbitrary cutoff for binding activity. *FGF*, Fibroblast growth factor; *FHF*, fibroblast growth factor homologous factors; *SS*, signal sequence. *Data is from Zhang et al., Journal of Biological Chemistry 281 (2006) 15694−15700 and Ornitz et al. Journal of Biological Chemistry 271 (1996) 15292−15297.*

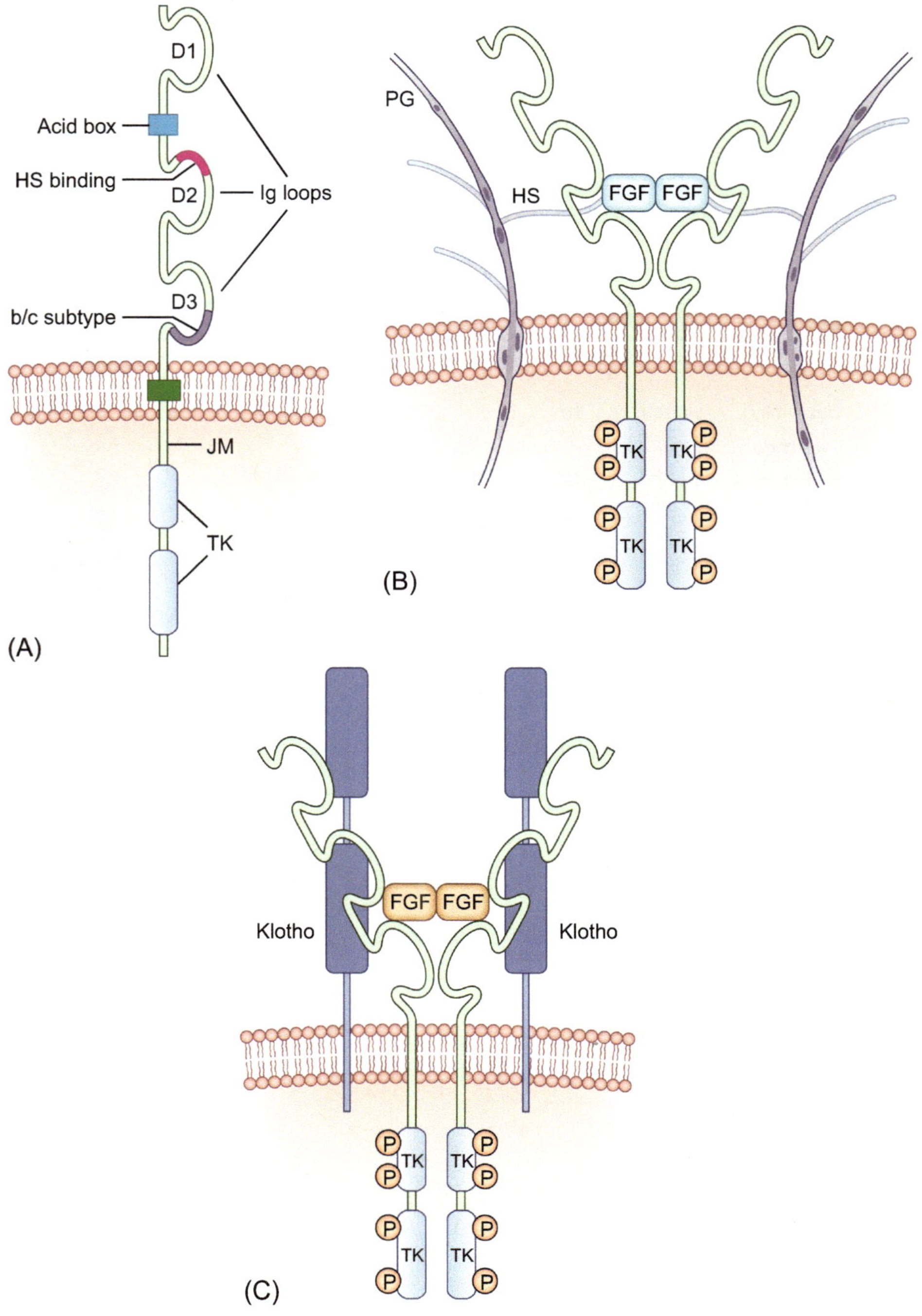

FIGURE 17.5 FGFR structure. (A) FGFR monomer. This diagram represents the common structural features of each of the four separately encoded FGFRs, FGFR1–FGFR4. The extracellular domain of the FGF monomer consists of three Ig-like loops, D1, D2, and D3. Loops D2 and D3 comprise the ligand-binding domain and are separated from loop D1 by a stretch of acidic amino acids referred to as the acid box. Alternative splicing of mRNA in loop D3 (*purple*) generates the −b and −c forms of FGFRS 1, 2, and 3 listed in Fig. 17.4B (there is an −a form that is a truncated secreted version of the receptor, not shown). To form a stable dimer upon ligand-binding (panel B), interaction between heparan sulfate and its binding site on the FGFR (*red*) is required. The FGF receptor has a single-pass transmembrane domain (*dark green*). The intracellular portion consists of a juxtamembrane region followed by a split TK in which the TK catalytic domain is interrupted by a nonfunctional amino acid sequence. (B) *FGF-HS-mediated dimerization of FGFR (paracrine FGF subfamilies 1,4,7,8,9)*. In this diagram, the 2:2:2 complex of FGF, FGFR, and HS is shown. HS is attached to membrane PG, ensuring its abundance in the extracellular membrane environment, and is required for high-affinity binding of FGF to its receptor and stabilization of the complex shown. (C) *FGF-klotho-mediated activation of FGFR (endocrine) subfamily 19*. FGFs that do not have a HS-binding site are released into the circulation and react with distant target cells that have both an FGFR and a membrane-bound protein called klotho. The details (e.g., dimerization) of klotho–FGFR interactions are not as well understood as those with heparan sulfate, but at least one molecule each of the ligand (FGF, *gold*), receptor, FGFR (*green*), and the coreceptor, klotho (*blue*), are required for activation of the receptor. *FGFR*, Fibroblast growth factor receptor; *HS*, heparan sulfate; *Ig*, immunoglobulin; *mRNA*, messenger RNA; *PG*, proteoglycan; *TK*, tyrosine kinase.

TABLE 17.2 Distribution of fibroblast growth factor receptors (FGFR).

FGFR	Cell/tissue[a]
FGFR/HS[b]	
FGFR-1b	Fibroblasts, endothelial, VSM[c]
FGFR-1c	Macrophages, hematopoietic progenitors
FGFR-2b	Epithelial
FGFR-2c	Fibroblasts, VSM, astrocytes, lymphocytes, hematopoietic progenitors
FGFR-3b	Epithelial, keratinocytes
FGFR-3c	Fibroblasts, monocytes, endothelial, hematopoietic progenitors
FGFR-4	Embryonic and multipotential stem cells
FGFR/Klotho[d]	
FGFR-1c	Gall bladder (19), adipocyte (21), kidney (23)
FGFR-3c	Kidney (23)
FGFR-4	Liver (19), kidney (23)

[a]*Partial list.*
[b]*HS, heparan sulfate; the receptor-binding members of FGF subfamilies 1,4,7,8,9 are listed later.*
[c]*VSM, vascular smooth muscle.*
[d]*The receptor-binding members of FGF subfamily 19 are listed later. The specific endocrine FGF ligand for these FGFRs is shown in parentheses.*

a single-pass transmembrane region, a juxtamembrane region followed by a split tyrosine kinase domain, that is, one in which the catalytic parts of the enzyme are separated by a run of noncatalytic amino acids.

HS is a chain of alternating sulfated glucuronic acid and *N*-acetyl glucosamine units and is attached to proteoglycan that is anchored to the cell membrane. The majority (the paracrine FGFs, subfamilies 1,4,7,8,9; Fig. 17.4) of FGFs require HS binding to specific sites on both ligand and receptor for receptor activation. The HS-binding site on the FGFR contains basic amino acids and is on the face of D2. The acid box, a stretch of about 20 amino acids with a predominance of aspartates, glutamates, and serines, along with the N-terminal loop, D1, interferes with the basic HS-binding site and, thus, acts as a restraining influence on FGF activation. This phenomenon is called autoinhibition and it is the removal of this restraint by ligand binding that contributes to the activation of the receptor.

Lacking functional HS-binding sites, members of the endocrine subfamily 19, FGFs 19, 21, and 23, are not retained in the HS-rich vicinity of the cells in which they are produced but rather are released into the circulation to be transported to distant target organs. To be responsive to one of these FGFs, a cell must have not only an FGFR but also a membrane protein called klotho, which acts as a coreceptor (Fig. 17.5C).

Klotho was originally discovered as an anti-aging protein in mice in 1997. Mice lacking the gene for this protein age rapidly and when it is overexpressed, aging is slowed. A few years later, disturbances in phosphate metabolism similar to some hyperphosphatemic diseases in humans (see Chapter 9: Calcium-Regulating Hormones: Vitamin D, Parathyroid Hormone, Calcitonin; Fibroblast Growth Factor-23) implicated klotho in FGF23/FGFR binding in the kidney. A second form of klotho, termed β-klotho (while α was added to the name of the original protein), was recognized as an FGFR coreceptor for binding of FGFs 19 and 21. Both forms of klotho are single-pass membrane proteins with a short intracellular domain. The extracellular domain consists of two tandem β-glucuronidase-like segments. Fig. 17.5A shows a schematic diagram of klotho in the context of one model for its interaction with the FGFR.

2 FGF/FGFR signaling

The signaling pathways of the activated FGFR are shown in Fig. 17.6. In two pathways the first step is phosphorylation and the activation of the FGFR substrate protein, FSR2α, which is localized in the cell membrane. Activated FSR2α binds and activates nearby growth factor receptor bound-2 (GRB2) which recruits one or both of two adaptor proteins. Son of seven less (SOS) leads to the activation of the mitogen-activated protein kinase (MAPK) pathway as seen in Fig. 17.3 and elsewhere in this book. The outcome of altered gene expression resulting from this pathway is increased cell proliferation as

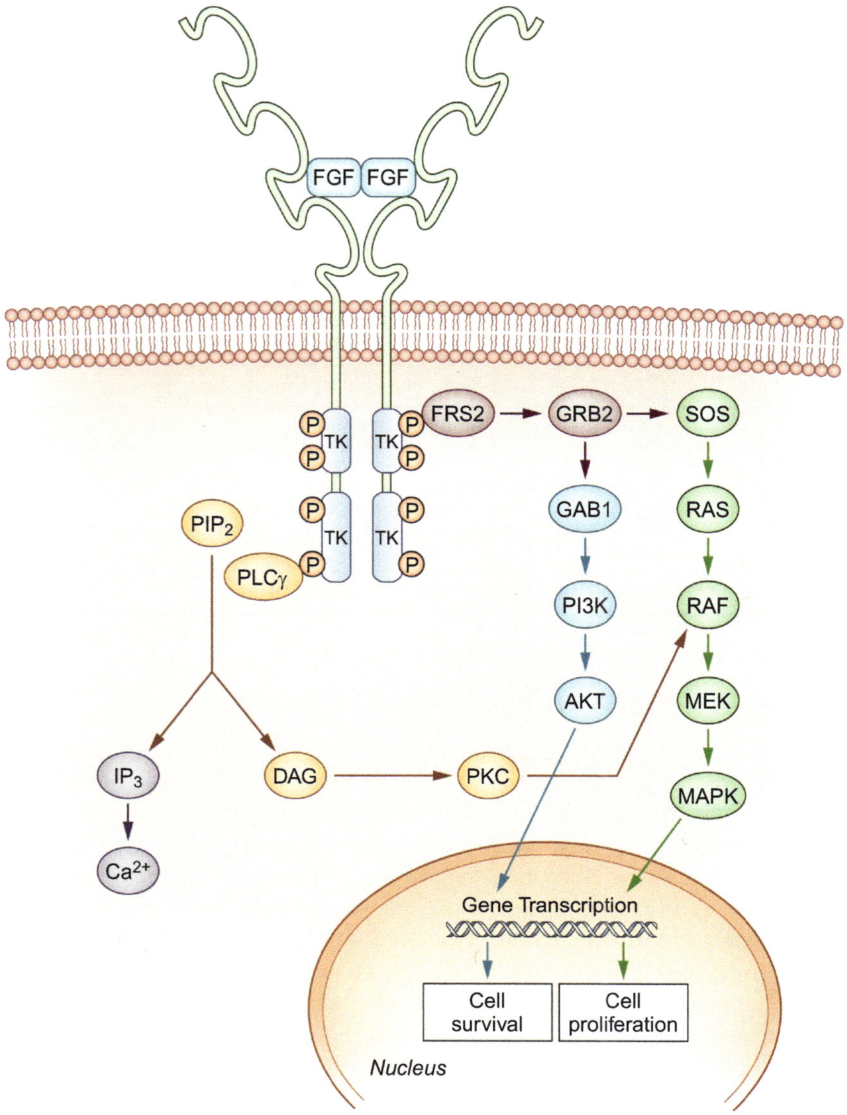

FIGURE 17.6 FGFR signaling. A major target for FGFR tyrosine kinase activity is FRS2α (FGFR substrate 2α; FRS2) which is associated with the kinase and is activated by tyrosine phosphorylation. Activated FRS2α binds GRB2, which then recruits either the guanine nucleotide exchange protein SOS or GAB1 (GRB2-associated binding protein), activating the MAPK pathway or PI3 kinase-mediated AKT activation pathway, respectively. In addition, the FGFR tyrosine kinase activation of PLCγ results in the release of DAG and IP_3, activating protein kinase C which phosphorylates and activates RAF, adding reinforcement to this pathway. *DAG*, Diacylglycerol; *FGFR*, Fibroblast growth factor receptor; *GRB2*, growth factor receptor bound-2; *IP_3*, inositol triphosphate; *MAPK*, mitogen-activated protein kinase; *PLCγ*, phospholipase Cγ; *RAF*, rapidly accelerated fibrosarcoma; *SOS*, son of seven less.

well as, in some cases, cell migration and differentiation. If the adaptor protein GAB1 is activated by GRB2, the PI3 kinase/protein kinase B pathway is activated promoting cell survival by inhibiting apoptosis. Other pathways may also participate in FGF signaling. For example, activation of the phospholipase C cascade leads to, among other things, amplification of the MAPK pathway by providing for the rat sarcoma oncogene-independent phosphorylation of rapidly accelerated fibrosarcoma. In some cells the Janus kinase/STAT pathway (not shown) is also activated by FGFR.

3 FGF biological activity

As mentioned in the introduction to this chapter, a catalog of the biological actions of the FGFs in all their cell types, both embryonic and adult, is not possible in this chapter. However, there are two aspects of FGF biological activity upon which some comment is appropriate.

The endocrine FGFs have, in the adult organism, specific physiological roles as do the other hormones discussed through this book. For example, the crucial actions of FGF23, interacting with FGFR1c and α- klotho, in the control of phosphate and, therefore, bone metabolism, have already been discussed in Chapter 9, Calcium-Regulating Hormones: Vitamin D, Parathyroid Hormone, Calcitonin; Fibroblast Growth Factor-23 (Fig. 9.17). FGF19, the main receptor for which is FGFR4 with β-klotho, is part of a negative feedback loop regulating bile acid synthesis following a meal. FGF21 binding to FGFR1c and β-klotho provides, under fasting conditions, a signal from the liver to the adipose tissue to increase fatty acid release and oxidation, gluconeogenesis, ketogenesis.

The second aspect of the biological activities of the FGF family of growth factors to emphasize is not unique to this family. This is that amplification of their normal signaling activity can and does lead to unrestrained growth of the cells in which this occurs. This can be due to a number of events, described more in Section VII, including mutations in the receptor which render them constitutively active. For example, one naturally occurring mutation in FGFR1, leading to the absence of IgG loop D1, behaves in such a way.

IV Platelet-derived growth factors

A Structure of PDGFs

The ability of a factor derived from platelets to stimulate the growth of fibroblasts, arterial smooth muscle cells, and glial cells in culture was discovered in the early 1970s. Over the next decade, two forms of this platelet-derived growth factor, PDGF-A and PDGF-B, were characterized. In the early 2000s, two more forms, PDGF-C and PDGF-D, were identified through genetic and protein biochemical approaches. Genes for the two receptors for PDGFs, PDGFRα, and PDGFRβ were characterized just prior to this time. Each of the PDGF forms and the two receptors are encoded by separate genes.

Fig. 17.7 shows in schematic form the structural components of the PDGFs. In addition to a 20-amino-acid signaling sequence, each of the four PDGFs contains a highly conserved core region that has eight strictly conserved cysteine residues. Six of these participate in a stability-promoting structure termed a cystine-knot fold, in which one disulfide bridge passes through a loop formed by two other disulfide bridges. The remaining two cysteines form disulfide bridges between two protomers to form homodimers (AA, BB, CC, or DD) or, when both protomers are synthesized in the same cell, one heterodimer, AB. PDGF-C and PDGF-C do not appear to form heterodimers. PDGF dimerization occurs prior to secretion from the cell.

Two features distinguish the A and B forms of PDGF from the C and D forms. The former have a carboxyl-terminal tail consisting of positively charged amino acids, promoting the interaction of the secreted factors with negatively charged pericellular molecules such as HS, as described earlier for FGF. This retention sequence keeps the growth factors in the vicinity of the cells in which they will carry out their paracrine functions. PDGF-A can be alternatively spliced so that it lacks this carboxyl-terminal tail, in which case the secreted homodimer is fully soluble when it leaves the cell. PDGFs-C and -D lack this sequence, but have a longer prosequence, containing a region (structural motif of ~110 residues; see legend to Fig. 17.7) which may fulfill the same pericellular retention function.

The prosequences of dimers containing PDGF-A and/or -B are removed at a specific cleavage site during secretion, whereas PDGFs-C and -D are secreted as latent forms with an intact prosequence that is then cleaved in the extracellular matrix. For PDGFs A, B, and D (and perhaps C as well), the cleaved prosequences remain associated with the active peptide to maintain it in a latent form until the dimer binds to its cognate receptor.

B PDGF receptors and signaling

1 The PDGF receptors, PDGFRα and PDGFRβ

Separate genes encode the two PDGFRs, α and β. As shown in Fig. 17.8A, the extracellular domain of the protomer consists of five Ig domains. D1 and D2 actually have a long linker between them, allowing the two domains to form a single structural module with a highly

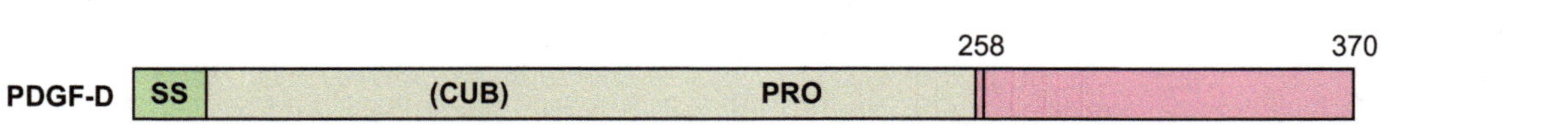

FIGURE 17.7 PDGFs. The schematic structures of the four PDGF-A, -B, -C, and -D, are shown. They contain a common conserved sequence (*pink*) of 95–112 amino acids which is the cystine-knot peptide that binds to the receptor. This portion of the protein contains three disulfide bonds that form a loop through which one of the three disulfide bridges passes, giving the peptide its stability and propensity to form dimers. Each PDGF also has a secretory sequence (*green*) and a prosequence (*gray*). The difference in the length of PDGFs A and B compared to C and D is largely accounted for by the difference in the length of this prosequence, with C and D, containing the CUB (an acronym combining features of three peptide families) sequence. The double line between the prosequence and the conserved sequence indicates the cleavage between the two to release the active peptide. Finally, PDGFs A and B have a carboxyl-terminal tail (*blue*), sometimes referred to as a R sequence, consisting of positively charged amino acids that facilitate the binding of the released form to extracellular heparan sulfate. *CUB*, structural motif of ~110 residues; *PDGF*, Platelet-derived growth factor; *R*, retention.

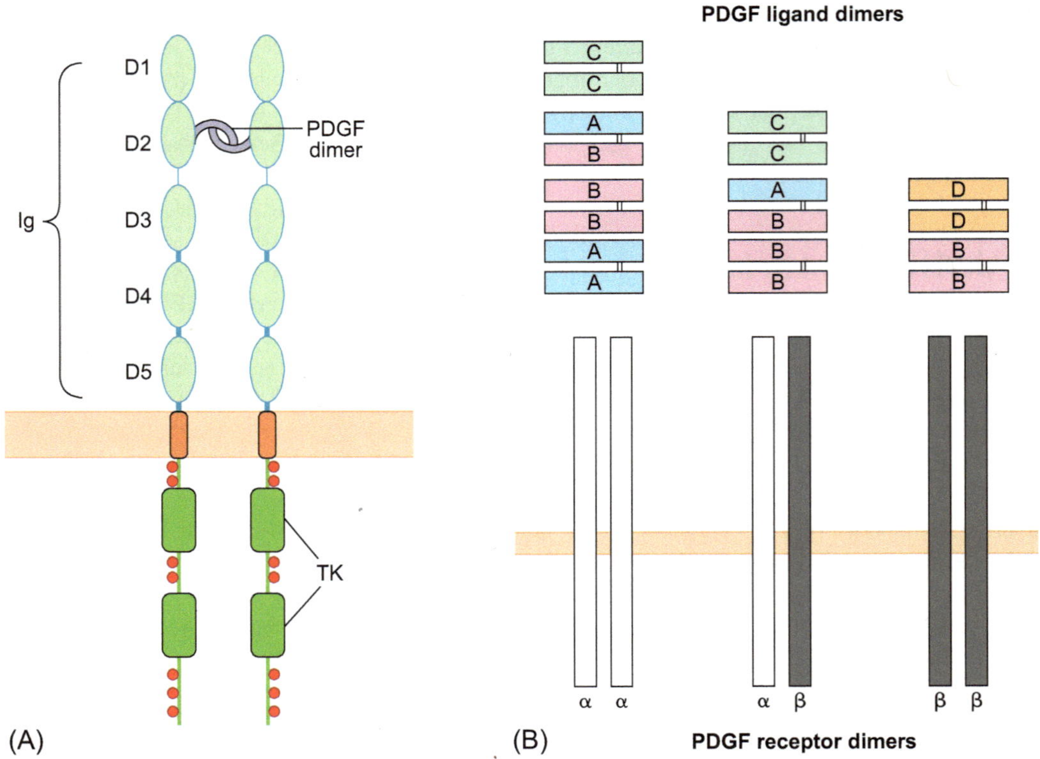

FIGURE 17.8 The PDGF receptor and ligand interactions. (A) Structural features of the α and β forms of the PDGF receptor. The PDGF receptor consists of an extracellular domain containing five Ig loops, D1–D5 (*green ovals*), a transmembrane domain of about 25 amino acids, and an intracellular domain with a split tyrosine kinase (see Fig. 17.5A). The binding of dimerized ligand (*purple*) promotes receptor protomer dimerization. Both homodimers and a heterodimer occur. Ligand binding takes place mostly on the hydrophobic surface of D2, which is closely associated with D1. The changes in D3 conformation brought about by ligand binding are passed through D4 and D5 by the more rigid D3–D4 and D4–D5 hinge regions (heavy bars) to the intracellular JM region and the tyrosine kinase moieties. (B) *PDGF ligand–receptor interactions*. The tendency of each of the four PDGF homodimers and one heterodimer (AB) to bind to each of the three forms of the PDGF receptor is shown. *Ig*, Immunoglobulin; *JM*, juxtamembrane; *PDGF*, platelet-derived growth factor.

N-glycosylated hydrophilic N-terminal region to protect against nonspecific hydrophobic interactions at the D2 ligand-binding site. Binding of the PDGF ligand dimer to two receptor protomers brings about receptor dimerization. The functional forms PDGFRs are either a homodimer, αα or ββ, or heterodimer, αβ (Fig. 17.8B). Structural studies indicate that while the linker between D2 and D3 is quite flexible, allowing the conformation of D3 to change significantly in response to ligand binding, the D3–D4 and D4–D5 linkers are rigid. This helps the change in conformation signal occurring in D3 upon ligand binding to be conveyed through the two remaining Ig domains and the plasma membrane to the intracellular signaling portion of the receptor. The intracellular portions of the PDGFRs have a juxtamembrane region, a split tyrosine kinase domain, and a cytoplasmic tail, which is phosphorylated upon activation of the receptor.

Fig. 17.8B shows the PDGF dimer/receptor dimer pairings that occur most often in vivo and in cells in tissue culture. The PDGFRα homodimer binds three (AA, BB, and CC) of the four homodimers with high affinity, whereas the PDGFβ homodimer binds the remaining homodimer (DD) as well as BB. The less restrictive amino acid composition of the ligand-binding site of PDGFRα relative to PDGFRβ accounts for the greater promiscuity of the α homodimer. Similarly, PDGF-B is more flexible than the other PDGFs due to the less restrictive amino acids in its binding interface. Thus BB homodimer can bind to all three forms of the PDGFR.

2 PDGFR signaling

Since there are five structurally distinct PDGF ligands that signal differentially through three forms of the PDGFR, it is clear that cell context is of great importance in understanding the intracellular signaling of any one ligand–receptor interaction. Add to this the secretion of PDGF-CC and PDGF-DD in latent forms that must be proteolytically activated in the extracellular matrix of a given cell, and the potential for heterogeneity in cell responses is expanded even further. As shown in Fig. 17.9, there are ample docking sites available among the 11 tyrosine residues on PDGFRα and 13 on PDGFRβ that are phosphorylated upon receptor activation. The

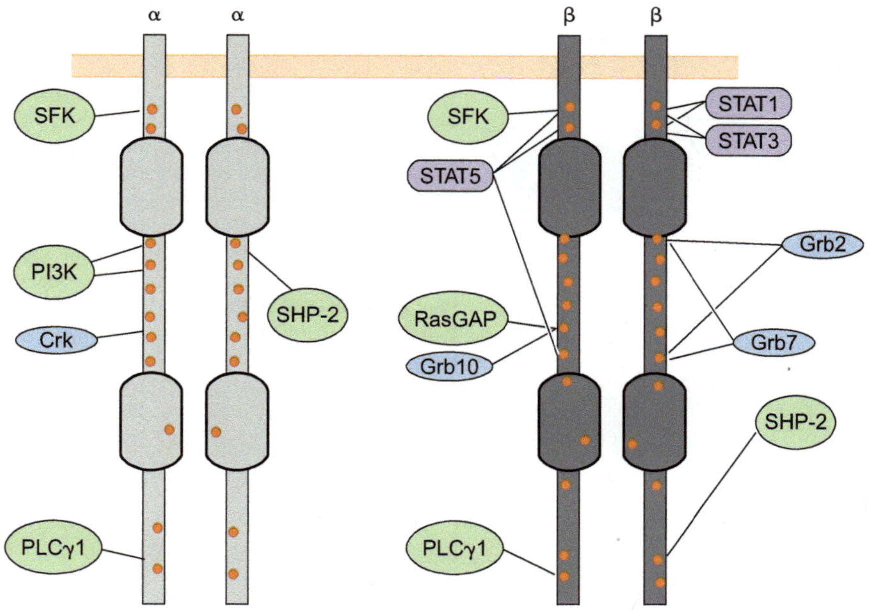

FIGURE 17.9 Proximal signaling events at the PDGF receptor. The dimerized PDGF contains numerous tyrosines that can undergo phosphorylation when the receptor is activated. These are shown by orange circles, 11 on each α protomer and 13 on each β protomer. Each of the sites interacts, as indicated by a solid line, specifically with one or more proteins to initiate the intracellular events that will lead to the biological response. These proteins may be enzymes (*green*), transcription factors (*purple*), or adaptor proteins (*blue*). *Crk*, CT10 regulator of kinase; *GRB*, growth factor receptor bound; *PDGF*, platelet-derived growth factor; *PI3K*, phosphatidylinositol-3′-kinase; *PLCγ − 1*, phospholipase Cγ − 1; *RasGAP*, Ras GTPase activating protein; *SFK*, Src family kinases; *SHP-2*, SH2 domain-containing protein tyrosine phosphatase-2; *STAT*, signal transducer and activator of transcription.

figure shows the specific sites to which various proteins are able to bind with high affinity. These proteins include enzymes such as phospholipase C (PLC)γ − 1 and PI3-kinase; members of the STAT family of transcription factors; and members of the GRB family of adaptor proteins. Most of the associations between these proteins and their sites on the receptor occur through SH2 (Src-homology 2) domains on the binding proteins. It is important to note that the exact proteins that are bound to the activated receptor in any particular cell, and, therefore, the pathways initiated, will depend on the availability of the proteins and, most likely, the particular ligand–receptor complex present in the cell.

Among the outcomes of PDGF signaling are proliferation, chemotaxis, survival in the form of protection from apoptosis, and, in some cells, differentiation. The next examples will indicate that the response of a given cell to a particular ligand depends to a great extent on which PDGFR(s) are present as the three forms are distinct from one another in their binding and signaling properties. In cells with the ββ receptor, GRB2, through the recruitment of Ras and SOS (see Fig. 17.6), initiates the MAPK pathway to stimulate proliferation. RasGAP interaction with the β PDGFR stimulates Ras GTPase activity, rendering Ras inactive and thereby countering the Ras-mediated stimulation of proliferation. In contrast, GRB2 interaction with the αα receptor does not lead to Ras so other pathways, possibly PI3 kinase, account for increased proliferation in response to PDGF in cells with the αα PDGFR. Thus the proliferative response of a given cell to PDGF will depend on the receptor types present and the balance between positive and negative influences on the process.

The ability of α and β PDGFRs to stimulate chemotaxis also differs; PDGFR ββ and PDGFRαβ potently stimulate chemotaxis, whereas the ability of PDGFRαα to do so is open to question. Both PI3 kinase and PLCγ1 mediate pathways that contribute to the chemotactic response which involves marked cytoskeletal changes and cell adhesion properties. This role of PDGF in chemotaxis, bringing new cells to a particular site, is crucial to its role in capillary formation during embryonic development and in wound healing.

V Insulin-like growth factors

A Structure of insulin-like growth factor 1 and insulin-like growth factor 2

The primary sequence of insulin-like growth factor (IGF)-1 is shown in Fig. 3.20, in which the similarity of its structure to that of insulin is illustrated. Fig. 17.10 shows the sequence of IGF-2 which has the same overall structure as IGF-1 but is 3 amino acids shorter than IGF-1. The two IGFs share nearly 70% amino acid identity and the same placement of disulfide bonds but differ in their binding properties to the receptors that mediate their actions.

B Insulin and IGF receptors and signaling

1 IGF and insulin receptors

Fig. 17.11 shows the structural features of the receptors for insulin (IR-A and IR-B) and for IGF-1 (IGF-1R) and IGF-2 (IGF-2R). We will first focus on the insulin (*blue*) and IGF-1 (*green*) receptors. The genes for IR (22 exons)

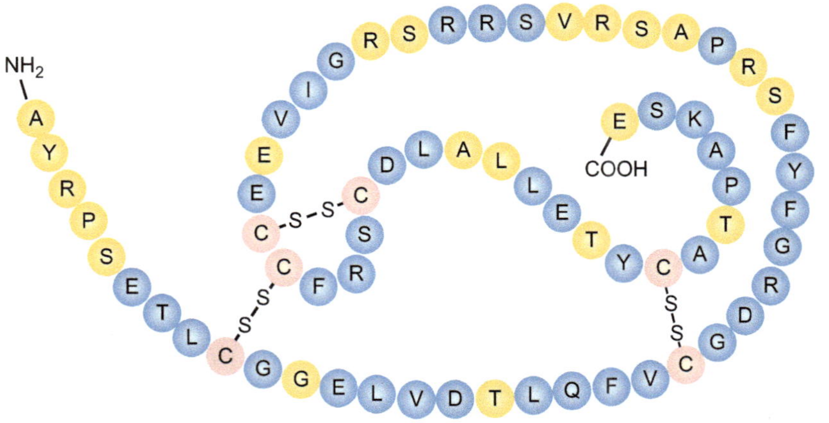

FIGURE 17.10 Structure of IGF-2. The amino acid sequence of IGF-2 is shown; see Fig. 3.20 for the structure of IGF-1 which is compared to that of proinsulin. Here the (*blue*) circles are those that are identical between IGF-1 and IGF-2. Yellow residues are those that differ between the two peptides. The distance between the N-terminal and the first disulfide bond is greater by three amino acids in IGF-2 than in IGF-1, and the distance between the second and third disulfide bonds is smaller by four amino acids. Finally, the C-terminal piece following the third disulfide bond is shorter by 2 amino acids in IGF-2 and the peptide overall is 67 amino acids compared to 70 in IGF-1. *IGF*, Insulin-like growth factor.

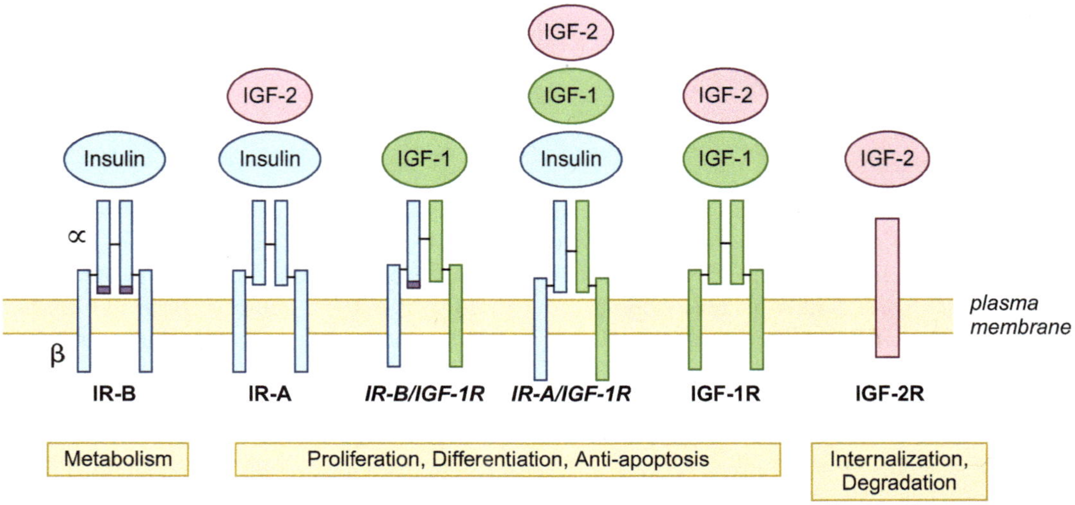

FIGURE 17.11 Receptors for insulin and IGFs. Each hemireceptor (consisting of one α and one β subunit) is encoded by a separate gene (insulin, *blue*; IGF-1, *green*; IGF-2, pink). On the left side of the figure is the B form of the IR-B (*blue*; see Chapter 6: Pancreatic Hormones: Insulin and Glucagon) which mediates most of the metabolic actions of insulin and does not bind either IGF. Of the 22 exons comprising the gene for the insulin receptor, exon 11, which encodes a 12-amino-acid segment at the carboxyl end of the α subunit (*purple*) of IR-B, is skipped in IR-A. The 21-exon gene for the IGF-1 receptor also lacks this exon. The absence of the 12 amino acids allows these receptors to bind IGF. Above each receptor, including two hybrids, are listed the ligands that bind to it with relatively high affinity (see Table 17.3). The main consequences of ligand binding to the various receptors are shown in the boxes below the receptors. The receptor for IGF-2 (*pink*) is a mannose-6-phosphate receptor which binds proteins tagged with this lysosomal breakdown signal. This receptor also binds IGF-2, bringing about its internalization and breakdown. *IGF*, Insulin-like growth factor; *IR*, insulin receptor.

and IGF-1R (21 exons) are related through a duplication of a 21-exon gene early in vertebrate evolution. Each gene encodes a polypeptide that is processed into the α and β subunits of the receptor. The extracellular domain contains the ligand-binding region; there is a single transmembrane segment of 22 or 23 amino acids, and an intracellular tyrosine kinase domain. The IR and IGF-1R share about 50% and 80% homology in the ligand-binding and tyrosine kinase domains, respectively.

In mammals a small exon, #11, has been incorporated into the gene for the IR. This exon encodes a 12-amino-acid sequence that occurs at the carboxyl terminus of the α subunit (*purple segment* in Fig. 17.11), the presence of which prohibits the binding of both IGFs. This exon can be and is skipped, on a cell-specific basis, during the processing of the messenger RNA (mRNA) for IR, generating IR-A which lacks the amino acids encoded by exon 11 and binds both insulin and IGF-2. This renders mammalian cells capable of responding only or primarily to the metabolic effects of insulin (IR-B) or to the growth-promoting effects one of the IGFs (IR-A). For example, the liver, skeletal muscle, and adipose tissue contain 80%, 60%, and 60%, respectively, IR-B, which binds only insulin. These tissues, along with the kidney, are the predominant sites of IR-B expression in the adult, whereas IR-A is ubiquitous. IR-A is also the major form expressed in fetal tissues and in certain cancerous tissues.

Some cells and tissues exhibit hybrid receptors as shown in Fig. 17.11. The ability of the various receptor forms to bind insulin, IGF-1, or IGF-2 are also depicted. A semiquantitative comparison of these affinities is presented in Table 17.3.

The receptor for IGF-2 (IGF-2R) is also known as a cation-independent mannose-6-phosphate receptor. In addition to binding extracellular IGF-2, it binds mannose-6-phosphate-tagged proteins in the *trans*-golgi network that are destined for lysosomal breakdown. Bound IGF-2 is also transported to lysosomes. Thus the function of this protein, in the context of the IGFs, is to remove IGF-2 from the pericellular environment and, thus, attenuates its intracellular signaling potential through IR-A or IGF-1R. The binding site on IGF-2 for the IGF-2R is different from the one used to bind the IR-A and IGF-1Rs.

The bottom portion of Fig. 17.11 depicts the biological consequences of activating the different receptor forms. IR-B mediates the metabolic effects of insulin; the IGF-2R mediates the removal of its ligand. The remaining receptors, in the central portion of the figure, bring about, depending on cell context, changes in pathways that lead to the cell's growth and survival.

TABLE 17.3 **Ligand binding by insulin receptor (IR)-A, IR-B, and insulin-like growth factor (IGF)-1R.**

	Relative binding[a]		
	IGF1	IGF-2	Insulin
IR-A	−−	++	+++
IR-B	−−	+	+++
IGF-1 R	+++	+++	−−

[a]*The entries show binding affinities that are very strong (+++), strong (++), weak but detectable (+), and undetectable or nearly so (− −).*
Source: Based on Table 17.1 from A. Belfiore et al., Endocrine Reviews 30 (2009) 586–623.

2 IR and IGF-R signaling

The signaling pathways utilized by the activated IRs and IGF-1R begin in the typical way for receptor tyrosine kinase receptors with autophosphorylation of the cytoplasmic portion of the receptor and recruitment of effector and adaptor proteins such as IRS-1 (insulin receptor substrate). These provide docking sites SH2 effector proteins such as the regulatory p85 subunit of PI3 kinase or the adaptor protein GRB2. These events lead to the activation of the PI3 kinase pathway or the Ras-mediated activation of the MAPK pathway as shown in Figures 6–6b for insulin and 17.3 and 17.6 for translocase of translation and FGF, respectively. Evidence has accumulated that the predominant pathway and its physiological outcome depend on (1) the receptor initiating the pathway, that is, IR-B, IR-A; or IGF-1R; (2) the ligand binding to the receptor, especially whether insulin or IGF-2 is the ligand for IR-B; and the cell context, particularly other mitogenic signals the cell is subject to. It is this variability in the precise signaling pathways used that leads to the different constellation of physiological outcomes depicted at the bottom of Fig. 17.11.

C IGF-binding proteins

The majority (approximately 98%) of circulating IGFs are bound to one of a six-membered family of IGF-binding proteins, IGFBPs. Of these, IGFBP-3 is the most abundant and, therefore, binds most of the circulating IGFs. Table 17.4 lists some of the characteristics of these proteins, each encoded by a separate gene. Fig. 17.12 shows the general structural features of this family of proteins. The IGFBPs have three domains of approximately the same size, the N- and C-domain and the L (linker)-domain. Highly conserved disulfide bonds in these cysteine-rich proteins occur within the two terminal domains giving these parts of the molecule the structural rigidity necessary to form, together, a high-affinity ligand-binding domain for one or both IGFs (see Table 17.4). The mid-region or L-domains of the

TABLE 17.4 **Insulin-like growth factor-binding proteins (IGFBPs).**

	Chromosome	No. of amino acids	No. of cysteines	IGF relative affinity[a]	Effect on IGF activity
IGFBP-1	7	234	18	IGF1=IGF2	↑ and/or ↓
IGFBP-2	2	289	18	IGF1 < IGF2	↓
IGFBP-3	7	264	18	IGF1=IGF2	↑ and/or ↓
IGFBP-4	17	237	20	IGF1=IGF2	↓
IGFBP-5	5	252	18	IGF1 < IGF2	↑
IGFBP-6	12	216	20	IGF1 ≪ IGF2	↓

[a] *< means two- to sixfold difference; ≪ means 100-fold difference.*

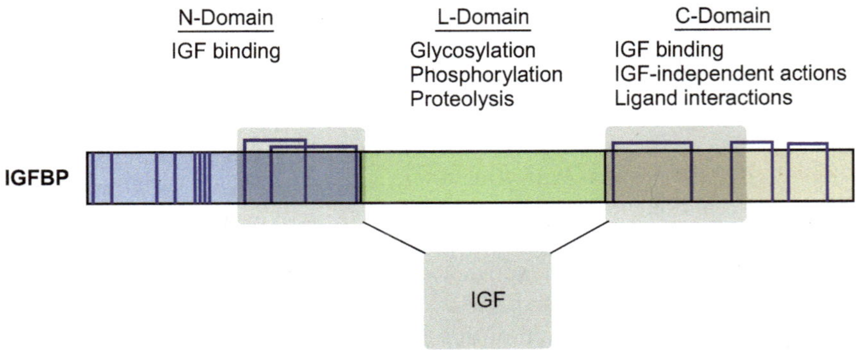

FIGURE 17.12 Structure of IGFBPs. The 6 members of the IGFBP family range from 216 to 289 amino acids long (see Table 17.4). They are divided into three domains of about the same size. IGFB requires both the N- and C-domains as shown by the shaded areas in each of these two domains. There are 12 conserved cysteine residues (*purple lines*) in the N-domain and 6 in the C-domain. These form disulfide bonds, some of which are depicted by the horizontal lines above the rectangle representing the peptide. IGFBP6 lacks two of the cysteines in the N-terminal portion of the N-domain but retains the disulfide bonds in the ligand-binding region. The L (linker)-domain (*green*) is not conserved among the members of the family. It undergoes considerable posttranslational modification as indicated. The C-domain, in addition to contributing to IGFB, also has interactions with several other proteins in addition to IGF. It also appears to be the site of IGF-independent activities. *IGFBPs*, Insulin-like growth factor binding protein.

immunoglobulin binding proteins share less than 15% similarity with each other. Considerable posttranslational modification of the residues within this region, such as glycosylation and phosphorylation, occurs. These alterations in the proteins probably affect their stability and susceptibility to proteolysis and, therefore, their half-life in the circulation, but have little or no effect on ligand binding. There is no known function of this region beyond serving as a hinge between the N- and C-domains and thereby promoting maximum ligand binding.

Through the interactions of their C-domains with extracellular proteins such as proteoglycans, IGFBPs can either inhibit, by restraining receptor binding, or enhance, by raising the effective pericellular concentration of the IGF ligand, the actions of the IGFs (see Table 17.4). In addition, IGFBPs appear to have IGF-independent actions, attributed to certain protein-binding sites, such as those for caveolin and collagen (cell surface association) and fibrinogen and plasminogen (wound healing). A nuclear localization signal and interactions with nuclear receptors such as vitamin D receptor, retinoic acid receptor α, and peroxisome proliferator-activated receptor γ suggest the possibility of the involvement of IGFBPs with nuclear functions, including gene transcription.

VI Transforming growth factor β

The structurally defined TGFβ superfamily consists of TGFβ itself, activins, inhibins (which antagonize the effects of activins), bone morphogenetic proteins (BMPs), anti-Mullerian hormone, nodal (and its antagonist lefty), and growth and differentiation factors. In this section the focus will be on TGFβ that plays an important homeostatic role in wound healing and controlling the immune system. TGFβ is secreted into the extracellular matrix by numerous cell types, including endothelial cells, macrophages, platelets, and fibroblasts and its receptors are nearly ubiquitous.

A Structure and secretion of TGFβ

Three forms of TGFβ are found in mammals, encoded by three separate genes. TGFβ − 1, TGFβ − 2, and TGFβ − 3 have distinctive physical and biological properties but share the same general structure and are synthesized and secreted by the same pathway. These are shown in Fig. 17.13. Pro-TGFβ1 is a peptide of 390 amino acids (TGFβ2 and TGFβ − 3 are 24 and 22 amino acids longer, respectively) of which the first 23 comprise the signal sequence directing it to the proper site for further processing. Proteolytic processing results in the cleavage of the remaining transcript into the latency-associated peptide, LAP, and the mature monomer of TGFβ. Four intramolecular disulfide bonds give this molecule its characteristic cysteine knot structure and a dimer is formed through the formation of an interchain disulfide bond near the carboxyl termini of the monomers. LAP also forms a homodimer with the formation of two adjacent disulfide bonds. While still within the cell, these two dimers associate with each other to form the small latency complex, as shown in Fig. 17.13B. In this complex, stabilized by noncovalent interactions, TGFβ is surrounded by LAP and unavailable to interact with any receptors it might encounter when secreted; that is, its activity is latent. As this complex is secreted, a third protein, the latent TGFβ binding protein (LTBP), is added forming the large latent complex. The N-terminal region of LTBP binds to the extracellular matrix of cells surrounding the site of TGFβ

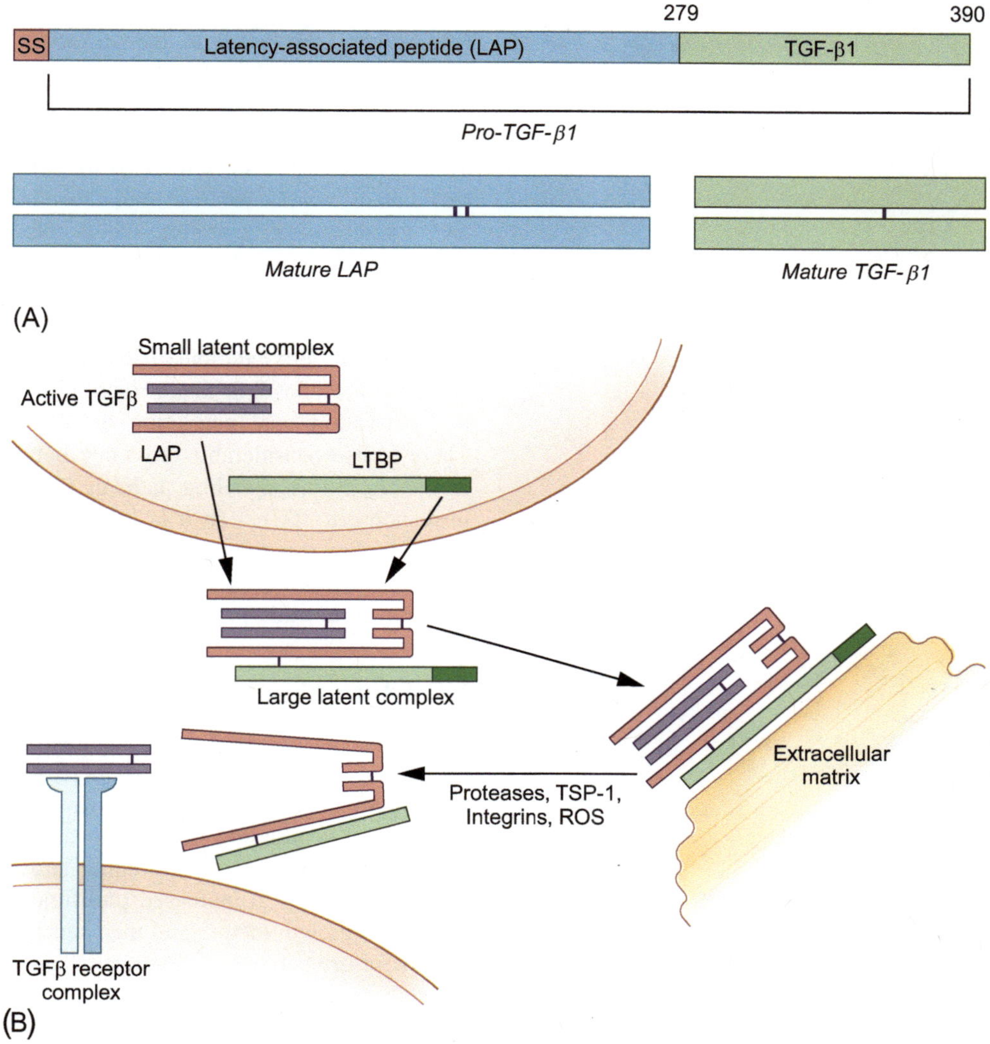

FIGURE 17.13 Formation and activation of TGFβ. (A) Processing of pro-TGFβ. The primary transcript of TGFβ − 1 is 390 amino acids long. The signal peptide directs it to the RER where it is cleaved by the proteolytic enzyme furin into the 112 amino acid TGFβ − 1 (*green*) and a 257-amino-acid peptide, known as LAP (*blue*). Dimers of each of the peptides are formed through one (TGFβ − 1) or two (LAP) adjacent disulfide bonds. The pathways for TGFβ − 2 and TGFβ − 3 are the same as those for TGFβ − 1. (B) *Formation of latent complexes, secretion, and activation of TGFβ.* While still in the RER, the small latent complex is formed, sequestering TGFβ (*purple*) within the LAP (*red*). This complex is secreted along with another peptide, the LTBP (*green*), forming the extracellular large latent complex. The LTBP binds to the extracellular matrix of the cell of origin or nearby cells, particularly at its carboxyl terminal, forming a reservoir of inactive TGFβ. Depending on the cell types and regulatory events involved, the latent complex can be opened, releasing active TGFβ to bind to specific cell receptors (Fig. 17.14). *LAP*, Latency-associated peptide; *LTBP*, latent TGFβ-binding protein; *RER*, rough endoplasmic reticulum; *ROS*, reactive oxygen species (from irradiation); *TGF*, transforming growth factor; *TSP*, thrombospondin-1.

biosynthesis. This serves as a storage form of TGFβ so that the regulation of the availability of active hormone occurs at the step of its release from the latent complex.

Several mechanisms exist to accomplish this release of TGFβ, depending on the target cells and the regulatory events to which they are subject. These include various proteolytic events, reactive oxygen species produced in response to radiation, and interaction of the complex with extracellular matrix proteins such as thrombospondin or cell surface integrins that can open the complex releasing TGFβ. Other signaling molecules, such as BMPs, can also play a specific role in releasing and, therefore, activating TGFβ at its site of action. This regulation of the activation of TGFβ is often the most important step, compared to, for example, the rate of transcription, at which the amounts of active growth factor are modulated.

B TGFβ receptors and signaling

When active dimerized TGFβ is released from the large latent complex through one of the regulatory mechanisms mentioned earlier, it binds to a dimer of a constitutively

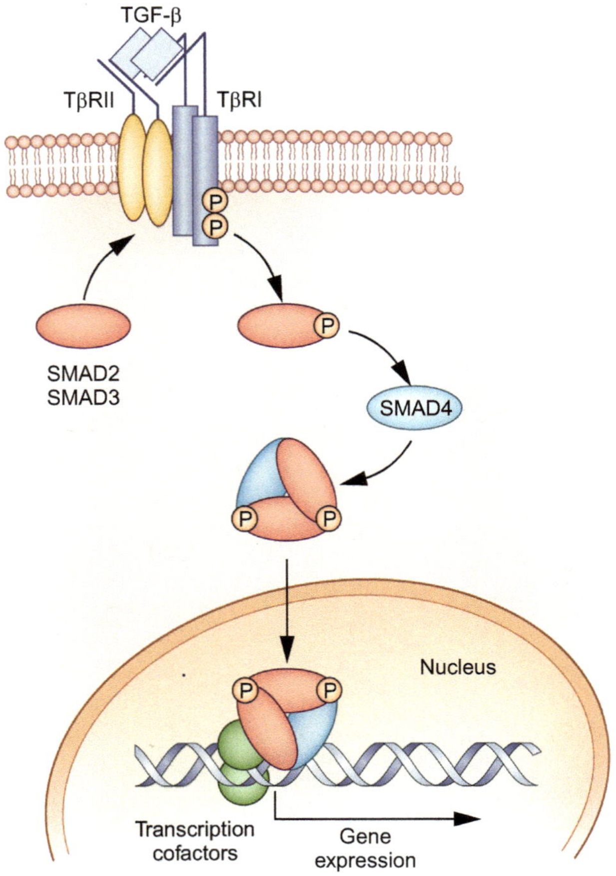

FIGURE 17.14 TGFβ signaling. TGFβ, like other members of its family, interacts sequentially with two different types of receptor, TGFβRI and TGFβRII. The type 1 receptor, ALK-5, is one of seven ALK receptor proteins that serve as the Type I receptors for other members of the TGFβ family of growth factors (see Table 17.5). Dimerized active TGFβ binds to a dimer of TGFβRII (*yellow*), which recruits a dimer of TGFβRI (*blue*); the heterotetrameric complex is stabilized by the bound ligand. The constitutively active serine/threonine kinase of the Type II receptor phosphorylates the Type I receptor, which phosphorylates Smads 2 and 3, two of the regulatory Smads. These are joined by a coregulating Smad, Smad 4, in a heterotrimer which enters the nucleus and interacts with transcriptional coactivators and corepressors to modulate, along with other transcription factors, the rate of gene transcription. *TGF*, Transforming growth factor; *ALK*, activin receptor–like kinase.

active serine/threonine kinase receptor known as TGFβ receptor Type II or TβRII. This entity recruits a second receptor dimer, TβRI, which is phosphorylated by TβRII (also known as ALK5; see legend to Fig. 17.14 and Table 17.5). The resulting heterotetramer, stabilized by the ligand, is shown in Fig. 17.14. Phosphorylated and activated TβRI phosphorylates Smads 2 and 3, which combine with Smad 4 to form a heterotrimer that enters the nucleus and interacts with corepressor or coactivation proteins in the transcription complex of specific genes to alter the pattern of gene expression of the cell. Table 17.5 lists some of the individual proteins that form the signaling complexes that participate in signaling by members of the TGFβ family of growth factors.

There are eight Smad proteins in vertebrates. Smads 1, 2, 3, 5, and 8 are phosphorylated and regulated by the activated receptors for members of the TGFβ family as described earlier and are referred to as R-Smads. Smad 4 participates in translocation into the nucleus for binding to a transcriptional complex and is referred to as a co-Smad. Smads 6 and 7 compete with Smad 4 in this function and are, therefore, known as I (inhibitory) Smads.

TGFβ is also capable of stimulating non-Smad signaling pathways. For example, TβRI can phosphorylate tyrosines of ShcA leading to the recruitment of GRB2/SOS, Ras activation, and activation of the MAPK pathway. There is also considerable cross talk between downstream mediators in one signaling pathway with those in another in response to TGFβ, so it is important to recognize that while the Smad pathway is central to the actions of TGFβ and other members of this family, it does not operate in isolation from other pathways.

VII Clinical aspects

Cancer is characterized by the unregulated autonomous growth of cells. Uncontrolled increased production of active growth factors can be a cause of this condition, as can the overexpression of the cognate membrane receptors for the growth factor. In either case the downstream pathways leading to increased proliferation are activated. Thus both growth factors and their receptors have become targets for drugs to rein in the uncontrolled growth of cancer cells.

Table 17.6 lists some of the drugs that have been developed to target specific growth factors or their receptors. It should be noted that this is just a limited illustrative list of such drugs, of which there are several-fold more in various stages of development. The purpose of the illustration is to underscore the importance of understanding the structures of the growth factors and their receptors; the pathways by which growth factors are synthesized and activated from latent forms; the mechanisms of receptor activation; and the downstream signaling pathways.

Because the majority of human epithelial cancers overexpress EGF or one of its receptors, the EGFRs, EGFR and HER2, have been targets of monoclonal antibody drugs (denoted by the syllable *mab* in the name). Particularly in breast cancer, but in others as well, it is standard to test the cancer cells for overexpression of HER2, which is active in the absence of ligand and is the preferred dimerization partner of the family, and, if positive, to treat with a drug of the kind listed in Table 17.6.

Another common target is the inhibition of the tyrosine kinase activity of growth factor receptors. Examples

TABLE 17.5 **Transforming growth factor (TGF)β family: typical signaling components.a**

Ligand	Type I receptor[b]	Type II receptor[c]	R-smad[d]
TGFβ	ALK-5 (TβRI)	TβRII	Smad 2/3
Activins	ALK-4 (ActRIB)	ActRIIA, ActRIIB	Smad 2/3
BMP 9, 10	ALK-1 (Ser/Thr kinase R3)	BMPRII, ActRIIA	Smad 1/5/8
GDF 5, 6, 7	ALK-2 (ActRI)	BMPRII, ActRIIB,	Smad 1/5/8
Nodal	ALK-7 (ActR1C)	ActRIIA, ActRIIB	Smad 2/3

[a]*This list is for illustration purposes and is not intended to be exhaustive.*
[b]*There are seven ALK (activin receptor-like kinase) of which all except ALK-6 are found in humans. These act as Type I receptors for the TGFβ family of growth factors. Beneath the ALK designation listed in the table is the other name by which the protein is known. ActR, activin receptor.*
[c]*There are five proteins that serve as Type II receptors for the TGFβ family of growth factors, of which four are listed here. TβR, TGFβ receptor; ActR (A and B), activin receptor; BMPR, bone morphogenetic protein receptor.*
[d]*The receptor-regulated Smads, R-Smads that are activated by the given receptor pair, are shown.*

TABLE 17.6 **Cancer drugs targeting growth factors and their receptors.**

Cancer	Drug	Target
Breast	Trastuzumab	HER2: JM region
	Pertuzumab	HER2: dimerization region
	Lapitinib	EGFR/HER2 TK
	Cixutumumab	IGFR1
CMLa	Imatinib	BCR-Ablb/PDGFR TK
Colorectal	Cetuximab/Erbutix	HER2
Endometrial	FP-1039	FGF1, − 2, − 4
Esophageal/stomach	Matuzumab	EGFR
	AZD4547	FGFR-TK *Phase II*
Glioblastoma	Trabedersen	TGFβ − 2 mRNA
	LY2157299	TβRI TK
Head/neck	Nimotuzumab	EFGR
Lung	Gefitinib	EGFR-TK
Pancreatic	Erlotinib	EGFR-TK

EGFR, Epidermal growth factor receptor; *HER*, human epidermal growth factor receptor; *IGFR*, Insulin-like growth factor receptor; *JM*, juxtamembrane; *mRNA*, messenger RNA; *PDGFR*, platelet-derived growth factor receptor; *TK*, tyrosine kinase.
[a]*CML, Chronic myelogenous leukemia.*
[b]*BCR-Abl codes for a constitutive TK found in cancer cells.*

of these for the EGF and FGF families are shown in Table 17.6; the names of these drugs often end in the syllable *ib*. Inhibition of a specific growth factor with an antisense oligonucleotide has been achieved in the case of TGFβ − 2 mRNA in the treatment of glioblastoma.

The development and use of these important and promising drugs is hampered by the very complexity of the systems that are targeted. For example, since the growth factors have normal physiological functions, interference with these can lead to intolerable side effects. In some cases, such as TGFβ, the window of treatment may be limited because the target has growth-controlling properties early in the oncogenic process but promotes metastatic disease later in the process. And, for reasons that are not understood, patients may become refractory to a given treatment whereupon the metastatic process continues. As our understanding of the growth factors, their receptors, and their roles in particular neoplasms deepens, these obstacles should be surmountable.

Further reading

Epidermal growth factor family

T. Bavaro, et al., Design of epidermal growth factor immobilization on 3D biocompatable scaffolds to promote tissue repair and regeneration, Scientific Reports 11 (2021) 2629. Available from: https://doi.org/10.1038/s41598-021-81905-1.

S. Park, et al., Comparison of epidermal growth factor receptor tyrosine kinase inhibitors for patients with lung adenocarcinoma harboring deficient epidermal growth factor receptor mutation types, BMC Cancer 21 (2021) 52. Available from: https://doi.org/10.1186/s12885-020-07765-6.

Fibroblast growth factor family

A. Asiri, et al., Epidermal and fibroblast growth factors incorporated polyvinyl alcohol electrospun nanofibers as biological dressing scaffold, Scientific Reports 11 (5634) (2021). Available from: https://doi.org/10.1038/s41598-021-85149-x.

C. Dorsoe, et al., Roles of Fibroblast growth factors and their therapeutic potential in treatment of ischemic stroke, Frontiers in Pharmacology 12 (2021) 671131. Available from: https://doi.org/10.3389/fphar.2021.671131.

H.R. Ferguson, M.P. Smith, C. Francavilla, Fibroblast growth factor receptors (EGFRs) and noncanonical partners in cancer signaling, Cells 10 (2021). Available from: https://doi.org/10.3390/cells10051201. 100.1201.

Platelet-derived growth factors

A. Freeman, et al., MicroRNA 219−5p inhibits alveolarization by reducing platelet-derived growth factor-alpha, Respiration Research 22 (2021) 57. Available from: https://doi.org/10.1186/s12031-021-01654-7.

T. Li, T. Guo, H. Liu, H. Jiang, Y. Wang, Platelet-derived growth factor-BB mediates pancreatic cancer malignancy via regulation of the Hippo/Yes-associated protein signaling pathway, Oncology Reports 45 (2020) 83−94. Available from: https://doi.org/10.3892/or.2020.7859.

L. Tavelli, et al., Recombinant human platelet-derived growth factor: a systematic review of clinical findings in oral regenerative procedures, JDR Clinical and Translational Research 6 (2020). Available from: https://doi.org/10.1177/2390084420921853.

Insulin-like growth factor binding proteins

A. Bharwaj, S.V. Shrivastav, Insulin-like growth factor binding protein-3 binds to histone 3, International Journal of Molecular Sciences 2021 (2021) 407. Available from: https://doi.org/10.3390/ijms22010407.

S.T. Ferreira, Brain insulin, insulin-like peptide 1 signaling in Alzheimer's disease, Journal of Neuroendocrinology 33 (2021). Available from: https://doi.org/10.1111/jne.12959.

B.J. Guercio, et al., IGF-binding proteins, adiponectin, and survival in metastatic colorectal cancer: results from CALGB (Alliance)/SWOG 80495, JNCI Cancer Spectrum 5 (2021). Available from: https://doi.org/10.1093/jncics/pkaa074. pkaa074.

C. Talia, L. Connolly, P.A. Fowler, The insulin-like growth factor system: a target for endocrine disruptors, Environment International 147 (2021) 106311. Available from: https://doi.org/10.1016/j.envint.2020.106311.

Transforming growth factor β

J. Chen, et al., Targeting transforming growth factor-β signaling for enhanced cancer chemotherapy, Theranostics 11 (2021) 1345−1363.

C. Guevara, F.C. Ortiz, Glial-derived transforming growth factor β (TGF-β1) a key factor in multiple sclerosis neuroinflammation, Neural Regeneration Research 16 (2021) 510−511.

C. Stolfi, E. Troncone, I. Marafini, G. Montelione, Role of TGF-beta and Smad7 in gut inflammation, fibrosis and cancer, Biomolecules 11 (2021) 17. Available from: https://doi.org/10.3390/biom11010017.

Growth factors and cancer

J.-M. Bao, et al., SPARC is a key mediator of TGF-β-induced renal cancer metastasis, Journal of Cellular Physiology 236 (2020). Available from: https://doi.org/10.1002/jcp.29975.

M. Gugnoni, et al., Linc 00941 is a novel transforming growth factor β target that primes papillary thyroid cancer metastatic behavior by regulating the expression of cadherin 6, Thyroid 31 (2021). Available from: https://doi.org/10.1089/thy.2020.0001.

K.E. Huang, et al., Salinomycin suppresses TGF-β1- induced EMT by down-regulating MMP-2 and MMP-9 via the AMPK/SIRT1 pathway in non-small cell lung cancer, International Journal of Medical Sciences 18 (2020) 715−728.

K.A. Thies, et al., Stromal platelet-derived growth factor receptor-β signaling promotes breast cancer metastasis in the brain, Cancer Research 81 (2021). Available from: https://doi.org/10.1158/0008-5472.CAN-19-3731.

Appendix 1

Compilation of many known hormones in higher mammals and humans

Trivial name and class	Abbreviation	Source	Action
Amino acid–derived hormones			
Adrenals			
Epinephrine (adrenaline)	EP	Adrenal medulla (CNS)	Glycogenolysis in liver, increases blood pressure
Adrenomedullin (52 aa)	AM	Cardiovascular system	Fall in blood pressure via dilation of vascular beds
Intestine			
Histamine		Gut, CNS, mast cells, many tissues	Gastric secretion; may affect CNS
Fetal growth factor 23 (251 aa protein)	FGF-23	Is secreted by osteocytes (bone cells)	Regulates blood phosphate concentration via actions on the kidney which increases phosphate excretion
Nervous system			
Acetylcholine		Neurons	Variety of activities in nervous system, innervates adrenal medulla
Dopamine (also believed to be PIF)		CNS	Inhibits PRL release (and other actions)
γ-Aminobutyric acid	GABA	CNS	Neurotransmitter; inhibits release of CRF and PRL
Norepinephrine	NEP	CNS neurons	Neurotransmitter; increases blood pressure
Serotonin		CNS neurons, gut	Affects smooth muscles, nerves, stimulates release of GH, TSH, ACTH (CRF), and inhibits LH release
Pineal gland			
Acetylserotonin			Affects GH release from anterior pituitary
Melatonin			Inhibits GH release from anterior pituitary; affects reproductive functions
Thyroid gland			
Thyroxine and triiodothyronine	T4 and T3	Thyroid gland	Increases oxidation rates in tissues
Fatty acid (arachidonic acid)-derived hormones			
Blood			
Prostacyclin	PGI	Vascular endothelium, blood	Prevents aggregation of platelets
Thromboxane B_2	TXB_2	Platelets	Metabolite of TXA_2

Trivial name and class	Abbreviation	Source	Action
Lung			
Leukotriene C_4	LTC_4		Long-acting bronchoconstrictor
Leukotriene D_4	LTD_4		Long-acting bronchoconstrictor
Leukotriene B_4	LTB_4		
Leukotriene A_4	LTA_4		Intermediate en route to LTC_4/LTD_4
Lipoxin A	LPA		Inflammatory
Lipoxin B	LPB		Inflammatory
Prostaglandin E_1+E_2	PGE_1 or PGE_2	Wide variety of cells	Stimulates cyclic AMP
Prostaglandin $F_1\alpha$+$F_2\alpha$	$PGF_1\alpha$ or $PGF_2\alpha$	Wide variety of cells	Active in dissolution of *corpus luteum*, ovulation+parturition contractions
Prostaglandin A_2	PGA_2	Kidney	Hypotensive effect
Thromboxane A_2	TXA_2	Platelets, lung, etc.	Causes platelet aggregation
Polypeptides/proteins			
Adipose tissue			
Leptin (167 aa)	–	Fat cells	Loss of fat; satiety factor
Adrenals			
Met-enkephalin+leu-enkephalin		Adrenal medulla and CNS cells	Analgesic actions in CNS; other unknown effects
Blood			
Angiotensin II (8aa)	ANGII	Blood, lungs, brain, many tissues	*Zona glomerulosa* cells of adrenal cortex to stimulate synthesis+release of aldosterone
Bradykinin (9 aa)	BK	Plasma, gut, other tissues	Vasodilator; lowers blood pressure
Bone marrow			
Granulocyte-macrophage colony stimulatory	GM-CSF	Bone marrow	Stimulates production of granulocytes and macrophages
Granulocyte-colony stimulatory factor	G-CSF	Bone marrow	Stimulates production of granulocytes
Macrophage-colony stimulatory factor	M-CSF	Bone marrow	Stimulates production of macrophages
Heart			
Atrial natriuretic factor (also known as atriopeptin) (28 aa)	ANF	Atria	Blood pressure lowering; stimulates renal sodium excretion; increases GFR and urine volume
Hypothalamus			
Corticotropic-releasing hormone	CRH	Hypothalamus	Releases ACTH+β-endorphin in anterior pituitary
Gonadotropic-releasing hormone	GnRH	Hypothalamus, distributed in CNS; milk, gonadal cells—containing GnHR receptors	Releases PSH and LH in anterior pituitary
Growth hormone release inhibiting hormone (somatostatin)	GIH	Hypothalamus, extrahypothalamic brain, spinal cord, pancreas, stomach, and intestine	Inhibits release of GH and TSH in anterior pituitary; regulates pancreatic hormones

Trivial name and class	Abbreviation	Source	Action
Growth hormone–releasing hormone	GRH		Releases GH in anterior pituitary
Melanotropin release–inhibiting factor	MIF		Prevents release of MSH in anterior pituitary, probably not in man
Melanotropin releasing factor	MRF		Releases MSH in anterior pituitary probably not in man
Neurotensin		Hypothalamus, intestine (mucosa)	May have neurotransmitter actions; in pharmacological amounts, has several effects on gut
Prolactin-releasing hormone	PRH or TRH		Releases PRL in anterior pituitary
Prolactin release–inhibiting hormone	PIF		Inhibits prolactin release
Substance P	SP	Hypothalamus and CNS, intestine	Transmits pain and other functions, increases smooth muscle contractions of GI tract
Thyrotropin-releasing hormone	TRH	Hypothalamus, extrahypothalamic brain, spinal cord, and brain stem	Release TSH and PRL in anterior pituitary
Gastrointestinal			
Adrenocorticotropic hormone (ACTH)	ACTH	Pituitary gland	↑ Cortisol release
α-Melanocyte-stimulating hormone	α-MSH	Intermediary pituitary	↑ Melanin release
Beta-endorphin		Anterior/intermediate pituitary	↓ Intestinal transit, analgesic action in CNS
Bombesin (14 aa)		Skin of toad (nerves +endocrine cells)	Hypothermic hormone; increases gastrin+gastric acid secretion; many other actions
Cholecystokinin (pancreozymin)	CCK		Stimulates gallbladder contraction+bile flow; enhances secretion of pancreatic enzymes
Kisspeptin (10 aa)	KISS	The brain's secretion is regulated by estradiol and testosterone	Could be vital in controlling the time of puberty
Enteroglucagon (different from pancreatic glucagon, glucagon-like immunoreactivity)	GLI	Gut, L cells of ileum and colon, brain	Increases insulin release
Ghrelin (a 28 aa)	GHR	Produced by fundus of the stomach	Stimulates appetite and food intake by actions in the brain
Gastrin (27 aa)	GAS	G cells in midpyloric, glands in stomach antrum	Increases secretion of gastric acid+pepsin+many other effects
Gastrin inhibitory polypeptide	GIP		↓ Gastric secretion, ↑ intestinal secretion
Gastrin releasing peptide	GRP		↑ Release gastrin
Glicentin (64 aa)		Secreted by intestinal L cells	↑ Hepatic glucose, ↓ acid secretion
Growth hormone–releasing hormone	GRH		↑ Release of growth hormone
Motilin (22 amino acid peptide)		Duodenum (jejunum), pineal, pituitary	Acts on GI tract to alter motility; stimulates contraction of fundus+antrum+decreases gastric emptying

Trivial name and class	Abbreviation	Source	Action
Neuromedin B (36 aa)	NB		Contraction of gastrointestinal smooth muscle
Neuromedin K	NK		
Neuropeptide Y	NPY		↑ Vasoconstriction
Neurotensin (13 aa)	NT		↓ Acid secretion, ↓ gastric motor; its activity is implicated in the regulation of luteinizing hormone
Oxyntomodulin (37 aa)	OM	Is secreted by fundic cells	↑ Insulin release, ↓ acid secretion
			It suppresses appetite
Pancreastatin (44 aa)			↓ Islet somatostatin release
Pancreatic polypeptide (36 aa)	PP		↓ Pancreatic secretion
Peptide YY	PYY		↓ Pancreatic secretion, ↓ gallbladder contraction
Secretin (27 aa)		Duodenum when pH of its contents is less than 4.5	Stimulates pancreatic acinar cells to release bicarbonate+water which are transported to duodenum to elevate pH
Somatostatin (14 and 28 aa)			↓ Acid secretion, ↓ pancreatic function
Substance P (11 aa)			Contraction gastrointestinal smooth muscle
Vasointestinal peptide (28 aa)	VIP	GI tract, hypothalamus, + elsewhere	Neurotransmitters in peripheral autonomic nervous system; relaxes smooth muscles of circulation; increases secretion of water and electrolytes from pancreas and gut
Kidney			
Erythropoietin (166 aa)	EPO	Kidney	Acts on bone marrow to induce terminal differentiation+initiation of hemoglobin synthesis
Lung			
Eosinophil chemotactic factor of anaphylaxis		Lung mast cells	After release, selective chemoattractants for eosinophils
Ovaries			
Relaxin		*Corpus luteum*	Increases during gestation (may inhibit myometrial contractions)
Pancreas			
Glucagon (29 aa)		A cells	Glycogenolysis in liver; increases cyclic AMP
Insulin (51 aa; 2 peptides)		B cells	Glucose utilization in liver; promotes synthesis of glycogen
Pancreatic polypeptide (36 aa)	PP	Pancreatic islets (peripheral cells of)	Has a number of effects on gut in pharmacological amounts
Proinsulin		B cells	Precursor to insulin
Pituitary			
Adrenocorticotrophic hormone	ACTH	Adenohypophysis	Stimulates synthesis and release of cortisol and dehydroepiandrosterone from adrenal cortex
Arg-vasopressin	AVP or ADH	Posterior pituitary	Increases water reabsorption in kidney
β-Endorphin		*Pars intermedia* and anterior pituitary	Analgesic actions in CNS

Trivial name and class	Abbreviation	Source	Action
Fibroblast growth factor	FGF		Stimulates proliferation of cells derived from endoderm and mesoderm in the presence of serum
Acid FGF	aFGF		
Basic FGF	bFGF		
Follicle-stimulating hormone	FSH	Adenohypophysis	Stimulates the development of ovarian follicle and secretion of estrogen; stimulates seminal tubules and spermatogenesis
Growth hormone (somatomammotropin or somatotropin)	GH	Adenohypophysis	Somatic cell growth mediated by somatomedins, hyperglycemia, liver steroid metabolism, bone sulfation reactions
Lipotropin	LPH	Adenohypophysis	Fat mobilization; source of opioid peptides
Luteinizing hormone (formerly "interstitial cell regulating hormone" in reference to male)	LH	Adenohypophysis	Stimulates Leydig (interstitial) cell development in male+production of testosterone; stimulates corpus luteum and its production of progesterone in female
Melanocyte-stimulating hormone	MSH	Adenohypophysis *pars intermedia*	CNS functions (e.g., in memory retention) and skin-darkening reaction
Oxytocin (9 aa)		Posterior pituitary, hypothalamus	Lactating mammary gland, milk letdown; uterine contraction at parturition
Parathyroid hormone (84 aa)	PTH	Parathyroid glands	Stimulates bone resorption; elevates serum Ca^{2+}, stimulates HPO=excretion by the kidney
Prolactin (198 aa)	PRL	Adenohypophysis	Synthesis of milk constituents in mammary gland; stimulates testosterone production; secondary growth hormone effects in liver (e.g., as hyperglycemic acid); mammary gland secretory cell differentiation
Thyroid-stimulating hormone	TSH	Adenohypophysis	Stimulates thyroid gland follicles to secrete thyroid hormone
Placenta			
Human chorionic gonadotropin	hCG		LH-like functions, maintains progesterone productivity during pregnancy
Human placental lactogen	hPL		Acts like PRL and like GH because of large amount of hPL produced
Transforming growth factor α	TGFα		Binds to EGF receptor
Transforming growth factor β	TGFβ		Confers transformed phenotype or normal cells
Salivary gland			
Epidermal growth factor (formerly urogastrone)	EGF		Stimulates proliferation of cells of ectodermal and mesodermal origin with serum; inhibits gastric secretion
Nerve growth factor	NGF		Stimulates neuronal growth
Skin			
Alytesin		Amphibian skin	Stimulates gastric acid secretion
Cerulein		Frog skin	Similar to CCK+gastrins
Litorin (9 or 4 aa)		Amphibian skin	Stimulates gastric secretion
Ranatensin		Amphibian skin	Stimulates gastric secretion

Trivial name and class	Abbreviation	Source	Action
Submaxillary gland			
Nerve growth factor	NGF	Saliva gland	Differentiation and growth of embryonic dorsal root ganglia
Testes			
Antimüllerian hormone		Fetal Sertoli cells of the testes	Mediates involution of the Müllerian ducts
Inhibin (a dipeptide)		Seminiferous tubule (and ovary)	Negative feedback inhibitors of FSH secretion from anterior pituitary
Thymus			
Thymic humoral factor	THF		Activates adenylate cyclase in thymus+spleen cells
Thymopoietin I and II, α-thymosin			Stimulates phagocytes; stimulates differentiation of precursors into immune competent T cells
Thyroid parathyroid gland(s)			
Calcitonin (32 aa)	CT	Parafollicular C cells of thyroid gland	Lowers serum calcium
Parathyroid hormone (84 aa)	PTH	Parathyroid glands	
Vascular endothelium eindothelins	E-1, E-2 E-3	Endothelial cells	Vasoconstriction
Steroid hormones			
Adrenals			
Aldosterone and 11-deoxycorticosterone	DOC	*Zona glomerulosa* of adrenal cortex	Salt retention in kidney
Cortisol (hydrocortisone)		*Zona fasciculata* and *zona reticularis* of adrenal cortex	Antistress hormone; carbohydrate metabolism; circulating glucose increased; liver glycogen increased; depresses immune system; antiinflammatory agent
Dehydroepiandrosterone	DHEA	*Zona reticularis* of adrenal cortex	Weak androgen; major secretion of fetal adrenal cortex; can be converted to estrogen; may have other unknown actions
Kidney			
1α,25-Dihydroxyvitamin D_3 or also 1α,25-Dihydroxycholeacalciferol	1,25D	Proximal tubule of the kidney	Stimulates intestinal Ca^{2+} absorption, stimulates a Ca^{2+}-binding protein in a variety of tissues, especially in intestine, kidney, and many other responses
24R,25-Dihydroxyvitamin D_3 or 24R,25-dihydroxycholecalciferol	24,25D		Has receptor in chondrocytes
Ovaries			
17β-Estradiol and estriol		Ovarian follicle (corpus luteum)	Uterine endometrium development, female tissues
Progesterone		Corpus luteum, placenta (ovarian follicle)	Breast development; uterine endometrium development
Testes			
Dihydrotestosterone	DHT	Seminiferous tubule+other male tissues (e.g., prostate)	Conversion product of testosterone which binds to androgen receptor
Testosterone	T	Leydig cells (interstitial cells of testis) (adrenal)	Spermatogenesis/male characteristics
Other			
Nitric oxide	NO	Vascular endothelium	Vasodilation

Trivial name and class	Abbreviation	Source	Action
Insect hormones			
Ecdysone (steroid)			
Juvenile hormone (hydrocarbon terpenoid)	JH		Stimulates molting and controls molting
Plant hormones			
Auxins (indoleacetic acid)		All higher plants and some lower plants	Stimulates extension growth and cell division in cambium
Gibberellins (diterpenoid cells)		Widespread plant regulators	Stimulates extension growth
Kinins (N-substituted)			Promotes cell division

More recent assay results for many hormones are available via the database that the National Institute of Health's National Library of Medicine maintains Consider utilizing the following URL: http//http://www.nlm.nih.gov/medlineplus/encyclopedia.html.
This table is composed of many of the known hormones of higher animals and man; in addition, the key insect hormones and plant hormones are listed at the end of the table. The first level presents hormones by structure (amino acid–derived hormones, fatty acid–derived hormones, polypeptide/protein hormones, steroid hormones, insect hormones, and plant hormones). Each of these categories is then subdivided, where appropriate, by organ; finally, the individual hormones in a category are entered alphabetically. (↑) Indicates an increase; (↓) indicates a decrease. The blood concentrations of many hormones are available at the NIH.
S denotes serum, P plasma, and U urine.
Source: Abstracted from C.D. Jordan, J.G. Flood, M. LaPosata, K.B. Lewandrowski, Normal reference laboratory values, New England Journal of Medicine 327 (1992) 718–724.

Appendix 2

Human blood concentrations of major hormones

Hormone	Fluid[a]	Units
Aldosterone		
Standing (normal-salt diet)	S, P	4–31 ng/dL
Recumbent (normal-salt diet)	S, P	<16 ng/dL
Normal-salt diet (100–180 meq of sodium)	U	6–25 μg/day
Low-salt diet (10 meq of sodium)	U	17–44 μg/day
High-salt diet	U	0–6 μg/day
Androstenedione	S	60–260 ng/dL
Antidiuretic hormone (arginine vasopressin)	P	1.0–13.3 pg/mL
Calcitonin	S	
Female		0–20 pg/mL
Male		0–28 pg/mL
Catecholamines		
Dopamine	U	65–400 μg/day
	P	0–30 pg/mL
Epinephrine	U	1.7–22.4 μg/day
Supine	P	0–110 pg/mL
Standing	P	0–140 pg/mL
Norepinephrine	U	12.1–85.5 μg/day
Supine	P	70–750 pg/mL
Standing	P	200–1700 pg/mL
Chorionic gonadotropin (hCG) (nonpregnant)	S	<10 mIU/mL
Corticotropin (ACTH)	P	6.0–76.0 pg/mL
Cortisol	P	
Fasting, 8:00 a.m.–12:00 noon		5.0–25.0 μg/dL
12:00 noon–8:00 p.m.		5.0–15.0 μg/dL
8:00 p.m.–8:00 a.m.		0.0–10.0 μg/dL
Cortisol, free	U	20–70 μg/day
C peptide	S	0.30–3.70 μg/L
11-Deoxycortisol (after metyrapone)	P	>75 μg/dL
Erythropoietin	S	<19 mU/mL
Estradiol	S, P	

Hormone	Fluid[a]	Units
Female		
Premenopausal adult		23–361 pg/mL
Postmenopausal		<30 pg/mL
Prepubertal		<20 pg/mL
Male		<50 pg/mL
Gastrin	P	0–200 pg/mL
Growth hormone	P	2.0–6.0 ng/mL
Hemoglobin A_{lc}	P	3.8%–6.4%
Homovanillic acid	U	0.0–15.0 mg/day
17-Hydroxycorticosteroids	U	
Female		2.0–6.0 mg/day
Male		3.0–10.0 mg/day
5-Hydroxyindoleacetic acid (lower in women than in men)	U	2–9 mg/day
17-Hydroxyprogesterone	S	
Female		
Prepubertal		0.20–0.54 μg/L
Follicular		0.02–0.80 μg/L
Luteal		0.90–3.04 μg/L
Postmenopausal		<0.45 μg/L
Male		
Prepubertal		0.12–0.30 μg/L
Adult		0.20–1.80 μg/L
25-Hydroxyvitamin D_3	S	8–55 ng/mL
1α,25-Dihydroxyvitamin D_3	S	16–42 pg/mL
Insulin	S	0–29 μU/mL
Male		5.0–23.0 mg/day
17-Ketogenic steroids	U	
Female		3.0–15.0 mg/day
17-Ketosteroids	U	
Female and male ≤10 years old		0.1–3.0 mg/day
Female and male 11–14 years old		2.0–7.0 mg/day
Female ≥15 years old		5.0–15.0 mg/day
Male ≥15 years old		9.0–22.0 mg/day
Metanephrines, total	U	0.0–0.90 mg/day
Parathyroid hormone	P	10–60 pg/mL
Parathyroid-related protein	P	<1.5 pmol/L
Pregnanediol	U	
Female		0.2–6.0 mg/day
Follicular phase		0.1–1.3 mg/day
Luteal phase		1.2–9.5 mg/day
Pregnancy		Gestation period dependent

Hormone	Fluid[a]	Units
Male		0.2–1.2 mg/day
Pregnanetriol	U	0.5–2.0 mg/day
Prolactin	S	
Female		0–15 ng/mL
Male		0–10 ng/mL
Renin activity	P	
Normal salt intake		
Recumbent 6 h		0.5–1.6 ng/mL/h
Upright 4 h		1.9–3.6 ng/mL/h
Low salt intake		
Recumbent 6 h		2.2–4.4 ng/mL/h
Upright 4 h		4.0–8.1 ng/mL/h
Upright 4 h, with diuretic		6.8–15.0 ng/mL/h
Somatomedin C	P	
Female		
Preadolescent		60.8–724.5 ng/mL
Adolescent		112.5–450.0 ng/mL
Adult		141.8–389.3 ng/mL
Male		
Preadolescent		65.5–841.5 ng/mL
Adolescent		83.3–378.0 ng/mL
Adult		54.0–328.5 ng/mL
Testosterone, total, morning sample	P	
Female		20–90 ng/dL
Male, adult		300–1100 ng/dL
Testosterone, unbound, morning sample	P	
Female, adult		0.09–1.29 ng/dL
Male, adult		3.06–24.0 ng/dL
Thyroglobulin	S	0–60 ng/mL
Thyroid hormone–binding index		0.83–1.17
Thyroid-stimulating hormone	S	0.5–5.0 μU/mL
Thyroxine free	S	0.8–2.7 ng/dL
Thyroxine-binding globulin	S	Age and sex dependent
Thyroxine free index		4.6–11.2
Thyroxine, total (T_4)	S	4–12 μg/dL
Triiodothyronine, total (T_3)	S	75–195 ng/dL
Vanillylmandelic acid (VMA)	U	1.4–6.5 mg/day

More recent assay results for many hormones are available via the database that the National Institute of Health's National Library of Medicine maintains. Consider utilizing the following URL: http://www.nlm.nih.gov/medlineplus/encyclopedia.html.

[a] S denotes serum, P plasma, and U urine.

Source: Abstracted from C.D. Jordan, J.G. Flood, M. LaPosata, K. B. Lewandrowski, Normal reference laboratory values, New England Journal of Medicine 327 (1992) 718–724.

Appendix 3

Clinically relevant endocrine disorders

Disease	Description
Acromegaly	Inappropriate and continued secretion of growth hormone by a tumor of pituitary cells which leads to soft tissue swelling and hypertrophy of the skeletal extremities (usually in the third or fourth decade).
Addison's disease	Adrenocortical insufficiency resulting from a deficient production of glucocorticoids and/or mineralocorticoids due to a destruction of the adrenal cortex.
Bartter's syndrome	Characterized by increased angiotensin, renin, aldosterone, secretion, hypokalemia, alkalosis, and hyperplasia of the renal juxtaglomerular cells, but with normal blood pressure; the disease, which is possibly an autosomal-dominant disorder, usually appears in late infancy or early childhood; the primary defect may be a lesion in chloride reabsorption in the kidney loop of Henle.
Celiac disease	An intestinal malabsorption disorder occurring in some individuals who may lack an enzyme necessary for the hydrolysis of *N*-glutamyl peptides in the small intestine. As a consequence, the affected individual is intolerant of some proteins, usually those derived from wheat, oats, barley, or rye; the disease is also referred to as gluten-sensitive enteropathy.
Chiari–Frommel syndrome	The occurrence of galactorrhea (nonnursing-related lactation) and amenorrhea (the absence of expected menstrual periods) during the postpartum period; the disease may be due to the presence of a prolactin-secreting tumor.
Cretinism	Characterized by a permanent neurological and skeletal retardation and results from an inadequate output of thyroid hormone during uterine and neonatal life; may be caused by iodine deficiency, thyroid hypoplasia, genetic enzyme defects, or excessive maternal intake of goitrogens.
Cushing's disease	Hypercortisolism resulting from the presence of small pituitary tumors that secrete adrenocorticotropic hormone leading to excess production of cortisol by the adrenals.
Cushing's syndrome	The circumstance of glucocorticoid excess without specification of the specific etiology; it may result from endogenous causes but is more commonly iatrogenic.
Diabetes insipidus	A deficient secretion of vasopressin that is manifested clinically as *diabetes insipidus;* it is a disorder characterized by the excretion of an increased volume of dilute urine.
Diabetes mellitus	A disease characterized by a chronic disorder of intermediary metabolism due to a relative lack of insulin which is characterized by hyperglycemia in both the postprandial and fasting state (see also types I and II *diabetes mellitus*).
Diabetes mellitus (insulin-dependent or type I diabetes)	The form of diabetes which appears in the second and third decade of life and is characterized by a destruction of the pancreas B cells; this form of the disease is normally treated with daily administration of insulin.
Diabetes mellitus (insulin-independent or type II diabetes)	The form of diabetes arising after the fourth decade, usually in obese individuals; this form of the disease does not normally require treatment with insulin.
Empty sella syndrome	Empty sella is a term that describes *sellae* that fill with air during pneumoencephalography. It is frequently associated with the flattening of the pituitary gland. The etiology of the disease is unknown; the pituitary function is usually normal.
Fanconi syndrome	A renal tubular defect in the absorption of a variety of substances, including H_2O, phosphate, sodium, bicarbonate, and amino acids; frequently an osteomalacic bone disease and a distal tubular acidosis may accompany the disease.
Feminization	Feminization of males, usually manifested by the enlargement of the breasts (gynecomastia) which can be attributed to an increase in estrogen levels relative to the prevailing androgen levels.

Disease	Description
Froehlich's syndrome	A condition usually caused by craniopharyngioma (a tumor of the hypothalamus) which results in a combination of obesity and hypogonadism; sometimes termed adiposogenital dystrophy.
Galactorrhea	The persistent discharge from the breast of a fluid that resembles milk and that occurs in the absence of parturition or else persists postpartum (4–6 months) after the cessation of nursing.
Gigantism	This condition appears in the first year of life and is characterized by a rapid weight and height gain; affected children usually have a large head and mental retardation; to date no specific endocrine abnormalities have been detected.
Goiter	Goiter may be defined as a thyroid gland that is twice its normal size; endemic goiter is the major thyroid disease throughout the world. Goiter is frequently associated with a dietary iodine deficiency; in instances of sporadic goiter, it may occur as a consequence of a congenital defect in thyroid hormone synthesis.
Grave's disease	An autoimmune disease characterized by the presence in serum of a LATS that is an antibody for the receptor for thyroid stimulating hormone. Grave's disease is the most common cause of thyrotoxicosis.
Gynecomastia	Abnormal breast enlargement that may occur in males during puberty.
Hartnup's disease	An intestinal transport disorder. The condition may be diagnosed by the massive urinary excretion of monoamino-monocarboxylic amino acids; frequently there are pellagra-like rashes after exposure to sunlight as well as attacks of cerebellar ataxia.
Hermaphroditism	True hermaphroditism is defined as the presence of both testicular and ovarian tissue in the same individual; pseudohermaphroditism is a discrepancy between gonadal and somatic sex.
Hirsutism	An increase in facial hair in women which is beyond that cosmetically acceptable; this condition may be associated with a number of masculinizing disorders, including Cushing's syndrome, congenital adrenal hyperplasia, and polycystic ovary syndrome.
Hyperaldosteronism	An inappropriate secretion of aldosterone. It can occur as a primary adrenal problem (e.g., adrenal tumor) or can be secondary to other metabolic derangements that stimulate its release; it is often characterized by inappropriately high levels of plasma renin
Hyperparathyroidism	Inappropriately high secretion of parathyroid hormone (PTH) leading to hypercalcemia. Frequently associated with hyperparathyroidism is a metabolic bone disease characterized by excessive bone calcium reabsorption; frequently attributable to an adenoma of the parathyroid gland.
Hypoparathyroidism	Inappropriately low secretion of PTH leading to hypocalcemia; the disease is either idiopathic or iatrogenically induced.
Hypophosphatasia	An autosomal recessive trait characterized by elevated serum and urine inorganic pyrophosphate, a low serum alkaline phosphatase, and frequently hypercalcemia; may be related to a dysfunction of the osteoblasts.
Klinefelter's syndrome	Typically characterized by male hypogonadism; the presence of extra X chromosomes is likely the fundamental underlying etiological factor. It is characterized by varying degrees of decreased Leydig cell function and seminiferous tubule failure.
Milk-alkali syndrome	Affected subjects have hypercalcemia, nephrocalcinosis, soft tissue calcification, renal impairment, alkalosis, and hyperphosphatemia; the syndrome can result as a consequence of an excessive dietary intake of milk and other absorbable alkali (e.g., Na_2CO_3 or $NaHCO_3$); it is uncommon today.
Myxedema	Hypothyroidism clinically manifested by the presence of a mucinous edema; the disease may appear at any time throughout life and is attributable to disorders of the thyroid gland or to pituitary insufficiency.
Nelson's syndrome	A pituitary adenoma occurring in 10% of patients with Cushing's disease; afflicted subjects have a severe skin pigmentation.
Osteomalacia	A bone disease in adults characterized by a failure of the skeletal osteoid to calcify; it is usually caused by an absence of adequate access to vitamin D.
Polycystic ovary syndrome	A complex of varying symptoms ranging from amenorrhea to anovulatory bleeding often associated with obesity and hirsutism. The term denotes an absence of ovulation in association with continuous stimulation of the ovary by disproportionately high levels of luteinizing hormone.
Pseudohypoparathyroidism	A familial disorder characterized by hypocalcemia, increased circulating levels of PTH and a peripheral unresponsiveness to the hormone; afflicted individuals frequently are of short stature, with mental retardation and short metacarpals and/or metatarsals.
Rickets	A failure in the child of the skeletal osteoid to calcify; it is usually caused by an absence of adequate amounts of vitamin D; it is characterized by a bowing of the femur, tibia, and fibulas.
Turner's syndrome	A condition present in females with a 45, XO chromosome pattern (i.e., complete absence of the X chromosome). The XO individual is typically short with a thick neck and trunk and no obvious secondary sex characteristics.

Disease	Description
Waterhouse–Friderichsen syndrome	Acute adrenal insufficiency resulting from severe systemic infection by *meningococcus* characterized by a high fever, meningeal irritation, and vascular collapse.
Werner's syndrome	Multiple endocrine neoplasia, type 1, caused by an autosomal recessive inheritance. It is often characterized by a severe testicular atrophy and a mild insulin-resistant diabetes.
Zollinger–Ellison syndrome	Tumors of the pancreas which result in excessive secretion of gastrin; the afflicted subject has recurrent duodenal ulcers and diarrhea caused by hypersecretion of gastric acid.

LATS, Long-acting thyroid stimulator.
This list was abstracted from the United States National Library of Medicine—Medical Subject Headings—Tree Structures—1996, pp. 294–298. United States Department of Commerce, Washington, DC, 1996. The diseases were included in the National Library of Medicine table on the basis of frequency of publication of papers about the given disease topics.

Appendix 4

The genetic code

First position (5′ end)	Second position				Third position (3′ end)
	U	**C**	**A**	**G**	
	Phe	Ser	Tyr	Cys	U
U	Phe	Ser	Tyr	Cys	C
	Leu	Ser	stop	stop	A
	Leu	Ser	stop	Trp	G
	Leu	Pro	His	Arg	U
C	Leu	Pro	His	Arg	C
	Leu	Pro	Gln	Arg	A
	Leu	Pro	Gln	Arg	G
	Ile	Thr	Asn	Ser	U
A	Ile	Thr	Asn	Ser	C
	Ile	Thr	Lys	Arg	A
	Met	Thr	Lys	Arg	G
	Val	Ala	Asp	Gly	U
G	Val	Ala	Asp	Gly	C
	Val	Ala	Glu	Gly	A
	Val	Ala	Glu	Gly	G

The genetic code is expressed in terms of triplet sequences as they would appear in a natural messenger RNA molecule. The sequence of each codon is read from left to right, starting with the 5′ end and going toward the 3′ end. The codons UAA, UAG, and UGA are "stop" signals that result in chain termination and are used to indicate the end of a polypeptide chain. No separate codon is required to start the polypeptide chain, since the first amino acid is always methionine (Met).

Appendix 5

Amino acid abbreviations

Amfino acid	Three letter	Single letter
Alanine	Ala	A
Arginine	Arg	R
Asparagine	Asn	N
Aspartic acid	Asp	D
Cysteine	Cys	C
Glutamic acid	Glu	E
Glutamine	Gln	Q
Glycine	Gly	G
Histidine	His	H
Isoleucine	Ile	I
Leucine	Leu	L
Lysine	Lys	K
Methionine	Met	M
Phenylalanine	Phe	F
Proline	Pro	P
Serine	Ser	S
Threonine	Thr	T
Tryptophan	Trp	W
Tyrosine	Tyr	Y
Valine	Val	V

Appendix 6

Units of measurement in biological systems

The Systemé International d'Unités (SI units), which is now generally accepted as the universal scientific system of measure, is based on the meter and kilogram (m, kg) rather than the centimeter and gram of the older cgs system. It has seven basic units of measurement:

Quantity	SI unit	Symbol
Amount of substance	Mole	mol
Mass	Kilogram	kg
Time	Second	s
Length	Meter	m
Electric current	Ampere	A
Thermodynamic temperature	Kelvin	K
Light intensity	Candela	cd

Several important units employed in biological systems which are derived from SI units include the following:

Quantity to be measured	Definition in terms of basic units	SI unit	Symbol
Volume	10^{-3} m^3	Liter	L
Frequency	s^{-1}	Hertz	Hz
Pressure	kg/(m s^2)	Pascal	Pa
Energy[a]	kg/(m^2 s)	Joule	J

For a complete description of SI units, see the IUPAC Manual of Symbols and Terminology for Physicochemical Quantities and Units (1970). Pure Applied Chemistry 21, 3–44.
[a] *1 cal is the energy required to heat 1 cm^3 of water from 14.5°C to 15.5°C. One calorie, which is equivalent to 1000 cal or 1.0 kcal, is the standard unit employed nutritionally. To convert calories into joules, multiply by the conversion factor of 4.185.*

Other units employed in biological systems include the following:

Unit	Abbreviation	Comment
Mole	mol	
Becquerel[a]	Bq	Units of radioactivity
Curie	Ci	
Calorie	cal	
Kilocalorie	kcal or Cal	
Svedberg	S	(1 S=10^{-13} s)

[a] *1 Bq=1 dps or 60 dpm, while 1 Ci=3.7 × 10^{10} Bq.*

The following prefixes and symbols are used to indicate multiples or decimal fractions of the SI units.

Prefix	Factor	Symbol
Exa	10^{18}	E
Peta	10^{15}	P
Tera	10^{12}	T
Giga	10^{9}	G
Mega	10^{6}	M
Kilo	10^{3}	k
Hector	10^{2}	h
Deca	10^{1}	da
Deci	10^{-1}	d
Centi	10^{-2}	c
Milli	10^{-3}	m
Micro	10^{-6}	μ
Nano	10^{-9}	n
Pico	10^{-12}	p
Femto	10^{-15}	f
Atto	10^{-18}	a

Index

Note: Page numbers followed by "*f*" and "*t*" refer to figures and tables, respectively.

H

O

P

R

S

T